T0418829

METAL OXIDES FOR NON-VOLATILE MEMORY

The Metal Oxides Book Series Edited by Ghenadii Korotcenkov

Forthcoming Titles

- Palladium Oxides Material Properties, Synthesis and Processing Methods, and Applications, Alexander M. Samoylov, Vasily N. Popov, 9780128192238
- Metal Oxides for Non-volatile Memory, Panagiotis Dimitrakis, Ilia Valov, Stefan Tappertzhofen, 9780128146293
- Metal Oxide Nanostructured Phosphors, H. Nagabhushana, Daruka Prasad, S.C. Sharma, 9780128118528
- Nanostructured Zinc Oxide, Kamlendra Awasthi, 9780128189009
- Multifunctional Piezoelectric Oxide Nanostructures, Sang-Jae Kim, Nagamalleswara Rao Alluri, Yuvasree Purusothaman, 9780128193327
- Transparent Conductive Oxides, Mirela Petruta Suchea, Petronela Pascariu, Emmanouel Koudoumas, 9780128206317
- Metal oxide-based nanofibers and their applications, Vincenzo Esposito, Debora Marani, 9780128206294
- Metal-oxides for Biomedical and Biosensor Applications, Kunal Mondal, 9780128230336
- Metal Oxide-Carbon Hybrid Materials, Muhammad Akram, Rafaqat Hussain, Faheem K Butt, 9780128226940
- Metal Oxide-based heterostructures, Naveen Kumar, Bernabe Mari Soucase, 9780323852418
- Metal Oxides and Related Solids for Electrocatalytic Water Splitting, Junlei Qi, 9780323857352
- Advances in Metal Oxides and Their Composites for Emerging Applications, Sagar Delekar, 9780323857055
- Metallic Glasses and Their Oxidation, Xinyun Wang, Mao Zhang, 9780323909976
- Solution Methods for Metal Oxide Nanostructures, Rajaram S. Mane, Vijaykumar Jadhav, Abdullah M. Al-Enizi, 9780128243534
- Metal Oxide Defects, Vijay Kumar, Sudipta Som, Vishal Sharma, Hendrik Swart, 9780323855884
- Renewable Polymers and Polymer-Metal Oxide Composites, Sajjad Haider, Adnan Haider, 9780323851558
- Metal Oxides for Optoelectronics and Optics-based Medical Applications, Suresh Sagadevan, Jiban Podder, Faruq Mohammad, 9780323858243
- Graphene Oxide-Metal Oxide and Other Graphene Oxide-Based Composites in Photocatalysis and Electrocatalysis, Jiaguo Yu, Liuyang Zhang, Panyong Kuang, 9780128245262

Published Titles

- Metal Oxides in Nanocomposite-Based Electrochemical Sensors for Toxic Chemicals, Alagarsamy Pandikumar, Perumal Rameshkumar, 9780128207277
- Metal Oxide-Based Nanostructured Electrocatalysts for Fuel Cells, Electrolyzers, and Metal-Air Batteries, Teko Napporn, Yaovi Holade, 9780128184967
- Titanium Dioxide (TiO2) and its applications, Leonardo Palmisano, Francesco Parrino, 9780128199602
- Solution Processed Metal Oxide Thin Films for Electronic Applications, Zheng Cui, 9780128149300
- Metal Oxide Powder Technologies, Yarub Al-Douri, 9780128175057
- Colloidal Metal Oxide Nanoparticles, Sabu Thomas, Anu Tresa Sunny, Prajitha V, 9780128133576
- Cerium Oxide, Salvatore Scire, Leonardo Palmisano, 9780128156612
- Tin Oxide Materials, Marcelo Ornaghi Orlandi, 9780128159248
- Metal Oxide Glass Nanocomposites, Sanjib Bhattacharya, 9780128174586
- Gas Sensors Based on Conducting Metal Oxides, Nicolae Barsan, Klaus Schierbaum, 9780128112243
- Metal Oxides in Energy Technologies, Yuping Wu, 9780128111673
- Metal Oxide Nanostructures, Daniela Nunes, Lidia Santos, Ana Pimentel, Pedro Barquinha, Luis Pereira, Elvira Fortunato, Rodrigo Martins, 9780128115121
- Gallium Oxide, Stephen Pearton, Fan Ren, Michael Mastro, 9780128145210
- Metal Oxide-Based Photocatalysis, Adriana Zaleska-Medynska, 9780128116340
- Metal Oxides in Heterogeneous Catalysis, Jacques C. Vedrine, 9780128116319
- Magnetic, Ferroelectric, and Multiferroic Metal Oxides, Biljana Stojanovic, 9780128111802
- Iron Oxide Nanoparticles for Biomedical Applications, Sophie Laurent, Morteza Mahmoudi, 9780081019252
- The Future of Semiconductor Oxides in Next-Generation Solar Cells, Monica Lira-Cantu, 9780128111659
- Metal Oxide-Based Thin Film Structures, Nini Pryds, Vincenzo Esposito, 9780128111666
- Metal Oxides in Supercapacitors, Deepak Dubal, Pedro Gomez-Romero, 9780128111697
- Transition Metal Oxide Thin Film-Based Chromogenics and Devices, Pandurang Ashrit, 9780081018996

METAL OXIDES FOR NON-VOLATILE MEMORY

Materials, Technology and Applications

Edited by

PANAGIOTIS DIMITRAKIS
Institute of Nanoscience and Nanotechnology, NCSR "Demokritos", Athens, Greece

ILIA VALOV
Peter Grünberg Institute 7, Jülich Research Centre, Jülich, Germany

STEFAN TAPPERTZHOFEN
Chair for Micro- and Nanoelectronics, Department of Electrical Engineering and Information Technology, TU Dortmund University, Dortmund, Germany

Series Editor

GHENADII KOROTCENKOV
Department of Physics and Engineering, Moldova State University, Chisinau, Republic of Moldova

ELSEVIER

Elsevier
Radarweg 29, PO Box 211, 1000 AE Amsterdam, Netherlands
The Boulevard, Langford Lane, Kidlington, Oxford OX5 1GB, United Kingdom
50 Hampshire Street, 5th Floor, Cambridge, MA 02139, United States

ISBN: 978-0-12-814629-3

For information on all Elsevier publications
visit our website at https://www.elsevier.com/books-and-journals

Publisher: Matthew Deans
Acquisitions Editor: Kayla Dos Santos
Editorial Project Manager: Rafael Guilherme Trombaco
Production Project Manager: Stalin Viswanathan
Cover Designer: Miles Hitchen

Typeset by STRAIVE, India

Contents

4. Technology and neuromorphic functionality of magnetron-sputtered memristive devices

A.N. Mikhaylov, M.N. Koryazhkina, D.S. Korolev, A.I. Belov, E.V. Okulich, V.I. Okulich, I.N. Antonov, R.A. Shuisky, D.V. Guseinov, K.V. Sidorenko, M.E. Shenina, E.G. Gryaznov, S.V. Tikhov, D.O. Filatov, D.A. Pavlov, D.I. Tetelbaum, O.N. Gorshkov, A.V. Emelyanov, K.E. Nikiruy, V.V. Rylkov, V.A. Demin, and B. Spagnolo

5. Metalorganic chemical vapor deposition of aluminum oxides: A paradigm on the process-structure-properties relationship

Constantin Vahlas and Brigitte Caussat

6. MOx materials by ALD method

Elena Cianci and Sabina Spiga

7. Nano-composite MOx materials for NVMs

C. Bonafos, L. Khomenkhova, F. Gourbilleau, E. Talbot, A. Slaoui, M. Carrada, S. Schamm-Chardon, P. Dimitrakis, and P. Normand

8. MOx in ferroelectric memories

Stefan Slesazeck, Halid Mulaosmanovic, Michael Hoffmann, Uwe Schroeder, Thomas Mikolajick, and Benjamin Max

9. "Metal oxides in magnetic memories": Current status and future perspectives

Andreas Kaidatzis, Georgios Giannopoulos, and Dimitris Niarchos

10. Correlated transition metal oxides and chalcogenides for Mott memories and neuromorphic applications

Laurent Cario, Julien Tranchant, Benoit Corraze, and Etienne Janod

11. The effect of external stimuli on the performance of memristive oxides

Yang Li, Dennis Valbjørn Christensen, Simone Sanna, Vincenzo Esposito, and Nini Pryds

Contributors

I.N. Antonov Lobachevsky State University of Nizhny Novgorod, Nizhny Novgorod, Russia

A.I. Belov Lobachevsky State University of Nizhny Novgorod, Nizhny Novgorod, Russia

C. Bonafos CEMES-CNRS, University of Toulouse, CNRS, Toulouse, France

S. Brivio CNR—IMM, Unit of Agrate Brianza, Agrate Brianza, Italy

Laurent Cario Institut des Matériaux Jean Rouxel (IMN), Nantes, France

M. Carrada CEMES-CNRS, University of Toulouse, CNRS, Toulouse, France

Brigitte Caussat LGC-CNRS, Toulouse, France

Georgios Ch. Sirakoulis Department of Electrical and Computer Engineering, Democritus University of Thrace, Xanthi, Greece

Dennis Valbjørn Christensen Department of Energy Conversion and Storage, Technical University of Denmark (DTU), Fysikvej, Kongens Lyngby, Denmark

Elena Cianci CNR-IMM, Unit of Agrate Brianza, Agrate Brianza (MB), Italy

Benoit Corraze Institut des Matériaux Jean Rouxel (IMN), Nantes, France

V.A. Demin National Research Center "Kurchatov Institute", Moscow, Russia

P. Dimitrakis National Center for Scientific Research "Demokritos", Institute of Nanoscience and Nanotechnology, Attiki, Greece

A.V. Emelyanov National Research Center "Kurchatov Institute", Moscow, Russia

Manuel Escudero Department of Electronic Engineering, Universitat Politècnica de Catalunya, Barcelona, Spain

Vincenzo Esposito Department of Energy Conversion and Storage, Technical University of Denmark (DTU), Fysikvej, Kongens Lyngby, Denmark

D.O. Filatov Lobachevsky State University of Nizhny Novgorod, Nizhny Novgorod, Russia

Georgios Giannopoulos Institute of Nanoscience and Nanotechnology, NCSR "Demokritos", Athens, Greece; Department of Physics and Astronomy, University College London, London, United Kingdom

O.N. Gorshkov Lobachevsky State University of Nizhny Novgorod, Nizhny Novgorod, Russia

F. Gourbilleau CIMAP Normandie University, ENSICAEN, UNICAEN, CEA, CNRS, Caen, France

E.G. Gryaznov Lobachevsky State University of Nizhny Novgorod, Nizhny Novgorod, Russia

D.V. Guseinov Lobachevsky State University of Nizhny Novgorod, Nizhny Novgorod, Russia

Michael Hoffmann NaMLab gGmbh, Dresden, Germany

Etienne Janod Institut des Matériaux Jean Rouxel (IMN), Nantes, France

Andreas Kaidatzis Institute of Nanoscience and Nanotechnology, NCSR "Demokritos", Athens, Greece

L. Khomenkhova CIMAP Normandie University, ENSICAEN, UNICAEN, CEA, CNRS, Caen, France; V. Lashkaryov Institute of Semiconductor Physics of National Academy of Sciences of Ukraine, Ukraine and National University "Kyiv-Mohyla Academy", Kyiv, Ukraine

D.S. Korolev Lobachevsky State University of Nizhny Novgorod, Nizhny Novgorod, Russia

M.N. Koryazhkina Lobachevsky State University of Nizhny Novgorod, Nizhny Novgorod, Russia

Yang Li Department of Energy Conversion and Storage, Technical University of Denmark (DTU), Fysikvej, Kongens Lyngby, Denmark

Qi Liu Frontier Institute of Chip and System, Fudan University, Shanghai, China

Benjamin Max TU Dresden, IHM, Dresden, Germany

A.N. Mikhaylov Lobachevsky State University of Nizhny Novgorod, Nizhny Novgorod, Russia

Thomas Mikolajick NaMLab gGmbh; TU Dresden, IHM, Dresden, Germany

Halid Mulaosmanovic NaMLab gGmbH, Dresden, Germany

Dimitris Niarchos Institute of Nanoscience and Nanotechnology, NCSR "Demokritos", Athens, Greece

K.E. Nikiruy National Research Center "Kurchatov Institute", Moscow, Russia

P. Normand National Center for Scientific Research "Demokritos", Institute of Nanoscience and Nanotechnology, Attiki, Greece

E.V. Okulich Lobachevsky State University of Nizhny Novgorod, Nizhny Novgorod, Russia

V.I. Okulich Lobachevsky State University of Nizhny Novgorod, Nizhny Novgorod, Russia

D.A. Pavlov Lobachevsky State University of Nizhny Novgorod, Nizhny Novgorod, Russia

Nini Pryds Department of Energy Conversion and Storage, Technical University of Denmark (DTU), Fysikvej, Kongens Lyngby, Denmark

Krishnaswamy Ramkumar Infineon Technologies, San Jose, CA, United States

Antonio Rubio Department of Electronic Engineering, Universitat Politècnica de Catalunya, Barcelona, Spain

V.V. Rylkov National Research Center "Kurchatov Institute", Moscow, Russia

Simone Sanna Department of Energy Conversion and Storage, Technical University of Denmark (DTU), Fysikvej, Kongens Lyngby, Denmark

S. Schamm-Chardon CEMES-CNRS, University of Toulouse, CNRS, Toulouse, France

Uwe Schroeder NaMLab gGmbH, Dresden, Germany

M.E. Shenina Lobachevsky State University of Nizhny Novgorod, Nizhny Novgorod, Russia

R.A. Shuisky Lobachevsky State University of Nizhny Novgorod, Nizhny Novgorod, Russia

K.V. Sidorenko Lobachevsky State University of Nizhny Novgorod, Nizhny Novgorod, Russia

A. Slaoui ICube, CNRS and University of Strasbourg, Strasbourg, France

Stefan Slesazeck NaMLab gGmbH, Dresden, Germany

B. Spagnolo Lobachevsky State University of Nizhny Novgorod, Nizhny Novgorod, Russia; University of Palermo and The National Interuniversity Consortium for the Physical Sciences of Matter, Palermo, Italy

Sabina Spiga CNR-IMM, Unit of Agrate Brianza, Agrate Brianza (MB), Italy

E. Talbot Normandie Univ, UNIROUEN, INSA Rouen, CNRS, Groupe de Physique des Matériaux, Rouen, France

Stefan Tappertzhofen Chair for Micro- and Nanoelectronics, Department of Electrical Engineering and Information Technology, TU Dortmund University, Dortmund, Germany

D.I. Tetelbaum Lobachevsky State University of Nizhny Novgorod, Nizhny Novgorod, Russia

S.V. Tikhov Lobachevsky State University of Nizhny Novgorod, Nizhny Novgorod, Russia

Julien Tranchant Institut des Matériaux Jean Rouxel (IMN), Nantes, France

Constantin Vahlas CIRIMAT-CNRS, Toulouse, France

E. Vianello CEA-LETI, Université Grenoble Alpes, Grenoble, France

Ioannis Vourkas Department of Electronic Engineering, Universidad Técnica Federico Santa María, Valparaíso, Chile

Xiaolong Zhao School of Microelectronics, University of Science and Technology of China, Hefei, China

Series editor biography

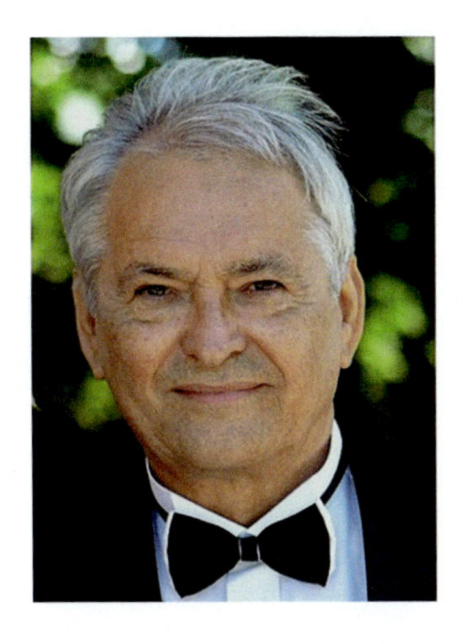

Ghenadii Korotcenkov received his PhD in physics and technology of semiconductor materials and devices in 1976 and his Doctor of Science degree (Doc. Hab.) in physics of semiconductors and dielectrics in 1990. He has more than 45 years of experience as a teacher and scientific researcher. Long a time he was a leader of gas sensor group and the manager of various national and international scientific and engineering projects carried out in the Laboratory of Micro- and Optoelectronics, Technical University of Moldova, Chisinau, Moldova. In 2007–08 he was an invited scientist at Korea Institute of Energy Research (Daejeon). After that, until 2017, Dr. G. Korotcenkov was a research professor in the School of Materials Science and Engineering at Gwangju Institute of Science and Technology (GIST), Korea. Currently, G. Korotcenkov is a chief scientific researcher at Moldova State University, Chisinau, Moldova. His present scientific interests starting from 1995 include material sciences, focusing on metal oxide film deposition and characterization, surface science, thermoelectric conversion, and design of physical and chemical sensors, including thin-film gas sensors.

G. Korotcenkov is the author or editor of 39 books and special issues, including 11-volume *Chemical Sensors* series published by Momentum Press; 15-volume *Chemical Sensors* series published by Harbin Institute of Technology Press, China; 3-volume *Porous Silicon: From Formation to Application* issue published by CRC Press; 2 volumes *Handbook of Gas Sensor Materials* published by Springer; and 3-volume *Handbook of Humidity Measurements* published by CRC Press. Currently, he is the book series editor of *Metal Oxides* published by Elsevier.

G. Korotcenkov is the author and coauthor of more than 650 scientific publications, including 31 review papers, 38 book chapters, and more than 200 peer-reviewed articles published in scientific journals (h-factor = 41 (Web of Science), h = 42 (Scopus), and h = 56 (Google scholar citation)). He holds 18 patents to his credit. He has presented more than 250 reports in both national and international conferences, including 17 invited talks.

G. Korotcenkov, as a cochair or a member of program, scientific and steering committees, participated in the organization of more than 30 international scientific conferences. He is a member of editorial boards in five scientific international journals. His name and activities have been listed by many biographical publications including *Who's Who*. His research activities are honored by the Honorary Diploma of the Government of the Republic of Moldova (2020); Award of the Academy of Sciences of Moldova (2019); Award of the Supreme Council of Science and Advanced Technology of the Republic of Moldova (2004); the Prize of the Presidents of the Ukrainian, Belarus, and Moldovan Academies of Sciences (2003); Senior Research Excellence Award of Technical University of Moldova (2001, 2003, 2005); and the National Youth Prize of the Republic of Moldova in the field of science and technology (1980); among others. G. Korotcenkov also received a fellowship from the International Research Exchange Board (IREX, USA, 1998), Brain Korea 21 Program (2008–12), and BrainPool Program (Korea, 2015–17).

Preface to the series

Synthesis, study, and application of metal oxides are the most rapidly progressing areas of science and technology. Metal oxides are one of the most ubiquitous compound groups on Earth and are available in a large variety of chemical compositions, atomic structures, and crystalline shapes. In addition, they are known to possess unique functionalities that are absent or inferior in other solid materials. In particular, metal oxides represent an assorted and appealing class of materials, exhibiting a full spectrum of electronic properties—from insulating to semiconducting, metallic, and superconducting. Moreover, almost all the known effects, including superconductivity, thermoelectric effects, photoelectrical effects, luminescence, and magnetism, can be observed in metal oxides. Therefore, metal oxides have emerged as an important class of multifunctional materials with a rich collection of properties, which have great potential for numerous device applications. Availability of a wide variety of metal oxides with different electrophysical, optical, and chemical characteristics; their high thermal and temporal stabilities; and their ability to function in harsh environments make them highly suitable materials for designing transparent electrodes, high-mobility transistors, gas sensors, actuators, acoustical transducers, photovoltaic and photonic devices, photocatalysts and heterogeneous catalysts, solid-state coolers, high-frequency and micromechanical devices, energy harvesting and storage devices, and nonvolatile memories, among many others in the electronics, energy, and health sectors. In these devices, metal oxides can be successfully used as sensing or active layers, substrates, electrodes, promoters, structure modifiers, membranes, or fibers, i.e., they can be used as both active and passive components.

Among other advantages of metal oxides are their low fabrication cost and robustness in practical applications. Furthermore, metal oxides can be prepared in various forms such as ceramics, thick films, and thin films. At that for thin film deposition can be used deposition techniques that are compatible with standard microelectronic technology. This factor is extremely important for large-scale production because the microelectronic approach ensures low cost for mass production, offers the possibility of manufacturing devices on a chip, and guarantees good reproducibility. Various metal oxide nanostructures including nanowires, nanotubes, nanofibers, core–shell structures, and hollow nanostructures can also be synthesized. As such, the field of metal oxide-nanostructured morphologies (e.g., nanowires, nanorods, and nanotubes) has become one of the most active research areas within the nanoscience community.

The ability to both create a variety of metal oxide-based composites and synthesize various multicomponent compounds significantly expands the range of properties that metal oxide-based materials can offer, making them a truly versatile multifunctional material for widespread use. As it is known, small changes in their chemical composition and atomic structure result in a spectacular variation in the properties and behavior of metal oxides. Current advances in synthesizing and characterizing techniques reveal numerous new functions of metal oxides.

Taking into account the importance of metal oxides in the progress of microelectronics, optoelectronics, photonics, energy conversion, sensors, and catalysis, a large number of books devoted to this class of materials have been published. However, one should note that some books from this list are too general, some are collections of various original works without any generalizations, and yet others were published many years ago. However, during the past decade, great progress has been made in the synthesis as well as in the structural, physical, and chemical characterization and application of metal oxides in various devices, and a large number of papers have been published on metal oxides. In addition, till now, many important topics related to the study and application of metal oxides have not been discussed. To remedy this situation, we decided to generalize and systematize the results of research in this direction and to publish a series of books devoted to metal oxides.

The proposed book series "Metal Oxides" is the first of its kind devoted exclusively to metal oxides. We believe that combining books on metal oxides in a series could help readers search for the required information on the subject. In particular, we hope that the books from our series, with its clear specialization by content, will provide interdisciplinary discussion for various oxide materials with a wide range of topics, from material synthesis and deposition to characterization, processing, and then to device fabrications and applications. This book series was prepared by a team of highly qualified experts, which guarantees its high quality.

I hope that our books will be useful and easy to navigate. I also hope that readers will consider this "Metal Oxides" book series an encyclopedia of metal oxides that will help them understand the present status of metal oxides, to estimate the role of multifunctional metal oxides in the design of advanced devices, and then, based on observed knowledge, to formulate new goals for further research.

This book series is intended for scientists and researchers, working or planning to work in the field of materials related to metal oxides, i.e., scientists and researchers whose activities are related to electronics, optoelectronics, energy, catalysis, sensors, electrical engineering, ceramics, biomedical designs, etc. I believe that the book series will also be interesting for practicing engineers or project managers in industries and national laboratories involved in designing metal oxide-based devices, helping them with the process and in selecting optimal metal oxide for specific applications. With many references to the vast resource of recently published literature on the subject, this book series will serve as a significant and insightful source of valuable information, providing scientists and engineers with new insights for understanding and improving existing metal oxide-based devices and for designing new metal oxide-based materials with new and unexpected properties.

I believe that the book series would be extremely helpful for university students, postdocs, and professors. The structure of these books offers a basis for courses in the field of materials science, chemical engineering, electronics, electrical engineering, optoelectronics, energy technologies, environmental control, and many others. Graduate students would also find the book series extremely useful in their research and understanding the synthesis of metal oxides and the study and applications of these multifunctional materials in various devices. We are sure that each of these audiences will find information useful for their activity.

Finally, I thank all contributing authors and book editors involved in the creation of these books. I am thankful that they agreed to participate in this project and for their efforts in the preparation of these books. Without their participation, this project would not have been possible. I also express my gratitude to Elsevier for giving us the opportunity to publish this series. I especially thank all the teams of the editorial office at Elsevier for their patience during the development of this project and for encouraging us during the various stages of preparation.

Ghenadii Korotcenkov

CHAPTER

1

Introduction to non-volatile memory

Stefan Tappertzhofen

Chair for Micro- and Nanoelectronics, Department of Electrical Engineering and Information Technology, TU Dortmund University, Dortmund, Germany

1.1 Introduction and history

In computers based on the *von Neumann* architecture, logic information is stored in memory cells and separately processed in logic gates (Fig. 1.1). One can categorize between volatile and non-volatile memories (NVM). In volatile memories, information is kept as long as power is supplied, whereas non-volatile memories can store information without being powered. The type of access is another way for memory classification. Sequential access in rotating magnetic disks, or more historically, magnetic tape memories, usually allows for storing information with high density, while random access memories (RAMs) typically allow for fast read and write access. While this book focusses on metal oxides for non-volatile memories, the most important volatile RAMs, namely static RAM (SRAM) and dynamic RAM (DRAM) are also briefly introduced [1]. Afterwards, an overview of Flash as the dominating state-of-the-art non-volatile memory technology is given. Novel concepts for non-volatile memories with particular focus on those based on metal oxides are finally summarized.

The advent of first integrated circuits in by Jack Kilby [2], Robert Noyce [3] and Kurt Lehovec [4] in 1957–1959, respectively, and advances in microelectronics in the beginning of the 1960s were not only important milestones for early information technology but also paved the way to one of the most venturesome journeys in human mankind: the race to the moon. The Apollo Guidance Computer [5] (AGC) consisted of a RAM with not more than 4 kB capacity and 74 kB read-only-memory (ROM) space. Today's mobile phones have a RAM capacity that is more

Metal Oxides for Non-volatile Memory
https://doi.org/10.1016/B978-0-12-814629-3.00001-5

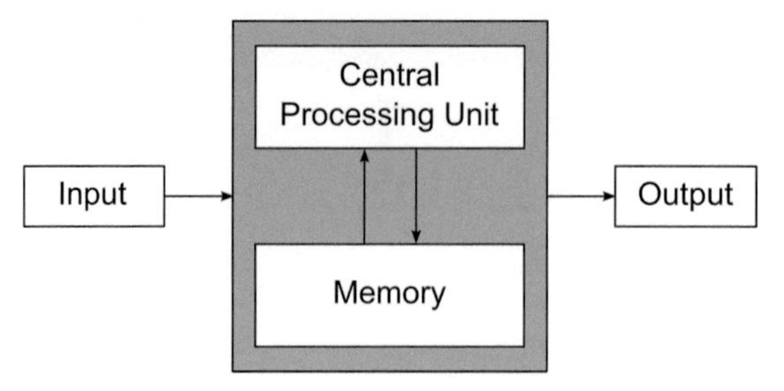

FIG. 1.1 *Von Neumann* architecture in which memory- and logic-operation are separated.

than 10-million times larger than the AGC-RAM. By that time, in the 1960s and 1970s, non-volatile magnetic-core memories [6] were the dominating RAM technology and had to be wired by hand. Disadvantages of magnetic-core memories were their destructive and power-inefficient read-access and integration limitations. Therefore, solid-state memory concepts, including Nb_2O_5-based resistive switches [7], were studied in the 1960s and 1970. Nevertheless, these early demonstrations did not reach the stability and operation performance of magnetic-core memories, and—more importantly—for space-flights and defence (as arguably the most important driving forces for innovation). The benefit of magnetic-core memory was their high radiation tolerance compared to charge-based solid-state memories.

When computer became affordable as mass-market consumer products, charge-based volatile memories as well as magnetic disks and tape memories as non-volatile memories began to be dominating. An important milestone was the discovery of the magnetoresistance effects with large resistance response that is now used in magnetic disk memories [8,9]. The general memory-organization in computers based on *von Neuman* architectures did not change much in the last few decades. For fast read- and write-access SRAM-cells are arranged as close as possible to logic gates in form of cache-memories. SRAM is faster and more expansive than dense DRAM, which is used as main memory. Magnetic disks and tapes are used for mass storage with capacities reaching some TB and more today. With the advent of Flash non-volatile memories (NAND-Flash was invented by F. Masuoka [10]), charge-based solid-state memories are becoming the dominating memory technology mass-storage. While magnetic disks still offer a higher memory density, Flash allows for random access with shorter access times (especially for fragmented data) and ultra-low power consumption [11,12] ($<$fJ/bit, with circuit overhead $\sim$100 pJ/bit), which makes Flash attractive for mobile applications, mobile phones, tablets and laptops. Due to the advances of Flash-technology in the 1990s to 2000s, research on resistive switches stepped out of sight. However, in the 2000s scaling of Flash was considered to

TABLE 1.1 Comparison of state-of-the-art memory technology for today's computer and typical performance data for resistive switches [11,18].

	SRAM	DRAM	NAND-Flash	Magnetic disks	Resistive switches
Energy/bit (fJ)	0.5	5	0.02–10	10^9	~1–10^3 [19]
Read/write access (ns)	0.1–0.3	10	10^5	10^6	<ns [20] to 100 ns [21]
Density (F^2)	140	6–12	1–4	2/3	4
Retention	As long as voltage is applied	~60 ms, refresh required	Years	Years	Years
Endurance	$>10^{16}$	$>10^{16}$	$>10^4$	$>10^4$	10^{12} [22]

Note, multilevel storage, circuit overhead and data transfer is not considered and it is difficult to compare between different device technologies. F is the feature size, that is the length/distance of the smallest feature that can be prepared with a given technology node.

become a technical problem and rediscovery and research on alternative technologies, including resistive switching [13–15] and in general memristive devices [16] as well as magnetic memories [17] started.

An overview of state-of-the-art memory technology compared to metal-oxide resistive switches as one example of an emerging memory device is listed in Table 1.1. While SRAM and DRAM are fast but volatile, non-volatile Flash is about four times of magnitude in time slower. The different access times of DRAM and Flash result in a latency-gap that limits the performance of modern computers for applications where huge amounts of data are processes, including simulation, databases, deep learning and artificial intelligence [23]. A new type of non-volatile memories, so-called storage class memories (SCM) [24–26], has been suggested to fill this latency-gap (Fig. 1.2). SCMs like Intel® Optane™ are also known as Data Center Persistent Memory Modules (DCPMMs) [27–30]. SCMs can store relatively large amount of data buffered between main memory and mass storage memories like Flash and magnetic disks. They may allow for relatively fast access for data being frequency read and programmed.

Several requirements must be fulfilled by a non-volatile memory device for SCM-operation, in particular: the read/write access may be faster than 1 μs, ideally as fast as 100–300 ns or below, and the endurance must be 10^8 cycles or more [31]. Obviously from Table 1.1, NAND-Flash does not fulfill these requirements and even NOR-Flash, which is faster (~μs) but more power-consuming (~100 pJ/bit) is not fast enough, and both Flash topologies show a low endurance. In contrast, resistive switches and magnetic RAMs have been considered as potential candidates for storage class memories with access-times in the order

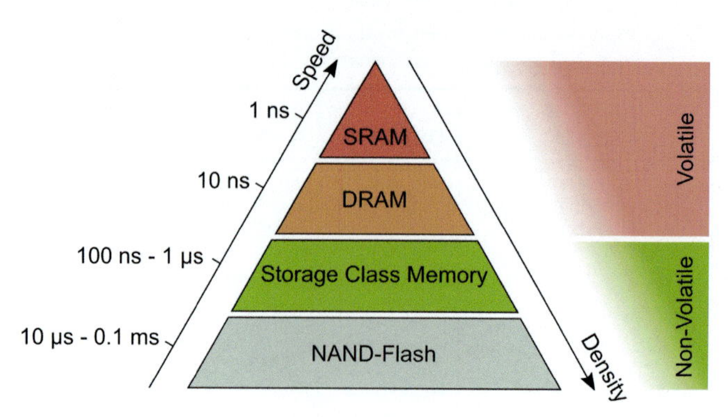

FIG. 1.2 Memory hierarchy including volatile SRAM and DRAM, and non-volatile storage class memories and NAND-Flash.

of ~100 ns or much lower, high endurance and relatively low power-consumption (compared to NOR-Flash).

1.1.1 Outline of this work

Fabrication methods and materials for metal-oxide memories are introduced in Chapters 4 (magnetron-sputtering), 5 (chemical vapor deposition), 6 (atomic layer deposition), 7 (nanocomposites), 12 (2D-materials). Important phenomena that are exploited for memory operation are discussed in Chapters 2 and 11 (resistive switching), 3 (charge trapping), 8 (ferroelectric memories), 9 (magnetoresistive memories), and 10 (Mott-memories). An overview of important applications, including those beyond traditional memory operation is given in Chapters 13 (logic applications) and 14 (neuromorphics). The following section introduces into state-of-the-art Flash non-volatile memories and later novel concepts for non-volatile memories with particular focus on metal-oxide based systems are briefly discussed.

1.2 Flash non-volatile memory

Flash is today the by-far dominating non-volatile random access memory technology. The core-element of Flash is the floating-gate transistor [32], which is also used in erasable programmable read-only memory (EPROM) and electrically erasable programmable read-only memory (EEPROM). The term "read-only memory" may be confusing in this case. Both EPROMs and EEPROMs can be electrically programmed. In case of EPROMs, ultra-violet light is required for erase, while EEPROMs can be electrically erased. At first glance, Flash is similar to EEPROM optimized

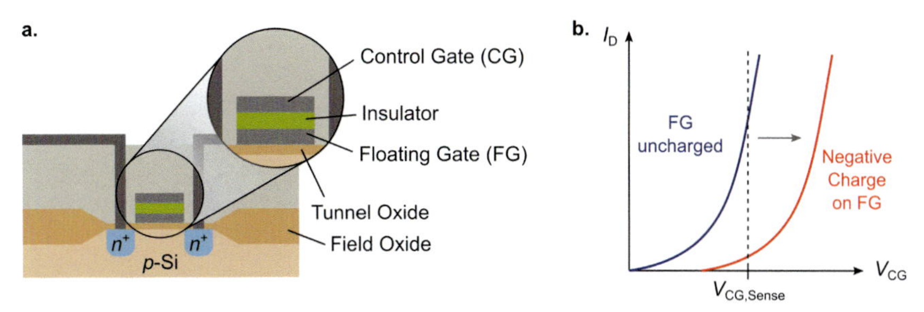

FIG. 1.3 (A) Schematic of a cross-sectional view of a floating-gate transistor fabricated in LOCOS-technology [33]. (B) Transfer characteristics (control-gate voltage V_{CG} vs. drain-current I_D) of a floating-gate transistor for a programmed (negative charge on FG) and erased (FG uncharged) state.

for high-speed and density. A cross-sectional view of a floating-gate-transistor is shown in Fig. 1.3A.

By applying appropriate voltages between the control gate (CG), bulk, source and drain, electrons are injected from the channel into the floating gate (FG). This process is called programming or write-operation. Accumulation of charge in the FG results in a shift of the transfer characteristics (Fig. 1.3B). When a sense voltage $V_{CG} = V_{CG,Sense}$ is applied to CG a relatively low or relatively high drain current is detected depending on the drain-source-voltage V_{DS} and the charge on the floating-gate. That allows to encode at least two different logic states, 0 and 1, respectively.

In addition to a 1-bit-storage capacity, by control of the charge on the FG, intermediate states can be programmed, which enables to encode multiple bit in one floating-gate transistor (multilevel cell, MLC) [34]. State-of-the-art are nowadays 4-bit Flash cells. That means, by keeping the footprint F^2 of a memory cell constant (with the feature size F) the information-density could be increased by a factor of 4 using MLC-technology. A 4-bit MLC encodes $2^4 = 16$ different logic levels. Further increase of the information density is sophisticated due to noise margins and charge fluctuations must be also considered. Today, the size of Flash memory cells results in an effective charge stored on a FG in the order of only some 10^2–10^3 electrons [35]. Thus, only limited freedom is available for further intermediate logic levels.

1.2.1 Programming- and erase-mechanism

Programming and erase in Flash is achieved by hot-carrier injection and Fowler-Nordheim tunneling (FN, also known as field electron emission) depending on the topology (NOR- vs. NAND-Flash). Simplified cross-sectional band-diagrams for hot-electron injection and Fowler-Nordheim tunneling are shown in Fig. 1.4A and B, respectively. The tunnel oxide

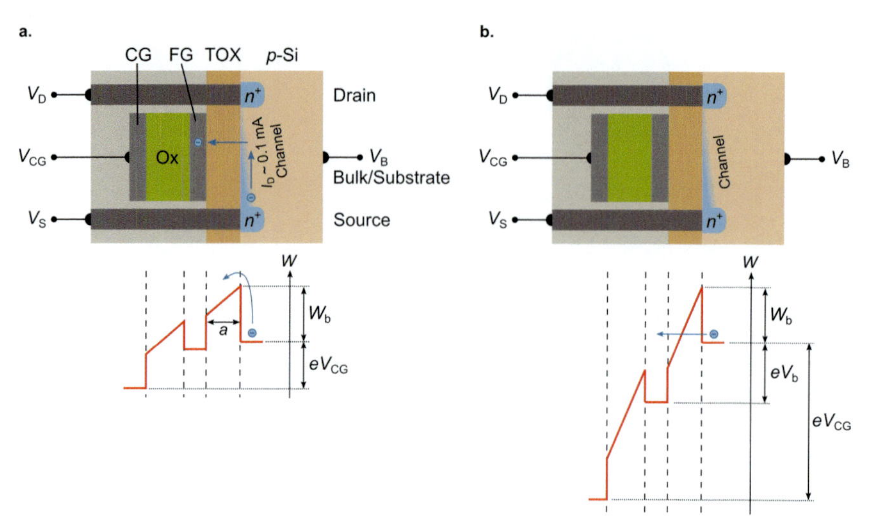

FIG. 1.4 Programming sequence for floating-gate transistors and corresponding energy (W) diagrams: (A) hot-electron injection and (B) Fowler-Nordheim tunneling. CG and FG denote the control- and floating-gate, OX is the oxide between CG and FG, TOX the tunnel oxide, p-Si is p-doped silicon, and the n^+-regions are the n-doped source and drain regions of the transistor, respectively. V_D, V_S, V_B, and V_{GC} are the drain-, source-, bulk- and gate-control voltages, respectively.

(TOX) with thickness a (in the order of some nanometer) forms an energy barrier ($W_b \sim 2\ldots4\,\text{eV}$) between the FG and channel/substrate. For hot-electron injection, electrons in the channel between source and drain are accelerated by applying a sufficiently high drain-source-voltage $V_{DS} = V_D - V_S$. The hot-carrier injection current density J can be calculated by the barrier-dependent current density J_0 and injection probability P:

$$J = \underbrace{\frac{4\pi e m {k_B}^2 T^2}{h^3}}_{=J_0} \cdot \underbrace{\exp\left(-\frac{W_b}{k_B T}\right)}_{=P} \tag{1.1}$$

At relatively high drain currents ($I_D \sim 0.1\,\text{mA}$) some electrons gain enough kinetic energy to overcome the energy barrier. A sufficiently high CG-voltage $V_{CG} = 6\text{–}12\,\text{V}$ forms a trapezoid energy-barrier (Fig. 1.4A) and attracts these electrons, which eventually charge the floating-gate. Higher CG-voltages ($V_{CG} \sim 15\ldots20\,\text{V}$) lead to a triangular shape of the energy-barrier and the effective distance is shorter than the barrier-thickness a (see Fig. 1.4B). Now, electrons have a relatively high probability to tunnel through the energy barrier into the floating-gate by Fowler-Nordheim tunneling. Note, in contrast to hot-electron injection, a large drain-current is not needed for Fowler-Nordheim tunneling. The FN-tunneling current density J reads:

$$J = \underbrace{\frac{e^3}{8\pi h W_b} \bullet \frac{{V_b}^2}{a^2}}_{=J_0} \bullet \underbrace{\exp\left(-\frac{4a\sqrt{2m}{W_b}^{\frac{3}{2}}}{3\hbar e V_b}\right)}_{=P} \quad (1.2)$$

Here, V_b is the voltage drop between the transistor-channel and the floating-gate, and m the effective mass of the charge carriers. A tunneling process based on Fowler-Nordheim tunneling can be identified by plotting $\log(J/V) = f(\sqrt{V})$. In comparison for Schottky-emission one finds $\log(J) = f\left(V^{\frac{1}{4}}\right)$ and $J \sim T^2$.

Hot-electron injection allows for relatively fast programming of the floating-gate transistor within some tens of μs and below. However, hot-electron injection is an energy−/power-inefficient programming method (typ. 100–1000 pJ/bit) since a high drain-current is required and the injection efficiency is only about 1:10^5–1:10^6 (i.e. 1 out of $\eta = 10^5$ to 10^6 electrons is injected into the FG). For $N = 1000$ electrons that will be stored on the FG and a drain-voltage of V_D=6 V a write energy of approximately 1 nJ is required:

$$W_{\text{write}} \approx \eta e N V_D \approx 10^{-9}\,\text{J} \quad (1.3)$$

Moreover, this process only allows to be used for programming and cannot be used to erase the floating-gate, since electrons on the FG cannot gain sufficiently high kinetic energies. In contrast, Fowler-Nordheim tunneling can be used both for programming and erase. The process is extremely energy-efficient (~ 10 fJ/bit or even less) but slower compared to hot-electron injection (~100 μs vs. 10 μs). Both programming methods result in significant tunnel oxide degradation over time which limits the programming-endurance to some 10^4 to 10^5 cycles.

1.2.2 NOR- and NAND-Flash

Two types of Flash-topologies are mainly used: NOR- and NAND-flash. Fig. 1.5 depicts the NOR-topology. With regard to the horizontal middle-electrode line (i.e. the connection of all drain-contacts) the circuitry corresponds to a parallel (thus NOR-) connection of floating-gate cells.

Each floating-gate cell can be individually selected for read- (Fig. 1.5A) and write- (Fig. 1.5B) operation. In contrast to EEPROMs, all cells in an array are usually erased simultaneously (Fig. 1.5B), hence the name

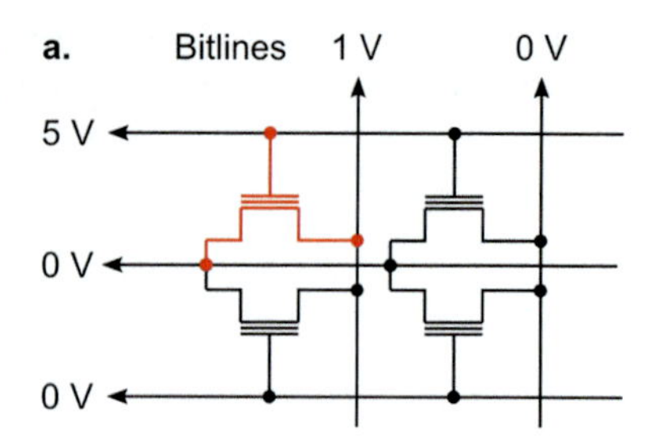

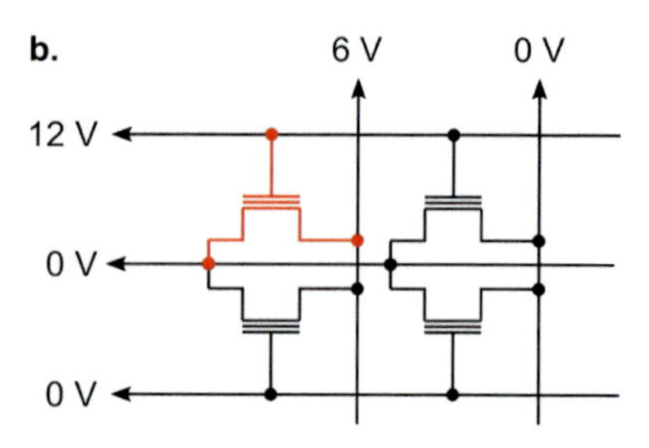

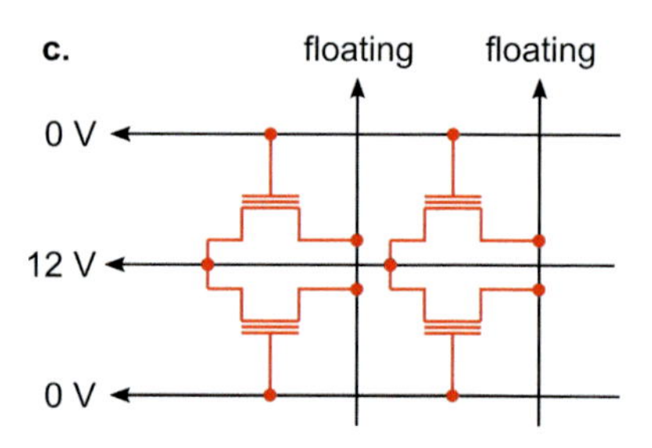

FIG. 1.5 NOR-Flash topology and examples of applied array voltages for (A) read, (B) write, and (C) array-erase. The selected cell(s) is/are highlighted in red, respectively.

Flash-memory. For read-operation the control-gate of the selected floating-gate cell is activated with e.g. a sense wordline voltage of 5V, while a drain-source-voltage of e.g. 1V (bitline) is applied. This corresponds to a relatively low or high drain-current depending on the floating-gate charge (see Fig. 1.3B). Write-access is similar to the read-operation but now the voltage-drop between drain-source channel and control-gate is larger. Write-operation is performed by hot-carrier injection (Eq. 1.1). Typical voltages are $\approx$ 5–6V for the drain-source- bitline-voltage and 12V for the wordline voltage.

The electrons on the floating-gate cannot be released by hot-carrier injection. Therefore, the erase-operation is based on Fowler-Nordheim tunneling. Since charge transport by FN tunneling in floating-gate transistors is typically slower than by hot-carrier injection, the whole array (block) is

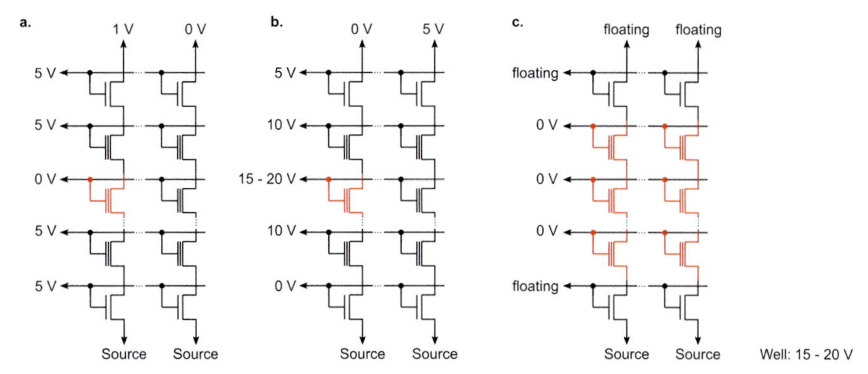

FIG. 1.6 NAND-Flash topology and examples of applied array voltages for (A) read, (B) write, and (C) array-erase. The selected cell(s) is/are highlighted in red, respectively. Note, for erase-operation the well-voltage is set to 15–20 V (in all other cases 0 V).

erased simultaneously. Here, no significant drain-current is needed and the bitlines remain floating, while a relatively large voltage drop between channel/substrate and control-gate is required for FN-tunneling.

The NAND-Flash topology (Fig. 1.6) corresponds to a series (and thus NAND-) connection of floating-gate transistors. Similar to NOR-Flash, individual cells can be selected for read- and write-operation, while the entire array is erased in parallel. However, due to the series connection of the floating-gate cells, the unselected cells within one bitline need to be conductive to drive a drain-current that is only determined by the stored floating-charge in the selected cell. Additionally, the floating-gate-cells are sandwiched between selector-field-effect-transistors.

For read-operation (see Fig. 1.6A), a bitline voltage of e.g. 1 V is applied. A control-gate-voltage of 0 V is applied to the selected memory cell. All other (unselected) floating-gate-cells and the selector-transistors are operated in ON-state by applying a voltage of e.g. 5 V to the control-gate terminals. In this case, the not-selected floating-gate-cells can drive a sufficiently high drain-current. The drain-current that is sensed to read the logic state is then only determined by the floating-gate induced shift of the transfer characteristic (Fig. 1.3B) of the selected cell. Write-operation is performed by Fowler-Nordheim tunneling (Fig. 1.6B). Here, a medium voltage of 10 V is applied to the not-selected floating-gate cells, while a high-voltage of typ. 15–20 V is applied to the selected cell. This configuration results in tunneling of electrons from the substrate (well) only into the selected floating-gate. To prevent programming of the cells in the unselected bitline a sufficiently high voltage of 5 V or larger is applied to the bitline. This allows the substrate to be shielded from the channel. Therefore, the bottom selector-transistors need to be in OFF-state, otherwise the vertical current through the unselected floating-gate transistors would result in a voltage drop across the drain- and source-terminals of each floating-gate cell, and hence, effectively reduce the shielding from the substrate. For erase-operation (Fig. 1.6C), the voltage-drop for tunneling is reversed compared to the write-operation (now, the well-voltage is typically 15–20 V). The gate-terminals of the selector-transistors are floating to avoid damage of their gate-oxides during erase-operation.

The main advantage of the NAND-topology is the high memory density compared to the NOR-configuration. NAND-Flash requires a memory footprint of 1–4 F^2/cell [18] while NOR-Flash has a larger footprint of 8–10 F^2/cell [36]. Another advantage is that conventional NAND-Flash is programmed solely by energy-efficient Fowler-Nordheim tunneling. Despite the high voltages required for programming, Fowler-Nordheim tunneling allows for an ultra-low programming energy of some fJ/bit. In contrast, write-operation for NOR-Flash is typically performed by hot-carrier injection, which results in a programming energy of ~100 pJ/bit to nJ/bit. However, Fowler-Nordheim tunneling is a relatively slow process with programming speeds in the order of 10^{-4} s, while NOR-Flash can be operated much faster (1 μs or below). Today NAND-Flash is the dominating topology for mass-storage, while NOR-Flash is used for niche products such as embedded applications.

1.2.3 Performance and scaling issues

An important property for non-volatile memories is the retention, that is how long information can be non-volatilely stored. The general retention requirement for memories is 10 years (=$3.2 \cdot 10^8$ s). The floating-gate and channel form a capacity C that is charged with N electrons. This charging results in a built-in voltage V_b' across the tunnel-oxide. To prevent a significant tunneling current, the built-in voltage needs to be well below the tunneling barrier energy, e.g. $V_b' < W_b/2e$. With a typical cell-size of $A = 20\,\text{nm} \cdot 20\,\text{nm}$ and a TOX thickness of $a = 5\,\text{nm}$ C is

$$C = \varepsilon_0 \varepsilon_r \frac{A}{a} \approx 3 \cdot 10^{-18}\,\text{F} \tag{1.4}$$

ε_0 and ε_r are the vacuum and relative permittivity (for SiO_2 $\varepsilon_r \approx 4$). With C the number N of electrons stored on the floating-gate can be calculated:

$$N \leq \frac{CV_b'}{e} = \frac{CW_b}{2e^2} \approx 27 \tag{1.5}$$

The retention time is determined by the leakage current, which results in a loss of stored electrons on the floating gate. Heisenberg's uncertainty principle $\Delta p \Delta x \geq \hbar/2$ with the momentum $\Delta p = mv = \sqrt{2mW_b}$ defines the theoretical thickness-limit $a_{min} \gg \Delta x \geq \hbar/2\sqrt{2mW_b}$ of the TOX-layer. In general, a thin oxide between the FG and channel results in a rectangular energy barrier through which electrons can tunnel with a current density:

$$J_T = \underbrace{\frac{e^2 V_b' \sqrt{2mW_b}}{ah^2}}_{=J_0} \bullet \underbrace{\exp\left(-\frac{2a\sqrt{2m}}{\hbar}\sqrt{W_b}\right)}_{=P} \text{ for } eV_b' \rightarrow 0 \text{ and } eV_b' \ll W_b \quad (1.6)$$

However, in this case the electrons stored on the floating-gate result in a voltage drop $eV_b' > 0$. Consequently, the energetic profile of the barrier is bend by V_b' and the leakage current can be expressed by tunneling through a trapezoidal barrier (for $eV_b' < W_b$, e.g. $eV_b' = 0.5 \bullet W_b$):

$$J_T = \underbrace{\frac{e^2 V_b' \sqrt{2m\left(W_b - \frac{eV_b'}{2}\right)}}{ah^2}}_{=J_0} \bullet \underbrace{\exp\left(-\frac{2a\sqrt{2m}}{\hbar}\sqrt{W_b - \frac{eV_b'}{2}}\right)}_{=P} \quad (1.7)$$

The retention time can be approximated by $t_r = Ne/I_T$. $I_T = A \bullet J_T$ is the total tunneling current. Thus, the current needs to be below 10^{-26} A for a retention time of 10 years. Note, in this case we assume that all electrons are released from the floating-gate. Evidently, for practical reasons only a fraction of N is allowed to be discharged, hence limiting the total tunneling current well below 10^{-27} A in our case. For SiO_2 ($W_b = 3$ eV, and $m = 0.5\, m_e$, with m_e the invariant electron mass) $I_T \ll 10^{-34}$ A, and for HfO_2 ($W_b = 1.5$ eV, $\varepsilon_r \approx 25$, $m = 0.15\, m_e$, and here $N \approx 83$) the tunneling current is $I_T \approx 10^{-21}$ A. Thus, HfO_2 does not fulfill the retention requirements for Flash-devices. High-*k* alternatives to SiO_2 are Al_2O_3 and Si_3N_4 which fulfill the retention requirements.

From Eq. (1.5) it is obvious that in ultra-small devices a small number of electrons is stored on the floating gate. This is critical especially for multi-level storage as fluctuation of even a small (single-digit) number of electrons can affect the logic information stored in a floating-gate-transistor. This so-called few-electron phenomena (Fig. 1.7) is one of the limitations for the cell-size of Flash memory [35]. Other short-channel limitations are also important, such as dopant-variability, drain-induced-barrier-lowering [37], and field-effect control of the channel, that limit the integration density of planar Flash memory technology.

In order to mitigate scaling limitations, vertical and three-dimensional Flash structures are part of current research. Some concepts are already in an even quite mature stage of mass-production. Vertical Flash allows to effectively increase the memory density (bits/μm^2) by stacking several memory cells on top of each other [38]. In contrast to NOR-Flash, the series connection of NAND Flash can be easily transferred into a vertical structure. A schematic circuitry of a vertical Flash device is shown in Fig. 1.8A. Several vertical Flash concepts have been reported in literature, including

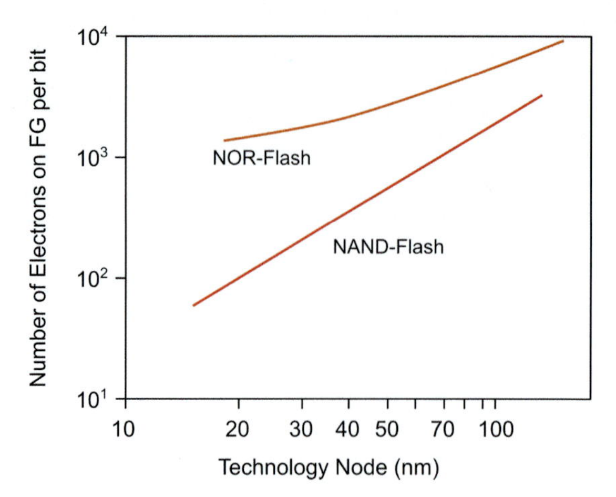

FIG. 1.7 Few electron phenomena for NOR- and NAND-Flash. The number of electrons stored on the floating-gate per bit decrease by further miniaturization. The trend lines are based on Refs. [11, 35].

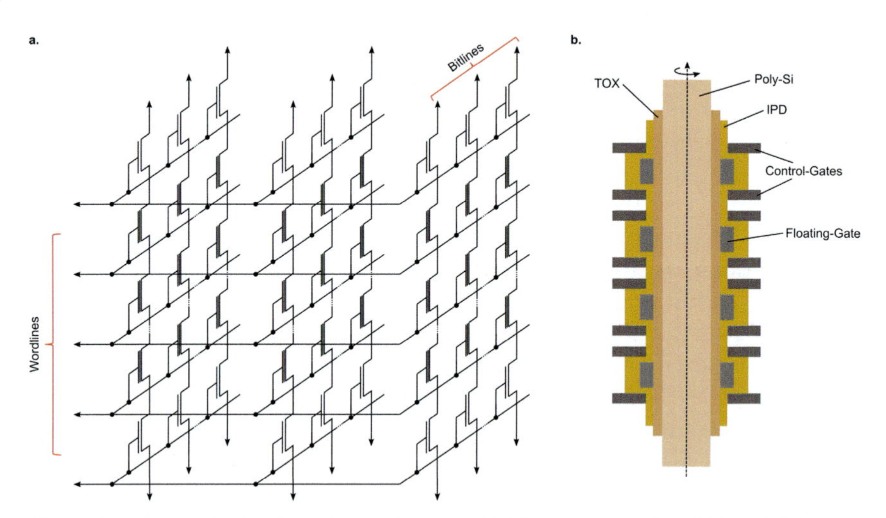

FIG. 1.8 Schematic circuitry of a vertical NAND-Flash topology [39] (A), and (B) cross-sectional view of a vertical dual control gate structure with surrounding floating-gate [40] (DC-SF Flash).

Terabit Cell Array Transistor Flash [41] (TCAT), Pipe-shaped Bit Cost Scalable (PBiCS) Flash [42], and Multi-Layered Vertical Gate NAND Flash [43]. Whang et al. demonstrated a vertical dual control gate structure with surrounding floating-gates (DC-SF) [40]. An illustration of the DC-SF Flash implementation is shown in the cross-sectional schematic in Fig. 1.8B. Note, selector-transistors are not shown for simplification. The floating-gate transistors have gate-all-around [44] (GAA) structures that allow ideal control of the field-effect. Each floating-gate cell is controlled by two control-gates, respectively. The memory cells are separated by a so-called interpoly-dielectric (IPD). The IPD also separates the control-gates from the floating gates. Since the transistor performance requirements (in particular the current density and charge velocity) for Flash are different compared to CMOS logic, polycrystalline Silicon (Poly-Si) can be used as channel without additional doping. A tunnel oxide such as Si_3N_4 or SiO_2 is used to separate the floating-gates from the Poly-Si nanowire channel. The Flash device can be operated in multi-level mode with a capacity of 2–4 bit/cell. Up to 128 cells can be stacked along the Poly-Si channel giving a total 2 Tb capacity.

1.3 Novel concepts for non-volatile memories

In the 2000s concerns about further Flash scaling triggered intensified research on alternative memory architectures. Though vertical integration of NAND-Flash relaxed scaling challenges, Flash technology still shows low endurance, slow write/erase-access and high energy consumption (in case of NOR-Flash). Flash also requires high programming voltages, which results in considerable circuitry overhead and additional energy consumption. Therefore, novel concepts for non-volatile memory (including ferroelectric [45], resistive [46], and magnetoresistive [47] devices) are part of ongoing research in academia and industry. The following section gives an overview of two emerging candidates for future non-volatile memory: resistive switches (RRAMs) and spintronic/magnetoresistive-devices (MRAMs). Alternative concepts such as ferroelectric and Mott-memories are covered in Chapters 8 and 10. Some devices based on the phenomena described in this sections were recently introduced for consumer products, including Panasonic's RRAM-based MN101 microcontroller. Further details on RRAMs are given in Chapters 2, 4, 11, and 14. A detailed introduction into MRAM-technology can be found in Chapter 9.

1.3.1 Resistive switches

Resistive switching random access memories are based on simple two-terminal metal/insulator/metal (MIM) structures [46]. By applying appropriate voltage pulses (SET-operation) between the two terminals the resistance of the device can be switched between at least a high resistive (HRS) and low resistive (LRS) state, respectively. Similar to Flash, intermediate resistance levels can be programmed, which allows to store multiple bit [48] in a single RRAM cell. The device can be switched to HRS during RESET-operation. For unipolar devices [49], SET and RESET have the sample polarity, whereas for bipolar devices, SET and RESET have different voltage polarities. Among the most important working principles are electrochemical metallization [50] (ECM, also known as conductive bridge RAM, CBRAM) and valence change mechanism [51] (VCM, also known as OxRAM). Devices based on either ECM or VCM are both bipolar devices. Their current-voltage hysteresis is shown in Fig. 1.9A for ECM-, and Fig. 1.9B for VCM-cells.

ECM-devices are based on an electrochemically active electrode such as Ag or Cu, an insulator, and an inert counter electrode like Pt. Metal oxides like Ta_2O_5 [52], dichalcogenides such as Ag:GeSe [53], SiO_2 [48], and iodides like AgI [54] are typically used as insulators. During SET-operation (Fig. 1.9a A, B), metal cations (e.g. Ag^+) are formed by anodic oxidation at the active electrode and migrate into the insulator. At the counter-electrode, these ions are reduced and electric-field enhancement results in formation of a thin metal filament (Fig. 1.9a C), which eventually bridges the two electrodes (LRS, Fig. 1.9a D). When the voltage polarity is reversed, RESET takes place, and the filament is (partly [55]) dissolved (HRS, Fig. 1.9a E). By tuning the current level during SET, the filament diameter and contact resistance is controlled, and the device resistance can be set between intermediate resistance levels.

In case of VCM-cells, high (e.g. Pt) and low (e.g. Zr) work function electrodes are used, respectively. The insulator is typically a transition metal oxide such as ZrO_x [1], Ta_2O_5 [56], HfO_2 [57] or TiO_2 [58]. During SET, electric-field driven relocation of oxygen vacancies takes place (Fig. 1.9a A–B), which results in growth of a nanoscale oxygen-deficient (thus almost metallic) path between the two electrodes. This electronically conductive path (plug) is most likely not forming a direct metallic contact to the high work function electrode. Instead, a Schottky-barrier is assumed to be formed. Movement of oxygen vacancies back and forth in this so-called disc could tune the Schottky-barrier's height/width [1]. This process decreases the overall device resistance (Fig. 1.9a C). During RESET, oxygen vacancies from the disc are removed and the change of the Schottky-barrier (thicker and/or higher) results in increase of the device resistance (Fig. 1.9a D).

1.3.1.1 Memristive devices

In 1971, L. Chua developed the theoretical concept of so-called memristors and memristive systems [59]. As can been seen in Fig. 1.10, by binary combination of the four basic physical quantities (blue): voltage V, current I, charge q and magnetic flux linkage Φ, three fundamental passive circuit elements (light-yellow), namely resistance R, capacitance C, and inductance L are defined. From symmetry considerations, a fourth, missing circuit element linking charge and magnetic flux is missing. L. Chua

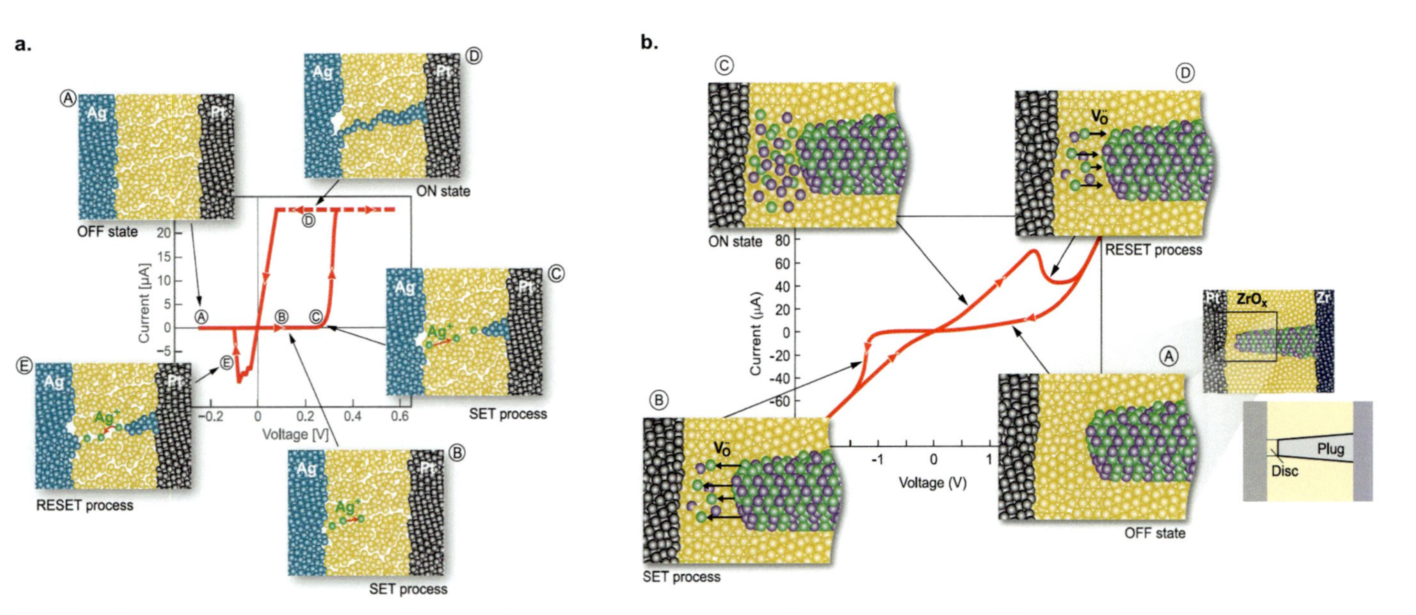

FIG. 1.9 Current/voltage-curves and schematics of the working principle of resistively switching of (a) ECM- and (b) VCM-devices, respectively. *Reprinted with permission from R. Waser, Nanoelectronics and Information Technology, 3rd ed. Wiley-VCH, 2012. Copyright 2012 John Wiley & Sons, Inc.*

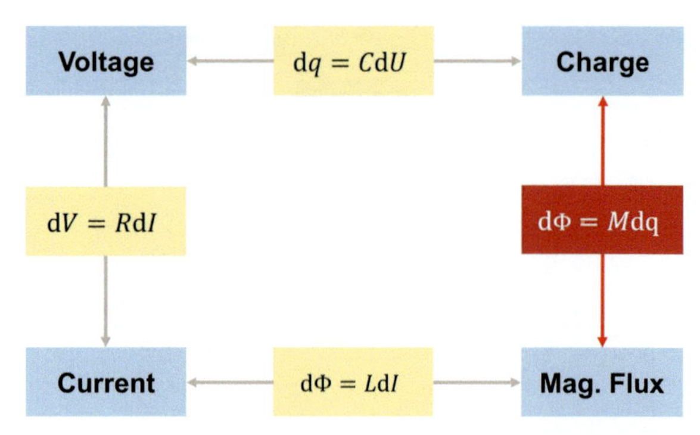

FIG. 1.10 The three fundamental passive circuit elements: resistance R, capacitance C, inductance L, and a new forth device, the so-called memristor with memristance M.

introduced a new passive circuit element (red), the memristor that can be described by a memristance M.

The memristance links the derivative of the magnetic flux linkage and charge [59]:

$$M(q(t)) = \frac{\mathrm{d}\Phi(t)}{\mathrm{d}q(t)} = \frac{\frac{\mathrm{d}\Phi}{\mathrm{d}t}}{\frac{\mathrm{d}q}{\mathrm{d}t}} = \frac{V(t)}{I(t)} \tag{1.8}$$

In the following, Φ and q are defined mathematically as magnetic flux linkage and charge without any further physical interpretation for simplification. We assume a sinusoidal current $I(t) = A\sin\omega t$ (with ω being the current's angular frequency, and A an arbitrary factor) and a cubic relation between Φ and q, $\Phi \sim 1 + \frac{1}{3}q^3$ [60]. With the charge $q = \int I(t)\ \mathrm{d}t$ the charge and magnetic flux linkage reads:

$$\begin{aligned} &\to q(t) = \frac{A}{\omega}(1 - \cos\omega t) \\ &\to \Phi(t) \sim \frac{A}{\omega}(1 - \cos\omega t) \cdot \left[1 + \frac{A^2}{3\omega^2}(1 - \cos\omega t)^2\right] \end{aligned} \tag{1.9}$$

The relation between voltage and magnetic flux linkage is $\Phi(t) \sim \int V(t)$ $\mathrm{d}t$. Therefore, a frequency-dependent voltage/current hysteresis is found:

$$\to V(t) \sim A\left[1 + \frac{A^2}{\omega^2}(1 - \cos\omega t)^2\right] \sin\omega t = I(t) \bullet \left[1 + \frac{A^2}{\omega^2}(1 - \cos\omega t)^2\right] \tag{1.10}$$

Such an I/V-hysteresis can be demonstrated by resistive switches (Fig. 1.9). Research on resistive switches started in the 1960s [7,61] and intensified in the 2000s [13,62,63]. In 2008, HP labs reported that memristor-functionality can be demonstrated using a TiO_2-based resistive switch [16]. It is noteworthy that at the same time, application of resistive switches for neuromorphics [64–66], such as spike timing dependent plasticity (STDP) [67], started, which paved the way to a new research field beyond classical memories. Such applications were already discussed in the context of memristors by L. Chua in 1971 [59,68]. Today, the term memristor or in more general memristive system [69] is synonymously used for resistive switches [60,70] despite concerns [71,72] that memristors would violate the Landau-principle for pure memory operation [73,74]. Moreover, memristors show the signature of a nanobattery effect that makes these devices essentially active (see section below) [75]. This effect depends on the ambient [76] and the nanobattery may limit the retention of practical devices [77]. However, it could be also used for beyond-memory applications such as bio-chemical sensors [78].

1.3.1.2 Nanoionic effects

The processes responsible for the resistance transition are of electrochemical nature [46] on the nanoscale and are termed *nanoionic effects*. They include cation/anion drift [79], (partial-) redox reactions [80], involvement of countercharges [76,81–85], nucleation and charge-transfer effects [86], and phenomena that are similar to those in (nano-) batteries [75]. In case of VCM-devices, the migration of oxygen-vacancies is additionally accelerated by Joule heating [87]. This chapter focusses on ECM-devices, a detailed description of VCM-devices in given in Chapter 2.

A simple electrode/electrolyte/electrode system is depicted in Fig. 1.11A, where the anode and cathode are made of silver, and the electrolyte is a crystalline solid silver ion conductor (e.g. silver iodide, AgI). AgI is here used as a model material system as the defect concentration (Ag^+-concentration) is almost constant during operation. Therefore, diffusion rate effects due to a concentration change of mobile ions do not need to be considered. At the electrode/electrolyte interface a double layer is formed with a thickness of typically some nm. It is composed of a rigid thin Helmholtz plane and a diffuse layer within the electrolyte (Fig. 1.11B). When a positive voltage is applied between the anode and cathode, mobile silver cations Ag^+ are formed at the anode (anodic oxidation) and are reduced at the cathode. Note, for nanoscale devices such as ECM- and VCM devices, the diffuse layer of both electrodes overlap, and the behavior is considerably different to macroscopic devices with large bulk electrolytes.

In this case the anodic reaction obeys

$$\mathrm{Me} \rightarrow \mathrm{Me}^{z+} + z\mathrm{e}^- \tag{1.11}$$

where Me = Ag or Cu and z is the charge number of the dominating ion species ($z = 1$ for Ag [88,89], and $z = 1$ or 2 for Cu [88]). The reduction of cations at the cathode is:

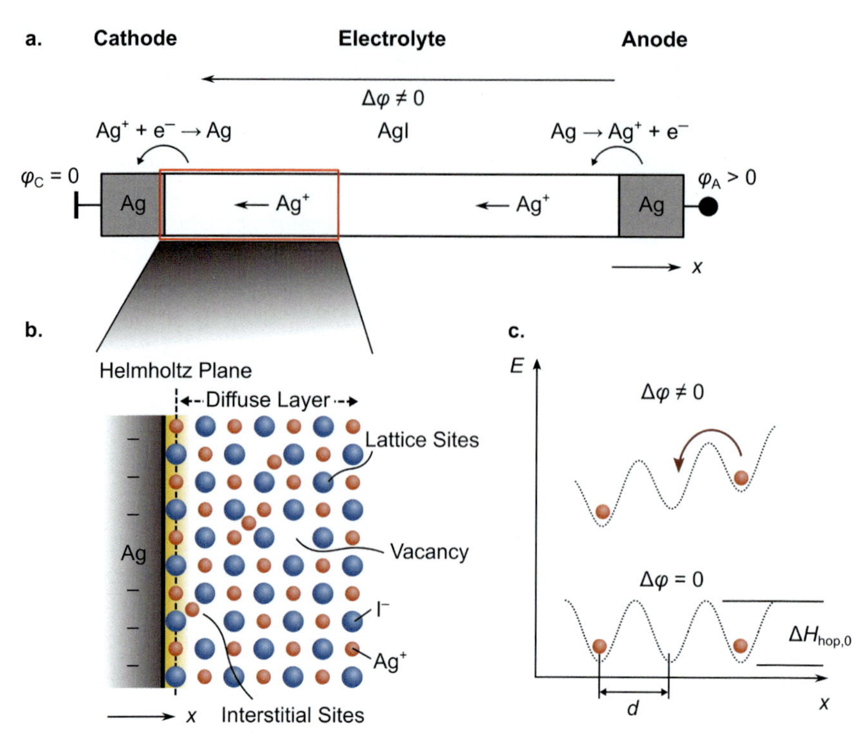

FIG. 1.11 Ion conduction in solid electrolytes. (A) A simple Ag/AgI/Ag system is considered here with $\Delta\varphi = \varphi_A - \varphi_C \neq 0$ (φ_A, φ_C are the cathode and anode overpotential, respectively). (B) Schematic of the electrode/electrolyte interface for AgI (zoom into interface from (A)). At the interface a double layer is formed in the electrolyte. Lattice sites, vacancies and interstitial sites are shown exemplarily. (C) Hopping migration of a cation (red particle) assuming a periodic energy barrier ΔH_{hop} (dotted line) for $\Delta\varphi = 0$ and $\Delta\varphi \neq 0$ and a field-accelerated process with $kT \ll ze\,\Delta\varphi$.

$$Me^{z+} + ze^- \rightarrow Me \tag{1.12}$$

In a simplified picture, the injection of cations into the electrolyte would require presence of vacancies nearby the anode interface. In a defect-free electrolyte the current driving these redox reactions would thus immediately diminish. However, in practice, ions can be incorporated on interstitial position. In this case no vacancies are required. Moreover, a small fraction of Ag atoms is always slightly displaced from their rigid lattice sites and occupy interstitial positions. Within the electrolyte ions can in principle move between I. unoccupied lattice sites (vacancies), II. interstitial lattice sites, or III. from interstitial lattice sites to lattice sites. Cation migration via interstitial sites (with $[Me_I^\bullet]$ being the cation concentration on interstitial sites and $[V_{Me}']$ the concentration of cation vacancies, i.e. according to Frenkel disorder [90] $[V_{Me}'] = [Me_I^\bullet]$) is energetically more favorable than migrating from a lattice site to a vacancy. The hopping process is thermally activated with an (average) energy barrier $\Delta H_{\text{hop},0}$ and a characteristic (average) hopping distance d. It can be described by a Mott-Gurney model [91]. By applying a potential-gradient $\Delta\varphi \neq 0$ (Fig. 1.11C), the effective energy barrier ΔH_{hop} decreases:

$$\Delta H_{\text{hop}} = \Delta H_{\text{hop},0} - \alpha_{\text{hop}} e \Delta\varphi \tag{1.13}$$

Here, α_{hop} is a barrier lowering coefficient [92]. Eq. (1.13) applies for high electrical fields. For low electrical fields, a diffusion-process is driven by a concentration gradient between the two electrodes. In a crystalline system such as AgI the hopping distance and energy barrier may be similar for all hopping-sites, but in some other materials such as amorphous SiO_2 the hopping process depends on the morphology and defects (like dangling bonds) on the nanoscale. It is challenging to experimentally determine d but it will be in the order of interatomic distances. The hopping distance is therefore a fit parameter in simulation models and a good measure to judge how realistic the model is.

For some ion conductors such as AgI, the degree of defect disorder can abruptly increase above a transition temperature (for AgI a transition from γ-AgI to α-AgI is observed at $T = 420\,\text{K}$) [93] and the cation sub-lattice is in a quasi-liquid phase. This results in a superionic conductivity of these materials at higher temperatures, reaching conductivity values of some liquid electrolytes. This superionic phase can be even observed at room temperature when the lattice is disturbed in ternary systems (e.g. in $RbAg_4I_5$) [94].

Due to the electrostatic force the concentration of mobile charges is reduced in the immediate vicinity of a charge carrier with same polarity. This is described by the Debye-Hückel theory [95] and the Debye length (for mobile species)

$$\lambda_D = \sqrt{\frac{\varepsilon_r \varepsilon_0 R T}{2 F^2 z^2 c_\infty}} \tag{1.14}$$

is a measure how far this charge screening persists. Here, R is the ideal gas constant, T is the absolute temperature, F is the Faraday constant and c_∞ is the mobile charge concentration of the bulk. λ_D is typically in the order of a nanometer, e.g. for AgI $\lambda_D = 1.33\,\text{nm}$ with only $[\text{V}_{\text{Ag}}^-] = [\text{Ag}_\text{I}^+] \approx 0.007\,\text{mmol/cm}$ [3] (i.e. about 0.06% of all Ag^+ ions are mobile and contributing to the ionic conduction) [90]. Some salts like silver chloride show rather long Debye lengths due to their extremely low mobile defect concentration (for AgCl $\lambda_D = 24.5\,\text{nm}$ with $[\text{Ag}]/[V_{\text{Ag}}^-] = 7 \cdot 10^{-7}$, $[V_{\text{Ag}}^-] \approx 14\,\text{nmol/cm}^3$ and $\varepsilon_r = 7$) [90].

For silver iodide, charge screening of the cations is provided by the iodine ions. In contrast to Ag^+, I^- ions are immobile but the charge screening length can be still estimated by Eq. (1.14) since AgI is stoichiometric. In oxide-based switching materials, the electrochemical conditions may be described by a special case of immobile charge carries and depleted mobile charges. In this case the charge screening length may be expressed by a Mott-Schottky approximation. Thus, the extension into the electrolyte can be in the order of some tens of nanometers, which is the thickness of most resistive switching materials [96,97]. In that case the ion-conduction can be significantly increased due to a pure nano-size effect [97,98]. The CaF_2/BaF_2 superlattice is an example for a well-studied fast fluorine ion conductor at room temperature [99].

AgI is a classical macroscopic electrolyte and mainly used as a model material system. For practical reasons, oxides like SiO_2, Ta_2O_5 or Al_2O_3 and also higher dichalcogenides Ag-GeS [100] or Ag-GeSe [101] are used for ECM-devices. These material systems are insulators in the macroscopic scale but become electrolytes or mixed ionic-electronic conductors on the nanoscale, e.g. by exploiting fast diffusion along grains, dangling bonds, interfaces or by overlapping of the diffuse layers (Fig. 1.11).

1.3.1.3 *Switching kinetics and energy consumption*

Similar phenomena as the electrochemical effects in ECM-cells can be observed in VCM-cells. In addition, both cation- and anion-conduction in VCM-cells may take place depending on the operation conditions [79]. The aforementioned effects are typically very sensitive to the applied voltage (including overpotentials that drop within the device across interfaces), the film and electrode morphology, stoichiometry and (local) temperature. For ECM-cells operation with ultra-low currents down to pA have been demonstrated [53]. Therefore, for ECM-cells Joule-heating is potentially not required for resistance transition in relatively short timescales ($\ll$ms). In contrast, for VCM-cells switching assisted by ultra-fast localized temperature gradients (with rise-times in the order of ps) have been discussed in literature [20,87]. In general, these effects result in a strong non-linear switching kinetics, that is, the switching time depends exponentially on the applied voltage pulse amplitude. For ECM-cells, switching between 10^5 s down to 10^{-9} s have been reported by changing the voltage amplitude between <0.25 and 5 V for several material systems (Fig. 1.12) [21]. These non-linear switching kinetics are advantageous, because unintentional switching at low voltage bias is suppressed (long resistance transition times) while the devices allows for fast switching at moderate voltages. Several processes can determine the shape and slope(s) of the switching kinetics, including filament nucleation effects [86,102,103], charge transfer and ion-drift [89,104].

Nowadays, resistive switches can be scaled down to ultra-low volumes with dimensions <10 nm [105]. The signature of quantum conductance levels even indicate a scalability to an almost atomic level [93,106,107]. The switching energy of typically some pJ is larger than for NAND-Flash. However, a theoretical study projects a minimum switching energy down

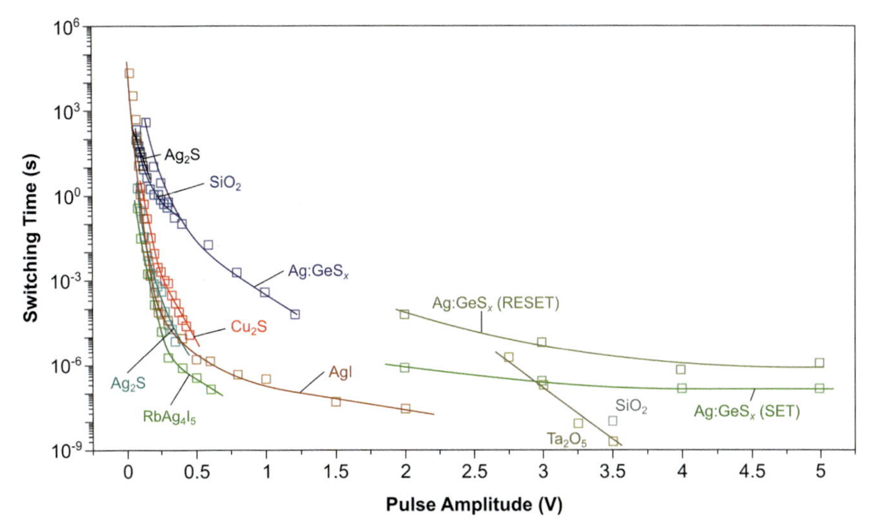

FIG. 1.12 Switching kinetics of several ECM-type resistive switches. *Redrawn from W. Chen, S. Tappertzhofen, H.J. Barnaby, M.N. Kozicki, SiO_2 based conductive bridging random access memory, J. Electroceram. (2017). doi:10.1007/s10832-017-0070-5 with permission from the Springer Nature.*

to some fJ, which is comparable to charging of a CMOS-gate [19]. In contrast to Flash, fast switching below 10 ns with high endurance up to 10^{12} cycles has been demonstrated [20,56,86,108–110], making resistive switches ideal candidates for storage class memories.

1.3.2 Magnetoresistive random access memories

Magnetoresistive RAM (MRAM) devices make use of magnetic effects and properties, in particular spin-based scattering and spin-polarization of electric current, for information processing and memory operation. Magnetoresistance effects are an ongoing part of research in the context of spintronics. Of technical importance are the giant magnetoresistance (GMR) and tunnel magnetoresistance (TMR) effects. The GMR-effect has been independently discovered in magnetic heterostructures and superlattices by Albert Fert [9] and Peter Grünberg [8] in 1988 and 1989. Today, magnetic hard disk drives are based on the GMR-effect. The TMR-effect was discovered by Michel Jullière in 1975 but did not attract high attention until the 2000s due to a relatively small resistance change at cryogenic conditions [111].

1.3.2.1 Giant magnetoresistance

The GMR-effect can be observed in a heterostructure such as Fe/Cr/Fe (Fig. 1.13A) composed of two ferromagnetic layers (Fe) separated by a thin (~1 nm) non-magnetic layer (Cr). The thickness of the individual layers is

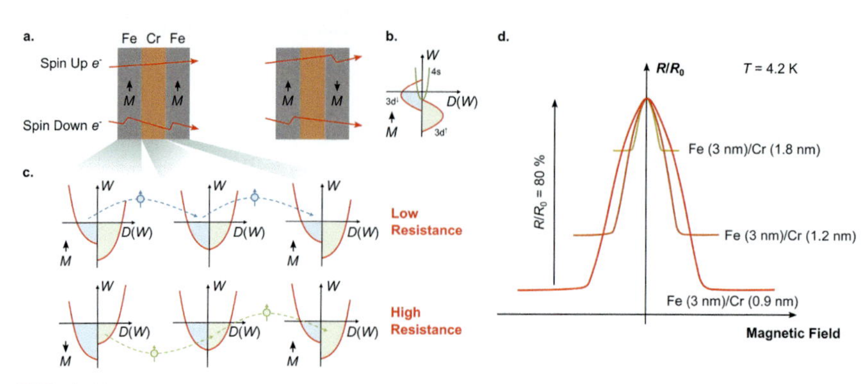

FIG. 1.13 Description of the giant magnetoresistance effect. (A) Cross-section of a Fe/Cr/Fe spin-valve with different magnetizations. (B) Splitting of the density of states in a ferromagnet. (C) Electronic transport (low and high resistance) explained by splitting of the d-orbitals. (D) Resistance vs. external magnetic field characteristics of a GMR-device. *Redrawn from A. Fert, The present and the future of spintronics, Thin Solid Films 517 (2008) 2–5 with permission from Elsevier B.V.*

critical and a breakthrough that allowed to fabricate such thin layers with high quality was the molecular beam epitaxy (MBE). The working principle is as follows: the electric current is composed of both Spin Up and Spin Down electrons. When passing a ferromagnetic layer with magnetization in a certain direction (Fe) spin-depended scattering takes places. In the example of Fig. 1.13A, Spin Up electrons are relatively less scattered compared to Spin Down electrons (as indicated by red arrows). If the magnetization of both Fe layers would be in the opposite direction, Spin Up electrons would be more scattered than Spin Down electrons. In both configurations, i.e. when the magnetization in both layers is in parallel, a relatively low resistance is measured. In contrast, when the magnetization in both Fe-layers is antiparallel (opposite directions), the electrons are scattered either in the right or left Fe-layer resulting in an overall relatively high resistance.

The spin-dependent scattering results from splitting of the density of states $D(W)$ (Fig. 1.13B), where W is the energy. Due to their lower effective mass, electric current is driven mainly by *s*- and *p*-orbital electrons. In ferromagnets such as Fe, electrons occupy free states in the *d*-orbital after scattering. Note, the spin-polarization is not affected by the scattering process. The electrons in the *d*-orbital have larger effective mass. In a ferromagnet, the dispersion relation $E_\uparrow(k) = E(k) - In_\uparrow/N$ of Spin Up $n_\uparrow$, and $E_\downarrow(k) = E(k) - In_\downarrow/N$ of Spin Down $n_\downarrow$ electrons differ from the dispersion relation of spinless electrons $E(k)$. N is the number of lattice sites and I the Stoner parameter [112]. It follows that the *d*-orbital is split for Spin Up and Spin Down electrons. For majority electrons, that are electrons with a spin in the same direction as the magnetization in Fig. 1.13B, the Fermi-level is in the *sp*-orbital. In contrast, for minority electrons, the Fermi-level is in a hybridized *spd*-orbital with a higher density of states, and thus, higher scattering probability. By combination of a Fe/Cr/Fe layer stack, this phenomenon can be used to design a so-called spin-valve. In case of a parallel magnetization of both Fe-layers (Fig. 1.13C, top), majority electrons can pass the spin-valve with relatively low scattering. In contrast, for an antiparallel magnetization of both Fe-layers (Fig. 1.13C, bottom) the high density of state at the Fermi-level results a high scattering probability.

The layer-structure in Fig. 1.13A is the simplest case of a spin-valve. Here, the non-magnetic layer ensures that the magnetization (orientation) in both Fe-layers can be different. For example, the magnetization in the left Fe layer could be fixed (hard magnetic, sometimes called references layer) while the magnetization of the right Fe-layer (soft magnetic, sometimes called storage layer) could be relatively easily changed by an external magnetic field. Thus, the resistance of the devices depends on the orientation of the soft magnetic layer. The corresponding resistance vs. external magnetic field is depicted in Fig. 1.13D for various Cr-thicknesses. In case of hard disks, bits are encoded by different magnetic orientations. These bits result in a localized pattern of an external magnetic field that can manipulate the magnetic orientation of the soft magnetic layer. In this case, the GMR spin-valve is used as a magnetic sensor to read the locally magnetically encoded information.

1.3.2.2 Tunnel magnetoresistance

While the GMR-effect is of high importance for magnetic hard disk drives, the tunnel magnetoresistance effect is used for MRAMs. Here, the core-element is a magnetic tunneling junction (MTJ). The qualitative description of the TMR-effect is very similar in a sense to the GMR-effect. In contrast to the GMR-effect, where the density of states splitting resulted in different scattering probabilities, for the TMR-effect the number of free states that can be occupied after tunneling is critical. The number of free states and therefore the tunneling probability depends again on the *d*-orbital splitting. The tunneling probability is high when the magnetization in both ferromagnetic layers is in parallel, otherwise, the probability is low for an antiparallel alignment.

A schematic of a conventional MRAM memory array is illustrated in Fig. 1.14A. The actual memory cell is composed of an MTJ, a selector-transistor, and a bit-, digit- and word-line. The selector-transistor is controlled by the word-line. The digit-line is required for write-operation. The logic states are encoded by different resistance levels, which can be programmed by an external magnetic field (Fig. 1.14B). The resistance hysteresis ensures non-volatility and prevents from unintentional state distribution by smaller magnetic fields of neighboring cells. For read-operation (Fig. 1.14C) the selector-transistor is active and the resistance

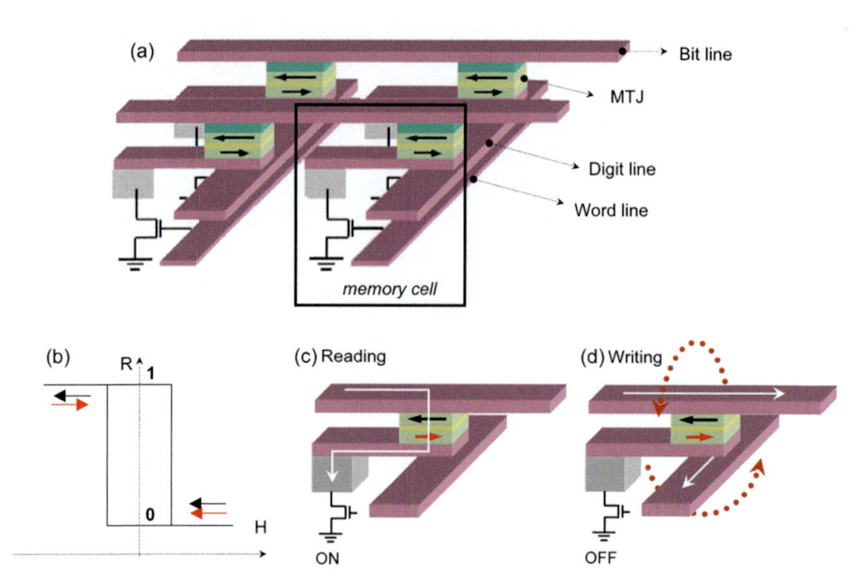

FIG. 1.14 (A) Schematic of a conventional MRAM. (B) Resistance vs. external magnetic field H hysteresis to encode two logic states "0" and "1," respectively. (C) Reading and (D) writing operation. *Reprinted from I.L. Prejbeanu, M. Kerekes, R.C. Sousa, H. Sibuet, O. Redon, B. Dieny, J.P. Nozières, Thermally assisted MRAM, J. Phys. Condens. Matter 19 (2007) 0–23 with permission from IOP Publishing.*

of the device is probed by driving a small current through the MTJ. For write-operation, an external magnetic field is required, which is generated by driving orthogonal currents through the bit-line and digit-lines. Now the selector-transistor is switched off. The orthogonal arrangement of the external magnetic field is used for a selective write-access to an individual memory cell without disturbing neighboring cells.

This concept has significant drawbacks. At first, the digit-line results in a large memory cell footprint of 20–30 F^2. Due to uniformity-challenges, those devices could not be scaled below 0.1 μm^2 [113,114]. Another issue is related to the stability and power efficiency. As a consequence of the ongoing miniaturization, the energy between two stable magnetizations must be relatively high compared to the thermal energy for non-volatile operation. This results in large write-current to ensure sufficiently high magnetic fields for programming. Hence, further miniaturization results in an increase of the write-currents up to 20 mA for a technology node of 100 nm [113].

A significant improvement was the invention of a thermally assisted write-operation. Here an MTJ with a small so-called blocking temperature is used, that is, the field to switch the MTJ is decreased by heating in a temperature range of up to 200 °C. A sufficiently high current driven across the junction heats a ferromagnetic storage layer above the Curie-temperature (Joule heating). Upon cooling, the magnetization is switched to encode the logic state. The difficulty is to be able to easily switch the magnetization at a still relatively moderate temperature of 200 °C while keeping the logic state stable in standby operation. This is technically achieved by pinning the magnetization of the storage layer (e.g. CoFe) with an antiferromagnetic layer (e.g. IrMn) [113]. The pinning fixes the magnetization of the ferromagnetic layer in standby. After heating above the blocking temperature of the antiferromagnetic layer, the storage layer can be switched. The resulting switching energy is significantly smaller compared to conventional MRAM.

1.3.2.3 Spin-transfer torque MRAM (STT-MRAM)

A further improvement was the development of Spin-Transfer Torque MRAM. STT-MRAM does not require a digit-line and in a 1T1R (1 transistor, 1 resistor) configuration the footprint is as low as 6 F^2. This is identical to other competing technologies such as 1T1R RRAM. A simplified schematic of the write-operation for writing a logic "0" and "1" is shown in Fig. 1.15A (including the selector-transistor). The basic idea of STT-MRAM is to switch the magnetization of a storage layer in a MTJ purely by changing the polarity of an appropriate switching current.

To switch the magnetization the current needs to be spin-polarized. This can be achieved when the current passes a relatively thick magnetic layer with a fixed magnetization. This spin-polarized current can switch the orientation of the magnetization of a thin ferromagnetic layer. Therefore, besides a thin storage layer stack with switchable magnetization an STT-MRAM cell (Fig. 1.15B) consists of two thick polarization layers with opposite magnetic orientations. Depending on the polarity of the switching current the current is either Spin Up or Spin Down polarized [115]. One of these polarization layers can be also used as a reference layer with fixed magnetization to provide the TMR-effect, that is, the resistance of the memory cell depends on the orientation of the storage layer in respect to the reference layer. The coupling is achieved using MgO-layers to provide a large magnetoresistance [116].

STT-MRAM can be operated with switching current densities in the order of 10^6 A/cm^2 (i.e. some tens of μA for nanoscale devices) [1]. This results in a switching energy of some pJ/bit or below [117]. STT-MRAM shows a remarkably high endurance of up to 10^{15} cycles [18], and programming times as low as 1–10 ns [117]. The endurance is mainly limited by voltage stress of the MTJ with a breakdown voltage of approximately 1 V. A challenge is the integration of the complex material stack (Fig. 1.15B) including materials like Fe, Co, and Pd in semiconductor processes. STT-MRAM can be easily scaled down to 45 nm. However, thermal stability for non-volatile operation is a concern in case of further scaling. This could be

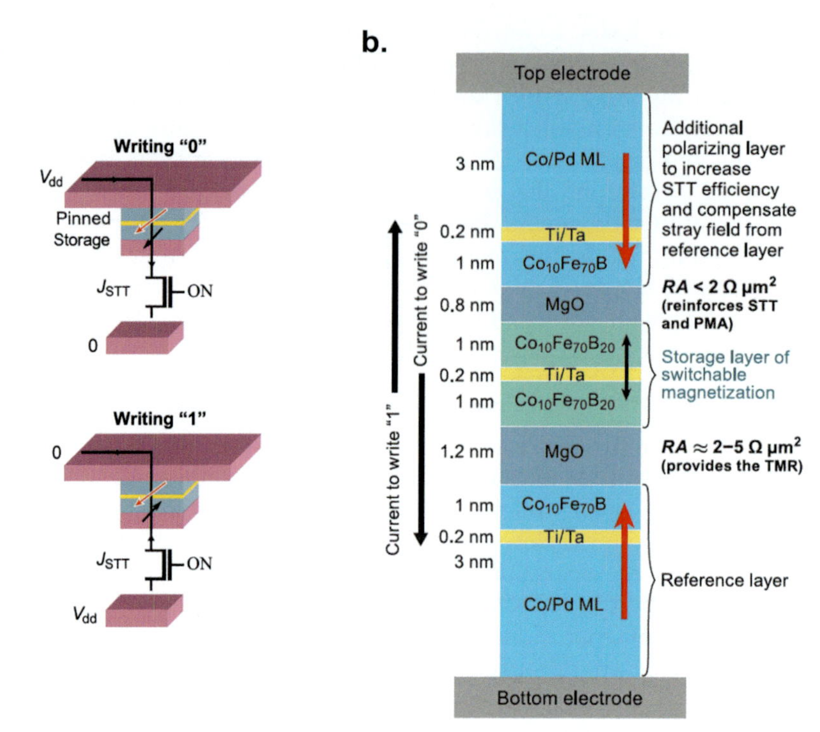

FIG. 1.15 Concept of a Spin-Transfer Torque MRAM. (A) Simplified schematic of the write-operation for STT-MRAM. (B) Layer stack of an STT-MRAM device (without selector-transistor). *Reprinted with permission from R. Waser, Nanoelectronics and Information Technology, 3rd ed., Wiley-VCH, 2012. Copyright 2012 John Wiley & Sons, Inc.*

mitigated by implementation of combined spin-transfer torque and thermally assisted write-operation as described above [115].

Acknowledgements

The author would like to thank I. Valov, D. Merten, A. Gumprich and R. Ahlmann for comments and proof-reading

References

[1] R. Waser, Nanoelectronics and Information Technology, third ed., Wiley-VCH, 2012.
[2] J. Kilby, Miniaturized Electronic Circuits, US Pat. US3029366A, 1959.
[3] R. Noyce, Semiconductor Circuit Complex Having Isolation Means, US Pat. US3150299A, 1959.
[4] L. Kurt, Multiple Semiconductor Assembly, US Pat. US3029366A, 1959.
[5] C. Brady, D. Apollo, Guidance and navigation electronics, IEEE Trans. Aerosp. AS-3 (1965) 354–362.
[6] E. Pugh, Ferrite core memories that shaped an industry, IEEE Trans. Magn. 20 (1984) 1499–1502.

[7] W.R. Hiatt, T.W. Hickmott, Bistable switching in niobium oxide diodes, Appl. Phys. Lett. 6 (1965) 106–108.

[8] G. Binasch, P. Grünberg, F. Saurenbach, W. Zinn, Enhanced magnetoresistance in layered magnetic structures with antiferromagnetic interlayer exchange, Phys. Rev. B 39 (1989) 4828–4830.

[9] M.N. Baibich, J.M. Broto, A. Fert, F.N. Van Dau, F. Petroff, P. Etienne, G. Creuzet, A. Friederich, J. Chazelas, Giant magnetoresistance of (001)Fe/(001)Cr magnetic superlattices, Phys. Rev. Lett. 61 (1988) 2472–2475.

[10] F. Masuoka, M. Momodom, Y. Iwata, R. Shirota, New ultra high density EPROM and flash EEPROM with NAND structure cell, in: 1987 International Electron Devices Meeting, IRE, 1987, pp. 552–555, https://doi.org/10.1109/IEDM.1987.191485.

[11] ITRS, Edition 2013, 2013. http://www.itrs.net/.

[12] C.S. Yang, D.S. Shang, N. Liu, E.J. Fuller, S. Agrawal, A.A. Talin, Y.Q. Li, B.G. Shen, Y. Sun, All-solid-state synaptic transistor with ultralow conductance for neuromorphic computing, Adv. Funct. Mater. 28 (2018) 1–10.

[13] M.N. Kozicki, M. Park, M. Mitkova, Nanoscale memory elements based on solid-state electrolytes, IEEE Trans. Nanotechnol. 4 (2005) 331–338.

[14] M. Mitkova, M. Kozicki, Silver incorporation in Ge-Se glasses used in programmable metallization cell devices, J. Non-Cryst. Solids 299 (2002) 1023–1027.

[15] K. Tsunoda, Y. Fukuzumi, J.R. Jameson, Z. Wang, P.B. Griffin, Y. Nishi, Bipolar resistive switching in polycrystalline TiO[sub 2] films, Appl. Phys. Lett. 90 (2007) 113501.

[16] D.B. Strukov, G.S. Snider, D.R. Stewart, R.S. Williams, The missing memristor found, Nature 453 (2008) 80–83.

[17] Y. Fujisaki, Review of emerging new solid-state non-volatile memories, Jpn. J. Appl. Phys. 52 (2013), 040001.

[18] J.J. Yang, D.B. Strukov, D.R. Stewart, Memristive devices for computing, Nat. Nanotechnol. 8 (2012) 13–24.

[19] D. Ielmini, Filamentary-switching model in RRAM for time, energy and scaling projections, in: 2011 International Electron Devices Meeting, IEEE, 2011, pp. 17.2.1–17.2.4, https://doi.org/10.1109/IEDM.2011.6131571.

[20] M.D. Pickett, R.S. Williams, Sub-100 fJ and sub-nanosecond thermally driven threshold switching in niobium oxide crosspoint nanodevices, Nanotechnology 23 (2012) 215202.

[21] W. Chen, S. Tappertzhofen, H.J. Barnaby, M.N. Kozicki, SiO_2 based conductive bridging random access memory, J. Electroceram. (2017), https://doi.org/10.1007/s10832-017-0070-5.

[22] M.-J. Lee, C.B. Lee, D. Lee, S.R. Lee, M. Chang, J.H. Hur, Y.-B. Kim, C.-J. Kim, D.H. Seo, S. Seo, et al., A fast, high-endurance and scalable non-volatile memory device made from asymmetric Ta_2O(5-x)/TaO(2-x) bilayer structures, Nat. Mater. 10 (2011) 625–630.

[23] A. Fumarola, P. Narayanan, L.L. Sanches, S. Sidler, J. Jang, K. Moon, R.M. Shelby, H. Hwang, G.W. Burr, Accelerating machine learning with non-volatile memory: exploring device and circuit tradeoffs, in: 2016 IEEE Int. Conf. Rebooting Comput. ICRC 2016—Conf. Proc, 2016, https://doi.org/10.1109/ICRC.2016.7738684.

[24] R.F. Freitas, W.W. Wilcke, Storage-class memory: the next storage system technology, IBM J. Res. Dev. 52 (2008) 439–447.

[25] G.W. Burr, B.N. Kurdi, J.C. Scott, C.H. Lam, K. Gopalakrishnan, R.S. Shenoy, Overview of candidate device technologies for storage-class memory, IBM J. Res. Dev. 52 (2008) 449–464.

[26] C.H. Lam, Storage class memory, in: ICSICT-2010–2010 10th IEEE Int. Conf. Solid-State Integr. Circuit Technol. Proc, 2010, pp. 1080–1083, https://doi.org/10.1109/ICSICT.2010.5667551.

[27] M. Weiland, B. Homölle, Usage scenarios for byte-addressable persistent memory in high-performance and data intensive computing, J. Comput. Sci. Technol. 36 (2021) 110–122.

[28] I.B. Peng, M.B. Gokhale, E.W. Green, System evaluation of the Intel optane byte-addressable NVM, in: Proceedings of the International Symposium on Memory Systems, ACM, 2019, pp. 304–315, https://doi.org/10.1145/3357526.3357568.
[29] K. Ishimaru, Future of non-volatile memory—from storage to computing, in: 2019 IEEE International Electron Devices Meeting (IEDM) 2019-Decem, IEEE, 2019, pp. 1.3.1–1.3.6.
[30] T. Mason, T. Dimitra Doudali, M. Seltzer, A. Gavrilovska, Unexpected performance of Intel® Optane™ DC persistent memory, IEEE Comput. Archit. Lett. 19 (2020) 55–58.
[31] G.W. Burr, M.J. BrightSky, A. Sebastian, H.Y. Cheng, J.Y. Wu, S. Kim, N.E. Sosa, N. Papandreou, H.L. Lung, H. Pozidis, et al., Recent progress in phase-change memory technology, IEEE J. Emerging Sel. Top. Circuits Syst. 6 (2016) 146–162.
[32] R. Bez, E. Camerlenghi, A. Modelli, A. Visconti, Introduction to flash memory, Proc. IEEE 91 (2003) 489–502.
[33] S. Kal, Isolation technology in monolithic integrated circuits: an overview, IETE Tech. Rev. 11 (1994) 91–103.
[34] R. Micheloni, L. Crippa, Multi-bit NAND Flash Memories for Ultra High Density Storage Devices. Advances in Non-Volatile Memory and Storage Technology, Woodhead Publishing Limited, 2014, https://doi.org/10.1533/9780857098092.1.75.
[35] G. Molas, D. Deleruyelle, B. De Salvo, G. Ghibaudo, M. Gely, S. Jacob, D. Lafond, S. Delconibus, Impact of few electron phenomena on floating-gate memory reliability, in: IEDM Technical Digest. IEEE International Electron Devices Meeting, 2004, IEEE, 2004, pp. 877–880, https://doi.org/10.1109/IEDM.2004.1419320.
[36] H. Veendrick, Bits on Chips, Springer International Publishing, 2019, https://doi.org/10.1007/978-3-319-76096-4.
[37] M. Jang, Scalability of Schottky barrier metal-oxide-semiconductor transistors, Nano Converg. 3 (2016).
[38] A. Spinelli, C. Compagnoni, A. Lacaita, Reliability of NAND flash memories: planar cells and emerging issues in 3D devices, Computers 6 (2017) 16.
[39] P.Y. Du, H.T. Lue, Y.H. Shih, K.Y. Hsieh, C.Y. Lu, Overview of 3D NAND flash and progress of split-page 3D vertical gate (3DVG) NAND architecture, in: Proc.—2014 IEEE 12th Int. Conf. Solid-State Integr. Circuit Technol. ICSICT 2014, 2014, pp. 24–27, https://doi.org/10.1109/ICSICT.2014.7021429.
[40] S.J. Whang, K.H. Lee, D.G. Shin, B.Y. Kim, M.S. Kim, J.H. Bin, J.H. Han, S.J. Kim, B.M. Lee, Y.K. Jung, et al., Novel 3-dimensional dual control-gate with surrounding floating-gate (DC-SF) NAND flash cell for 1Tb file storage application, in: Tech. Dig.—Int. Electron Devices Meet. IEDM, 2010, pp. 668–671, https://doi.org/10.1109/IEDM.2010.5703447.
[41] J. Jang, H.S. Kim, W. Cho, H. Cho, J. Kim, I.S. Sun, Y. Jang, J.H. Jeong, B.K. Son, W.K. Dong, et al., Vertical cell array using TCAT (terabit cell array transistor) technology for ultra high density NAND flash memory, Dig. Tech. Pap. - Symp. VLSI Technol. (2009) 192–193.
[42] R. Katsumata, M. Kito, Y. Fukuzumi, M. Kido, H. Tanaka, Y. Komori, M. Ishiduki, J. Matsunami, T. Fujiwara, Y. Nagata, et al., Pipe-shaped BiCS flash memory with 16 stacked layers and multi-level-cell operation for ultra high density storage devices, Dig. Tech. Pap. - Symp. VLSI Technol. (2009) 136–137.
[43] K. Wonjoo, C. Sangmoo, S. Junghun, L. Taehee, P. Chulmin, K. Hyoungsoo, J. Juhwan, Y. Inkyong, P. Yoondong, Multi-layered vertical gate NAND flash overcoming stacking limit for terabit density storage, Dig. Tech. Pap. - Symp. VLSI Technol. (2009) 188–189.
[44] D. Nagy, G. Indalecio, A.J. Garcia-Loureiro, M.A. Elmessary, K. Kalna, N. Seoane, FinFET versus gate-all-around nanowire FET: performance, scaling, and variability, IEEE J. Electron Devices Soc. 6 (2018) 332–340.
[45] Y. Fujisaki, Overview of emerging semiconductor non-volatile memories, IEICE Electron. Express 9 (2012) 908–925.

[46] R. Waser, R. Dittmann, G. Staikov, K. Szot, Redox-based resistive switching memories-nanoionic mechanisms, prospects, and challenges, Adv. Mater. 21 (2009) 2632–2663.
[47] J. Akerman, F. Masuoka, M. Asano, H. Iwahashi, T. Komuro, M. Johnson, L. Savtchenko, B.N. Engel, N.D. Rizzo, M.F. Deherrera, et al., Applied physics. Toward a universal memory, Science 308 (2005) 508–510.
[48] C. Schindler, S.C.P. Thermadam, R. Waser, M.N. Kozicki, Bipolar and unipolar resistive switching in Cu-doped SiO_2, IEEE Trans. Electron Devices 54 (2007) 2762–2768.
[49] D. Ielmini, R. Bruchhaus, R. Waser, Thermochemical resistive switching: materials, mechanisms, and scaling projections, Phase Transit. 84 (2011) 570–602.
[50] I. Valov, M.N. Kozicki, Cation-based resistance change memory, J. Phys. D Appl. Phys. 46 (2013), 074005.
[51] R. Waser, M. Aono, Nanoionics-based resistive switching memories, Nat. Mater. 6 (2007) 833–840.
[52] T. Tsuruoka, T. Hasegawa, I. Valov, R. Waser, M. Aono, Rate-limiting processes in the fast SET operation of a gapless-type Cu-Ta_2O_5 atomic switch, AIP Adv. 3 (2013), 032114.
[53] C. Schindler, M. Meier, R. Waser, M.N. Kozicki, Resistive switching in Ag-Ge-Se with extremely low write currents, in: 2007 Non-Volatile Memory Technology Symposium, IEEE, 2007, pp. 82–85, https://doi.org/10.1109/NVMT.2007.4389953.
[54] X.F. Liang, Y. Chen, L. Shi, J. Lin, J. Yin, Z.G. Liu, Resistive switching and memory effects of AgI thin film, J. Phys. D Appl. Phys. 40 (2007) 4767–4770.
[55] S. Tappertzhofen, S. Hofmann, Embedded nanoparticle dynamics and their influence on switching behaviour of resistive memory devices, Nanoscale 9 (2017) 17494–17504.
[56] A. Prakash, D. Jana, S. Maikap, TaOx-based resistive switching memories: prospective and challenges, Nanoscale Res. Lett. 8 (2013) 418.
[57] Y.-M. Kim, J.-S. Lee, Reproducible resistance switching characteristics of hafnium oxide-based nonvolatile memory devices, J. Appl. Phys. 104 (2008) 114115.
[58] F. Lentz, B. Roesgen, V. Rana, D.J. Wouters, R. Waser, Current compliance-dependent nonlinearity in TiO_2 ReRAM, IEEE Electron Device Lett. 34 (2013) 996–998.
[59] L.O. Chua, Memristor—the missing circuit element, IEEE Trans. Circuit Theory 18 (1971) 507–519.
[60] L.O. Chua, The fourth element, Proc. IEEE 100 (2012) 1920–1927.
[61] K.L. Chopra, Current-controlled negative resistance in thin niobium oxide films, Proc. IEEE 51 (1963) 941–942.
[62] R. Symanczyk, M. Balakrishnan, C. Gopalan, T. Happ, M. Kozicki, M. Kund, T. Mikolajick, M. Mitkova, M. Park, C.-U. Pinnow, et al., Electrical characterization of solid state ionic memory elements, in: Proceedings of the Non-Volatile Memory Technology Symposium, 2003, pp. 17–21.
[63] K. Szot, W. Speier, G. Bihlmayer, R. Waser, Switching the electrical resistance of individual dislocations in single-crystalline $SrTiO_3$, Nat. Mater. 5 (2006) 312–320.
[64] T. Ohno, T. Hasegawa, T. Tsuruoka, K. Terabe, J.K. Gimzewski, M. Aono, Short-term plasticity and long-term potentiation mimicked in single inorganic synapses, Nat. Mater. 10 (2011) 591–595.
[65] M. Hu, C.E. Graves, C. Li, Y. Li, N. Ge, E. Montgomery, N. Davila, H. Jiang, R.S. Williams, J.J. Yang, et al., Memristor-based analog computation and neural network classification with a dot product engine, Adv. Mater. 30 (2018) 1–10.
[66] M.A. Zidan, J.P. Strachan, W.D. Lu, The future of electronics based on memristive systems, Nat. Electron. 1 (2018) 22–29.
[67] X. Zhang, S. Liu, X. Zhao, F. Wu, Q. Wu, W. Wang, R. Cao, Y. Fang, H. Lv, S. Long, et al., Emulating short-term and long-term plasticity of bio-synapse based on Cu/a-Si/Pt memristor, IEEE Electron Device Lett. 38 (2017) 1208–1211.
[68] L. Chua, V. Sbitnev, H. Kim, Hodgkin-Huxley axon is made of memristors, Int. J. Bifurcation Chaos 22 (2012) 1230011.
[69] L.O. Chua, S.M. Kang, Memristive devices and systems, Proc. IEEE 64 (1976) 209–223.

[70] L. Chua, Resistance switching memories are memristors, Appl. Phys. A Mater. Sci. Process. 102 (2011) 765–783.
[71] S. Vongehr, Missing the memristor, Adv. Sci. Lett. 17 (2012) 285–290.
[72] P. Meuffels, H. Schroeder, Comment on "exponential ionic drift: fast switching and low volatility of thin-film memristors" by D.B. Strukov and R.S. Williams in Appl. Phys. A (2009) 94: 515–519, Appl. Phys. A Mater. Sci. Process. 105 (2011) 65–67.
[73] S. Vongehr, X. Meng, The missing memristor has not been found, Sci. Rep. 5 (2015) 1–7.
[74] I. Abraham, The case for rejecting the memristor as a fundamental circuit element, Sci. Rep. 8 (2018) 1–9.
[75] I. Valov, E. Linn, S. Tappertzhofen, S. Schmelzer, J. van den Hurk, F. Lentz, R. Waser, Nanobatteries in redox-based resistive switches require extension of memristor theory, Nat. Commun. 4 (2013) 1771.
[76] S. Tappertzhofen, I. Valov, T. Tsuruoka, T. Hasegawa, R. Waser, M. Aono, Generic relevance of counter charges for cation-based nanoscale resistive switching memories, ACS Nano 7 (2013) 6396–6402.
[77] S. Tappertzhofen, E. Linn, U. Bottger, R. Waser, I. Valov, Nanobattery effect in RRAMs—implications on device stability and endurance, IEEE Electron Device Lett. 35 (2014) 208–210.
[78] I. Tzouvadaki, N. Aliakbarinodehi, G. De Micheli, S. Carrara, The memristive effect as a novelty in drug monitoring, Nanoscale 9 (2017) 9676–9684.
[79] A. Wedig, M. Luebben, D.-Y. Cho, M. Moors, K. Skaja, V. Rana, T. Hasegawa, K.K. Adepalli, B. Yildiz, R. Waser, et al., Nanoscale cation motion in TaOx, HfOx and TiOx memristive systems, Nat. Nanotechnol. 11 (2015) 67–74.
[80] S. Tappertzhofen, H. Mündelein, I. Valov, R. Waser, Nanoionic transport and electrochemical reactions in resistively switching silicon dioxide, Nanoscale 4 (2012) 3040.
[81] S.M. Yang, E. Strelcov, M.P. Paranthaman, A. Tselev, T.W. Noh, S.V. Kalinin, Humidity effect on nanoscale electrochemistry in solid silver ion conductors and the dual nature of its locality, Nano Lett. 15 (2015) 1062–1069.
[82] T. Tsuruoka, I. Valov, C. Mannequin, T. Hasegawa, R. Waser, M. Aono, Humidity effects on the redox reactions and ionic transport in a $Cu/Ta_2O_5/Pt$ atomic switch structure, Jpn. J. Appl. Phys. 55 (2016) 06GJ09.
[83] F. Messerschmitt, M. Kubicek, J.L.M. Rupp, How does moisture affect the physical property of memristance for anionic-electronic resistive switching memories? Adv. Funct. Mater. 25 (2015) 5117–5125.
[84] I. Valov, T. Tsuruoka, Effects of moisture and redox reactions in VCM and ECM resistive switching memories, J. Phys. D Appl. Phys. 51 (2018).
[85] S. Tappertzhofen, R. Waser, I. Valov, Impact of the counter-electrode material on redox processes in resistive switching memories, ChemElectroChem 1 (2014) 1287–1292.
[86] S. Menzel, S. Tappertzhofen, R. Waser, I. Valov, Switching kinetics of electrochemical metallization memory cells, Phys. Chem. Chem. Phys. 15 (2013) 6945–6952.
[87] S. Menzel, M. Waters, A. Marchewka, U. Böttger, R. Dittmann, R. Waser, Origin of the ultra-nonlinear switching kinetics in oxide-based resistive switches, Adv. Funct. Mater. 21 (2011) 4487–4492.
[88] D.-Y. Cho, S. Tappertzhofen, R. Waser, I. Valov, Bond nature of active metal ions in SiO_2-based electrochemical metallization memory cells, Nanoscale 5 (2013) 1781–1784.
[89] D.-Y. Cho, I. Valov, J. van den Hurk, S. Tappertzhofen, R. Waser, Direct observation of charge transfer in solid electrolyte for electrochemical metallization memory, Adv. Mater. 24 (2012) 4552–4556.
[90] N. Dudney, Enhanced ionic conductivity in composite electrolytes, Solid State Ionics 28–30 (1988) 1065–1072.
[91] N.F. Mott, R.W. Gurney, Electronic Processes in Ionic Crystals, Oxford University Press, 1948.

[92] D. Ielmini, Modeling the universal set/Reset characteristics of bipolar RRAM by field- and temperature-driven filament growth, IEEE Trans. Electron Devices 58 (2011) 4309–4317.
[93] S. Tappertzhofen, I. Valov, R. Waser, Quantum conductance and switching kinetics of AgI-based microcrossbar cells, Nanotechnology 23 (2012) 145703.
[94] I. Valov, I. Sapezanskaia, A. Nayak, T. Tsuruoka, T. Bredow, T. Hasegawa, G. Staikov, M. Aono, R. Waser, Atomically controlled electrochemical nucleation at superionic solid electrolyte surfaces, Nat. Mater. 11 (2012) 530–535.
[95] P. Debye, E. Hückel, The theory of electrolytes. I. Lowering of freezing point and related phenomena, Phys. Z. 24 (1923) 185–206.
[96] J. Maier, Defect chemistry and ionic conductivity in thin films, Solid State Ionics 23 (1987) 59–67.
[97] J. Maier, Defect chemistry and ion transport in nanostructured materials. Part II. Aspects of nanoionics, Solid State Ionics 157 (2003) 327–334.
[98] J. Maier, Nanoionics: ion transport and electrochemical storage in confined systems, Nat. Mater. 4 (2005) 805–815.
[99] N. Sata, K. Eberman, K. Eberl, J. Maier, Mesoscopic fast ion conduction in nanometre-scale planar heterostructures, Nature 408 (2000) 946–949.
[100] J. van den Hurk, A.-C. Dippel, D.-Y. Cho, J. Straquadine, U. Breuer, P. Walter, R. Waser, I. Valov, Physical origins and suppression of Ag dissolution in GeS x-based ECM cells, Phys. Chem. Chem. Phys. 16 (2014) 18217.
[101] C. Schindler, I. Valov, R. Waser, Faradaic currents during electroforming of resistively switching Ag-Ge-Se type electrochemical metallization memory cells, Phys. Chem. Chem. Phys. 11 (2009) 5974.
[102] S. Menzel, P. Kaupmann, R. Waser, Understanding filamentary growth in electrochemical metallization memory cells using kinetic Monte Carlo simulations, Nanoscale 7 (2015) 12673–12681.
[103] C. Schindler, G. Staikov, R. Waser, Electrode kinetics of Cu–SiO_2-based resistive switching cells: overcoming the voltage-time dilemma of electrochemical metallization memories, Appl. Phys. Lett. 94 (2009), 072109.
[104] I. Valov, R. Waser, J.R. Jameson, M.N. Kozicki, Electrochemical metallization memories—fundamentals, applications, prospects, Nanotechnology 22 (2011) 289502.
[105] S. Pi, C. Li, H. Jiang, W. Xia, H. Xin, J.J. Yang, Q. Xia, Memristor crossbar arrays with 6-nm half-pitch and 2-nm critical dimension, Nat. Nanotechnol. 14 (2019) 35–39.
[106] T. Tsuruoka, T. Hasegawa, K. Terabe, M. Aono, Conductance quantization and synaptic behavior in a Ta_2O_5-based atomic switch, Nanotechnology 23 (2012) 435705.
[107] Y. Li, S. Long, Y. Liu, C. Hu, J. Teng, Q. Liu, H. Lv, J. Suñé, M. Liu, Conductance quantization in resistive random access memory, Nanoscale Res. Lett. 10 (2015).
[108] V. Havel, K. Fleck, B. Rösgen, V. Rana, S. Menzel, U. Böttger, R. Waser, Ultrafast switching in Ta_2O_5-based resistive memories, in: Proc. Silicon Nanoelectron. Workshop, 2016, pp. P1–15, https://doi.org/10.1109/SNW.2016.7577995.
[109] X.-D. Huang, Y. Li, H.-Y. Li, K.-H. Xue, X. Wang, X.-S. Miao, Forming-free, fast, uniform, and high endurance resistive switching from cryogenic to high temperatures in W/AlO_x/Al_2O_3/Pt bilayer memristor, IEEE Electron Device Lett. 41 (2020) 549–552.
[110] J.J. Yang, M.-X. Zhang, J.P. Strachan, F. Miao, M.D. Pickett, R.D. Kelley, G. Medeiros-Ribeiro, R.S. Williams, High switching endurance in TaO[sub x] memristive devices, Appl. Phys. Lett. 97 (2010) 232102.
[111] M. Julliere, Tunneling between ferromagnetic films, Phys. Lett. A 54 (1975) 225–226.
[112] E.C. Stoner, Collective electron ferronmagnetism, Proc. R. Soc. Lond. Ser. A Math. Phys. Sci. 165 (1938) 372–414.
[113] I.L. Prejbeanu, M. Kerekes, R.C. Sousa, H. Sibuet, O. Redon, B. Dieny, J.P. Nozières, Thermally assisted MRAM, J. Phys. Condens. Matter 19 (2007). 0–23.

[114] J.W. Lau, J.M. Shaw, Magnetic nanostructures for advanced technologies: fabrication, metrology and challenges, J. Phys. D Appl. Phys. 44 (2011).
[115] A.V. Khvalkovskiy, D. Apalkov, S. Watts, R. Chepulskii, R.S. Beach, A. Ong, X. Tang, A. Driskill-Smith, W.H. Butler, P.B. Visscher, et al., Basic principles of STT-MRAM cell operation in memory arrays, J. Phys. D Appl. Phys. 46 (2013) 074001.
[116] W.H. Butler, X. Zhang, T.C. Schulthess, Spin-dependent tunneling conductance of Fe|MgO|Fe sandwiches, Phys. Rev. B 63 (2001) 1–12.
[117] K.L. Wang, J.G. Alzate, P. Khalili Amiri, Low-power non-volatile spintronic memory: STT-RAM and beyond, J. Phys. D Appl. Phys. 46 (2013) 074003.

CHAPTER

2

Resistive switching in metal-oxide memristive materials and devices

A.N. Mikhaylov[a], M.N. Koryazhkina[a], D.S. Korolev[a], A.I. Belov[a], E.V. Okulich[a], V.I. Okulich[a], I.N. Antonov[a], R.A. Shuisky[a], D.V. Guseinov[a], K. V. Sidorenko[a], M.E. Shenina[a], E.G. Gryaznov[a], S. V. Tikhov[a], D.O. Filatov[a], D.A. Pavlov[a], D.I. Tetelbaum[a], O.N. Gorshkov[a], and B. Spagnolo[a,b]

[a]Lobachevsky State University of Nizhny Novgorod, Nizhny Novgorod, Russia, [b]University of Palermo and The National Interuniversity Consortium for the Physical Sciences of Matter, Palermo, Italy

2.1 Mechanisms of resistive switching in metal-oxide memristive materials and devices

At the beginning of the 21st century, the effect of resistive switching (memristive effect) has been a subject of intensive research due to the fact that, according to available forecasts, it can be used to create the new generation of information storage and logic devices [1,2], and also perspective neuromorphic and quantum computing systems [3–6], which are approaching the capabilities of the human brain. The special interest in non-volatile resistive memory (Resistive Random-Access Memory, RRAM) is due to a number of its advantages, such as unique scalability and simplicity of manufacturing technology, as well as fast and energy efficient operation. The structure of the RRAM cell is a thin (usually less

https://doi.org/10.1016/B978-0-12-814629-3.00002-7

than 100nm) insulator layer sandwiched between the layers (electrodes) of well-conducting materials. Particularly attractive are the so-called metal-oxide memristive devices based on metal/insulator/metal (MIM) and metal/insulator/semiconductor (MIS) structures, which are characterized by good compatibility with complementary metal/oxide/semiconductor (CMOS) technology.

The operation of RRAM is based on the switching of a memory cell from a high resistance state (HRS) to a low resistance state (LRS) and vice versa with the application of voltage of certain amplitude and polarity. When this voltage is removed, the resistive state is preserved. From this we can conclude that RRAM is a non-volatile memory. The process of transition from HRS to LRS is called SET, and the reverse process—RESET.

2.1.1 Classification of resistive switching

It is adopted to classify memristive devices either by the type of current-voltage characteristics (*I-V* curves), or by electroforming method with the formation of conductive channels in an insulator material—filaments. Let us consider primarily the first type of classification shown schematically in Fig. 2.1 [7]. There are different modes of resistive switching (RS): unipolar RS, when HRS and LRS are formed at any but the same polarity of applied voltage; bipolar RS, when these states are formed with different polarities of voltage; threshold RS, when LRS is obtained at some threshold voltage and spontaneously goes into HRS after the voltage is removed (the latter type of RS is volatile).

In turn, the bipolar RS is divided into two subtypes—F8 (figure-of-eight) and cF8 (counter-figure-of-eight) differing in the sequence of transitions from HRS to LRS and back when the polarity of *I-V* curves changes (the names F8 and cF8 are given because the *I-V* curves have the type of horizontal eight figure). Fig. 2.1 also shows the *I-V* curves obtained with current limitation and without current limitation (current compliance, CC).

Another classification of memristive devices is related to the nature of conducting filaments: the electrochemical formation of a metal phase inside an insulator from electrode atoms (extrinsic type) or due to atoms of the insulator itself (intrinsic type). The second type is characteristic of memristive devices based on oxides with high mobility of oxygen ions (vacancies). Typically these oxides are represented by the transition metal oxides (TMO), although cation mobility can also be used in the TMO-based memristive devices [8]. Since in a number of oxides, the metal atoms in the filament region change their valence state, the RS mechanism in such memristive materials can be considered based on the effect of changing valence (valence-change mechanism, VCM [9,10]), although this "chemical" point of view does not always reflect the real processes in

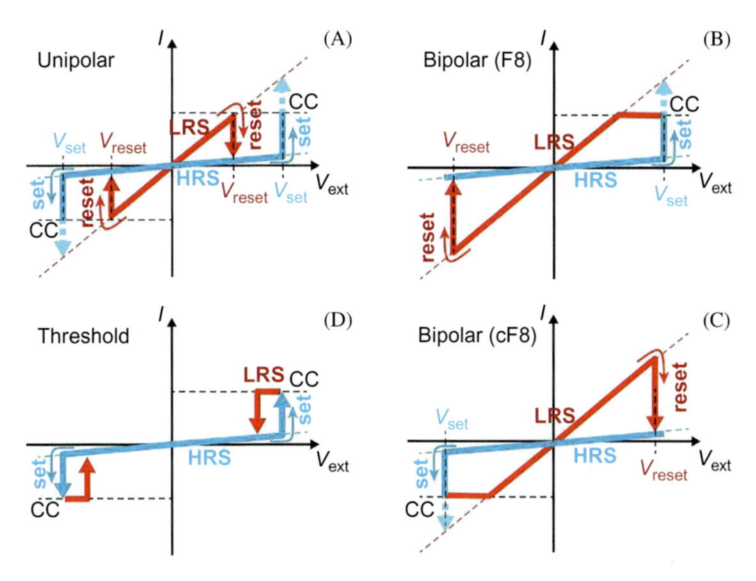

FIG. 2.1 (A) Schematic of a typical unipolar (or nonpolar) *I-V* curve, which is symmetric with respect to the polarity of an external voltage. Thus, SET and RESET can occur with a single polarity. To prevent a permanent dielectric breakdown, a compliance current is needed. In all observed experiments, the resistance change during SET and RESET was discontinuous. With a few exceptions, the RESET voltage was generally smaller than the SET voltage. (B) Schematic of a typical figure-of-eight (F8) bipolar *I-V* curve, which is asymmetric with respect to the polarity of an external voltage. Therefore, SET and RESET both occur with different polarities. In some cases, a compliance current is not needed and the resistance change during SET and RESET is not sudden, but gradual. (C) Schematic of a typical counter-figure-of-eight (cF8) bipolar *I-V* curve: the SET and RESET occur at a different polarity compared with the F8 case. (D) Schematic of a typical (unipolar) threshold *I-V* curve, which is symmetric with respect to the polarity of an external voltage. The LRS is maintained only when a bias is applied [7].

the memristive materials. Another (more general) name of the switching mechanism in metal-oxide devices is the RS of anionic type, since the movement of vacancies essentially means jumping anions (oxygen atoms), unlike the devices of first type, in which the filament is formed due to the drift of metal atoms (cations). This name is also suitable for those cases when oxygen vacancies are neutral and immobile, and their concentration changes due to the escape of oxygen atoms from the lattice sites and their return to the sites (analogous to the redox-reaction). A typical example is the MIM or MIS structure based on oxides with a predominantly covalent bond type, in particular SiO_x, which, by virtue of the common mechanism of anionic type, will also be referred to as metal-oxide memristive material in this chapter.

2.1.2 Operation mechanisms of memristive devices based on TMO and SiO_x

Let us briefly review the existing ideas about the operation mechanisms of memristive devices based on TMO and SiO_x. In the following, we will agree for definiteness in the notation of memristive devices (e.g., Pt/HfO_2/TiN) to indicate on the left the material of top electrode (TE) and on the right—the material of bottom electrode (BE). The voltage sign corresponds to the potential of TE relative to the potential of BE.

The oxides of such metals as Hf, Ta, Ti, Zr, Ni, as well as double- and multilayered insulators (oxides of various metals or oxides of the same metal but with different crystalline structure and different stoichiometry [7]) can be used as TMO. As for silicon oxide, the deposited films of SiO_x with a large oxygen deficiency are most often used ($x \leq 1.7$) [11–15] and less often with $x \approx 2$ [16–18]. In the case of VCM, the nature of switching can be either unipolar or bipolar. It depends on whether TE and BE are made from one or from different materials.

In most cases, in order to implement RS, metal-oxide memristive structures are preliminarily subjected to electroforming (EF) by applying a certain threshold voltage. It should be noted that for some oxide materials, as well as in the case of very thin layers (less than 4 nm), this operation is not required. In the process of EF, the filament is formed inside the oxide with a diameter of the order of units or tens of nm, or a group of filaments can be formed. The shape and size of filaments are little studied and differ not only for various types of memristive devices, but even within the same device. They are usually depicted either as thin cords with a constant or variable cross-section, or in a percolating filamentary shape [7]. For the TMO-based memristors, it is unambiguously assumed that the filaments are areas with a high concentration of oxygen vacancies. According to the literature data, the situation for silicon oxide based devices is not unambiguous: the filaments are areas with a high concentration of oxygen vacancies [19] or thin cords consisting either of silicon [20] or silicon nano-inclusions close to each other [21]. Such a representation provides low resistivity of the filament material and a conductivity of the structure is close to ohmic one. The mechanisms of filament formation during EF are described in many papers, for example, in the review [7]. In details, these mechanisms are different for different authors, but a common feature for them is the key role of oxygen vacancies. The authors also discuss the source of such vacancies. For silicon oxide and some TMOs (e.g., stoichiometric HfO_2), it is assumed that vacancies are absent in the initial oxides and their generation at EF occurs due to the breaking of chemical bonds in local areas near the electric field concentrators adjacent to one of the electrodes [22]. Another source (reservoir) of vacancies can be an oxygen-depleted insulator layer adjacent to one of the electrodes, which is either specially created in the process of fabrication of memristive devices (e.g., the TaO_x layer in the $Pd/Ta_2O_5/TaO_x/Pd$ structure [23]), or such a layer is formed unintentionally at the stage of the deposition of oxide film. A combined case is also possible: oxygen vacancies are uniformly distributed in the insulator and their generation is not needed [24]. Schematic representation of the VCM type filamentary mechanism [9] and standard approaches to its implementation [25] is shown in Fig. 2.2.

The electrode and insulator interface usually is not ideal: there are nanoscale irregularities or nanoareas with an increased concentration of defects such as grain boundaries, clusters of impurity atoms, etc. When voltage is applied to the device, these areas serve as electric field concentrators, and this is where the formation of filament nucleus begins. After the nucleus formation, it is extended (possibly with branching and/or increase in diameter) with appropriate polarity and the magnitude of applied voltage until it reaches the opposite electrode or approaches the tunneling distance. In cases where the mobility of vacancies is too low at room temperature with increasing voltage, a "soft breakdown" may occur, at which a current flow channel is formed. The magnitude of this current is considered to be sufficient for a local increase in temperature due to the Joule heating and the associated increase in the mobility of vacancies. Anyway, spatial redistribution of vacancies occurs in VCM-based devices at EF. In general case, this is accomplished as a result of three processes: (a) thermophoresis (Soret process), in which oxygen ions move in the opposite direction of the temperature gradient; (b) diffusion

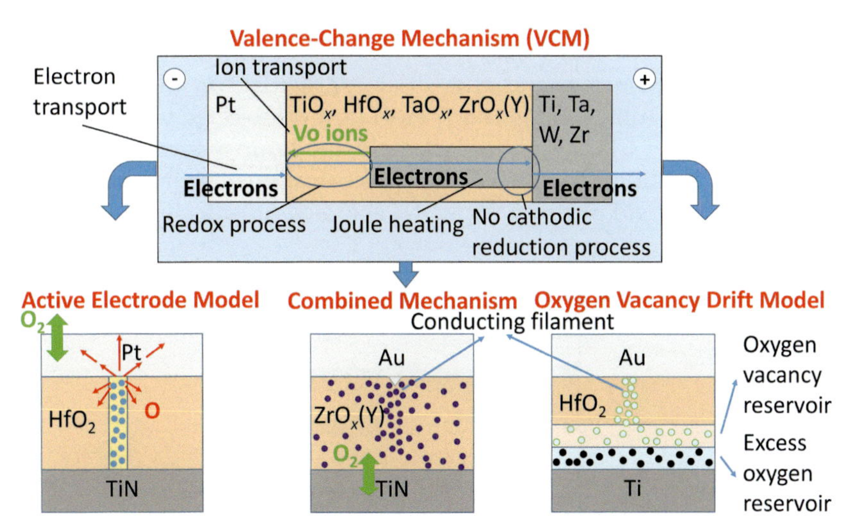

FIG. 2.2 Schematic representation of the main processes responsible for RS of anionic type (VCM).

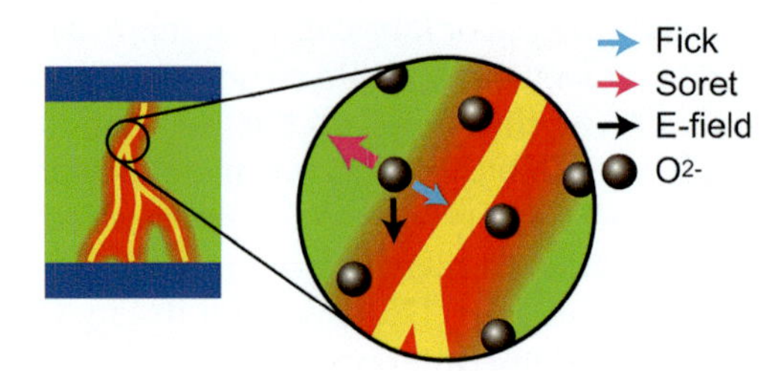

FIG. 2.3 A large electrical current (*yellow line*) due to the soft dielectric breakdown produces Joule heat, which gives the temperature gradient (*red region*) around the filament. The electric field, temperature gradient, and concentration gradient simultaneously lead to the three microscopic forces (electric, Soret, and Fick) that move the oxygen anion or vacancies [7].

(Fick process)—movement in the opposite direction of the concentration gradient; (c) drift in electric field that has two components—horizontal (parallel to the electrodes) and vertical. These three processes can occur simultaneously as shown in Fig. 2.3 [7,26,27].

In some cases, the filament formation is described in terms of a thermal-field local phase transition [28]. This process differs from the above mentioned one in the following: as a result of Joule heating, a cord is locally formed, which is another oxide phase with a lower resistivity.

There is a number of studies aimed at the direct confirmation of existence of filaments by visualizing them *in situ* using transmission electron microscopy (TEM) or scanning transmission electron microscopy (STEM) [29–32].

In [33–35], the composition and structure of columnar magnetron yttria-stabilized zirconia (YSZ) films used in VCM memristive devices were investigated before and after EF. In particular, for the structure of Au (40 nm)/Zr (3 nm)/YSZ (40 nm)/TiN (25 nm) subjected to EF, an image was obtained by STEM in the dark field mode, in which the contrast depends on the atomic number Z and concentration of chemical elements localized in the field of study (Fig. 2.4). The Z-contrast technique is based

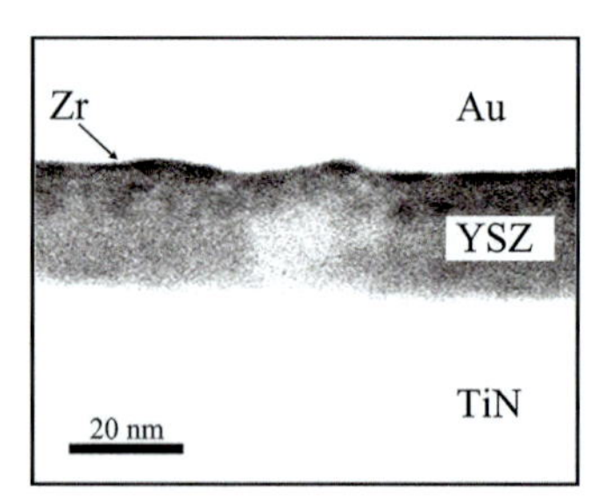

FIG. 2.4 Cross-sectional image showing the filament area in YSZ film after EF.

on the presence of a direct relationship between the composition of the structure under study and the dark-field contrast of its STEM images and usually used to increase the spatial resolution of elemental analysis.

Bright contrast at the top and bottom of the image corresponds to the layers containing Au and Ti heavy metals, respectively. *Z*-contrast features associated with changes in the composition of oxide film and/or the density of material during EF were observed in the center of the YSZ layer. Such areas were absent in the samples that did not undergo EF.

The chemical composition of such areas was directly determined using the method of energy-dispersive X-ray (EDX) spectroscopy. The concentration profiles were measured in directions parallel to the plane of electrodes. Before EF, the distribution of elements was uniform within the experimental error. After EF, a significant local redistribution of zirconium and oxygen concentrations in the region of grain boundaries was observed—increase in Zr concentration and decrease in O concentration (Fig. 2.5). The obtained results indicate that it is the grain boundary that is the most probable region for the formation of filaments.

In order to establish the mechanisms of electron and ion transport in the YSZ-based MIM devices, a set of independent electrophysical methods was used in [36] including measurements of dynamic current-voltage characteristics (dynamic *I-V* curves) in the temperature range of 77–600 K, small-signal characteristics in the frequency range of 10^3–10^7 Hz, depolarization currents and quasi-static capacitance. As a result, the value of one of the fundamental parameters of ion transport in YSZ nanocrystalline films was obtained—the activation energy of oxygen ion migration (0.50–0.55 eV), which is important for understanding the mechanisms of filament growth/rupture during EF and RS. Table 2.1 shows the specific values of the activation energy and the concentration of oxygen ions. The obtained value is ~2 times lower than the values reported in the literature for bulk single-crystal YSZ in the temperature range of 400–900 °C, which is associated with the predominant migration

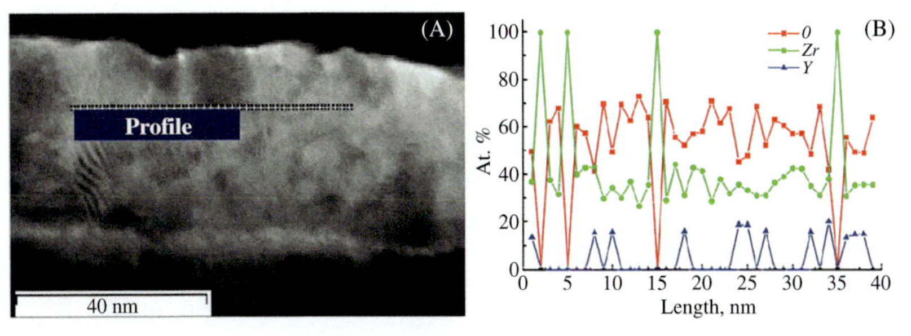

FIG. 2.5 (A) TEM image and (B) the EDX distribution profiles of Zr (*green circle*), Y (*blue triangle*) and O (*red square*) in the memristive device based on YSZ after EF.

TABLE 2.1 Parameters of oxygen ions in columnar magnetron-sputtered YSZ films.

	Parameters of oxygen ions	
Measurement technique	**Activation energy, eV**	**Ion concentration at 500 K, cm^{-3}**
Dynamic *I-V* curves	0.55	2.7×10^{19}
Thermal depolarization	0.55	2.6×10^{19}
Quasi-static capacitance	0.50–0.53	–

of oxygen ions along grain boundaries in columnar magnetron-sputtered films.

RS becomes possible after the EF process in those structures, for which it was necessary. After EF, the device is usually transferred from LRS to HRS (RESET process). Under the action of strong electric field, as well as an elevated temperature associated with Joule heating, part of the filament is oxidized due to the filling of oxygen vacancies with oxygen atoms moving toward the filament. As a result, the filament becomes thinner, and high-resistance gaps may appear in it. Mostly this occurs near one of the electrodes.

In HRS, the electron transport (current flow) is carried out by one of the following mechanisms known from the theory of contact phenomena and electrical conductivity in semiconductors and insulators: (1) space-charge-limited current (SCLC); (2) currents through the Schottky barrier; (3) trap-assisted tunneling; (4) Poole-Frenkel emission. Other electron transport mechanisms are possible, as well as combinations of different mechanisms. In the mechanism (1), vacancies change their charge by capturing carriers, thereby ensuring the space charge and the nonlinear dependence of the current on voltage. In the mechanism (2), the height and width of the Schottky barrier are determined by the concentration of vacancies and their charge, which also ensures the nonlinear dependence of the current on voltage. The mechanism (3) is associated with vacancies playing the role of traps for carriers. In the mechanism (4), vacancies create energy levels in the band gap of the insulator and exchange electrons with the conduction band due to thermal ionization and the return capture of carriers from the band. It should be noted that, in the LRS, the current through filament is usually associated with electron transfer through the conduction band or through the impurity band (defect band).

The processes at the electrode-insulator interface play an important role in EF and RS. Electrodes can be of two types—chemically active (for example, Ti, Ta, TiN, Zr) and inert (Pt and Au). The active electrodes during the EF and SET processes can absorb oxygen atoms moving to the electrode under the action of electric field (filament gaps are eliminated), and during the RESET process they can return the oxygen atoms to the insulator layer, contributing to the formation of gaps in filament. These processes were described, in particular, for memristive devices based on ZrO_2 [37] and Ta_2O_5 [38]. Since the active electrodes facilitate not only EF, but also the switching process, a layer of another metal is often introduced into the region between the main electrode and insulator. Thus, in [39], an underlayer of active material (Zr) was injected between the inert HfO_x/Pt contact. It should be noted that the role of electrodes as oxygen reservoirs can be judged by the difference between the free energy of the switching oxide and the oxide of metal electrode: the metal is active if its oxidation leads to a decrease in the free energy of the system.

The topography of the interface between insulator and electrodes strongly influences on the processes of EF and RS. By intentionally changing this topography, one can vary the parameters of memristive device in the proper direction. Thus, for the Pt/ZrO_2/Pt device [37], the Au nanoclusters were formed on the BE serving as electric field concentrators, which reduced the variation of switching parameters. Metallic nanoclusters formed inside an insulator can also serve as electric field concentrators [40].

Another way to control the switching parameters is the doping of insulator layers with impurities that affect the oxygen exchange between insulator and electrodes, the diffusion coefficients of vacancies and their formation energy, the mechanism of current flow through the interface between the electrode and insulator, as well as the crystallization temperature when using amorphous insulators (thus on the retention and endurance of memristive devices) [18,37,41].

One more factor affecting the switching mechanism is the geometric (dimensional) factor. Devices based on ultrathin (less than 10 nm) insulators, as already mentioned, most often do not require EF, whereas this operation is necessary for thicker insulator [7]. In thin insulators, those local regions, which serve as the filament nuclei in thick insulators, already penetrate through the entire thickness of the insulator film.

The mechanism of EF and RS in metal-oxide memristive devices based on silicon oxide requires separate consideration [14,16–18]. Unlike most types of the TMO-based memristive devices, the SiO_x layers contain practically no vacancies with a concentration sufficient to form a filament and, accordingly, RS. Vacancies and interstitial oxygen ions are formed when EF voltage is applied due to the breaking of chemical bonds under the action of strong electric field. However, vacancies and oxygen ions in SiO_x have low mobility (even taking into account Joule heating). The local injection of carriers in SiO_x from the electrode (in the defect-rich intercolumnar regions) leads to a sharp decrease in the barriers for the formation of vacancies and/or migration of oxygen ions [42]. After this, the EF and RS become possible (see also the section on simulation).

An important feature of switching processes typical of memristive devices based on both TMO and SiO_x is the random (stochastic) character of the physico-chemical processes responsible for switching. Stochasticity is most pronounced in the parameters of RS (V_{SET}, V_{RESET}, I_{HRS}, I_{LRS}) with repeated measurement of the *I-V* characteristics for even one and the same device. As a rule, the *I-V* curve varies from cycle-to-cycle, but the required difference between resistive states can be maintained. The number of cycles, for which the difference between resistive states acceptable from the performance viewpoint is maintained, is called endurance. The variation in the parameters of RS from cycle to cycle can be explained by the fact that EF produces not one, but many filaments. Such an assumption is quite reasonable: if the filaments are formed at the locations of the electric field concentrators, in particular, at the interface irregularities, then a large number of such sites are located on area of the order of 1 μm^2, each of which in principle can initiate the formation of a filament. The filaments differ from each other in size, shape and configuration of oxygen vacancies. On each cycle after EF, only one filament "works" with minimal resistance. The uniqueness of the "working" filament (or a bundle of close to each other filaments acting in each individual cycle) results from the experimentally observed independence of the RS parameters from the device area if this area is not too small. Let us suppose that a bipolar memristor operates in current compliance (CC) mode for the polarity in which the SET process is carried out. At a certain current value (not reaching the CC value), the "burning out" of this filament occurs due to Joule heating. In CC mode, this causes an increase in the voltage drop on the device and leads to the "re-growth" of another filament, which becomes effective in the next cycle. Since its parameters are somewhat different from the parameters of first filament, the *I-V* curve for the next cycle may also differ from the *I-V* curve for the first cycle and so on. Another reason leading to the stochasticity of the process in memristive devices may be a noticeable fluctuation of the number of oxygen vacancies located in the filament gap with a small area. The latter determines the current in the HRS, as well as the voltage required to fill the vacancies in the RESET process. It is also possible to combine these two scenarios.

The variation of RS parameters is most pronounced when studying the role of CC in the process of EF [37,43–46]. The variation measured in CC mode is usually higher than that measured without compliance. This is due to the fact that "loose" filaments are formed at CC. At the same time, there is a very small number of vacancies in the gap region of the filament; therefore, fluctuations of this amount from cycle-to-cycle are large.

The variation of RS parameters is undesirable from the viewpoint of using memristive structures as memory devices, as well as non-volatile "storages" of weights in artificial neural networks, when an unambiguous, essentially programmed change in weight (resistance of the memristive device) is required in response to external influence. Therefore, numerous works were aimed at reducing the variation of RS parameters by engineering insulator (insulators) materials and electrode materials. For example, the report [38] presents the results of experimental study of the influence of W, Ta, Ti, and Hf electrodes on the parameters of RS of a memristive device based on tantalum oxide. These metals were chosen due to a significant difference in the free energy of formation of oxides with different oxygen content (in particular, TaO_2 and Ta_2O_5). Memristive devices with Ti and Hf electrodes, which have negative formation energy, showed a faster degradation of the RESET process in the endurance study, which is associated with effective accumulation of oxygen vacancies in the Ta_2O_5 layer. This leads to an irreversible transition of the structure to LRS. The Ta and W oxide formation energy is positive, which made it possible to obtain rather stable switching and reduce the stochastic variation of RS parameters.

An example of effective materials engineering approach to solving the problem of reducing the RS variations is given in [47], where multilayer oxide structures were used, in which ultrathin Al_2O_3 layers are embedded between the TiO_2 layers. The results show that Al_2O_3 layers, acting as reservoirs of oxygen vacancies, improve the stability of RS parameters. In [48], it was also shown that the RRAM crossbar arrays with multilayer $TiN/[AlO_y/HfO_x]_m/Pt$ (number of layers $m=10$) structure shows excellent device performances such as a large resistance window (>80), uniform switching voltage, high switching endurance ($>10^7$ cycles), high-disturbance immunity ($>10^9$ cycles), and good retention ($>10^5$ s at 150 °C).

These results indicate the importance of finding the best combinations of insulator materials and electrodes in order to provide reproducible RS parameters in metal-oxide memristive devices.

In conclusion, it should be noted that, despite the very large number of studies of the resistive switching mechanisms in metal-oxide memristive devices, a complete clarity has not yet been achieved on this issue. In this direction, additional research is needed with the use of an arsenal of modern experimental (including local) and computational tools, e.g., computer simulation.

2.2 Local analysis of resistive switching of anionic type

By using a variety of Scanning Probe Microscopy (SPM) techniques, one can not only understand the RS mechanisms more deeply, but also carry out the nanometer scale studies to detect individual filaments.

SPM has been proven to be a very informative method for studying the RS phenomena in thin insulator films. The following two SPM methods are applied to study RS in insulator films: Scanning Tunneling Microscopy/Spectroscopy (STM/STS) [49,50] and Atomic Force Microscopy (AFM) utilizing conductive AFM probe usually called Conductive AFM (CAFM) [51]. The latter CAFM method has been demonstrated to be a powerful tool for studying the RS phenomena at the nanometer scale [52].

2.2.1 Conductive atomic force microscopy

Originally, CAFM was invented to study the local electrical properties of solid surfaces with the nanometer scale spatial resolution [53]. In this method, the electric current through the contact of a conductive AFM probe to the sample surface I_t is measured. A bias voltage V_g is applied between the AFM probe and the sample (Fig. 2.6). The sample surface is scanned in contact AFM mode. Along with measuring the surface topography of the sample $z(x, y)$ (here x and y are the probe coordinates in the sample surface plane and z is the arbitrary height of the sample surface at the point x, y), the map of probe current $I_t(x, y)$ (hereafter referred to as the current image) is recorded. Also, the *I-V* curves of the probe-to-sample contact $I_t(V_g)$ can be measured.

Spreading Resistance Microscopy (SRM) is a kind of CAFM [54]. In this method, the experiment is arranged in such a way that the probe current I_t is limited by the spreading resistance of the sample. It is essential to provide the resistance of the contact between the CAFM probe and the sample surface to be negligible. In this case, I_t is determined by the specific resistivity of a small volume of the sample material just under the probe-to-sample contact ($\sim D_p^3$, where D_p is the size of contact area). Thus, SRM allows mapping the local specific resistivity of the sample surface with lateral spatial resolution $\sim D_p$, when all the voltage in the circuit falls on the spreading resistance.

On the contrary, when studying the insulator films on conductive substrates by CAFM, almost whole bias voltage V_g drops on the insulator film,

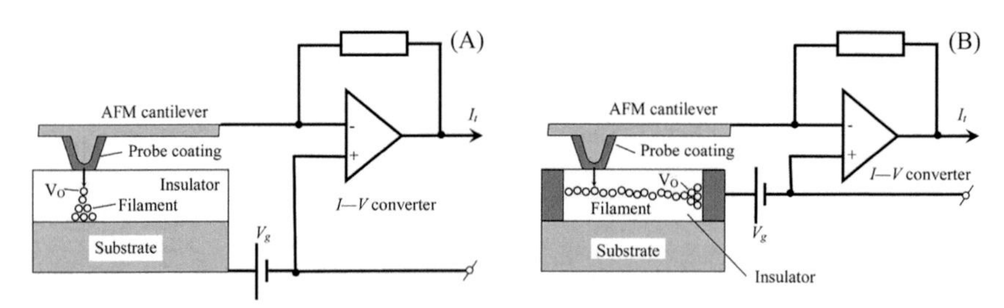

FIG. 2.6 (A) Transverse and (B) lateral scheme for the study of resistive switching in thin insulator films by the AFM method with a conducting probe.

whereas the spreading resistance of the substrate should be minimized. This method allows, in particular, tunnel spectroscopy of the insulator films (at appropriate film thickness) as well as of the localized electronic states inside the films due to some defects, nanoinclusions, etc. Therefore, this kind of CAFM is often called *Tunneling AFM*. The main fields of application of this method include the investigation of local electrical properties of the ultrathin gate insulators in the MIS stacks for modern field-effect transistors (FET) [55,56].

The opportunity of studying the RS by CAFM was demonstrated in [57] when investigating thin $SrTiO_3$ films. The following two configurations of the experimental setup are employed usually in the studies of RS in the thin insulator films by SPM:

(1) Transverse scheme (Fig. 2.6A). In this scheme, the objects of investigation are thin insulator films deposited onto a conductive substrate. A gap voltage V_g is applied between the STM or CAFM tip and the substrate, and the electric current between the tip and the substrate I_t is measured. In another measuring mode, the *I-V* curves of the probe-to-sample contact are measured. The transverse scheme allows studying the local transverse electron transport in the insulator films, including tracing the EF and RS dynamics, visualizing the filaments, studying the electron transport via individual filaments [8], and even obtaining the three-dimensional images of individual filaments inside the insulator films utilizing the tomography technique [58].

(2) Lateral scheme (Fig. 2.6B). In this scheme, the objects under investigation are the insulator films deposited onto insulating substrates with metal electrodes preliminary formed on the substrate surfaces. The RS takes place in the film volume inside the narrow gaps between the electrodes formed by high-definition lithography (electron beam lithography, synchrotron radiation X-ray lithography, focused ion beams, etc.). The filaments are growing through the film volume inside the gap between the electrodes from one electrode to another under the driving voltage applied between the electrodes. The film surface in the gap above the filaments is scanned by CAFM or STM tip. The gap voltage V_g is applied between the SPM tip and one of the electrodes. The probe current I_t is flowing between the tip and the electrode through the film volume. The lateral scheme allows the in-plane visualization of filaments inside the gaps, revealing the longitudinal structure of the filament along its axis, and tracing the transformation dynamics of the filament shape during growth. This scheme was applied also for the *in situ* real-time imaging the filament growth and rapture by TEM [49].

In general, STM can provide a better spatial resolution in the current images than CAFM (down to the atomic scale) due to the "last atom" effect [59]. However, there is some advantage of CAFM in studying the RS in thin insulator films, which consists in the following. The STM/STS investigations always imply a measurable electric current flowing through the sample in order to maintain a stable feedback in every point of the STM scan using the tip current channel. This requirement imposes rather strict limitations of the electrical properties of the insulator films under study. Namely, the film should have high enough electron conductivity in the HRS in order to ensure high enough tip current (at least ~1 pA for ordinary STM and ~10–100 fA for the low current versions) to maintain the feedback in every point of the film surface at reasonable values of $|V_g| < |V_{SET}, V_{RESET}|$ in order not to modify the electrical properties of the investigated film during examination. This requirement should be satisfied in every point of the STM scan: if even a single very small not enough conductive plasma occurs within the scan area, the experiment will end with crushing the tip.

In the transverse measurement scheme, there is an opportunity to maintain the feedback utilizing the direct electron tunneling between the STM tip and conductive substrate through the insulator film. However, utilizing this mechanism imposes very strict limitations onto the film thickness (actually, no more than few nanometers, subject to the electronic properties of the film material, namely, to the height of the energy barrier between the Fermi level in the tip/substrate material and the conduction band edge in the insulator film) since the film should be tunnel transparent at a reasonable gap voltage (again, the condition $|V_g| < |V_{SET}, V_{RESET}|$ should be satisfied).

On the contrary, in CAFM the feedback channel (the force sensor) and the tip current I_t measurement are separated from each other. This method allows studying the samples with nonconductive areas on the insulator film surface. This advantage of CAFM is especially valuable when the investigated insulator films are nonconductive in the initial state (i.e., prior to EF).

Another advantage of CAFM consists in the following. At present, scaling is a serious problem in the development of RRAM. The RS mechanisms in macroscopic RRAM cells (with the electrodes of several tens and hundred micrometers in size) were found to differ from the ones in the nanometer-sized cells [60]. The difference consists in the number of filaments under the electrodes. In the macroscopic memristive cells, a large number of filaments can be covered by electrodes (up to 10^3–10^4) [61]. On the other hand, typical sizes of the contact area between the AFM probe tip and the sample surface D_p can be estimated from the solution of Hertz problem [62] to be ~10 nm at reasonable loading forces ~1 nN employed in the CAFM experiments [63]. Thus, typical values of D_p are of the same order of magnitude as the expected sizes of the RRAM cells in future RRAM devices [64]. Under an electrode of such size, no more than one filament can occur only. So far, the CAFM probe contact to an insulator film on a conductive substrate can be considered as a virtual device (a nanometer-sized RRAM cell), which appears to be a good model system for studying the RS mechanisms in emerging RRAM devices.

2.2.2 Investigations of individual filaments by CAFM

There are two main approaches to the visualization of individual filaments in the insulator films by CAFM utilizing the transverse scheme:

(1) AFM tip-induced RS. In this approach, the filaments are formed in the initially uniform insulator films first by applying an appropriate voltage V_{EF} between the CAFM tip and the conductive substrate [65]. EF can be performed either at the points selected on the insulator film surface or by scanning a selected area on the film surface. Then, the film surface around the modified area is scanned at $|V_g| < |V_{SET}, V_{RESET}|$ in order to avoid further modification of the already formed filaments (some analog of read procedure in real memristive devices). So far, in this approach, CAFM plays a dual role: it is applied both for forming the filaments and for the visualization of the formed filaments.

(2) In the second approach, the memristor cells of the ordinary design are fabricated, and EF is performed by applying an appropriate voltage between the cell electrodes. Then the cells are switched either into LRS or into HRS. Then, the TE are removed by selective etching, and the bare surface is examined by CAFM. This approach has been applied for the first time in [66] to study RS in $Al/TiO_2/Ru$ memristive devices. First, the HRS and LRS of the TiO_2 films were prepared at the device level. Then, the Al top electrodes were removed by selective wet etching in nitric acid, and the surface topography and current images of the TiO_2 film surfaces under the Al electrodes were examined by CAFM. The surface density of the conductive spots (current channels) in LRS was found to be larger than in HRS. Also, the values of probe current I_t through the individual filaments in HRS was found to be higher than in LRS.

In [67], the mercury droplets sprayed onto the NiO/Pt film surface were used as the liquid metal top electrodes. After EF and switching to HRS and LRS for the NiO films, the Hg droplets were washed away,

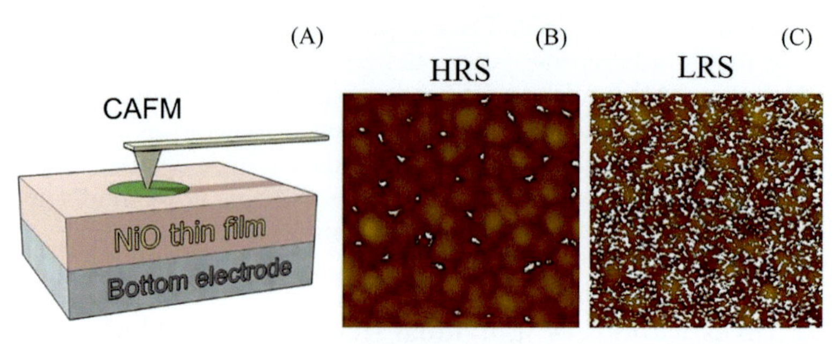

FIG. 2.7 (A) CAFM measurements for HRS or LRS after the removal of Hg top electrode. (B) The CAFM image of NiO thin film for HRS corresponding to 100 switching cycles. The bias voltage of 0.1 V was used for the CAFM measurement. (C) The CAFM image of NiO thin film for LRS corresponding to 100 switching cycles.

and the bare NiO film surfaces were examined by CAFM (Fig. 2.7A). Again, the surface density of filaments in LRS was found to be much larger than in HRS (Fig. 2.7B and C) confirming the results of [66]. Besides, by comparing the topographic and current images of the bare NiO film surface, the authors of [67] conclude that the investigated NiO films have the nanocrystalline structure and that conductive filaments were formed preferentially at the grain boundaries. Later, the same conclusions were drawn from the results of investigations of local RS in the polycrystalline HfO_2 films by CAFM [68,69]. The defects of crystal structure (first of all, the vacancies) are well known to concentrate at the grain boundaries in the polycrystalline materials, so the preferential formation of filaments at the grain boundaries seems to be natural.

Also, by using CAFM, the conductance quantization in electron transport through individual filaments has been observed in the course of investigations of local AFM tip-induced RS in HfO_x films [70]. The manifestation of conductance quantization in the *I-V* curves of the CAFM probe contact to individual filaments was interpreted within the framework of the quantum point contact (QPC) concept [71]. Namely, the minimum possible thickness of a conductive filament is a diameter of single oxygen vacancy V_O, i.e., the filament can be treated as a chain of oxygen vacancies. If a single V_O in this chain is filled by an oxygen ion O^{2-}, such a system can be treated as a QPC. The quantization of the memristor conductance ascribed to the formation of QPC's inside thin filaments has been observed first in the memristive structures with the macroscopic contacts [11,72]. Application of CAFM makes the observation of conductance quantization during the RS easier even at room temperature since it allows measuring the *I-V* curves of single filaments.

Below, the original data on RS in YSZ films studied by CAFM are discussed.

2.2.3 Investigations of resistive switching in YSZ films by CAFM

2.2.3.1 *Sample preparation*

Thin (3-10 nm) YSZ films for the investigation of RS by CAFM were deposited onto the n^+-type Si substrates covered by native oxide SiO_2 (~2–3 nm) with or without Au (40 nm)/Cr (10 nm) sublayers. The YSZ films were formed by the radio frequency (RF) magnetron sputtering deposition in $Ar:O_2$ gas mixture (50:50 mol%) from the pressed $ZrO_2+Y_2O_3$ mixture powder targets using Torr International MagSputt 3G2 and 2G1-1G2-EB4-TH1 vacuum systems for thin film deposition. The gas pressure inside the vacuum chamber was $\sim 10^{-2}$ Torr. The molar fraction of the stabilizing oxide Y_2O_3 in the target material was 0.12. The substrate temperature T_g was 300 °C.

2.2.3.2 *CAFM measurements technique*

The CAFM experiments were carried out in Ultra High Vacuum (UHV) using Omicron UHV AFM/STM LF1 installed in Omicron Multiprobe RM UHV system. The base residual gas pressure inside the AFM/STM chamber was $\sim 10^{-10}$ Torr. The YSZ film surface was scanned by the conductive AFM probe in Contact Mode. Silicon NT-MDT NSG01 AFM probes with Pt coating and DCP11 probes with diamond-like coating were used. The values of the probe tip curvature radius R_p for the NSG01 and DCP11 probes were 35 nm and 70 nm, respectively (according to the vendor's specifications). The bias voltage V_g was applied to the Si substrate relative to the CAFM probe, which was grounded virtually. The *I-V* curves of the probe-to-sample contact were measured in every point of a selected area in the AFM scan that allowed excluding the errors in the probe positioning onto the individual filaments due to the creep of the piezoelectric ceramics.

The RS effect in YSZ films was studied by recording the cyclic *I-V* curves of the probe-to-sample contact. The bias voltage between the CAFM probe and the sample V_g was swept forward from V_{min} to V_{max} ($|V_{min}|<|V_{RESET}|$ and $|V_{max}|>|V_{SET}|$, where V_{SET} is the threshold voltage for switching from HRS to LRS and V_{RESET} is the threshold for switching back from LRS to HRS) and backward from V_{max} to V_{min}, and so forth. Prior to RS investigation, the EF was performed by scanning a selected area on the sample surface (one or several times, if necessary) at $V_g=V_{EF}$, $|V_{EF}|>|V_{SET}, V_{RESET}|$.

In another measuring mode, a selected area on the sample surface was scanned in Contact Mode with $|V_g|>|V_{SET}|$ in order to switch the YSZ film within this area into the LRS. In order to switch the film back into HRS, the respective area was scanned again at $|V_g|>|V_{RESET}|$. The results of switching was examined by acquiring the current images $I_t(x, y)$ $|V_g|<|V_{SET}, V_{RESET}|$.

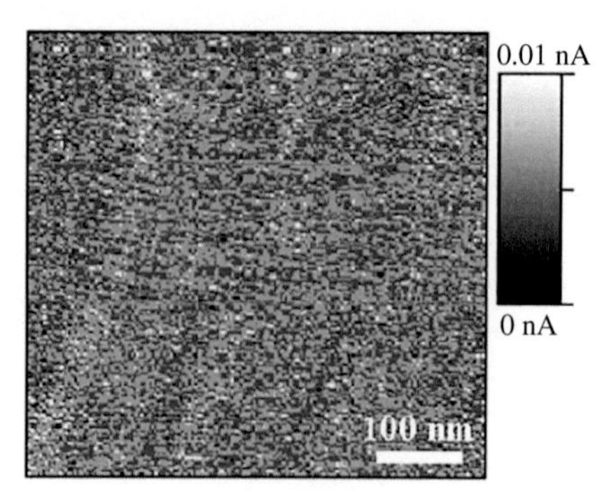

FIG. 2.8 Current image of initial (as-grown) YSZ film on the SiO_2/Si substrate ($V_g = 6\,V$).

2.2.3.3 *Effect of annealing on the CAFM images of YSZ films*

The as-grown YSZ films (~4 nm in thickness) on the SiO_2/Si substrate were not conductive at $|V_g| \leq 6\,V$: the values of I_t averaged over the scan area did not exceed the level of ~2 pA (Fig. 2.8) that was comparable to the intrinsic noise level of the STM preamplifier (*I-V* converter) of Omicron UHV AFM/STM LF1.

The annealing of sample in UHV at $T_A = 300\,°C$ for 1 h has not resulted in the essential change in the YSZ film conductivity (Fig. 2.9).

Further annealing at $T_A = 500\,°C$ for 1 h resulted in a slight increase in the probe current (the maximum value of I_t in Fig. 2.10B was ~30 pA). There was some correlation between the topography image (Fig. 2.10A) and the current one (Fig. 2.10B): the probe current I_t in the places corresponding to the hillocks in the topography image (Fig. 2.10A) was lower than the STM preamplifier detection threshold. The annealing likely resulted in the crystallization of the initially amorphous YSZ film (the

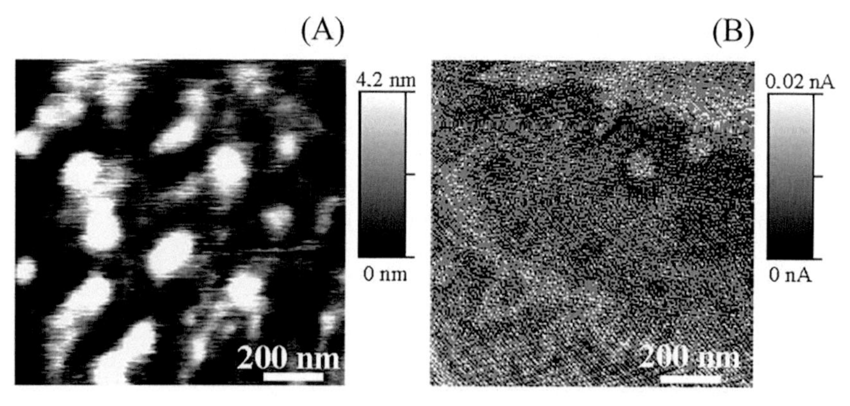

FIG. 2.9 (A) Topography and (B) current image of YSZ film on the SiO_2/Si substrate after annealing in UHV at $T_A = 300\,°C$ for 1 h ($V_g = 6\,V$).

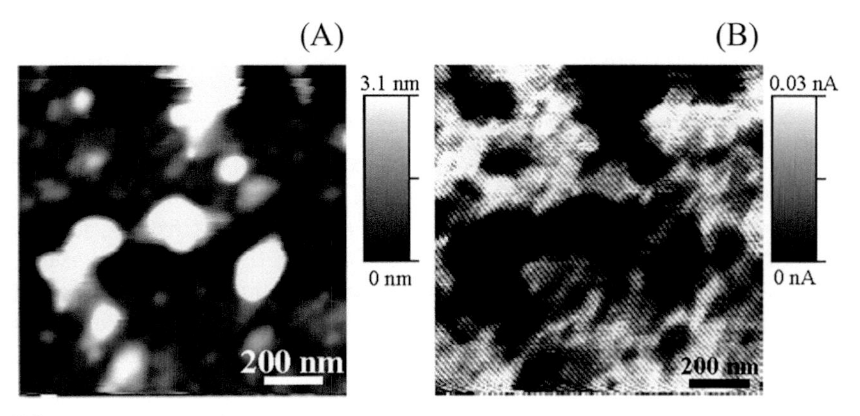

FIG. 2.10 (A) Topography and (B) current image of YSZ film on the SiO_2/Si substrate after annealing in UHV at $T_A = 500\,°C$ for 1 h ($V_g = 6\,V$).

YSZ nanocrystals are manifested as the hillocks in the topography images in Figs. 2.9A and 2.10A), and the current between the CAFM probe and the substrate flows through the grain boundaries only. This result agrees with the results obtained earlier in [67–69].

Typical *I-V* curve of the CAFM probe contact to the YSZ (3 nm)/SiO_2/Si structure surface obtained after EF by double scanning at $V_{EF} = 6\,V$ is shown in Fig. 2.11A [73]. Well expressed hysteresis typical of the bipolar RS was observed. Such *I-V* curves of the CAFM probe contact to the insulator films on conductive substrates have been reported in the literature extensively (see, for example, Refs. [51,52]) and are attributed usually to

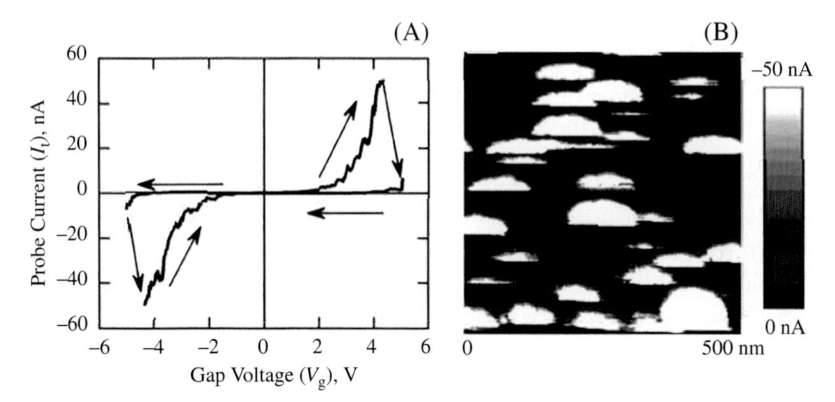

FIG. 2.11 (A) Typical *I-V* curve of CAFM probe contact to YSZ film on the SiO_2/Si substrate after annealing in UHV at $T_A = 500\,°C$ for 1 h. (B) CAFM image of the same film recorded at $V_g = -6\,V$ [73]. *Reproduced under Creative Commons Attribution 4.0 International License.*

the AFM tip-induced bipolar RS in the insulator films under the electric field between the CAFM probe and the substrate.

2.2.3.4 *Imaging of individual filaments by CAFM*

Fig. 2.11B shows the CAFM image of the YSZ film on the SiO_2/Si substrate after annealing in UHV at $T_A = 500\,°C$ for 1 h [73]. The CAFM image was acquired at $V_g = -6\,V$ ($|V_g| > V_{SET} \approx 5\,V$). The current channels (places of increased probe current I_t) observed in the CAFM image were attributed to the electric current flowing through the individual filaments formed in the course of preliminary scanning. The sizes of current channels (~100 nm) are consistent with the values of R_p (≈70 nm). Earlier, the sizes of current images of the localized electron states inside the tunnel transparent insulator films (metal nanoclusters (NCs), point defects, defect clusters, etc.) have been shown to be determined by the size and shape of the probe-to-sample contact area and not to depend on the defect size [74]. The same should be true in the case of imaging the filaments in insulator films. This effect is a manifestation of well-known convolution effect [75], which, in turn, is a particular manifestation of the general principle theory of measurement claiming the result of any measurement to be a convolution of the object function $f(x, y)$ and the apparatus one $g(x, y)$. In the case of current images of the conductive filaments in insulator films, the apparatus function $g(x, y)$ represents the shape of contact area of the CAFM probe tip to the investigated insulator film. Since the size of contact area D_p exceeds essentially the size of filament tip (nearest to the film surface), which can be as low as the size of single V_O into the ultimate limit, the object function of the filament $f(x, y)$ can be approximated by Dirac delta-function $\delta(x, y)$. So far, $I_t(x, y) = f(x, y) \times g(x, y) \approx \delta(x, y) \times g(x, y) \to g(x, y)$. In other words, the current image of a filament obtained by scanning by a CAFM probe of relatively large value of D_p (with respect to the filament tip size) represents rather an image of the tip-to-sample contact area obtained by a δ-function like filament.

At the same time, the current channels in Fig. 2.11B had a half-moon shape whereas the AFM probe tips are known to be round in general. This disagreement was explained assuming that the filaments can grow during scanning [73]. Initially, the film was not conductive. Once a filament has grown through the film, I_t increases. So far, the current images in Fig. 2.11B can be interpreted as the images of a half of the contact area of the CAFM probe to the sample surface imaged by the emerged filaments. The surface density of current channels is increased after the second scanning the same area that supports the suggestion on forming new filaments while scanning.

Most likely, the filaments form preferentially in the points, where some defects are located inside the film. These defects concentrate the electric field between the probe and the sample promoting the filament growth.

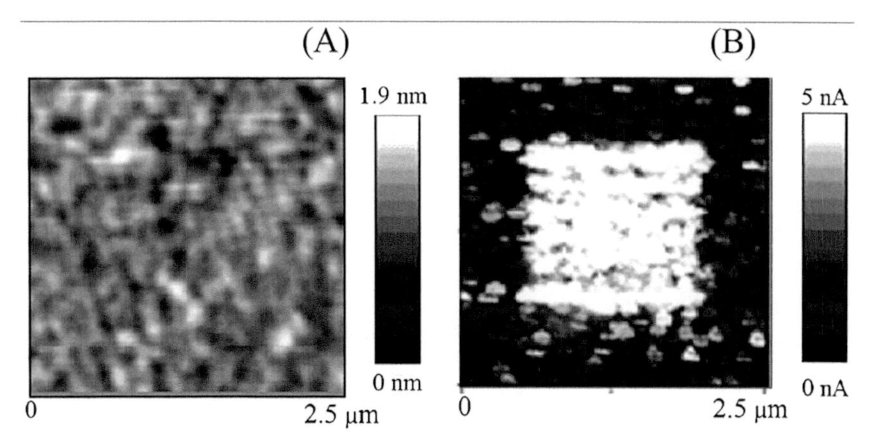

FIG. 2.12 (A) AFM image of $Au/Cr/SiO_2/Si$ stack. (B) CAFM image (recorded at $V_g=2V$) of YSZ film deposited onto $Au/Cr/SiO_2/Si$ stack and annealed in UHV at $T_A=300°C$ for 1 h. The square area $1.5 \times 1.5 \mu m^2$ in the center was modified by scanning two times at $V_g=-2V$. *Reproduced under Creative Commons Attribution 4.0 International License.*

In order to confirm the above suggestion, the YSZ films (4 nm in thickness) deposited onto Au (40 nm)/Cr (10 nm) sublayers on the SiO_2 (3 nm)/Si substrates were studied [73]. The AFM image of an $Au/Cr/SiO_2/Si$ sublayer is shown in Fig. 2.12A. The Au surface had an islanded structure typical for the thin Au films deposited in vacuum.

The CAFM image of the YSZ film deposited onto the $Au/Cr/SiO_2/Si$ stack and annealed in UHV at $T_A=300°C$ for 1 h is shown in Fig. 2.12B. In the square area of $1.5 \times 1.5 \mu m^2$ in size (in the center of the CAFM scan), preliminary EF was performed by scanning two times at $V_g=-2V$. A fair correlation between the Au surface topography in Fig. 2.12A and the CAFM image in Fig. 2.12B indicate the Au grains to promote the filament forming. Earlier, the promotion of filament growth by the substrate roughness was reported in [76,77]. The authors of the cited works suggested the hillocks on the substrate to act as the electric field concentrators stimulating the nucleation of filaments. This assumption is illustrated in Fig. 2.13. The results of the present study confirm the suggestions in the early works by direct experimental observations at the nanometer scale.

2.2.3.5 Quantum size effects in the electron transport via individual filaments

Besides the butterfly-type hysteresis loops in the *I-V* curves of the CAFM probe-to-sample contact typical of the bipolar RS (Fig. 2.11A), the *I-V* curves with the negative differential resistance (NDR) have been observed (Fig. 2.14A). Earlier, the NDR has been observed in the *I-V*

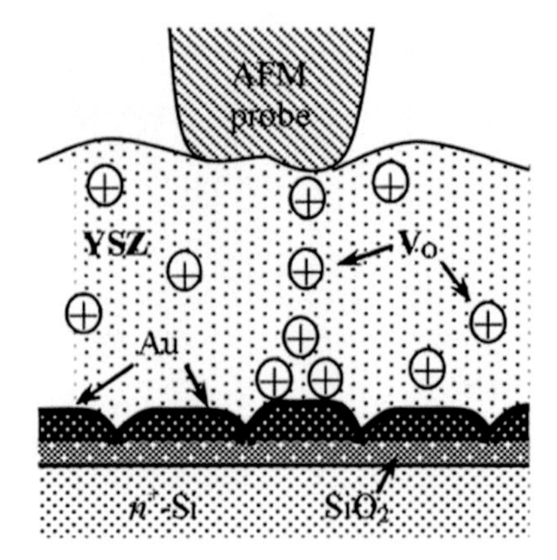

FIG. 2.13 A qualitative schematic of a filament growth stimulated by an Au island on the substrate (not to scale) [73]. *Reproduced under Creative Commons Attribution 4.0 International License.*

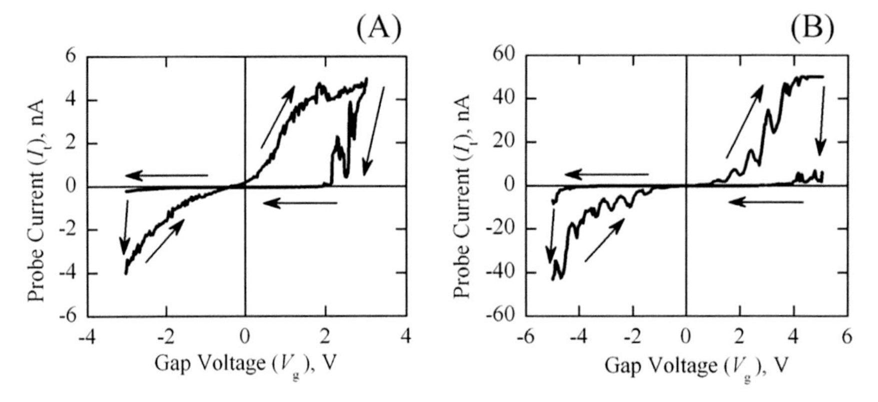

FIG. 2.14 (A) The examples of *I-V* curves of the AFM probe contact to the YSZ (4nm)/Au/Cr/SiO_2/Si sample and (B) to the YSZ (5nm)/SiO_2/Si sample [73]. *Reproduced under Creative Commons Attribution 4.0 International License.*

curves of the CAFM probe contacts to the tunnel-transparent SiO_2 [78] and YSZ [79] films with embedded Au NC arrays and were attributed to the resonant electron tunneling between the CAFM probe and conductive Si substrate via the quantum-confined electron states in the ultrafine (~1 nm in size) Au NCs.

The absence of any hysteresis in the *I-V* curves reported in [78,79] points to the absence of RS compared with the results obtained in [73] (Fig. 2.14A). The RS was absent in [78] because the investigated films were made of stoichiometric SiO_2, whereas the oxygen vacancies are necessary for the manifestation of RS. On the other hand, the limits of V_g sweep in [79] were limited intentionally in order to avoid the RS ($|V_g| < |V_{SET}, V_{RESET}|$). On the contrary, when measuring the cyclic *I-V* curves presented in Fig. 2.14 [73], the limits of the bias voltage sweep were selected $|V_g| > |V_{SET}, V_{RESET}|$ that resulted in the appearance of hysteresis in the *I-V* curves due to bipolar RS.

The NDR in the *I-V* curves of the CAFM probe contact to the uniform $YSZ/SiO_2/Si$ films was attributed to the resonant electron tunneling via localized energy states of some quantum systems with discrete electron energy spectrum embedded into the filaments [73]. These quantum systems could be associated with small enough (~1 nm in size) Zr NCs emerged during annealing in UHV. Annealing promotes the diffusion of O^{2-} ions from the inner volume of YSZ film to its surface followed by the escape out of the film into UHV. The residual excess Zr atoms may coagulate into the Zr NCs [80]. These Zr NCs formed inside the YSZ film during annealing in UHV are most likely the defects promoting the filament growth in the $YSZ/SiO_2/Si$ samples.

On the other hand, such localized electronic states may originate from a V_O (or several ones coalesced in a vacancy cluster) separated from the other parts of a filament by the oxygen vacancies filled by O^{2-} ions as shown in Fig. 2.13. The vacancy clusters can form in the course of the RESET process due to the nonuniform diffusion of oxygen vacancies out of the filaments in the essentially nonequilibrium conditions [81] (e.g., due to overheating the filaments by Joule heat released when passing too high reverse currents through the filaments at the initial stages of RESET process (so called current overshoot) if the current limitation is absent). Earlier, the V_O filled with O^{2-} ion in a vacancy chain (filament) had been considered as a QPC [71]. In [73], the QPC model was developed further: a double QPC model was proposed to explain the NDR in the *I-V* curves of the filaments.

Another possible candidate for such a localized state could be the Y^{3+} ion inside the filament [73]. Among various types of *I-V* curves, ones with multiple equidistant peaks were observed (Fig. 2.14B). Such features are typical of the tunneling through so called "colored" atom [59] and could be attributed to the tunneling via the fine structure electron states in 4f shell of Y^{3+} ion in ZrO_2 lattice.

The presented results demonstrate CAFM to be a powerful method for studying the microscopic mechanisms of RS in the ultrathin metal-oxide films at the nanometer scale. Particularly, applying CAFM allows the real-time tracing the filament emerging and growth, associating the filament nucleation with the substrate roughness, and studying the electron transport via individual filaments. The NDR regions observed in the *I-V* curves of the CAFM probe contact to individual filaments were attributed to the resonant electron tunneling via the localized electronic states in the filaments related to the Zr NCs or to the Y^{3+} ions embedded into the filaments.

2.3 Multiscale simulation of resistive switching in metal-oxide memristive devices

To develop a technology of memristive devices of VCM type based on oxide materials (e.g., NiO, TiO_2, ZrO_2, HfO_2, GeO_x, WO_x, SiO_x) it is important to understand and predict the details of the RS mechanism in these devices. One of the generally accepted ways to achieve this goal is the theoretical calculation or computer simulation. As indicated in the previous paragraphs, the operation parameters of most devices of this type depend on the structure of filaments, as well as the processes occurring in the respective local areas of thin films when the control voltage and/or current changes.

Existing models of operation of memristive devices include consideration of such factors as the effect of temperature, ion conductivity, electronic processes, as well as phenomena at the electrode/insulator interfaces [82]. In this case, an important role in the models is assigned to taking into account the defects present in insulators, which are introduced during the process of their fabrication (in some cases intentionally) or during the EF process.

Many variants of models were proposed in [83] to be classified as follows: (1) "electronic scale"—taking into account the capture of carriers into electronic levels of traps; (2) "atomic scale"—taking into account the migration of ions and point defects of various nature in the force fields of electrical, concentration and thermal nature; (3) "metal-insulator transition"—considering the metal-insulator transitions with changes in atomic structure and their influence on energy barriers for charge carriers; (4) "thermochemical reaction"—taking into account the changes in microstructure in the active layers of memristive devices.

According to the used mathematical apparatus and calculation algorithm, the approaches to multiscale description and modeling of transport processes and statistical phenomena during RS can be divided into the following types: (1) *phenomenological*, in which the mentioned processes are described by a system of differential equations that take into account the internal electric field, drift/diffusion of oxygen ions/vacancies, carrier transfer in view of their statistics, and (2) *atomistic*, including kinetic Monte Carlo method and molecular dynamics, allowing virtually to carry out the "direct observation" of the movement of atoms. It should be noted that in all of the pointed above approaches the use of certain structural, energetic or electrochemical parameters, which can be determined in the experiment, is required. These parameters include the electronic energy barriers, trap parameters, energies of defects formation and migration. There is also a large class of publications devoted to the calculation of these parameters based on the first-principles methods such as density functional theory (DFT). Let us consider successively these approaches in more detail.

2.3.1 Phenomenological approach

One of the first phenomenological models of RS in metal-oxide memristive devices was proposed in [84]. It was based on the analysis of X-ray photoelectron spectroscopy data for ZnO films. This model includes the statements, which were also used in later theoretical works (with some variations):

(1) Conductivity in the filament area in both HRS and LRS is realized as a result of the motion of electrons by the trap-mediated mechanism. The traps are positively charged vacancies ($V_O^{\bullet\bullet}$), which are formed during the movement of oxygen atoms from the sites to interstitials.
(2) Switching between HRS and LRS is associated, respectively, with the rupture and reduction (partial oxidation) of the filament consisting of a material enriched with oxygen vacancies.
(3) During the switching to LRS (SET process), oxygen vacancies are generated, due to which filaments are formed, ensuring the effect of current carriers (electrons) flowing through the insulator.
(4) The transition from LRS to HRS (RESET process) when changing the polarity of the voltage is associated with a local decrease in the concentration of oxygen vacancies in the filament due to their recombination with O^{2-} ions, which leads to its rupture at a certain value of voltage.
(5) The reason for the observed disappearance of RS (degradation of memristive device) with multiple switching is the formation of an interstitial O^{2-} ion deficiency in the vicinity of the filament.

The described mechanism is schematically shown in Fig. 2.15.

In the calculations based on this model, an equation is used that describes the kinetics of oxygen vacancies formation depending on the magnitude of external electric field strength. It is assumed that the resistance of memristive cell in LRS is inversely related to the concentration of oxygen vacancies in the filament, and in HRS it is determined by the length of the insulator region with a small concentration of vacancies separating the filament from the corresponding electrode (Fig. 2.15). In the general case, a system of differential equations describing the motion of oxygen ions and vacancies in an electric field was solved numerically taking into account various factors: the influence of defects on the internal

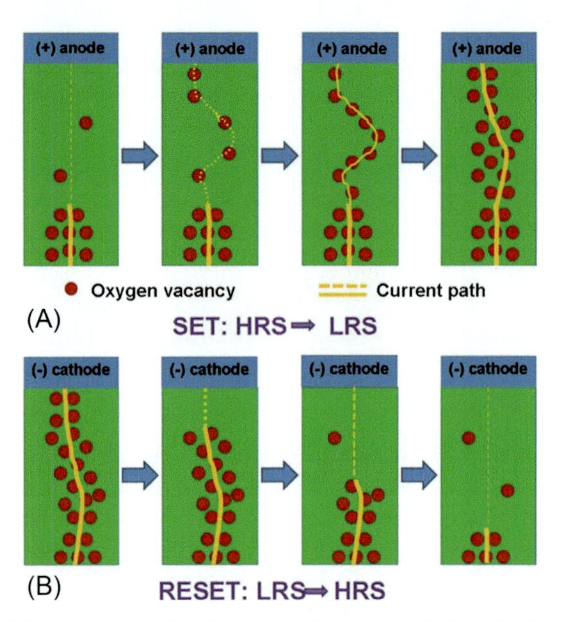

FIG. 2.15 Schematic view of the bipolar switching physical model describing the switching behavior. (A) SET process under the positive voltage bias. (B) RESET process under the negative voltage bias. The *arrows* indicate the increased trend of the applied voltage bias [82].

electric field, the effects of carrier tunneling through the Schottky barrier at the metal/insulator interface and traps created by vacancies, heating the filament region. Other examples are presented in the reports [85–89]. To obtain a quantitative agreement with the experimental data, the following dependence of the drift velocity ν on electric field is introduced in [89]:

$$\nu = f \cdot a \cdot exp\left(-\frac{U_a}{k_B T}\right) \cdot sinh\left(\frac{qE_\nu a}{2k_B T}\right), \tag{2.1}$$

where f—the oscillation frequency of atoms, a—the jump length, U_a—the activation energy of jump, q—the vacancy charge. E_ν is the electric field strength acting on the vacancy related to the applied electric field E as $E_\nu = (1+\varepsilon F)E$, where ε—the dielectric constant of a substance, F—the Lorentz factor. The relation for E_ν takes into account the influence of dielectric polarization by an external field on the magnitude of the force acting on a point charge.

In various models of memristive effect with the participation of oxygen vacancies, the latter are considered either as traps that provide tunnel transport of carriers injected into the insulator, or as agents affecting the electric field and carrier concentration in the oxide. In the latter case, the simulation consists in solving a system of equations that describe local changes in the electric field strength, concentrations of free charge carriers, the drift of vacancies in electric field, and the transfer of electrons through the filament.

In most reports on phenomenological modeling, the SET and RESET switching processes are associated with a change in the filament structure. Thus, in [90,91], when discussing experimental data for a HfO_2-based memristive device, a model was proposed in which the switching is due to thinning and breaking of the vacancy filament taking into account the local Joule heating the material by the flowing current. To this end, the heat conductivity equation is added to the system of differential equations describing the movement of vacancies and charge carriers. It is shown that heating significantly influences on the diameter of filament, which increases during the SET process and decreases during the RESET process until the appearance of a gap. The maximum value of resistance corresponds to the maximum length of this gap. The calculations [91] showed that the temperature for the $TiN/HfO_2/TiN$ system in the central part of the filament can reach 600 K. In the framework of this model, the authors obtain the experimentally observed dependence of the *I-V* curves on the sweep speed with a fairly good agreement between theoretical calculations and experimental results.

A similar approach is implemented in [92]. In this work, the SET and RESET processes for the HfO_x-based memristive devices are modeled. The dynamics of change in the filament diameter with the change in the magnitude and polarity of external voltage is considered. To calculate the current, the concept of "nanoscale filament resistance" is introduced. Despite the relative simplicity of the model, it allows the authors to explain the main features of the SET and RESET processes, for example, the dependence of the filament resistance and current in the RESET process on the CC during the preceding SET process. It is found that such characteristics as the device resistance and current at SET and RESET switching are practically independent of the energy barrier for vacancy jump in the range of 0.9–1.5 eV. This is due to the presence of negative feedback, due to which the self-organization of the filament resistance dynamics takes place.

In [93], the RESET process in the $Pt/Ta/Ta_2O_5/Pt$ structure is modeled in order to find the conditions for obtaining a gradual switching from HRS to LRS. It is assumed that the concentration of vacancies of $\sim 10^{20}\,cm^{-3}$ inside the filament with a radius of 10 nm, and outside of it, gradually decreases to $\sim 10^{16}\,cm^{-3}$. At the same time, at the Pt electrode, the concentration of vacancies in the filament is increased and reaches values of $10^{21}\,cm^{-3}$. To calculate the current, it is necessary to determine the potential barriers for charge carriers at the electrode/insulator interfaces. Outside the filament, the heights of the barriers are assumed to be equal to their nominal values for the Pt/TaO_x and TaO_x/Ta contacts (2.3 and 0.8 eV, respectively). This effectively blocks the electron current flowing through this region and allows current to flow only through the filament. It is assumed that in the filament region, the barrier height is 0.3 eV at the Pt/TaO_x interface and 0.075 eV at the TaO_x/Ta interface, which corresponds to the Schottky barrier type contact and the ohmic contact, respectively. When solving the equations of heat conductivity and electron transport, the stationary conditions are used. When considering the processes of transport of vacancies and their recombination with oxygen, the time-dependent equations are solved. The transport of electrons from the electrodes, which can change the charge state of vacancies, is also taken into account. The distribution of vacancy concentration, current and temperature along the filament are presented, depending on the amplitude of the applied voltage pulse.

As mentioned above, one of the most important problems for practical applications of memristive devices is the stochastic variation of RS parameters. In [94], on the example of the $Au/Ti/SiO_x/Mo$ structure, it was shown that in some cases the phenomenological approach is able to satisfactorily describe the instability of the *I-V* curves without using microscopic approaches. This approach is particularly attractive in those cases when the microscopic physical aspects of the electron transport problem are hard or impossible to implement in detail. Its advantage is also the relatively low requirements for computer resources.

Recently, to describe the behavior of memristive devices, so-called stochastic models have been used, which relate to the phenomenological approach and allow explicitly taking into account the internal fluctuations of the system [95]. Stochastic models are based on at least two equations: an ohmic-type relationship and a first-order differential equation with a noise source (the first order Langevin equation, known as the model of overdamped Brownian motion in the field of force).

2.3.2 Atomistic approach

It is generally accepted that the processes of atomic reconstruction and its kinetics are responsible for the processes of EF, RS and degradation of memristive devices. The phenomenological approaches with the above mentioned advantages do not allow taking all these processes into account with a sufficient degree of detail and therefore do not always lead to success. Simulation of processes by the kinetic Monte Carlo (kMC) method allows solving this problem more reliable.

The kMC is based on a simulation of random hopping processes of vacancies or atoms using a generator of random numbers that specifies the probabilities of a particular elementary event at each time step. Thus, in [96], the formation of filaments in the $Pt/TiO_2/Pt$ device is modeled. The motion of charged vacancies in TiO_2 occurs due to their jumps, while the external electric field lowers the potential barrier for the jump. The authors calculated the density of electron states in the oxide band gap depending on the concentration of vacancies and demonstrated the presence of a transition from an insulating state to a conducting state at a certain concentration.

In the report [97], the EF processes in the $TiN/HfO_2/TiN$ device are considered. By the AFM method, it is established that there are grain boundaries in the HfO_2 film, in one of which the filament grows as a result of EF. A numerical simulation of the EF process was carried out. It was assumed that electron tunneling transfer occurs through oxygen vacancy traps. The current flow and the associated heating lead to the formation of new vacancies—electron traps. Comparison of the calculation results with the experimental *I-V* curves shows good agreement.

In [35], the kMC simulation of EF, SET, and RESET processes for the Au/YSZ/TiN/Ti memristive device with the experimentally implemented geometrical and compositional parameters is carried out. The simulation is performed under the assumption that the above mentioned processes are determined by the transport of oxygen vacancies leading to the growth of filament from the initial nucleus located on one of the electrodes. It is shown that, in a system of 10^4 filaments, the statistical variation of switching voltages during repeated cycling can lead to the experimentally observed gradual nature of the RESET switching process due to the random selection of the filament system at each successive cycle. Subsequently, the calculation algorithm for this device used in [35] was modified in order to take into account more rigorously the kMC processes not only in the atomic but also in the electronic subsystem. Simple description of the algorithm and the original results are shown below.

The calculation is carried out in the three-dimensional (3D) space. A cubic lattice with the sites corresponding to the positions of oxygen atoms in cubic stabilized zirconia is considered. The Cartesian coordinate axes (x, y, z) are oriented parallel to the $\langle 100 \rangle$, $\langle 010 \rangle$ and $\langle 001 \rangle$ axes, so that the z axis is directed perpendicular to the device layers. The model YSZ volume (Fig. 2.16) is a cylinder with a diameter of 30 nm and a length of 40 nm located between the TE (Au) and BE (TiN). As in real EF experiments, the negative potential V on TE is changed linearly with time from zero to some maximum value V_{max}. The same modes of positive voltage sweep are set for the RESET process.

Let's first describe the simulation of EF process. According to the literature, the barrier for jumping of oxygen vacancies E_m in the absence of electric field is chosen to be 1.1 eV [98], although, as noted above, this barrier can be significantly lower for the migration of vacancies along grain

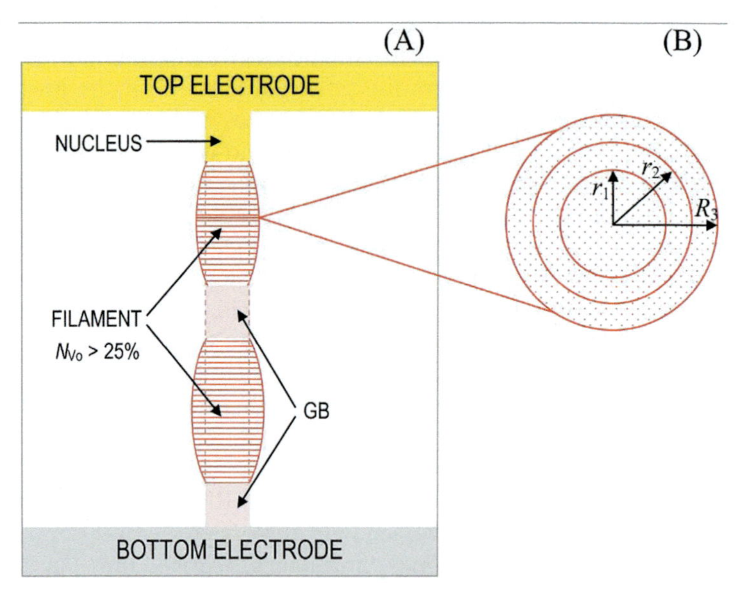

FIG. 2.16 Schematic representation of the model volume (cross-section) at the intermediate stage of (A) forming and (B) splitting the cross section of the filament into disks.

boundaries in magnetron-sputtered films. The application of external electric field to the device causes an asymmetric change in the potential barrier and, as a result, changes the probability W of the vacancy jump:

$$W = W_0 \cdot exp\left(-\frac{E_m - E_{bias}}{k_B T}\right), \tag{2.2}$$

where W_0—the Debye frequency ($\sim 10^{13}\,\mathrm{s}^{-1}$), E_{bias}—the barrier change due to electric field. Thus, when an external field is applied under the conditions of sufficiently high temperature, there is a redistribution (drift) of vacancies, leading to the formation of local low-resistance regions (filaments) with an increased concentration of vacancies and, accordingly, a change in the total resistance of the insulator layer. The Eq. (2.2) automatically describes three key processes responsible for the movement of vacancies, namely, the drift under the action of electric field, diffusion and thermophoresis (Soret effect).

The simplest estimates show that a barrier of 1.1 eV at room temperature and real fields ($\leq 10^8\,\mathrm{V/m}$) can be overcome by vacancy only in a time of the order of $10^5\,\mathrm{s}$, whereas the experimental EF times are no more than a few milliseconds. This implies that at such barrier the EF process requires heating, which apparently occurs immediately after applying the voltage and precedes the beginning of the EF itself, i.e., the formation of filaments. Such heating can occur due to the flow of current through the areas with increased conductivity present in the original insulator films. For example, in [99], it was assumed that the pathways associated with grain boundaries exist in YSZ. When the field is applied, the heating current flowing along these boundaries can appear when the threshold voltage is exceeded. In our experiments with the YSZ-based memristive devices, the EF stage usually begins at voltages of about 3 V. In the model, the resistance of the boundary is chosen so that when the voltage reaches 3 V, the current flow is sufficient to heat up, leading to the start of the vacancy structure transformation in the insulator.

The grain boundary is modeled as a cylinder with a diameter of 1 nm with a specific resistance of $5 \times 10^{-4}\,\Omega$cm. In fact, the shape of the border is unknown; in the case of typical films obtained by magnetron sputtering, which have, as a rule, columnar morphology, the border can be, for example, a junction of three crystallites penetrating the film, which is consistent with the TEM data [34]. According to these data, some initial inhomogeneities are on the TE (Au) in the form of bumps facing the insulator in the areas of grain boundaries. In the calculations, such a bump is considered as a filament nucleus with the shape of a metallic (equipotential with respect to Au) cylinder with a diameter of 1 nm and a height of 5 nm. It is located in the center of the top surface of the model volume.

When an external negative potential is applied to TE, a nonuniform electric field appears in the YSZ film, which, at a sufficiently high temperature, causes a decrease in the potential barrier for vacancy jumps and their movement mainly in the direction of the nucleus. The initial concentration of oxygen vacancies is assumed to be 6%—in accordance with a given concentration of Y_2O_3 used in the structure (12 mol%), and their initial distribution is assumed to be random and statistically homogeneous. Hereinafter, the percentage of oxygen vacancies corresponds to the ratio of the number of unoccupied oxygen positions to the total number of oxygen positions.

In our model, the growth of filament over the EF time is carried out by increasing the length and width of the initial nucleus. This is associated with the increase in concentration of oxygen vacancies (N_{Vo}) near the nucleus to a certain critical value, which is assumed to be 25% of the concentration of oxygen atoms in ZrO_2. This value of critical concentration is taken in connection with the results of ab initio calculations performed in [100].

The following approach was used to determine the geometrical parameters of the filament. For simplicity, an assumption was taken that the filament has cylindrical symmetry, i.e., it represents a rotation figure (Fig. 2.16). At each time step, the region of the model volume occupied by the filament is determined as the locus of points, in which the average concentration of vacancies exceeds 25%, and through which, as assumed in the model, the entire current flows.

The average concentration of vacancies N_{Vo} is determined within each disk, into which the region adjacent to the central axis is divided (Fig. 2.16B). The disk thickness is taken equal to 0.25 nm. The filament radius $r(z)$ at the depth z is determined as follows. At first, the disk radius is assumed to be r_1—half the distance between the oxygen atoms, and the average concentration of oxygen vacancies $(N_{Vo})_1$ in this disk is determined, then the radius $r_2 = 2r_1$ is taken, and the value $(N_{Vo})_2$ in the disk with such a radius is calculated, then with $r_3 = 3r_1$, an so on. In such a way a sequence of numbers $(N_{Vo})_1$, $(N_{Vo})_2$, ... $(N_{Vo})_n$ is received. The filament radius $r_i(z)$ is chosen from the condition that the value $(N_{Vo})_i \geq 25\%$ at $r \leq r_i$, and this value is less than 25% at $r \geq r_i$. Further, it is assumed that at each moment of time all the current flows through the filament, the resistivity of which is considered to be an order of magnitude smaller than in the material of the boundary. In those areas where the filament has not been formed at this time, the current flows through the material of the boundary. Thus, the current I is calculated by the formula:

$$I = \frac{V}{\sum R_i + R_b} \tag{2.3}$$

where R_i—is the resistance of each disk with $N_{Vo} \geq 25\%$, and R_b—is the resistance of those parts of the boundary that were not included in the filament body at this stage.

At each time step, the electric field distribution is calculated by solving the Poisson equation. The calculation is performed without taking into account the fields created by the charges of vacancies and ions of the substance. The fields created by vacancies and ions can be neglected for the following reasons. First, taking into account these fields in the presence of a concentration of vacancies in the central part of the model volume leads to such strong internal fields repulsing the vacancies located outside the concentration that the latter cannot participate in the formation of filament, i.e., such an assumption will lead to the insolvency of filament model. Second, it is not clear what charges should be attributed to vacancies and ions, taking into account the nonstationarity of the simulated processes, as well as the uncertainty of the true distribution of electron clouds in the structure. Third, the experimentally determined barrier for jumping over a vacancy to some extent already takes into account the influence of the fields of neighboring ions and vacancies. Fourth, the paper [101] has shown that the main role in the processes of EF and RS is played by the external electric field.

In the simulation, the potential distribution on lateral surface of the region occupied by the filament is calculated based on the magnitude of the flowing current and the resistance of each disk. At each stage of the filament growth, the temperature distribution in the model volume is calculated by solving the stationary heat conduction equation. For this, the Joule heating is calculated when the current flows through the filament, i.e., the local power P_i, emitted by the current inside each disk is determined as:

$$P_i = V_i^2 / R_i \tag{2.4}$$

The assumption of stationary conditions is based on the fact that the setting time of a stationary temperature is less than nanosecond, as shown by the estimates using the Fourier formula, while the characteristic calculated times of a "settled" life of a vacancy are in the order of microsecond. At the boundary of the model volume, the temperature is assumed to be 300 K.

The model considers the inflow and outflow of vacancies from the TiN electrode, which ensure a constant concentration of vacancies in the boundary layer between the insulator and TiN electrode. In this case, in the presence of a shift of the entire vacancy array in the direction of the field to the opposite electrode (Au) at $V < 0$, no depleted vacancy area forms near the bottom electrode (unlike the previous version of the calculation [35]).

The filament growth is stopped at a time equal to the voltage sweep duration from $V = 0$ to the maximum absolute value of V at a sweep rate of 200 V/s. Using the known values of the voltage drop and resistance at each stage of the calculation, the total current is calculated, which allows us to build *I-V* curves in real time.

Similarly, the calculation procedures are carried out for the RESET and SET processes, taking into account the corresponding voltage signs. For the RESET process, the initial distribution of vacancies is taken as received at the end of EF (or at the end of the preceding SET stage), and for the SET process—at the end of the preceding RESET stage.

Fig. 2.17 shows typical vacancy distributions in the oxide layer of memristive device after EF (SET) and RESET. It is seen that, during EF, vacancies move to the central axis, that is, to the location of the grain boundary.

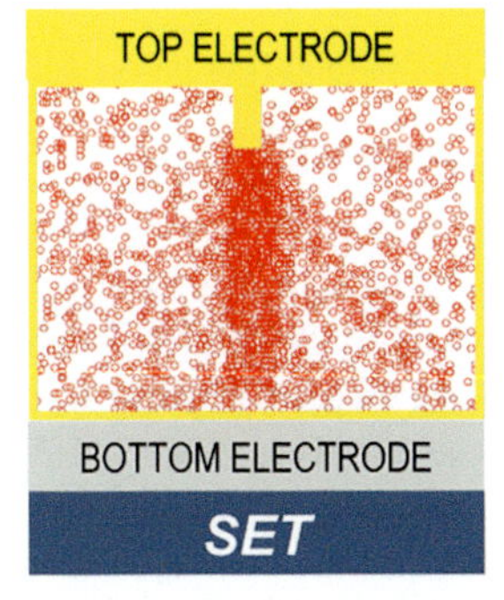

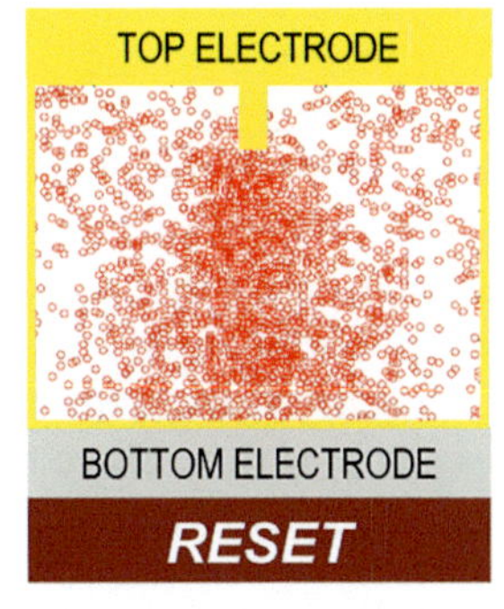

FIG. 2.17 The distribution of oxygen vacancies in the model volume in the original structure after SET and RESET processes for the first switching cycle.

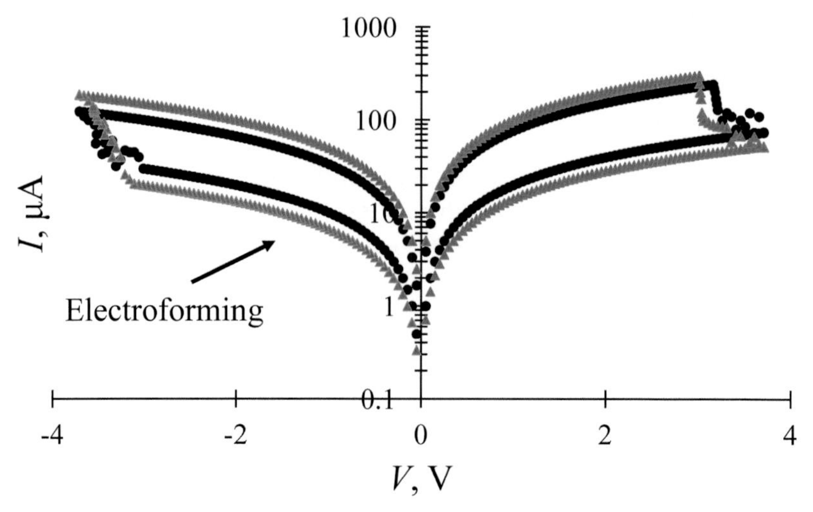

FIG. 2.18 Simulated *I-V* curves for the YSZ-based memristive device.

When switching to the HRS (RESET process), the concentration of vacancies near the central axis decreased, and when switching back to the LRS (SET process), they again move to the center.

Fig. 2.18 shows the simulated *I-V* curves at the EF, two RESET cycles, and one SET cycle. It is interesting that, as in real experiments, there is a cycle-to-cycle variation in the *I-V* curve, which is associated with the stochastic nature of RS.

A comparison of the simulated and measured *I-V* curves (taking into account their device-to-device and cycle-to-cycle variation) shows their good agreement. This indicates that the model satisfactorily reflects the physical processes during the EF and RS of memristive devices based on YSZ.

It should be noted that the Monte Carlo simulation, although it does not obviate the need to use fitting parameters, is more "physical" and more "flexible" as compared to the purely phenomenological approach used in several papers (e.g., Ref. [92]).

In the kMC method, the parameters of materials corresponding to the equilibrium thermodynamic properties are usually used. With its help, it is rather difficult to take into account the changes in the energy parameters of the processes under study, which are caused by the evolution of atomic structure, especially during the fast processes and in time and space at a nonconstant temperature. Since the molecular dynamics (MD) method is based on the numerical integration of the equations of motion of individual atoms, it can take into account such factors and be used as a fairly universal method of considering the behavior of a substance under external influences, in particular, to study the dynamics of equilibrium and nonequilibrium, including the fast processes [102]. However, its practical application is limited by the capabilities of the computer tools used, allowing for a reasonable time only to track processes in relatively small systems and only for short periods of time (from a few nanoseconds to tens of nanoseconds) even with the use of modern cluster computing systems. Nevertheless, the small size (nanometer scale) of the memristive devices and the high speed of the processes in them facilitate this task and allow us to hope for a sufficiently adequate use of the MD method.

As an example, in [103], the MD method is used to simulate an electrochemical memristive cell of Cu/amorphous SiO_2/Cu. Although the electrochemical switching mechanism is not discussed in this chapter, these results are well suited for the demonstration of molecular dynamics capabilities. In the calculations, the classical potential ReaxFF [104] is used, which was developed to describe chemical reactions in systems with different phase states, from solid to gaseous. Its characteristic feature is that the parameters are selected using the results of both ab initio calculations and experimental data. This potential provides a fairly correct description of changes in interatomic interactions with various rearrangements of the atomic structure (including those associated with rearrangement of the electron density in the interatomic space).

In [103], the structure of Cu/amorphous SiO_2/Cu with an area of $7 \times 7\,nm^2$ and an insulator thickness between 1 and 4 nm (the total cell thickness was 6 nm including electrodes). Despite the relatively small size, the use of first-principle methods for describing the processes of filament formation and switching is impossible because of the large size and time scales of the structure and process. Therefore, the MD method using the ReaxFF interatomic interaction potential is used for the simulation, which allows one to take into account the dynamic change of the charge state (electronegativity) of the atoms participating in the simulation. A special procedure is used to simulate a time-varying potential difference applied to the structure. According to the authors, their approach allows taking into account all possible atomic processes involved in the operation of electrochemical memristive devices, including electrochemical decay (electrochemical dissolution), ion transport, nucleation and growth of moving particles, as well as the changes in electric field around the developing asperities and filaments. However, as the authors note, they do not consider the stages of electric current passage and therefore do not take into account the effects associated with the Joule heating of the device. Nevertheless, their study allows the consideration in many details of the EF and RS processes in the simulated memristive cell. The possibility of its ultrafast switching and formation of copper clusters is proved. The typical snapshots during the switching process are shown in Fig. 2.19.

Memristive devices based on silicon oxide are of particular interest, since their fabrication is most well compatible with traditional methods

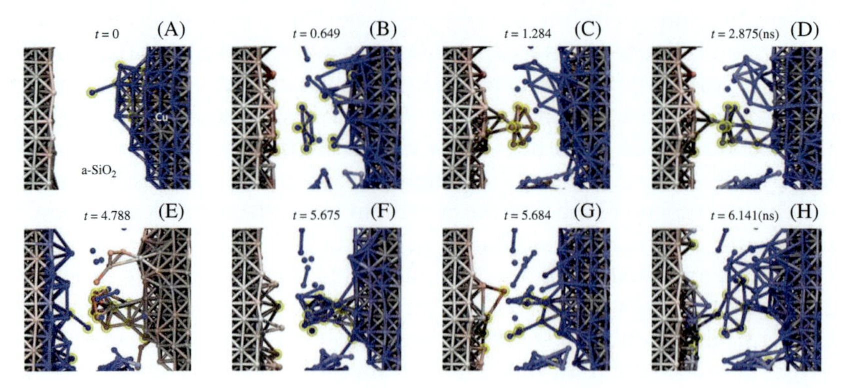

FIG. 2.19 Atomic mechanisms of filament formation and dissolution. (A–H) Snapshots zooming in at the filament during the EF/RESET/SET phases. (D) The *yellow spheres* follow the constituent atoms of the first filament established during the EF phase. In all snapshots, color represents atomic charges, ranging from *red* (−0.3e) to *blue* (+0.3e), and amorphous SiO_2 has been hidden for clarity [103].

of silicon microelectronics technology. At the same time, the mechanisms of EF and RS in them have their own specifics (as compared to TMO-based VCM devices). In [105–107], the most complete study of the properties of SiO_x-based memristive devices demonstrating the VCM type RS is reported. The proposed model of memristive effect is generally similar to that for TMO. The current in the film is provided by the transfer of charge carriers by the trap-mediated mechanism, while the traps are created by oxygen vacancies. The transfer of oxygen ions occurs due to their migration by analogy with the interstitial mechanism for crystals and is characterized by a small value of the energy barrier. The key points in this model are assumptions, first, on the facilitated escape of the oxygen ion from the site under the influence of external electric field and, second, on the determining influence of the Fermi level on the concentration of charged defects. On the basis of these provisions, a design scheme is developed and implemented for 3D simulation of RS for a silica-based memristor [106]. The model is universal in the sense that it includes almost all factors that can affect the movement of ions and charge carriers, such as changes in the temperature field and various electron transfer mechanisms. The calculation scheme is presented in Fig. 2.20.

It should be noted that this approach requires some calibration for comparison with experimental data. In particular, it is necessary to take into account a large number of electronic transport mechanisms that include trap-to-trap tunneling, electrode-trap tunneling, Schottky emission, Fowler-Nordheim tunneling, emission from traps to the conduction band, tunneling from trap to the conduction band, and direct tunneling.

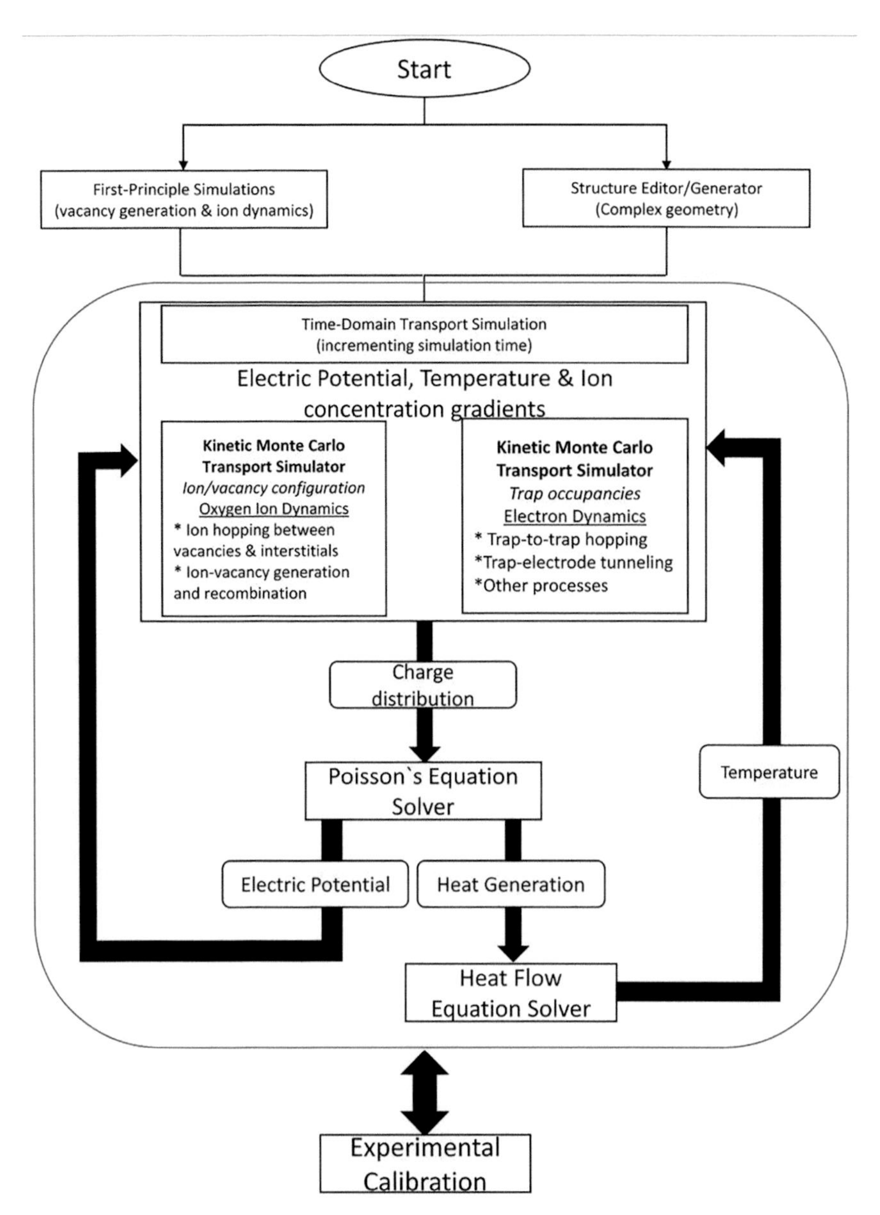

FIG. 2.20 The simulator coupling self-consistently ion and electron transport to the electric field and temperature distributions. *Redrawn from [106].*

Results of calculating the *I-V* curves in this model are shown in [105]. According to the authors, the simulation is in satisfactory agreement with experiment.

In any approach, modeling requires the use of certain parameters of elementary processes. In the case of nanostructures, the experimental determination of these parameters is difficult. Therefore, as mentioned above, the methods of their determination based on theoretical calculations are of particular importance. Activities in this direction for the films of amorphous SiO_2 began long time ago [108]. In recent studies [109,110], the parameters of the formation and migration of vacancies and interstitial atoms are calculated using the first-principle DFT method. This calculation method is applied to SiO_2 clusters, the atomic coordinates of which are determined as a result of bringing them to an equilibrium state by the MD method using the ReaxFF potential. In all reports on the MD simulation of amorphous SiO_2, the initial positions of the atoms are determined according to the following scheme. The crystal cluster (SiO_2) is heated to a high temperature (about 5000 K), then kept at this temperature for some time (~1 ps), then cooled to 0 K. The most interesting results of [109,110] are the detection for amorphous SiO_2 of the variation in the energy parameters of the formation of defects in a fairly wide range depending on the local environment and the electron density distribution accompanying their motion in the oxide.

The dependence of the parameters of atomic processes in an amorphous state on the position of an atom greatly complicates the modeling of memristive effect using the kMC method and requires proper consideration of this circumstance. Strictly speaking, such an account is also required for crystalline oxides, since filaments in them are usually formed in areas with increased disorder, such as grain boundaries [97]. In addition, even in a nominally perfect crystal, with a high concentration of oxygen vacancies characteristic of the filament and the surrounding areas, significant fluctuations in the energy parameters are possible, which are usually not taken into account. This disadvantage can be partially compensated by adjusting the parameters according to the experiment. In the future, as the efficiency of computing systems increases, it will be possible to perform a rather strict simulation.

In particular, in one of the first works devoted to the study of memristive devices [111], the transition of the amorphous TiO_2 phase to one of the crystalline polymorphic modifications of titanium dioxide (anatase) was observed.

In order to verify the effect of increased concentration of oxygen vacancies on the local structure of amorphous SiO_2, the MD simulation of this situation was carried out on the Lobachevsky high-power computing cluster [112]. The structure of amorphous SiO_2 is obtained by the standard method of heating to melting and subsequent rapid cooling [110]. The size of the simulated cluster is $60 \times 60 \times 60\,\text{Å}^3$. It includes 11,616 Si and O nodes. Then, oxygen nodes (created oxygen vacancies) are randomly removed in the central area of the cluster $14 \times 14 \times 14\,\text{Å}^3$ to provide the percentage of oxygen in accordance with the stoichiometric formula

$SiO_{1.7}$. The cluster constructed in this way is minimized (bringing the system to a minimum of potential energy) at temperatures of 0 K and 900 K for 10^8 time steps with duration of 10^{-4} ps, which corresponds to the total physical process time ≈ 10 ns. The calculation is carried out using the LAMMPS program [113] in the thermostat mode NVT. These commands perform time integration on Nose-Hoover style non-Hamiltonian equations of motion which are designed to generate positions and velocities sampled from the canonical (NVT), isothermal-isobaric (NPT), and isenthalpic (NPH) ensembles. This updates the position and velocity for atoms in the group each time step. To maintain the desired temperature, the temp/rescale algorithm is used. To describe the interatomic interaction, one of the modifications of the Tersoff potential [114] is used.

Then a cluster point is selected, around which a maximum local density of the introduced oxygen vacancies is created in a cube of limited size (due to the fact that their placement is random). In this case, it is a cube with dimensions of $7 \times 7 \times 7\,\text{Å}^3$, containing 9 vacancies out of a total of 24. It is assumed that the effect of vacancy defects will manifest itself in this area as much as possible. For the region found in this way, the Radial Distribution Functions (RDF) of the mutual distribution of silicon nodes (Si-Si) are calculated. Calculations are carried out for both defect-free and defect-containing structures of silicon dioxide at temperatures of 0 K and 900 K (in memristive devices, this temperature is reached during the SET process). To obtain static averaged atomic coordinates, their averaging is performed according to the data of the last 10,000 time steps. The results are presented in Figs. 2.21 and 2.22.

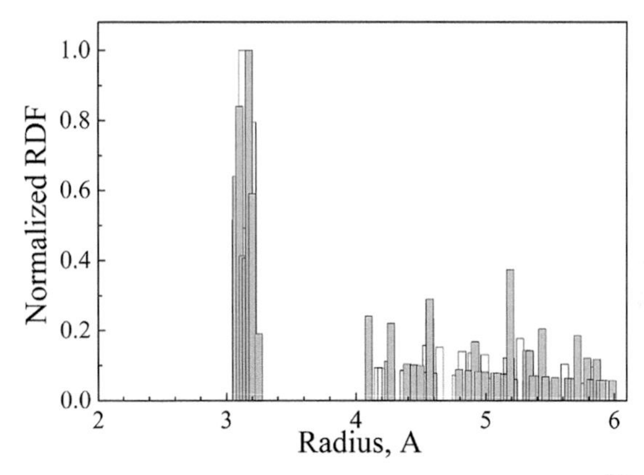

FIG. 2.21 Normalized RDF of silicon nodes in the simulated $7 \times 7 \times 7\,\text{Å}^3$ volume without vacancies and with 24 vacancies at a temperature of 0 K (*white* and *gray* columns, respectively).

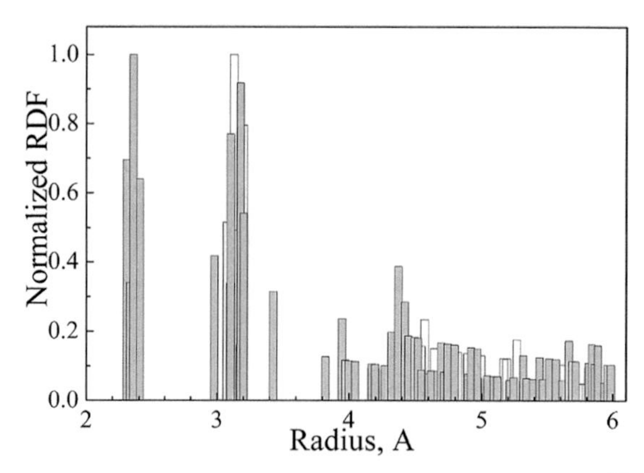

FIG. 2.22 Normalized RDF of silicon nodes in the simulated $7 \times 7 \times 7 \text{Å}^3$ volume without vacancies at a temperature of 0K and with 24 vacancies at a temperature of 900K (*white* and *gray* columns, respectively).

It is clearly seen that in the volume under study the relative positions of silicon nodes have changed significantly. Comparison with the interatomic distance of crystalline Si equal to 2.34Å suggests that the observed process may be the initial stage of the formation of Si nanoclusters.

The observed time scale, for which these changes occur, gives reason to believe that these processes are also probable in real memristive devices based on silicon oxide. If this is indeed the case, then the thermodynamic parameters of these kinetic processes can hardly be considered constant, which may be one of the reasons for the stochastic behavior of memristive devices. It can also be assumed that the accumulation of structural transformations, the initial stage of which is observed during modeling, may be the cause of the degradation of memristive devices with time.

2.4 Conclusions

The provided analysis of extensive literature data shows that significant progress has been made in understanding the specific microscopic mechanisms of resistive switching in metal-oxide memristive materials and devices based on the valence-change mechanism (anionic mechanism). The corresponding processes of ion and electron transport under the action of electric field and Joule heating have been studied both experimentally (using local microscopic techniques) and theoretically using a wide range of simulation approaches from the first-principle and atomistic to phenomenological. However, even detailed knowledge of the physicochemical mechanisms of growth, local destruction and restoration of filaments in metal-oxide structures obtained using high performance computing does not allow describing correctly the complex response of memristive devices, which is generally determined by a stochastic filament ensemble.

Acknowledgments

This work was supported by the Government of the Russian Federation through the Agreement No. 074-02-2018-330 (2). The local analysis of resistive switching phenomenon was carried out at the shared research facilities of Research and Educational Center for Physics of Solid State Nanostructures at Lobachevsky University.

References

[1] D. Ielmini, R. Waser (Eds.), Resistive Switching: From Fundamentals of Nanoionic Redox Processes to Memristive Device Applications, Wiley-VCH, Weinheim, 2016.
[2] I. Vourkas, G.C. Sirakoulis, Emerging memristor-based logic circuit design approaches: a review, IEEE Circuits Syst. Mag. 16 (2016) 15–30.
[3] D. Kuzum, S. Yu, H-S PhilipWong, synaptic electronics: materials, devices and applications, Nanotechnology 24 (2013) 1–22.
[4] S.D. Ha, S. Ramanathan, Adaptive oxide electronics: a review, J. Appl. Phys. 110 (2011). 071101-1-071101-20.
[5] A. Thomas, Memristor-based neural networks, J. Phys. D Appl. Phys. 46 (2013) 1–12.
[6] P. Pfeiffer, I.L. Egusquiza, M. Di Ventra, M. Sanz, E. Solano, Quantum memristors, Sci. Rep. 6 (2016) 1–6.
[7] J.S. Lee, S. Lee, T.W. Noh, Resistive switching phenomena: a review of statistical physics approaches, Appl. Phys. Rev. 2 (2015). 031303-1-031303-57.
[8] A. Wedig, M. Luebben, D.-Y. Cho, M. Moors, K. Skaja, V. Rana, et al., Nanoscale cation motion in TaO_x, HfO_x and TiO_x memristive systems, Nat. Nanotechnol. 11 (2016) 67–74.
[9] R. Waser, R. Dittmann, G. Staikov, K. Szot, Redox-based resistive switching memories—nanoionic mechanisms, prospects, and challenges, Adv. Mater. 21 (2009) 2632–2663.
[10] R. Waser, R. Dittmann, S. Menzel, T. Noll, Introduction to new memory paradigms: memristive phenomena and neuromorphic applications, Faraday Discuss. 213 (2019) 11–27.
[11] A. Mehonic, A. Vrajitoarea, S. Cueff, S. Hudziak, H. Howe, C. Labbe, et al., Quantum conductance in silicon oxide resistive memory devices, Sci. Rep. 3 (2013) 1–8.
[12] A. Mehonic, M. Buckwell, L. Montesi, L. Garnett, S. Hudziak, S. Fearn, et al., Structural changes and conductance thresholds in metal-free intrinsic SiO_x resistive random access memory, J. Appl. Phys. 117 (2015). 124505-1-124505-8.
[13] A. Mehonic, M.S. Munde, W.H. Ng, M. Buckwell, L. Montesi, M. Bosman, et al., Intrinsic resistance switching in amorphous silicon oxide for high performance SiO_x ReRAM devices, Microelectron. Eng. 178 (2017) 98–103.
[14] A. Mehonic, M. Buckwell, L. Montesi, M.S. Munde, D. Gao, S. Hudziak, et al., Nanoscale transformations in metastable, amorphous, silicon-rich silica, Adv. Mater. 28 (2016) 7486–7493.
[15] A. Mehonic, T. Gerard, A.J. Kenyon, Light-activated resistance switching in SiO_x RRAM devices, Appl. Phys. Lett. 111 (2017). 233502-1-233502-5.
[16] A.N. Mikhaylov, A.I. Belov, D.V. Guseinov, D.S. Korolev, I.N. Antonov, D.V. Efimovykh, et al., Bipolar resistive switching and charge transport in silicon oxide memristor, Mater. Sci. Eng. B 194 (2015) 48–54.
[17] C.M.M. Rosário, O.N. Gorshkov, A.P. Kasatkin, I.A. Antonov, D.S. Korolev, A.N. Mikhaylov, et al., Resistive switching and impedance spectroscopy in SiO_x-based metal-oxide-metal trilayers down to helium temperatures, Vacuum 122 (2015) 293–299.

[18] K.C. Chang, T.C. Chang, T.M. Tsai, R. Zhang, Y.C. Hung, Y.E. Syu, et al., Physical and chemical mechanisms in oxide-based resistance random access memory, Nanoscale Res. Lett. 10 (2015) 1–27.
[19] A. Mehonic, S. Cueff, M. Wojdak, S. Hudziak, C. Labbe, R. Rizk, et al., Electrically tailored resistance switching in silicon oxide, Nanotechnology 23 (2012) 1–9.
[20] R. Yamaguchi, S. Sato, Y. Omura, Roles of chemical stoichiometry and hot electrons in realizing the stable resistive transition of sputter-deposited silicon oxide films, Jap. J. Appl. Phys. 56 (2017). 041301-1-041301-6.
[21] A.A. Gismatulin, G.N. Kamaev, Electrophysical properties of Si/SiO_2 nanostructures fabricated by direct bonding, Tech. Phys. Lett. 42 (2016) 590–593.
[22] J. McPherson, J.Y. Kim, A. Shanware, H. Mogul, Thermochemical description of dielectric breakdown in high dielectric constant materials, Appl. Phys. Lett. 82 (2003) 2121–2123.
[23] S. Kim, S.H. Choi, W. Lu, Comprehensive physical model of dynamic resistive switching in an oxide memristor, ACS Nano 8 (2014) 2369–2376.
[24] D.V. Guseinov, D.S. Korolev, A.I. Belov, E.V. Okulich, V.I. Okulich, D.I. Tetelbaum, et al., Flexible Monte-Carlo approach to simulate electroforming and resistive switching in filamentary metal-oxide memristive devices, Model. Simul. Mater. Sci. Eng. 28 (2019) 1–16.
[25] D. Wouters, Resistive switching materials and devices for future memory applications, in: 43rd IEEE Semiconductor Interface Specialists Conference, 2012.
[26] D.B. Strukov, F. Alibart, R. Stanley Williams, Thermophoresis/diffusion as a plausible mechanism for unipolar resistive switching in metal–oxide–metal memristors, Appl. Phys. A 107 (2012) 509–518.
[27] J.J. Yang, D.B. Strukov, D.R. Stewart, Memristive devices for computing, Nat. Nanotechnol. 8 (2013) 13–24.
[28] H. Okamoto, O-Ti (oxygen-titanium), J. Phase Equilib. 22 (2001) 515.
[29] D.H. Kwon, K.M. Kim, J.H. Jang, J.M. Jeon, M.H. Lee, G.H. Kim, et al., Atomic structure of conducting nanofilaments in TiO_2 resistive switching memory, Nat. Nanotechnol. 5 (2010) 148–153.
[30] S.H. Jeon, W.J. Son, B.H. Park, S. Han, Multiscale simulation on electromigration of the oxygen vacancies in metal oxides, Appl. Phys. A Mater. Sci. Process. 102 (2011) 909–914.
[31] K. Fujiwara, T. Nemoto, M.J. Rozenberg, Y. Nakamura, H. Takagi, Resistance switching and formation of a conductive bridge in metal/binary oxide/metal structure for memory devices, Jap. J. Appl. Phys. 47 (2008) 6266–6271.
[32] M. Sowinska, T. Bertaud, D. Walczyk, S. Thiess, M.A. Schubert, M. Lukosius, et al., Hard X-ray photoelectron spectroscopy study of the electroforming in Ti/HfO_2-based resistive switching structures, Appl. Phys. Lett. 100 (2012). 233509-1-233509-5.
[33] N.V. Malekhonova, Profiling of the Composition of Heteronanostructures by Z-Contrast and X-Ray Energy Dispersive Spectroscopy: dis, PhD thesis/Natalya Viktorovna Malekhonova, UNN, 2016, p. 118.
[34] A.N. Mikhaylov, E.G. Gryaznov, A.I. Belov, D.S. Korolev, A.N. Sharapov, D.V. Guseinov, et al., Field- and irradiation-induced phenomena in memristive nanomaterials, Phys. Status Solidi C 13 (2016) 870–881.
[35] D.V. Guseinov, D.I. Tetelbaum, A.N. Mikhaylov, A.I. Belov, M.E. Shenina, D.S. Korolev, et al., Filamentary model of bipolar resistive switching in capacitor-like memristive nanostructures on the basis of yttria-stabilised zirconia, Int. J. Nanotechnol. 14 (2017) 604–617.
[36] D. Filatov, S. Tikhov, O. Gorshkov, I. Antonov, M. Koryazhkina, A. Morozov, Ion migration polarization in the yttria stabilized zirconia based metal-oxide-metal and metaloxide-semiconductor stacks for resistive memory, Adv. Condens. Matter Phys. 2018 (2018) 1–8.
[37] D. Panda, T.Y. Tseng, Growth, dielectric properties, and memory device applications of ZrO_2 thin films, Thin Solid Films 531 (2013) 1–20.

[38] W. Kim, S. Menzel, D.J. Wouters, Y. Guo, J. Robertson, B. Roesgen, et al., Impact of oxygen exchange reaction at the ohmic interface in Ta_2O_5-based ReRAM devices, Nanoscale 8 (2016) 17774–17781.

[39] J. Lee, E.M. Bourim, W. Lee, J. Park, M. Jo, S. Jung, et al., Effect of ZrO_x/HfO_x bilayer structure on switching uniformity and reliability in nonvolatile memory applications, Appl. Phys. Lett. 97 (2010). 172105-1-172105-3.

[40] M.C. Wu, T.H. Wu, T.Y. Tseng, Robust unipolar resistive switching of Co nano-dots embedded ZrO_2 thin film memories and their switching mechanism, J. Appl. Phys. 111 (2012). 014505-1-014505-6.

[41] Y.Y. Chen, R. Roelofs, A. Redolfi, R. Degraeve, D. Crotti, A. Fantini, et al., Tailoring switching and endurance/retention reliability characteristics of HfO_2/Hf RRAM with Ti, Al, Si dopants, in: VLSI Technology: Digest of Technical Papers, 2014.

[42] M.S. Munde, A. Mehonic, W.H. Ng, M. Buckwell, L. Montesi, M. Bosman, et al., Intrinsic resistance switching in amorphous silicon suboxides: the role of columnar microstructure, Sci. Rep. 7 (2017) 1–7.

[43] S. Ambrogio, S. Balatti, A. Cubeta, A. Calderoni, N. Ramaswamy, D. Ielmini, Statistical fluctuations in HfO_x resistive-switching memory: part I—set/reset variability, IEEE Trans. Electron Devices 61 (2014) 2912–2919.

[44] S. Balatti, S. Ambrogio, D.C. Gilmer, D. Ielmini, Set variability and failure induced by complementary switching in bipolar RRAM, IEEE Electron Device Lett. 34 (2013) 861–863.

[45] A. Fantini, L. Goux, R. Degraeve, D.J. Wouters, N. Raghavan, G. Kar, et al., Intrinsic switching variability in HfO_2 RRAM, in: 5th IEEE International Memory Workshop, 2013.

[46] D.S. Korolev, A.I. Belov, E.V. Okulich, V.I. Okulich, I.N. Antonov, E.G. Gryaznov, et al., Manipulation of resistive state of silicon oxide memristor by means of current limitation during electroforming, Superlattice. Microst. 122 (2018) 371–376.

[47] M. Trapatseli, S. Cortese, A. Serb, A. Khiat, T. Prodromakis, Impact of ultra-thin Al_2O_{3-y} layers on TiO_{2-x} ReRAM switching characteristics, J. Appl. Phys. 121 (2017). 184505-1-184505-8.

[48] H. Runze, H. Peng, Z. Yudi, C. Zhe, L. Lifeng, L. Xiaoyan, et al., Demonstration of logic operations in high-performance RRAM crossbar array fabricated by atomic layer deposition technique, Nanoscale Res. Lett. 12 (2017) 1–6.

[49] Y. Yang, Y. Takahashi, A. Tsurumaki-Fukuchi, M. Arita, M. Moors, M. Buckwell, et al., Probing electrochemistry at the nanoscale: in situ TEM and STM characterizations of conducting filaments in memristive devices, J. Electrochem. 39 (2017) 1–21.

[50] Y.L. Chen, J. Wang, C.M. Xiong, R.F. Dou, J.Y. Yang, J.C. Nie, Scanning tunneling microscopy/spectroscopy studies of resistive switching in Nb-doped $SrTiO_3$, J. Appl. Phys. 112 (2012). 023703-1-023703-5.

[51] M. Lanza, A review on resistive switching in high-k dielectrics: a nanoscale point of view using conductive atomic force microscope, Materials 7 (2014) 2155–2182.

[52] M. Lanza (Ed.), Conductive Atomic Force Microscopy: Applications in Nanomaterials, Wiley-VCH, Weinheim, 2017.

[53] P. de Wolf, E. Brazel, A. Erickson, Electrical characterization of semiconductor materials and devices using scanning probe microscopy, Mater. Sci. Semicond. Process. 4 (2001) 71–76.

[54] P. Eyben, M. Xu, N. Duhayon, T. Clarysse, S. Callewaert, W. Vandervorst, Scanning spreading resistance microscopy and spectroscopy for routine and quantitative two-dimensional carrier profiling, J. Vac. Sci. Technol. B 20 (2002) 471–478.

[55] X. Blasco, J. Pétry, M. Nafría, X. Aymerich, O. Richard, W. Vandervorst, C-AFM characterization of the dependence of $HfAlO_x$ electrical behavior on post-deposition annealing temperature, Microelectron. Eng. 72 (2004) 191–197.

[56] J. Pétry, W. Vandervorst, X. Blasco, Effect of N_2 anneal on thin HfO_2 layers studied by conductive atomic force microscopy, Microelectron. Eng. 72 (2004) 174–179.

[57] K. Szot, R. Dittmann, W. Speier, R. Waser, Nanoscale resistive switching in $SrTiO_3$ thin films, Phys. Status Solidi 1 (2007) R86–R88.
[58] U. Celano, L. Goux, A. Belmonte, K. Opsomer, A. Franquet, A. Schulze, et al., Three-dimensional observation of the conductive filament in nanoscaled resistive memory devices, Nano Lett. 14 (2014) 2401–2406.
[59] G. Binnig, H. Rohrer, In touch with atoms, Rev. Mod. Phys. 71 (1999) S324–S330.
[60] Y. Hou, U. Celano, L. Goux, L. Liu, A. Fantini, R. Degraeve, et al., Sub-10nm low current resistive switching behavior in hafnium oxide stack, Appl. Phys. Lett. 108 (2016). 123106-1-123106-3.
[61] M. Rogala, G. Bihlmayer, W. Speier, Z. Klusek, C. Rodenbücher, K. Szot, Resistive switching of a quasi-homogeneous distribution of filaments generated at heat-treated TiO_2(110)-surfaces, Adv. Funct. Mater. 25 (2015) 6382–6389.
[62] M.H. Sadd, Elasticity: Theory, Applications, and Numerics, second ed., Academic Press, New York, 2009.
[63] D.O. Filatov, D.A. Antonov, O.N. Gorshkov, A.P. Kasatkin, D.A. Pavlov, V.N. Trushin, I.A. Antonov, M.E. Shenina, Investigation of resistive switching in the nanocomposite zirconia films by tunneling atomic force microscopy, in: H. Yang (Ed.), Atomic Force Microscopy (AFM): Principles, Modes of Operation and Limitations, Nova Science, New York, 2014, pp. 335–355.
[64] D.S. Jeong, R. Thomas, R.S. Katiyar, J.F. Scott, H. Kohlstedt, A. Petraru, et al., Emerging memories: resistive switching mechanisms and current status, Rep. Prog. Phys. 75 (2012) 1–31.
[65] C. Yoshida, K. Kinoshita, T. Yamasaki, Y. Sugiyama, Direct observation of oxygen movement during resistance switching in NiO/Pt film, Appl. Phys. Lett. 93 (2008). 042106-1-042106-3.
[66] B.J. Choi, D.S. Jeong, S.K. Kim, C. Rohde, S. Choi, J.H. Oh, et al., Resistive switching mechanism of TiO_2 thin films grown by atomic-layer deposition, J. Appl. Phys. 98 (2005). 033715-1-033715-10.
[67] J.Y. Son, Y.H. Shin, Direct observation of conducting filaments on resistive switching of NiO thin films, Appl. Phys. Lett. 92 (2008). 222106-1-222106-3.
[68] V. Iglesias, M. Porti, M. Nafría, X. Aymerich, P. Dudek, T. Schroeder, et al., Correlation between the nanoscale electrical and morphological properties of crystallized hafnium oxide-based metal oxide semiconductor structures, Appl. Phys. Lett. 97 (2011). 262906-1-262906-3.
[69] M. Lanza, G. Bersuker, M. Porti, E. Miranda, M. Nafría, X. Aymerich, Resistive switching in hafnium dioxide layers: local phenomenon at grain boundaries, Appl. Phys. Lett. 101 (2012). 193502-1-193502-5.
[70] Y. Hou, U. Celano, L. Goux, L. Liu, R. Degraeve, Y. Cheng, et al., Multimode resistive switching in nanoscale hafnium oxide stack as studied by atomic force microscopy, Appl. Phys. Lett. 109 (2016). 023508 1 023508 5.
[71] S. Long, X. Lian, C. Cagli, X. Cartoixà, R. Rurali, E. Miranda, et al., Quantum-size effects in hafnium-oxide resistive switching, Appl. Phys. Lett. 102 (2013). 183505-1-183505-4.
[72] S. Tapertzhofen, I. Valov, R. Waser, Quantum conductance and switching kinetics of AgI-based microcrossbar cells, Nanotechnology 23 (2012) 1–6.
[73] D. Filatov, D. Antonov, I. Antonov, A. Kasatkin, O. Gorshkov, Resistive switching in stabilized zirconia films studied by conductive atomic force microscopy, J. Mater. Sci. Chem. Eng. 5 (2017) 8–14.
[74] M.A. Lapshina, D.O. Filatov, D.A. Antonov, Current image formation in combined STM/AFM of metal nanoclusters in dielectric films, J. Surf. Invest.: X-Ray, Synchrotron Neutron Tech. 2 (2008) 616–619.
[75] A.A. Bukharaev, N.V. Berdunov, D.V. Ovchinnikov, K.M. Salikhov, Three-dimensional probe and surface reconstruction for atomic force microscopy using deconvolution algorithm, Scanning Microsc. 12 (1998) 225–234.

[76] D.-Y. Lee, S.-Y. Wang, T.-Y. Tseng, Ti-induced recovery phenomenon of resistive switching in ZrO_2 thin films, J. Electrochem. Soc. 157 (2010) G166–G169.
[77] H.Y. Jeong, J.Y. Lee, S.-Y. Choi, Direct observation of microscopic change induced by oxygen vacancy drift in amorphous TiO_2 thin films, Appl. Phys. Lett. 97 (2010). 042109-1-042109-3.
[78] D.O. Filatov, M.A. Lapshina, D.A. Antonov, O.N. Gorshkov, A.V. Zenkevich, Y.Y. Lebedinskii, Resonant tunnelling through individual au nanoclusters embedded in ultrathin SiO_2 films studied by tunnelling AFM, J. Phys. Conf. Ser. 245 (2010) 1–4.
[79] O.N. Gorshkov, D.O. Filatov, D.A. Antonov, I.N. Antonov, M.E. Shenina, D.A. Pavlov, An oscillator based on a single Au nanoclusters, J. Appl. Phys. 121 (2017). 014308-1-014308-6.
[80] O.N. Gorshkov, D.A. Pavlov, V.N. Trushin, I.N. Antonov, M.E. Shenina, A.I. Bobrov, et al., Peculiarities in the formation of gold nanoparticles by ion implantation in stabilized zirconia, Tech. Phys. Lett. 38 (2012) 185–187.
[81] Y. Yang, P. Gao, L. Li, X. Pan, S. Tappertzhofen, S. Choi, et al., Electrochemical dynamics of nanoscale metallic inclusions in dielectrics, Nat. Commun. 5 (2014) 1–9.
[82] L. Liu, B. Chen, B. Gao, F. Zhang, Y. Chen, X. Liu, et al., Engineering oxide resistive switching materials for memristive device application, Appl. Phys. A Mater. Sci. Process. 102 (2011) 991–996.
[83] L. Zhu, J. Zhou, Z. Guo, Z. Sun, An overview of materials issues in resistive random access memory, J. Mater. 1 (2015) 285–295.
[84] N. Xu, L. Liu, X. Sun, X. Liu, D. Han, Y. Wang, et al., Characteristics and mechanism of conduction/set process in TiN/ZnO/Pt resistance switching random-access memories, Appl. Phys. Lett. 92 (2008). 232112-1-232112-3.
[85] D.S. Jeong, H. Schroeder, R. Waser, Mechanism for bipolar switching in a Pt/TiO_2/Pt resistive switching cell, Phys. Rev. B 79 (2009). 195317-1-195317-10.
[86] S. Larentis, F. Nardi, S. Balatti, D.C. Gilmer, D. Ielmini, Resistive switching by voltage-driven ion migration in bipolar RRAM—part II: modeling, IEEE Trans. Electron Devices 59 (2012) 2468–2475.
[87] M. Noman, W. Jiang, P.A. Salvador, M. Skowronski, J.A. Bain, Computational investigations into the operating window for memristive devices based on homogeneous ionic motion, Appl. Phys. A Mater. Sci. Process. 102 (2011) 877–883.
[88] N. Hashem, S. Das, Switching-time analysis of binary-oxide memristors via a nonlinear model, Appl. Phys. Lett. 100 (2012). 262106-1-262106-3.
[89] H. Abunahla, B. Mohammad, D. Homouz, C.J. Okelly, Modeling valance change memristor device: oxide thickness, material type, and temperature effects, IEEE Trans. Circuits Syst. Regul. Pap. 63 (2016) 2139–2148.
[90] P. Huang, X.Y. Liu, W.H. Li, Y.X. Deng, B. Chen, Y. Lu, et al., A physical based analytic model of RRAM operation for circuit simulation, in: International Electron Devices Meeting, 2012.
[91] S. Ambrogio, S. Balatti, D.C. Gilmer, D. Ielmini, Analytical modeling of oxide-based bipolar resistive memories and complementary resistive switches, IEEE Trans. Electron Devices 61 (2014) 2378–2386.
[92] D. Ielmini, Modeling the universal set/reset characteristics of bipolar RRAM by field– and temperature—driven filament growth, IEEE Trans. Electron Devices 58 (2011) 4309–4317.
[93] A. Marchewka, B. Roesgen, K. Skaja, H. Du, C. Jia, J. Mayer, et al., Nanoionic resistive switching memories: on the physical nature or the dynamic reset process, Adv. Electron. Mater. 2 (2015) 1–13.
[94] E. Miranda, A. Mehonic, W.H. Ng, A.J. Kenyon, Simulation of cycle-to-cycle instabilities in SiO_x-based ReRAM devices using a self-correlated process with long-term variation, IEEE Electron Device Lett. 40 (2019) 28–31.

[95] N.V. Agudov, A.V. Safonov, A.V. Krichigin, A.A. Kharcheva, A.A. Dubkov, D. Valenti, et al., Nonstationary distributions and relaxation times in a stochastic model of memristor, J. Stat. Mech: Theory Exp. 2020 (2020) 1–23.
[96] D. Li, M. Li, F. Zahid, J. Wang, H. Guo, Oxygen vacancy filament formation in TiO_2: a kinetic Monte Carlo study, J. Appl. Phys. 112 (2012). 073512-1-073512-7.
[97] G. Bersuker, D.C. Gilmer, D. Veksler, P. Kirsch, L. Vandelli, A. Padovani, et al., Metal oxide resistive memory switching mechanism based on conductive filament properties, J. Appl. Phys. 110 (2011). 124518-1-124518-12.
[98] Oxygen Ion Conductivity in Yttria Stabilized Zirconia. http://electronicstructure.wikidot.com/predicting-the-ionic-conductivity-of-ysz-from-ab-initio-calc>. (accessed 11.03.16).
[99] I. Karkkanen, A. Shkabko, M. Heillila, M. Vehkamaki, J. Niinisto, N. Aslam, et al., Impedance spectroscopy study of the unipolar and bipolar resistive switching states of atomic layer deposited polycrystalline ZrO_2 thin films, Phys. Status Solidi A 212 (2015) 1–16.
[100] K.-H. Xue, P. Blaise, L.R. Fonseca, Y. Nishi, Prediction of semimetallic tetragonal Hf_2O_3 and Zr_2O_3 from first principles, Phys. Rev. Lett. 110 (2013). 065502-1-065502-5.
[101] P. Meuffels, H. Schroeder, Comment on "Exponential ionic drift: fast switching and low volatility of thin-films memristors" by D.B. Strukov and R.S. Williams in Appl. Phys. A (2009) 94:515-519, Appl. Phys. A Mater. Sci. Process. 105 (2011) 65–67.
[102] G.E. Norman, V.V. Stegailov, Stochastic theory of the classical molecular dynamics method, Math. Models Comput. Simul. 5 (2013) 305–333.
[103] N. Onofrio, D. Guzman, A. Strachan, Atomic origin of ultrafast resistance switching in nanoscale electrometallization cells, Nat. Mater. 14 (2015) 440–446.
[104] A.C.T. van Duin, V.S. Bryantsev, M.S. Diallo, W.A. Goddard, O. Rahaman, D.J. Doren, et al., Development and validation of a ReaxFF reactive force field for cu cation/water interactions and copper metal/metal oxide/metal hydroxide condensed phases, J. Phys. Chem. A 114 (2010) 9507–9514.
[105] A. Mehonic, Resistive Switching in Silicon-Rich Silicon Oxide: Dis, Doctoral thesis/ Adnan Mehonic, UCL, 2014, p. 160.
[106] T. Sadi, L. Wang, D. Gao, A. Mehonic, L. Montesi, M. Buckwell, et al., Advanced physical modeling of SiO_x resistive random access memories, in: International Conference on Simulation of Semiconductor Processes and Devices, 2016.
[107] T. Sadi, A. Mehonic, L. Montesi, M. Buckwell, A. Kenyon, A. Asenov, Investigation of resistance switching in SiO_x RRAM cells using a 3D multi-scale kinetic Monte Carlo simulator, J. Phys. Condens. Matter 30 (2018) 1–17.
[108] P.V. Sushko, S. Mukhopadhyay, A.S. Mysovsky, V.B. Sulimov, A. Taga, A.L. Shluger, Structure and properties of defects in amorphous silica: new insights from embedded cluster calculations, J. Phys. Condens. Matter 17 (2005) S2115–S2140.
[109] D.Z. Gao, A.M. El-Sayed, A.L. Shluger, A mechanism for Frenkel defect creation in amorphous SiO_2 facilitated by electron injection, Nanotechnology 27 (2016) 1–7.
[110] M.S. Munde, D.Z. Gao, A.L. Shluger, Diffusion and aggregation of oxygen vacancies in amorphous silica, J. Phys. Condens. Matter 29 (2017) 1–10.
[111] J.P. Strachan, D.B. Strukov, J. Borghetti, J.J. Yang, G. Medeiros-Ribeiro, R.S. Williams, The switching location of a bipolar memristor: chemical, thermal and structural mapping, Nanotechnology 22 (2011) 1–6.
[112] Volga Research and Education Center for Supercomputing Technologies. http://hpc-education.unn.ru/ru/>. (accessed 06.03.19).
[113] S. Plimpton, Fast parallel algorithms for short-range molecular dynamics, J. Comp. Physiol. 117 (1995) 1–19.
[114] S. Munetoha, T. Motooka, K. Moriguchib, A. Shintanic, Interatomic potential for Si–O systems using Tersoff parameterization, Comput. Mater. Sci. 39 (2007) 334–339.

CHAPTER

3

Charge trapping NVMs with metal oxides in the memory stack

Krishnaswamy Ramkumar

Infineon Technologies, San Jose, CA, United States

3.1 Introduction

With the explosive growth of consumer electronic gadgets such as smart phones, tablets, medical devices and electronic games and the proliferation of electronics into automobiles, the need for non-volatile memories (NVMs) has been growing rapidly in recent years. To meet the needs of these applications, the performance and capacity of NVMs for data and code storage have been improving dramatically during this period. While the need for larger and larger memory capacity is obvious, the industrial and automotive electronics has demanded more and more robust data retention performance over an ever-widening range of temperature. This coupled with the need for low-cost solutions is challenging the semiconductor memory technology. Among all available NVM options, charge trap (CT) memory seems to offer solutions in an ever-widening range of applications. While the oxide-nitride-oxide (ONO) dielectric has been the CT memory stack all these years, new materials are now being explored for the memory stack. The key features of the CT memory stack containing metal oxides and their dependence on material characteristics are discussed in the following sections.

3.2 History of charge trap memory devices

The early CT non-volatile (NV) memory device conceived was the SONOS (silicon-oxide-nitride-oxide-silicon) FET with its structure being very similar to that of a conventional MOS (metal-oxide-semiconductor) FET. The only significant difference is the gate dielectric is replaced by

Metal Oxides for Non-volatile Memory
https://doi.org/10.1016/B978-0-12-814629-3.00003-9

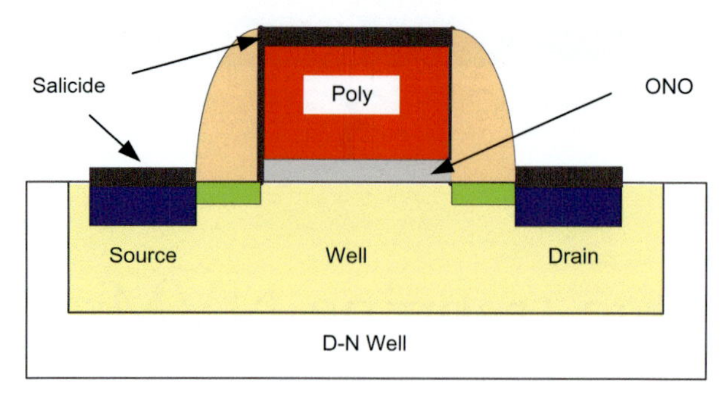

FIG. 3.1 Schematic cross-section of a typical SONOS FET.

an ONO dielectric which acts as the charge storage dielectric stack. In the ONO stack, the nitride layer is the charge storage layer. This device has been studied extensively in the last three decades to understand the carrier transport, trapping and de-trapping mechanisms in the nitride layer. Based on the understanding from these studies, the SONOS FET has been optimized significantly for robust performance and reliability.

A SONOS FET fabrication requires very few changes to the basic CMOS process. The only significant process change is the formation of the ONO dielectric as the gate dielectric of the SONOS FET. Schematic cross section of a typical SONOS FET is shown in Fig. 3.1.

The early SONOS memories required very high (>20 V) program/erase voltages to get significant tunneling currents because of the difficulty in scaling down the thicknesses of the layers of the ONO stack. With the availability of better film deposition and metrology equipment, the ONO stack thickness could be scaled down with robust process control in recent years and SONOS memories capable of program/erase at lower voltages were developed. Additionally, extensive research on charge trap devices yielded novel methods of program and erase that allowed low voltage operation [1]. With this feature, SONOS memories have carved a niche in the NVM market, for both standalone and embedded NVMs.

3.3 SONOS memory devices

The charge trap NV memory stack in a SONOS memory device is shown in Fig. 3.2 as a ONO dielectric made up of a tunnel oxide, a nitride and a blocking oxide. In more general terms, these three dielectric layers can be referred to as

(1) The tunnel oxide layer, used for injecting charges from substrate into the trap layer and to prevent charges stored in the trap layer from being lost to the substrate
(2) The charge trapping nitride layer which stores charge by trapping the mobile carriers in traps located in the band gap

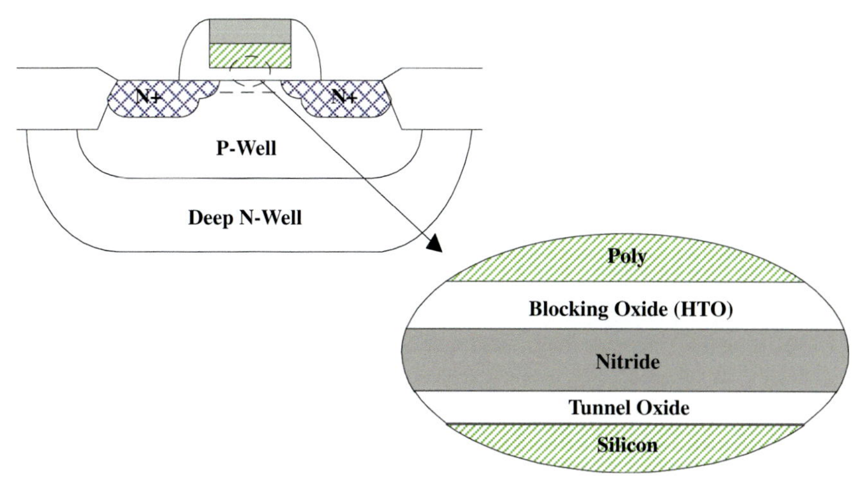

FIG. 3.2 Schematic of SONOS FET showing layers of ONO stack.

(3) The blocking layer which keeps the trapped charge in the trap layer isolated from the gate.

The ONO with some modifications has been the dominant charge trap memory stack so far. However, since the next generation CMOS technologies will have Hafnium or other metal oxides (such as HfSiON or Al_2O_3) in the baseline flow, it is imperative that charge trap memory stacks with metal oxides will be the focus of research and development in the coming years. In fact, such memory stacks have already been studied widely in the last decade. The progress made in this effort will be discussed in subsequent sections. Hereafter, the memory stack will be referred to as CT stack because it is not necessarily ONO.

3.3.1 Traps in the CT NV memory stack

It is obvious that the performance and reliability of the CT memory devices are strongly dependent on the density, distribution and energy levels of the traps in the trapping layer. Studies on trap energy levels and trapping-de-trapping mechanisms over the last three decades have shown that the location of traps in the energy band gap of the trapping layer can vary depending on the material of this layer and the method of deposition [2]. With the increased focus on the use of high-K dielectrics (generally metal oxides) for gate dielectric of CMOS FETs, several studies have focused on the impact of incorporating the high-K materials as charge trapping layer in a CT memory device on the traps in that layer. The high-K materials such as HfO_2 or HfAlON with smaller band gap make the trap levels close to the band edges much deeper than in conventional SONOS stacks. With such deeper traps, the charge retention in the trap layer can be significantly better [3,4].

3.3.2 Program and erase of CT memory device

In a charge trap memory device (which is basically a MOSFET), there are multiple methods by which charge carriers from the channel can be injected into the trapping layer to be trapped.

(1) By applying a high electric field between the gate and the channel, sufficient band bending can be created in the gate stack to enable injection of electrons or holes from the channel into the trap layer through the tunnel oxide by Fowler-Nordheim (FN) tunneling.

(2) By turning ON the MOSFET with an appropriate gate voltage and applying a high enough voltage to the drain. Hot carriers can be generated in the channel and a fraction of the hot carriers will get injected across the tunnel oxide (hot carrier injection, HCI) into the trap layer to get trapped there.

The key difference between these two methods is that for FN tunneling, significant band bending must be achieved with the applied gate voltage and this means that the tunnel oxide must be very thin (<20 Å) whereas with HCI, the carrier injection is less dependent on the tunnel oxide thickness and hence it can be much thicker (30–50 Å). This difference has a big impact on the charge retention, making the HCI based CT memory device to be more robust because of less charge loss through the tunnel oxide. Another difference is that in FN tunneling, carrier injection occurs uniformly over the entire length of the channel whereas in HCI, most of the carrier injection occurs near the drain where the charge carriers have the maximum energy.

The energy band diagram of the gate stack with a simple ONO gate dielectric in FN tunneling based device with no bias voltage applied is shown in Fig. 3.3A.

On application of a voltage to the gate with respect to the channel, band-bending occurs (Fig. 3.3B). In the programing condition, with

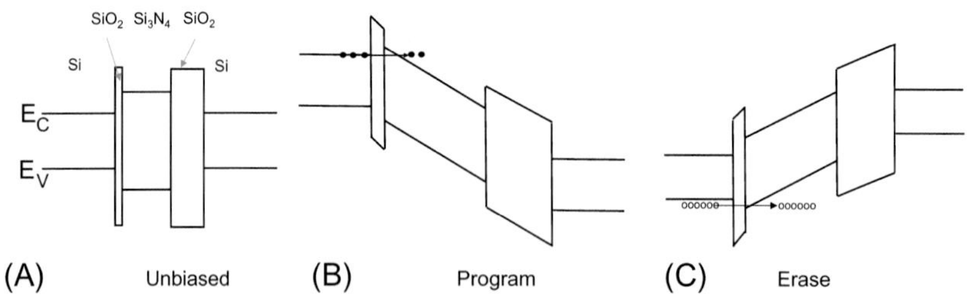

FIG. 3.3 Band diagram of SONOS device for different operating conditions of (A) no bias (B) programmed and (C) erased.

positive bias applied to the gate, electrons from the channel tunnel through the tunnel oxide by FN tunneling into the conduction band of the nitride. Many of them get trapped by the deep level traps in the nitride while the remaining move through the nitride, tunnel through the blocking oxide and get collected by the gate. As long as the gate bias is present, more and more electrons get injected into and trapped in the nitride thereby building up the trapped charge and hence changing the threshold voltage, V_t of the transistor. The change in $V_{t,}$ will saturate only when the built-in electric field due to the stored charge becomes large enough to reduce the FN tunneling from the channel or induce injection of holes from gate through the blocking oxide. A very similar process takes place in the erasing condition, with the negative bias applied to the gate, except that holes are the ones that get trapped and hence change the device V_t. As the stored charge changes, depending on the initial channel doping (enhancement or depletion type), V_t, also changes. This is typically shown as a variation of V_t in the program state (VTP) and erase state (VTE) with the pulse-width of the program or erase pulse. A typical variation of V_t for a depletion mode SONOS FET is shown in Fig. 3.4. The difference between VTP and VTE is referred to as "V_t window."

The saturation of the V_t can be seen for both program and erase states, although it is more pronounced for the erase state. This is because the back injection of electrons from the gate in the erase state is more pronounced than the hole injection in the program state because of the energy barrier difference.

As mentioned earlier, the injection of carriers from the channel into the nitride can also be achieved by hot carrier injection which is also referred to as channel hot electron injection (CHE). As explained earlier, the

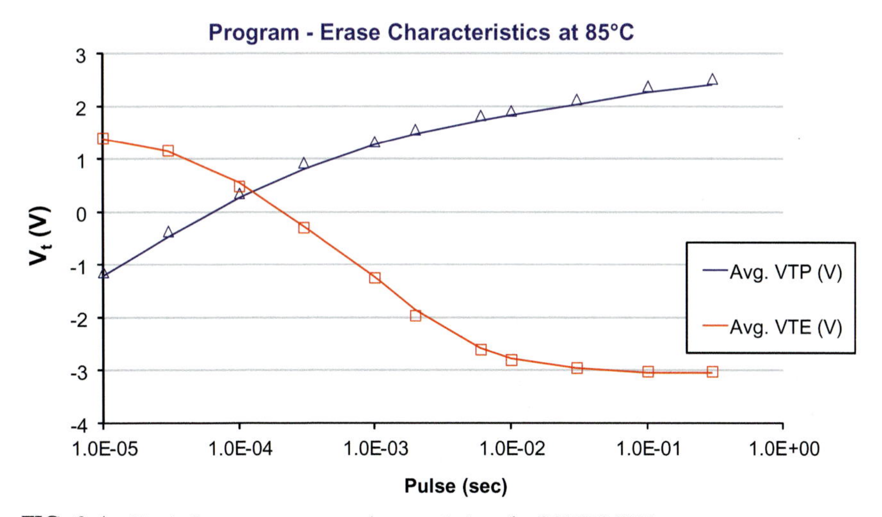

FIG. 3.4 Typical program—erase characteristics of a SONOS FET.

accelerated carriers become highly energetic near the drain end and a fraction of them can cross the tunneling barrier into the trapping layer and get trapped there leading to a change in the V_t of the transistor. In a NMOS device, programming is done by hot electron injection while erasing is done by hot hole injection or hole band to band tunneling. Unlike in FN tunneling, there is a large channel current which can supply a copious amount of charge carriers for injection into the nitride in the case of hot carrier injection. Therefore, significantly faster program/erase is achieved in the case of hot carrier injection (μs) as compared to FN tunneling (ms).

3.4 CT memory cell reliability

In any type of charge trap memory array, it is expected that, ideally, the charge stored in the memory cell is not lost or changed over a long period of time (typically 10–20 years) irrespective of what goes on in the ambient and in neighboring bit cells. However, in reality, the state of the bit cell does not stay constant because of several phenomena that affect the stored charge. The phenomenon that changes the stored charge over time, even with no bias applied, due to loss of the charge from the trap layer is referred to as "retention." The change in stored charge that occurs due to repeated programming and erasing is referred to as "endurance." The impact of these effects is that they slowly shift the threshold voltages of the cell in the program or erase states and beyond some point their sensing becomes problematic. These phenomena will be discussed briefly below.

3.4.1 SONOS endurance

Endurance of a NVM cell refers to the wear out that occurs in the NVM device due to repeated cycling between "program" and "erase" states. The mechanism of the wear out is typically interface state generation due to transfer of high energy charge carriers across the silicon-tunnel dielectric and gate-blocking dielectric interfaces. The interface state density increases with the number of cycles and hence the threshold voltages also change with the number of cycles. Since the interface states disturb the charge injection into the trap layer, the charge trapping can get reduced with cycling and this will result in a decrease of the threshold voltage during program or erase. Such a decrease could lead to a gradual closing of the V_t window by further cycling. Endurance performance of SONOS is generally very good as shown in Fig. 3.5. Typically, there is a minimal shift of V_t even after 10^6 cycles.

This is because the tunnel oxide in SONOS is much thinner and the charge carriers tunnel through it easily thereby causing little damage to the interface. This results in lower density of interface states generated

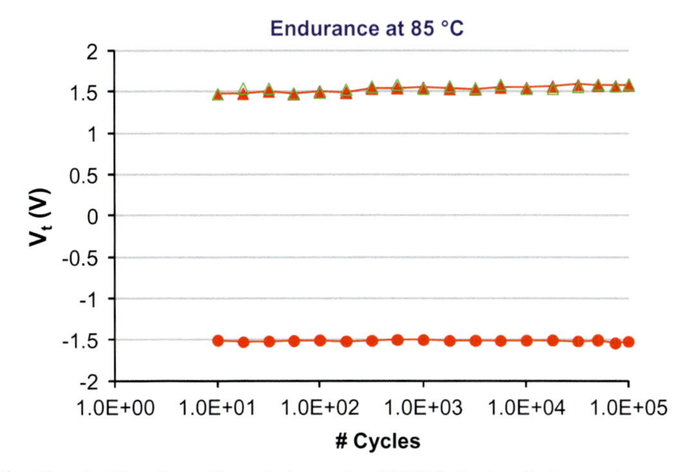

FIG. 3.5 Threshold voltage V_t variations of a SONOS device during endurance cycling.

during cycling and hence very small V_t shifts. The interface state generation in the blocking oxide—gate interface, due to the damage caused by cycling is typically responsible for V_t shift during cycling. The injection of carriers from the gate to the trap layer caused by the erase saturation can lead to charging of these interface states, and hence, a shift of V_t. A solution is to increase the blocking oxide layer thickness, which dramatically reduces carrier injection from gate to the trap layer. However, this increases the thickness of the CT stack which can have negative impact on other electrical parameters.

In another type of endurance effect, the charging up of the existing interface states at the Si-tunnel oxide interface can occur without damage to the interface (less degradation in the charge injection). This effect usually shifts both the Program V_t and erase V_t in the same direction, depending on the nature of the interface charge. This is common in devices that have a thick tunnel oxide. However, in SONOS devices that use FN tunneling for both program and erase, there is little interface charging because the electrons and holes can tunnel into and out of the nitride. Therefore, there is almost no shifting of V_t due to interface charging. The observed V_t shift is mainly due to interface damage which tends to close the V_t window.

3.4.2 Data retention

Data Retention in the NVM cell is the most critical specification because it indicates how long the cell can retain the data written into it. The retention is specified as the number of years at a given ambient temperature for which the state of the memory cell can be read without any ambiguity. The data retention issue arises because the charge stored in the device is slowly

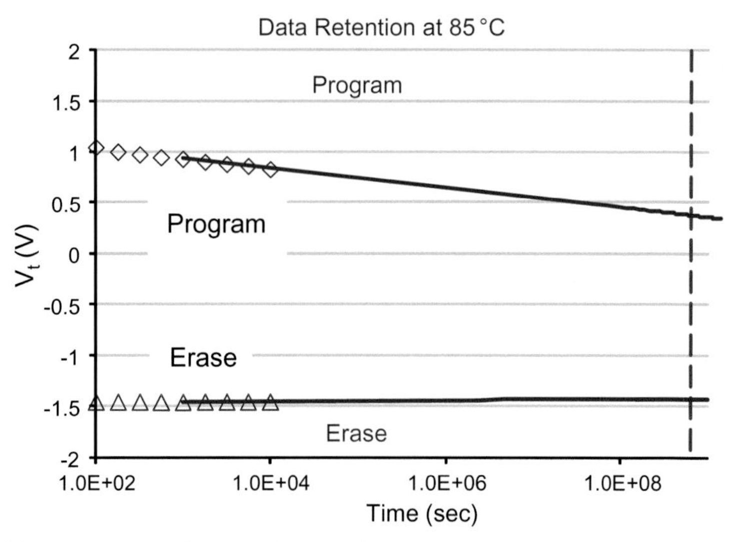

FIG. 3.6 V_t variation of a SONOS device during retention bake.

lost over time. Higher ambient temperature accelerates the charge loss. In the SONOS memory device, the charge loss from the charge trap layer, which is silicon nitride in most devices, can be due mechanisms such as trap assisted tunneling in the nitride and tunneling through the thin tunnel oxide.

The retention performance is characterized by measuring the variation of program and erase threshold voltages with time at an elevated temperature. A typical retention characteristic (worst-case scenario at 85°C) of a SONOS memory array is shown in Fig. 3.6.

As shown in Fig. 3.6, the program and erase threshold voltages decay with time due to the trapped charge being lost. The rate of decay depends on the temperature during the retention testing. The slopes of the program and erase V_t determine the retention performance of the memory. All methods to optimize the CT stack are aimed at reducing these decay rates to enable better retention performance of the memory.

3.5 New materials for charge trap memory stack—Metal oxides

It is clear from the previous sections that the CT stack is the most critical component of any CT memory array and needs the most amount of optimization to get the best endurance and retention performance from the memory. A significant amount of research has been done to change the material or stoichiometry of three layers of the stack to improve performance and reliability without compromising on any other parameter. For the first three or four decades, the focus was on SiO_2, Si_3N_4 and SiON because these were the materials acceptable to the semiconductor industry. Once these options were fully exhausted, the attention turned to the use of other materials. In the last two decades, oxides and nitrides of a variety of elements have been explored for the CT stack. Of these, metal oxides have been a dominant category studied widely. The primary reason for the choice of metal oxides is the compatibility with semiconductor processing. For instance, metal oxides that have been explored for gate dielectrics of CMOS FETs were obvious choices for the CT stack because they have already been studied and optimized for processes such as deposition and patterning. Hence, oxides such as Al_2O_3, La_2O_3, HfO_2 and ZrO_2 have been widely studied from the point of their suitability for the layers of the CT stack and novel stacks have been proposed to address the weaknesses of conventional ONO-stack.

3.5.1 Why metal oxides in CT memory stack?

The conventional ONO stack, although simple to fabricate in high volume manufacturing, suffers from insufficient retention, especially when FN tunneling is used for program/erase. The energy band diagram of the ONO stack is quite straightforward and the only knob for retention improvement is the energy level of the traps. There is a limited amount of optimization of the trap layer that can be done to maximize the effect of traps. However, the introduction of a metal oxide in the CT stack alters the energy band diagram of the stack significantly and this can have a big impact on the stack reliability. Due to much higher dielectric constant than SiO_2 or SiN, many of the metal oxides can be candidates for the next generation CT stack. Metal oxides in small amounts (like doping) can also give rise to deep level traps that can really enhance the retention. These possibilities are illustrated below.

The simplest modification to the ONO stack involves the replacement of SiO_2 blocking oxide by another dielectric with a higher dielectric constant, such as Al_2O_3. In fact, this became a reality when deposition technology for Al_2O_3 became mature and suitable for volume manufacturing. The energy band diagram of the conventional ONO stack is compared in Fig. 3.7 with a CT stack containing metal oxide with higher dielectric constant, Al_2O_3, acting as the blocking oxide [5].

As can be seen clearly, the band bending and hence the electric field in the metal oxide (Al_2O_3) is lower than in SiO_2 of the conventional ONO stack. This will reduce the carrier back-injection from the gate to the nitride since the tunneling distance for these carriers becomes larger, as shown in the figure. This will greatly reduce the erase saturation of the memory device which should improve the endurance performance.

High-K blocking oxide directly in contact with the trapping layer has been found to degrade the reliability performance of the CT stack. Because of this, generally, a thin SiO_2 layer is sandwiched between the trap layer

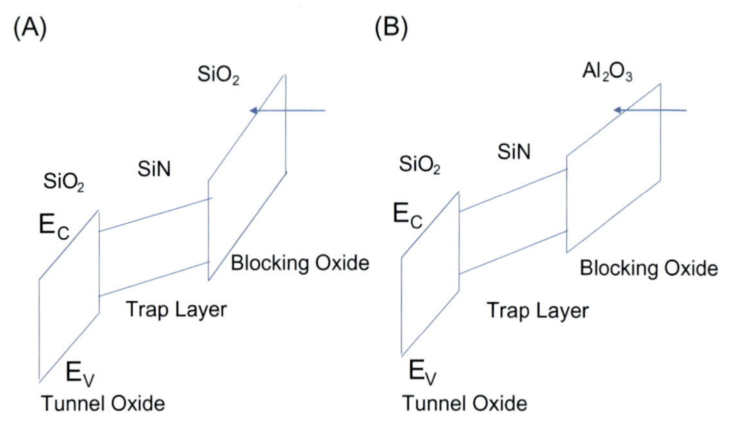

FIG. 3.7 Energy band diagrams of CT stacks with (A) SiO_2 and (B) Al_2O_3 blocking oxides [5].

and the high-K dielectric layer. This will be discussed in more detail in Section 3.5.2.1.

Metal oxide with high dielectric constant can also replace the trap layer provided it has high enough density of traps. The energy band diagram of this type of stack is shown in Fig. 3.8 [6].

The reason to use a metal oxide instead of the conventional SiN for the trap layer is mainly for the higher trap density and for the deep energy level (E_t) of the traps. As already explained, metal oxides such as HfO_2 and HfSiON are already used as gate dielectric for CMOS FETs and it will be relatively easy to adopt these materials for CT memory devices as well.

Substitution of SiO_2 by a metal oxide as tunnel oxide is also an option, when the metal oxide has a significantly higher dielectric constant than SiO_2 and is trap free (both in bulk and interface with substrate). The higher dielectric constant enables the use of a physically thick layer of the metal oxide for the tunnel oxide and a physically thicker tunnel dielectric can

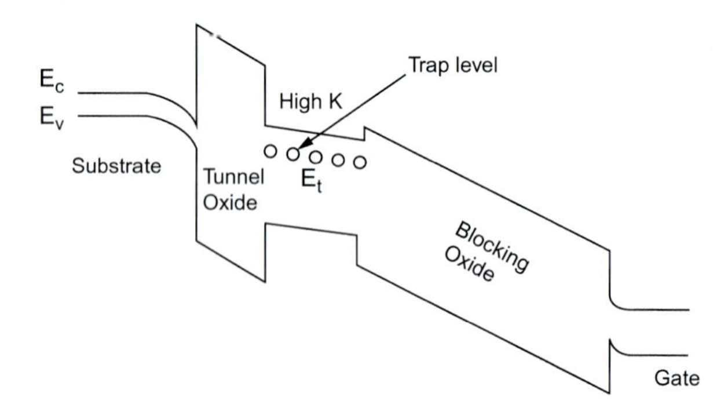

FIG. 3.8 Energy band diagram of a CT stack with high-K trap layer.

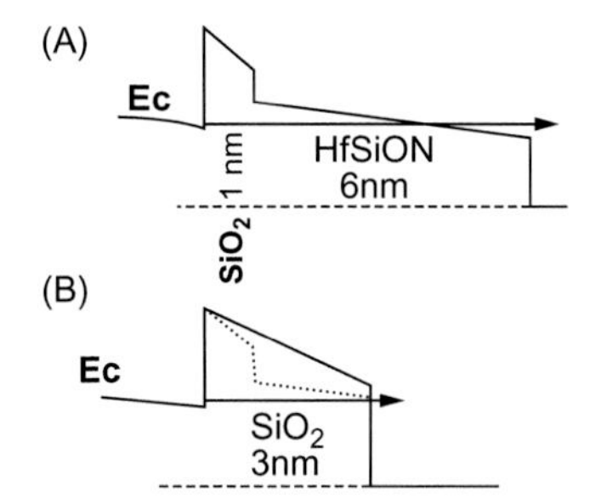

FIG. 3.9 Energy barrier at the Si-tunnel oxide interface (A) a high-K tunnel oxide, and (B) SiO_2.

greatly improve the retention performance of the CT memory device. Since the electrical thickness of the tunnel dielectric is small, there will be minimal impact on the FN injection from the substrate to the CT layer. The band diagram of a CT stack with a metal oxide as part of tunnel oxide is shown in Fig. 3.9 for the case of HfSiON [7].

As depicted in Fig. 3.9, the lower band gap of the high-K dielectric and a thinner SiO_2 enables easier injection of carriers out of the tunnel dielectric on to the trapping layer. This ensures robust program/erase performance of the device. During retention, the entire thickness of the tunnel dielectric (SiO_2+high-K) comes into play and this greatly reduces the charge loss from the trap layer to the substrate. Hence replacing the SiO_2 tunnel oxide by SiO_2/high-K sandwich can greatly improve retention performance. However, the quality of the high-K dielectric becomes critical for this type of stack. If the high-K dielectric or its interfaces with the SiO_2 or the trapping layer has too many interface traps, which are essentially shallow traps, then they will trap a significant number of carriers but lose them quickly thereby degrading the overall retention performance of the CT memory device. Having a thin SiO_2 between the substrate and the high-K-material eliminates the problem at one of the interfaces. Making the high-K layer trap free will also help improve reliability.

The above discussion shows that metal oxides can potentially replace any of the three layers of the CT stack. However, the trap density in this material plays a crucial role in determining the best location for the layer.

There is a wide variety of metal oxides that have been explored for different layers of the CT stack and the key ones will be discussed below.

3.5.2 Charge trap memory stack with high-K metal oxides

Logic CMOS process flows at 28 nm and beyond are already primarily based on the high-K metal gate option in which a Hafnium based oxide ($K>20$) is used as the gate dielectric for the high-performance core devices and metals with appropriate work functions are used as gates. Therefore, it is imperative that when CT memories are made or embedded into 28 nm technology and beyond, CT stacks with high-K dielectrics will need to be developed and optimized for performance and reliability. In fact, HfO_2 may become the most widely used metal oxides in future CT memory devices. Therefore, it is critical to understand the impact, positive or negative, of high-K dielectric on the CT stack performance and reliability. Depending on how the SONOS device is integrated, the high-K dielectric can replace or be a part of different layers of the SONOS stack.

3.5.2.1 *High-K dielectric for blocking layer*

In a conventional SONOS device, the simplest memory stack would have the metal oxide based high-K metal oxide as part of the blocking layer. Replacing SiO_2 blocking oxide by a high-K dielectric changes the energy bands as shown in Fig. 3.10.

When SiO_2 is replaced by high-K blocking oxide, the band bending and electric field (E-field) in that layer decreases while the band bending, and E-field increase in the trap layer and tunnel oxide. This can improve the FN injection from the substrate to the trap layer which will increase the program/erase speed. Alternatively, it allows the use of a thicker tunnel oxide to get the same band bending as in the case of SiO_2 blocking oxide. This can significantly reduce the charge loss during retention. Both approaches are used in memories. Also, a thicker blocking oxide reduces back tunneling from gate to the trap layer. One of the early studies on this topic was reported by Duurenn et al. [8]. In this study, the HfSiON was incorporated as part of the blocking layer and its impact on the program-erase behavior was examined. The threshold voltage variation during program-erase clearly indicates that the high-K dielectric prevents the erase saturation seen in the pure ONO stack (Fig. 3.11).

In the study by Tan et al., CT stacks with Al_2O_3 and HfO_2 as blocking oxide were compared [9]. The results showed that HfO_2 can degrade the barrier for trapped charge in the trap layer, as shown in Fig. 3.12.

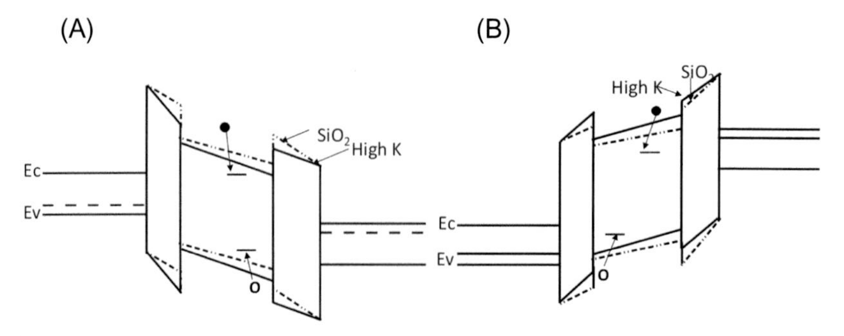

FIG. 3.10 Energy band diagram of CT stack with SiO_2 and high-K blocking layer for (A) program state and (B) erase state.

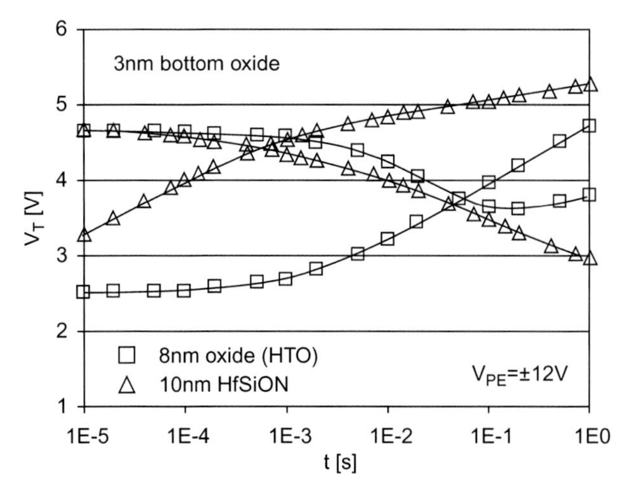

FIG. 3.11 Threshold voltage variation of a SONOS device with SiO_2 and HfSiON blocking oxide [8].

As is clear from the band diagram in Fig. 3.12, the barrier for carriers from the gate to the trap layer is much lower for HfO_2 as compared to Al_2O_3. This will cause greater charge loss in the case of HfO_2 blocking layer and degrade the retention as shown in Fig. 3.13.

However, because of the much higher dielectric constant of HfO_2, a thick layer of HfO_2 can be used for the blocking dielectric to get the same electrical thickness as Al_2O_3. Thicker blocking oxide will dramatically lower the tunneling of carriers from the gate to the trap layer. A similar finding has been reported by Tang et al. [10]. Their study found that HfAlO blocking dielectric shows better retention than HfO_2 but not as good as Al_2O_3.

Another approach to improve retention with HfO_2 is to sandwich a SiO_2 layer in between the trap layer and the HfO_2. Lai et al. [11] published a detailed study on the impact of a buffer oxide layer in between the blocking high-K layer and the nitride layer of the ONO stack. It was found that

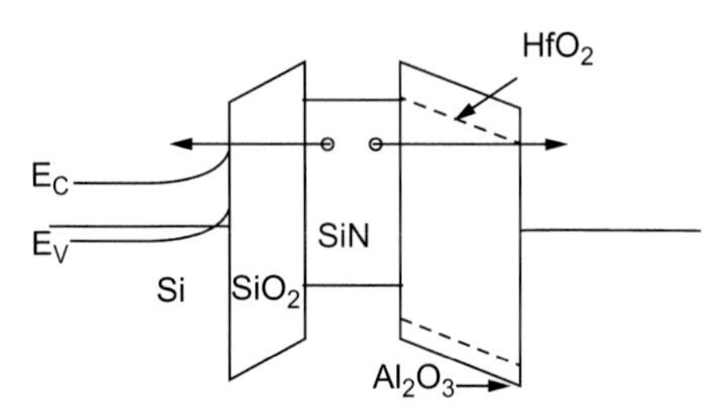

FIG. 3.12 Energy band diagram of CT stack during retention using Al_2O_3 and HfO_2 blocking dielectrics.

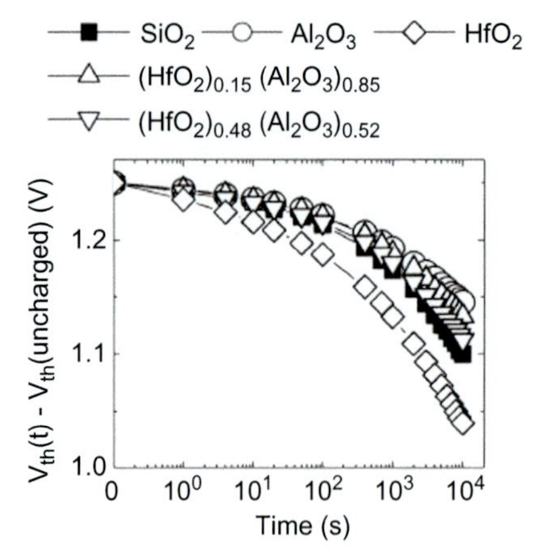

FIG. 3.13 Retention performance of CT stacks with different metal oxide blocking layers [9].

with the high-K layer (HfO_2 or Al_2O_3) directly deposited on the nitride, the retention performance is severely degraded, as shown in Fig. 3.14.

However, when an oxide layer is sandwiched between the high-K layer, the retention is much improved. This has been attributed to the reduction in charge loss due to the presence of the oxide layer.

The key message from all these studies is that using HfO_2 as the blocking layer, although helpful to prevent erase saturation can degrade the retention unless the thickness is optimized and a buffer layer of SiO_2 is inserted between the trap layer and the HfO_2 layer.

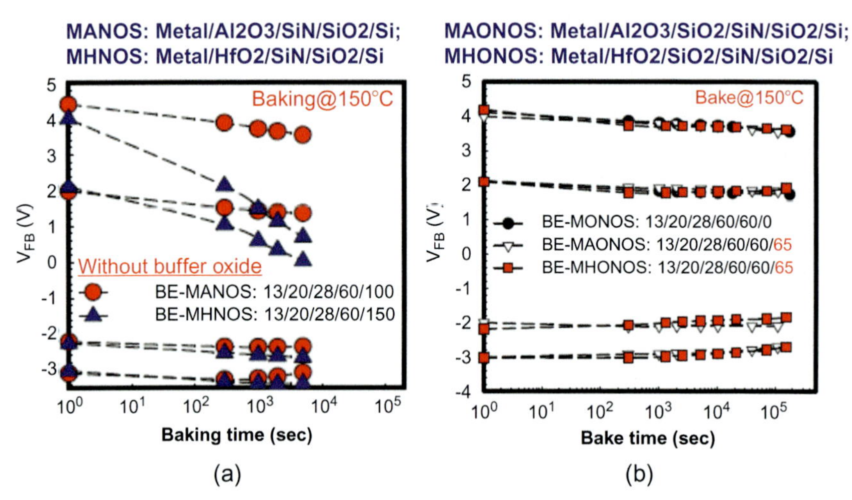

FIG. 3.14 Impact of buffer oxide on retention of CT stacks with metal oxide blocking layer [11].

3.5.2.2 *High-K dielectric for tunneling layer*

Another approach for advanced CT stack is to use a metal oxide based high-K dielectric as part of the tunneling layer. The key motivation for this is to replace the thin tunneling SiO_2 layer with a much thicker high-K oxide layer which can preserve the same programming efficiency as SiO_2 tunnel oxide but can improve the data retention significantly by making the physical barrier for charge loss much thicker. As an alternate approach, one can use the same thickness of high-K layer as SiO_2 but take advantage of the impact of the high-K on the hole tunneling from the substrate to the trap layer on account of the lower band offset of the high-K layer. This enables a deeper erase with the high-K tunneling layer as shown in Fig. 3.15 [8].

HfSiON as part of the tunneling layer was investigated by Molas et al. [7]. The CT stack studied in this research is shown in Fig. 3.16. With the HfSiON as part of the tunneling layer, the leakage through that layer gets reduced dramatically as shown in Fig. 3.17 which compares the leakage current density J_G in a SiO_2 tunneling layer vs SiO_2-HfSiON layer.

With a band gap engineered CT stack having HfSiON as part of the tunneling layer a significant improvement in endurance was observed, as shown in Fig. 3.18.

A study on CT stack with high-K dielectrics for both the bottom and top layers [12] showed superior erase characteristics of the memory stack. The adaption of such stacks by the industry is more dependent on the complexity of process integration than the benefit for the reliability of the stack.

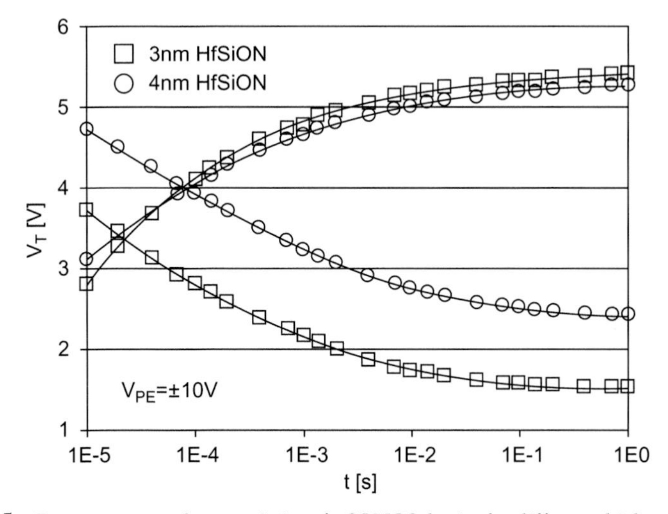

FIG. 3.15 Program-erase characteristics of a SONOS device for different thicknesses of the high-K dielectric tunneling layer [8].

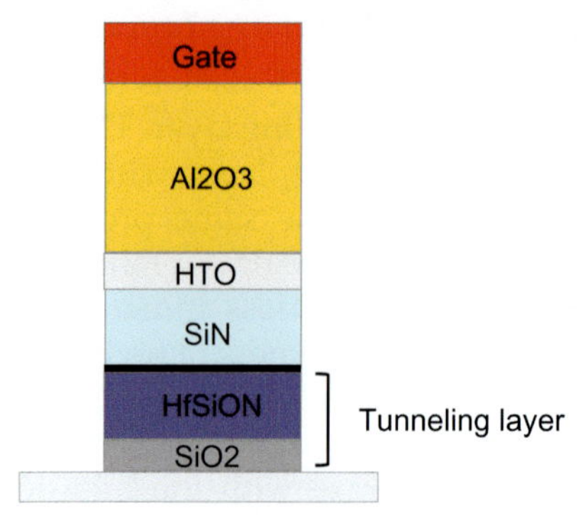

FIG. 3.16 CT stack with HfSiON as part of the tunneling layer.

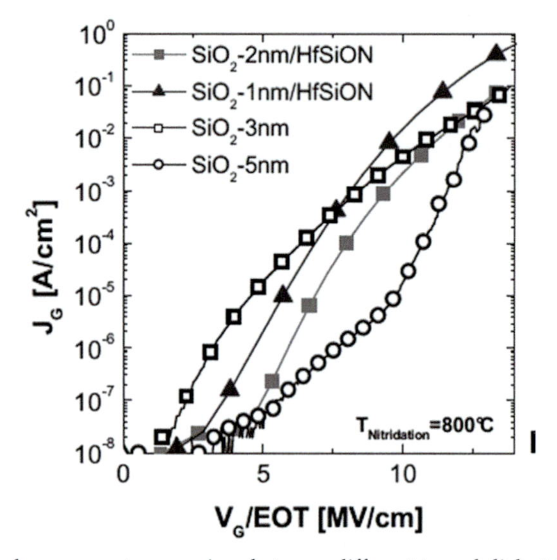

FIG. 3.17 Leakage current comparison between different tunnel dielectrics [7].

3.5.2.3 *High-K dielectric for trap layer*

Another option that has been evaluated is the use of high-K dielectric as the trapping layer in the SONOS stack. Here the choice of the material and processing is directed toward generating a trap rich layer. Techniques such as Nitrogen implantation are adopted to make the high-K layer suitable. When the materials and deposition tools become more readily available, it may become feasible to have an all high-K layer charge trap memory stack.

Replacing the SiN trap layer by a high-K metal oxide trap layer changes the energy band diagram significantly and that can improve the reliability. The band diagram of a stack using HfAlO as the trap layer, in program and erase states is shown below (Fig. 3.19) and compared with that of a standard ONO stack [9].

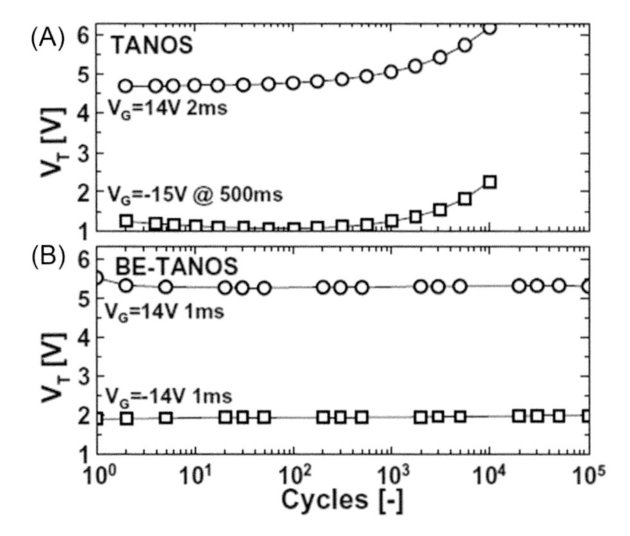

FIG. 3.18 Comparison of endurance between (A) TANOS and (B) TANOS with HfSiON showing improvement in endurance [7].

The figure shows the bands being flat in the unbiased condition of the CT stack. With bias applied during program or erase, the bands bend. Because the band shifts between the SiN trap layer and the HfAlO trap layer, the band bending due to an applied electric field is also different as shown in Fig. 3.20. The consequence of this is a significant change in both the FN injection during program-erase and the charge loss during retention. Tunneling distance for electron injection into the trap layer increases while it increases for hole injection.

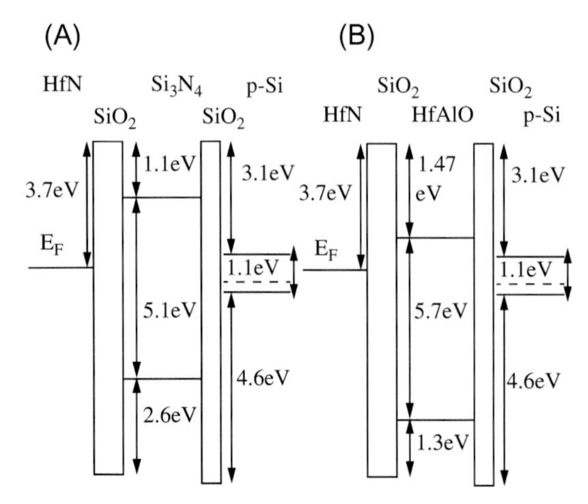

FIG. 3.19 Energy band diagram comparison between CT stacks with (A) SiN and (B) HfAlO trap layers [9].

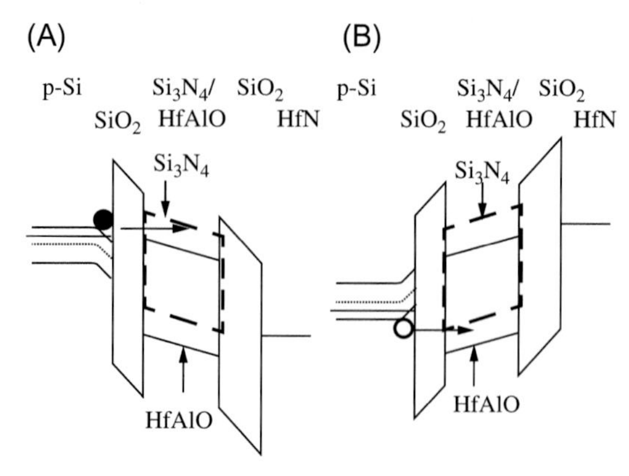

FIG. 3.20 Band bending in CT stacks with SiN and HfAlO trap layers during (A) program and (B) erase [9].

Another example is a CT stack that uses HfON as the trap layer. Yang et [13] have studied this CT stack in detail. According to the findings of this study, by increasing the N content in $Hf_{1-x-y}N_xO_y$ both the memory window and high temperature retention improved. The authors examined several metal oxides from the point of the energy bands with respect to Si and SiO_2 and determined that HfON had the best band offset relative to SiO_2 (dependent on N concentration added to HfO_2) which could result in deep trap levels that result in very robust retention (Fig. 3.21).

They further explored varying the concentration of N in HfON and found that $Hf_{0.3}N_{0.2}O_{0.5}$ shows the best window and low retention decay (Fig. 3.22) in a CT stack that used HfLaON as the blocking dielectric.

A study published by Buckley et al. [6] evaluated a variety of Hf based oxides for the trap layer of SONOS. This study concludes that the crystallization of the high-K layer after deposition holds the key to its success as a trap layer. The results show the more crystalline the high-K layer the higher is the trapping capability. In addition to deeper traps, the study also found the smaller trap cross section in the Hf based high-K dielectric is the reason of the superior retention performance (Fig. 3.23).

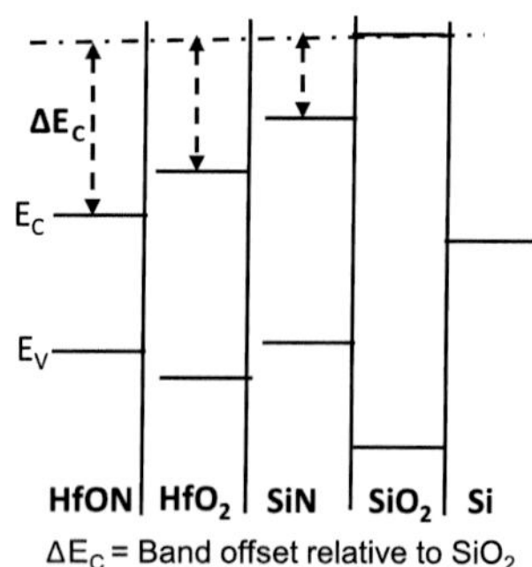

FIG. 3.21 Energy band diagrams of different metal oxides showing band offsets [13].

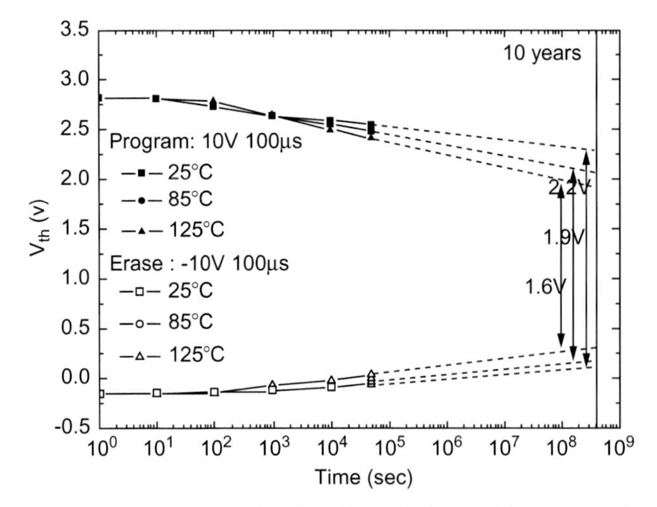

FIG. 3.22 Retention performance of a HfLaON/$Hf_{0.3}N_{0.2}O_{0.5}$ CT stack at different temperatures [13].

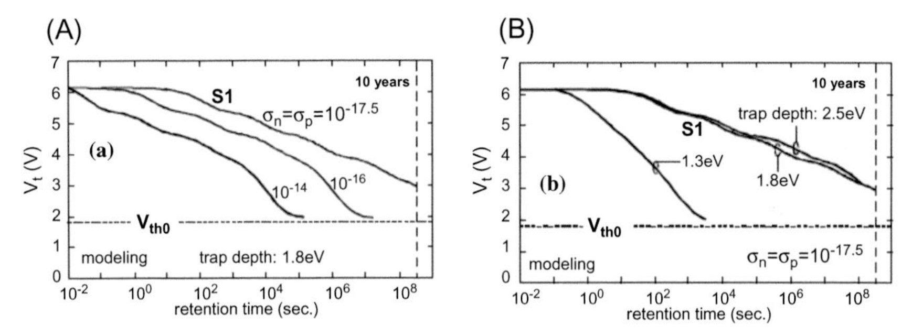

FIG. 3.23 Dependence of retention on trap cross section and trap energy level in SONOS with HfSiON trap layer [6].

In contrast, another study by Park et al. [14] showed that the retention performance of a CT stack with HfO_2 as the trapping layer has worse retention performance than the stack with SiN as the trapping layer. The authors attribute this to the presence of a large number of interface traps in the case of HfO_2. To generate deep traps, which are desirable for good retention, this study proposes implanting N_2 into the HfO_2 layer. Significant improvement in retention is reported because of the N_2 implant, as shown in Fig. 3.24.

Another interesting concept evaluated with high-K dielectrics for CT layer is to combine two materials. The idea here is to achieve band gap engineering by mixing the two materials as part of the trap layer. Jie et al. [15] studied a CT stack which had a composition modulated $(HfO_2)_x(Al_2O_3)_{1-x}$. To achieve this, they used atomic layer deposition

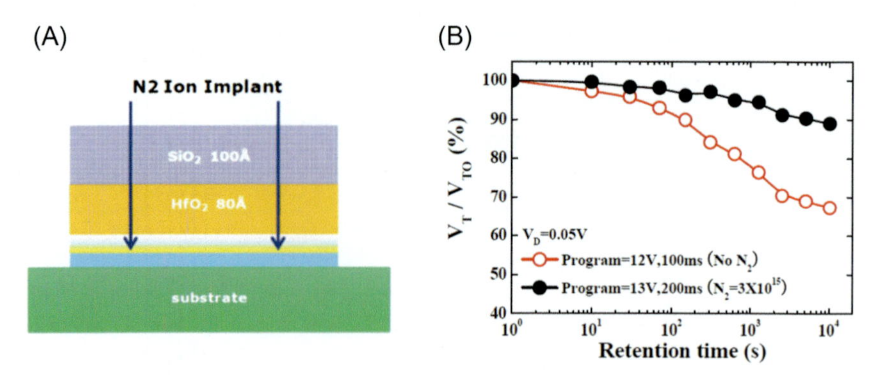

FIG. 3.24 Nitrogen implant into CT stack with high-K dielectric—(A) schematic and (B) impact on retention [14].

(ALD) with precursors $HfCl_4$ and $Al(CH_3)_3$. The percentage of Al was first increased and then decreased by controlling the number of ALD cycles. For comparison, they also looked at CT layers made of a homogeneous $(HfO_2)_{0.2}(Al_2O_3)_{0.8}$(HA) and pure HfO_2 (HfO_2). The Al composition and the corresponding energy band structure in the three cases was examined. A modulation of the band diagram was clearly seen when the composition is modulated across the thickness of the CT layer. This has a significant impact on the reliability of the stack because of the large band shifts at the two interfaces of the CT layer which minimizes charge loss from the layer. A reduction in charge loss is clearly seen due to the composition modulation.

Instead of this rather complicated method of modifying the composition, other studies have evaluated doping HfO_2 film with Al_2O_3 [16] or introducing a homogenous film of $(HfO_2)_{0.8}(Al_2O_3)_{0.2}$. It was found that doping of Al_2O_3 in HfO_2 improved the program-erase speed of the NVM device because of the band offset between the tunnel layer and the trapping layer.

The use of high-K metal oxides as blocking layers and the trap layer has also been explored. Lai et al. [17] investigated a TaN-HfAlO-HfON-SiO_2-Si stack and found that it could be programmed and erased at voltage as low as ±8 V and the stack showed good retention performance (Fig. 3.25).

Mirror Bit™ memories which can store 2 bits in a cell are widely used in standalone NVM memories. These memories generally use the conventional ONO stack for the CT stack. However, some studies have been reported on Mirror Bit™ memory stack which uses a high-K metal oxide as the trap layer. Zhang and Yoo [18] studied a SiO_2-HfAlO-SiO_2 stack and saw the advantage of high-speed program-erase and robust data retention.

Some uncommon metal oxides such as ZrO_2 and Y_2O_3 have also been explored for the trap layer. Detailed reviews on these have been published

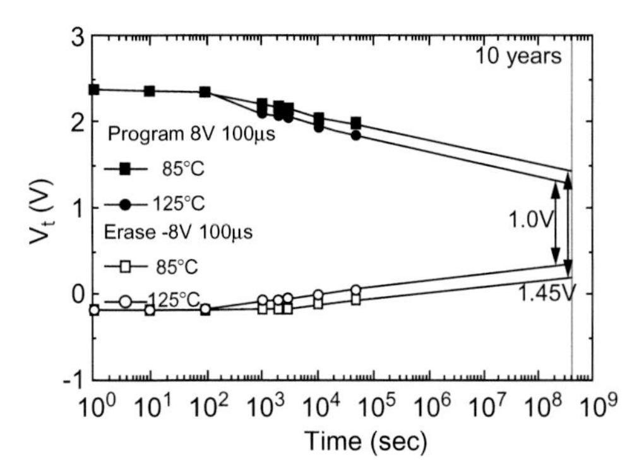

FIG. 3.25 Retention of CT stack with high-K metal oxides for blocking and trap layers [17].

by Zhao et al. [19] and Y.H. Wu [20]. However, these materials are not available in high volume manufacturing and therefore not the first choice for semiconductor devices.

3.5.3 TANOS with high-K metal oxide dielectric

TANOS stands for TaN-Al_2O_3-silicon nitride-silicon dioxide-silicon. This is also a three-layer charge trap memory stack but in this case the Al_2O_3 replaces the conventional SiO_2 for the blocking oxide. The high dielectric constant of this blocking oxide reduces the electric field across this layer, and hence, the field across the tunnel oxide will increase. This means the tunnel oxide thickness can be increased to reduce the electric field to the relaxed condition. A thicker tunnel oxide improves the data retention (un-biased condition) performance of the cell dramatically. As explained earlier, a high-K dielectric for the blocking oxide reduces the injection from the gate to the trap layer. The energy bad diagram of this stack is shown in Fig. 3.26A.

Robust retention and endurance performance have been demonstrated in TANOS at 63nm technology node, as shown in Fig. 3.26B [21]. TANOS stacks have been studied widely to improve its performance. Some of the improvements suggested are discussed below.

3.5.3.1 Enhancement of TANOS stacks

Park et al. [22] studied a TANOS stack in which the nitride layer was doped with La. This was done by sandwiching a La_2O_3 layer in the TANOS stack as shown in Fig. 3.27 and then annealing the stack so that mixing can occur and La can get into the nitride layer.

Other oxides such as Ga_2O_3, Al_2O_3 and ZrO_2 were also studied. It was found that the while La and Gd doping of LPCVD SIN CT layer improved the retention performance, AL and Zr doping actually degraded the

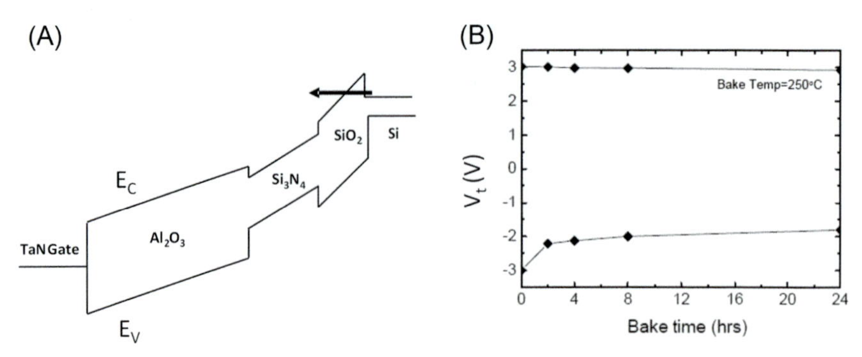

FIG. 3.26 TANOS device showing (A) energy band diagram and (B) retention characteristics [21].

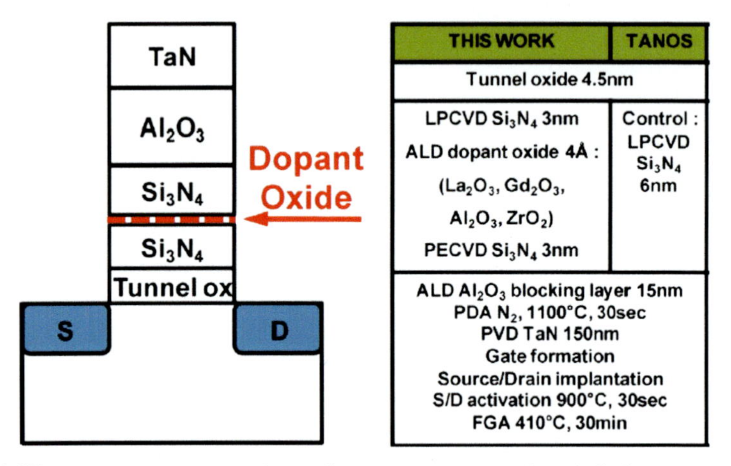

FIG. 3.27 TANOS structure with trap layer containing metal oxide [22].

retention. La-doped trap layer showed the best retention performance (Fig. 3.28). The superior retention is attributed to the deep traps generated due to the doping by La_2O_3. The thickness of the La_2O_3 layer is critical and needs to be about 4 Å. Beyond 6 Å, the charge loss, expressed as change in threshold voltage V_t, increases significantly as shown in Fig. 3.29.

3.5.3.2 Band gap engineered SONOS (BE-SONOS) with high-K dielectric

Replacing the SiO_2 tunnel oxide by a stack (band engineering) such as ONO or OXO (X being HfO_2 or Al_2O_3) is also another option for improving the TANOS performance. With optimized stack (shown in Fig. 3.30), significant increase in memory window has been achieved [23].

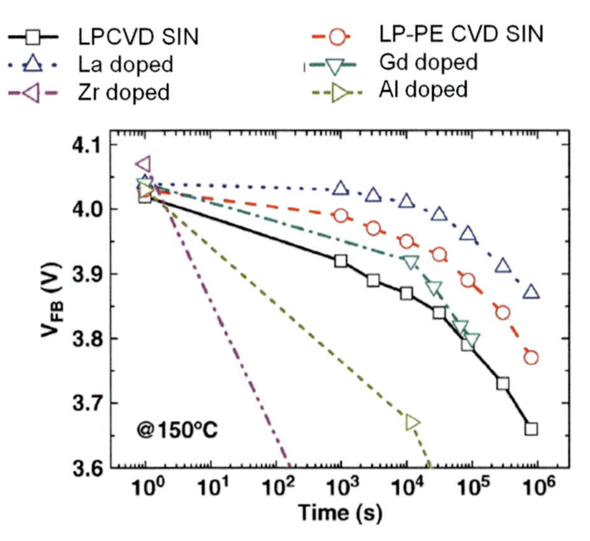

FIG. 3.28 Retention performance of SONOS, expressed as variation of flat band voltage V_{FB}, with several dopants (La, Zt, Gd, Al) in nitride layer [22].

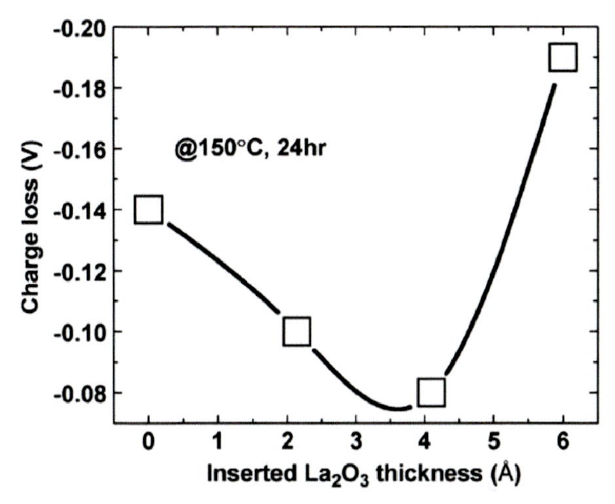

FIG. 3.29 Dependence of charge loss on thickness of La_2O_3 layer [22].

A 400% increase of the V_t margin after retention bake was obtained when OXO was SiO_2-Al_2O_3 stack (Fig. 3.31). The improvement was 300% when the OXO was an ONO stack. The improvement is due to the band offset caused by the introduction of the stack instead of the simple SiO_2.

The high-K dielectric layer will reduce the injection from the gate significantly, as discussed earlier. Lai et al. [24] also compared a SONOS stack

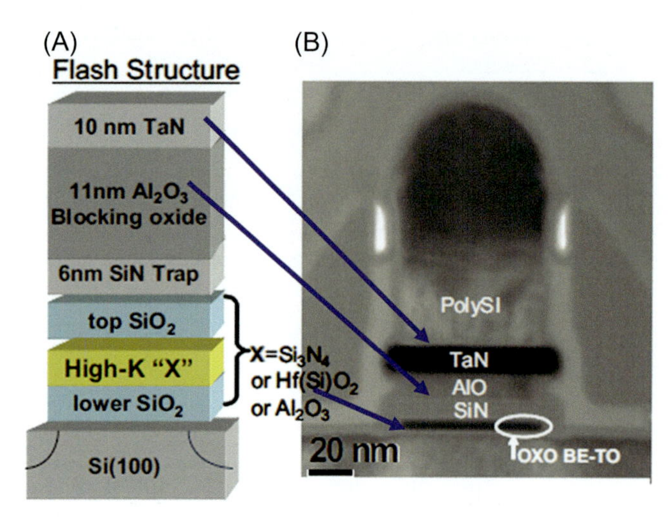

FIG. 3.30 Band gap engineered CT stack with high-K dielectric: (A) schematic and (B) TEM cross section [23].

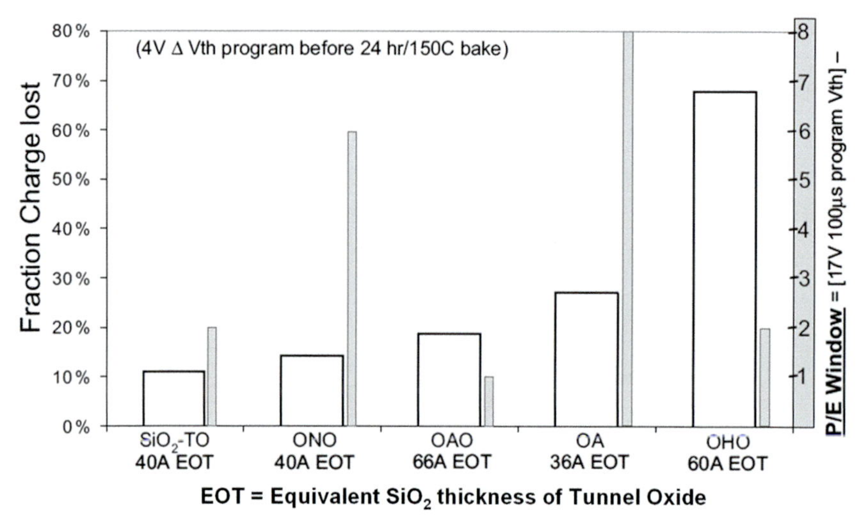

FIG. 3.31 V_t window and charge loss comparison between different CT stacks [23].

with a stack in which the blocking SiO_2 was replaced by Al_2O_3 and a third stack in which in addition to blocking dielectric, the SiO_2 tunnel oxide was also replaced by an ONO stack, as shown in Fig. 3.32.

Al_2O_3 has a relatively high dielectric constant which can greatly reduce the electric field in the blocking dielectric during the erase. In addition, Al_2O_3 also possesses sufficiently large electron barrier height (~2.8 eV

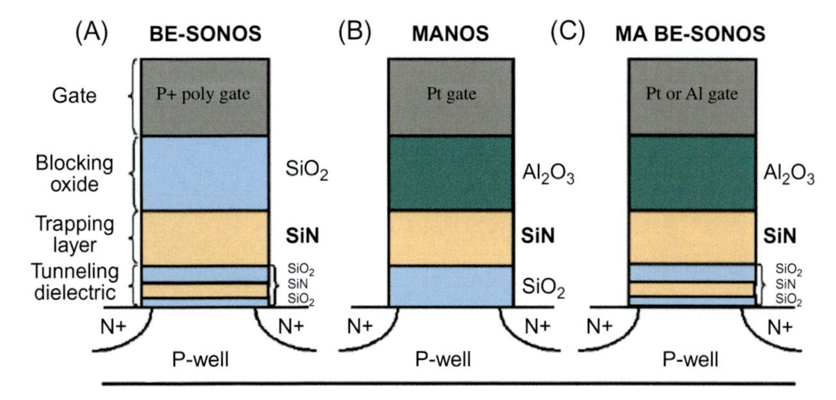

FIG. 3.32 CT stacks with high-K blocking oxide compared to SiO_2 blocking oxide and band gap engineered tunneling layer compared to SiO_2 tunneling layer [24].

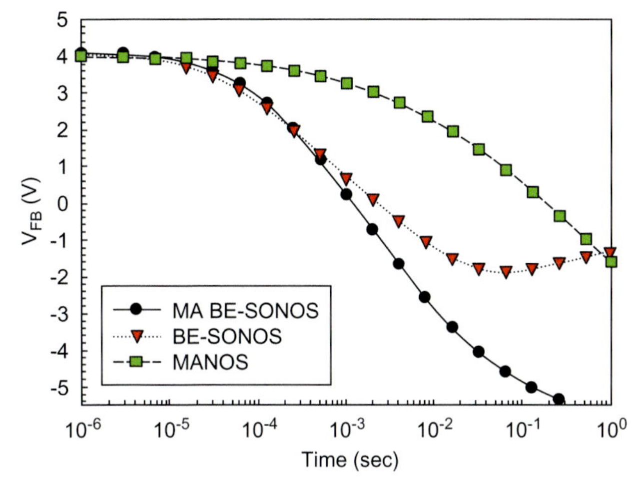

FIG. 3.33 Erase characteristics for CT stacks with SiO_2 and Al_2O_3 blocking oxide showing the impact of Band Gap Engineering [24].

band offset with Si). These factors eliminate back injection of carriers from gate to trap layer during erase and can greatly reduce the erase saturation (Fig. 3.33). Use of a high work function metal as gate instead of polysilicon helps further because of the larger potential barrier for carriers in the gate.

3.5.4 Metal oxides in FINFET based CT memory device

It is widely accepted that FinFET will be the core device for high performance logic circuits in 21 nm technology and beyond. The advantages of the FinFET architecture for device drive current and leakage have been

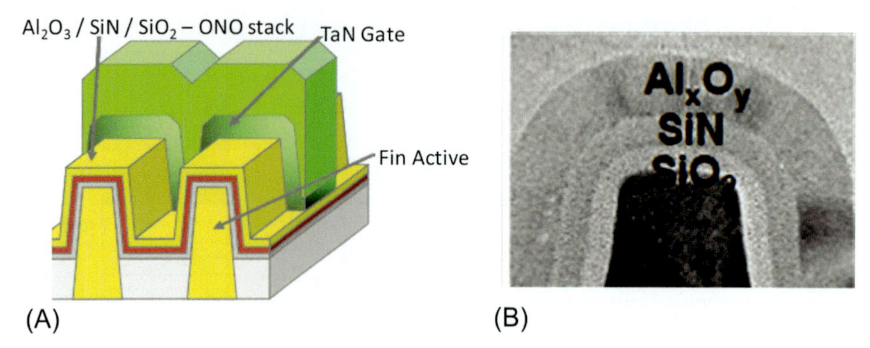

FIG. 3.34 FinFET with high-K dielectric in CT stack—(A) schematic and (B) TEM cross section of CT stack on the Fin [25].

studied in detail for CMOS FETs in advanced platforms. When SONOS memory is embedded into such a logic platform, the FinFET architecture and process is available to be used for SONOS devices also. It is interesting to determine whether the FinFET architecture helps CT devices. The impact of FinFET architecture on CT memory devices has been studied quite extensively in the last 10 years. Most of the studies have been on FinFETs on bulk silicon, not SOI. It is found that the electric field gets significantly enhanced in the curved part of the fin. The enhancement substantially increases the carrier injection from the channel into the trap layer of the charge trap memory stack. On account of the field enhancement, the program and erase speeds are significantly increased in a FinFET. Since SONOS FinFET is a new device architecture, introduction of a metal oxide into this device is a bound to happen once FinFET technology goes mainstream. Such a CT device would look as shown in Fig. 3.34 [25].

It has been shown that the electric field enhancement factor (FE) in the curved part of the fin can be twice as high compared to the flat region, depending on the width and this substantially increases the carrier injection from the channel into the trap layer of the charge trap memory stack. Because of the different electric fields in the flat and curved portions of the fin, the carrier injection will also be non-uniform. Sung et al. have shown that with a FinFET SONOS device with high-K blocking dielectric (Al_2O_3), cell current which is 3× higher than in a planar device can be achieved [26]. Typical programming behavior of a FinFET based CT device with Al_2O_3 as the blocking oxide is compared with a reference device in Fig. 3.35.

Clearly the FinFET structure enhances the threshold voltage during programming and erasing due to the electric field enhancement. Reliability of this type of device is also very similar or better than the conventional device (Fig. 3.36).

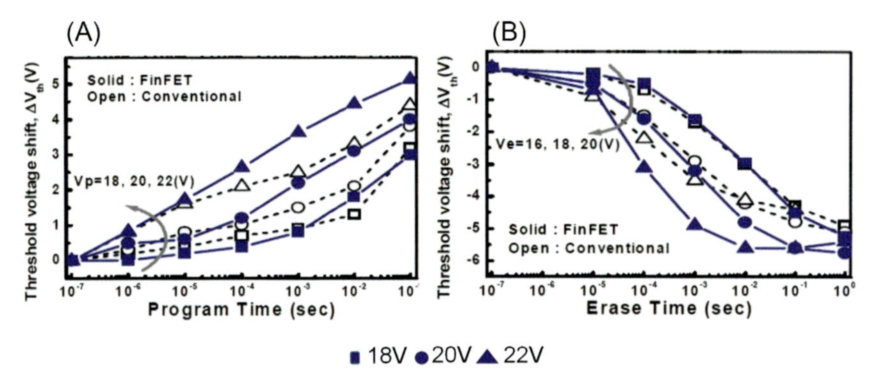

FIG. 3.35 Threshold voltage variations during (A) program and (B) erase at three different voltages for conventional *(open)* and FinFET *(solid)* for CT device with Al_2O_3 as blocking oxide [26].

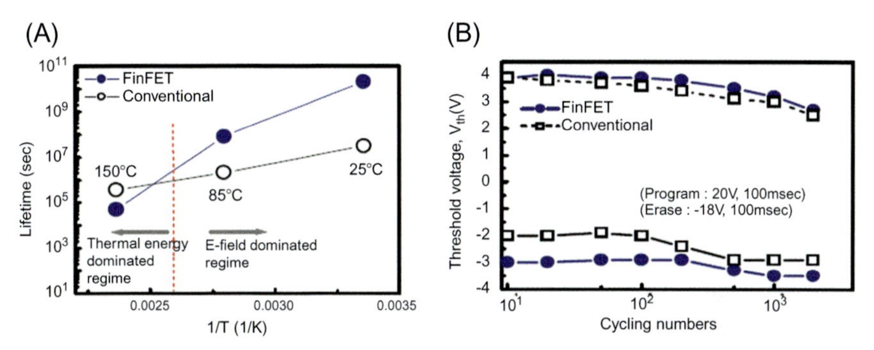

FIG. 3.36 Comparison of FinFET and conventional CT device for (A) retention and (B) endurance for CT device with Al_2O_3 as blocking oxide [26].

These results basically show that with the semiconductor industry adopting the FinFET architecture for the CMOS logic devices to gain performance improvements, the CT memory devices in the embedded NVM can also adopt this architecture for the CT devices, for process commonality, without losing in performance. There may be some benefits to the performance from the field enhancement effect in the Fins. The CT memory stack in the FinFET will most likely have high-K dielectric such as Al_2O_3 or HfO_2 because such stacks will be mainstream for CMOS devices by the time the FinFETs become part of the CT memories. The band gap engineered SONOS stack, with an ONO stack acting as the tunneling layer, can be implemented in the FinFET SONOS device also. The same is true for the TANOS stack.

Thus, when the FinFETs are adopted widely for NVM devices, metal oxides will smoothly get incorporated in them from the beginning and any integration issues will be addressed.

References

[1] M. White, et al., A low voltage SONOS nonvolatile memory technology, in: IEEE Trans. Components, Packaging and Manufacturing—Part A, 20, 1997, p. 190.
[2] V.J. Kapoor, et al., Charge storage and distribution in the nitride layer of Metal-Nitride-Oxide Semiconductor structures, J. Appl. Phys. 52 (1981) 311.
[3] A. Chin, et al., Improved retention and cycling characteristics of MONOS memory using charge-trapping engineering, in: IEEE Proc. 16th IPFA, 2009.
[4] A. Chin, et al., Extremely low voltage and high speed deep trapping MONOS memory with good retention, in: ICSICT, 2006, p. 744.
[5] H.T. Lue, et al., A critical review of charge trapping NAND flash devices, in: ICSICT, 2008, p. 807.
[6] J. Buckley, et al., In-depth investigation of Hf-based high-k dielectrics as storage layer of charge-trap NVMs, in: IEDM, 2006.
[7] G. Molas, et al., Layered HfSiON—Based tunnel stacks for voltage reduction and improved reliability in TANOS memories, in: Int. Symp. VLSI Technology Systems and Applications, 2010, p. 56.
[8] M.V. Duuren, et al., Pushing the scaling limits of embedded non-volatile memories with high-k materials, in: ICICDT, 2006, p. 1.
[9] Y.N. Tan, et al., Hafnium aluminum oxide as charge storage and blocking-oxide layers in SONOS-type nonvolatile memory for high-speed operation, IEEE Trans. Electron Devices (2006) 654.
[10] Z. Tang, et al., Progress of High K dielectrics applicable to SONOS type nonvolatile semiconductor memories, Trans. Electr. Electron. Mater. (11) (2010) 155.
[11] S.C. Lai, et al., A study of barrier engineered Al2O3 and HfO2 high-K charge trapping devices (BE-MAONOS and BE-MHONOS) with optimal high-K thickness, in: Int. Memory Workshop, 2010, p. 1.
[12] R. Van Schaijk, et al., A novel SONOS memory with HfSiON/Si3N4/HfSiON stack for improved retention, in: NVSMW, 2006, p. 50.
[13] H.J. Yang, et al., Comparison of MONOS memory device integrity when using Hf1−-$x-y$NxOy trapping layers with different N compositions, IEEE Trans. Electron Devices (2008) 1417.
[14] J.G. Park, et al., Improvement of reliability characteristics using the N2 implantation in SOHOS flash memory, in: IEEE Nanotech Mat and Devi Conf, 2010, p. 364.
[15] T.Z. Jie, et al., Performance improvement of charge trap flash memory by using a composition-modulated high-k trapping layer, Chin. Phys. B (2013), 097701-1.
[16] Y.N. Tan, et al., High-K HfAlO charge trapping layer in SONOS-type nonvolatile memory device for high speed operation, in: Proc. IEDM, 2004, p. 889.
[17] C.H. Lai, et al., Very low voltage SiO_2/HfON/HfAlO/TaN memory with fast speed and good retention, in: Symp. VLSI Tech, 2006, p. 44.
[18] G. Zhang, W.J. Woo, Novel HfAlO charge trapping layer in SONOS type flash memory for multi-bit per cell operation, in: ISCICT, 2006, p. 781.
[19] C. Zhao, et al., Review on non-volatile memory with high K dielectrics: flash for generation beyond 32nm, Materials 7 (2014) 5117.
[20] Y.H. Wu, et al., Nonvolatile memory with nitrogen-stabilized cubic-phase ZrO2 as charge-trapping layer, IEEE Electron Device Lett. 31 (2010) 1008.
[21] C.H. Lee, et al., Charge trapping memory cell of TANOS (Si-Oxide-SiN-Al2O3-TaN) structure compatible to conventional NAND flash memory, in: NVSMW, 2006, p. 54.
[22] J.K. Park, et al., Lanthanum-oxide-doped nitride charge-trap layer for a TANOS memory device, IEEE Trans. Electron Devices (2011) 3314.
[23] D.C. Gilmer, et al., Band engineered tunnel oxides for improved TANOS-type flash program/erase with good retention and 100K cycle endurance, in: Int. Symp. VLSI Tech. Systems and Applications, 2009, p. p156.

[24] S.C. Lai, et al., MA BE-SONOS: a bandgap engineered SONOS using metal gate and Al2O3 blocking layer to overcome erase saturation, in: IEEE NVSMW, 2007, p. 88.
[25] S.H. Lee, et al., Improved post-cycling characteristic of FinFET NAND flash, in: IEDM, 2006, p. 1.
[26] S.K. Sung, et al., SONOS-type FinFET device using P+ Poly-Si gate and high-k blocking dielectric integrated on cell array and GSL/SSL for multi-gigabit NAND flash memory, in: IEEE VLSI Tech. Symp, 2006, p. 86.

CHAPTER

4

Technology and neuromorphic functionality of magnetron-sputtered memristive devices

A.N. Mikhaylov[a], *M.N. Koryazhkina*[a], *D.S. Korolev*[a], *A.I. Belov*[a], *E.V. Okulich*[a], *V.I. Okulich*[a], *I.N. Antonov*[a], *R.A. Shuisky*[a], *D.V. Guseinov*[a], *K.V. Sidorenko*[a], *M.E. Shenina*[a], *E.G. Gryaznov*[a], *S.V. Tikhov*[a], *D.O. Filatov*[a], *D.A. Pavlov*[a], *D.I. Tetelbaum*[a], *O.N. Gorshkov*[a], *A.V. Emelyanov*[b], *K.E. Nikiruy*[b], *V.V. Rylkov*[b], *V.A. Demin*[b], *and B. Spagnolo*[a,c]

[a]Lobachevsky State University of Nizhny Novgorod, Nizhny Novgorod, Russia, [b]National Research Center "Kurchatov Institute", Moscow, Russia, [c]University of Palermo and The National Interuniversity Consortium for the Physical Sciences of Matter, Palermo, Italy

4.1 Features of magnetron sputtering

Physical Vapor Deposition (PVD) methods (often called thin film processes) are well compatible with the current CMOS process and are most often used to fabricate memristive devices and arrays because they allow both insulator layers and electrodes to be formed in a wide range of thicknesses (up to the ultrathin ones). Common to all PVD methods is that they are based on atomistic deposition process, in which the material is vaporized from a solid or liquid source in the form of atoms or molecules transported as a vapor through vacuum or low pressure gaseous (or plasma) environment to the substrate, where it condenses [1]. The main categories

Metal Oxides for Non-volatile Memory
https://doi.org/10.1016/B978-0-12-814629-3.00004-0

of PVD processing are vacuum/thermal/electron-beam evaporation, sputter deposition, and ion plating.

The method of magnetron sputtering, as a kind of PVD, is especially attractive because it allows one to obtain metal-oxide structures with a characteristic heterogeneous structure and composition of oxide films, which play a key role in the electroforming (EF) and resistive switching (RS) processes [2,3]. The EF conditions depend on the device design or materials used, and often the EF step is not required to achieve RS [4,5]. This section discusses typical examples of the application of magnetron sputtering method to fabricate metal-oxide memristive devices and arrays with specified properties based on two different types of oxides: silicon oxide and yttria stabilized zirconium oxide (YSZ). Silicon oxides are intrinsic RS materials and seem to be most compatible with the current CMOS technology [6]. At the same time, zirconium oxide shows a high mobility of oxygen vacancies and its doping with stabilizing yttrium oxide enables additional control of the oxygen vacancy concentration, as well as the parameters of anionic RS type.

When fabricating memristive devices by magnetron sputtering, the deposition conditions are important because their change provides control over the composition, structure and thickness of the deposited layers. Such conditions include target composition, radio-frequency (RF) discharge power, substrate temperature, deposition pressure, O_2:Ar gas mixture ratio, deposition time (or growth rate).

The choice of magnetron sputtering conditions is based on the need to control the oxygen content in oxide film. For example, in the case of zirconium oxide, the oxygen concentration can be controlled by the doping with stabilizing yttrium oxide, and in the case of reactive sputtering to deposit silicon oxide films, by varying the parameters for sputtering of a silicon target—by adjusting, for example, the fraction of oxygen in the sputter ambient.

Table 4.1 shows typical conditions of magnetron sputtering required for the deposition of zirconium and silicon oxide layers of a given composition.

Combinations of materials for electrodes can be selected from a number of metals (Au, W, Ta, Ti and Zr) with different work functions and oxygen affinity. Table 4.2 shows typical conditions of magnetron sputtering used for the deposition of metal electrodes.

4.2 Performances and reproducibility of memristive devices

In this section, the RS and adaptive behavior of the resistive state in response to electrical stimulation of memristive devices are considered to demonstrate the performances and reproducibility of memristive

TABLE 4.1 Typical magnetron sputtering conditions required for the deposition of oxide layers.

Sputtered material	Layer thickness, nm	Process ambient O_2:Ar, %	Sputtered target	Initial/process pressure, Torr	Substrate temperature, °C	Growth rate, nm/s
ZrO_2	40	50:50	ZrO_2	$1 \times 10^{-6}/1.8 \times 10^{-2}$	300	0.016
$ZrO_2+Y_2O_3$ (12 mol%)	40	50:50	$ZrO_2+Y_2O_3$ (12 mol%)	$1 \times 10^{-6}/1.8 \times 10^{-2}$	300	0.011
SiO_x ($x \sim 2$)	40	92:8	Si	$1 \times 10^{-6}/3.2 \times 10^{-3}$	200	0.223

TABLE 4.2 Typical magnetron sputtering conditions required for the deposition of metal electrodes.

Sputtered material	Layer thickness, nm	Process ambient	Sputtered target	Initial/process pressure, Torr	Substrate temperature, °C	Growth rate, nm/s
Au	40	Ar	Au	$2\times10^{-6}/5\times10^{-3}$	200	0.40
W	8	Ar	W	$2\times10^{-6}/5\times10^{-3}$	200	0.13
Ti	40	Ar	Ti	$2\times10^{-6}/5\times10^{-3}$	200	4.00
Zr	8	Ar	Zr	$2\times10^{-6}/5\times10^{-3}$	200	0.16
Ta	8	Ar	Ta	$2\times10^{-6}/5\times10^{-3}$	200	0.17

devices. For this purpose, a comparison of electrical characteristics of SiO_x-based memristive devices subjected to EF without or with current compliance (CC) [7] is made. The bias polarity in all the cases considered in this chapter corresponds to the potential of Au top electrode (TE) relative to the grounded TiN bottom electrode (BE). The EF is performed during the first current-voltage sweep at the negative bias. It should be emphasized here that, the CC during EF affects the parameters of growing conductive path (or filament) ensembles and reduction-oxidation reactions resulting in a gradual character and a wide dynamic range of resistance change important for neuromorphic applications.

The ability of memristive structures to continuously change the resistance in dependence on electrical input signals is equivalent to the adaptive behavior (plasticity) of synapses as a key mechanism underlying the phenomena of learning and memory in neural networks. The realization of adaptive behavior of metal-oxide memristive devices [8] is based on a continuous spectrum of resistive states related to different configurations of filaments grown in the switching oxide. Gradual (multilevel) switching between the different resistive states has been already studied for a number of transition metal oxides (TMOs) [9–11] as well as for silicon oxide [12]. Each study was carried out for a given technique of EF. The EF conditions (e.g., CC) are crucial for the subsequent memristive behavior, because the range and sensitivity of a resistive state to electrical stimulation depends on the initial configuration of filaments and related kinetics of microscopic reduction-oxidation phenomena driven by the electric field and temperature.

The effect of CC applied during EF on the character and parameters of RS was investigated in several works for different compositions of materials in memristive structures [13,14]. For the memristive devices based on HfO_2 [15], it was reported that the character of RS (gradual or abrupt) in the processes of increase (SET) or decrease (RESET) of conductivity depends on the measurement conditions. Obviously, such peculiarities of a memristive device must depend on specific materials used in a thin film structure. The systematic and comparative study of the CC effect on the adaptive behavior of resistive state in CMOS-compatible devices based on SiO_2 [16] was reported in [7].

Typical *I–V* curves of EF realized in the range of 2–6 V of negative bias and subsequent bipolar RS of valence-change or anionic mechanism [16] are shown as bold curves in Fig. 4.1 for same oxide but different EF conditions. The cycle-to-cycle RS variation for both conditions is also shown by thin curves.

If no CC is used during EF, silicon filaments are grown in the oxide film at high current and by Joule heating. After EF, the device switches between the LRS and HRS in the so-called high-current mode. The corresponding RS is shown by the red *I–V* curves in Fig. 4.1 and characterized

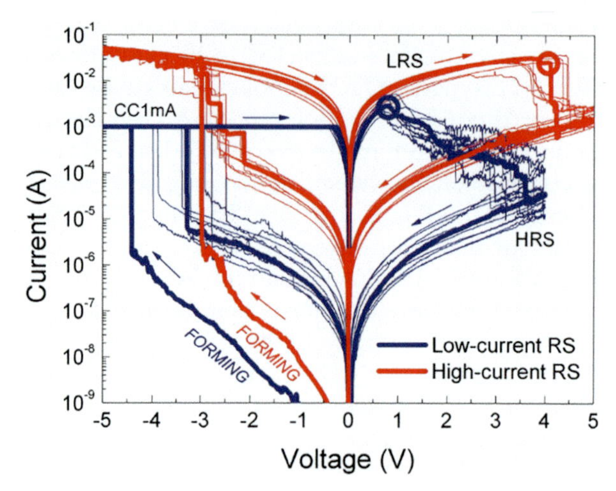

FIG. 4.1 *I–V* curves of memristive devices measured in cyclic sweeping mode with different conditions of EF: with *(blue)* or without *(red)* CC. *Bold lines* represent typical curves which are close to the averaged ones [7]. *Reproduced with permission from D.S. Korolev, et al., Manipulation of resistive state of silicon oxide memristor by means of current limitation during electroforming, Superlattice. Microst. 122 (2018) 371. Copyright 2018 Elsevier B.V.*

by both the gradual or abrupt stepwise current changes in the SET and RESET transitions at relatively high voltages (up to ±4V), which are poorly reproducible from cycle to cycle. The EF voltage also varies from device to device as will be discussed below.

If a CC is applied during EF, the abrupt SET transition is followed by a monotonous RESET process to the "deeper" HRS state with a much lower threshold voltage. This gradual RESET transition is reproducible and is much more appropriate for the demonstration of adaptive behavior compared with the stepwise SET transition in the high-current RS mode [17]. It should be noted that the observed deep HRS state cannot be achieved in the high-current mode even if CC is additionally applied during the SET transition. It should be also noted that all this behavior can depend on the "quality" (i.e., speed and overshoots) of the CC. That's why one may use physical CC based on external transistor [18] or special current-limiting circuits [19].

In addition to cycle-to-cycle RS variation, special attention should be paid to the wafer-scale device-to-device variation. Fig. 4.2 shows the variation of the EF curves and subsequent RS for eight memristive devices. The strong variation of the EF behavior is due to the stochastic nature of the filament growth. SiO_2 films obtained by magnetron sputtering on polycrystalline electrodes have a columnar structure, contain numerous intercolumnar boundaries and corresponding electrode roughness enabling potential sites for nucleation of filaments [3]. The dielectric

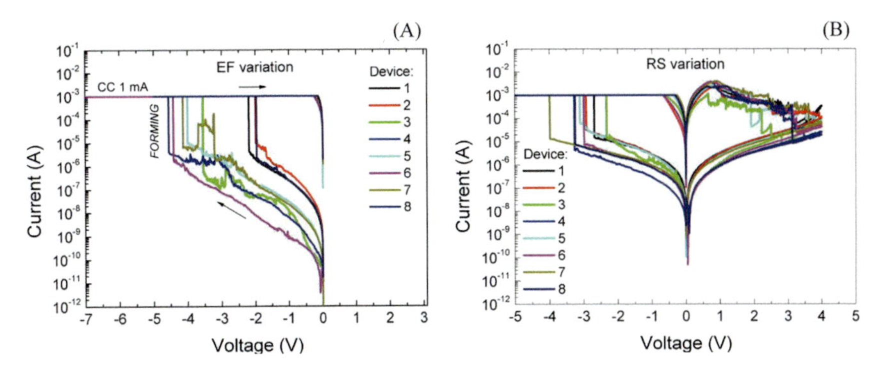

FIG. 4.2 Device-to-device variation of current-voltage characteristics for different memristive devices measured for EF with CC (A) and for subsequent RS (B) processes [7]. *Reproduced with permission from D.S. Korolev, et al., Manipulation of resistive state of silicon oxide memristor by means of current limitation during electroforming, Superlattice. Microst. 122 (2018) 371. Copyright 2018 Elsevier B.V.*

properties and the conditions for electric field concentration can differ significantly from site to site, which explains the large differences of the EF characteristics (Fig. 4.2A). The device-to-device variation of the resistive states is less pronounced after EF (Fig. 4.2B). It does not reflect the initial state of a memristive device and it is comparable with the cycle-to-cycle variation (Fig. 4.1).

Taking into account the observed RS variation, it can be concluded that the mentioned differences in the high-current and low-current RS modes have a systematic character and can be related to different conditions of the EF. The main differences are observed for the RESET process in the low-current and high-current RS modes. One can estimate the Joule heat released in the memristive structure at the beginning of RESET process (the corresponding points on LRS curves are shown by circles in Fig. 4.1). A typical device resistance is about 275 Ω in the low-current LRS, and it is about 135 Ω in the high-current LRS. The Joule power required for RESET in the low-current mode is 2 mW, and in the high-current mode it is 120 mW. Thus, the CC during EF leads to a twofold increase in the device LRS resistance and 60 times decrease in the power, at which the RESET occurs. It is reasonable to suppose that the characteristic filament diameter is in the order of a few nanometers. The solution of the heat conduction equation for a heat source with dimensions of 2 nm in diameter and 30 nm in height (located in oxide film with the non-zero heat conduction) and generating a power of the order of 2 mW (as in the low-current regime) leads to heating of a filament region to an absolute temperature that is five times higher than the melting point of SiO_2 (1900 K). To avoid such a result, an ensemble of at least 40 nanometric filaments should be considered to have about 50 μW of power per filament and an appropriate temperature of 600 K. If 60 times more power is applied to this ensemble of filaments (as in the high-current LRS), then a temperature incredibly higher than the melting point is obtained again.

Thus, the following conclusions can be drawn based on the measured *I–V* curves and simple temperature estimation: an ensemble of filaments is responsible for the current flow and RS. Probably, this ensemble is localized in a region of a small area, which follows from the independence of RS parameters of the device area up to a micrometer size [6]. When the EF mode is changed by using current limitation, different heating conditions result in either different composition and morphology of filaments, or different composition and structure of surrounding oxide matrix. This explains the differences in RS behavior for the high- and low-current modes, even if the total resistances appear to be similar for various resistive states.

The observed effect of CC applied during EF can be generally interpreted from the viewpoint of the RS model [6,16] based on the formation and partial oxidation of a silicon filament accompanied by the migration of interstitial O^{2-} ions produced during the breaking of Si-O-Si bonds. In the high-current RS mode, the high current through the filaments provides strong Joule heating of the filaments and in a local area adjacent to the filaments. As a result, the temperature gradient (the Soret effect) moves oxygen far away from the filaments. In the low-current RS mode, the temperature is smaller, and the effective distance of oxygen removal from the filaments is lower too. In this case, the RESET process associated with the filament oxidation near the electrode begins at lower voltages and leads to a more expressed oxidation, which causes the decrease in RESET voltage and increase in HRS resistance. The RESET process becomes gradual because the oxidation takes place in a wide range of voltages, whereas the "high-current" (high-temperature) RESET transition is observed in a narrow voltage range and has mainly an abrupt character.

This interpretation based on the peculiarities of redox reactions can be supported by the analysis of microscopic changes in the geometry of filaments depending on the conditions of EF. For this purpose, the filament can be considered as a number (bundle) of conductive paths with certain statistical distribution depending on the filament composition and geometry [20–22]. The individual conductive path can be a chain of oxygen vacancies (chain of silicon atoms) that behaves as a quantum wire with one-dimensional system of electron states. Thus, for the narrower part of a filament (quantum constriction), the conductance quantization effect according to the Landauer transmission approach can be demonstrated at room temperature [20]. The so-called quantum point contact model was used to describe the transition from "linear" conduction through a number of conductive paths in LRS (with the conductance represented as multiples of the fundamental unit $G_0 = 2e^2/h$) to the

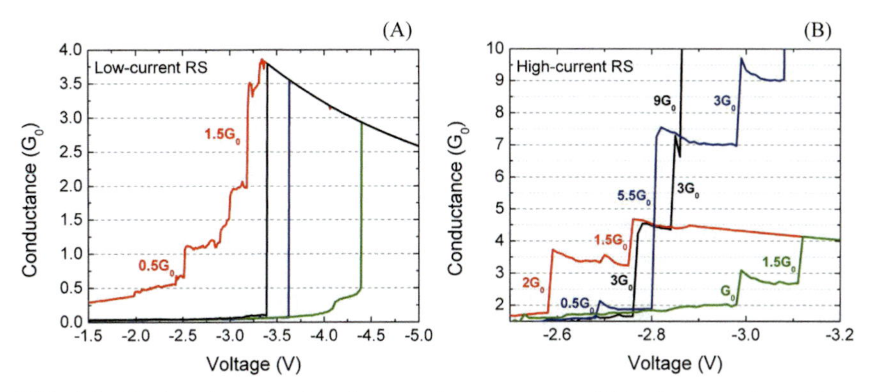

FIG. 4.3 Typical dependencies of the conductance on the SET voltage for the (A) low-current and (B) high-current RS modes [7]. *Reproduced with permission from D.S. Korolev, et al., Manipulation of resistive state of silicon oxide memristor by means of current limitation during electroforming, Superlattice. Microst. 122 (2018) 371. Copyright 2018 Elsevier B.V.*

nonlinear transport through the metallic filament spatial gaps in HRS with $G < G_0$ [22].

For the memristive device considered in this section, the characteristic conductance steps can be also observed and interpreted as related to the quantization effect. Fig. 4.3 shows typical dependencies of the conductance on the voltage applied during the SET transition. It can be seen that for both modes of RS, there are conductance steps with heights around $0.5\,G_0$ or higher. This is in accordance with the conclusion [20] about the origin of half-integer conductance quantization in the Si-based filaments. At the same time, the limitation of the current during EF has a significant effect on the manifestation of conductance quantization. In the high-current mode, the conductance steps occur almost at every SET transition up to the used CC. On the contrary, in the low-current regime, only rare steps of conductance are observed, while there is generally a sharp increase in the current to the value of CC applied in the SET process. This is due to the fact that, in the low-current mode, the conductivity of the device is very low before the SET process due to the "deep" oxidation in the preceding RESET process. As a result, the reconstruction of current paths begins at higher SET voltages, but earlier than the conductivity reaches the level of $0.5\,G_0$, which is necessary for the conductance quantization. It should be also noted that the deformed shapes of conductance steps are typical of the devices based on SiO_x, because, in general, the total conductivity should be deconvolved into parallel linear and nonlinear terms related to the filaments and surrounding semiconducting matrix, respectively [20].

The given microscopic interpretation of the observed effect of EF conditions can be supplemented by a statistical mechanism based on a complicated interplay of field- and temperature-driven phenomena in the filament ensemble [21]. In the high-current RS mode, high temperature results in formation of a filament ensemble that provides its low resistance and a large statistics of possible conductive paths. In this case, the RESET transition starts at a high temperature, and the heating allows dramatic changes in the filament configurations even for small changes in electric field (corresponding to the abrupt changes in resistive state). On the contrary, when the low-current RESET is realized, the heating is not enough for abrupt changes, and gradual RS behavior is provided by successive and delayed reconstruction of the filament ensemble driven by an electric field over a wide voltage range. Of course, such a complex stochastic nature of RS leads to the cycle-to-cycle fluctuations, because each switching cycle results in completely new atomic configurations in the local switching areas that are formed or destroyed by different switching currents and voltages.

The dependence of the resistance of a memristive device on the amplitude of RESET voltage pulses obtained in the gradual low-current RS mode is shown in Fig. 4.4. This dependence suggests the following technique of programming the resistance (weight) of a memristor for the application in traditional neural networks: the desired value of the resistance can be set by applying a programming pulse with an amplitude in the range of 1–5 V. To set the next resistance value (in any direction of resistance change, i.e., decrease or increase), the initializing pulse of opposite polarity with the amplitude of 5–8 V (which turns the device into initial resistance state) is applied before the programming pulse. The amplitude

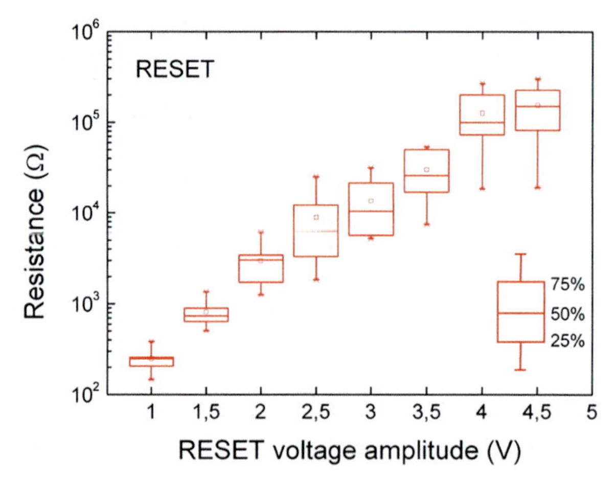

FIG. 4.4 The dependence of resistive state of memristive device programmed by the application of RESET voltage pulses (5 ms) on the pulse amplitude [7]. *Reproduced with permission from D.S. Korolev, et al., Manipulation of resistive state of silicon oxide memristor by means of current limitation during electroforming, Superlattice. Microst. 122 (2018) 371. Copyright 2018 Elsevier B.V.*

of a voltage pulse, which does not change the resistance, should be lower than ±0.5 V. The variations of resistive states obtained by multiple (100 times) programming sequences of each resistance value with 0.5 V voltage steps are also shown in Fig. 4.4. It is seen that the pulse-to-pulse variation of resistance increases with increasing the resistance of memristive device. It is evident that the observed variation of memristive states is caused by the mentioned stochastic nature of RS and should be taken into account in the design of artificial neural networks based on memristive devices. In particular, the random variation of resistance values may strongly affect the training process and accuracy of a small-size neural network, however, this stochastic effect will be mitigated with scaling (with the increase of weights number). The impact of the weights variation can also be minimized by increasing the number of neural network training steps.

Obviously, the observed complex stochastic behavior of memristive devices requires further analysis at a mesoscopic level [6] and is a significant problem for their application as a nonvolatile weight "storage" in traditional neural networks, when the unambiguous programmed weight change is required in response to a specified voltage level.

It should be emphasized that noise plays an important role both in determining the fundamental intrinsic properties of memristive devices, considered as a multistable nonlinear system, and in practical terms [23,24]. It is shown that the effective potential profile of memristive devices depends on the parameters of external white Gaussian noise and changes during exposure to noise. This indicates that the effect of noise on the memristive system is essentially multiplicative, that is, the noise signal causes a rearrangement of the filament's system and a corresponding change in the electrophysical properties of the memristive structure during exposure. Such a well-known phenomenon as stochastic resonance can be also observed in memristive devices, namely, the stabilized resistive switching is achieved, when optimal-level white Gaussian noise is superimposed on the sub-threshold sinusoidal driving signal [25]. This behavior is a clear manifestation of the constructive role of noise and fluctuations in nonequilibrium memristive systems.

4.3 Functionality of memristors as elements for neuromorphic systems

Despite the significant variation in the RS parameters at this stage of technology development, special techniques of adaptive programming can be used to demonstrate the simplest functionality of memristor as element for neuromorphic systems. An important condition for the integration of memristive devices into neuromorphic circuits is their implementation in the form of integrated arrays cross-point and cross-bar.

In this section, an array of cross-point memristive devices on the basis of an YSZ thin film is used in a prototype for spiking neural networks. The RS phenomena and the plasticity nature of such memristive devices are considered. Reproducible bipolar RS and precise tuning of resistive state are demonstrated and used to implement the plasticity rule according to the so-called spike-timing-dependent plasticity (STDP) mechanism that was shown to emerge naturally in memristive devices [26–30]. STDP learning is found to be dependent on the memristors initial resistive state value and discussed in terms of the finite conductance change in studied structures. Obtained results provide the foundation for the development of autonomous neuromorphic circuits with unsupervised learning.

Below, the arrays of microscale ($20 \times 20\,\mu m^2$) cross-point memristive devices either in paired form (overall 44 devices) or combined in a 16×16 cross-bar with a Au/Zr/YSZ/TiN/Ti structure on a silicon chip are considered fabricated according to previously developed special layouts [31,32] and mounted in a standard 64-pin package (Fig. 4.5).

The schematic view of the thin-film structure fabricated in cross-bar topology is shown in Fig. 4.6, together with a TEM cross-section image of the structure. Previously, it was shown in detail that the YSZ films obtained by magnetron sputtering have a columnar structure sandwiched between the polycrystalline TE and BE [34], and the grain boundaries in YSZ film correlate well with the roughness of the TE and play an important role for resistive switching [35].

The EF process was carried out by applying a negative voltage up to 20 V to the TE of pristine sample. This resulted in formation of filaments

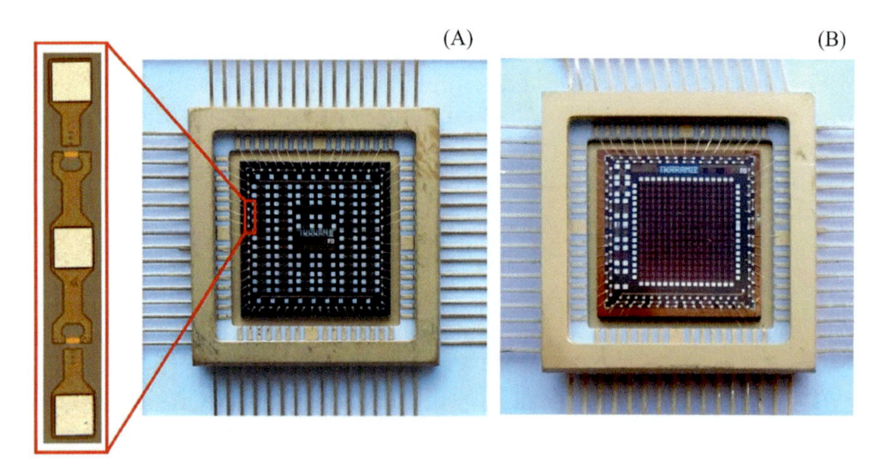

FIG. 4.5 Photographs of a fabricated microchips with (A) cross-point [33] and (B) cross-bar memristive devices based on YSZ. *Part (A) reproduced with permission from A.V. Emelyanov, et al., Yttria-stabilized zirconia cross-point memristive devices for neuromorphic applications, Microelectron. Eng. 215 (2019) 110988. Copyright 2019 Elsevier B.V.*

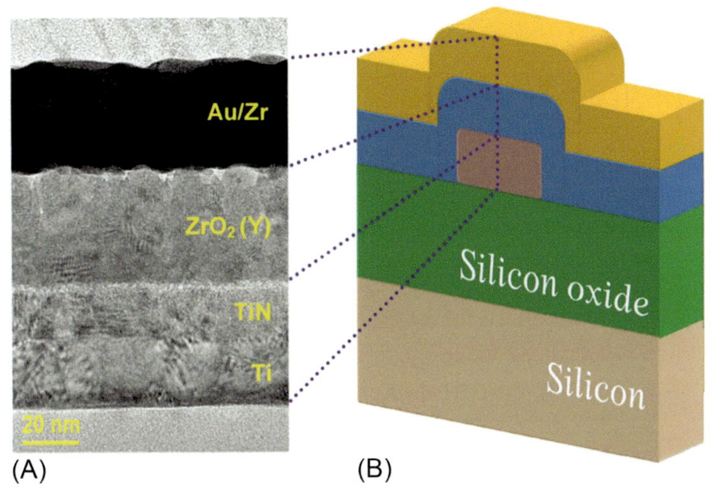

FIG. 4.6 (A) TEM image of a Au/Zr/YSZ/TiN/Ti memristive structure. (B) Corresponding schematic view of a Au/Zr/YSZ/TiN/Ti cross-bar memristive device [33]. *Reproduced with permission from A.V. Emelyanov, et al., Yttria-stabilized zirconia cross-point memristive devices for neuromorphic applications, Microelectron. Eng. 215 (2019) 110988. Copyright 2019 Elsevier B.V.*

and hence enabled the switching process. The initial resistive state was found to vary from device-to-device which is probably due to the mentioned effect of randomly distributed grain boundaries and the corresponding surface roughness of the YSZ film. For EF in case of a high-resistance initial state, the CC was set (in the range of 1–100 μA depending on the initial state) and the voltage sweep was stopped when the device current reached the CC value. It is worth noting that more than 90% of devices revealed RS after EF or first *I–V* sweeps irrespective of the initial state.

Fig. 4.7A presents typical *I–V* curves of a Au/Zr/YSZ/TiN/Ti memristive device showing a typical bipolar RS performance. Bipolar RS cycles demonstrate stable repetitiveness and a resistance window between the LRS and HRS states $R_{HRS}/R_{LRS} \approx 10^4$ at a read voltage $V_{read} = -0.5$ V. Each cycle was carried out by applying a voltage-sweep as follows: a positive RESET voltage to switch the sample to the HRS state ($0 \rightarrow V_{RESET} \rightarrow 0$) was applied; after that, a negative SET voltage to switch the sample to the LRS state was applied ($0 \rightarrow V_{SET} \rightarrow 0$). The voltage step was 0.1 V with a step-length of 50 ms. In general, LRS is characterized by a linear conduction mechanism for the investigated YSZ-based devices, whereas the HRS can be described by the Frenkel-Poole emission [36]. At the same time, there is almost a linear *I–V* dependence at the low voltage region used for reading the device resistive state (Figs. 4.7A and 4.8). This is a good

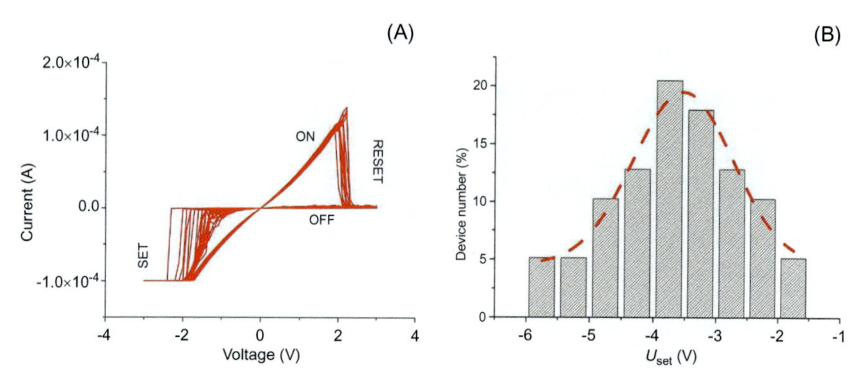

FIG. 4.7 (A) Typical *I–V* curves of a Au/Zr/YSZ/TiN/Ti memristive device. (B) Histogram for the device-to-device SET voltage distribution. The *dashed line* is a guide showing a Gaussian distribution [33]. *Reproduced with permission from A.V. Emelyanov, et al., Yttria-stabilized zirconia cross-point memristive devices for neuromorphic applications, Microelectron. Eng. 215 (2019) 110988. Copyright 2019 Elsevier B.V.*

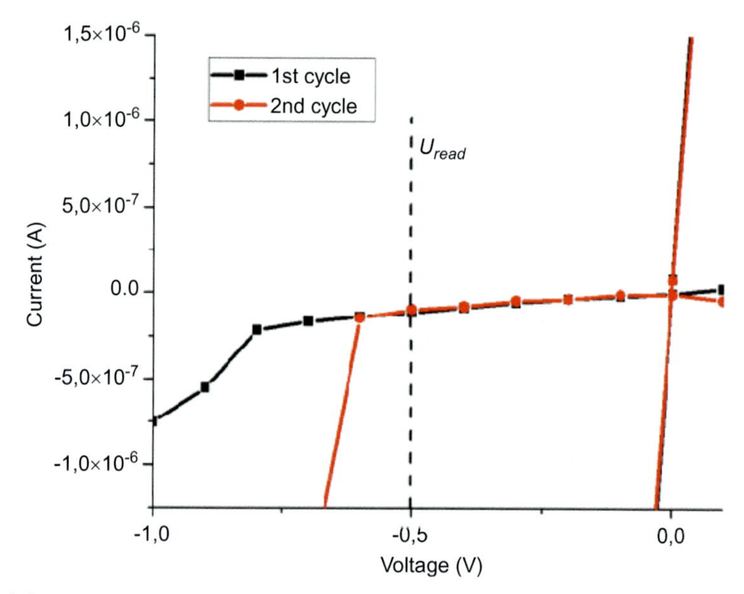

FIG. 4.8 Zoom into the low-voltage region of two cycles of typical *I–V* curves of a Au/Zr/YSZ/TiN/Ti memristive device. The read voltage $V_{read} = -0.5$ V was chosen to be within the linear region of *I–V* curves [33]. *Reproduced with permission from A.V. Emelyanov, et al., Yttria-stabilized zirconia cross-point memristive devices for neuromorphic applications, Microelectron. Eng. 215 (2019) 110988. Copyright 2019 Elsevier B.V.*

characteristic for a neuromorphic computing where constant weights are required to allow accurate learning and calculations.

The observed RS effect in an YSZ-based memristor can be explained by the reversible partial oxidation and recovery of filaments in the oxide film under electrical stress of different polarity [35]. The corresponding local reduction-oxidation phenomena are typical for anionic based RS devices and are often reduced to the change of valence state of cations in a TMO. In contrast to canonical filamentary models proposed for asymmetric device structures and related to one active electrode and/or built-in oxygen vacancy reservoir (see e.g., [37]), the present approach is based on the use of YSZ, which has a high concentration of mobile oxygen vacancies, and does not require the generation of oxygen vacancies in stoichiometric oxide or introduction of oxygen scavenging layer. The filament grows via the redistribution of oxygen vacancies with the initially uniform depth profile in YSZ film predefined by the yttrium doping level, as has been shown by kinetic Monte Carlo simulation [6,35] (see Chapter 3 for the approach and simulation results).

According to TEM data (Fig. 4.6A), there are initial inhomogeneities in the oxide at the interface with the TE and grain boundaries. Such inhomogeneities can serve as the filament nuclei. If a negative potential is applied to the TE, the inhomogeneous electric field occurs in the YSZ film, which leads to the increased concentration of oxygen vacancies around the nucleus and, as a consequence, the growth of metallic-like filaments [6]. Correspondingly, the device switches to the linear LRS. The RESET transition at the application of positive potential to the TE is determined by the escape of oxygen vacancies from the region of filament tip (a gap near the active BE). This results in a decreased vacancy concentration and conductivity of this region followed by an abrupt current drop. As a result, the device switches to the HRS with a nonlinear electron transport mechanism. If a negative potential is applied after the RESET, it leads to the SET transition. The concentration of vacancies inside and near the gap is increased, and the current grows up during this process. The device switches to the LRS again.

Of course, the growth and oxidation of filaments is strongly affected by the temperature due to Joule heating. Thus, the specific value of CC used during the SET process (100 μA) limits the number and size of filaments, and hence, the possible resistance window. The LRS is achieved by applying negative voltage bias—a process that is stochastic in nature. Fig. 4.7B shows the histogram for the device-to-device (spatial) SET voltage distribution extrapolated with the normal one with the mean value $\mu_V = -3.5\,\mathrm{V}$ and standard deviation $\sigma_V = 1.6\,\mathrm{V}$. The obtained results show comparable device-to-device RS with a good set voltage variation ($\sigma_V/\mu_V = 45\%$). It is worth noting that such spatial variation is preferable for hardware-intrinsic security primitives (e.g., physical unclonable functions) [38].

As the performance of many types of neuromorphic circuits (especially those using error-back-propagation learning algorithms—perceptrons, convolutional neural networks, etc.) directly depends on how accurately the resistive state can be set [38–41], the following is an algorithm of precision tuning of state for the memristive devices based on YSZ [33]. For this purpose, an active feedback scheme was used, i.e., iterative write and read pulses were applied to converge to a certain desired resistive state of the device as it was proposed in [42,43]. The distinctive feature of the YSZ-based memristive device in comparison with that used in [42,43] is a much bigger R_{HRS}/R_{LRS} ratio, and hence, more abrupt SET process (small voltage change values can lead to the resistance drop over several orders of magnitude). In order to tolerate this feature, the algorithm used in [42,43] has been modified and is based on RS during the *I–V* cycle measurement. The algorithm consists of two steps: writing, i.e., applying a voltage pulse (the pulse magnitude is selected according to current *I–V* iteration), and reading at voltage of −0.5 V (an example of pulse sequence is shown in Fig. 4.9). If the memristor's resistive state at the current iteration equals to the predefined one with selected accuracy, then the algorithm stops, else the *I–V* cycle is continued or repeated. The writing steps followed the voltage sweep when measuring the *I–V* curves; therefore, the switching with the used algorithm was similar to that in the measurement of standard *I–V* curves and reading steps did not affect the resistive state of the device. In this way, it is possible to avoid sticking the memristor in one state due to a good repeatability of cycles. On the other hand, applying few cycles of *I–*

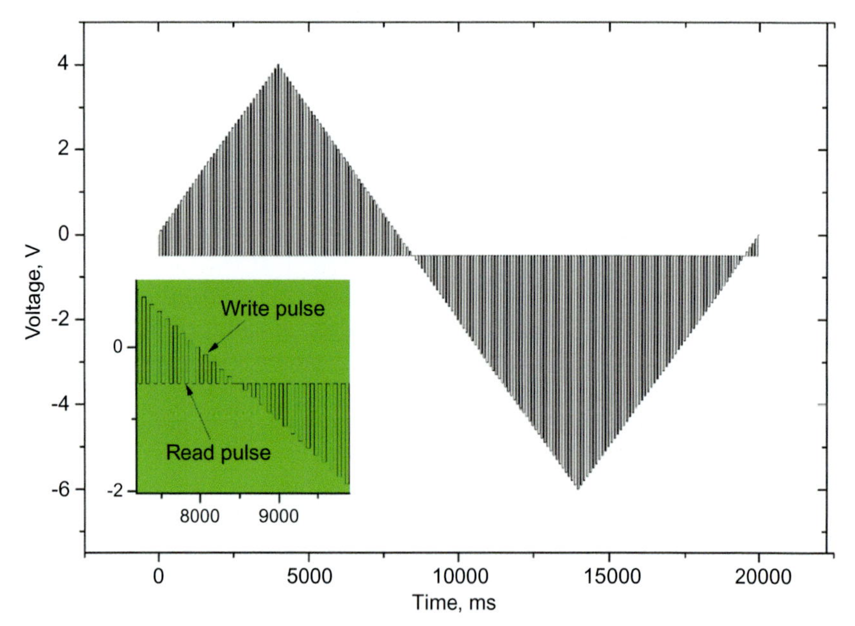

FIG. 4.9 One cycle of pulses applied to the memristor during switching by the developed algorithm. Write pulses were gradually changed with the step $|\Delta V| = 0.1$ V, read pulses were $V_{read} = -0.5$ V [33]. *Reproduced with permission from A.V. Emelyanov, et al., Yttria-stabilized zirconia cross-point memristive devices for neuromorphic applications, Microelectron. Eng. 215 (2019) 110988. Copyright 2019 Elsevier B.V.*

V iterations ensures reaching the required state for writing voltages close to the switching voltage value. On the average 10–15 of I–V cycles were needed to set the required resistance with the 5% precision. If the desired resistance values were not reached within 30 cycles, the device was assumed as defective.

By applying the above described algorithm to YSZ-based memristive devices, more than eight levels of resistance states or three bits of digital precision were demonstrated in a resistance range of 10^3–10^7 Ω (Fig. 4.10). The resolution precision could be further increased up to 5–8 bits, i.e., 32–256 resistive states [38]. However, the more challenging goal was to investigate the possibility of memristor resistive state change due to the STDP mechanism.

For STDP implementation, the BE of an YSZ-based memristive structure was assigned as a presynaptic input and the TE was considered as a postsynaptic input. Identical potential pulses for the pre- and postsynaptic spikes were used as the corresponding signals (Fig. 4.11A). The amplitudes of spikes were chosen to be 1 V, so that the spike itself could not lead to a conductivity change of the memristor. At the same time, if two spikes are specifically summed up (as illustrated in Fig. 4.11B), the maximum voltage drop across a memristive device could be 2 V and the minimum one could be −2 V, which is within the switching range of the memristive device based on YSZ. The pulse duration was

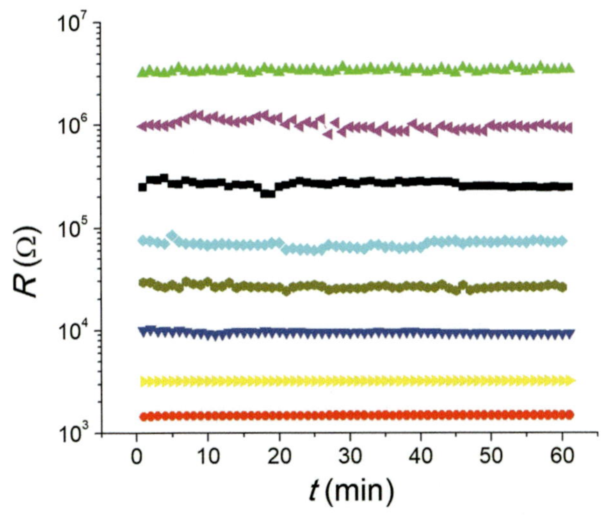

FIG. 4.10 Room-temperature retention of eight different resistive states of a Au/Zr/YSZ/TiN/Ti memristive device [33]. *Reproduced with permission from A.V. Emelyanov, et al., Yttria-stabilized zirconia cross-point memristive devices for neuromorphic applications, Microelectron. Eng. 215 (2019) 110988. Copyright 2019 Elsevier B.V.*

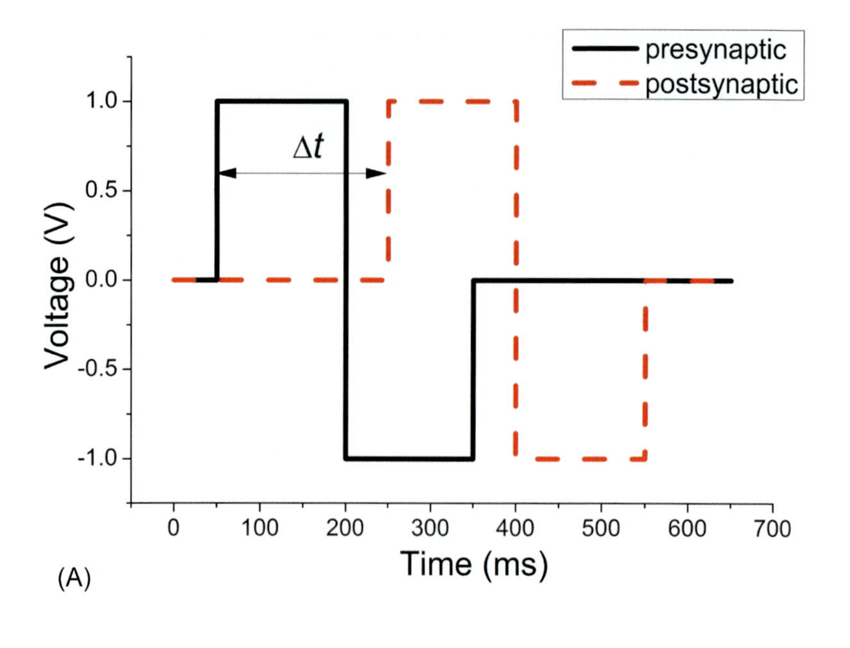

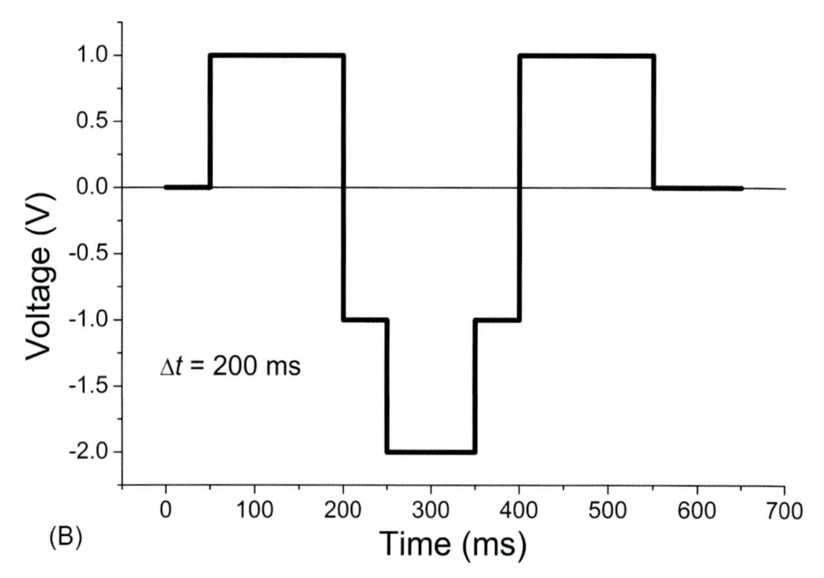

FIG. 4.11 (A) Shapes of presynaptic *(solid line)* and postsynaptic *(dashed line)* potential pulses. (B) Resulting voltage across the memristive device for the specific value of $\Delta t = 200\,\text{ms}$ [33]. *Reproduced with permission from A.V. Emelyanov, et al., Yttria-stabilized zirconia cross-point memristive devices for neuromorphic applications, Microelectron. Eng. 215 (2019) 110988. Copyright 2019 Elsevier B.V.*

300 ms with the discretization of 50 ms. Postsynaptic pulses were applied after presynaptic pulses with a varied delay time Δt. The pulse's shapes and the definition of Δt are shown in Fig. 4.11A. The applied voltage across the memristor during the measurement (the difference between post- and pre-synaptic potentials) for a delay time $\Delta t = 200$ ms is shown in Fig. 4.11B.

The device conductance G is regarded as the synaptic weight, and then its change (ΔG) is equivalent to the synaptic weight change. Conductance values were measured at V_{read} before and after the pre- and post-synaptic pulse sequence. Weight changes were considered as $\Delta G = G_f - G_i$, where G_f and G_i are the final and initial conductance values, respectively. Thus, the determined weight change dependence on the delay time (STDP window) averaged over 10 samples for three different initial states is shown in Fig. 4.12.

As can be seen from Fig. 4.12 the experimental results obey the STDP rule observed in biological systems, especially for inhibitory synapses [44]. The synaptic potentiation ($\Delta G > 0$) was achieved for $\Delta t > 0$, and the synaptic depression ($\Delta G < 0$) was achieved for $\Delta t < 0$. As it follows from simulation results in [45] the STDP window shape depends on the pre- and post-synaptic spike shapes: e.g., for triangle pulses STDP window would follow hyperbolic-like function [26], for square pulses it would follow square-like function (as in the considered case). It should be noted that

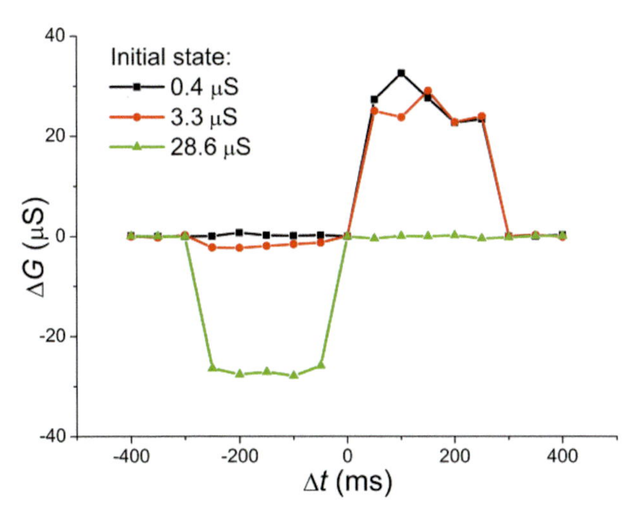

FIG. 4.12 STDP window for a memristive device based on YSZ—weight (conductance) changes for different delay Δt values and different initial conductive states: 0.4 μS *(black squares)*, 3.3 μS *(red dots)* and 28.6 μS *(green triangles)* [33]. *Reproduced with permission from A.V. Emelyanov, et al., Yttria-stabilized zirconia cross-point memristive devices for neuromorphic applications, Microelectron. Eng. 215 (2019) 110988. Copyright 2019 Elsevier B.V.*

STDP learning is found to depend on the G_i value. If the memristor state is close to LRS, then its synaptic weight would likely depress than potentiate (as for the 28.6 μS state in Fig. 4.12), and vice versa (0.4 μS state in Fig. 4.12). For the intermediate state, the weight could both depress and potentiate (relative conductance change $\Delta G/G_i$ is varied from −70% (for negative Δt) to +800% (for positive Δt)—red dots in Fig. 4.12). This could be explained by taking into account the finite conductance change of the studied memristors. Moreover, such G_i-dependent STDP behavior may be expected for many types of memristive devices with saturating switching dynamics [46,47].

It should be emphasized here that the experimental STDP window is in a good agreement with the data available for biological synapses and could provide a basis for unsupervised learning [26,48]. Therefore, the arrays of memristive devices can be used for hardware implementation of biologically inspired spiking neural networks.

4.4 Conclusions

The obtained good background in the technology of fabrication of metal-oxide devices and arrays demonstrates excellent compatibility with the standard CMOS process and provides the possibility of hybridization of memristive electronics and analog-digital circuits. The latter, in turn, is necessary to create primitives for small sized neuromorphic systems, including spike neural networks based on local Hebbian learning (on the example of the STDP model). At the same time, the functionality of hardware memristive systems is hampered by the insufficient reproducibility of device parameters due to the stochastic nature of the resistive switching effect.

The mentioned problems are traditionally solved with the use of adaptive programming of the resistive state or by tuning materials and interface properties in memristive structures but this requires either complex peripheral circuits or advanced technological processes. A very promising approach to modeling and controlling the response of memristive devices is the macrosystem approach, in which the memristor is considered as a complex multistable system driven by noise and fluctuations. This approach will ensure the development of compact physics-based models of memristors [49]. On the one hand, such models would most accurately describe the stochastic behavior of memristive devices and their imperfections (for example, the variability of parameters). On the other hand, such simulation would not require large computational resources and could be implemented by computer-aided design tools for the use in design of large spiking neural networks. The unique scalability of memristors and their "live" dynamic response to complex spike-like signals are beneficial for brain-like neural network architectures with self-organization of synaptic connections.

Acknowledgments

This work was supported by the Government of the Russian Federation through the Agreement No. 074-02-2018-330 (2). The part related to the effect of electroforming conditions on adaptive behavior of memristive devices was supported by the grant of the Russian Science Foundation (16-19-00144).

References

[1] D.M. Mattox, Handbook of Physical Vapor Deposition (PVD) Processing: Film Formation, Adhesion, Surface Preparation and Contamination Control, second ed., Noyes Publications, New Jersey, USA., 1998.

[2] A. Mehonic, M.S. Munde, W.H. Ng, M. Buckwell, L. Montesi, M. Bosman, et al., Intrinsic resistance switching in amorphous silicon oxide for high performance SiO_x ReRAM devices, Microelectron. Eng. 178 (2017) 98–103.

[3] M.S. Munde, A. Mehonic, W.H. Ng, M. Buckwell, L. Montesi, M. Bosman, et al., Intrinsic resistance switching in amorphous silicon suboxides: the role of columnar microstructure, Sci. Rep. 7 (2017) 9274.

[4] D.S. Jeong, B.K. Cheong, H. Kohlstedt, $Pt/Ti/Al_2O_3/Al$ tunnel junctions exhibiting electroforming-free bipolar resistive switching behavior, Solid State Electron. 63 (2011) 1–4.

[5] S. Tappertzhofen, I. Valov, R. Waser, Quantum conductance and switching kinetics of AgI-based microcrossbar cells, Nanotechnology 23 (2012) 145703.

[6] A.N. Mikhaylov, E.G. Gryaznov, A.I. Belov, D.S. Korolev, A.N. Sharapov, D.V. Guseinov, et al., Field- and irradiation-induced phenomena in memristive nanomaterials, Phys. Status Solidi C 13 (2016) 870–881.

[7] D.S. Korolev, A.I. Belov, E.V. Okulich, V.I. Okulich, I.N. Antonov, E.G. Gryaznov, et al., Manipulation of resistive state of silicon oxide memristor by means of current limitation during electroforming, Superlattice. Microst. 122 (2018) 371–376.

[8] S.D. Ha, S. Ramanathan, Adaptive oxide electronics: a review, J. Appl. Phys. 110 (2011), 071101.

[9] M.-C. Wu, W.-Y. Jang, C.-H. Lin, T.-Y. Tseng, A study on low-power, nanosecond operation and multilevel bipolar resistance switching in $Ti/ZrO_2/Pt$ nonvolatile memory with 1T1R architecture, Semicond. Sci. Technol. 27 (2012), 065010.

[10] L. Zhao, H.-Y. Chen, S.-C. Wu, Z. Jiang, S. Yu, T.-H. Hou, et al., Multi-level control of conductive nano-filament evolution in HfO_2 ReRAM by pulse-train operations, Nanoscale 6 (2014) 5698–5702.

[11] S.R. Lee, Y.-B. Kim, M. Chang, K.M. Kim, C.B. Lee, J.H. Hur, et al., Multi-level switching of triple-layered TaO_x RRAM with excellent reliability for storage class memory, in: Symposium on VLSI Technology, 2012.

[12] A. Mehonic, M. Buckwell, L. Montesi, L. Garnett, S. Hudziak, S. Fearn, et al., Structural changes and conductance thresholds in metal-free intrinsic SiO_x resistive random access memory, J. Appl. Phys. 117 (2015) 124505.

[13] S. Kim, B.-G. Park, Nonlinear and multilevel resistive switching memory in $Ni/Si_3N_4/Al_2O_3/TiN$ structures, Appl. Phys. Lett. 108 (2016) 212103.

[14] M. Sowinska, T. Bertaud, D. Walczyk, S. Thiess, P. Calka, L. Alff, et al., In-operando hard X-ray photoelectron spectroscopy study on the impact of current compliance and

switching cycles on oxygen and carbon defects in resistive switching Ti/HfO_2/TiN cells, J. Appl. Phys. 115 (2014) 204509.
[15] S. Brivio, E. Covi, A. Serb, T. Prodromakis, M. Fanciulli, S. Spiga, Experimental study of gradual/abrupt dynamics of HfO_2-based memristive devices, Appl. Phys. Lett. 109 (2016) 133504.
[16] A.N. Mikhaylov, A.I. Belov, D.V. Guseinov, D.S. Korolev, I.N. Antonov, D.V. Efimovykh, et al., Bipolar resistive switching and charge transport in silicon oxide memristor, Mater. Sci. Eng., B 194 (2015) 48–54.
[17] D.S. Korolev, A.N. Mikhaylov, A.I. Belov, V.A. Sergeev, I.N. Antonov, O.N. Gorshkov, et al., Adaptive behaviour of silicon oxide memristive nanostructures, J. Phys. Conf. Ser. 741 (2016) 012161.
[18] A. Mikhaylov, A. Belov, D. Korolev, I. Antonov, V. Kotomina, A. Kotina, et al., Multilayer metal-oxide memristive device with stabilized resistive switching, Adv. Mater. Technol. 5 (2020) 1900607.
[19] T. Hennen, E. Wichmann, A. Elias, J. Lille, O. Mosendz, R. Waser, et al., Current-Limiting Amplifier for High Speed Measurement of Resistive Switching Data, Arxiv preprint, 2021. arXiv:2102.05770 [physics.ins-det].
[20] A. Mehonic, A. Vrajitoarea, S. Cueff, S. Hudziak, H. Howe, C. Labbe, et al., Quantum conductance in silicon oxide resistive memory devices, Sci. Rep. 3 (2013) 2708.
[21] D. Ielmini, Modeling the universal set/reset characteristics of bipolar RRAM by field– and temperature—driven filament growth, IEEE Trans. Electron Devices 58 (2011) 4309–4317.
[22] X. Lian, X. Cartoixà, E. Miranda, L. Perniola, R. Rurali, S. Long, et al., Multi-scale quantum point contact model for filamentary conduction in resistive random access memories devices, J. Appl. Phys. 115 (2014) 244507.
[23] D.O. Filatov, D.V. Vrzheshch, O.V. Tabakov, A.S. Novikov, A.I. Belov, I.N. Antonov, et al., Noise-induced resistive switching in a memristor based on $ZrO_2(Y)/Ta_2O_5$ stack, J. Stat. Mech: Theory Exp. 2019 (2019) 124026.
[24] D.O. Filatov, A.S. Novikov, V.N. Baranova, D.A. Antonov, A.V. Kruglov, I.N. Antonov, et al., Experimental investigations of local stochastic resistive switching in yttria stabilized zirconia film on a conductive substrate, J. Stat. Mech: Theory Exp. 2020 (2020) 024005.
[25] A.N. Mikhaylov, D.V. Guseinov, A.I. Belov, D.S. Korolev, V.A. Shishmakova, M.N. Koryazhkina, et al., Stochastic resonance in a metal-oxide memristive device, Chaos, Solitons Fractals 144 (2021) 110723.
[26] M. Prezioso, F. Merrikh-Bayat, B. Hoskins, K. Likharev, D. Strukov, Self-adaptive spike-time-dependent plasticity of metal-oxide memristors, Sci. Rep. 6 (2016) 21331.
[27] A. Serb, J. Bill, A. Khiat, R. Berdan, R. Legenstein, T. Prodromakis, Unsupervised learning in probabilistic neural networks with multi-state metal-oxide memristive synapses, Nat. Commun. 7 (2016) 12611.
[28] C.-C. Hsieh, A. Roy, Y.-F. Chang, D. Shahrjerdi, S.K. Banerjee, A sub-1-volt analog metal oxide memristive-based synaptic device with large conductance change for energy-efficient spike-based computing systems, Appl. Phys. Lett. 109 (2016) 223501.
[29] Y. Matveyev, K. Egorov, A. Markeev, A. Zenkevich, Resistive switching and synaptic properties of fully atomic layer deposition grown TiN/HfO_2/TiN devices, J. Appl. Phys. 117 (2015), 044901.
[30] D.A. Lapkin, A.V. Emelyanov, V.A. Demin, T.S. Berzina, V.V. Erokhin, Spike-timing-dependent plasticity of polyaniline-based memristive element, Microelectron. Eng. 185-186 (2018) 43–47.
[31] E.G. Gryaznov, I.N. Antonov, A.I. Belov, A.Y. Kotina, V.E. Kotomina, Y.K. Makarychev, A.N. Mikhaylov, D.O. Filatov, A.N. Sharapov, O.N. Gorshkov, The Topology of the Test Crystal With Elements of Non-Volatile, Repeatedly Programmable

Resistive Memory, RU. Patent 2016630152 filed November 30, 2016, and issued January 21, 2017.
[32] E.G. Gryaznov, I.N. Antonov, A.Y. Kotina, V.E. Kotomina, A.N. Mikhaylov, A.N. Sharapov, O.N. Gorshkov (Eds.), Topology of a Test Crystal With a Matrix of Memristive Microdevices, 2018. RU. Patent 2018630123 filed July 12, 2018, and issued August 08.
[33] A.V. Emelyanov, K.E. Nikiruy, V.A. Demin, V.V. Rylkov, A.I. Belov, D.S. Korolev, et al., Yttria-stabilized zirconia cross-point memristive devices for neuromorphic applications, Microelectron. Eng. 215 (2019) 110988.
[34] O.N. Gorshkov, A.N. Mikhaylov, A.P. Kasatkin, S.V. Tikhov, D.O. Filatov, D.A. Pavlov, et al., Resistive switching in the Au/Zr/ZrO_2–Y_2O_3/TiN/Ti memristive devices deposited by magnetron sputtering, J. Phys. Conf. Ser. 741 (2016) 012174.
[35] D.V. Guseinov, D.I. Tetelbaum, A.N. Mikhaylov, A.I. Belov, M.E. Shenina, D.S. Korolev, et al., Filamentary model of bipolar resistive switching in capacitor-like memristive nanostructures on the basis of yttria-stabilised zirconia, Int. J. Nanotechnol. 14 (2017) 604–617.
[36] S.A. Gerasimova, A.N. Mikhaylov, A.I. Belov, D.S. Korolev, A.V. Lebedeva, O.N. Gorshkov, et al., Design of memristive interface between electronic neurons, AIP Conf. Proc. 1959 (2018), 090005.
[37] D. Wouters, Resistive switching materials and devices for future memory applications, in: 43rd IEEE Semiconductor Interface Specialists Conference, 2012.
[38] C. Li, M. Hu, Y. Li, H. Jiang, N. Ge, E. Montgomery, et al., Analogue signal and image processing with large memristor crossbars, Nat. Electron. 1 (2018) 52–59.
[39] H. Nili, G.C. Adam, B. Hoskins, M. Prezioso, J. Kim, M.R. Mahmoodi, et al., Hardware-intrinsic security primitives enabled by analogue state and nonlinear conductance variations in integrated memristors, Nat. Electron. 1 (2018) 197–202.
[40] A.V. Emelyanov, D.A. Lapkin, V.A. Demin, V.V. Erokhin, S. Battistoni, G. Baldi, et al., First steps towards the realization of a double layer perceptron based on organicmemristive devices, AIP Adv. 6 (2016) 111301.
[41] S. Battistoni, V. Erokhin, S. Iannotta, Organic memristive devices for perception applications, J. Phys. D Appl. Phys. 51 (2018) 284002.
[42] F. Alibart, L. Gao, B.D. Hoskins, D.B. Strukov, High precision tuning of state for memristive devices by adaptable variation-tolerant algorithm, Nanotechnology 23 (2012), 075201.
[43] K.E. Nikiruy, A.V. Emelyanov, V.A. Demin, V.V. Rylkov, A.V. Sitnikov, P.K. Kashkarov, A precise algorithm of memristor switching to a state with preset resistance, Tech. Phys. Lett. 44 (2018) 416–419.
[44] G. Hennequin, E.J. Agnes, T.P. Vogels, Inhibitory plasticity: balance, control, and codependence, Annu. Rev. Neurosci. 40 (2017) 557–579.
[45] S. Saïghi, C.G. Mayr, T. Serrano-Gotarredona, H. Schmidt, G. Lecerf, J. Tomas, et al., Plasticity in memristive devices for spiking neural networks, Front. Neurosci. 9 (2015) 51.
[46] Z. Wang, S. Joshi, S. Savel'ev, W. Song, R. Midya, Y. Li, et al., Fully memristive neural networks for pattern classification with unsupervised learning, Nat. Electron. 1 (2018) 137–145.
[47] M. Prezioso, F. Merrikh-Bayat, B.D. Hoskins, G.C. Adam, K.K. Likharev, D.B. Strukov, Training and operation of an integrated neuromorphic network based on metal-oxide memristors, Nature 521 (2015) 61–64.
[48] S. Boyn, J. Grollier, G. Lecerf, B. Xu, N. Locatelli, S. Fusil, et al., Learning through ferroelectric domain dynamicsin solid-state synapses, Nat. Commun. 8 (2017) 14736.
[49] N.V. Agudov, A.V. Safonov, A.V. Krichigin, A.A. Kharcheva, A.A. Dubkov, D. Valenti, et al., Nonstationary distributions and relaxation times in a stochastic model of memristor, J. Stat. Mech: Theory Exp. 2020 (2020) 024003.

CHAPTER

5

Metalorganic chemical vapor deposition of aluminum oxides: A paradigm on the process-structure-properties relationship

Constantin Vahlas[a] *and Brigitte Caussat*[b]
[a]CIRIMAT-CNRS, Toulouse, France, [b]LGC-CNRS, Toulouse, France

5.1 Introduction

The stiffness, hardness and chemical inertia of amorphous aluminum oxide (Al_2O_3) thin films, combined with their optical transmittance make them a key player in several applications including microelectronics, optics, catalysis, protection against corrosion and high temperature oxidation. In particular, it was reported that, in contrast to doped polycrystalline hafnium dioxide HfO_2, amorphous Al_2O_3 films possess the advantages of often low process temperature and operating frequencies up to ~GHz, which enable multi-gate non-volatile field-effect transistor (NVFET) with nanometerscale fin pitch [1–3]. Furthermore, combination of Al_2O_3 with other oxides such as HfO_2 results in high-*k* dielectric layers that exhibit high thermal stability, large band gap and high barrier for oxygen diffusion [4], while the thickness of such films can be assessed toward improving the ferroelectric performance of the device [5]. These advantages allow fabricating non-volatile resistive memory devices on thermally sensitive supports such as paper [6] or polyethylene naphthalate [3].

Overall, the number of reports on the preparation of alumina films is significant using either physical vapor deposition (PVD) or chemical (atomic layer deposition, ALD, chemical vapor deposition, CVD) techniques. A practical way to produce amorphous alumina films is metal-

Metal Oxides for Non-volatile Memory
https://doi.org/10.1016/B978-0-12-814629-3.00005-2

organic CVD (MOCVD) [7–12], in which a controlled amount of metal-organic precursor carried over a surface decomposes and reacts to form a conformal condensate thin film.

MOCVD is a potentially attractive technique for processing of such coatings, especially on complex-in-shape temperature sensitive substrates. It allows depositing thin films on heated substrates, by the transport in the reactor of gaseous reactive molecules and then the occurrence of gas phase and surface chemical reactions leading to the formation of the film. CVD processes in general are of industrial interest offering a variety of coating compositions and morphologies to answer severe technological specifications, such as anti-corrosion or catalytic conformal deposits on complex surfaces, under the form of either thin films or nanomaterials (particles, tubes, wires). To overcome drawbacks of thermal incompatibility with the substrate and risks of manipulations of corrosive effluents, MOCVD processes have been developed at relatively low temperature involving liquid or solid metalorganic precursors. Composition, nature of deposited phases, stoichiometry, crystallinity and microstructure of the material can be adjusted by finely tuning the MOCVD operating parameters. However, MOCVD processes are rather difficult to develop and optimize, due to the complicated decomposition mechanisms of thermally sensitive molecular compounds in the gas phase and on the growing surface. The chemistry is often hard to catch due to the difficulty to implement non-intrusive metrology for the in situ monitoring of the process. Process numerical modeling allows correlating the local behavior of the MOCVD reactor parameters like the gas flow, temperature, concentration and deposition rate profiles with the reactor geometry and deposition parameters. When available, such models are a valuable tool to optimize a MOCVD process in terms of film thickness and composition, in particular on complex substrates, like micro-patterned silicon wafers, porous membranes or foams or hollow objects like vials. However, the bottleneck is often the knowledge of the chemical reactions and of the associated kinetic laws leading to the film formation.

Because of the previously mentioned well-known technological interest of alumina thin films in various domains, including catalysis, corrosion protection, and electrical insulation, insight into the influence of the MOCVD process on the atomic structure of the ceramic is a valuable key to process and property optimization. Indeed, the development of materials with such superior properties is strongly linked to the understanding of their atomic structure dictated by the fabrication process. This has been illustrated for amorphous alumina films concerning, for example, the corrosion resistance measured by electrochemical impedance spectroscopy [7] or their superior mechanical properties (hardness and Young's modulus) [12]. It is therefore crucial to obtain comprehensive insights regarding the influence of a synthesis process on the material composition and atomic structure.

The methodology used to develop a set of chemical reactions and kinetic laws is presented in a first part, together with the capability of a numerical model to help optimizing a MOCVD process on complex substrates. The particular case of an amorphous aluminum oxide deposit on glass vials is described.

The second part is dedicated to the experimental analysis of the interplay existing between the amorphous alumina deposition conditions, the microstructure and local atomic organization of the thin films and their barrier properties against oxidation or biofouling.

5.2 Process kinetic modeling and simulation of the MOCVD of metal oxides: The case of Al_2O_3 films

CVD processes are subjected to a strong interplay between the complex reactive transport phenomena occurring inside the reactor and the film characteristics in terms of thickness, composition and even morphology and microstructure [13]. With the increasingly severe specifications regarding film properties, thickness and composition uniformities on always more complex 3D substrates, the empirical development of CVD processes is no more efficient. This is particularly true in the fields of microelectronics, photovoltaics (microelectromechanical systems) MEMS and thermal barrier and anticorrosion coatings [13–15].

Computational Fluid Dynamics (CFD) models of CVD reactors associated with the appropriate kinetics laws of gas phase and surface chemical reactions allow to calculate the local profiles of gas flow, temperature, species concentration and deposition rates, as a function of the reactor and substrate geometry and of the deposition parameters. This corresponds to a purely continuous approach at a macroscopic scale. When combined with experimental studies, such models allow accurately analyzing the interplay between the reactive transport phenomena and the film thickness and composition, in order de develop and optimize new CVD processes with high capabilities.

The governing equations that describe the transport phenomena occurring inside a CVD reactor include the conservation of mass, momentum, energy and chemical species, associated with the kinetic laws of the homogeneous and heterogeneous chemical reactions. These equations are implemented in standard CFD codes such as Comsol, Fluent/Ansys, CFX among others [13–15]. However, the kinetic laws are rarely known for CVD processes and their establishment is the real bottleneck of this field. This is particularly true for MOCVD processes involving complex metalorganic precursor molecules, which are decomposed in the gas phase in numerous unknown by-products themselves likely to contribute to the formation of the film.

The methodology used to develop these kinetic laws is empirical. It combines (i) literature data on the reaction mechanisms and on deposition experiments, (ii) dedicated experiments providing the deposition rate and the chemical composition of the film and when possible, the gas phase composition at the exit of the reactor measured by gas chromatography-mass spectrometry (GC-MS) or other analytical techniques, for various operating conditions, and (iii) simulation work using apparent homogeneous and heterogeneous chemical reactions and kinetic laws, developed and optimized by trial and error from (i) and (ii). These apparent reactions represent in a simplified way the complex and numerous real reactions, which are difficult to precisely determine. It must be emphasized that, as they are only global and apparent, they are only valid for the studied range of deposition conditions.

The capability of such approach is detailed below for the MOCVD of amorphous Al_2O_3 films deposited at moderate temperature (350–500 °C) and low pressure from aluminum tri-isopropoxide (ATI) diluted in N_2, first on planar [16] and then on complex substrates [17]. The CFD code used is first presented with its main assumptions and boundary conditions. Then, a bibliographic analysis is performed about the available chemical mechanisms and kinetic data, followed by the description of the development of the kinetic model and its validation using experimental results obtained on planar substrates. Finally, the model is applied for the optimization of an original MOCVD reactor dedicated to coat the internal surface of a complex part, namely a glass vial.

5.2.1 Model characteristics

The CFD code Fluent 16.2 (Ansys) was used for the works reported in this chapter. Fluent is a pressure-based, implicit Reynolds Averaged Navier Stokes solver that employs a cell-centered finite volume scheme having second-order spatial accuracy. It discretizes any computational domain into elemental control volumes, and permits the use of quadrilateral or hexahedral, triangular or tetrahedral and hybrid meshes. The following assumptions were made:

- Steady state conditions,
- Laminar gas flow (Reynolds number lower than 1000),
- Compressible ideal gas (Mach number of max. 1.4),
- No consideration of the precursor in the evaluation of the physical properties of the inlet gas, and no effect of the heats of reaction, due to the high precursor dilution.

The associated boundary conditions are the following:

- At the gas inlet, a flat profile was imposed on the gas velocity, and the mass flow rate and species mass fractions were fixed in accordance with the process conditions.

- At the symmetry axis and at the exit, classical Danckwerts conditions (diffusive flux densities equal to zero), were applied for gas velocity and mass fractions.
- On the walls and substrate surfaces, a classical no-slip condition was used for the gas velocity; the mass flux density of each species was assumed to be equal to the corresponding heterogeneous reaction rate.
- At the exit, the total pressure was fixed at the experimental value.

The physical properties of the gaseous species and the multicomponent diffusion coefficients were calculated using the kinetic theory of gases and the Chapman-Enskog theory, respectively [18]. Lennard-Jones parameters for N_2, O_2 and H_2O were taken from the FLUENT database. The Lennard Jones parameters for C_6H_{12} and C_3H_6 were taken from Bird et al. [18]. These parameters are unknown for ATI and were considered the same as those of n-decane, this molecule having almost the same number of carbon atoms. The ATI molar fraction being low, a small error on these values did not affect the simulation results.

5.2.2 Bibliographic analysis on the chemical mechanisms and kinetic laws

Some groups worked on the Al_2O_3 MOCVD process from ATI. Kawase et al. [19] developed a CVD rate analysis method by using non-isothermal tubular reactors of small diameter (1.6–5 mm) for alumina deposition from ATI. Since the gas composition can be kept practically constant in the thin tubular reactor, the deposition rate can be measured simultaneously at various temperatures [20]. Working at temperatures higher than 800 °C and at 4 kPa of total pressure, they found that gas phase reactions control the deposition-rate and that the ATI decomposition rate is first order with respect to the ATI concentration. They mention that Barybin and Tomilin [21] reported the following overall reaction:

$$2Al(i-OC_3H_7)_3 \rightarrow Al_2O_3 + 3C_3H_7OH + 3C_3H_6 \quad (5.R1)$$

and that Pauer et al. [22] reported:

$$2Al(i-OC_3H_7)_3 \rightarrow Al_2O_3 + 3\,(C_3H_7)_2O \quad (5.R2)$$

$$(C_3H_7)_2O \rightarrow 2C_3H_6 + H_2O \quad (5.R3)$$

Hofman et al. [23] performed mass spectroscopic experiments at the exit of a hot wall CVD reactor operating at 400 Pa between 220 °C and 450 °C. The authors found that the deposition kinetics is a second order process, explained by the fact that the rate-limiting step is dehydration of the deposited AlOOH into Al_2O_3. They also showed that deposition by-products are propene, water and 2-propanol, in agreement with Pauer et al. [22].

Blittensdorf et al. [24] used a stagnation point cold wall CVD reactor to investigate the alumina deposition rate from ATI/O_2/Ar mixtures between 350 °C and 1080 °C and 5–25 kPa total pressure. The aim was to minimize gas phase reactions in order to mainly analyze the heterogeneous reactions. They found a deposition rate independent of temperature at 10 kPa. They attributed this behavior to the fact that the process was operated in the transport-limited regime, meaning that the transition temperature between a surface to a transport limited regime occurs at temperatures below 400 °C. This result is in agreement with that of Saraie et al. [25] who found that the temperature of the transition regime increased from 270 °C to 350 °C with the ATI concentration in the inlet gas. These authors [25] experimentally studied the influence of various atmospheres (N_2, N_2/O_2, N_2/H_2, N_2/H_2O) on the deposition rate of alumina from ATI in a hot wall reactor between 200 °C and 350 °C at a total pressure 2 kPa and proposed an activation energy of 80 kJ/mol. Kim et al. [26] studied the deposition of alumina on membranes in a hot wall tubular CVD reactor using ATI/N_2/O_2 mixtures between 350 °C and 400 °C and found a linear temperature dependence of the deposition rate.

The discrepancy of literature results shows that the involved chemical mechanisms strongly depend on the geometry of the reactor, in particular on its surface on volume ratio and on the adopted operating conditions, especially on the deposition temperature and on the partial pressure of ATI. More largely, this bibliographic survey reveals that the low pressure MOCVD process for the deposition of alumina from ATI in the temperature range 350–500 °C has only been fragmentarily investigated. To the best of the authors' knowledge, no local kinetic modeling of this CVD process was available before our work published in 2011 [16] summarized below.

5.2.3 Kinetic model development and validation

Based on the literature analysis summarized in the previous section, the following apparent chemical heterogeneous reaction was considered:

$$[Al(OC_3H_7)_3]_{4(g)} \rightarrow 2Al_2O_{3(s)} + 12C_3H_{6(g)} + 6H_2O_{(g)} \qquad (5.R4)$$

This reaction represents the complete set of unknown homogeneous and heterogeneous chemical reactions involved. Hofman et al. [23] showed that the alumina deposition rate R_{Di}, from ATI can be represented by the following apparent kinetic law:

$$R_{Di} = k_0 \exp(-E_a/RT)[ATI]n \qquad (5.1)$$

where k_0 is a pre-exponential constant, E_a (J/mol) an activation energy, T (K) the temperature and n the apparent reaction order (with R_{Di} in kg/m^2/s).

In order to determine the values of these kinetic parameters for the range of operating conditions of interest, dedicated experiments were performed in a horizontal hot wall tubular CVD reactor heated by a Trans Temp transparent furnace. More information can be found in [16]. As received ATI (Acros Organics) was vaporized using a heated bubbler in which pure nitrogen (Air Products) was sent to carry ATI to the deposition zone. The films were deposited on 15 mm × 10 mm or on 8 mm × 10 mm silicon substrates. They were weighed before and after the alumina deposition using a Sartorius balance (±10 μg) to determine the deposition rate of the films. The substrate holder was an horizontal aluminum alloy plate with dimensions 1 mm × 358 mm × 19 mm. Seven substrates were used per run, located between 8 and 36 cm from the reactor inlet. Four runs were performed using the same deposition conditions except the bubbler temperature and the deposition duration, as detailed in Table 5.1. The total pressure was of 658 Pa and the total flow rate of 653 sccm in all runs.

Fig. 5.1 presents the axial evolution of the temperature profile measured along the symmetry axis of the reactor as well as the alumina deposition rates deduced from substrates weight difference after and before deposition. The furnace and reactor characteristics led to a temperature profile presenting an isothermal zone close to 490 °C between 0.15 and 0.25 m from the reactor inlet. Fig. 5.1 also reveals a strong influence of the bubbler temperature; i.e., of the concentration of ATI, on the alumina deposition rate: a 10 degrees increase of the bubbler temperature results in a high increase of the average deposition rate and in a decrease of the deposit thickness uniformity. This is due to a peak in deposition rate existing at the beginning of the isothermal zone and then to a depletion in ATI.

Based on the kinetic equation (5.1), the results of experiments E_1, E_2 and E_3 were used to fit the three parameters k_0, E_a and n. The fitting was performed by virtually dividing the tubular CVD reactor into a series of perfectly mixed cylindrical reactors of 1 cm in length. The deposition rate was assumed to be uniform in each compartment i. An average value deduced from the weighing of each sample was considered for the substrate and for

TABLE 5.1 Investigated operating conditions.

Run	$T_{bubbler}$ (°C)	Deposition duration (min)	Q_{ATI} (sccm)	ATI molar fraction (10^{-2})	ATI mass fraction (10^{-2})
E_1	100	36	1.54	0.24	1.69
E_2	110	21	3.40	0.52	3.68
E_3	90	60	0.70	0.11	0.77
E_4	80	175	0.31	0.05	0.34

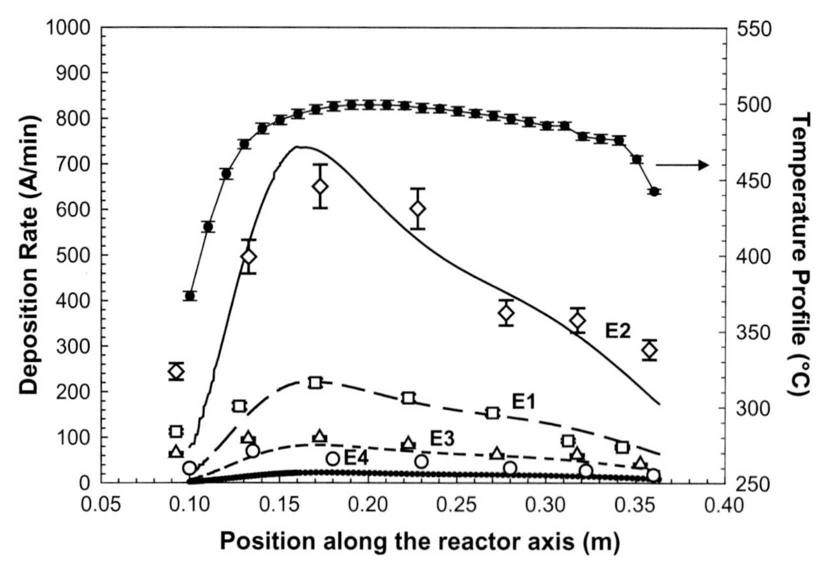

FIG. 5.1 Experimental axial evolution of the reactor temperature *(full circles)* and comparison between the experimental (E1: *square*, E2: *diamond*, E3: *triangle*, E4: *circle*) and calculated (E1: *long dash*, E2: *solid line*, E3: *medium dash*, E4: *dotted*) deposition rates along the reactor axis for runs E1, E2, E3, E4 [16].

the reactor walls. In these conditions, the following mass balance was written for ATI:

$$Q_{i-1}\,y_{i-1} - Q_i\,y_i = \pm R_{Di}\,S_i \tag{5.2}$$

where Q_i ($kg\,s^{-1}$) is the total gas mass flow rate, S_i is the deposition surface and y_i is the mass fraction of ATI in the ith compartment. Q_i was calculated along the CVD reactor by considering the mass of alumina deposited. The inlet mass fraction of ATI at the first compartment was considered equal to that provided by the bubbler at the reactor entrance. By minimizing the error between the experimentally determined deposition rate and the calculated one R_{Di} for the three experiments, the following values of the kinetic parameters k_0, E_a and n were obtained: $k_0 = 1.5 \times 10^6\,kg\,m^{2.5}/(mol^{1.5}s)$, $E_a = 78\,kJ/mol$, $n = 1.5$.

This value of E_a corresponds to the main limiting phenomena involved during ATI decomposition and alumina deposition for the conditions tested. As detailed in Section 5.2.2, Saraie et al. [25] obtained a very close value of 80 kJ/mol by operating between 200 °C and 350 °C at 2000 Pa, with dry N_2 as carrier gas. The apparent order of 1.5 proposed in this work reflects the dependence on ATI concentration of the dominant homogeneous and heterogeneous mechanisms existing in this range of operating conditions. It can be compared with the value of 1 adopted by Kawase et al. [19] between 830 °C and 1160 °C and with the value of 2 found by Hofman et al. [23] at temperatures slightly lower than the present ones.

This kinetic law was implemented into the CFD code Fluent and the results obtained in terms of deposition rate along the reactor axis are compared with the experimental results in Fig. 5.1. The average relative error for the three experiments is of 18%. This error can be attributed to several factors, including the uncertainty of the inlet flow rate of ATI, due to that of the ATI saturation pressure law and the use of a simplified apparent kinetic law. The model was used to simulate the deposition rate profile for run E4 (not used for the establishment of the kinetic data). The obtained profile together with the experimental deposition rates are also reported in Fig. 5.1. A satisfactory agreement is observed, which validates as a first approach the kinetic model for the investigated range of operating conditions. This is the reason why the model has then been applied to another reactor operating in the same range of deposition conditions, but treating a complex substrate.

5.2.4 Optimization of an original reactor by process simulation

As detailed elsewhere [17], a direct liquid injection (DLI) MOCVD process was developed to deposit alumina thin films from ATI on the internal surface of glass vials. The term "DLI-MOCVD" refers to reactors, which use liquid delivery units to feed the deposition zone in reactants [27]. This technology allows performing reliable, accurate and repeatable vaporization of liquid and dissolved solid compounds and precursors since it is much easier to regulate precisely a liquid flow rate than that of powders. The aim was to form an anti-corrosion barrier against aggressive liquids which could be stored into the vials. The uniform deposition of thin films inside the inner surface of a closed geometry is a non-trivial target since the complex gas flow strongly interacts with the species local concentration profiles and then with the local deposition rates. Developing such a process from scratch purely empirically would have been time and cost consuming. We then decided to combine coating experiments and process simulation to develop and optimize the process.

The DLI-MOCVD reactor is schematically represented in Fig. 5.2. The glass vial to coat (4.25 cm in diameter, 7.3 cm in height) is put on a substrate holder in an inductively heated vertical quartz tube (5.5 cm in diameter, 33.5 cm in height).

All experiments were performed at a temperature of 480 °C on the external surface of the central part of the vial bottom. Fig. 5.3 shows the experimentally measured thermal profiles on the inner walls of the vial for various reactor configurations, as it will be detailed below. ATI was dissolved in cyclohexane and then the solution was atomized and vaporized in a Kemstream Vapbox 500® DLI instrument. This reactive gas phase was diluted with pure N_2 and introduced into the reactor through a SS nozzle. The total gas flow rate was of 585 sccm and the inlet ATI molar

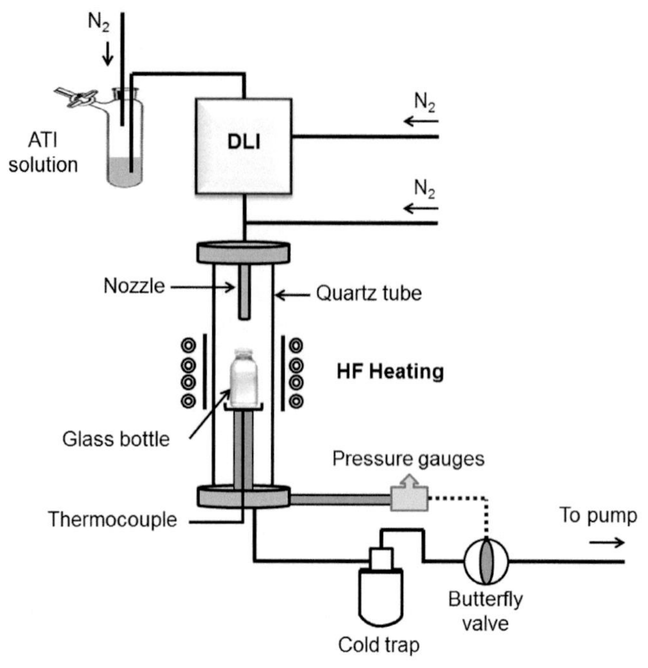

FIG. 5.2 Schematic of the DLI-MOCVD reactor for the Al_2O_3 deposition inside a vial [17].

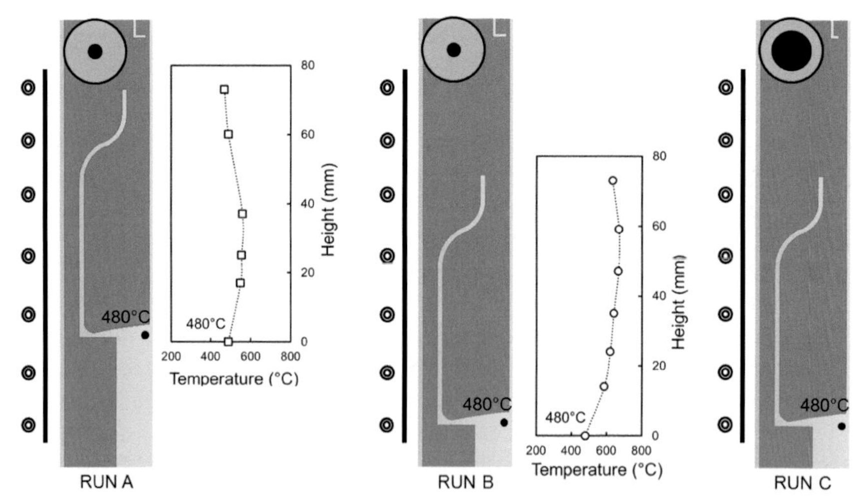

FIG. 5.3 Schematic representation of the reactor for RUNS A, B and C with relative positions of the bottle, the heating tube and the nozzle, associated with the experimental thermal profiles. The *circles* on the top represent bottom views of the nozzle outlet with the feeding hole in *black* [17].

fraction was fixed at 1.7×10^{-3}. The total pressure was of 666 Pa, close to the value used in the tubular reactor.

After deposition, the deposit thickness profile was measured by scanning electron microscopy (SEM) on a LEO 435VP instrument after fracture of the vials.

A first set of runs called Run A was performed with an arbitrary fixed distance between the nozzle and the vial of 15 mm and with a 2 mm inner nozzle diameter at its outlet. The experimental thickness profile along the internal walls of the vial is presented in Fig. 5.4A. The thickness on the lateral walls (height of bottle from 20 to 55 mm) ranged between 600 and 1000 nm. An over-thickness was observed at the lower part (up to 5600 nm from 0 to 20 mm) and at the entrance (up to 1300 nm from 55 to 70 mm). Visually, the thickest zone appeared as a powdery white layer while the other parts remained optically transparent. Cross-section SEM analyses of the deposit (not shown) revealed that that the powdery zone was composed of spherical particles aggregated on the film surface, whereas the other zones were dense and smooth. The simulated thickness profile using the kinetic law developed in the previous section is shown in Fig. 5.4A. A good agreement can be observed between the experiment and the model only in the regions where the deposit is smooth and dense. This can be explained by the fact that a different deposition mechanism probably exists for the powdery zone, which is not represented by the model.

In order to better understand the mechanisms involved, the local gas velocity profiles are plotted in Fig. 5.5 RUN A into the reactor and at the bottom of the vial in Fig. 5.6 RUN A. The scales of velocities respectively higher than 100 and 150 m/s are saturated for a better representation of the flow. At the nozzle outlet, the gas velocity is very high (470 m/s) due to its narrow section. This creates an impinging jet perpendicular to the vial bottom surface with high gradients and a sharp change of direction after hitting the bottom surface. When zooming at the bottom corner (Fig. 5.6 RUN A), a recirculation loop is present. The high residence time of the gas in this loop could explain the formation of the white powdery deposit by gas phase nucleation. Fig. 5.7 presents the ATI molar fraction into the reactor. The ATI concentration is not uniform into the vial and especially near its inner walls. The zones of highest ATI concentration correspond logically to the regions with the thickest deposits (neck and bottom).

New simulations were then performed aiming to decrease the velocity of the impinging jet in order (i) to limit the recirculation zone leading to the powdery deposit and (ii) to have more uniform ATI local concentrations near the inner walls of the vial, so more uniform deposit thicknesses. After several trials, this was achieved (Run B) by increasing the distance between the nozzle and the vial from 15 to 40 mm. Experimentally, the

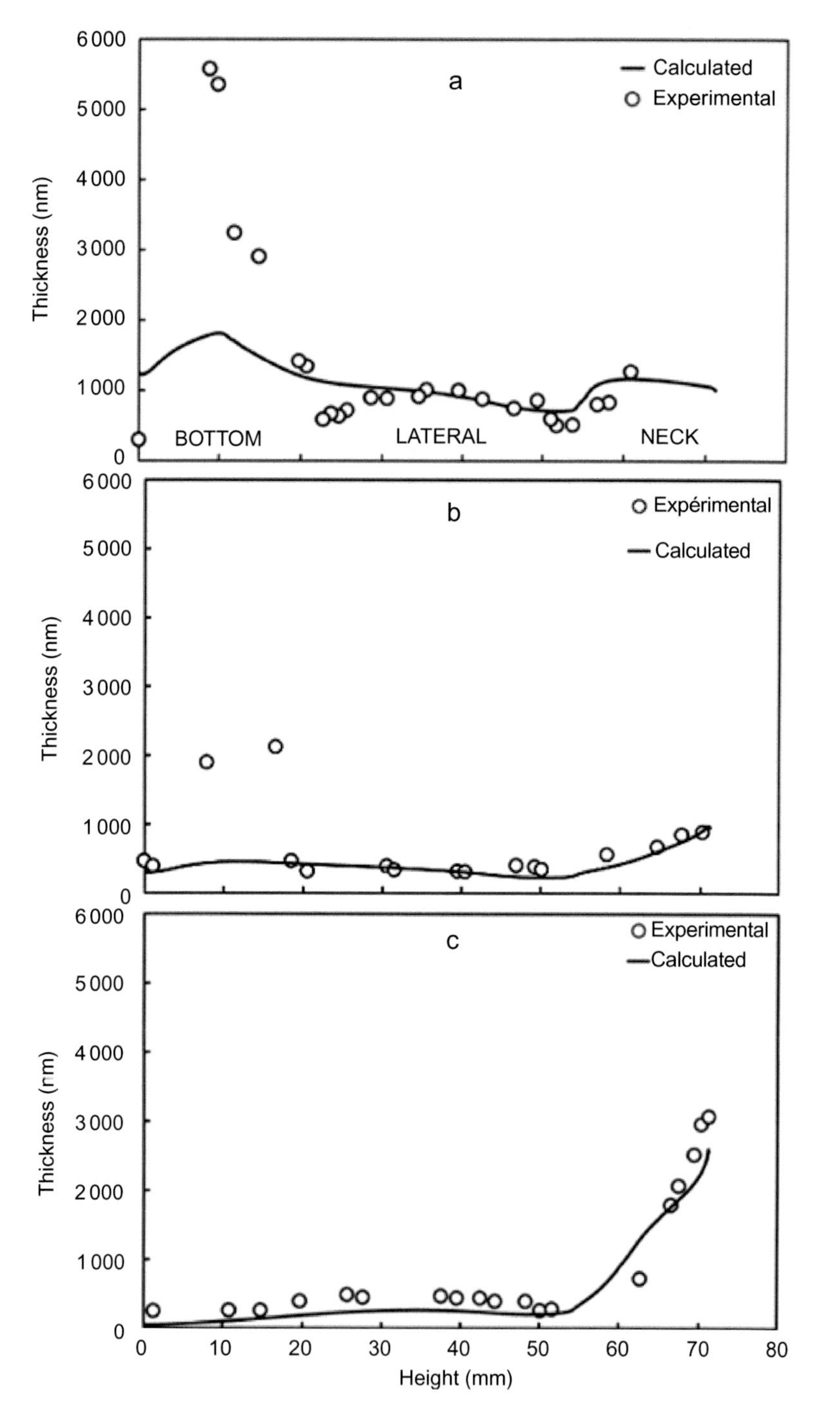

FIG. 5.4 Comparison between the experimental and the calculated thickness along the vial height for (A) RUN A, (B) RUN B, (C) RUN C. *Adapted from P.-L. Etchepare, H. Vergnes, D. Samélor, D. Sadowski, B. Caussat, C. Vahlas, Modelling of a DLI MOCVD process to coat by alumina the inner surface of bottles, Surf. Coat. Technol. 275 (2015) 167–175.*

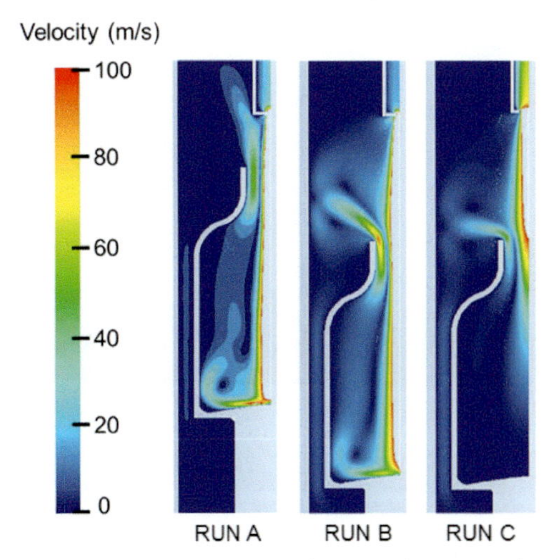

FIG. 5.5 Simulated profiles of the local gas velocity in the reactor for the three runs [17].

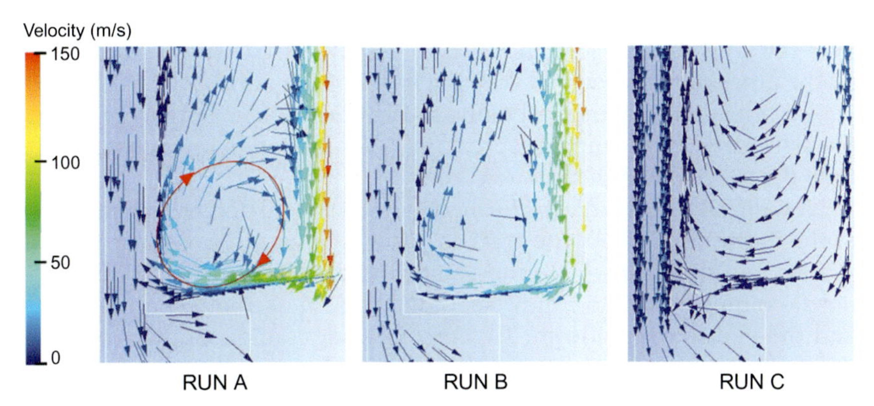

FIG. 5.6 Simulated profiles of the local gas velocity for the three runs at the bottom of the vial [17].

setup was modified accordingly and the vial was then translated toward the bottom part of the heating system, leading to a new measured thermal profile (Fig. 5.3B). The corresponding local gas velocity profiles calculated into the reactor and at the bottom of the vial are detailed in Fig. 5.5 RUN B and Fig. 5.6 RUN B respectively. The phenomena of impinging jet and of recirculation still exist but are attenuated. Fig. 5.7 RUN B shows that the local ATI molar fraction profile is more uniform near the vial walls. However, the corresponding experimental results indicated that a white powdery thicker deposit was still present in the bottom corner of the vial, this

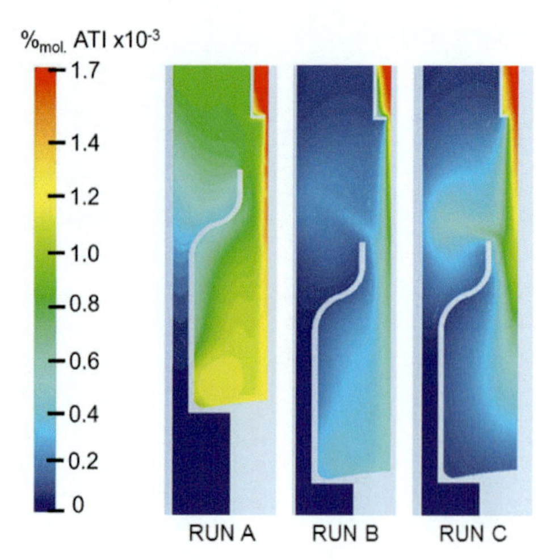

FIG. 5.7 Simulated profiles of the local ATI molar fraction in the reactor for the three runs [17].

over-thickness can be seen in Fig. 5.4B. A transparent dense layer coated all the other parts of the vial. The thickness at the neck and on the lateral walls of the vial was uniform while a slight over thickness was present at the vial entrance. The model reproduces well this behavior, except another time the powdery high thickness zone.

New simulations were performed in order to still decrease the velocity of the impinging jet. As the total gas flow rate was imposed by the DLI system, the simplest way found was to increase the inlet diameter of the nozzle outlet from 2 to 5 mm, keeping the distance between the nozzle and the vial fixed at 40 mm. This new reactor configuration called Run C can be seen in Fig. 5.3. Fig. 5.5 RUN C indicates that the velocity profile inside the vial is different from Runs A and B. An impinging jet is no more present inside the vial, even if the inlet flow is mainly axial. The gas velocities are also much lower, not exceeding 100 m/s. Consequently, Fig. 5.6 RUN C shows that the gas flow reaches the bottom part of the vial smoothly and no recirculation loop is present. The ATI concentration profile inside the vial (Fig. 5.7 RUN C) is more uniform near the inner walls, but the convective transport being decreased, the local ATI concentrations are lower. The corresponding calculated thickness profile is presented in Fig. 5.4C. From 0 to 55 mm, the thickness is quite uniform, ranging from 250 to 450 nm and is logically lower than for Runs A and B. An over-thickness is present on the neck walls till 3000 nm due to the higher ATI concentration existing in this zone. The calculated profile is in good agreement with the corresponding experimental thickness profile.

5.3 Local coordination affects properties: The case of amorphous Al_2O_3 barrier coatings

Following Aboaf's pioneering works [28], ATI is one of the most commonly used, single source precursors for the deposition of alumina films. It presents significant advantages since it is stable in air, only slowly reactive with water, and it is cheap. Once melt, it remains liquid below the melting temperature, its vapors can thus be transported to the process chamber by simply bubbling through a carrier gas. MOCVD of Al_2O_3 from ATI at moderate deposition temperature (T_d) was first used to coat thermally sensitive semiconductor materials with isolating films in transistor devices. Films of Al_2O_3 were deposited in the range 350–500 °C [28–34]. The intrinsic corrosion resistance [35], and the use of such films for the protection against corrosion of stainless steel [7] and of titanium alloys [36] were investigated. Most of these works were more pragmatic approaches to tune film properties rather than systematic investigations on the correlation between experimental parameters and film composition and properties. Compositional uncertainties are frequently reported for the films prepared at the lowest T_d. More fundamental approaches were tentatively carried out with films grown in hot-wall reactors, by Morssinkhof [23,37,38], by Yoshikawa et al. [39] and by Vahlas et al. [16,40,41]

However, ATI does not meet all the criteria required for an ideal MOCVD precursor. Its shelf-life, polymerization, and the condensation behavior of its tetramer during thermal aging are problematic [41] and lead to worsening film properties as well as to tedious reactor maintenance due to line pollution and injector blockage. Besides ATI, other volatile aluminum compounds have been proposed: aluminum tri-(2-ethylhexanoate) [42], aluminum tri-(sec-butoxide) [43–45], aluminum (ethylacetoacetate)-di-isopropoxide [26], aluminum triacetylacetonate [46–48], dimethyl, diethyl and di-isobutyl aluminum acetylacetonate [49], and dimethyl aluminum isopropoxide (DMAI) [10,11,50–56]. The latter consists of an Al atom linked to two methyl and one isopropyl groups, and has an intermediate structure between ATI and trimethyl aluminum (TMA). Unlike TMA, the remaining isopropyl group in DMAI stabilizes the molecule, which is non-pyrophoric. Moreover, the molecule presents a significantly higher vapor pressure than ATI, it is liquid at room temperature, has a longer shelf life and is not prone to polymerization. The use of ATI and DMAI for the CVD of alumina will be taken as example of the impact of the precursor chemistry on the characteristics and the properties of the obtained films.

The generation of precursor vapors and their transport to the process chamber through the sublimation and the vaporization of solid and liquid precursors, respectively, presents a priori simplicity and cost effectiveness. Nevertheless, maintaining of temperature sensitive precursors such as ATI at the vaporization temperature for a long period impacts the stability of the molecule and subsequently the coating quality [41]. In addition, it is difficult to access and control the generated reactive gas mass flow rate. This poorly controlled situation leads to non-reproducible operation, especially in cases where low activation energy prevails in the entire temperature range of interest, resulting in mass transport limited process. The direct liquid injection (DLI) technology allows overcoming these drawbacks with the controlled atomization and vaporization of pure liquid precursors or of solid ones dissolved in an appropriate carrier solvent [57–60]. Both vaporization and DLI technologies will be used hereafter to feed the process chamber with the alumina precursors.

5.3.1 Microstructure of amorphous alumina films

Fig. 5.8 presents cross-sectional SEM images of Al_2O_3 CVD films, deposited on Si wafers at 5 Torr from ATI (left column) and from DMAI (in the presence of O_2, right), with the precursor molecules shown on top. The films are processed either by vaporization of the precursor (up, from ATI only) or from DLI (lower row), with the schematic of a bubbler and the photograph of a DLI equipment shown on the left of the upper and lower rows, respectively. Process temperature is indicated in each case. The AFM 3D topography image of the surface of the film processed from DMAI is also presented. Its scale spans from 0 to 15 nm. Films diffract neither X-rays nor electrons, with a characteristic electron diffraction pattern shown in the center of the figure. They are dense and present pinholes-free, highly smooth surfaces, witnessed by low roughness values, such as 1.35 nm and 1.71 nm for the arithmetic and rms, respectively, of the DLI DMAI/O_2 sample. The only exception concerns the film processed from ATI by DLI, which is still smooth but slightly porous with nanometer cavities over the entire height. This particular morphology is attributed to the presence of the solvent, in this case cyclohexane in the DLI process.

The O/Al ratio of such films was determined by electron probe for microanalysis (EMPA) using internal standards. It is presented in Fig. 5.2 as a function of T_d and is compared to the ratios expected for AlOOH and Al_2O_3 references. Their carbon content was determined by X-ray photoelectron spectroscopy (XPS).

Films deposited from both evaporated and DLI ATI present O/Al ratios of 2.12 and ca. 1.7 when deposited at 360 °C and 420 °C, respectively. These values correspond to compositions close to AlOOH (atomic ratio 2.0) and partially hydroxylated $AlO_{1+x}(OH)_{1-2x}$ films. Films deposited between 470 °C and 650 °C show O/Al ratios close to, though slightly higher than 1.5, in good agreement with the formula Al_2O_3. The presence

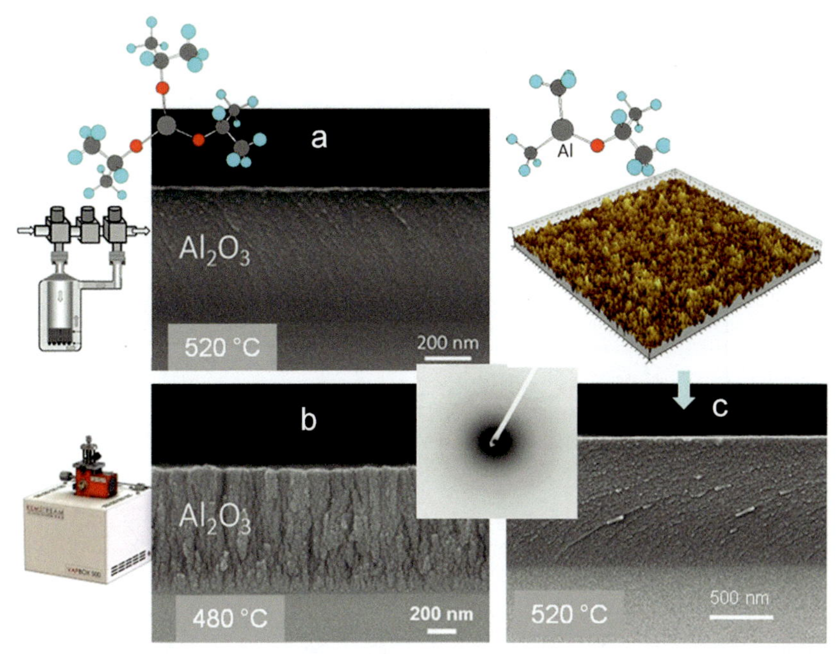

FIG. 5.8 Cross-sectional SEM micrographs of Al_2O_3 CVD films deposited on Si wafers from ATI (A, B, left column) and from DMAI (C, right column), with the precursor molecules shown on top. The films are processed either by sublimation of the precursor (upper row) or from DLI (lower row). Temperatures refer to process conditions. The atomic force microscopy (AFM) 3D topography image is that of the surface of the film processed from DMAI with the two horizontal scales spanning from 0 to 2 μm and the vertical scale spanning from 0 to 15 nm. The electron diffraction pattern shown in the center of the figure is characteristic of all samples. *Adapted from P.-L. Etchepare, L. Baggetto, H. Vergnes, D. Samélor, D. Sadowski, B. Caussat, C. Vahlas, Amorphous alumina barrier coatings on glass: MOCVD process and hydrothermal aging, Adv. Mater. Interfaces 3(8) (2016) 1600014; L. Baggetto, J. Esvan, C. Charvillat, D. Samélor, H. Vergnes, B. Caussat, A. Gleizes, C. Vahlas, Alumina thin films prepared by direct liquid injection chemical vapor deposition of dimethylaluminum isopropoxide: a process-structure investigation, Phys. Status Solidi C 12(7) (2015) 989–995.*

of hydroxylated aluminas at low T_d and the decrease of the concentration of hydroxyl groups with increasing T_d is confirmed by the O1s binding energy shown in the lower part of Fig. 5.9. At higher T_d (700 °C), the O/Al ratio decreases to 1.44. The C content is below 1 at.% for all T_d below 650 °C, while a small concentration of ca. 4 at.% C is detected by XPS. The XPS probed C1s binding energy reveals that this carbon is almost exclusively related to aliphatic moieties, and most likely results from carbon residues of ATI isopropoxide ligands trapped in the film.

The films deposited from DMAI at 150 °C and 200 °C in the presence of H_2O vapor show O/Al atomic ratios of 1.81 and 1.67, respectively. These values are contained between those of AlOOH and Al_2O_3 (1.5), thus

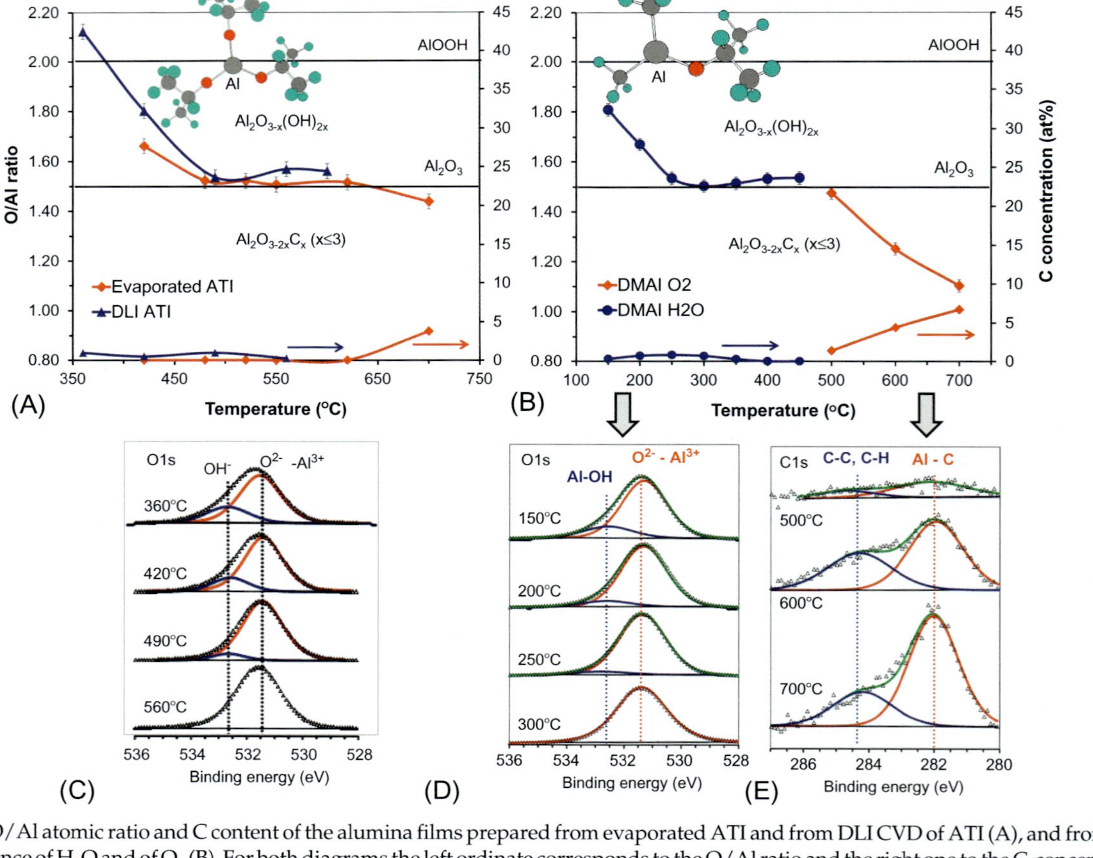

FIG. 5.9 *Top*: O/Al atomic ratio and C content of the alumina films prepared from evaporated ATI and from DLI CVD of ATI (A), and from DLI CVD of DMAI in the presence of H_2O and of O_2 (B). For both diagrams the left ordinate corresponds to the O/Al ratio and the right one to the C-concentration (at.%). *Bottom*: High resolution X-ray photoelectron O1s et C1s spectra of alumina films deposited from ATI in the range 360–560 °C (C), from DMAI + H_2O in the range 150–450 °C (D) and from DMAI + O_2 in the range 500–700 °C (E). *Adapted from L. Baggetto, C. Charvillat, J. Esvan, Y. Thébault, D. Samélor, H. Vergnes, B. Caussat, A. Gleizes, C. Vahlas, A process-structure investigation of aluminum oxide and oxycarbide thin films prepared by direct liquid injection chemical vapor deposition of dimethylaluminum isopropoxide (DMAI), Chem. Vap. Depos. 21 (2015) 343–351; P.-L. Etchepare, L. Baggetto, H. Vergnes, D. Samélor, D. Sadowski, B. Caussat, C. Vahlas, Amorphous alumina barrier coatings on glass: MOCVD process and hydrothermal aging, Adv. Mater. Interfaces 3(8) (2016) 1600014.*

revealing partially hydroxylated alumina $AlO_{1+x}(OH)_{1-2x}$. The O/Al ratio of films prepared at 250 °C and above is in good agreement with the formula Al_2O_3. This result is confirmed by the evolution of the O1s binding energy which, above this T_d reveals the presence of O—Al bonds, exclusively. It is concluded that stoichiometric alumina films can be processed in this way for the protection and surface functionalization of thermally sensitive parts. The C content of such films is below 1 at.%. The film prepared at 500 °C in the presence of O_2 has an O/Al ratio of 1.48, which may be ascribed to the formula Al_2O_3. The films prepared at 600 °C and 700 °C have O/Al ratios of 1.25 and 1.10, well below 1.50. The decrease in O/Al ratio is clearly related to an increase in the C concentration, up to about 6.5 at.% in the film prepared at 700 °C, as has been also evidenced by Schmidt et al. during the preparation of alumina films from sublimed DMAI in this T_d range [55,61]. This suggests the formation of the aluminum carbide Al_4C_3 and/or of aluminum oxycarbides, e.g., Al_4O_4C or Al_2OC [39–42], along with Al_2O_3, to lead to the generic composition of $Al_2O_xC_{3/2-x/2}$. The formation of oxycarbides can be thought to result from the partial substitution of one C^{4-} anion for two O^{2-} anions, leading to the formulas Al_4O_4C, Al_2OC, and ultimately Al_4C_3, as also illustrated by the strong C—Al bond contribution in the C1s binding energies at 600 °C and above [62].

The above resumed compositional and property variations of the alumina films are indicative of modifications of the amorphous structure upon changing T_d. We showed that for the films of general formula $AlO_{1+x}(OH)_{1-2x}$, structural modifications are expected to accompany dehydroxylation from AlOOH to Al_2O_3 with increasing the T_d up to 250 °C for the DMAI process and up to 450 °C for the ATI one. $\{^1H\}^{27}Al$ rotational echo double resonance (REDOR) nuclear magnetic resonance (NMR) experiments performed on films obtained from vaporized ATI confirm this trend [63]. The question arises as to possible further structural modifications in Al_2O_3 for the films grown at higher T_d. This question is investigated for both systems by very high-field (20 T) ^{27}Al NMR and the main results are resumed in the next paragraphs.

The left part of Fig. 5.10 illustrates the 2D ^{27}Al 3-quantum magic-angle spinning nuclear magnetic resonance (3QMAS NMR) spectra of alumina films processed from vaporized ATI at six different T_d [63]. The use of very high magnetic field allows high spectral resolution and yields a clear identification of the tetra-, penta-, and hexacoordinated aluminum sites ($^{[n]}Al$ with $n=4$, 5, or 6). Their line shapes are characteristic of a typical amorphous structure as seen in alumina-containing glasses. The percentage of each $^{[n]}Al$ site was calculated from these and from the corresponding 1D ^{27}Al MAS spectra. Its evolution with T_d is drown in the upper left part of the figure. The respective percentages of tetra- and hexacoordinated sites vary with opposite trends. The $^{[4]}Al$ content increases from 33%

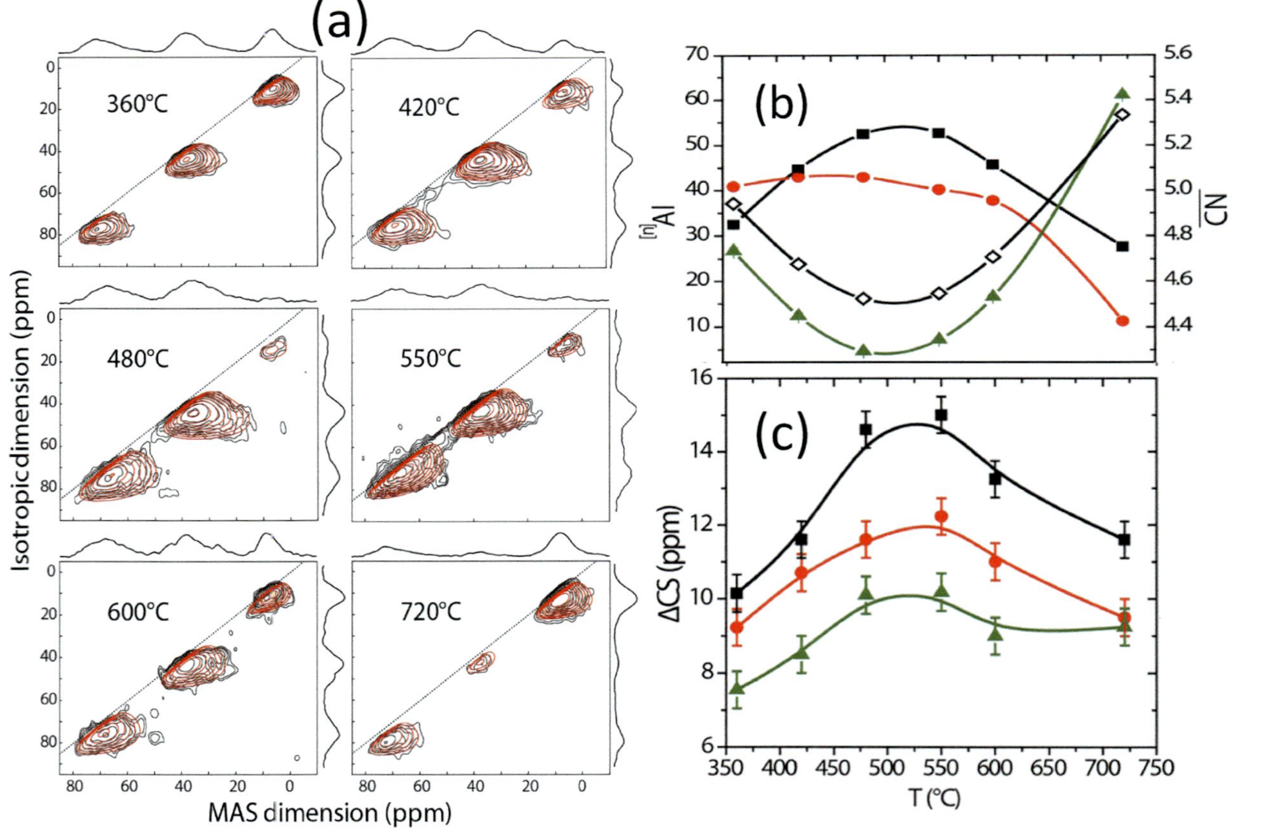

FIG. 5.10 (A): Two-dimensional ^{27}Al 3QMAS spectra of the films deposited at different temperatures. *Black and red lines* refer to experimental and simulated spectra, respectively. (B): Percentages of the $^{[n]}$Al environments and (C): their corresponding chemical shift distributions (ΔCS), as a function of T_d. *Black squares*, *red circles*, and *green triangles* refer to $^{[4]}$Al, $^{[5]}$Al, and $^{[6]}$Al sites, respectively. *Open diamonds* in (B) correspond to the averaged coordination number calculated from the formula $\overline{CN} = [4\ (^{[4]}\text{Al atom \%}) + 5\ (^{[5]}\text{Al atom \%}) + 6\ (^{[6]}\text{Al atom \%})]/100$. *All lines* are a guide to the eye. *Adapted from V. Sarou-Kanian, A.N. Gleizes, P. Florian, D. Samelor, D. Massiot, C. Vahlas, Temperature-dependent 4-, 5- and 6-fold coordination of aluminum in MOCVD-grown amorphous alumina films: a very high field Al-27-NMR study, J. Phys. Chem. C 117(42) (2013) 21965–21971.*

for $T_d = 360\,°C$ up to 53% for $Td = 480\,°C$; at the same time, the $^{[6]}Al$ content decreases from 26% to 5%. For $T_d = 550\,°C$, the respective percentages have values that are nearly the same as those for 480 °C, suggesting that the maximum of $^{[4]}Al$ sites and the minimum of $^{[6]}Al$ sites occur at ~515 °C. From 360 °C to 515 °C, the $^{[5]}Al$ content is at its highest with values between 40% and 43%. It can be noticed that the $^{[n]}Al$ percentages measured for $T_d = 480\,°C$ (and 550 °C) compare well with those measured by Lee et al. for two aluminum oxide films prepared by PVD and ALD [64,65]. In the 550–720 °C range, the $^{[6]}Al$ content increases to 59% and the $^{[4]}Al$ and $^{[5]}Al$ contents decrease to 29% and 12%, respectively. The lower right diagram of Fig. 5.3 shows the evolution of chemical shift distribution (ΔCS) of each $^{[n]}Al$ environment. ΔCS is maximal for a T_d more or less half-way between 480 °C and 550 °C for each coordination number. This suggests that the utmost structural disorder corresponds to the highest rates of $^{[4]}Al$ and $^{[5]}Al$ sites.

A similar, nuclear magnetic-resonance spectroscopy (NMR) investigation was applied to the films processed from DMAI [66]. As an example, the left part of Fig. 5.11 presents 1D ^{27}Al MAS NMR spectra of alumina films prepared from DMAI + O_2. All spectra consist of 3 or 4 overlapping broad lines each corresponding to a specific Al coordination, characteristic of amorphous alumina-containing materials. The spectral reconstructions evidence $^{[4]}Al$ (AlO_4), $^{[5]}Al$ (AlO_5) and $^{[6]}Al$ (AlO_6) coordination units in all films. Furthermore, an additional broad line centered at around 90 ppm is detected for the films processed at 600 °C and 700 °C, which suggests the presence of a carbon containing aluminum environment. The reported in the Fig. 5.2D 3QMAS spectrum for the 700 °C sample confirms the presence in the film of these four Al environments. The diagram on the right of Fig. 5.11 presents the distribution, as a function of T_d, of the concentrations (at.%) of the four Al environments in films processed from DMAI plus H_2O (up to 300 °C) and from DMAI plus O_2 (from 500 °C and above). The $^{[4]}Al$ and $^{[5]}Al$ concentrations increase(decrease) with increasing T_d for films processed in the presence of $H_2O(O_2)$. All corresponding $^{[n]}Al$ values evolve in a way which allows assuming either a saturation or a maximum of $^{[4]}Al$ plus $^{[5]}Al$ coordination environments in the T_d range 300–500 °C. In this T_d range the concentration of the $^{[6]}Al$ coordination environments reaches a remarkably low value of less than 5 at.%. The carbon containing aluminum environment is null at T_d less or equal 500 °C; it strongly increases at higher T_d and approaches 30 at.% at 700 °C.

The impact of the distribution of the $^{[n]}Al$ environments of films processed from vaporized ATI on different properties of such films is presented in Fig. 5.12. The bottom diagram of the figure resumes the compositional and structural information of the films and their evolution as a function of T_d. The middle diagram presents the corresponding evolution of their hardness and of the Young modulus, determined from

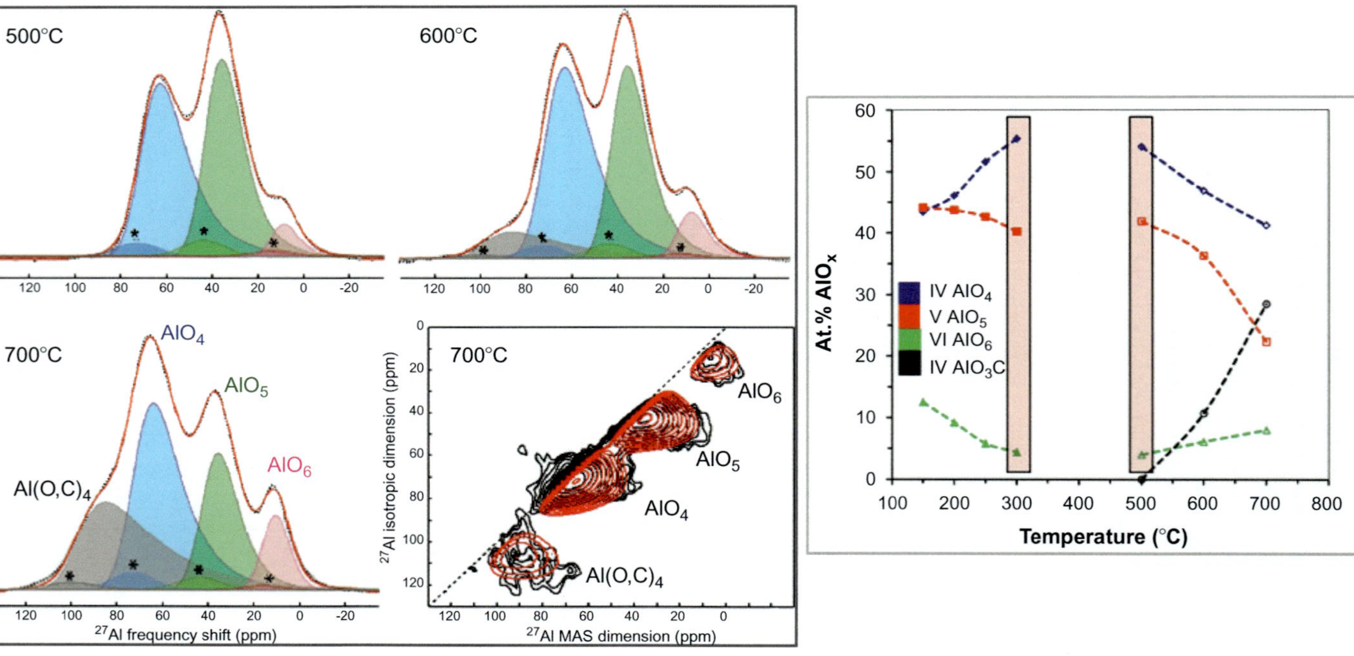

FIG. 5.11 *Left*: Experimental *(dashed black lines)* and reconstructed *(full red lines)* ^{27}Al MAS NMR spectra of alumina films prepared from DMAI plus O_2 from 500 °C to 700 °C. The 2D 3QMAS for 700 °C is shown as an example. The AlO_4 *(blue)*, AlO_5 *(green)*, AlO_6 *(pink)* and $Al(O,C)_4$ *(gray)* coordination environments are fitted with their respective MAS-related sidebands (*) which correspond to the bands $n=0$ of the external transitions. *Right*: Distribution, as a function of T_d, of the AlO_4 *(blue diamonds)*, AlO_5 *(red squares)*, AlO_6 *(green triangles)*, and $Al(O,C)_4$ *(black circles)* concentrations (at.%) in films processed from DMAI plus H_2O *(full symbols)* and from DMAI plus O_2 *(open symbols)*. *Pink bars* reveal investigated process temperatures resulting in films with maximum AlO_4 plus AlO_5 coordination environments. *Adapted from L. Baggetto, V. Sarou-Kanian, P. Florian, A.N. Gleizes, D. Massiot, C. Vahlas, Atomic scale structure of amorphous aluminum oxyhydroxide, oxide and oxycarbide films probed by very high field 27Al nuclear magnetic resonance, Phys. Chem. Chem. Phys. 19 (2017) 8101–8110.*

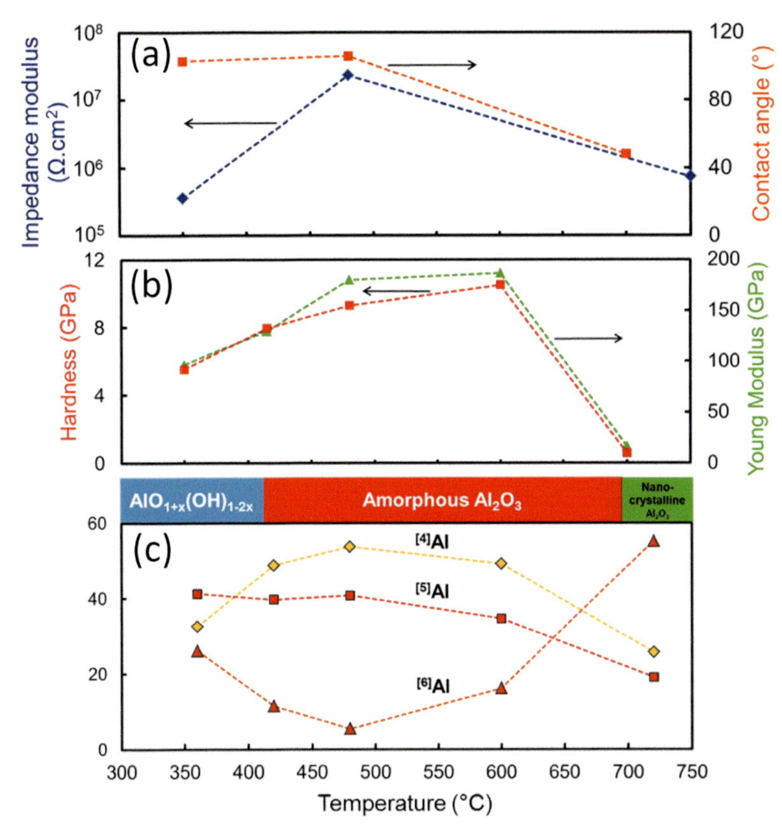

FIG. 5.12 Evolution of the impedance modulus and the contact angle with water (A), of the hardness and the Young modulus (B) as a function of deposition temperature for alumina films processed from evaporated ATI. (C) Corresponding evolution of the concentration of the $^{[4]}$Al, $^{[5]}$Al, and $^{[6]}$Al sites and of the composition and the crystallinity of the films. *Adapted from Y. Balcaen, N. Radutoiu, J. Alexis, J.D. Béguin, L. Lacroix, D. Samelor, C. Vahlas, Mechanical and barrier properties of MOCVD processed alumina coatings on TA6V titanium alloy, Surf. Coat. Technol. 206(7) (2011) 1684–1690; D. Samélor, A.M. Lazar, M. Aufray, C. Tendero, L. Lacroix, J.D. Béguin, B. Caussat, H. Vergnes, J. Alexis, D. Poquillon, N. Pébère, A. Gleizes, C. Vahlas, Amorphous alumina coatings : processing, structure and remarkable barrier properties, J. Nanosci. Nanotechnol. 11(9) (2011) 8387–8391; D. Samélor, M. Aufray, L. Lacroix, Y. Balcaen, J. Alexis, H. Vergnes, D. Poquillon, J.D. Beguin, N. Pébère, S. Marcelin, B. Caussat, C. Vahlas, Mechanical and surface properties of chemical vapour deposited protective aluminium oxide films on TA6V alloy, Adv. Sci. Tech. 66 (2010) 66–73.*

nanoindentation experiments on alumina films deposited on Ti_6Al_4V titanium alloy [12]. The hardness increases with increasing T_d up to 480 °C, where it reaches a maximum value of 10.8 ± 0.8 GPa. Similar trends are obtained for the modulus, which increases from 92 ± 8 GPa to 155 ± 6 GPa with increasing T_d from 350 °C to 480 °C. The value of both parameters vanishes at T_d 750 °C, because at this temperature, the film becomes nano-crystallized and thus it presents limited cohesion and high roughness. The upper diagram presents the evolution, with increasing T_d of the impedance modulus, obtained after 1h of immersion in a 0.1 NaCl solution, for films deposited on Ti6242 titanium alloy [36]. The film deposited at 480°C stands out by improving the corrosion resistance of nearly two orders of magnitude in comparison with the uncoated sample, whose impedance modulus is $1.3 \times 10^5\,\Omega\,cm^2$. Contrastingly, films processed at 350°C and at 750°C slightly improve the corrosion resistance of the bare alloy. The evolution of the contact angle with pure water is also presented in this diagram [67]. It slightly increases from $103 \pm 1°$ at 350°C to $106 \pm 2°$ at 480°C, these values illustrating the hydrophobic nature of the amorphous aluminum oxyhydroxides and alumina surfaces. It decreases to $48 \pm 3°$at 700°C, revealing the hydrophilic nature of the surface of the nanocrystalline film.

5.3.2 Application examples of barrier function of amorphous alumina films

Such alumina films find applications in a large number of key enabling technologies. Two examples will be sketched hereafter to illustrate this utility.

The first concerns the protection of lightweight titanium alloys against high temperature oxidation. We have already provided selected data from this research in the previous paragraphs to illustrate intrinsic characteristics of the alumina films. Here, we focus on their capacity to durably protect such alloys in operating conditions, which simulate the operation of specific parts of aero turbines. The scope of the work is to extend the implementation of Ti alloys to hotter parts of the turbine, in replacement of stainless steel. The limited resistance of these alloys to high temperature oxidation hinders this implementation. It is due to the rapid formation on the surface of titanium alloys in contact with the air, of a very stable, nonprotective titanium oxide layer, and also, to the ingress of oxygen at interstitial position in the alpha phase of these alloys [68,69].

Alumina films were deposited on all faces of a near-alpha titanium alloy coupons Ti6242S from both ATI and DMAI following the previously described CVD processes. Fig. 5.13 presents parabolic rate constants k_p for the oxidation of the Ti6242S alloy, bare and coated with ca. 500nm thick amorphous alumina processed from DMAI at 300°C (DMAI300) and at 500°C (DMAI500), and from vaporized ATI at 520°C (ATI520). The values were calculated according to the equation: $k_p = [(\Delta m_f/S)^2 - (\Delta m_i/S)^2]/(t_f - t_i)$, where Δm_i and Δm_f are respectively the mass gain of a sample with free surface S, at the beginning, t_i and at the end, t_f of each temperature dwell. The diagram shows that the three coatings DMAI500, DMAI300

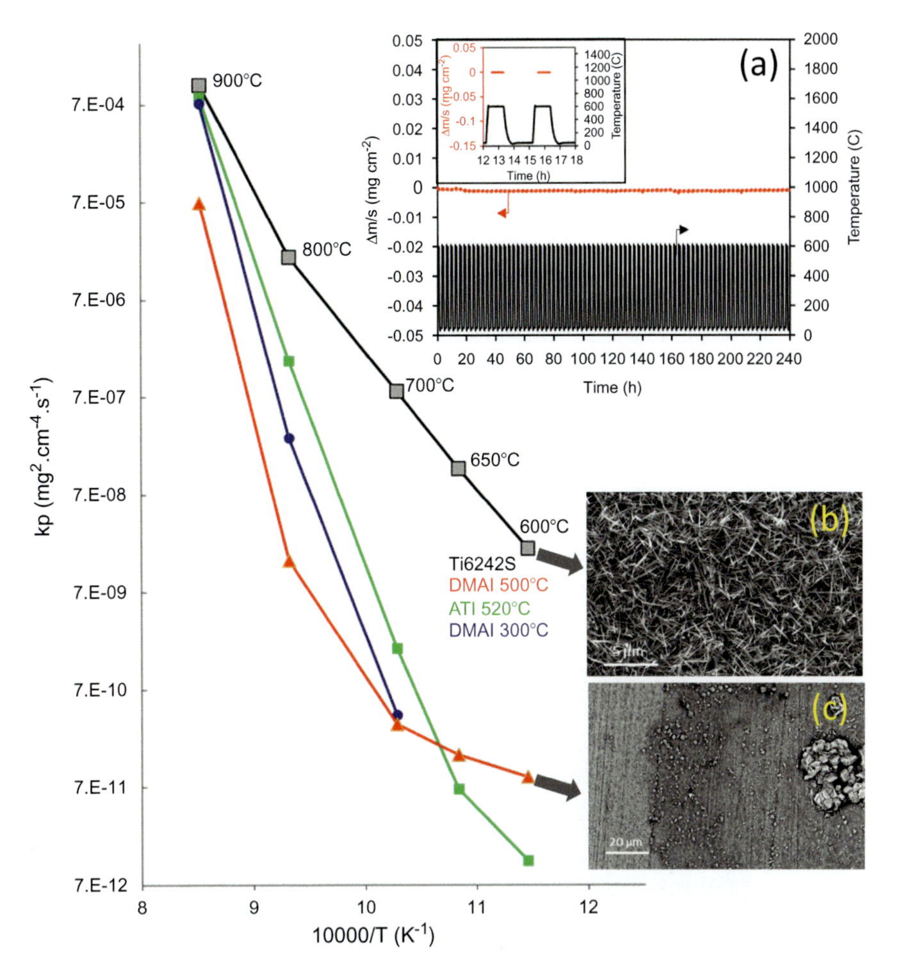

FIG. 5.13 Parabolic rate constants for the oxidation of the Ti6242S alloy, bare and coated with amorphous alumina processed from DMAI at 300 °C (DMAI300) and at 500 °C (DMAI500), and from vaporized ATI at 520 °C (ATI520). Results are compared with the kinetics of the development of gamma, theta and alpha alumina on NiAl, from [70]. (A): Evolution of the mass gain per unit area of the DMAI300 sample, and of the temperature variation (cycling) as a function of time. The insert is a zoom on two cycles, detailing the temperature profile. (B) and (C): Surface SEM micrographs of two samples after 5000 h of isothermal oxidation at 600 °C: bare Ti6242S (B), and DMAI500 coated coupon (C). *Adapted from D. Samelor, L. Baggetto, R. Laloo, V. Turq, A.N. Gleizes, T. Duguet, D. Monceau, C. Vahlas, Efficient, durable protection of the Ti6242S titanium alloy against high-temperature oxidation through MOCVD processed amorphous alumina coatings, J. Mater. Sci. 55(11) (2020) 4883–4895.*

and ATI520 (where the codes correspond to alumina deposition from DMAI at 550 °C and at 300 °C, and from vaporized ATI at 520 °C, respectively), are efficient for oxidation protection during the short term (10–20 h) thermogravimetric experiment, between 600 °C and 700 °C. They are less efficient at 800 °C except for the DMAI500 one, and they are inefficient at 900 °C. This first set of TGA experiments was completed by long term oxidation annealing in laboratory air at 600 °C, which is close to the maximum temperature of use of such alloys. The weight gain of the DMAI500 coated coupon after 5000 h is ca. 0.20 mg cm^{-2}, to be compared with 1.63 mg cm^{-2} for the bare alloy. The surface micrographs of the two samples are presented in the bottom right side of Fig. 5.13. The whole surface of the bare alloy is covered with thin, few-micrometers-long needles. This morphology corresponds to the well-known acicular rutile that forms when Ti6242S is exposed to oxidation treatments [71]. The surface of the DMAI500 coated coupon presents a smaller number and thinner needles than those observed on the oxidized Ti6242S. The formation of a few excrescences on the surface, one of which is shown in the left bottom part is due to scarce defects of the coating, probably created from homogeneous nucleation during deposition. These results confirm the excellent oxidation protection conferred to Ti6242S by the 500 nm thick amorphous Al_2O_3 films.

Fig. 5.14 shows SEM images in backscattered electrons mode of cross-sections obtained by focused ion beam (FIB), and corresponding electron

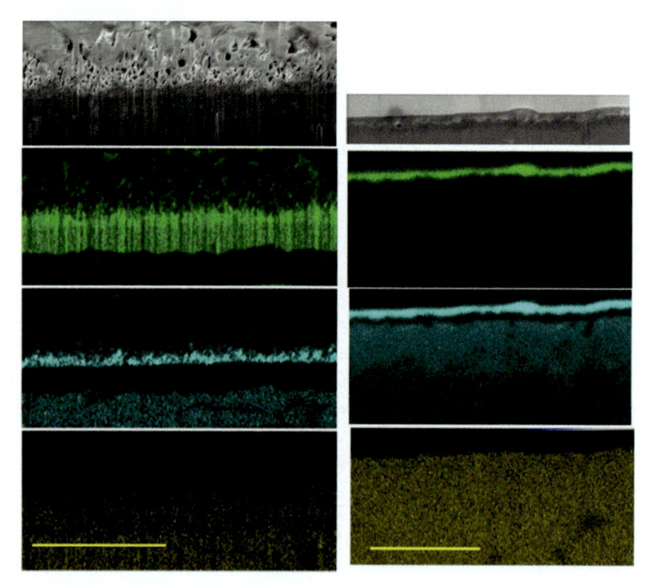

FIG. 5.14 SEM cross-sections and EDS mapping of O, Al, Ti elements from top to the botom, of two samples after 5000 h of isothermal oxidation at 600 °C: bare Ti6242S (left column), and DMAI500 coated coupon (right column). The scale bars correspond to 10 μm (left) and 2.5 μm (right) *Adapted from D. Samelor, L. Baggetto, R. Laloo, V. Turq, A.N. Gleizes, T. Duguet, D. Monceau, C. Vahlas, Efficient, durable protection of the Ti6242S titanium alloy against high-temperature oxidation through MOCVD processed amorphous alumina coatings, J. Mater. Sci. 55(11) (2020) 4883–4895.*

dispersive spectroscopy (EDS) maps of a bare Ti6242S alloy (micrographs A–D) and a coated coupon (E–H) after oxidation at 600 °C during 5000 h. The bare Ti6242S cross-section after oxidation (A) can be divided into three layers. The top 1.6 μm-thick layer corresponds to the needles observed on the surface, with a great density of large open pores. The intermediate layer is 2 μm-thick, with a high density of small pores. The deepest region corresponds to the bulk part of the sample with the known microstructure of the alloy. The EDS mapping (B–D) reveals that a (Al, Ti) mixed oxide is formed on the top layer of the acicular rutile TiO_2 with a clear preferential segregation of Al from the bulk to the free surface. In contrast, the intermediate layer mainly contains Ti and O. After oxidation of the coated coupon, its microstructure and chemical composition are strongly modified. The cross sectional SEM micrograph (E) shows the presence of pores in the vicinity of the initial alumina/Ti6242S interface. In this porous region, the Al_2O_3 coating has been replaced by a mixed Ti, O, Al layer. The thickness of the initial alumina protective film has been reduced by half at the end of the oxidation. Finally, the behavior of the coated coupons was also evaluated under thermal cycling conditions. Indeed, in such conditions spallation of the ceramic coating on the metallic substrate may rapidly occur resulting in limited durability of the material. The tests are performed with a TGA apparatus which allows thermal cycling with continuous recording of the mass [72]. The sensitivity on mass variation of the instrument is 0.1 μg, allowing for detection of tiny spallation events. Eighty 1 h cycles were performed between 50 °C and 600 °C, under flowing synthetic air. The top right insert of Fig. 5.13 presents the evolution of the mass gain per unit area of the DMAI300 coated sample, and of the temperature variation as a function of time. No mass gain or mass loss could be detected for this sample at the microgram scale. Despite the fact that this test is rather short compared with the usual service periods in turbine applications, it is still a valuable indicator of the excellent adherence of the alumina layer and of the relatively low level of elastic strain energy in the system during thermal cycling. These results confirm the efficiency of the protection of the alloy by the alumina coating. The considered CVD processes are transferable and can be used for the surface treatment of parts of aeroturbines with complex shapes.

The second example of the use of alumina to meet diverse technological challenges concerns the environmental monitoring. Here, alumina is considered as the matrix component of a nanocomposite film. The work is motivated by the increased needs for environmental monitoring and consequently for robust and reliable sensors. It aims at increasing the service time of optical sensors immersed in riverine waters by decreasing the development of biofouling on their surface. In this aim, nanocomposite coatings composed of metallic silver, Ag nanoparticles (NPs) embedded in alumina are co-deposited on sensor glass windows [73]. The CVD process consists in simultaneously introducing in the process chamber vaporized ATI and direct liquid injected silver pivalate $(CH_3)_3CC(O)OAg$ in the form of a solution in a 90/10 vol. mesytylene/dipropylamine solvent [74]. Co-deposition was performed at 450°C and 5 Torr. The left part of Fig. 5.15 presents a bright field TEM cross section micrograph of such a coating (15a). It is composed of a 70 nm thick Al_2O_3 sub-layer (2) by the Si substrate (1), followed by a 260 nm thick layer consisting of Ag NPs (black dots) embedded in Al_2O_3 (3). The external part of the coating is composed of a ca. 75 nm thick Al_2O_3 sub-layer (4). In Fig. 5.15B one can observe the Ag NPs. Their size distribution is monodisperse centered at 6 nm. Fig. 5.15C shows a high resolution TEM micrograph of one Ag NP and the insert in this figure reveals the fcc compact stacking of Ag, perfectly oriented in the [111] direction, with a lattice spacing of 0.290 nm between the (110) planes. The right part of Fig. 5.15 presents a phase-contrast image of a $400 \times 400\,nm^2$ surface area. A two scale microstructure, one with ca. 90 nm features size, on the top of which are smaller features, attributed to Ag NPs, whose size does not exceed 10 nm. The quadratic roughness (Sq) of the surface is 5.0 ± 0.5 nm, averaged over 10 images, to be compared with the slightly lower Sq of the bare substrate, 0.8 ± 0.2 nm. Such moderate increase of the roughness shouldn't affect the sensor operation and will be taken into account during its calibration phase.

Fig. 5.16 summarizes the results of the evaluation of the two targeted functional properties of such films, namely their optical transparency and protection against biofouling. The left part of the figure shows optical transmittance spectra of four nanocomposite films of increasing thickness from 200 nm (Q4) to 300 nm (Q4). The four dotted lines point on the wavelengths used by the sensor. All four spectra consist of a transmittance band peaking at a fixed position of 315 nm and a broad absorption band covering the vis-NIR spectral domain, which contains several interference fringes. The fixed transmittance band results from the minimum of the imaginary part of the refractive index of bulk Ag k_{Ag} [75] due to Ag interband transitions from 3.83 eV (323 nm) to higher energies [76]. The transmittance band intensity gradually increases from Q1 to Q4, due to the decrease of the film thickness. Thickness increase results in the increase of the absorption since the imaginary part of the resulting effective refractive index of the medium composed of Ag NPs ($k_{Ag} \neq 0$) and Al_2O_3 ($k_{alumina} = 0$) host media is non null. Besides, the gradual displacement of the interference fringes toward longer wavelengths is also characteristic of the film thickness increase. The absorption band originates from surface plasmon resonance of Ag NPs which is known to occur between 350 and 800 nm in Al_2O_3 environments [77,78]. The transmittance at the four wavelengths of interest for sample Q4 is between 0.5 and 0.7. This degradation of the transmittance can be reduced through appropriate tuning of the

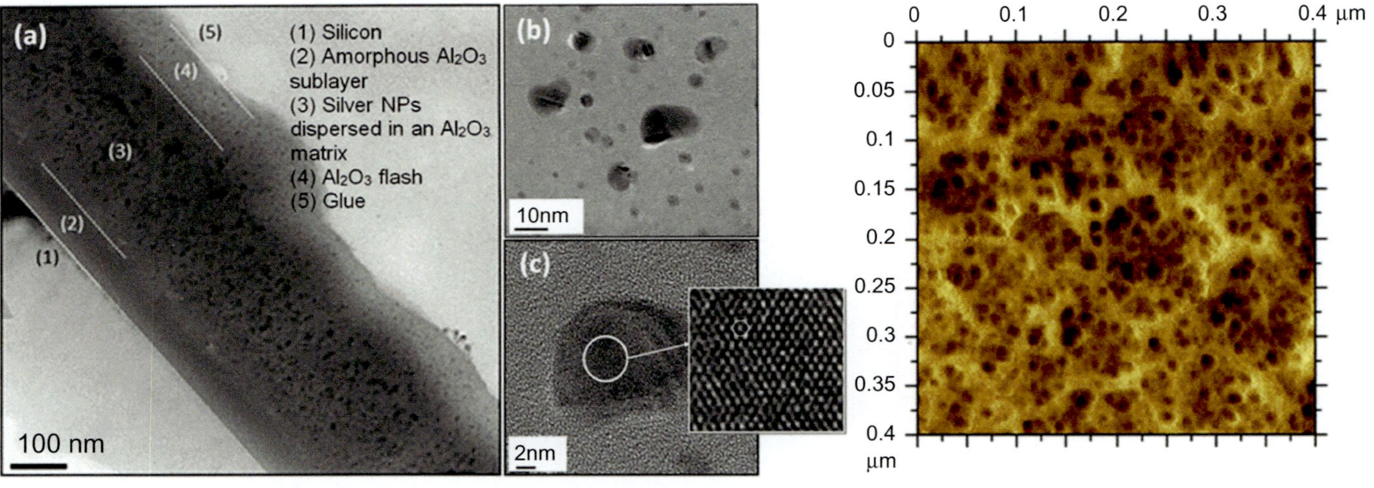

FIG. 5.15 *Left*: (A) Bright field TEM micrograph of a cross section of a representative amorphous Al_2O_3/Ag NPs composite coating with an alumina rich top-layer, deposited on Si. (B) and (C) High-resolution TEM micrographs illustrating Ag NPs dispersed in Al_2O_3. The insert in (C) highlights the [111] orientation of an fcc-crystallized Ag NP. The spacing between the (110) reticular planes is 0.290nm. *Right*: Atomic force microscopy phase-shift mapping that highlights the contrast between Ag NPs and the Al_2O_3 matrix. *Adapted from C. Tendero, A.M. Lazar, D. Samelor, O. Debieu, V. Constantoudis, G. Papavieros, A. Villeneuve, C. Vahlas, Nanocomposite thin film of Ag nanoparticles embedded in amorphous Al2O3 on optical sensors windows: synthesis, characterization and targeted application towards transparency and anti-biofouling, Surf. Coat. Technol. 328 (2017) 371–377.*

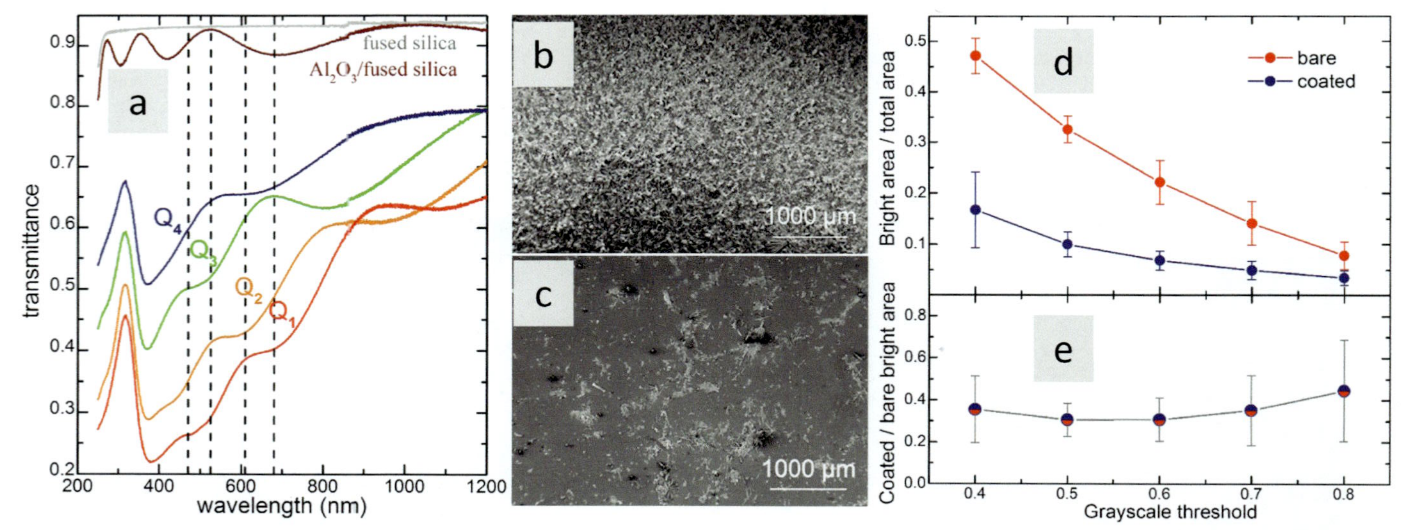

FIG. 5.16 *Left* (A): Transmittance spectra of four as processed samples composed of Ag NPs doped alumina films co-deposited from ATI and Ag pivalate at 450°C on commercial sensors glass windows. The thickness of the nanocomposite coatings decreases from Q1 to Q4. Fused silica window and a-Al_2O_3 deposited on fused silica are shown for comparison. The four *dotted lines* point on the wavelengths used by the sensor. *Middle*: SEM surface micrographs of bare (B) and coated (C) glass samples after 1-week immersion. *Right*: Bright area over total area ratios of bare and coated glass samples for a wide range of grayscale thresholds used in the analysis of SEM images (D), coated over bare bright area ratio as a function of the grayscale threefold (E). The error bars indicate the uncertainty of measurements in different images. *Adapted from C. Tendero, A.M. Lazar, D. Samelor, O. Debieu, V. Constantoudis, G. Papavieros, A. Villeneuve, C. Vahlas, Nanocomposite thin film of Ag nanoparticles embedded in amorphous Al2O3 on optical sensors windows: synthesis, characterization and targeted application towards transparency and anti-biofouling, Surf. Coat. Technol. 328 (2017) 371–377.*

characteristics of the coating, namely its thickness, and the density and size of the Ag NPs, while maintaining the anti-biofouling efficiency.

The middle part of Fig. 5.16 presents two SEM surface micrographs of bare (Fig. 5.16A) and coated (Fig. 5.16B) glass surfaces after immersion. The observed surfaces confirm visual observations of the samples, namely the complete coverage by microorganisms of the surface of the untreated sample in contrast to the limited, and localized development of biofouling on the surface of the coated one. Quantitative assessment of the biofouling fraction was made by measuring the fraction of bright areas of SEM micrographs of four bare and the four (Q1–Q4) coated samples by using the using nanoTOPO_SEM™ software (Nanometrisis). The results of the analysis are shown on the top right part of Fig. 5.16 where the biofouling fraction is plotted for various grayscale thresholds (0.4–0.7) used to define the bright areas. In addition to an expected decrease of measured biofouling vs grayscale threshold, one can notice that, for all grayscale thresholds, the biofouling fraction of coated samples is systematically and unambiguously lower than that of bare samples. To quantify this decrease, we plot the ratio of biofouling of coated to the biofouling of bare samples for four grayscale thresholds, shown in the bottom right diagram of the figure. The value of this ratio is included between 0.30 ± 0.10 and 0.35 ± 0.17; it does not monotonically vary with the value of the grayscale threshold. It is concluded that CVD nanocomposite thin films of Ag NPs embedded in Al_2O_3 are efficient in decreasing biofouling on glass windows.

5.4 Concluding remarks

Applied to aluminum oxide thin films, this chapter illustrates the capabilities of the MOCVD process to meet severe requirements of various high-tech industrial fields like microelectronics, photovoltaics, MEMS or anticorrosion barrier layers, to deposit thin films on complex substrates.

Two different and complementary aspects are treated. First, process modeling allows analyzing the interplay between the reactive transport phenomena and the film thickness and composition, in order de develop and optimize new MOCVD processes able to conformally coat complex-in-shape substrates. The second part demonstrates how dedicated multi-criteria experimental studies can correlate the film microstructure and local atomic order to key properties like protection against oxidation and biofouling.

These two generic approaches can be applied to MOCVD thin films of oxides, metals, carbides, nitrides, to coat a great variety of substrates, including patterned wafers, foams, porous media like membranes and even powders.

Acknowledgements

The authors acknowledge A. Gleizes, D. Samelor, D. Sadowski, H. Vergnes, P.L. Etchepare, S. Ponton, C. Tendero, D. Monceau, T. Duguet, V. Turq, R. Laloo for their valuable contribution to the works reported in this chapter.

References

[1] Y. Peng, G.Q. Han, F.N. Liu, W.W. Xiao, Y. Liu, N. Zhong, C.G. Duan, Z. Feng, H. Dong, Y. Hao, Ferroelectric-like behavior originating from oxygen vacancy dipoles in amorphous film for non-volatile memory, Nanoscale Res. Lett. 15 (1) (2020) 134.

[2] H. Ryu, H.N. Wu, F.B. Rao, W.J. Zhu, Ferroelectric tunneling junctions based on aluminum oxide/zirconium-doped hafnium oxide for neuromorphic computing, Sci. Rep. 9 (2019) 20383.

[3] M. Kaltenbrunner, P. Stadler, R. Schwodiauer, A.W. Hassel, N.S. Sariciftci, S. Bauer, Anodized aluminum oxide thin films for room-temperature-processed, flexible, low-voltage organic non-volatile memory elements with excellent charge retention, Adv. Mater. 23 (42) (2011) 4892–4896.

[4] J.P.B. Silva, F.L. Faita, K. Kamakshi, K.C. Sekhar, J.A. Moreira, A. Almeida, M. Pereira, A.A. Pasa, M.J.M. Gomes, Enhanced resistive switching characteristics in Pt/BaTiO3/ITO structures through insertion of HfO2:Al2O3 (HAO) dielectric thin layer, Sci. Rep. 7 (2017) 46350.

[5] J.P.B. Silva, K.C. Sekhar, K. Veltruska, V. Matolín, R. Negrea, C. Ghica, M. Oliveira, J.A. Moreira, M.A.C. Castro Pereira, M.J.M. Gomes, HfO2-Al2O3 dielectric layer for a performing metalferroelectric-insulator-semiconductor structure with ferroelectric 0.5Ba(Zr0.2Ti0.8)O3-0.5(Ba0.7Ca0.3)TiO3 thin film, ACS Appl. Electron. Mater. 2 (9) (2020) 2780–2787.

[6] J. Jang, Y. Song, K. Cho, Y. Kim, W. Lee, D. Yoo, S. Chung, T. Lee, Non-volatile aluminum oxide resistive memory devices on a wrapping paper substrate, Flex. Print. Electron. 1 (3) (2016), 034001.

[7] A.-M. Lazar, W.P. Yespica, S. Marcelin, N. Pébère, D. Samélor, C. Tendero, C. Vahlas, Corrosion protection of 304L stainless steel by chemical vapour deposited alumina coatings, Corros. Sci. 81 (2014) 125–131.

[8] S. Krumdieck, S. Davies, C.M. Bishop, T. Kemmitt, J.V. Kennedy, Al2O3 coatings on stainless steel using pulsed-pressure MOCVD, Surf. Coat. Technol. 230 (2013) 208–212.

[9] F. Wiest, V. Capodieci, O. Blank, M. Gutsche, J. Schulze, I. Eisele, J. Matusche, U.I. Schmidt, Conformal aluminum oxide coating of high aspect ratio structures using metalorganic chemical vapor deposition, Thin Solid Films 496 (2) (2006) 240–246.

[10] D. Barreca, G.A. Battiston, R. Gerbasi, E. Tondello, Al2O3 thin films from aluminium dimethylisopropoxide by metal-organic chemical vapour deposition, J. Mater. Chem. 10 (9) (2000) 2127–2130.

[11] G.A. Battiston, R. Gerbasi, Aluminum dimethylisopropoxide decomposition and the growth of dense alumina thin films at low temperature, Chem. Vap. Depos. 8 (5) (2002) 193–195.

[12] Y. Balcaen, N. Radutoiu, J. Alexis, J.D. Béguin, L. Lacroix, D. Samelor, C. Vahlas, Mechanical and barrier properties of MOCVD processed alumina coatings on TA6V titanium alloy, Surf. Coat. Technol. 206 (7) (2011) 1684–1690.

[13] C.R. Kleijn, R. Dorsman, K.J. Kuijlaars, M. Okkerse, H. van Santen, Multi-scale modeling of chemical vapor deposition processes for thin film technology, J. Cryst. Growth 303 (1) (2007) 362–380.

[14] N. Cheimarios, G. Kokkoris, A.G. Boudouvis, A multi-parallel multiscale computational framework for chemical vapor deposition processes, J. Comput. Sci. 15 (2016) 81–85.

[15] S. Ponton, H. Vergnes, D. Samelor, D. Sadowski, C. Vahlas, B. Caussat, Development of a kinetic model for the moderate temperature chemical vapor deposition of SiO2 films from tetraethyl orthosilicate and oxygen, AICHE J. 64 (11) (2018) 3958–3966.

[16] H. Vergnes, D. Samelor, A. Gleizes, C. Vahlas, B. Caussat, Local kinetic modeling of aluminium oxide metal organic chemical vapor deposition from aluminium triisopropoxide, Chem. Vap. Depos. 17 (7–9) (2011) 181–185.

[17] P.-L. Etchepare, H. Vergnes, D. Samélor, D. Sadowski, B. Caussat, C. Vahlas, Modelling of a DLI MOCVD process to coat by alumina the inner surface of bottles, Surf. Coat. Technol. 275 (2015) 167–175.

[18] R.B. Bird, W.E. Stewart, E.N. Lightfoot, Transport Phenomena, second ed., John Wiley & Sons, 2007.

[19] M. Kawase, Y. Ikuta, T. Tago, T. Masuda, K. Hashimoto, Modeling of a thermal gradient chemical vapor infiltration process for production of silicon carbide whisker alumina composite, Chem. Eng. Sci. 49 (24A) (1994) 4861–4870.

[20] M. Kawase, K. Miura, Rate analysis of chemical vapor deposition by use of the thin tubular reactor, Thin Solid Films 498 (1–2) (2006) 25–29.

[21] A.A. Barybin, V.I. Tomilin, Influence of conditions of pyrolysis of triisopropoxyaluminum on certain properties of Al2O3 films, J. Appl. Chem. USSR 49 (8) (1976) 1723–1726.

[22] G. Pauer, H. Altena, B. Lux, AI2O3 CVD with organic AI-donors, Int. J. Refract. Met. Hard Mater. (**1986**, *Sept.*) 165–170.

[23] R. Hofman, R.W.J. Morssinkhof, T. Fransen, J.G.F. Westheim, P.J. Gellings, Thin alumina and silica films by chemical vapor deposition (CVD), Mater. Manuf. Process. 8 (3) (1993) 315–329.

[24] S. Blittersdorf, N. Bahlawane, K. Kohse-Höinghaus, B. Atakan, J. Müller, Chemical vapor deposition of Al2O3 thin films using aluminium tri-isopropoxide, Chem. Vap. Depos. 9 (4) (2003) 194–198.

[25] J. Saraie, K. Ono, S. Takeuchi, Effect of various atmospheres on the reduced-pressure CVD of Al_2O_3 thin films at low temperatures, J. Electrochem. Soc. 136 (10) (1989) 3139–3141.

[26] J.H. Kim, G.J. Choi, J.K. Lee, S.J. Sim, Y.D. Kim, Y.S. Cho, A novel precursor for the sol gel and CVD methods to prepare alumina permselective membranes, J. Mater. Sci. 33 (5) (1998) 1253–1262.

[27] V. Astié, C. Millon, J.-M. Decams, A. Bartasyte, Direct liquid injection chemical vapor deposition, in: P. Mandracci (Ed.), Chemical Vapor Deposition for Nanotechnology, IntechOpen, London, UK, 2019, pp. 29–51.

[28] J.A. Aboaf, Deposition and properties of aluminum oxide obtained by pyrolytic decomposition of an aluminum alkoxide, J. Electrochem. Soc. 114 (9) (1967) 948–952.

[29] M.T. Duffy, J.E. Carnes, D. Richman, Dielectric and interface properties of pyrolytic aluminum oxide films on silicon substrates, Metall. Trans. A. 2 (1971) 667–672.

[30] M. Okamura, T. Kobayashi, Improved interface in inversion-type INP-misfet by vapor etching technique, Jpn. J. Appl. Phys. 19 (11) (1980) 2151–2156.

[31] J. Fournier, W. Desisto, R. Brusasco, M. Sosnowski, R. Kershaw, J. Baglio, K. Dwight, A. Wold, Preparation and characterization of thin-films of alumina by metal-organic chemical vapor-deposition, Mater. Res. Bull. 23 (1) (1988) 31–36.

[32] K. Tanaka, H. Takahashi, S. Kuniyoshi, H. Ohki, Electrical-properties of CVD Al2O3-GaAs MIS capacitors, Solid State Electron. 23 (10) (1980) 1093–1094.

[33] M.M. Sovar, D. Samélor, A.N. Gleizes, P. Alphonse, S. Perisanu, C. Vahlas, Protective alumina coatings by low temperature metalorganic chemical vapour deposition, Adv. Mat. Res. 23 (2007) 245–248.

[34] A. Gleizes, M.M. Sovar, D. Samélor, C. Vahlas, Low temperature MOCVD-processed alumina coatings, Adv. Sci. Tech. 45 (2006) 1184–1193.

[35] N. Hara, S. Nagata, N. Akao, K. Sugimoto, Formation of Al2O3-Ta2O5 double-oxide thin films by low-pressure MOCVD and evaluation of their corrosion resistances in acid and alkali solutions, J. Electrochem. Soc. 146 (2) (1999) 510–516.

[36] G. Boisier, M. Raciulete, D. Samélor, N. Pébère, A.N. Gleizes, C. Vahlas, Electrochemical behavior of chemical vapor deposited protective aluminium oxide coatings on Ti6242 titanium alloy, Electrochem. Solid St. 11 (10) (2008) C55–C57.
[37] R.W.J. Morssinkhof, T. Fransen, M.M.D. Heusinkveld, P.J. Gellings, Mechanistic aspects of the deposition of thin alumina films deposited by MOCVD, Mater. Res. Soc. Symp. Proc. 168 (1990) 125.
[38] V.A.C. Haanappel, H.D. van Corbach, R. Hofman, R.W.J. Morssinkhof, T. Fransen, P.J. Gellings, Formation of thin oxide films by metal-organic chemical vapor deposition, High Temp. Mater. Processes 15 (4) (1996) 245–262.
[39] N. Yoshikawa, S. Takamura, S. Taniguchi, A. Kikuchi, MOCVD kinetics and morphologies of Al2O3 deposits using aluminium tri-isopropoxide (ATI) precursor, Trans. Mater. Res. Soc. Jpn 24 (1999) 151.
[40] A. Gleizes, C. Vahlas, M.M. Sovar, D. Samélor, M.C. Lafont, CVD-fabricated aluminum oxide coatings from aluminum tri-iso-propoxide: correlation between processing conditions and composition, Chem. Vap. Depos. 13 (2007) 23–29.
[41] M.M. Sovar, D. Samélor, A.N. Gleizes, C. Vahlas, Aluminium tri-isopropoxide: shelf life, transport properties, and decomposition kinetics for the low temperature processing of aluminium oxide-based coatings, Surf. Coat. Technol. 201 (2007) 9159–9162.
[42] T. Maruyama, T. Nakai, Aluminum oxide thin films prepared by chemical vapor deposition from aluminum 2-ethylhexanoate, Appl. Phys. Lett. 58 (19) (1991) 2079–2080.
[43] V.A.C. Haanappel, J.B. Rem, H.D. Vancorbach, T. Fransen, P.J. Gellings, Properties of alumina films prepared by metal-organic chemical vapor deposition at atmospheric pressure in the presence of small amounts of water, Surf. Coat. Technol. 72 (1–2) (1995) 1–12.
[44] V.A.C. Haanappel, D.V.D. Vendel, H.S.C. Metselaar, H.D.V. Corbach, T. Fransen, P.J. Gellings, The mechanical properties of thin alumina films deposited by metal-organic chemical vapour deposition, Thin Solid Films 254 (1995) 153–163.
[45] D.-H. Kuo, B.-Y. Cheung, R.-J. Wu, Growth and properties of alumina films obtained by low-pressure metal-organic chemical vapor deposition, Thin Solid Films 398–399 (2001) 35–40.
[46] O.B. Ajayi, M.S. Akanni, J.N. Lambi, H.D. Burrows, O. Osasona, B. Podor, Preparation and optical characterization of pyrolytically deposited thin films of some metal oxides, Thin Solid Films 138 (1) (1986) 91–95.
[47] O.B. Ajayi, M.S. Akanni, J.N. Lambi, C. Jeynes, J.F. Watts, Compositional studies of various metal oxide coatings on glass, Thin Solid Films 185 (1) (1990) 123–136.
[48] M. Pulver, W. Nemetz, G. Wahl, CVD of ZrO2, Al2O3 and Y2O3 from metalorganic compounds in different reactors, Surf. Coat. Technol. 125 (1–3) (2000) 400–406.
[49] G.A. Battiston, G. Carta, G. Cavinato, R. Gerbasi, M. Porchia, G. Rossetto, MOCVD of Al2O3 films using new dialkylaluminum acetylacetonate precursors: growth kinetics and process yields, Chem. Vap. Depos. 7 (2) (2001) 69–74.
[50] W. Koh, S.-J. Ku, Y. Kim, Chemical vapor deposition of A1203 films using highly volatile single sources, Thin Solid Films 304 (1–2) (1997) 222–224.
[51] S.Y. Lee, B. Luo, Y. Sun, J.M. White, Y. Kim, Thermal decomposition of dimethylaluminum isopropoxide on Si(100), Appl. Surf. Sci. 222 (1–4) (2004) 234–242.
[52] M. Natali, G. Carta, V. Rigato, G. Rossetto, G. Salmaso, P. Zanella, Chemical, morphological and nano-mechanical characterizations of Al2O3 thin films deposited by metal organic chemical vapour deposition on AISI 304 stainless steel, Electrochim. Acta 50 (23) (2005) 4615–4620.
[53] F. Guidi, G. Moretti, G. Carta, M. Natali, G. Rossetto, Z. Pierino, G. Salmaso, V. Rigato, Electrochemical anticorrosion performance evaluation of Al2O3 coatings deposited by MOCVD on an industrial brass substrate, Electrochim. Acta 50 (23) (2005) 4609–4614.
[54] G. Carta, M. Casarin, N. El Habra, M. Natali, G. Rossetto, C. Sada, E. Tondello, P. Zanella, MOCVD deposition of CoAl2O4 films, Electrochim. Acta 50 (23) (2005) 4592–4599.

[55] B.W. Schmidt, B.R. Rogers, C.K. Gren, T.P. Hanusa, Carbon incorporation in chemical vapor deposited aluminum oxide films, Thin Solid Films 518 (14) (2010) 3658–3663.
[56] L. Baggetto, C. Charvillat, J. Esvan, Y. Thébault, D. Samélor, H. Vergnes, B. Caussat, A. Gleizes, C. Vahlas, A process-structure investigation of aluminum oxide and oxycarbide thin films prepared by direct liquid injection chemical vapor deposition of dimethylaluminum isopropoxide (DMAI), Chem. Vap. Depos. 21 (2015) 343–351.
[57] M. Manin, S. Thollon, F. Emieux, G. Berthome, M. Pons, H. Guillon, Deposition of MgO thin film by liquid pulsed injection MOCVD, Surf. Coat. Technol. 200 (5–6) (2005) 1424–1429.
[58] J. Mungkalasiri, L. Bedel, F. Emieux, J. Doré, F.N.R. Renaud, F. Maury, DLI-CVD of TiO2–Cu antibacterial thin films: growth and characterization, Surf. Coat. Technol. 204 (6–7) (2009) 887–892.
[59] P.-L. Etchepare, H. Vergnes, D. Samélor, D. Sadowski, C. Brasme, B. Caussat, C. Vahlas, Amorphous alumina coatings on glass bottles using direct liquid injection MOCVD for packaging applications, Adv. Sci. Tech. 91 (2014) 117–122.
[60] C. Vahlas, B. Caussat, W.L. Gladfelter, F. Senocq, E.J. Gladfelter, Liquid and solid precursor delivery systems in gas phase processes, Recent Pat. Complex Metal. Alloys 8 (2015) 91–108.
[61] B.W. Schmidt, B.R. Rogers, W.J. Sweet Iii, C.K. Gren, T.P. Hanusa, Deposition of alumina from dimethylaluminum isopropoxide, J. Eur. Ceram. Soc. 30 (11) (2010) 2301–2304.
[62] L. Baggetto, J. Esvan, C. Charvillat, D. Samélor, H. Vergnes, B. Caussat, A. Gleizes, C. Vahlas, Alumina thin films prepared by direct liquid injection chemical vapor deposition of dimethylaluminum isopropoxide: a process-structure investigation, Phys. Status Solidi C 12 (7) (2015) 989–995.
[63] V. Sarou-Kanian, A.N. Gleizes, P. Florian, D. Samelor, D. Massiot, C. Vahlas, Temperature-dependent 4-, 5- and 6-fold coordination of aluminum in MOCVD-grown amorphous alumina films: a very high field Al-27-NMR study, J. Phys. Chem. C 117 (42) (2013) 21965–21971.
[64] S.K. Lee, S.B. Lee, S.Y. Park, Y.S. Yi, C.W. Ahn, Structure of amorphous aluminum oxide, Phys. Rev. Lett. 103 (2009), 095501.
[65] S.K. Lee, S.Y. Park, Y.S. Yi, J. Moon, Structure and disorder in amorphous alumina thin films: insights from high-resolution solid-state NMR, J. Phys. Chem. C 114 (32) (2010) 13890–13894.
[66] L. Baggetto, V. Sarou-Kanian, P. Florian, A.N. Gleizes, D. Massiot, C. Vahlas, Atomic scale structure of amorphous aluminum oxyhydroxide, oxide and oxycarbide films probed by very high field 27Al nuclear magnetic resonance, Phys. Chem. Chem. Phys. 19 (2017) 8101–8110.
[67] D. Samélor, A.M. Lazar, M. Aufray, C. Tendero, L. Lacroix, J.D. Béguin, B. Caussat, H. Vergnes, J. Alexis, D. Poquillon, N. Pébère, A. Gleizes, C. Vahlas, Amorphous alumina coatings : processing, structure and remarkable barrier properties, J. Nanosci. Nanotechnol. 11 (9) (2011) 8387–8391.
[68] G. Lütjering, J.C. Williams, Titanium, Springer-Verlag, Berlin, Heidelberg, 2007.
[69] R.N. Shenoy, J. Unnam, R.K. Clark, Oxidation and embrittlement of Ti-6Al-2Sn-4Zr-2Mo alloy, Oxid. Met. 26 (1–2) (1986) 105–124.
[70] M.W. Brumm, H.J. Grabke, The oxidation behaviour of NiAl—I .Phase transformations in the alumina scale during oxidation of NiAl and NiAl-Cr alloys, Corros. Sci. 33 (11) (1992) 1677–1690.
[71] C. Dupressoire, A. Rouaix-Vande Put, P. Emile, C. Archambeau-Mirguet, R. Peraldi, D. Monceau, Effect of nitrogen on the kinetics of oxide scale growth and of oxygen dissolution in the Ti6242S titanium-based alloy, Oxid. Met. *87* (3–4) (**2017**) 343–353.
[72] D. Monceau, D. Poquillon, Continuous thermogravimetry under cyclic conditions, Oxid. Met. 61 (1–2) (2004) 143–163.
[73] C. Tendero, A.M. Lazar, D. Samelor, O. Debieu, V. Constantoudis, G. Papavieros, A. Villeneuve, C. Vahlas, Nanocomposite thin film of ag nanoparticles embedded in

amorphous Al2O3 on optical sensors windows: synthesis, characterization and targeted application towards transparency and anti-biofouling, Surf. Coat. Technol. 328 (2017) 371–377.

[74] J. Mungkalasiri, L. Bedel, F. Emieux, J. Dore, F.N.R. Renaud, C. Sarantopoulos, F. Maury, CVD elaboration of nanostructured TiO_2-ag thin films with efficient antibacterial properties, Chem. Vap. Depos. 16 (2010) 35–41.

[75] X.K. Meng, S.C. Tang, S. Vongehr, A review on diverse silver nanostructures, *J. Mater. Sci.* Technol. 26 (6) (2010) 487–522.

[76] M. Fox, Introduction to nanophotonics, Sergey V. Gaponenko, Contemp. Phys. 52 (3) (2011) 257.

[77] S. Camelio, E. Vandenhecke, S. Rousselet, D. Babonneau, Optimization of growth and ordering of ag nanoparticle arrays on ripple patterned alumina surfaces for strong plasmonic coupling, Nanotechnology 25 (2014), 035706.

[78] A. Esteban-Cubillo, C. Dıaz, A. Fernandez, L.A. Dıaz, C. Pecharroman, Silver nanoparticles supported on α-, η-, and δ-alumina, J. Eur. Ceram. Soc. 26 (2006) 1–7.

CHAPTER

6

MOx materials by ALD method

Elena Cianci and Sabina Spiga
CNR-IMM, Unit of Agrate Brianza, Agrate Brianza (MB), Italy

6.1 Introduction

Atomic layer deposition (ALD) is a thin film growth technique belonging to the chemical vapor phase processing methods, with the key advantage of growing films with atomic level thickness control and optimal conformality on three-dimensional (3D) topographies. ALD was initially proposed in works on molecular layering in the 1960s in the Soviet Union and then developed in the 1970s in Finland with the name of ALE, Atomic Layer Epitaxy [1,2], for application in electroluminescent flat panel displays [3–5]. A breakthrough for ALD technology began in the 1990s driven by the semiconductor industry, leading to the integration of ALD high permittivity (high-k) oxides as gate dielectrics in metal-oxide field effect transistors as the industry standard in 2007 [6–8]. Since then, the number of ALD materials [9–18], applications [19–22], and equipment [23] has been largely increased so far that the ALD inventor Tuomo Suntola was awarded in May 2018 by the biennial Millennium Technology Prize [24], a technology equivalent of the Nobel Prizes for the sciences, for his ALD innovation.

The ever scaling memory technologies have been largely impacted by ALD [25–27], thanks to its capability of unique accuracy in thickness and composition control, high thickness uniformity over large areas and optimal step coverage over high aspect ratio structures. For instance, in Flash non-volatile memory devices, both for floating gate (FG) and charge trap (CT) storage approaches, ALD oxides have been investigated as interpoly dielectrics and blocking oxide to improve gate to gate coupling and charge retention; in dynamic random access memory (DRAM) capacitors ALD high-k materials have been implemented on 3D structures to increase the capacitance density.

This chapter describes, in the first part, the fundamental aspects of atomic layer deposition, together with its advantageous features, and its

Metal Oxides for Non-volatile Memory
https://doi.org/10.1016/B978-0-12-814629-3.00006-4

critical process parameters. The second part of the chapter focuses on some ALD oxide processes, employed for the deposition of thin oxide films integrated in memory devices.

6.2 ALD fundamentals

6.2.1 Saturating self-limiting reactions

Atomic layer deposition is based on self-terminating surface reactions of gaseous precursors, a metal-containing one and a co-reactant (an oxidant agent for metal oxide deposition), separately delivered onto the substrate, as sketched in Fig. 6.1. The deposition process consists of discrete alternating pulsing of precursor and co-reactant separated by purging steps (Fig. 6.2A), differently from Chemical Vapor Deposition (CVD) where they flow together in the vapor phase over the substrate [28]. Each precursor reacts with the surface on available reactive sites; once all reactive surface sites are saturated, excess precursor cannot react anymore and is pumped away during the purging step in inert gas, usually nitrogen or argon, together with reaction by-products. The film is deposited in a bottom-up-like process, half-reaction by half-reaction in a cyclic manner. During each cycle a sub-monolayer of material, limited by both the reactive site density and precursor steric hindrance, is formed on the substrate in a self-controlled manner. In this sense, the film growth is self-limited,

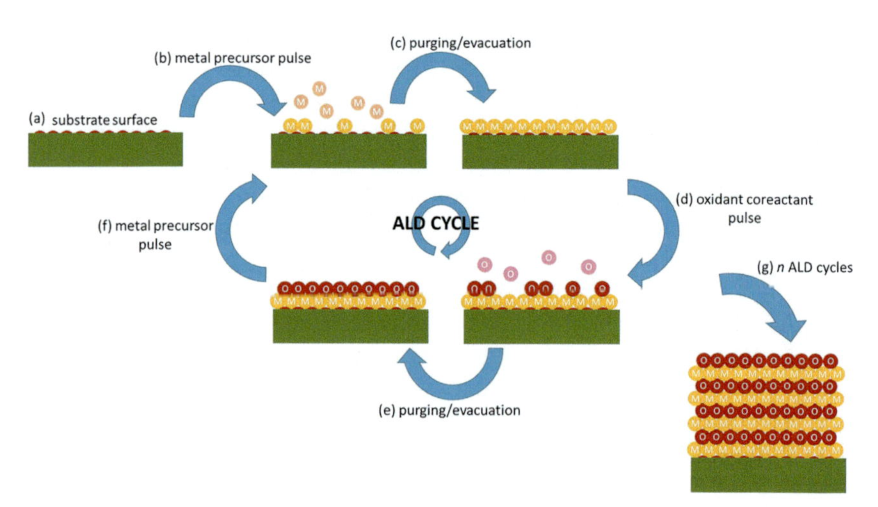

FIG. 6.1 Schematic of ALD deposition process steps for the synthesis of binary metal oxide. Metal precursor (initially on the substrate surface (A) and then on the film growing surface (B, F)) and oxidant co-reactant (D) are pulsed alternatively, separated by purging steps (C) and (E). The ALD cycle is repeated *n* times until the desired film thickness is achieved (G).

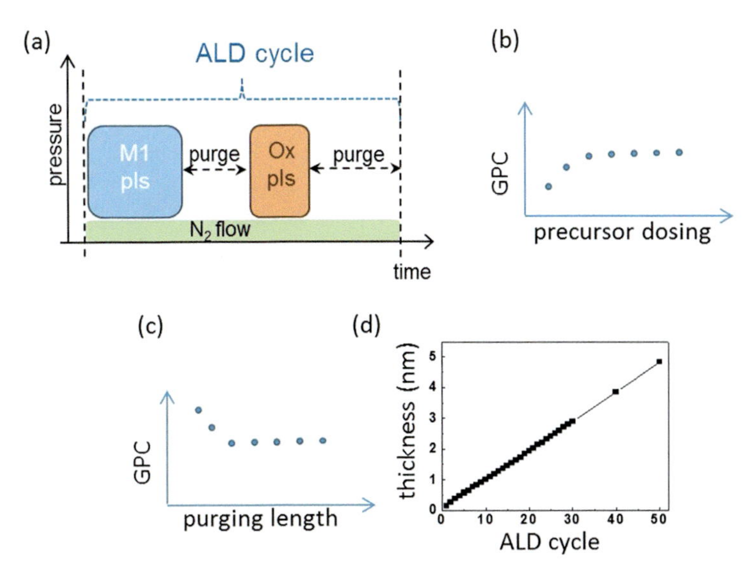

FIG. 6.2 (A) Schematic of an ALD cycle for a binary metal oxide in which precursor M1 and co-reactant Ox are pulsed separately by purging intervals, an inert gas flows through the reactor, both as carrier and as purging gas; GPC saturation curves vs precursor dosing (B) and purging length (C); (D) linear thickness increase vs number of ALD cycles indicating constant ALD GPC.

provided that enough precursor is dosed to the substrate saturating the surface. Saturation is one of the key process parameters for achieving true atomic layer growth: controlling the deposition rate as a function of precursor dosing, the growth rate per cycle (GPC) increases for increasing precursor dose until saturation is achieved (Fig. 6.2B), then it remains constant despite of longer precursor pulses. In the same way, the purging step between precursor and co-reactant pulses must be long enough to avoid their superposition and reaction in the vapor phase (Fig. 6.2C), that will cause CVD-like reactions resulting in faster GPC. The saturating self-limiting mechanism makes the growth rate linearly dependent only on the number of ALD cycles (Fig. 6.2D), not on the reactant flow or concentration. As a consequence, ALD ensures a very accurate control over the film thickness, uniformity over large surface areas, and excellent step coverage properties [20,29,30].

Another benefit coming from the monolayer or sub-monolayer-like deposition in ALD cycles is the precise control over material composition, resulting in high-density pin-hole free films provided that enough time is given for reactions to proceed to completion. In addition, combining multiple ALD processes (Fig. 6.3A), allows the deposition of both nanolaminates and alloys or doped materials, sketched in Fig. 6.3B and C [31,32] for further tailoring the material physical and chemical properties, and

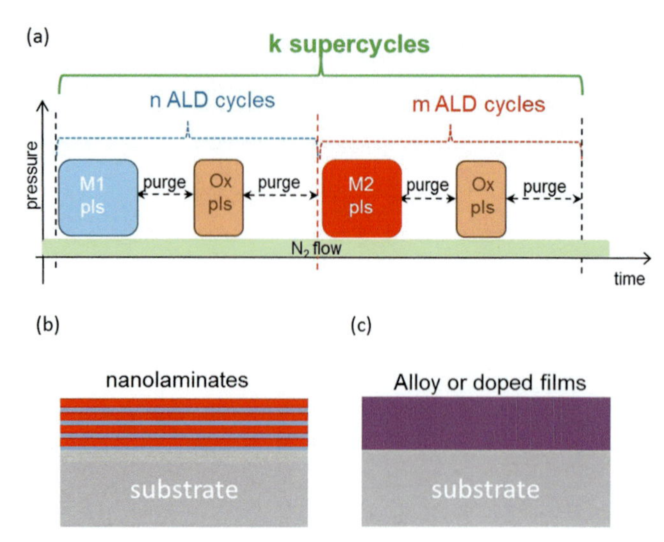

FIG. 6.3 (A) Schematic of a supercycle in which n ALD cycles for metal oxide $M1O_x$ are followed by m ALD cycles for metal oxide $M2O_x$, is repeated k times to deposit a nanolaminated film (B). Alternating $M1O_x$ and $M2O_x$ binary cycles can result in alloy films ($m=n=1$), while adjusting the n:m ratio, $M1O_x$ can be added as dopant into $M2O_x$ host film (C).

as a consequence, the functional properties toward the target application. Thanks to the cyclic nature of ALD processes, multicomponent or doped materials are generally grown employing supercycles in which ALD cycles for each oxide component are alternated. In principle ALD would allow precise control over concentration and distribution of dopant inside host oxide, by adjusting the binary oxides ALD cycle ratio n:m (Fig. 6.3). However, it must be taken into account that in an ALD supercycle, precursor chemisorption on the growing surface can be different from that in the growth of the binary oxides, so that alternating ALD cycles of dopant and host oxide in the appropriate cycle ratio, in order to tune the film composition, may not be straightforward and may even lead to dopant clustering or layering rather than homogeneous distribution in the film [33,34]. Furthermore the ALD windows of the binary oxides to be combined have to overlap at least partially to ensure ALD deposition of the composite film.

The main drawback accompanying the excellent control of film growth enabled by ALD is the low deposition rate, typically in the order of 1 Å/cycle or even below, where 1 cycle can last tens of seconds. This is not a big issue for application in devices where deposition of very thin layers is required, as in many memory devices. Otherwise new approaches to ALD have been explored in order to increase throughput, as the use of spatial ALD, where reactant exposure to the surface is performed moving the substrate in spatially separated regions where reactants flow continuously [35].

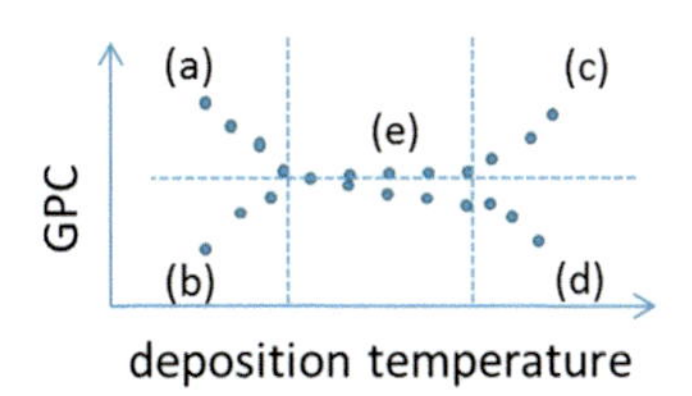

FIG. 6.4 ALD GPC may change as a function of deposition temperature, due to (A) precursor condensation on the substrate, (B) slow reaction kinetics, (C) precursor decomposition, (D) precursor desorption from the substrate. In (E) the ALD window is constant GPC or slightly decreases with reactive sites density, that is true ALD is performed.

6.2.2 Deposition temperature

Typical ALD process temperature varies from room temperature up to 400 °C, relatively low in respect to CVD process, making ALD compatible with thermal budgets typical of CMOS processes. The temperature, necessary to activate precursor reaction on the growing surface, can affect the amount of deposited material per cycle: it must be high enough to let surface reactions occur between gaseous precursors and functional groups on the surface, but not too high to cause precursor thermal decomposition, resulting in GPC increase (Fig. 6.4C) and often leading to poor uniformity and high impurity contents. Alternatively, too high deposition temperature can make precursor molecules desorb from the surface, resulting in GPC decrease (Fig. 6.4D). In the same way, uncontrolled growth can result from too low deposition temperature, as precursors may condense on the substrate surface, depositing larger amount of material respect to available reactive sites (Fig. 6.4A), or alternatively precursors may not have enough thermal energy to drive appropriate reactions causing lower deposition rate than expected (Fig. 6.4B). The temperature range between these limit conditions is called *ALD window*, in which the self-limiting feature of ALD growth is achieved and the growth rate stays constant. True ALD processes are performed at temperatures in the *ALD window*. Anyway, even in the *ALD window*, the GPC may be slightly dependent on temperature, as a consequence of reactive sites density variation on the substrate (Fig. 6.4E) [36].

To obtain the best possible film characteristics, the highest deposition temperature in the ALD window is chosen as often leading to denser films with low amounts of impurities.

6.2.3 Precursors

The precursor chemistry is of fundamental importance for ALD [37–45], being a chemical deposition process. The key requirements for ALD precursors are several. First of all, being ALD a vapor deposition process, precursors must be enough volatile to ensure high vapor pressure for efficient gas phase transport at the process temperature. This is why liquid and gaseous precursors are preferred to solid ones. Some precursors, as trimethylaluminum (TMA, $Al(CH_3)_3$), have high vapor pressure even at room temperature [46]; others must be heated to achieve the needed vapor pressure [47,48], and as a consequence their thermal stability over prolonged time becomes an issue. Typically, thermogravimetric analysis is used to determine the initial temperature when precursor volatilization starts and the residual mass after volatilization gives an indication of precursor decomposition [49,50]. Precursor thermal stability in the deposition temperature range is a fundamental requirement to avoid uncontrolled reactions, both in the gas phase and at the substrate surface, that mine the self-limiting key feature of ALD growth, leading to poor uniformity and high impurity content. On the other hand, precursor reactivity is required to drive ALD surface reactions and ALD growth itself. Furthermore, low toxicity, ease-to-handle, easy synthesis, and low cost are desirable.

Metal precursors mainly employed in oxide ALD are inorganic precursors as halides and metalorganic compounds, in which the metal ion can be bonded directly to carbon, as in alkyls and cyclopentadienyls, to oxygen as in alkoxides and β-diketonates or to nitrogen as in alkylamides and amidinates.

Together with metal precursors, co-reactants for oxide deposition are volatile oxygen sources, reactive enough at the deposition temperature to oxidize the metal precursor and to restore the initial surface groups. Water (H_2O) is used in most cases, even if it is difficult to purge away, especially at low temperature and its reactivity is often not sufficient. More reactive co-reactants are ozone (O_3) and oxygen (O_2) plasma, that can be employed in combination with precursors with high thermal stability and at lower deposition temperatures. H_2O_2 and many oxygen-containing organic molecules have also been reported as co-reactants for oxide deposition [9,51].

6.3 ALD of oxides for memory devices

ALD oxides have been widely investigated for memory applications. Requirements for oxide layers in various memory devices can be different, but the requests of very high quality films together with precise thickness control at the nanometric scale over high aspect ratio topography must be fulfilled for all applications, and ALD can be the technique-of-choice [25,26,52]. In the next paragraphs some of the most studied ALD oxides in the field of memory applications will be presented, starting from Al_2O_3 and oxides of the Group 4 metals (Zr, Hf, and Ti) both as binary compounds and doped or alloyed with other elements. Then ALD of Ta based oxides, NiO, and SiO_x are reported.

6.3.1 Atomic layer deposition of Al_2O_3

The ALD process for Al_2O_3 based on TMA and H_2O, as precursors for aluminum and oxygen sources respectively, is the most widely used and studied ALD process, so much to become a model system for ideal ALD [53–56], being TMA highly reactive to H_2O, exhibiting a wide ALD temperature window, from room temperature up to 400°C [36,53,57–61], in which the growth rate is 0.08–0.12nm/cycle. As-deposited Al_2O_3 films result amorphous, with some carbon contamination at temperature ≤150°C and hydrogen content, originating from hydroxyl groups entrapped in the films, decreasing with increasing deposition temperature [36,62,63] as shown in Fig. 6.5 for Al_2O_3 films deposited in the 100–250°C temperature range. After post-deposition annealing (PDA) at high temperature (≥900°C), Al_2O_3 films thickness shrinking, density increase (Fig. 6.6A), and crystallization (Fig. 6.6B), correspondently higher electrical performances are evidenced [9,64–69].

In the deposition of Al_2O_3 films by ALD other oxygen sources besides H_2O have been used in combination with TMA, such as O_3 and O_2 plasma. The advantage to use non aqueous co-reactants is motivated by the fact that they can be purged more efficiently than water, especially at low deposition temperature and on high aspect-ratio topography; O_3 and O_2 plasma high reactivity and different ALD reaction paths with TMA

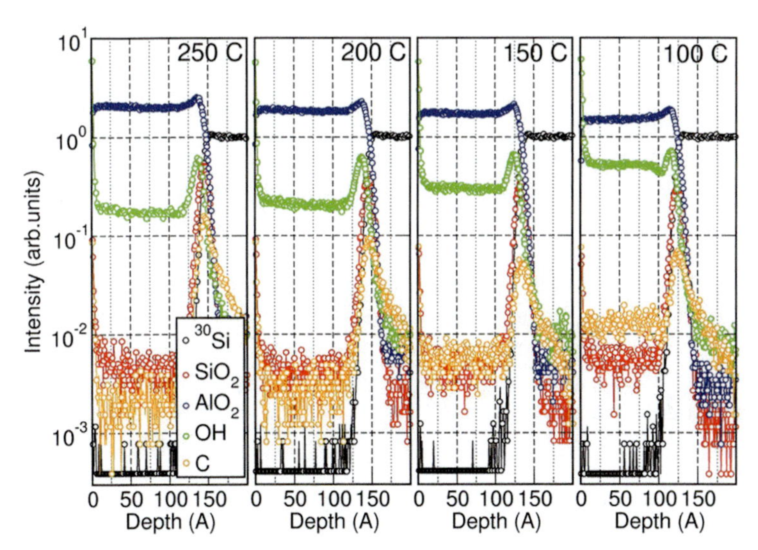

FIG. 6.5 ToF-SIMS (Time of Flight-Secondary Ion Mass Spectrometry) profiles of ALD Al_2O_3 deposited at temperature 100–250°C. The OH signal related to the hydrogen content in the film decreases as the deposition temperature increases, while the carbon content is the detection limit for $T_{dep} > 150$°C (by courtesy of Michele Perego, CNR-IMM, Agrate Brianza (MB), Italy).

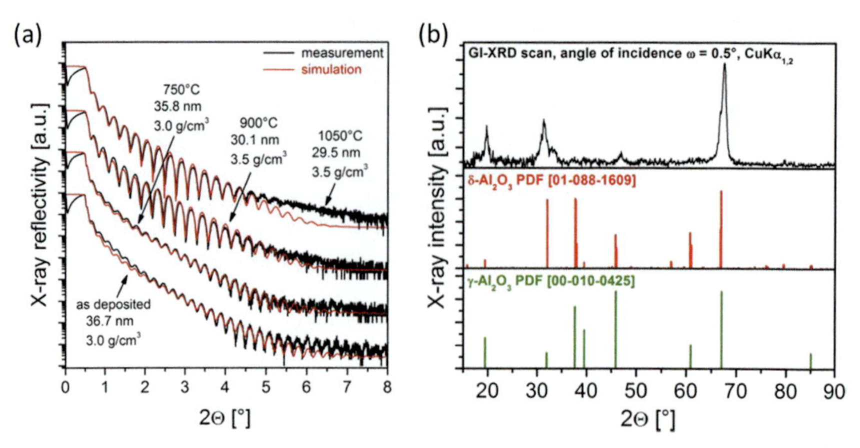

FIG. 6.6 (A) Measured X-ray reflectivity and fit curves for ALD AlOx films. Annealing temperatures and film density and thickness resulting from the fit are indicated. (B) GI-XRD (grazing-incidence X-ray diffraction) after PDA in N_2 at 1050°C and reference patterns for δAl_2O_3 and γAl_2O_3. *Reprinted from V. Cimalla, M. Baeumler, L. Kirste, M. Prescher, B. Christian, T. Passow, F. Benkhelifa, F. Bernhardt, G. Eichapfel, M. Himmerlich, S. Krischok, J. Pezoldt, Mater. Sci. Appl. 05 (2014) 628. Copyright © 2014 by authors and Scientific Research Publishing Inc. This work is licensed under the Creative Commons Attribution International License (CC BY). http://creativecommons.org/licenses/by/4.0/*

[70–73] can lead to different physical-chemical characteristics and electrical properties of the deposited Al_2O_3 films. Al_2O_3 films grown using O_2 plasma has been reported to have higher GPC, higher density and lower carbon and –OH contents than films deposited using H_2O or O_3 as oxidants [62,73]; moreover O_3-based ALD of Al_2O_3 films grown results in lower –OH impurity incorporation and in a reduction in the leakage current density [70,74–76]. As a consequence, Al_2O_3 films optimization, by tuning the ALD process, has been pursued for integration as interpoly dielectric in FG memory devices to replace the standard silicon oxide-nitride-oxide stack and as blocking oxide in CT memory device in combination with metal gates, achieving high dielectric breakdown fields, low trap density and a permittivity value of 9 [26,77–79]. Furthermore, ALD Al_2O_3 has been used in combination with other ALD oxides, as nanolaminates or as alloys or doped films in order to improve the final electrical performances of memory devices, as shown in the following paragraphs.

6.3.2 Atomic layer deposition of HfO_2 and ZrO_2

Intense research efforts have been devoted to the development of ALD processes for HfO_2 and ZrO_2 thin films since the 1990s, as high permittivity oxides, with a *k* value larger than Al_2O_3, for logic and memory devices, in order to keep pace with Moore's law scaling. Both HfO_2 and ZrO_2 are polymorphic materials: monoclinic, tetragonal and cubic phases are stable at different temperature conditions at atmospheric pressure, and their dielectric properties can vary with their crystallinity and crystallographic phases. In case of HfO_2 the dielectric constant strongly depends on its crystallographic phase, the cubic and tetragonal phases being predicted to have a much higher permittivity ($k \sim 29$ and ~ 70, respectively) than that of amorphous or monoclinic ones ($k \sim 16$) [80]. Furthermore, the orthorhombic phase is predicted at high pressure [81], which can induce ferroelectric properties in the material [82]. Similarly, for amorphous and monoclinic ZrO_2 a dielectric constant equal to 19.7 was estimated [83], while higher k values were calculated to be 36.8 and 46.6 for cubic and tetragonal phases, respectively [84]. In addition, higher k phases can be stabilized by doping incorporation of several elements, as it has been theoretically predicted and experimentally demonstrated, for both HfO_2 [31,85–91] and ZrO_2 [90,92–94]. Crystallinity and crystalline phases of ALD thin HfO_2 and ZrO_2 films have been demonstrated to be dependent on the film thickness and on the deposition process (including the employed precursors, deposition temperature, substrate, etc.). In the following, some ALD processes for the binary and doped HfO_2 and ZrO_2 will be reported, starting from some of the precursor available chemistries.

6.3.2.1 Chloride precursor-based ALD

$HfCl_4$ and $ZrCl_4$ were employed as metal precursors together with H_2O as oxidant in the earlier deposition processes, when ALD was still called ALE, and since then chlorides and water based processes have been the commercial ones for many years, especially for HfO_2. Halides precursors are characterized by high thermal stability, enabling the film growth over a large temperature range. Both ALD of ZrO_2 [95] and HfO_2 [96] was demonstrated on glass substrates at 500 °C, with growth rate of about 0.5 Å/cycle, and Cl and H contamination levels below 0.5 at.%. ZrO_2 films resulted mainly amorphous, becoming poorly crystallized in the tetragonal phase for thicknesses up to 50 nm and evolving in the monoclinic phases for thickness as large as 100 nm [97,98], while HfO_2 films showed larger crystallization grade, resulting amorphous for thickness below 15 nm, then fully polycrystalline, mainly in the monoclinic phase, for thickness larger than 40 nm.

Self-limited growth was demonstrated for HfO_2 in a large temperature window (180–600 °C) [99–101], even if the growth rate was observed to decrease for increasing temperature. Furthermore, the film critical thickness for crystallization, dominantly in the monoclinic phase, decreases with deposition temperature (25–30 nm at 300 °C, 8–10 nm at 400 °C) and the film density increases, as the Cl and H impurity content is reduced. $HfCl_4$ has been also used in combination with O_3 as oxidant, resulting in better crystallized and smoother films compared to those grown with H_2O, despite of the same deposition temperature of 300 °C [102].

Chloride precursors present some drawbacks: they are solid and have to be heated at high temperature (>130 °C) to volatilize, as a consequence particle contamination on the surface growing film can occur; furthermore, HCl reaction by-product can be corrosive against the substrate or the deposition equipment.

6.3.2.2 Alkylamide precursor-based ALD

Hf and Zr alkylamides, which contain metal-nitrogen bonds, have been introduced in order to replace the chloride based Hf and Zr precursors. They are volatile with high reactivity toward water and ozone. Using TEMAH ($Hf(NEtMe)_4$ where $Et=C_2H_5$ and $Me=CH_3$) and TEMAZ ($Zr(NEtMe)_4$), constant growth rates of 0.9–1.1 Å/cycle have been reported within variable ALD windows depending on the oxidant precursor and on the detailed process conditions, but generally with lower upper limits respect to chlorides: in the first reports [103,104] the maximum temperature for ALD deposition of ZrO_2 and HfO_2 using H_2O as co-reactant was 300 °C and 400 °C respectively; in more recent papers the upper limit of the ALD window is reported to be lower: for HfO_2 the available temperature range for self-limiting growth is 240–320 °C, 200–280 °C and 175–300 °C, using O_3, H_2O [105] and H_2O_2 [106] respectively, showing increasing growth rates and higher contamination levels for higher deposition temperature, ascribed to precursor decomposition. Furthermore, low temperature processes have been optimized using O_3 (T_{dep} = 175–225 °C) [107,108] and H_2O (80 °C) [109] for increasing the film density and lowering the impurity content. Low temperature deposited films result amorphous independently of their thickness, while films grown at 280 °C result crystalline in the monoclinic phase above the critical thickness of about 15 nm. Moreover, it has been reported that reducing the co-reactant oxidation strength by lowering the O_3 concentration, down to pure oxygen, modifies the crystalline phase of HfO_2 films deposited at 280 °C, allowing the stabilization of tetragonal (or cubic) phase upon annealing at 550 °C [110].

The film growth and crystallization behavior can depend on the underlying substrate as it was reported for ZrO_2 grown by TEMAZ in combination with O_3 on TiN, that is typically used as bottom electrode in DRAM or in RRAM, or on Si substrate covered by native oxide [111,112]. Higher growth rate for ZrO_2 at 275 °C on TiN respect to Si was measured and attributed the presence to more reaction sites provided by rough TiN layer. Films grow amorphous up to a critical thickness on both substrates (independently by the fact that TiN is crystalline while native oxide on Si is an amorphous interface). Then the film starts to crystallize in the tetragonal/cubic form. A dependence of the crystallization temperature on the substrate and on the film thickness was found for very thin films (<7.5 nm), which crystallize at higher temperature on TiN respect to Si, due to the different interfacial energy between the substrate and the film, becoming less important as the film thickness increases. The k value was observed to increase (from ~26 up to ~43) for increasing the deposition temperature, from 225 °C to 275 °C, corresponding to the film structural evolution from partially crystalline to fully crystalline with increasing tetragonal/cubic phases, both in Si/ZrO_2/TiN and TiN/ZrO_2/TiN stack structures [112,113]. Sandwiching an amorphous Al_2O_3 mono- or sub-monolayer, grown using TMA and O_3 at 300 °C, among two ZrO_2 thin films grown by TEMAZ and O_3, avoids film crystallization during ALD deposition, reducing surface roughness and defect density after crystallization upon PDA. Such ALD trilayer in a MIM configuration (ZAZ capacitor) with TiN electrodes [114–116] has been integrated in DRAM addressing the dielectric constant and leakage characteristics required by the device down to the design rule of ~20 nm. For sub-20 nm DRAM, other high k dielectrics in ZrO_2/high k/ZrO_2 configuration are under investigation (SrO, Y_2O_3) [117,118] to increase the total dielectric constant while maintaining the role of improving the leakage characteristics.

6.3.2.3 Cyclopentadienyl precursor-based ALD

Besides chlorides and alkylamides, cyclopentadienyl-based precursors have been widely investigated mainly for their thermal stability and low impurity contamination. In particular, HfD-CO4 and ZrD-O4 (respectively $(MeCp)_2Hf(Me)(OMe)$ and $(MeCp)_2Zr(Me)(OMe)$, where $Cp=C_5H_5$) have been employed in combination with H_2O and O_3, presenting 300–400 °C wide ALD window [119–124].

ZrO_2 films synthesized at 300 °C using H_2O result polycrystalline showing the co-presence of ZrO_2 cubic/tetragonal phases and monoclinic phase, while in films deposited using O_3 the monoclinic phase is drastically reduced and ZrO_2 films are almost completely stabilized in the cubic/tetragonal crystallographic phase [121,122,125]. As already pointed out for ALD Al_2O_3 films, the different ALD reactions occurring when using different oxidizing co-reactants [123,124] can affect the final film properties. For instance, the value of the dielectric constant, depending on the different ZrO_2 crystalline phases in the films, results larger for ZrO_2 films deposited using O_3 (~30) respect H_2O (~25), after PDA at 800 °C, corresponding to the non-detectable monoclinic phase in the O_3-based ZrO_2 films.

Both ZrO_2 and HfO_2 films start to crystallize during deposition at 300 °C after reaching a critical thickness, in the range of 6–8 nm, depending on the precursor and oxidant. The onset of crystallization can be monitored by the growth rate evolution, since the GPC of the film in the crystalline phase is larger than the GPC of the amorphous films. For instance, as shown in

Fig. 6.7A for HfO_2, where the GPC at 300 °C using the HfD-CO4 precursor with O_3 is reported, the change in the slope of the linear dependence is evident after about 130 ALD cycles (~8 nm thickness), increasing from 0.65 to 1.06 Å/cycle, corresponding to the crystallization onset. As-deposited HfO_2 films with thickness larger the critical value result crystalline in a mixture of monoclinic and cubic/tetragonal phases, while upon annealing, the most stable monoclinic component increases (Fig. 6.7B), corresponding to a k value of 16 for PDA at 900 °C [126,127].

The effect of inserting a doping element during the ALD growth of HfO_2 and ZrO_2 has been extensively investigated as a viable approach to engineer the thin film crystalline structure, stabilizing the metastable

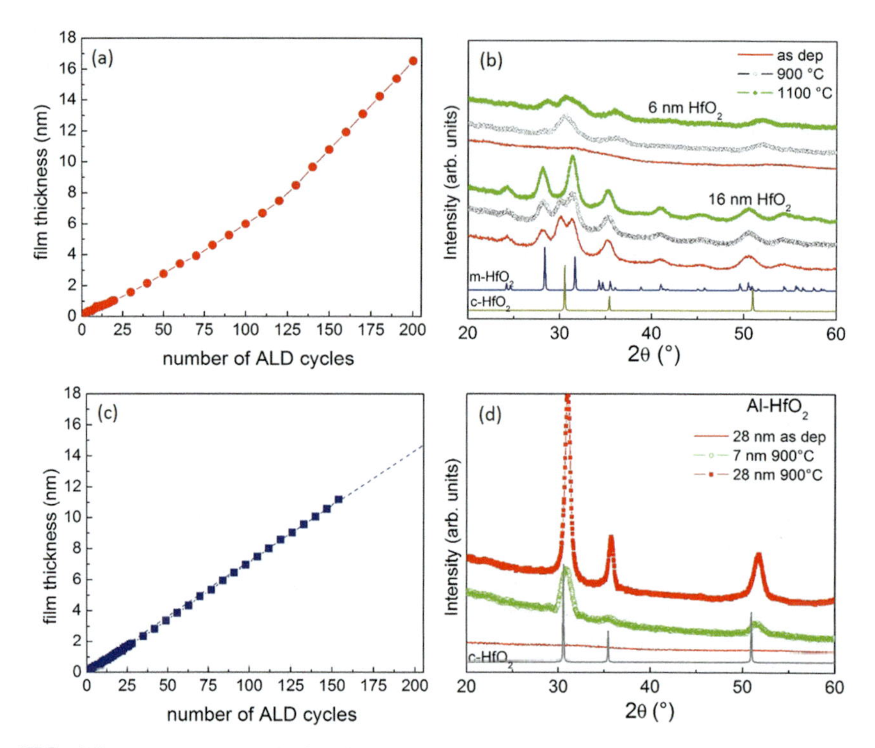

FIG. 6.7 (A) ALD GPC of HfO_2 film on SiO_2/Si deposited at 300 °C by HfD-CO4 and O_3; after a certain number of cycle, as a critical thickness is reached, the film starts to crystallize and the GPC increases; (B) GIXRD patterns of Si/SiO_2/HfO_2/Al_2O_3 stack as a function of PDA temperature, with reference patterns of monoclinic and cubic HfO_2; (C) ALD GPC of Al-HfO_2 films deposited at 300 °C by HfD-CO4, TMA and O_3 on SiO_2/Si; (D) GIXRD pattern of 7 and 28 nm thick Al-HfO_2 films on SiO_2/Si as-deposited and after PDA at 900 °C. *(A and B) Adapted from S. Spiga, F. Driussi, A. Lamperti, G. Congedo, O. Salicio, Appl. Phys. Express 5 (2012) 021102. ©2012 The Japan Society of Applied Physics. (C and D) Adapted with permission from S. Spiga, F. Driussi, G. Congedo, C. Wiemer, A. Lamperti, E. Cianci, ACS Appl. Nano Mater. 1 (2018) 4633–4641. © 2018, American Chemical Society.*

higher *k* crystalline phases [87,89,113,128–135]. Fig. 6.7C shows the GPC of Al-HfO_2 deposited at 300 °C by HfD-CO4, TMA and O_3 on SiO_2/Si, inserting an ALD TMA/O_3 cycle every 20 ALD HfO_2 cycles and resulting in 4.5% Al content in the film [136]. It is worth noticing that, differently from the GPC of pure HfO_2 deposited with the same precursors at the same deposition temperature, no GPC enhancement corresponding to crystallization onset is observed, suggesting that the film is growing in the amorphous phase. This is confirmed by GIXRD analysis (Fig. 6.7D), showing that as-deposited Al-HfO_2 films are amorphous for a large thickness range, while crystallization in the cubic/tetragonal phase is induced only by PDA at 900 °C, corresponding to a k value ~32 [136]. Increasing the Al content in ALD HfO_2 leads to the stabilization of the amorphous phase up to higher temperatures [137].

6.3.2.4 Application of Hf-based oxides in memory storage devices

The investigation of ALD HfO_2 in charge trapping memory devices as CT layer, both in the metal/high-*k*/SiO_2/Si (MHOS) [138,139] and in the metal/Al_2O_3/high-*k*/SiO_2/Si (MAHOS) [126,127,139–141] configurations, has brought to devices with better charge trap efficiency and larger memory windows than memory devices including silicon nitride CT layer. CT memory devices integrating ALD Al-HfO_2 charge trapping layer have been investigated and it has been demonstrated that, by suitably tuning the Al content in the Al-HfO_2 and the PDA process, the memory operation performance can be improved [142,143] and it is possible to achieve a CT layer with equivalent oxide thickness < 1 nm and good retention properties [136].

Hafnium-based oxides have also been studied for their application in ferroelectric memories (FeRAM); since in 2011 it has been demonstrated that ALD Si-doped HfO_2 shows remnant polarization correspondent to the film crystallization in the metastable orthorhombic non-centrosymmetric phase [82]. In fact, the structural origin of ferroelectricity in HfO_2-based thin films is attributed to the formation of a polar orthorhombic phase, however, the stabilization of this o-phase over other competitive phases in HfO_2-based thin films has been reported to be induced by many plausible factors, as doping, surface energy effect, capping layer effect, island coalescence, thermal expansion mismatch, and formation of oxygen vacancies [144]. Many dopant elements have been reported to induce ferroelectricity in HfO_2-based thin films. ALD Al-doped HfO_2 deposited by TEMAH, TMA and O_3, at 300 °C has shown ferroelectricity tunable by Al content in the film and thermal treatments and it has been integrated in deep trench MIM capacitors [145,146].

Hf-based oxides grown by ALD have been also exploited for application in RRAM devices, either as binary HfO_2 oxides or as Al-doped HfO_2. The Hf-based RRAM devices base their operation on redox reactions and electrochemical phenomena that allow the formation and dissolution of conductive filaments, shorting the two electrodes of a metal/oxide/metal stack [147,148]. While many material systems are under investigation for RRAM application, ALD HfO_2-based RRAMs are of particular interest due to their low power consumption, fast switching times, scalability down to nm scale [149,150] or even atomic level, and CMOS compatibility with the possibility to integrate the film in 3D architectures. Currently Hf-based RRAM are investigated either for non-volatile memory applications [150], and as synaptic element in neural networks [151].

HfO_2 layer in RRAM devices has been grown by several processes: $HfCl_4$ and H_2O [152–154], TEMAH and H_2O [155,156] and HfD-CO4 and H_2O [157]. A comparison of the precursor chemistry impact (halide vs metalorganic used at the same process conditions, not optimized for each single precursor) on the final RRAM device operation showed a dependence on the film amorphous/crystalline phase and carbon content: amorphous HfO_2 films with low carbon content resulted in a better stability of the resistance ratio during cycling and better inter-cell and intra-cell uniformity [158]. It is worth noticing that for RRAM devices also the electrode and reactions at the oxide/electrode interface play a significant role [159]. Among the most studied stacks, TiN/HfO_2/Ti/TiN [150,160] or TiN/HfO_2/Hf/TiN [154,161,162], where the Ti or the Hf layer acts as scavenging layer [163], have shown the best performances for device applications and integration in large memory array [160]. Anyway, also electrode combinations such as TiN/HfO_2/Pt have been studied [157,164–166].

The additional advantages of ALD for RRAM is the possibility to engineer the material properties by doping [167] and to address the material integration in 3D architectures [166,168,169], as shown in Fig. 6.8 where HfO_x thin film conformally coated the sidewall of a trench, forming a vertical RRAM between the TiN pillar electrode and Pt plane electrode.

6.3.3 Atomic layer deposition of TiO_2

ALD TiO_2 thin films have been widely investigated as high-k dielectrics for DRAM and resistive switching material in RRAM [170,171]. For DRAM application, the requirement of a very high-k capacitor oxide to be deposited with high conformality on high aspect-ratio structures made ALD TiO_2 thin films very attractive because one of the crystalline polymorphic phase in which TiO_2 can stabilize, the rutile one, presents a very high dielectric constant value (90 and 170 along the a-axis or the c-axis [172]). For RRAM application, the investigation of TiO_2 as resistive switching (RS) material has been kicked off by the large knowledge of the

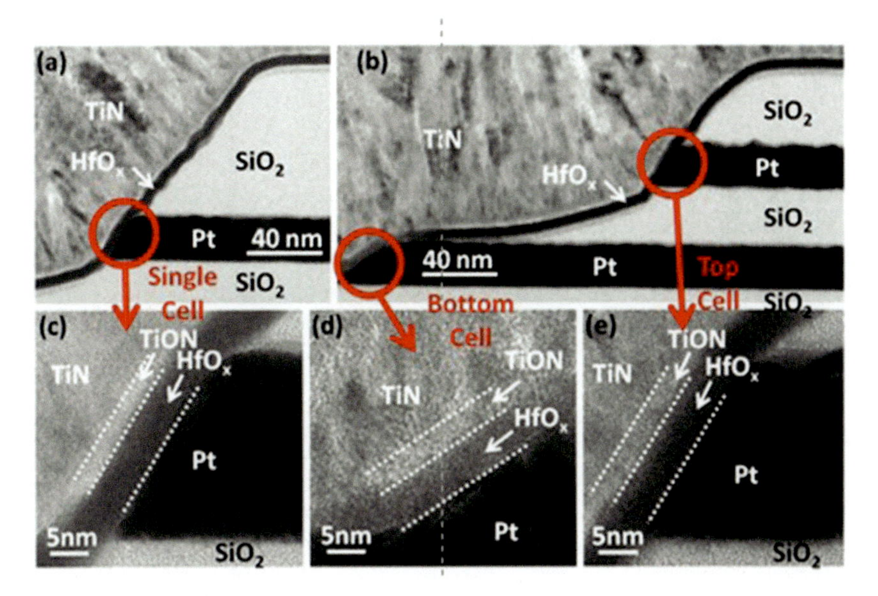

FIG. 6.8 Cross-section TEM of (A) the single-layer sample and (B) the double-layer sample; HR-TEM of (C) the cell in the single-layer sample, (D) the bottom cell, and (E) the top cell in the double-layer sample. The TiN pillar electrode, TiON interfacial layer, HfO_x resistive switching layer, and Pt plane electrode are clearly shown. *Reprinted with permission from S. Yu, H.-Y. Chen, B. Gao, J. Kang, H.-S.P. Wong, ACS Nano 7 (2013) 2320–2325. © 2013, American Chemical Society.*

material gained as high-*k* dielectrics for DRAM. Furthermore, TiO_2 shows both unipolar and bipolar resistive switching, allowing deep investigation of the underlying physical phenomena. For both applications, the 3D structure of DRAM and RRAM devices makes ALD the technique of choice for the high step coverage degree it offers.

Several ALD processes for TiO_2 thin films have been reported since early 1990s, following a similar evolution path as ALD HfO_2 and ZrO_2, starting from the use of halides precursors, followed by the development of processes employing alkoxide, alkylamide and heteroplectic precursors combined with H_2O, O_3 and O_2 plasma [173]. The most used halide precursor has been $TiCl_4$ that is stable up to high temperature (600 °C), while $Ti(iprox)_4$ ($iprox = OC_3H_7$) is the most investigated and employed Ti precursor among alkoxides [173–177] and TDMAT ($Ti(NMe_2)_4$) among the alkylamides [178–182], exhibiting similar ALD window with starting thermal decomposition above 250 °C.

The $TiCl_4$-H_2O based process has been investigated in a wide temperature range from 100 °C up to 600 °C [183], showing GPC values of around 0.5 Å/cycle. As-deposited TiO_2 films result in the amorphous phase at temperatures below 165 °C, in anatase polycrystalline structure at

165–350 °C, and in mixed rutile-anatase polycrystalline structure at temperatures above 350 °C [184,185]. Also for the other halide precursors, as they are stable at high temperature, the growth of rutile TiO_2 films has been reported at temperatures above 450 °C on Si (100) and at lower temperature (above 350 °C) on α-Al_2O_3 [186,187]. The effect of the substrate on the nucleation, growth and structure of ALD TiO_2 has shown to be relevant independently of the used Ti precursor since the early investigations on the growth process. The rutile structure of TiO_2 films has been obtained on ruthenium substrates at low temperature (250 °C) using $Ti(iprox)_4$ and O_3 [188] or TDMAT and O_2 plasma [178,179]: it was reported that a strong oxidizer can oxidize the substrate surface to a thin RuO_2 interfacial layer, which acts as an epitaxial seed having a TiO_2-lattice matched rutile structure. At support, it was shown that using a milder oxidizer as H_2O as co-reactant, TiO_2 crystallized in the anatase phase on Ru [189] and, when grown directly on RuO_2, using H_2O with $TiCl_4$ at 400 °C or $Ti(iprox)_4$ with O_3 at 250 °C, TiO_2 was in the rutile phase [188,190,191].

Furthermore, different nucleation and film grain size have been observed depending on the substrate. Films made of very large grains (more than 100 times larger than the film thickness) were grown on silicon with native oxide (Fig. 6.9A), while when grown on rough surface as TiN (Fig. 6.9B) or Ti [192], the film morphology reproduces that of the substrate both as roughness and grain size. It has been reported that the distribution of crystal nuclei on the substrate, governed by the surface energy, if sparse, can lead to the formation of extremely large grains in the as-grown TiO_2 [193].

For integration in DRAM capacitors, ALD processes for incorporation of Al atoms in TiO_2, in order to achieve sufficiently low leakage current, have been reported [170,176,194,195]. Al-doped TiO_2 (ATO) have been deposited using $Ti(iprox)_4$, TMA and O_3 at 250 °C on Ru electrode.

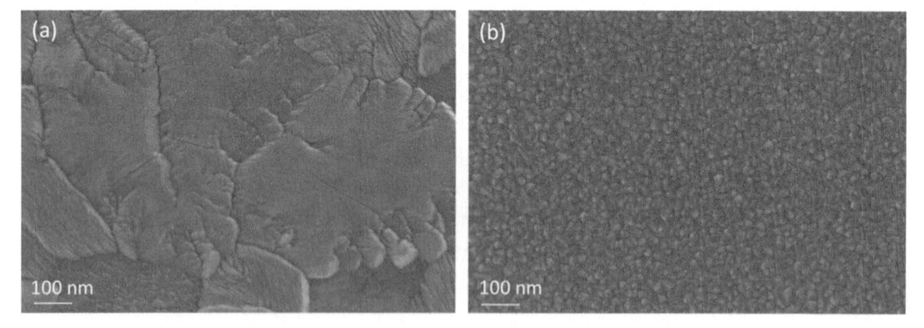

FIG. 6.9 SEM image of 30 nm thick TiO_2 films grown using $Ti(iprox)_4$ and O_3 at 295 °C on Si(100) with native oxide (A) and on TiN substrates (B). Very large grains (~100 times larger than the film thickness) are observed in films grown on Si while when grown on TiN the film follows the substrate morphology. In both cases films are in the anatase phase.

TMA/O_3 ALD cycles intercalating in TiO_2 growth resulted in uniform Al concentration in the film, which was increased by increasing the number of alumina to titania ALD cycle ratio, but not linearly due to reduced adsorption of Ti precursor on the growing surface after the Al incorporation. ATO films grown on Ru exhibited reduced leakage current [195] and very high dielectric constant ($k \sim 80$) corresponding to rutile structure, despite Al doping up to 6% Al/Al+Ti atomic ratio.

For application in RRAM devices [196], ALD TiO_2 film deposited by $Ti(iprox)_4$ and O_2 plasma at 250°C and sandwiched among Pt electrodes in MIM configuration has been one of the most investigated process. No critical dependence on TiO_2 structural phase on switching behavior has been evidenced [197–201]. Improved dispersion of resistive switching parameters has been reported also using a trilayer RS material $HfO_2/TiO_2/HfO_2$ deposited at 250°C by TEMAH, $TiCl_4$ and H_2O, with top Pt and bottom Pt or TiN electrodes [202]. TiO_2 is a material of choice to fabricate RRAM devices in a bi-layer configuration exhibiting uniform type of switching, for which the whole electrode area controls the conductance. These devices exploit interface effects in oxide-bilayers composed of a barrier layer (acting also as self-compliance layer) and a switching layer, e.g., a-Si(barrier)/TiO_2 [203] and Al_2O_3(barrier)/TiO_2 [204]. Moreover, these RRAM stacks display an analog control of the conductance, useful for implementation in synaptic devices for neuromorphic applications. In the TiN/a-Si/TiO_2/TiN device, the active stack consists of 8nm PVD a-Si and 8nm ALD TiO_2, crystallized in anatase phase [205], deposited at 250°C using $Ti(iprox)_4$ with O_3 and/or H_2O.

6.3.4 Atomic layer deposition of Ta_2O_5

Similar to TiO_2, tantalum oxide (Ta_2O_5) has received extensive attention because of its high dielectric constant (~25 for amorphous film and ~50 for crystalline film) for application as high-*k* dielectrics in the capacitive element of DRAM. Furthermore, Ta-based oxide has also entered the research field of RRAM application since 2012 when Ta_2O_5/TaO_x RS stack was integrated in a memory cell showing superior device performance (up to 10^{12} cycle endurance) in comparison to other transition metal oxides [206–208].

ALD of Ta_2O_5 thin films using several precursors including halides [206,209–212], alkoxides [213,214], and alkyl-amides and -imides [73,215,216] have been reported. ALD based on $TaCl_5$ and H_2O has been investigated since 1994 [217], resulting in amorphous films up to 250°C, in partially crystallized films, for thickness larger than 30nm, in δ-hexagonal phase at temperature above 300°C. For higher deposition temperature (>350°C) or in the case of thin films annealed at high temperature

(760 °C) Ta oxide films the orthorhombic structure is stabilized. For increasing deposition temperature or for increasing $TaCl_5$ precursor dose, etching processes appear to be competing against deposition, affecting growth rate and thickness uniformity over large areas. Ta_2O_5 integrated at 250 °C in a MOS capacitor with Pt top electrode, annealed in forming gas at 560 °C [210] showed a dielectric constant value of ~28.

PET ($Ta(OEt)_5$) has been used in combination with H_2O and O_3 up to 350 °C, resulting in amorphous films, almost stoichiometric in the deposition temperature range 250–325 °C, with hydrogen contamination decreasing with temperature. Deposited at 300 °C on TiN and annealed at 400 °C in N_2, Ta oxide films exhibit δ-Ta_2O_5 structure, with larger crystallization grade for films deposited using O_3 respect than H_2O. Correspondently, when incorporated in Ru/Ta_2O_5/TiN MIM configuration and annealed at 400 °C in N_2 and in forming gas, k values of 46 and 40 have been found for films deposited using O_3 respect than H_2O respectively [214].

PDMAT ($Ta(NMe_2)_5$) precursor with H_2O and O_2 plasma showed the ALD process window between 200 °C and 250 °C with 0.9 Å/cycle GPC and between 150 °C and 250 °C with 1.1 Å/cycle GPC respectively. Due to the high precursor reactivity, in the ALD process window, stoichiometric films are obtained with no detectable N and C impurity content. Grown on Si, films are amorphous for deposition temperature in the ALD window and crystallize in orthorhombic phase upon high temperature annealing (600–1000 °C).

Integration in RRAM storage devices has been reported for ALD Ta_2O_5 deposited using PET and H_2O at 300 °C in a TiN/Ta_2O_5/TiN configuration and in combination with Al_2O_3 layer in TiN/Ta_2O_5/Al_2O_3/TiN stack [218]. The formation of a thin interlayer of TaO_xN_y at TiN top electrode interface was evidenced while suppressed by the insertion of thin Al_2O_3 deposited by TMA and H_2O at 300 °C, resulting in a faster and more reliable RS behavior. ALD TaO_x/Ta_2O_5 oxide bilayers were also investigated. The control of film stoichiometry and oxygen concentration has been shown to be achieved by means of plasma ALD process parameters, in particular oxygen deficient TaO_x films were deposited by using a H_2/Ar plasma mixture tuning the H_2/Ar ratio in the plasma pulse and its duration [219,220].

6.3.5 Atomic layer deposition of NiO

Nickel oxide has been mainly researched as RS material since 2004, when Samsung reported on their investigation on RS of NiO thin films [221]. Precursors explored for ALD of NiO, summarized in Table 6.1, are typically others from the ones exploited for group IV and V metal oxides. β-diketonate compounds as $Ni(acac)_2$ (acac = acetylacetone,

TABLE 6.1 Precursors combination and process temperatures for ALD of NiO, with resulting growth rate and some film properties.

	Precursor temperature (°C)	Deposition temperature (°C)	Growth rate (Å/cycle)	Film characteristics	References
$Ni(acac)_2/O_3$ or O_3 $_+H_2O$	155	250	0.62, increased to 0.76	Polycrystalline	[222,223]
$Ni(tmhd)_2/O_3$	145	400		In ALD $LaNiO_3$	[224]
$Ni(tmhd)_2/H_2O$	–	200–275	0.2–0.4	Polycrystalline, (111) on-αAl_2O_3(001)	[225–227]
$Ni(Cp)_2/O_3$	40	250–300	0.12–0.08	Polycrystalline, (111) on Pt(111)	[228,229]
$Ni(Cp)_2/O_3$	90	270–330	0.22–03	Conformal growth in anodic alumina	[230]
$Ni(EtCp)_2/O_3$	80	200–300	0.1	Polycrystalline	[229,231]
$Ni(MeCp)_2/O_2$ plasma	50	150–200	0.48 on Pt 0.84 on W	Growth rate dependent on substrate	[232]
$Ni(dmamb)_2/H_2O$	80	130–150	1.28	Partially crystalline	[233,234]
$Ni(dmamb)_2/O_3$		140–175	0.25	Polycrystalline	[234]
$Ni(dmamp)_2/H_2O$	90	90–150	0.8	Amorphous	[235]

$C_5H_7O_2$) and $Ni(thd)_2$ (thd = 2,2,6,6-tetramethylheptane-3,5-dionate, $OCC(CH_3)_3CHCOC(CH_3)_3$), and cyclopentadienyl compounds as $Ni(Cp)_2$, $Ni(EtCp)_2$ and $Ni(MeCp)_2$, exhibit similar ALD windows at intermediate temperatures (<300 °C) and low reactivity against various oxidant co-reactants resulting in low growth rate and high carbon contamination. ALD NiO films are polycrystalline in the cubic phase with (111) preferential orientation when grown on (111)-Pt [228] or on $\alpha Al_2O_3(001)$ [225] independently of the chosen precursor combination.

More volatile and reactive Ni precursors are alkoxide based ones as $Ni(dmamb)_2$ (dmamb = 1-dimethylamino-2-methyl-2-butanolate, $OC(Me)(C_2H_5)CH_2N(Me)_2$), and $Ni(dmamp)_2$ (dmamp = 1-dimethylamino-2-methyl-2-propanolate, $-OCMe_2CH_2NMe_2$). They are liquid and very reactive against water at low temperature (<175 °C), resulting in amorphous or partially crystallized films deposited at high growth rate.

ALD NiO deposited using $Ni(dmamb)_2$ and H_2O at 140 °C in Pt/NiO/Pt configuration showed bistable resistive switching has been reported [236]. ALD NiO deposited using $Ni(Cp)_2$ and O_3 at 300 °C has been integrated as RS on several flat metal electrodes (Pt, Ni, W and TiN) [237,238]. Films grow conformally replicating the substrate roughness, they are cubic polycrystalline with density similar to bulk NiO. When deposited on W electrode, an interfacial WOx layer is detected and RS MIM devices with W bottom electrodes exhibit a larger resistance window with respect to TiN, Si, Ni and Pt electrodes. When grown on top of a tungsten pillar bottom electrode (Fig. 6.10A) to investigate the dependence of RS parameters on the electrode area scaling [239], direct observation of resistive switching through filamentary conduction has been evidenced by atomic force microscopy [240,241] (Fig. 6.10B).

6.3.6 Atomic layer deposition of SiO_2

Silicon oxide is one of the most important materials in the semiconductor industry, and its importance has gathered research efforts toward the investigation of ALD processes for applications where low thermal budgets, precise thickness control and high conformality on 3D structures are required. For instance, ALD SiO_2 has been investigated as blocking oxides in CT memory devices for 3D integration [136], and it has been exploited as sacrificial layer in DRAM capacitor fabrication process or for double patterning process [19,242].

Silicon chloride sources [243,244] and metalorganic precursors, such as BDMAS (bis-dimethylaminosilane, $SiH_2[N(Me)_2]_2$) [245], TDMAS (tris-dimethylaminosilane $SiH[N(Me)_2]_3$) [245–247], TEMASi (tetrakis-ethylmethylaminosilane, $Si[N(MeEt)]_4$), BDEAS (bis-diethylaminosilane $SiH_2[N(Et)_2]_2$) [248–250], DIPAS (di-isopropylamino silane, $H_3Si[N$

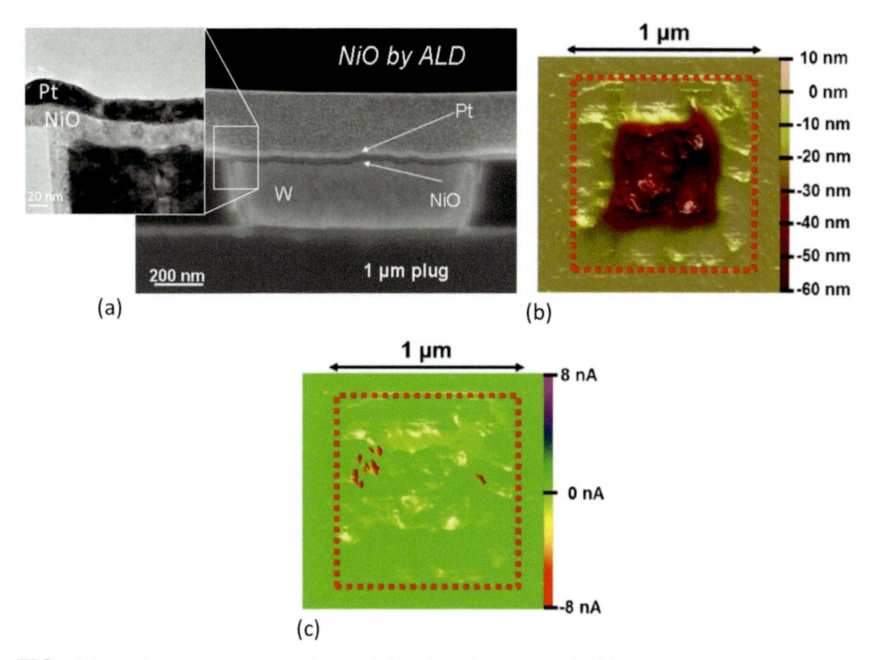

FIG. 6.10 (A) SEM cross-section and TEM in the onset of ALD-NiO based memory cells fabricated on W bottom plug with Pt top. (B) Topography map obtained by Conductive Atomic Force Microscopy (CAFM) scanning of the NiO surface, evidencing the roughness in the plug region, due to W dishing during the CMP step following the metal filling of the via hole, and reproduced by conformal ALD NiO, (C) Topography/current map after forming, showing surface topography with a light/dark scale and low/high currents as green/red color scale. *Red spots* reveal electrically-formed CFs at nanoscale. *Adapted from F. Nardi, D. Deleruyelle, S. Spiga, C. Muller, B. Bouteille, D. Ielmini, J. Appl. Phys. 112 (2012) 064310, with the permission of AIP Publishing.*

$(C_3H_7)_2]$) [251] have been studied for atomic layer deposition of SiO_2 and of silicates, in particular Hf-silicates.

Initial studies using silicon chloride and water vapor showed good film quality obtained at a relatively high temperature of 500–600 °C, with low deposition rate. In order to reduce the deposition temperature, avoiding corrosive byproduct of chloride based precursors, the attention of precursor designers focused on alkyaminosilane precursors, characterized by high vapor pressure and high reactivity to oxidizing co-reactants. The above listed precursors are all liquid, with high volatility even at room temperature, operating efficiently at low temperature (<320 °C). Optimization in ligand design is required for clean oxidization of the precursor without remnants of ligand incorporated in the film. For instance, bis-aminosilanes as BDEAS in combination with O_3 have been reported to yield to ALD of SiO_2 with higher growth rate, wider ALD window and low carbon impurities respect to tris-aminosilane as TDMAS and O_3 [247].

ALD SiO_2 deposited by BDEAS and O_3 at 300 °C has been investigated as blocking oxide in a CT cell on top of ALD Al-doped HfO_2 CT layer, resulting in good retention characteristics compared to ALD Al_2O_3 [136]. Recently SiO_2 has been gaining attention as RS material in RRAM, as it has been reported to offer several advantages compared to other competing technologies (higher stability, greater resistance contrast, the potential to minimize the requirement for cell selector elements) [252,253]. To the best of our knowledge, ALD SiO_2 has not yet been exploited by resistance-switching technologies, but thanks to ALD peculiarities, ease of process integration and controlled material properties can be foreseen as viable mean for the realization of SiO_x-based RRAM memory chips.

6.4 Conclusions

ALD appears to be a unique deposition technique for depositing thin films with exceptionally precise thickness and composition control on the nanometer scale over the 3D geometry. Such characteristics derive form the cyclic, self-saturating nature of the ALD process itself. Successful ALD precursors and processes have been developed for many metal oxides and for several applications. This chapter reviewed some ALD processes for metal oxides employed successfully in memory devices from FLASH and DRAM to emerging concepts including RRAM and FeRAM. The ALD processes are detailed for the following binary oxides: Al_2O_3, HfO_2, ZrO_2, NiO, TiO_2, Ta_2O_5 and SiO_2, together with some examples of their applications. Furthermore, the possibility of engineering these oxides by doping is discussed. ALD allows the choice of a large variety of precursors and low deposition temperature range, and enables the deposition of very thin and conformal films with a precise control of stoichiometry. All these features allow to select the best process to tailor on one hand the materials properties and on the other hand to make the metal oxide deposition process compatible with various integration strategies and substrates.

References

[1] G.N. Parsons, J.W. Elam, S.M. George, S. Haukka, H. Jeon, W.M.M.(.E.). Kessels, M. Leskelä, P. Poodt, M. Ritala, S.M. Rossnagel, J. Vac. Sci. Technol. A 31 (2013) 050818.
[2] R.L. Puurunen, Chem. Vap. Depos. 20 (2014) 332–344.
[3] T. Suntola, J. Antson, Method for Producing Compound Thin Films, US4058430A, 1977.
[4] T. Suntola, Mater. Sci. Rep. 4 (1989) 261–312.
[5] T. Suntola, Thin Solid Films 216 (1992) 84–89.
[6] K. Mistry, C. Allen, C. Auth, B. Beattie, D. Bergstrom, M. Bost, M. Brazier, M. Buehler, A. Cappellani, R. Chau, C.H. Choi, G. Ding, K. Fischer, T. Ghani, R.

Grover, W. Han, D. Hanken, M. Hattendorf, J. He, J. Hicks, R. Huessner, D. Ingerly, P. Jain, R. James, L. Jong, S. Joshi, C. Kenyon, K. Kuhn, K. Lee, H. Liu, J. Maiz, B. McIntyre, P. Moon, J. Neirynck, S. Pae, C. Parker, D. Parsons, C. Prasad, L. Pipes, M. Prince, P. Ranade, T. Reynolds, J. Sandford, L. Shifren, J. Sebastian, J. Seiple, D. Simon, S. Sivakumar, P. Smith, C. Thomas, T. Troeger, P. Vandervoorn, S. Williams, K. Zawadzki, 2007 IEEE Int Electron Devices Meet., 2007, pp. 247–250.

[7] K.J. Kuhn, IEEE Trans. Electron Devices 59 (2012) 1813–1828.

[8] C. Auth, C. Allen, A. Blattner, D. Bergstrom, M. Brazier, M. Bost, M. Buehler, V. Chikarmane, T. Ghani, T. Glassman, R. Grover, W. Han, D. Hanken, M. Hattendorf, P. Hentges, R. Heussner, J. Hicks, D. Ingerly, P. Jain, S. Jaloviar, R. James, D. Jones, J. Jopling, S. Joshi, C. Kenyon, H. Liu, R. McFadden, B. McIntyre, J. Neirynck, C. Parker, L. Pipes, I. Post, S. Pradhan, M. Prince, S. Ramey, T. Reynolds, J. Roesler, J. Sandford, J. Seiple, P. Smith, C. Thomas, D. Towner, T. Troeger, C. Weber, P. Yashar, K. Zawadzki, K. Mistry, 2012 Symp. VLSI Technol. VLSIT, 2012, pp. 131–132.

[9] V. Miikkulainen, M. Leskelä, M. Ritala, R.L. Puurunen, J. Appl. Phys. 113 (2013), 021301.

[10] J. Hämäläinen, M. Ritala, M. Leskelä, Chem. Mater. 26 (2014) 786–801.

[11] J. Päiväsaari, J. Niinistö, P. Myllymäki, C. Dezelah, C.H. Winter, M. Putkonen, M. Nieminen, L. Niinistö, in: Rare Earth Oxide Thin Films, Springer, Berlin, Heidelberg, n.d., pp. 15–32.

[12] B.S. Lim, A. Rahtu, R.G. Gordon, Nat. Mater. 2 (2003) 749–754.

[13] X. Meng, Y.-C. Byun, H.S. Kim, J.S. Lee, A.T. Lucero, L. Cheng, J. Kim, Materials 9 (2016).

[14] J. Musschoot, Q. Xie, D. Deduytsche, S. Van den Berghe, R.L. Van Meirhaeghe, C. Detavernier, Microelectron. Eng. 86 (2009) 72–77.

[15] M. Leskelä, V. Pore, T. Hatanpää, M. Heikkilä, M. Ritala, A. Schrott, S. Raoux, S. Rossnagel, ECS Trans. 25 (2009) 399–407.

[16] N.P. Dasgupta, X. Meng, J.W. Elam, A.B.F. Martinson, Acc. Chem. Res. 48 (2015) 341–348.

[17] M. Mattinen, T. Hatanpää, T. Sarnet, K. Mizohata, K. Meinander, P.J. King, L. Khriachtchev, J. Räisänen, M. Ritala, M. Leskelä, Adv. Mater. Interfaces 4 (n.d.) 1700123.

[18] J.J. Pyeon, S.H. Kim, D.S. Jeong, S.-H. Baek, C.-Y. Kang, J.-S. Kim, S.K. Kim, Nanoscale 8 (2016) 10792–10798.

[19] A.J.M. Mackus, A.A. Bol, W.M.M. Kessels, Nanoscale 6 (2014) 10941–10960.

[20] H. Kim, H.-B.-R. Lee, W.-J. Maeng, Thin Solid Films 517 (2009) 2563–2580.

[21] J.W. Elam, N.P. Dasgupta, F.B. Prinz, MRS Bull. 36 (2011) 899–906.

[22] C. Bae, H. Shin, K. Nielsch, MRS Bull. 36 (2011) 887–897.

[23] W.M.M.(.E.). Kessels, M. Putkonen, MRS Bull. 36 (2011) 907–913.

[24] https://taf.fi/millennium-technology-prize/winner-2018/.

[25] J. Niinistö, K. Kukli, M. Heikkilä, M. Ritala, M. Leskelä, Adv. Eng. Mater. 11 (2009) 223–234.

[26] K.H. Kuesters, M.F. Beug, U. Schroeder, N. Nagel, U. Bewersdorff, G. Dallmann, S. Jakschik, R. Knoefler, S. Kudelka, C. Ludwig, D. Manger, W. Mueller, A. Tilke, Adv. Eng. Mater. 11 (2009) 241–248.

[27] J.S. Lim, Y.K. Kim, S.J. Choi, J.H. Lee, Y.S. Kim, B.T. Lee, H.S. Park, Y.W. Park, S.I. Lee, 6th Int. Conf. VLSI CAD 1999 ICVC 99, 1999, pp. 506–509.

[28] K.L. Choy, Prog. Mater. Sci. 48 (2003) 57–170.

[29] K.-E. Elers, T. Blomberg, M. Peussa, B. Aitchison, S. Haukka, S. Marcus, Chem. Vap. Depos. 12 (n.d.) 13–24.

[30] C.W. Park, H.Y. Yu, U.H. Pi, S.-Y. Choi, Nanotechnology 16 (2005) 361.

[31] D.R.G. Mitchell, D.J. Attard, K.S. Finnie, G. Triani, C.J. Barbé, C. Depagne, J.R. Bartlett, Appl. Surf. Sci. 243 (2005) 265–277.
[32] D.H. Triyoso, R.I. Hegde, X.-D. Wang, M.W. Stoker, R. Rai, M.E. Ramon, B.E. White, P.J. Tobin, J. Electrochem. Soc. 153 (2006) G834–G839.
[33] T. Wang, J.G. Ekerdt, Chem. Mater. 23 (2011) 1679–1685.
[34] Y. Wu, S.E. Potts, P.M. Hermkens, H.C.M. Knoops, F. Roozeboom, W.M.M. Kessels, Chem. Mater. 25 (2013) 4619–4622.
[35] P. Poodt, D.C. Cameron, E. Dickey, S.M. George, V. Kuznetsov, G.N. Parsons, F. Roozeboom, G. Sundaram, A. Vermeer, J. Vac. Sci. Technol. A 30 (2011), 010802.
[36] M.D. Groner, F.H. Fabreguette, J.W. Elam, S.M. George, Chem. Mater. 16 (2004) 639–645.
[37] M. Putkonen, L. Niinistö, in: Precursor Chem. Adv. Mater., Springer, Berlin, Heidelberg, n.d., pp. 125–145.
[38] K. Bernal Ramos, M.J. Saly, Y.J. Chabal, Coord. Chem. Rev. 257 (2013) 3271–3281.
[39] R.J. Crutchley, Coord. Chem. Rev. 257 (2013) 3153.
[40] A. Devi, Coord. Chem. Rev. 257 (2013) 3332–3384.
[41] T. Hatanpää, M. Ritala, M. Leskelä, Coord. Chem. Rev. 257 (2013) 3297–3322.
[42] T.J. Knisley, L.C. Kalutarage, C.H. Winter, Coord. Chem. Rev. 257 (2013) 3222–3231.
[43] S.W. Lee, B.J. Choi, T. Eom, J.H. Han, S.K. Kim, S.J. Song, W. Lee, C.S. Hwang, Coord. Chem. Rev. 257 (2013) 3154–3176.
[44] F. Zaera, J. Mater. Chem. 18 (2008) 3521–3526.
[45] M.A. Malik, P. O'brien, Chem. Vap. Depos, Royal Society of Chemistry, Cambridge CB4 0WF, UK, 2008, pp. 207–271.
[46] S.E. Potts, G. Dingemans, C. Lachaud, W.M.M. Kessels, J. Vac. Sci. Technol. A 30 (2012), 021505.
[47] S.A. Rushworth, L.M. Smith, A.J. Kingsley, R. Odedra, R. Nickson, P. Hughes, Microelectron. Reliab. 45 (2005) 1000–1002.
[48] M. Fulem, K. Růžička, V. Růžička, E. Hulicius, T. Šimeček, K. Melichar, J. Pangrác, S.A. Rushworth, L.M. Smith, J. Cryst. Growth 248 (2003) 99–107.
[49] S. Rushworth, H. Davies, A. Kingsley, T. Leese, R. Odedra, Microelectron. Reliab. 47 (2007) 718–721.
[50] S. Rushworth, K. Coward, H. Davies, P. Heys, T. Leese, L. Kempster, R. Odedra, F. Song, P. Williams, Surf. Coat. Technol. 201 (2007) 9060–9065.
[51] K. Knapas, M. Ritala, Crit. Rev. Solid State Mater. Sci. 38 (2013) 167–202.
[52] R. Clark, Materials 7 (2014) 2913–2944.
[53] S.M. George, Chem. Rev. 110 (2010) 111–131.
[54] R.L. Puurunen, J. Appl. Phys. 97 (2005) 121301.
[55] A. Delabie, S. Sioncke, J. Rip, S. Van Elshocht, G. Pourtois, M. Mueller, B. Beckhoff, K. Pierloot, J. Vac. Sci. Technol. A 30 (2011) 01A127.
[56] S.D. Elliott, J.C. Greer, J. Mater. Chem. 14 (2004) 3246–3250.
[57] G. Seguini, E. Cianci, C. Wiemer, D. Saynova, J.A.M. van Roosmalen, M. Perego, Appl. Phys. Lett. 102 (2013) 131603.
[58] L. Lamagna, G. Scarel, M. Fanciulli, G. Pavia, J. Vac. Sci. Technol. A 27 (2009) 443–448.
[59] R.A. Wind, S.M. George, J. Phys. Chem. A 114 (2010) 1281–1289.
[60] A.W. Ott, J.W. Klaus, J.M. Johnson, S.M. George, Thin Solid Films 292 (1997) 135–144.
[61] M. Ritala, M. Leskelä, J.-P. Dekker, C. Mutsaers, P.J. Soininen, J. Skarp, Chem. Vap. Depos. 5 (1999) 7–9.
[62] J.L. van Hemmen, S.B.S. Heil, J.H. Klootwijk, F. Roozeboom, C.J. Hodson, M.C.M. van de Sanden, W.M.M. Kessels, J. Electrochem. Soc. 154 (2007) G165–G169.
[63] C. Guerra-Nuñez, M. Döbeli, J. Michler, I. Utke, Chem. Mater. 29 (2017) 8690–8703.

[64] P. Ericsson, S. Bengtsson, J. Skarp, Microelectron. Eng. 36 (1997) 91–94.
[65] L. Zhang, H.C. Jiang, C. Liu, J.W. Dong, P. Chow, J. Phys. D Appl. Phys. 40 (2007) 3707.
[66] S. Jakschik, U. Schroeder, T. Hecht, M. Gutsche, H. Seidl, J.W. Bartha, Thin Solid Films 425 (2003) 216–220.
[67] V. Cimalla, M. Baeumler, L. Kirste, M. Prescher, B. Christian, T. Passow, F. Benkhelifa, F. Bernhardt, G. Eichapfel, M. Himmerlich, S. Krischok, J. Pezoldt, Mater. Sci. Appl. 05 (2014) 628.
[68] J.M. Rafí, M. Zabala, O. Beldarrain, F. Campabadal, J. Electrochem. Soc. 158 (2011) G108–G114.
[69] M.D. Groner, J.W. Elam, F.H. Fabreguette, S.M. George, Thin Solid Films 413 (2002) 186–197.
[70] S.D. Elliott, G. Scarel, C. Wiemer, M. Fanciulli, G. Pavia, Chem. Mater. 18 (2006) 3764–3773.
[71] D.N. Goldstein, J.A. McCormick, S.M. George, J. Phys. Chem. C 112 (2008) 19530–19539.
[72] V.R. Rai, V. Vandalon, S. Agarwal, Langmuir 26 (2010) 13732–13735.
[73] S.B.S. Heil, J.L. van Hemmen, M.C.M. van de Sanden, W.M.M. Kessels, J. Appl. Phys. 103 (2008) 103302.
[74] J.B. Kim, D.R. Kwon, K. Chakrabarti, C. Lee, K.Y. Oh, J.H. Lee, J. Appl. Phys. 92 (2002) 6739–6742.
[75] S.-C. Ha, E. Choi, S.-H. Kim, J. Sung Roh, Thin Solid Films 476 (2005) 252–257.
[76] F. Campabadal, O. Beldarrain, M. Zabala, M.C. Acero, J.M. Rafí, Proc. 8th Span. Conf. Electron Devices CDE 2011, 2011, pp. 1–4.
[77] C. Lee, J. Choi, C. Kang, Y. Shin, J. Lee, J. Sel, J. Sim, S. Jeon, B. Choe, D. Bae, K. Park, K. Kim, 2006 Symp. VLSI Technol. 2006 Dig. Tech. Pap, 2006, pp. 21–22.
[78] C.-H. Lee, K.-C. Park, K. Kim, Appl. Phys. Lett. 87 (2005), 073510.
[79] M. Alessandri, R. Piagge, S. Alberici, E. Bellandi, M. Caniatti, G. Ghidini, A. Modelli, G. Pavia, E. Ravizza, A. Sebastiani, C. Wiemer, S. Spiga, M. Fanciulli, V. Fiorentini, E. Cadelano, G.M. Lopez, ECS Trans. 1 (2006) 91–105.
[80] X. Zhao, D. Vanderbilt, Phys. Rev. B 65 (2002) 233106.
[81] J. Kang, E.-C. Lee, K.J. Chang, Phys. Rev. B 68 (2003), 054106.
[82] T.S. Böscke, J. Müller, D. Bräuhaus, U. Schröder, U. Böttger, Appl. Phys. Lett. 99 (2011) 102903.
[83] X. Zhao, D. Ceresoli, D. Vanderbilt, Phys. Rev. B 71 (2005), 085107.
[84] X. Zhao, D. Vanderbilt, Phys. Rev. B 65 (2002), 075105.
[85] M.H. Park, T. Schenk, C.M. Fancher, E.D. Grimley, C. Zhou, C. Richter, J.M. LeBeau, J.L. Jones, T. Mikolajick, U. Schroeder, J. Mater. Chem. C 5 (2017) 4677–4690.
[86] C.-K. Lee, E. Cho, H.-S. Lee, C.S. Hwang, S. Han, Phys. Rev. B 78 (2008), 012102.
[87] C. Wiemer, L. Lamagna, S. Baldovino, M. Perego, S. Schamm-Chardon, P.E. Coulon, O. Salicio, G. Congedo, S. Spiga, M. Fanciulli, Appl. Phys. Lett. 96 (2010) 182901.
[88] P.R. Chalker, M. Werner, S. Romani, R.J. Potter, K. Black, H.C. Aspinall, A.C. Jones, C.Z. Zhao, S. Taylor, P.N. Heys, Appl. Phys. Lett. 93 (2008) 182911.
[89] E. Cianci, A. Molle, A. Lamperti, C. Wiemer, S. Spiga, M. Fanciulli, ACS Appl. Mater. Interfaces 6 (2014) 3455–3461.
[90] D. Fischer, A. Kersch, Appl. Phys. Lett. 92 (2008), 012908.
[91] J. Niinistö, K. Kukli, T. Sajavaara, M. Ritala, M. Leskelä, L. Oberbeck, J. Sundqvist, U. Schröder, Electrochem. Solid St. 12 (2009) G1–G4.
[92] D. Tsoutsou, L. Lamagna, S.N. Volkos, A. Molle, S. Baldovino, S. Schamm, P.E. Coulon, M. Fanciulli, Appl. Phys. Lett. 94 (2009), 053504.
[93] S. Spiga, R. Rao, L. Lamagna, C. Wiemer, G. Congedo, A. Lamperti, A. Molle, M. Fanciulli, F. Palma, F. Irrera, J. Appl. Phys. 112 (2012), 014107.
[94] A. Lamperti, L. Lamagna, G. Congedo, S. Spiga, J. Electrochem. Soc. 158 (2011) G221–G226.
[95] M. Ritala, M. Leskelä, Appl. Surf. Sci. 75 (1994) 333–340.

[96] M. Ritala, M. Leskelä, L. Niinistö, T. Prohaska, G. Friedbacher, M. Grasserbauer, Thin Solid Films 250 (1994) 72–80.
[97] E. Bonera, G. Scarel, M. Fanciulli, J. Non-Cryst, Solids 322 (2003) 105–110.
[98] K. Kukli, M. Ritala, J. Aarik, T. Uustare, M. Leskelä, J. Appl. Phys. 92 (2002) 1833–1840.
[99] J. Aarik, A. Aidla, A.-A. Kiisler, T. Uustare, V. Sammelselg, Thin Solid Films 340 (1999) 110–116.
[100] K. Kukli, J. Aarik, M. Ritala, T. Uustare, T. Sajavaara, J. Lu, J. Sundqvist, A. Aidla, L. Pung, A. Hårsta, M. Leskelä, J. Appl. Phys. 96 (2004) 5298–5307.
[101] G. Scarel, S. Spiga, C. Wiemer, G. Tallarida, S. Ferrari, M. Fanciulli, Mater. Sci. Eng., B 109 (2004) 11–16.
[102] H.B. Park, M. Cho, J. Park, S.W. Lee, C.S. Hwang, J.-P. Kim, J.-H. Lee, N.-I. Lee, H.-K. Kang, J.-C. Lee, S.-J. Oh, J. Appl. Phys. 94 (2003) 3641–3647.
[103] D.M. Hausmann, E. Kim, J. Becker, R.G. Gordon, Chem. Mater. 14 (2002) 4350–4358.
[104] J. Swerts, N. Peys, L. Nyns, A. Delabie, A. Franquet, J.W. Maes, S.V. Elshocht, S.D. Gendt, J. Electrochem. Soc. 157 (2010) G26–G31.
[105] S.Y. Lee, H.K. Kim, J.H. Lee, I.-H. Yu, J.-H. Lee, C.S. Hwang, J. Mater, Chem. C 2 (2014) 2558–2568.
[106] M.-J. Choi, H.-H. Park, D.S. Jeong, J.H. Kim, J.-S. Kim, S.K. Kim, Appl. Surf. Sci. 301 (2014) 451–455.
[107] M. Berdova, X. Liu, C. Wiemer, A. Lamperti, G. Tallarida, E. Cianci, M. Fanciulli, S. Franssila, J. Vac. Sci. Technol. A 34 (2016), 051510.
[108] M. Berdova, C. Wiemer, A. Lamperti, G. Tallarida, E. Cianci, L. Lamagna, S. Losa, S. Rossini, R. Somaschini, S. Gioveni, M. Fanciulli, S. Franssila, Appl. Surf. Sci. 368 (2016) 470–476.
[109] C. Richter, T. Schenk, U. Schroeder, T. Mikolajick, J. Vac. Sci. Technol. A 32 (2013) 01A117.
[110] D.-Y. Cho, H.S. Jung, I.-H. Yu, J.H. Yoon, H.K. Kim, S.Y. Lee, S.H. Jeon, S. Han, J.H. Kim, T.J. Park, B.-G. Park, C.S. Hwang, Chem. Mater. 24 (2012) 3534–3543.
[111] W. Weinreich, L. Wilde, J. Müller, J. Sundqvist, E. Erben, J. Heitmann, M. Lemberger, A.J. Bauer, J. Vac. Sci. Technol. A 31 (2013) 01A119.
[112] W. Weinreich, T. Tauchnitz, P. Polakowski, M. Drescher, S. Riedel, J. Sundqvist, K. Seidel, M. Shirazi, S.D. Elliott, S. Ohsiek, E. Erben, B. Trui, J. Vac. Sci. Technol. A 31 (2012) 01A123.
[113] J.-H. Kim, V.A. Ignatova, J. Heitmann, L. Oberbeck, J. Phys. D Appl. Phys. 41 (2008) 172005.
[114] W. Weinreich, A. Shariq, K. Seidel, J. Sundqvist, A. Paskaleva, M. Lemberger, A.J. Bauer, J. Vac. Sci. Technol. B 31 (2012) 01A109.
[115] H.J. Cho, Y.D. Kim, D.S. Park, E. Lee, C.H. Park, J.S. Jang, K.B. Lee, H.W. Kim, Y.J. Ki, I.-K. Han, Y.W. Song, Solid-State Electron. 51 (2007) 1529–1533.
[116] D. Martin, M. Grube, W. Weinreich, J. Müller, W.M. Weber, U. Schröder, H. Riechert, T. Mikolajick, J. Appl. Phys. 113 (2013) 194103.
[117] M. Pešić, S. Knebel, M. Geyer, S. Schmelzer, U. Böttger, N. Kolomiiets, V.V. Afanas'ev, K. Cho, C. Jung, J. Chang, H. Lim, T. Mikolajick, U. Schroeder, J. Appl. Phys. 119 (2016), 064101.
[118] T. Onaya, T. Nabatame, T. Sawada, K. Kurishima, N. Sawamoto, A. Ohi, T. Chikyow, A. Ogura, Meet. Abstr. MA2016-02, 2016, p. 2022.
[119] C.L. Dezelah, J. Niinistö, K. Kukli, F. Munnik, J. Lu, M. Ritala, M. Leskelä, L. Niinistö, Chem. Vap. Depos. 14 (2008) 358–365.
[120] P.R. Fischer, D. Pierreux, O. Rouault, J. Sirugue, P.M. Zagwijn, E. Tois, S. Haukka, ECS Trans. 16 (2008) 135–148.

[121] N. Nikolaou, P. Dimitrakis, P. Normand, V. Ioannou-Sougleridis, K. Giannakopoulos, K. Mergia, K. Kukli, J. Niinistö, M. Ritala, M. Leskelä, Solid-State Electron. 68 (2012) 38–47.

[122] J. Niinistö, K. Kukli, A. Tamm, M. Putkonen, C.L. Dezelah, L. Niinistö, J. Lu, F. Song, P. Williams, P.N. Heys, M. Ritala, M. Leskelä, J. Mater. Chem. 18 (2008) 3385–3390.

[123] J.W. Elam, M.J. Pellin, S.D. Elliott, A. Zydor, M.C. Faia, J.T. Hupp, Appl. Phys. Lett. 91 (2007) 253123.

[124] K. Knapas, M. Ritala, Chem. Mater. 20 (2008) 5698–5705.

[125] A. Lamperti, E. Cianci, O. Salicio, L. Lamagna, S. Spiga, M. Fanciulli, Surf. Interface Anal. 45 (2013) 390–393.

[126] S. Spiga, F. Driussi, A. Lamperti, G. Congedo, O. Salicio, Appl. Phys. Express 5 (2012), 021102.

[127] S. Spiga, G. Congedo, U. Russo, A. Lamperti, O. Salicio, F. Driussi, E. Vianello, 2010 Proc. Eur. Solid State Device Res Conf., 2010, pp. 408–411.

[128] A. Lamperti, E. Cianci, R. Ciprian, D. Sangalli, A. Debernardi, Thin Solid Films 533 (2013) 83–87.

[129] C. Adelmann, V. Sriramkumar, S. Van Elshocht, P. Lehnen, T. Conard, S. De Gendt, Appl. Phys. Lett. 91 (2007) 162902.

[130] T.J. Park, J.H. Kim, J.H. Jang, C.-K. Lee, K.D. Na, S.Y. Lee, H.-S. Jung, M. Kim, S. Han, C.-S. Hwang, Chem. Mater. 22 (2010) 4175–4184.

[131] P.K. Park, S.-W. Kang, Appl. Phys. Lett. 89 (2006) 192905.

[132] C. Wiemer, L. Lamagna, M. Fanciulli, Semicond. Sci. Technol. 27 (2012), 074013.

[133] P. Majumder, G. Jursich, C. Takoudis, J. Appl. Phys. 105 (2009) 104106.

[134] T. Wang, J.G. Ekerdt, Chem. Mater. 21 (2009) 3096–3101.

[135] H.K. Kim, H.-S. Jung, J.H. Jang, J. Park, T.J. Park, S.-H. Lee, C.S. Hwang, J. Appl. Phys. 110 (2011) 114107.

[136] S. Spiga, F. Driussi, G. Congedo, C. Wiemer, A. Lamperti, E. Cianci, ACS Appl. Nano Mater. 1 (2018) 4633–4641.

[137] T. Wang, J.G. Ekerdt, Chem. Mater. 23 (7) (2011) 1679–1685.

[138] H.-W. You, W.-J. Cho, Appl. Phys. Lett. 96 (2010), 093506.

[139] G. Chen, Z. Huo, L. Jin, D. Zhang, S. Zhao, Y. Han, S. Liu, M. Liu, Semicond. Sci. Technol. 29 (2014), 045019.

[140] S. Maikap, H.Y. Lee, T.-Y. Wang, P.-J. Tzeng, C.C. Wang, L.S. Lee, K.C. Liu, J.-R. Yang, M.-J. Tsai, Semicond. Sci. Technol. 22 (2007) 884.

[141] Y.-N. Tan, W. Chim, B.J. Cho, W.-K. Choi, IEEE Trans. Electron Devices 51 (2004) 1143–1147.

[142] Y.N. Tan, W.K. Chim, W.K. Choi, M.S. Joo, B.J. Cho, IEEE Trans. Electron Devices 53 (2006) 654–662.

[143] P. Tsai, K. Chang-Liao, C. Liu, T. Wang, P.J. Tzeng, C.H. Lin, L.S. Lee, M. Tsai, IEEE Electron Device Lett. 29 (2008) 265–268.

[144] Z. Fan, J. Chen, J. Wang, J. Adv. Dielectr. 06 (2016) 1630003.

[145] S. Mueller, J. Mueller, A. Singh, S. Riedel, J. Sundqvist, U. Schroeder, T. Mikolajick, Adv. Funct. Mater. 22 (2012) 2412–2417.

[146] P. Polakowski, S. Riedel, W. Weinreich, M. Rudolf, J. Sundqvist, K. Seidel, J. Muller, 2014 IEEE 6th Int. Mem. Workshop IMW, 2014, pp. 1–4.

[147] S. Brivio, G. Tallarida, E. Cianci, S. Spiga, Nanotechnology 25 (2014) 385705.

[148] S. Brivio, J. Frascaroli, S. Spiga, Nanotechnology 28 (2017) 395202.

[149] J. Frascaroli, S. Brivio, F. Ferrarese Lupi, G. Seguini, L. Boarino, M. Perego, S. Spiga, ACS Nano 9 (2015) 2518–2529.

[150] H.-Y. Chen, S. Brivio, C.-C. Chang, J. Frascaroli, T.-H. Hou, B. Hudec, M. Liu, H. Lv, G. Molas, J. Sohn, S. Spiga, V.M. Teja, E. Vianello, H.-S.P. Wong, J. Electroceram. 39 (2017) 21–38.

[151] E. Covi, S. Brivio, A. Serb, T. Prodromakis, M. Fanciulli, S. Spiga, Front. Neurosci. 10 (2016).
[152] L. Goux, P. Czarnecki, Y.Y. Chen, L. Pantisano, X.P. Wang, R. Degraeve, B. Govoreanu, M. Jurczak, D.J. Wouters, L. Altimime, Appl. Phys. Lett. 97 (2010) 243509.
[153] Y.Y. Chen, L. Goux, J. Swerts, M. Toeller, C. Adelmann, J. Kittl, M. Jurczak, G. Groeseneken, D.J. Wouters, IEEE Electron Device Lett. 33 (2012) 483–485.
[154] Y.Y. Chen, G. Pourtois, S. Clima, L. Goux, B. Govoreanu, A. Fantini, R. Degreave, G.S. Kar, G. Groeseneken, D.J. Wouters, M. Jurczak, ECS Trans. 50 (2013) 3–9.
[155] G. Niu, H.-D. Kim, R. Roelofs, E. Perez, M.A. Schubert, P. Zaumseil, I. Costina, C. Wenger, Sci. Rep. 6 (2016) 28155.
[156] K.V. Egorov, R.V. Kirtaev, Y.Y. Lebedinskii, A.M. Markeev, Y.A. Matveyev, O.M. Orlov, A.V. Zablotskiy, A.V. Zenkevich, Phys. Status Solidi A 212 (2015) 809–816.
[157] S. Brivio, J. Frascaroli, S. Spiga, Appl. Phys. Lett. 107 (2015), 023504.
[158] A. Grossi, E. Perez, C. Zambelli, P. Olivo, E. Miranda, R. Roelofs, J. Woodruff, P. Raisanen, W. Li, M. Givens, I. Costina, M.A. Schubert, C. Wenger, Sci. Rep. 8 (2018) 11160.
[159] L. Goux, X.P. Wang, Y.Y. Chen, L. Pantisano, N. Jossart, B. Govoreanu, J.A. Kittl, M. Jurczak, L. Altimime, D.J. Wouters, Electrochem. Solid St. 14 (2011) H244–H246.
[160] M. Azzaz, A. Benoist, E. Vianello, D. Garbin, E. Jalaguier, C. Cagli, C. Charpin, S. Bernasconi, S. Jeannot, T. Dewolf, G. Audoit, C. Guedj, S. Denorme, P. Candelier, C. Fenouillet-Beranger, L. Perniola, Solid-State Electron. 125 (2016) 182–188.
[161] Y.Y. Chen, R. Degraeve, B. Govoreanu, S. Clima, L. Goux, A. Fantini, G.S. Kar, D.J. Wouters, G. Groeseneken, M. Jurczak, IEEE Electron Device Lett. 34 (2013) 626–628.
[162] Y.Y. Chen, B. Govoreanu, L. Goux, R. Degraeve, A. Fantini, G.S. Kar, D.J. Wouters, G. Groeseneken, J.A. Kittl, M. Jurczak, L. Altimime, IEEE Trans. Electron Devices 59 (2012) 3243–3249.
[163] S. Clima, K. Sankaran, Y.Y. Chen, A. Fantini, U. Celano, A. Belmonte, L. Zhang, L. Goux, B. Govoreanu, R. Degraeve, D.J. Wouters, M. Jurczak, W. Vandervorst, S.D. Gendt, G. Pourtois, Phys. Status Solidi RRL 8 (2014) 501–511.
[164] Z. Jiang, Z. Wang, X. Zheng, S. Fong, S. Qin, H. Chen, C. Ahn, J. Cao, Y. Nishi, H.-P. Wong, 2016 IEEE Int. Electron Devices Meet. IEDM, 2016, pp. 21.3.1–21.3.4.
[165] Y. Matveyev, R. Kirtaev, A. Fetisova, S. Zakharchenko, D. Negrov, A. Zenkevich, Nanoscale Res. Lett. 11 (2016) 147.
[166] S. Yu, H.-Y. Chen, B. Gao, J. Kang, H.-S.P. Wong, ACS Nano 7 (2013) 2320–2325.
[167] A. Markeev, A. Chouprik, K. Egorov, Y. Lebedinskii, A. Zenkevich, O. Orlov, Microelectron. Eng. 109 (2013) 342–345.
[168] J.Y. Seok, S.J. Song, J.H. Yoon, K.J. Yoon, T.H. Park, D.E. Kwon, H. Lim, G.H. Kim, D.S. Jeong, C.S. Hwang, Adv. Funct. Mater. 24 (2014) 5316–5339.
[169] B. Hudec, I.-T. Wang, W.-L. Lai, C.-C. Chang, P. Jančovič, K. Fröhlich, M. Mičušík, M. Omastová, T.-H. Hou, J. Phys. D Appl. Phys. 49 (2016) 215102.
[170] S.K. Kim, K.M. Kim, D.S. Jeong, W. Jeon, K.J. Yoon, C.S. Hwang, J. Mater. Res. 28 (2013) 313–325.
[171] K. Fröhlich, Mater. Sci. Semicond. Process. 16 (2013) 1186–1195.
[172] U. Diebold, Surf. Sci. Rep. 48 (2003) 53–229.
[173] J.-P. Niemelä, G. Marin, M. Karppinen, Semicond. Sci. Technol. 32 (2017), 093005.
[174] J. Aarik, A. Aidla, T. Uustare, M. Ritala, M. Leskelä, Appl. Surf. Sci. 161 (2000) 385–395.
[175] Q. Xie, Y.-L. Jiang, C. Detavernier, D. Deduytsche, R.L. Van Meirhaeghe, G.-P. Ru, B.-Z. Li, X.-P. Qu, J. Appl. Phys. 102 (2007), 083521.
[176] S.K. Kim, S.W. Lee, J.H. Han, B. Lee, S. Han, C.S. Hwang, Adv. Funct. Mater. 20 (2010) 2989–3003.
[177] M. Ritala, M. Leskela, L. Niinisto, P. Haussalo, Chem. Mater. 5 (1993) 1174–1181.

[178] A. Chaker, P.D. Szkutnik, J. Pointet, P. Gonon, C. Vallée, A. Bsiesy, J. Appl. Phys. 120 (2016), 085315.
[179] J. Pointet, P. Gonon, L. Latu-Romain, A. Bsiesy, C. Vallée, J. Vac. Sci. Technol. A 32 (2013) 01A120.
[180] G.T. Lim, D.-H. Kim, Thin Solid Films 498 (2006) 254–258.
[181] C. Jin, B. Liu, Z. Lei, J. Sun, Nanoscale Res. Lett. 10 (2015) 95.
[182] M. Reiners, K. Xu, N. Aslam, A. Devi, R. Waser, S. Hoffmann-Eifert, Chem. Mater. 25 (2013) 2934–2943.
[183] E.-L. Lakomaa, S. Haukka, T. Suntola, Appl. Surf. Sci. 60–61 (1992) 742–748.
[184] J. Aarik, A. Aidla, T. Uustare, V. Sammelselg, J. Cryst. Growth 148 (1995) 268–275.
[185] W.-J. Lee, M.-H. Hon, J. Phys. Chem. C 114 (2010) 6917–6921.
[186] J. Aarik, A. Aidla, T. Uustare, K. Kukli, V. Sammelselg, M. Ritala, M. Leskelä, Appl. Surf. Sci. 193 (2002) 277–286.
[187] V. Pore, T. Kivelä, M. Ritala, M. Leskelä, Dalton Trans. 0 (2008) 6467–6474.
[188] J. Aarik, B. Hudec, K. Hušeková, R. Rammula, A. Kasikov, T. Arroval, T. Uustare, K. Fröhlich, Semicond. Sci. Technol. 27 (2012), 074007.
[189] W.D. Kim, G.W. Hwang, O.S. Kwon, S.K. Kim, M. Cho, D.S. Jeong, S.W. Lee, M.H. Seo, C.S. Hwang, Y.-S. Min, Y.J. Cho, J. Electrochem. Soc. 152 (2005) C552–C559.
[190] J. Aarik, T. Arroval, L. Aarik, R. Rammula, A. Kasikov, H. Mändar, B. Hudec, K. Hušeková, K. Fröhlich, J. Cryst. Growth 382 (2013) 61–66.
[191] T. Arroval, L. Aarik, R. Rammula, H. Mändar, J. Aarik, B. Hudec, K. Hušeková, K. Fröhlich, Phys. Status Solidi A 211 (2014) 425–432.
[192] A.P. Alekhin, S.A. Gudkova, A.M. Markeev, A.S. Mitiaev, A.A. Sigarev, V.F. Toknova, Appl. Surf. Sci. 257 (2010) 186–191.
[193] C.J. Cho, J.-Y. Kang, W.C. Lee, S.-H. Baek, J.-S. Kim, C.S. Hwang, S.K. Kim, Chem. Mater. 29 (2017) 2046–2054.
[194] S.K. Kim, G.J. Choi, J.H. Kim, C.S. Hwang, Chem. Mater. 20 (2008) 3723–3727.
[195] S.K. Kim, G.-J. Choi, S.Y. Lee, M. Seo, S.W. Lee, J.H. Han, H.-S. Ahn, S. Han, C.S. Hwang, Adv. Mater. 20 (2008) 1429–1435.
[196] D. Acharyya, A. Hazra, P. Bhattacharyya, Microelectron. Reliab. 54 (2014) 541–560.
[197] D.-H. Kwon, K.M. Kim, J.H. Jang, J.M. Jeon, M.H. Lee, G.H. Kim, X.-S. Li, G.-S. Park, B. Lee, S. Han, M. Kim, C.S. Hwang, Nat. Nanotechnol. 5 (2010) 148–153.
[198] K.M. Kim, S. Han, C.S. Hwang, Nanotechnology 23 (2012), 035201.
[199] B.J. Choi, D.S. Jeong, S.K. Kim, C. Rohde, S. Choi, J.H. Oh, H.J. Kim, C.S. Hwang, K. Szot, R. Waser, B. Reichenberg, S. Tiedke, J. Appl. Phys. 98 (2005), 033715.
[200] H.Y. Jeong, Y.I. Kim, J.Y. Lee, S.-Y. Choi, Nanotechnology 21 (2010) 115203.
[201] H.Y. Jeong, J.Y. Lee, S.-Y. Choi, Appl. Phys. Lett. 97 (2010), 042109.
[202] W. Zhang, J.-Z. Kong, Z.-Y. Cao, A.-D. Li, L.-G. Wang, L. Zhu, X. Li, Y.-Q. Cao, D. Wu, Nanoscale Res. Lett. 12 (2017) 393.
[203] B. Govoreanu, D. Crotti, S. Subhechha, L. Zhang, Y.Y. Chen, S. Clima, V. Paraschiv, H. Hody, C. Adelmann, M. Popovici, O. Richard, M. Jurczak, 2015 Symp. VLSI Technol. VLSI Technol, 2015, pp. T132–T133.
[204] B. Govoreanu, A. Redolfi, L. Zhang, C. Adelmann, M. Popovici, S. Clima, H. Hody, V. Paraschiv, I.P. Radu, A. Franquet, J. Liu, J. Swerts, O. Richard, H. Bender, L. Altimime, M. Jurczak, 2013 IEEE Int. Electron Devices Meet, 2013, pp. 10.2.1–10.2.4.
[205] M. Popovici, J. Swerts, K. Tomida, D. Radisic, M.-S. Kim, B. Kaczer, O. Richard, H. Bender, A. Delabie, A. Moussa, C. Vrancken, K. Opsomer, A. Franquet, M.A. Pawlak, M. Schaekers, L. Altimime, S.V. Elshocht, J.A. Kittl, Phys. Status Solidi RRL 5 (2011) 19–21.
[206] M.-J. Lee, C.B. Lee, D. Lee, S.R. Lee, M. Chang, J.H. Hur, Y.-B. Kim, C.-J. Kim, D.H. Seo, S. Seo, U.-I. Chung, I.-K. Yoo, K. Kim, Nat. Mater. 10 (2011) 625–630.
[207] J.J. Yang, M.-X. Zhang, J.P. Strachan, F. Miao, M.D. Pickett, R.D. Kelley, G. Medeiros-Ribeiro, R.S. Williams, Appl. Phys. Lett. 97 (2010) 232102.

[208] A. Prakash, D. Jana, S. Maikap, Nanoscale Res. Lett. 8 (2013) 418.
[209] C.W. Hill, G.J. Derderian, G. Sandhu, J. Electrochem. Soc. 152 (2005) G386–G390.
[210] C. Adelmann, A. Delabie, B. Schepers, L.N.J. Rodriguez, A. Franquet, T. Conard, K. Opsomer, I. Vaesen, A. Moussa, G. Pourtois, K. Pierloot, M. Caymax, S. Van Elshocht, Chem. Vap. Depos. 18 (2012) 225–238.
[211] K. Kukli, M. Ritala, R. Matero, M. Leskelä, J. Cryst. Growth 212 (2000) 459–468.
[212] K. Kukli, J. Aarik, A. Aidla, O. Kohan, T. Uustare, V. Sammelselg, Thin Solid Films 260 (1995) 135–142.
[213] K. Kukli, M. Ritala, M. Leskelä, J. Electrochem. Soc. 142 (1995) 1670–1675.
[214] M.-K. Kim, W.-H. Kim, T. Lee, H. Kim, Thin Solid Films 542 (2013) 71–75.
[215] W.J. Maeng, S.-J. Park, H. Kim, J. Vac. Sci. Technol. B 24 (2006) 2276–2281.
[216] T. Blanquart, V. Longo, J. Niinistö, M. Heikkilä, K. Kukli, M. Ritala, M. Leskelä, Semicond. Sci. Technol. 27 (2012), 074003.
[217] J. Aarik, A. Aidla, K. Kukli, T. Uustare, J. Cryst. Growth 144 (1994) 116–119.
[218] K.V. Egorov, Y.Y. Lebedinskii, A.M. Markeev, O.M. Orlov, Appl. Surf. Sci. 356 (2015) 454–459.
[219] K.V. Egorov, D.S. Kuzmichev, P.S. Chizhov, Y.Y. Lebedinskii, C.S. Hwang, A.M. Markeev, ACS Appl. Mater. Interfaces 9 (2017) 13286–13292.
[220] K.V. Egorov, D.S. Kuzmichev, A.A. Sigarev, D.I. Myakota, S.S. Zarubin, P.S. Chizov, T.-V. Perevalov, V.A. Gritsenko, C.S. Hwang, A.M. Markeev, J. Mater. Chem. C 6 (2018) 9667–9674.
[221] S. Seo, M.J. Lee, D.H. Seo, E.J. Jeoung, D.-S. Suh, Y.S. Joung, I.K. Yoo, I.R. Hwang, S.H. Kim, I.S. Byun, J.-S. Kim, J.S. Choi, B.H. Park, Appl. Phys. Lett. 85 (2004) 5655–5657.
[222] M. Utriainen, M. Kröger-Laukkanen, L. Niinistö, Mater. Sci. Eng., B 54 (1998) 98–103.
[223] M. Utriainen, M. Kröger-Laukkanen, L.-S. Johansson, L. Niinistö, Appl. Surf. Sci. 157 (2000) 151–158.
[224] H. Seim, H. Mölsä, M. Nieminen, H. Fjellvåg, L. Niinistö, J. Mater. Chem. 7 (1997) 449–454.
[225] E. Lindahl, J. Lu, M. Ottosson, J.-O. Carlsson, J. Cryst. Growth 311 (2009) 4082–4088.
[226] E. Lindahl, M. Ottosson, J.-O. Carlsson, Chem. Vap. Depos. 15 (2009) 186–191.
[227] E. Lindahl, M. Ottosson, J.-O. Carlsson, Surf. Coat. Technol. 205 (2010) 710–716.
[228] A. Lamperti, S. Spiga, H.L. Lu, C. Wiemer, M. Perego, E. Cianci, M. Alia, M. Fanciulli, Microelectron. Eng. 85 (2008) 2425–2429.
[229] H.L. Lu, G. Scarel, C. Wiemer, M. Perego, S. Spiga, M. Fanciulli, G. Pavia, J. Electrochem. Soc. 155 (2008) H807–H811.
[230] M. Daub, M. Knez, U. Goesele, K. Nielsch, J. Appl. Phys. 101 (2007) 09J111.
[231] H.L. Lu, G. Scarel, X.L. Li, M. Fanciulli, J. Cryst. Growth 310 (2008) 5464–5468.
[232] S.J. Song, S.W. Lee, G.H. Kim, J.Y. Seok, K.J. Yoon, J.H. Yoon, C.S. Hwang, J. Gatineau, C. Ko, Chem. Mater. 24 (2012) 4675–4685.
[233] M.-H. Ko, B. Shong, J.-H. Hwang, Ceram. Int. 44 (2018) 16342–16351.
[234] P.A. Premkumar, M. Toeller, C. Adelmann, J. Meersschaut, A. Franquet, O. Richard, H. Tielens, B. Brijs, A. Moussa, T. Conard, H. Bender, M. Schaekers, J.A. Kittl, M. Jurczak, S.V. Elshocht, Chem. Vap. Depos. 18 (2012) 61–69.
[235] T.S. Yang, W. Cho, M. Kim, K.-S. An, T.-M. Chung, C.G. Kim, Y. Kim, J. Vac. Sci. Technol. A 23 (2005) 1238–1243.
[236] Y.-H. You, B.-S. So, J.-H. Hwang, W. Cho, S.S. Lee, T.-M. Chung, C.G. Kim, K.-S. An, Appl. Phys. Lett. 89 (2006) 222105.
[237] S. Spiga, A. Lamperti, C. Wiemer, M. Perego, E. Cianci, G. Tallarida, H.L. Lu, M. Alia, F.-G. Volpe, M. Fanciulli, Microelectron. Eng. 85 (2008) 2414–2419.
[238] S. Spiga, A. Lamperti, E. Cianci, F.G. Volpe, M. Fanciulli, ECS Trans. 25 (2009) 411–425.
[239] D. Ielmini, S. Spiga, F. Nardi, C. Cagli, A. Lamperti, E. Cianci, M. Fanciulli, J. Appl. Phys. 109 (2011), 034506.

[240] D. Deleruyelle, C. Dumas, M. Carmona, C. Muller, S. Spiga, M. Fanciulli, Appl. Phys. Express 4 (2011), 051101.
[241] F. Nardi, D. Deleruyelle, S. Spiga, C. Muller, B. Bouteille, D. Ielmini, J. Appl. Phys. 112 (2012), 064310.
[242] H. Yaegashi, K. Oyama, A. Hara, S. Natori, S. Yamauchi, Adv. Resist Mater. Process. Technol. XXIX, International Society for Optics and Photonics, 2012, p. 83250B.
[243] J.W. Klaus, A.W. Ott, J.M. Johnson, S.M. George, Appl. Phys. Lett. 70 (1997) 1092–1094.
[244] S.-W. Lee, K. Park, B. Han, S.-H. Son, S.-K. Rha, C.-O. Park, W.-J. Lee, Electrochem. Solid St. 11 (2008) G23–G26.
[245] S. Kamiyama, T. Miura, Y. Nara, Thin Solid Films 515 (2006) 1517–1521.
[246] F. Hirose, Y. Kinoshita, S. Shibuya, Y. Narita, Y. Takahashi, H. Miya, K. Hirahara, Y. Kimura, M. Niwano, Thin Solid Films 519 (2010) 270–275.
[247] S. Ahn, Y. Kim, S. Kang, K. Im, H. Lim, J. Vac. Sci. Technol. A 35 (2016) 01B131.
[248] A. Kobayashi, N. Tsuji, A. Fukazawa, N. Kobayashi, Thin Solid Films 520 (2012) 3994–3998.
[249] S. Won, S. Suh, M.S. Huh, H.J. Kim, IEEE Electron Device Lett. 31 (2010) 857–859.
[250] G. Dingemans, C.A.A. van Helvoirt, D. Pierreux, W. Keuning, W.M.M. Kessels, J. Electrochem. Soc. 159 (2012) H277–H285.
[251] Y.-J. Choi, S.-M. Bae, J.-H. Kim, E.-H. Kim, H.-S. Hwang, J.-W. Park, H. Yang, E. Choi, J.-H. Hwang, Ceram. Int. 44 (2018) 1556–1565.
[252] A. Mehonic, A.L. Shluger, D. Gao, I. Valov, E. Miranda, D. Ielmini, A. Bricalli, E. Ambrosi, C. Li, J.J. Yang, Q. Xia, A.J. Kenyon, Adv. Mater. 30 (2018) 1801187.
[253] A. Bricalli, E. Ambrosi, M. Laudato, M. Maestro, R. Rodriguez, D. Ielmini, IEEE Trans. Electron Devices 65 (2018) 115–121.

CHAPTER

7

Nano-composite MOx materials for NVMs

C. Bonafos[a], L. Khomenkhova[b,c], F. Gourbilleau[b], E. Talbot[d], A. Slaoui[e], M. Carrada[a], S. Schamm-Chardon[a], P. Dimitrakis[f], and P. Normand[f]

[a]CEMES-CNRS, University of Toulouse, CNRS, Toulouse, France, [b]CIMAP Normandie University, ENSICAEN, UNICAEN, CEA, CNRS, Caen, France, [c]V. Lashkaryov Institute of Semiconductor Physics of National Academy of Sciences of Ukraine, Ukraine and National University "Kyiv-Mohyla Academy", Kyiv, Ukraine, [d]Normandie Univ, UNIROUEN, INSA Rouen, CNRS, Groupe de Physique des Matériaux, Rouen, France, [e]ICube, CNRS and University of Strasbourg, Strasbourg, France, [f]National Center for Scientific Research "Demokritos", Institute of Nanoscience and Nanotechnology, Attiki, Greece

7.1 Introduction

Conventional floating-gate flash memories continue to dominate the market for non-volatile memories (NVMs) even if suffering from serious scaling issues to follow the growing demand for increased information storage density, fast data transfer speed and low-voltage operation. These issues originate mainly from the scalability limitation of the tunnel dielectric (usually made of silicon dioxide) and the floating-gate interference effect between adjacent cells [1]. Several NVM approaches based on different concepts and/or technologies have been proposed as potential solutions to flash downscaling limitations. Among these approaches, the alternative of replacing the conventional floating gate (poly-Si layer) by

Metal Oxides for Non-volatile Memory
https://doi.org/10.1016/B978-0-12-814629-3.00007-6

laterally isolated floating nodes in the form of nanoparticles has retained considerable attention in the early 2000s. This floating gate concept has led to the emergence of the so-called nanocrystal memories (NCMs) which have the potential of operating at lower voltages and higher speeds compared to the conventional NVMs despite a low gate to NCs capacitive coupling. Basically, a NCM cell consists of a metal-oxide-semiconductor field effect transistor (MOSFET) with mono-disperse nanometer-scale crystals embedded within the gate dielectric. It associates the finite-size effects of NCs and the benefits (robustness and fault-tolerance) of a stored charge distribution. The floating gate made of NCs is sandwiched by a tunnel oxide (separating the trapping nodes and the channel device) and an inter-poly dielectric (separating the nodes and the gate). The last decade, significant advances have been made in NC fabrication (see, e.g., Refs. [2–5]) and prototype NCM-based products for low-power microcontroller applications have even been demonstrated [6]. However, the NCM technologies developed to date still face the concern of producing high-density of uniformly distributed size-homogeneous NCs and hence, cannot avoid fluctuations in device performance and fail to exploit size-dependence effects [2,6–8]. While more research is needed to overcome these limitations and other concerns related to the dielectric materials and device architecture [6,7], the NCM approach still competes with other charge storage (e.g., SONOS) [4] and non-charge storage (e.g., FeRAM, MRAM, PCRAM, ReRAM) NVMs alternatives [9,10] as a mid-term solution to Flash scaling.

Historically, NC memories using silicon nanocrystals (Si-NCs) embedded in SiO_2 have been first introduced and extensively studied [4,11–14]. Then, a lot of attention has been put on Ge-NCs [15–17], for which very good data retention has been obtained for holes because of the higher valence-band offset. More recently, metallic dots embedded in SiO_2 have been proposed as a way of obtaining a higher conduction-band offset [18–20]. Despite attractive results, this latter approach suffers from limitations in the thermal budget applied after dot fabrication, which may induce metal and oxide reactions or/and metal diffusion toward the silicon substrate and thereby, degrade the device performance.

For all these systems (Si, Ge or metallic NCs embedded in thin SiO_2 layers), charge transfer from the substrate to the NCs across a SiO_2 layer of thickness in the 3–5 nm range (thickness required for insuring 10-year data storage) is relatively slow and charging times typically above the μs range are necessary to obtain a functional memory window. In addition, the erasing and programming voltages remain at too high figures (10–14 V) for a number of applications (e.g., mobile phone, MP3 player, laptop, etc.) requiring low-voltage operation. This is rather a universal result, which does not depend on the NCs fabrication technique. Several solutions based on energy band engineering have therefore been examined for achieving NCMs with lower programming voltages and improved data retention. A promising alternative for increasing the performance of NCMs lays in dielectric engineering and especially in the use of NCs embedded in high permittivity (high-κ) gate dielectrics usually made of metal oxides (HfO_2 and Hf oxide based compounds, ZrO_2 and Zr oxide based compounds, Al_2O_3 etc.) or, but it is more prospective, rare earth oxides (as La_2O_3).

These high-κ gate dielectric films are widely studied as gate oxides in CMOS devices to suppress the unacceptable gate leakage current when the traditional SiO_2 gate oxide becomes ultrathin. In particular, after a decade-long search the semiconductor industry has converged on Hf-based oxides, such as HfO_2 or $HfSi_xO_y$, for the first generation CMOS products featuring high-κ gate dielectrics and metal gate electrodes [21]. The high-κ gate stack is made of a multi-layer structure, which includes, beside the deposited high-κ film, an amorphous SiO_2 layer at the interface with the Si substrate. This interfacial layer forms during routine high-κ film deposition due to the excess of oxygen needed to minimize the formation of oxygen vacancies in the case of metal–organic chemical vapor deposition (MOCVD), and the alternating H_2O cycles commonly used in atomic layer deposition (ALD) processing.

High-κ gate dielectrics can bring strong added value to NVMs, as exploiting the higher current densities that can be obtained at low voltages (thus requiring lower programming voltages) as well as the larger physical thickness of these materials, a feature that should lead to improved data retention time. In addition, using a metal oxide high-κ dielectric as inter-poly dielectric (IPD) material could reinforce the gate-coupling ratio. However, most high-k IPD materials with high Gate Coupling Ratio value generally exhibit a low band gap, which is not suitable for flash memory because its susceptibility to the gate injection [10]. Among the different metal oxides, HfO_2 and ZrO_2 present both a high dielectric constant (25 against 3.9 for silica) and a large band gap (>6 eV) meanwhile Al_2O_3 exhibits a moderate dielectric constant but a large bandgap (>8 eV against 9 eV for SiO_2). In addition, HfO_2 and Al_2O_3 have the advantage to be much less hygroscopic than most of the other high-κ layers and in particular comparing to rare earth oxides, hence leading to low moisture absorption reaction and penetration. This is of outmost importance for the stability of associated memory devices involving such layers on one hand and for the synthesis of pure semiconducting (Si, Ge) or metallic (except for Au) NCs embedded in these metal oxide matrices on the other hand.

While recent years have seen a decrease in research activity on NC memories, most of the task forces being focused on alternative solutions as ReRAM and PCRAM, the study of NVM with NCs as charge storage nodes involving metal oxide dielectrics (mainly HfO_2 and associated alloys or Al_2O_3) as tunnel oxide and/or inter-poly oxide is still active.

Most of them concern metal NPs as Au [22,23], Ag [24] or Pd [25] or metal oxide NCs as Gd_2O_3 [26], ZnO [27]. Finally, only a few recent papers concern Ge or Si quantum dots in metal oxide high-κ dielectrics for NVM [28,29].

Promising device results using metallic or semiconducting NCs involving HfO_2 or HfAlO dielectrics as tunnel and inter-poly oxides or as tunnel oxide only have been presented in [25,30–33]. Nevertheless, the attainable programming times remain in the 10–100 ms range and long retention times for NVM applications have not been demonstrated yet. In addition, these studies have failed to demonstrate that the device performances are clearly linked to the increase of the conduction-band offset. In such dielectric materials, deep traps are often at the origin of electron trapping due to a lack of control of the interfacial layer with silicon and there is still some room for their optimization.

In this chapter, we concentrate on Si- and Ge-NCs floating gate memories with thin metal oxide high-κ dielectric stacks as HfO_2 (and its associated silicate) and Al_2O_3. There are many technological approaches used for the production of high-κ metal oxide dielectric films including or not NCs, such as ALD [34,35], different chemical vapor deposition (CVD) techniques [36–38], sol-gel process [39,40], and pulsed laser deposition [41,42]. Despite its flexibility, only few investigations have explored the potentialities of radio-frequency magnetron sputtering (RF-MS) [29,43,44]. Recently the simplicity and utility of the RF-MS approach has been demonstrated for the production of fully Hf-based nanocrystal memory structures with large memory windows and enhanced dielectric properties in one deposition run [43]. Among the different technological routes explored the last few years for generating NCs in the gate oxide of MOS devices, two major techniques have been utilized, namely deposition in vacuum like low-pressure chemical vapor deposition (LPCVD, see, e.g., Refs. [5,6] and references therein), and ion beam synthesis (IBS) [45,46]. The latter technique has received substantial attention due to its flexibility, manufacturing advantages and full compatibility with standard CMOS technology. In particular, ultra-low energy IBS (ULE-IBS) has been proposed for the controlled synthesis of single planes of Si-NCs embedded in very thin (5–10 nm) oxide layers [47–49]. More recently, combined with nanometer sized block copolymer masks, ULE-IBS allowed the spatial arrangement of Si-NCs in a hexagonal network of 20 nm wide nanovolumes at controlled depth within 16 nm thick SiO_2 films [50].

In this chapter, we present a digest of the main materials science aspects of the controlled fabrication of 2D arrays of semi-conducting NCs (Si, Ge) in metal oxide dielectric layers by using the two above mentioned techniques, namely magnetron-sputtering and ULE-IBS. In particular, the involved phase separation processes as well as the main problems related to humidity penetration or controlled oxidation processes are among the discussed topics. The state of the art of different tools for structural characterization is also described. Finally, the associated electrical characterization is presented in order to evaluate the potential of such nanocomposite metal oxide layers for NC memory applications. It is worth to mention here that the same nanocomposite oxide layers have been recently used as active layers in RRAM memories. In these alternative NVM, the presence of NCs is shown to improve the variability of the switching cycles ([51] and references therein, [24,52–56]). Hence, all the advances and breakthroughs in the controlled synthesis of nanocomposite metal oxide layers also benefits to this more current application.

7.2 Experimental

7.2.1 Fabrication routes

7.2.1.1 Synthesis of nanocrystals embedded in metal oxide dielectrics by magnetron-sputtering

Magnetron sputtering (MS) belongs to the physical vapor deposition process. It is highly versatile and fully compatible to established CMOS technologies. Sputtering sources are equipped with magnetrons that utilize strong electric and magnetic fields to confine charged plasma particles close to the surface of the sputter target. In a magnetic field, electrons follow helical paths around magnetic field lines, undergoing more ionizing collisions with gaseous neutrals near the target surface. The sputter gas is typically argon. The sputtered atoms are neutrally charged and so are unaffected by the magnetic trap. Charge build-up on insulating targets can be avoided with the use of *RF magnetron sputtering* where the sign of the anode-cathode bias is varied at a high rate (commonly 13.56 MHz). RF-MS works well to produce highly insulating oxide films.

The main advantages of RF-MS are as follows: (i) low plasma impedance and thus high discharge currents from 1 to 100 A (depending on cathode length) at typical voltages around 500 V; (ii) deposition rates in the range from 1 to 10 nm/s; (iii) low thermal load to the substrate coating uniformity in the range of a few percent even for several meters long cathodes; (iv) easy to scale up; (v) dense and well adherent coatings; (vi) large variety of film materials available (nearly all metals, compounds, dielectrics); (vii) broadly tunable film properties. There are a lot of modifications of this method. However, all of them can be gathered in two main approaches (Fig. 7.1): (i) sputtering when the film is grown from the sputtering of single target (either simple or composite target); (ii) co-sputtering when the film is grown via simultaneous sputtering of several targets.

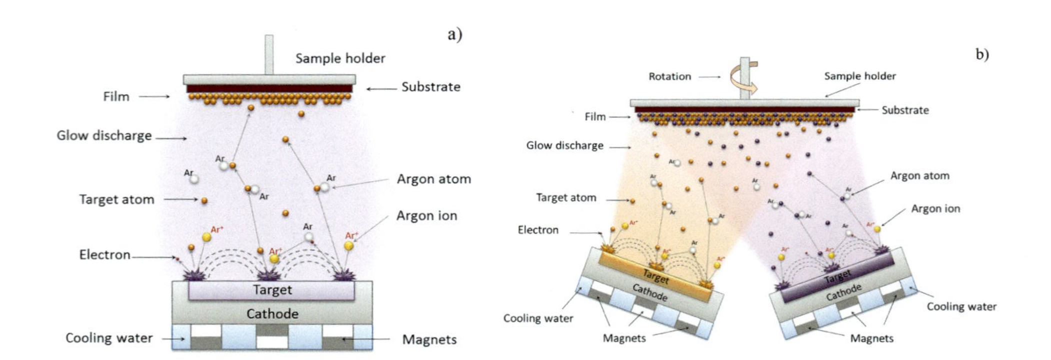

FIG. 7.1 A schematic illustration of magnetron sputtering from single cathode *(left image)* and confocal co-sputtering from two spaced-apart cathodes *(right image)*.

In the case of direct sputtering from single cathode, deposition can be performed on non-rotated substrate, but in this case the physical size of the target should exceed the size of the substrate (about 30%). Co-sputtering, being a more flexible approach, is usually performed on the rotated substrate to achieve homogeneous films. The deposition can be performed on the large substrate which size exceeds the sizes of the cathodes. Both techniques allow growing of both single layers and multi-layer structures. The deposition can be performed in the plasma stimulated by Ar ions only (so-called, standard deposition/sputtering) or in the mixture of Ar and N_2 (or O_2, or H_2) (reactive sputtering). The combination of both approaches permits a fine control of chemical composition and film microstructure to be achieved.

Nowadays, MS and RF-MS are used for the deposition of metallic, semiconductor and dielectric films. Besides, the combination of different targets in one deposition tool allows films with unique properties to be produced. For instance, this approach was used to grow fully high-κ based structure with embedded Si- or Ge-NCs [57,58].

It should be noted that most attention was paid to the fabrication of Si- and Ge-NCs in SiO_2 or Si_3N_4 matrices. It was shown that NCs formation is a two-step process. At the first step, the Si(Ge)-rich-SiO_2 (Si_3N_4) supersaturated solution is produced. The second step involves high-temperature annealing (usually, at 1050–1200 °C for Si-rich materials and 600–800 °C for Ge-rich ones in nitrogen) to favor NCs nucleation via phase separation process followed by the nuclei crystallization. The size distribution and spatial location of the Si-and Ge-NCs depend on both excess Si (or Ge) content and annealing conditions (temperature, duration, and environment).

The formation of supersaturated Si-rich solution by RF-MS can be achieved in several ways. For instance, for Si-SiO_2 systems this can be done: (i) by the sputtering of composed Si-SiO_2 target which is usually a SiO_2 target topped with calibrated Si chips; (ii) by the sputtering of pure SiO_2 target in mixed argon-hydrogen plasma, and (iii) by co-sputtering of pure Si and pure SiO_2 targets.

Despite the fact that RF-assisted-formation of Si or Ge-NCs in metal oxide films such as HfO_2 or ZrO_2 has not been often addressed, similar approaches as described here above can be used for this purpose. In this case, the Si (Ge) content in the films can be tuned via: (i) the variation of the amount of Si(Ge) chips topped on target surface; (ii) the partial pressure of Ar and H_2 gases and (iii) the RF power density (RFP) applied to the cathode(s).

Formation of Si and Ge-NCs in HfO_2 host via the phase separation process was recently reported. It was shown that non-stoichiometric $HfSiO_x$ or $HfGeO_x$ film is formed by the sputtering of HfO_2 (99.99%) target topped with calibrated Si(Ge) chips placed on the electron "race track." The Si(Ge) content can be tuned via variation of number of Si(Ge) chips, power applied on the cathode as well as by the sputtering of the target in mixed Ar-H_2 plasma. The use of hydrogen allows the variation of Si(Ge) content due to the ability of hydrogen to remove the oxygen from plasma and consequently, to reach higher Si(Ge) excess in the layers. A conventional furnace annealing in inert atmosphere at 900–1100 °C or 600–800 °C was used to form Si or Ge-NCs, respectively. The lower annealing temperatures for Ge-rich films are governed by the lower crystallization temperature of pure Ge (550–600 °C) as well as by the formation of volatile GeO (at about 850–900 °C).

The formation of NCs with uniform mean size and spherical shape is usually required. This demand can be hardly achieved for the composite layers when possibility of the coalescence of small NCs into large ones exists due to atom diffusion in the 3-D directions. In such case a multilayer approach is usually used. Formation of Ge-NCs in metal oxide host was successfully achieved as reported in Ref. [29,57,58].

7.2.1.2 Synthesis of nanocrystals embedded in metal oxide dielectrics by ULE-IBS

Ion implantation is material engineering technique used as ages for semiconductor doping. This technique has been diverted since two decades from its initial use to synthesize novel materials in the host matrix by implanting atoms at concentrations higher than the solid solubility level. In the traditional Ion Beam Synthesis (IBS) technique, controlled amounts of selected ions are implanted with energies of few tens or hundreds of keV (the choice in energy depends on the implanted species and the targeted application) into a dielectric matrix, which is subsequently annealed at a temperature sufficient for phase separation to occur. In general, such a process leads to a depth-distributed population of NCs centered at a depth close to the projected range of the implanted ions. As a rule of thumb, impurity concentrations larger than few at.%, at the projected range, are required to form isolated NCs through nucleation and growth phenomena, while concentrations of more than 20 at.% results in networks of connected particles obtained through spinodal decomposition of the supersaturated matrix [59].

There are, however, a couple of characteristics the NCs/matrix should possess. Firstly, the impurity should not be miscible in the matrix. If the matrix has to keep its intrinsic properties, the solid solubility limit of the implanted impurities should be as small as possible. In addition, the matrix should show only marginal chemical reactivity with the implanted species, so that the precipitation of a pure second phase is possible. In other words, all the reactions allowing the formation of the NCs phase should have a negative Gibbs free energy [60]. Finally, a low diffusivity of the implanted species at room temperature, although not mandatory, insures the final sizes and densities of NCs to be controlled only by the thermal treatment following the implantation step.

Phenomenologically, IBS can be divided into several stages, described below.

- *Accumulation and supersaturation:* after ions have been slowed down by electronic and nuclear stopping to thermal energy, they are incorporated into the target material at a depth, which depends on their acceleration energy. Supersaturation occurs when the concentration of implanted atoms exceeds the solubility limit.
- *Nucleation or early phase separation:* in the case of high dose implantation, nucleation of NPs can be encountered during implantation and/or during a post-implantation heat treatment. For a given ion-substrate system, the primary nucleation stage can roughly be classified with respect to the diffusion length L_{diff} of impurity monomers during implantation. Different scenarios can be envisaged by comparing L_{diff} with the two following characteristics distances as described in Ref. [61]:
  - R^*: the radius of the hypothetical sphere defined by $4/3\pi R^{*3}\overline{c}(\phi) \leq \overline{i^*}$ where $\overline{c}(\phi)$ denotes the averaged local concentration after a fluence ϕ has been implanted and i^* the number of atoms in the critical nucleus for nucleation.
  - ΔR_p corresponds to the width of the implanted profile. The ions species are indeed distributed into the matrix according to a Gaussian law around the mean projected range R_p with a mean square deviation given by ΔR_p.

(i) $L_{diff} < R^*$: This condition states that thermal fluctuation does not initiate nucleation during the implantation process. A thermal treatment is necessary for phase separation. This is the case for Si or Ge implantation in insulators (SiO_2 or in metal oxides) at room temperature [47,49]. Upon annealing thermal fluctuations could initiate phase separation through the (homogeneous) nucleation of precipitates.

(ii) $L_{diff} = \Delta R_p > R^*$: Nucleation starts during implantation when the monomer concentration exceeds the nucleation threshold. Due to the implantation profile, the first stable precipitates appear close to R_p. Since the concentration increase is proportional to the implanted profile, a larger supersaturation is reached in the center of the profile than in the tails. Accordingly, the critical NP size is supposed to increase from the center to the tails. In the course of implantation, the nucleation region spreads outside, whereas behind the nucleation front, the growth stage sets in. In this case, the NPs are formed essentially across the whole deposition range with a density roughly proportional to the impurity profile.

(iii) $L_{diff} >> \Delta R_p > R^*$: due to the high mobility of the impurities, during the ongoing implantation those deposited in the tails of the distribution either diffuse away from the implantation region or diffuse into the center, where the already formed NCs act as sinks. Hence, the resulting monomer fluxes prevent spreading of the nucleation region across the profile, because in the tails the critical monomer concentration will not be exceeded. Therefore, the nucleation region is restricted close to R_p.

(iv) The two last above-mentioned situations are usually met for metal implantation into insulators [62,63]. These processes are named "Ion Beam Direct Synthesis." In the case of low energy ion implantation, ΔR_p is narrow and the nucleation region is restricted to R_p (for both low diffusing species after post implantation annealing and for high diffusing species during the implantation process itself). Hence, while the fabrication of NCs by the "conventional" (i.e., at conventional ion energy) IBS technique results in the formation of a population of depth-distributed NCs, the ULE-IBS technique allows the fabrication of a single plane of NCs located at a fixed nanometer distance from the surface (see Fig. 7.2) that can nucleate either during the ion implantation process for high diffusion species or after further annealing at high temperature for low diffusing species.

In the results described in this chapter, the thin metal oxide layers that have been used as matrices for NCs implantation have been fabricated either by MOCVD for the Hf-based layers or by ALD for the Al_2O_3 layers. Si and Ge have been implanted at ultra-low energy (i. e. 1–5 keV) and high doses (some 10^{15} at/cm^2) and further annealed at high temperature (700–1050 °C) for some seconds to some minutes.

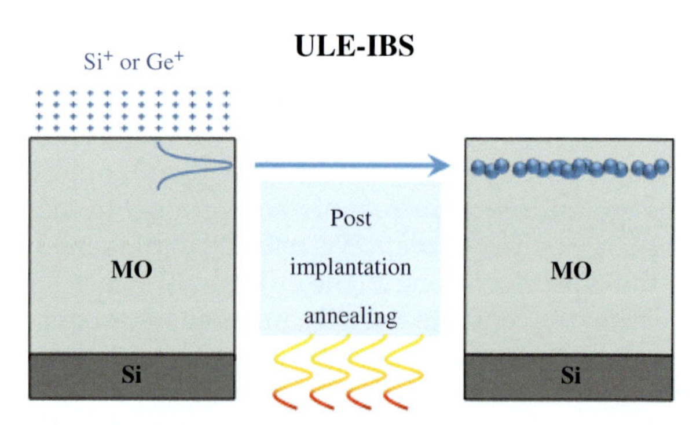

FIG. 7.2 Steps for the ion beam synthesis of Si and Ge nanocrystals in metal oxide layers by Ultra Low Energy Ion Implantation.

7.2.2 Characterization techniques

7.2.2.1 Transmission electron microscopy

The structural characterization of the nanocomposite oxide layer is crucial for optimizing the electrical properties of the produced structures. The relevant parameters are the distance of the NPs array to the electrodes, the NPs size distribution and inter-spacing, as well as the NPs and matrix crystallinity and surface occupancy. In particular, the NP density is a key factor as it would determine the number of charges that can be stored by the device and/or the number of charges stored per NP. Time-of-flight secondary ion mass spectrometry (TOF-SIMS) has been applied to detect the presence of Si-NPs in thin SiO_2 layers [64] but, transmission electron microscopy (TEM) techniques are yet the most powerful ones [65].

Nevertheless, evidencing and characterizing the formation of Si- and Ge-NCs embedded in metal oxide matrices (HfO_2, ZrO_2 and associated silicates, Al_2O_3 etc.) can be tricky. In classical bright field conditions, the Si (or Ge) rich regions show different contrasts with respect to the matrix. This type of imaging is sometimes used in the literature to claim the presence of Si- (or Ge-) NPs embedded in high-κ matrices but this is not completely reliable because Si oxide (or Ge oxide) as well would give similar contrast. These matrices are crystalline as most often also the NCs and the Si (Ge) interatomic distances are very close to those of the matrix so they cannot be easily discriminated by using HREM or electron diffraction, even if sometimes possible.

Chemical imaging techniques based on electron energy loss spectroscopy (EELS) as Energy Filtered Transmission Electron Microscopy are usually the most powerful techniques for imaging Si- and Ge-NCs embedded in dielectric matrices as silica or silicon nitride [65]. EELS analysis is performed in the low-energy-loss domain where plasmons, corresponding to plasma oscillations of valence electrons associated with each phase of the analyzed area, are the major signatures. Unfortunately, the contrast in EFTEM images obtained by filtering around the Si (or Ge) plasmon is poor when the NCs are embedded in Hf- or Zr-oxide based compounds because, contrary to the SiO_2 case, there is too much overlap between the Si (Ge) plasmon and the matrix low loss signature.

Scanning Transmission Electron Microscopy-High Angle Annular Dark Field (STEM-HHADF) imaging coupled to STEM-EDX analysis can also be used to image Si- or Ge-rich zones in a high-κ matrix [29] but one will not be able to discriminate between Si and SiO_2 or Ge and GeO_2.

Finally, evidencing presence of Si (or Ge)-rich domains in crystalline metal oxide matrices needs coupling HREM imaging to accurate chemical analysis by STEM-EELS in the core-loss region (edges), as it will be emphasized in the results presented hereafter. We will show that coupling

HREM imaging to accurate chemical analysis by STEM-EELS in the core-loss region (edges) is a relevant way to evidence the presence of Si (or Ge)-rich domains in crystalline metal oxide matrices.

For the studies described in this chapter, microscopy imaging was typically performed using a field emission TEM, FEI Tecnai™ F20 microscope operating at 200 kV, equipped with a spherical aberration corrector dedicated for high quality HREM images with an increased signal/noise ratio and nearly no delocalization effect at surfaces and interfaces. EFTEM imaging in the low-energy-loss domain was conducted with a TRIDIEM Gatan imaging filter attached to the microscope and coupled to a scanning stage for the STEM-EELS analyses. TEM lamellas transparent to electrons were prepared by mechanical grinding and Ar^+ ion milling.

7.2.2.2 Atom probe tomography

Atom Probe Tomography (APT) is a promising approach for materials' characterization when three-dimension reconstruction with atomic scale resolution and quantitative chemical characterization are required [66–69]. The implementation of femtosecond laser pulses [70] allowed enlarging the variety of materials to be studied, i.e., photonic materials, solar cells, magnetic semiconductor and nanoelectronic devices [71–77].

For the APT analysis, part of the sample is prepared as a tiny needle shape, using a focused ion beam annular milling procedure. To prevent the layer of interest from Ga damages and/or amorphization during sample processing, a 300 nm thick Pt layer was first deposited on the top of the sample. Then, films were ion-milled into sharp tips with an apex radius close to 30 nm. A low accelerating voltage (2 kV) was used for the final stage in order to avoid Ga implantation and sample amorphization. The APT apparatus used in this work is the CAMECA laser assisted Wide-Angle Tomographic Atom Probe (LaWaTAP). The experiments were performed with samples cooled down to 80 K, with a vacuum of $(2–3) \times 10^{-10}$ mbar in the analysis chamber and with ultra-violet ($\lambda = 343$ nm) femtosecond (350 fs) laser pulses. The laser energy was fixed at 50 nJ/pulse focused onto a $< 0.01\ mm^2$ spot.

To identify the clusters, the procedure described hereafter was applied. Each step of this identification consists in (a) the placement of a sphere (sampling volume) over one atom of the volume investigated and (b) the estimation of a local composition of these selected elements by counting atoms within this sphere. If the composition exceeds a given threshold, the atom at the center of the sphere is associated to a cluster. If the composition is lower than the threshold, the atom at the center of the sphere is belonging to the matrix. Then the sphere is moved to the next atom and this procedure is applied again to estimate the composition and to compare it with the threshold value. This approach was used for all the atoms of the volume to identify those belonging either to the clusters or to the matrix. We will show that this approach is valuable to identify phase separation in a HfSiOx matrix with formation of Si-NCs.

7.2.2.3 XRD

X-ray diffraction (XRD) is a powerful nondestructive technique for characterizing crystalline materials. It provides information on crystal structure, crystalline phase (polymorphs), preferred crystal orientation (texture), and other structural parameters, such as average grain size, crystallinity, strain, and crystal defects. XRD peaks are produced by constructive interference of a monochromatic beam of X-rays diffracted at specific angles from each set of lattice planes in a sample. The peak intensities are determined by the distribution of atoms within the lattice. Consequently, the X-ray diffraction pattern is the fingerprint of the periodic atomic arrangements in a given material. A search of the ICDD (International Centre for Diffraction Data) database of X-ray diffraction patterns enables the phase identification of a large variety of crystalline samples. The main applications of XRD analysis are: (a) identification/quantification of crystalline phase, (b) measurement of average crystallite size, strain, or micro-strain effects in bulk materials and thin films, (c) quantification of preferred orientation (texture) in thin films, multi-layer stacks, and manufactured parts, (d) determination of the ratio of crystalline to amorphous material in bulk materials and thin films.

Another XRD-related technique is X-ray reflectivity (XRR) that is widely used for the characterization of thin-films and multilayer structures. X-ray scattering at very small diffraction angles allows characterization of the electron density profiles of thin films down to a few nanometers thick. Using a simulation of the reflectivity pattern, coupled with layer thickness measurements by TEM, a highly accurate measurement of interface roughness, and layer density for crystalline and amorphous thin films and multilayers can be obtained. Besides, it allows fine control of the quality of multilayers, especially, annealed ones in terms of inter-diffusion and lost of periodicity.

The XRD and XRR analysis presented hereafter was performed using a Phillips XPERT HPD Pro device with Cu Kα radiation (0.154 nm) at a fixed grazing angle incidence of 0.5°. An asymmetric grazing geometry was chosen for the analysis of thin films to increase the volume of material interacting with X-ray beam, as well as to eliminate the contribution from an underlying substrate.

7.2.2.4 FTIR

This method is an effective analytical technique for quickly identifying the "chemical family" of a material. FTIR measures the absorbance of infrared light by a sample and generates a spectrum based on the functional groups in the material.

Typically, organic and polymeric compounds produce a "fingerprint" IR spectrum, which can be compared to extensive reference database and thereby, the unknown component's chemical family or actual identity may be determined. In the case of inorganic materials, FTIR spectra are used to a lesser degree. At the same time, they allow the information about microstructure transformation on composite materials sensitive to IR excitation to be also obtained in a quick and non-destructive way.

A noticeable example of the utility of FTIR method is the estimation of excess Si content in Si-rich-SiO_2 and Si-rich-Si_3N_4 composite materials, as well as the investigation of their microstructural evolution with annealing followed by the formation of Si nanocrystals [78,79].

The main principle of this method is based on the comparison of the parameters (such as peak position, full-width, and intensity) of a specific vibration band of the composite material with those of the pure dielectric host. This technique works well for the films thicker than 30 nm. In the case of ultrathin films, Attenuated Total Reflection (ATR-FTIR) technique is used. It holds an important place in the samples' characterization, allowing elimination of the contribution of an underlying substrate that is important for the films thinner 10 nm. However, referenced ATR-FTIR data for most high-κ materials are not so numerous [43,80]. Hereafter both techniques will be used for the analysis of samples properties. FTIR spectra were measured in the range of 460–4000 cm^{-1} by means of a Spectrum BX FTIR spectrometer (PerkinElmer Inc.) and a Nicolet Nexus FTIR spectrometer. The spectra were recorded in "transmission" mode at normal or Brewster (65°) incidence of excited light, using both atmospheric and Si substrate corrections. ATR-FTIR spectra were measured in the range 600–4000 cm^{-1} by means of a 60° Ge Smart Ark accessory inserted in a Nicolet Nexus spectrometer.

7.2.2.5 Electrical characterization

The memory properties of gate dielectrics with embedded NCs can be easily accessed using test MOS devices in the form of capacitors or field-effect transistors (FETs). These devices allow a fast evaluation of charge-storage characteristics such as programming/erasing (P/E) voltage and speed, endurance to P/E cycling and charge retention. In addition to the above, testing based on capacitor structures is helpful for extracting important basic parameters (e.g., breakdown voltage, interface state density, flat-band voltage hysteresis, leakage currents) and understanding the charging/discharging mechanisms,

7.2.3 Synthesis of nanocrystals embedded in metal oxides by radio frequency magnetron sputtering

7.2.3.1 Synthesis of Ge-NCs in HfO_2 host

This section reports on the synthesis and structural properties of Ge-NCs in HfO_2 host. The formation of such NCs in HfO_2-based films grown by RF-MS can be achieved via phase separation in $HfGeO_x$ solid solution. Hereafter, the properties of single $HfGeO_x$ films are considered as a function of the applied thermal treatment and their implementation in trilayer structures such as HfO_2/HfGeO/HfO_2 (Type A) and $HfSiO_x$/HfGeO/$HfSiO_x$ (Type B) are demonstrated. The schematic presentation of the structures is given in Figs. 7.3 and 7.4, respectively.

HfO_2 layers were sputtered from pure HfO_2 (99.9%) target, whereas $HfGeO_x$ and $HfSiO_x$ layers were grown from HfO_2 target topped by Ge or Si chips with a surface ratio of $R_{Ge} = R_{Si} = 12\%$. All the depositions were performed at 100 °C in pure argon plasma. More details about preparation of the films and structures can be found elsewhere [43,44,72,81,82]. The samples were annealed in a conventional furnace at different temperatures, $T_A = 500–1000\,°C$, during $t_A = 10–60\,min$ under a continuous nitrogen flow.

The evolution of structure properties with annealing was investigated by means of X-ray diffraction (XRD), Fourier Transmission Infrared spectroscopy (FTIR) and Raman scattering methods, as well as by high-resolution transmission electron microscopy (HREM). More details about these techniques are given here above.

To study the electrical properties of the structures, metal-insulator-semiconductor (MIS) capacitors with top electrodes of 150 μm in diameter

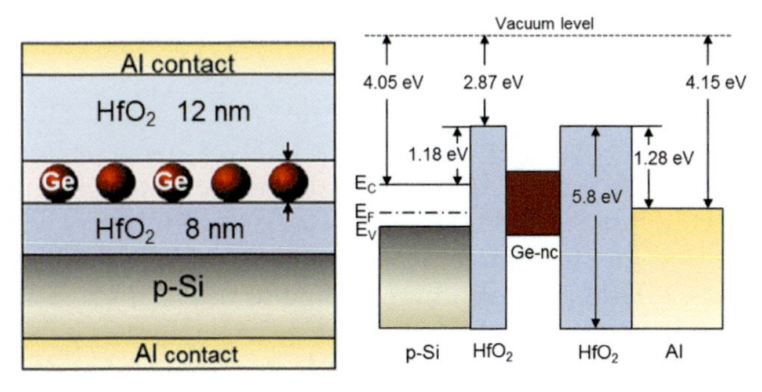

FIG. 7.3 Schematic presentation of the type A structures and corresponding flat band gap diagram.

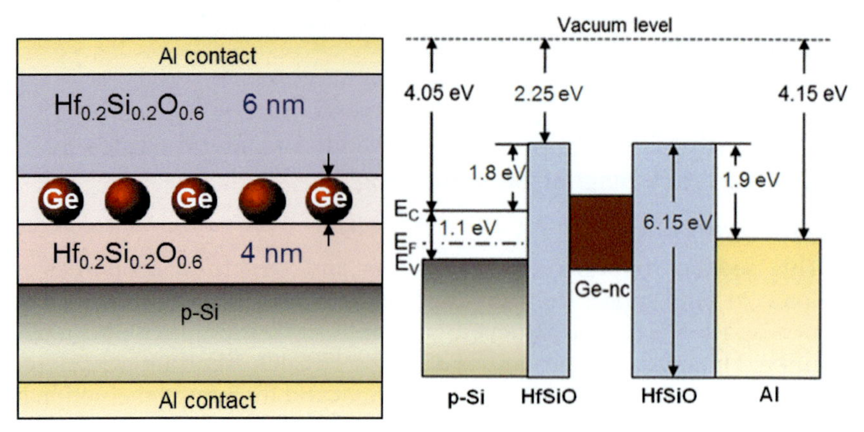

FIG. 7.4 Schematic presentation of the type B structures and corresponding flat band gap diagram.

were fabricated by Al thermal evaporation through a shadow mask. Al rear-side contact was formed after scratching the back surface of the samples. Following Al deposition, the MIS structures were annealed at 400 °C for 20 min in forming gas. In some cases, prior to metallization, an additional annealing in forming gas at 400 °C for 60 min was also performed to passivate the dangling bonds, if any. The electrical properties of the samples were studied through capacitance-voltage (C-V) measurements at different frequencies in the 10 kHz–1 MHz range using a HP 4294A LCR meter.

Properties of $HfGeO_x$ layers versus annealing treatment

The evolution of film microstructure was investigated by means of FTIR spectroscopy. The spectra recorded under Brewster angle incidence of excitation light for these films are presented in Fig. 7.5. It is worth to point that tabled FTIR data for such non-stoichiometric materials are not numerous. For the IR band identification the literature data proposed for HfO_2, GeO_2 and $HfGeO_4$ [58,83–87] were used.

FTIR spectra of as-deposited and annealed at $T_A < 800$ °C $HfGeO_x$ films (Fig. 7.5A) present a broad band in the 460–1000 cm^{-1} spectral range with a shoulder at about 820 cm^{-1}. This latter is due to Ge incorporation in a hafnia host. This IR band is shifted toward lower wavenumbers with respect to that observed for stoichiometric Ge-O-Ge one at 870 cm^{-1}. This can be attributed to Ge-Hf ion substitution leading to the formation of Ge-O-Hf bonds [58,87].

An annealing at $T_A < 600$ °C did not lead to a prominent microstructural transformation of $HfGeO_x$ films as it is demonstrated by the coincidence of the FTIR spectra for samples annealed at $T_A = 500$ °C (Fig. 7.5A). This is an

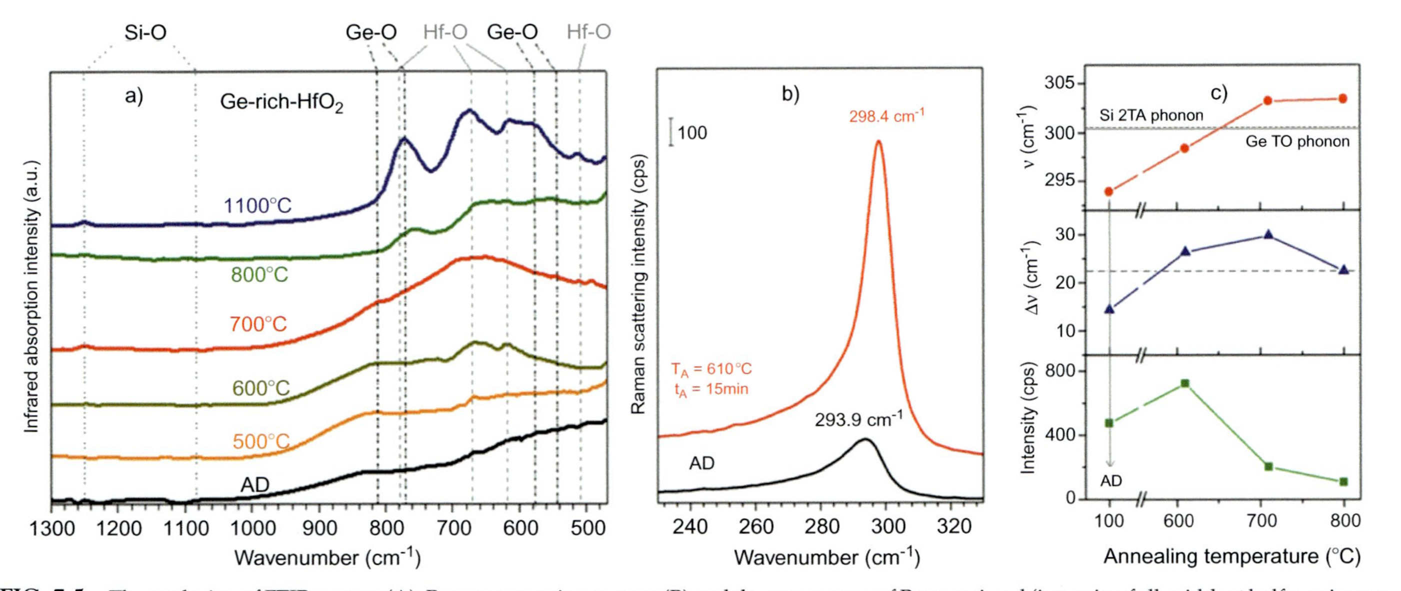

FIG. 7.5 The evolution of FTIR spectra (A), Raman scattering spectra (B) and the parameters of Raman signal (intensity, full-width at half maximum and spectral position) (C) of $HfGeO_x$ film with annealing temperature. FTIR spectra were recorded under Brewster angle incidence of excitation light. Annealing time is 15 min. *From L. Khomenkova, X. Portier, M. Carrada, C. Bonafos, B. S. Sahu, A. Slaoui, F. Gourbilleau, Ge-doped Hafnia-based dielectrics for non-volatile memory applications, ECS Trans. 45(3) (2012) 331–344.*

evidence of chemical and microstructural stability of the films. Further T_A increase up to 700°C results in the increase of the intensity of the 820 cm^{-1} vibration band as well as in the appearance and increase in intensity of the 680 cm^{-1} one. The latter is due to Hf—O bonds. Since the composition of the films remains unchangeable, one can conclude that a phase separation appears in the films upon annealing.

FTIR spectra of the samples annealed at $T_A = 700$°C show an intensity increase of the 680 cm^{-1} vibration band, whereas the shoulder at 820 cm^{-1} is also visible (Fig. 7.5A). Since the first band is due to Hf—O bonds, its increased contribution to FTIR spectra means a significant phase separation in $HfGeO_x$ films. Comparison with pure HfO_2 films annealed at similar conditions reveals that the HfO_2 phase is rather tetragonal (or cubic) than monoclinic.

Annealing at higher T_A induces the appearance of a 780 cm^{-1} band in $HfGeO_x$ samples (Fig. 7.5A). This band can be assigned to both monoclinic HfO_2 and pure GeO_2 phases. For the latter, the presence of the 770 cm^{-1} vibration band was demonstrated for the GeO_2 materials enriched in oxygen impurities [87]. Moreover, for $T_A = 1100$°C, the 770–780 cm^{-1} band is due to monoclinic HfO_2 phase (Fig. 7.5A).

Information on the phase separation process in $HfGeO_x$ films can be obtained based on thermodynamic considerations. The Gibbs free energy variations for the formation of GeO_2 and HfO_2 phases are $\Delta G_{GeO_2} = -111.8$ kcal/mol and $\Delta G_{HfO_2} = -260.1$ kcal/mol at 300 K, respectively. Thus, the formation of HfO_2 is preferable. It was found also that solid reaction of $GeO_2 + Hf \rightarrow HfO_2 + Ge$ has negative energy difference, and, consequently, the mixture of pure Ge and HfO_2 has the lower Gibbs free energy. In this regard, annealing of $HfGeO_x$ films has to result in Hf oxidation, and Ge atoms agglomeration into NCs in the HfO_2 matrix. One can conclude that phase separation in $HfGeO_x$ films at moderate temperatures occurs via formation of Ge-NCs embedded in HfO_2 matrix.

To confirm the formation of Ge-NCs, Raman scattering spectra were recorded. The asymmetric and broad first-order Ge phonon band peaked at about 293.9 cm^{-1} with full-width at half-maximum (FWHM) $\Delta\nu = 18$ cm^{-1} was observed for the as-deposited samples. An annealing treatment at $T_A = 600$°C leads to a "blue" shift of the Raman signal up to 298.4 cm^{-1} accompanied by an increase of the signal intensity and the $\Delta\nu$ decrease (Fig. 7.5B and C). Further T_A increase results in a further "blue" shift of the peak position up to 303.4 cm^{-1} and its broadening, as well as a significant decrease of its intensity (Fig. 7.4C). The dramatic decrease of the Raman signal for $T_A = 800$°C (Fig. 7.5B) that can be explained by the Ge out-diffusion via GeO formation.

The estimation of Ge-NCs sizes from the $\Delta\nu$ values using the methods presented in Refs. [88, 89] showed that the samples annealed at $T_A = 600$°C contain Ge-NCs with sizes of about 5 nm ($\Delta\nu = 14$ cm^{-1}). However, Raman peak position, observed at $298.4\,cm^{-1}$, corresponds to Ge-NCs size of about 18 nm. This discrepancy suggests the presence of compressive stresses for Ge-NCs in the HfO_2 matrix.

Evolution of trilayer structures with annealing treatment

The analysis of structural properties of trilayer stacks was performed by means of XRD. The identification of all XRD peaks was performed using the data of different HfO_2 crystalline phases, $HfSiO_4$, GeO_2 and pure Ge. The XRD data for tetragonal (t-HfO_2) and monoclinic (m-HfO_2) pure HfO_2 phases deposited by RF-MS shown in Fig. 7.6 are in good agreement with the standard XRD data (Fig. 7.6).

The XRD patterns for as-deposited structures are characterized by a broad band width in the range of $2\Theta = 27$–$37°$ (Fig. 7.6) that gives evidence of the amorphous nature of these structures. Annealing results in the appearance of some XRD peaks. For type A samples annealed at 600 °C, peaks at $2\Theta = 28.5°$ and $34.2°$ emerge (Fig. 7.6). The XRD peak at $2\Theta = 30.6°$ is due to the $HfGeO_x$ layer in which t-HfO_2 phase is formed upon annealing. The highest intensity of this XRD peak is an additional argument for the presence of t-HfO_2 in the $HfGeO_x$ layer.

XRD peaks observed at $2\Theta = 28.5°$ and $34.2°$ for type A samples annealed at $T_A = 600$ °C can be due to m-HfO_2 phase. Further increasing of T_A favors more prominent crystallization of m-HfO_2 phase that was evidenced by its significant contribution in XRD patterns for $T_A = 800$ °C (not shown here) [90].

It is interesting to note that the XRD pattern of type B samples annealed at 600 °C (Fig. 7.6A, curve 4) is similar to that observed for the as-deposited one (Fig. 7.6A, curve 3), pointing to a good stability of the structure upon such annealing. When type B samples were annealed at 800 °C, the formation of t-HfO_2 phase was observed (Fig. 7.6A, curve 5). It can be assumed that this phase is located mainly in the $HfGeO_x$ layer. Such a statement is based on the fact the $HfSiO_x$ layers remain amorphous at $T_A = 800$ °C as it was proved by TEM study for similar films [44]. Thus the formation of a tetragonal HfO_2 phase originates from the $HfGeO_x$ layer due to phase separation process.

To confirm the formation of Ge-NCs due to phase separation in $HfGeO_x$ layers—type A sample annealed at 615 °C for 15 min was analyzed by TEM and EELS methods. High-resolution TEM observations of cross-sections (Fig. 7.7) showed the bright regions corresponding to Ge-NCs in $HfGeO_x$ layer. The HfO_2 phase was found to be tetragonal as confirmed by the electron diffraction pattern (Fig. 7.7A, inset). Besides, the formation of a SiO_x interfacial layer (IL) upon annealing was detected. This can affect the electrical properties of the structures via fast trap formation. Furthermore, an appearance of oxygen vacancies in the nearest tetragonal HfO_2

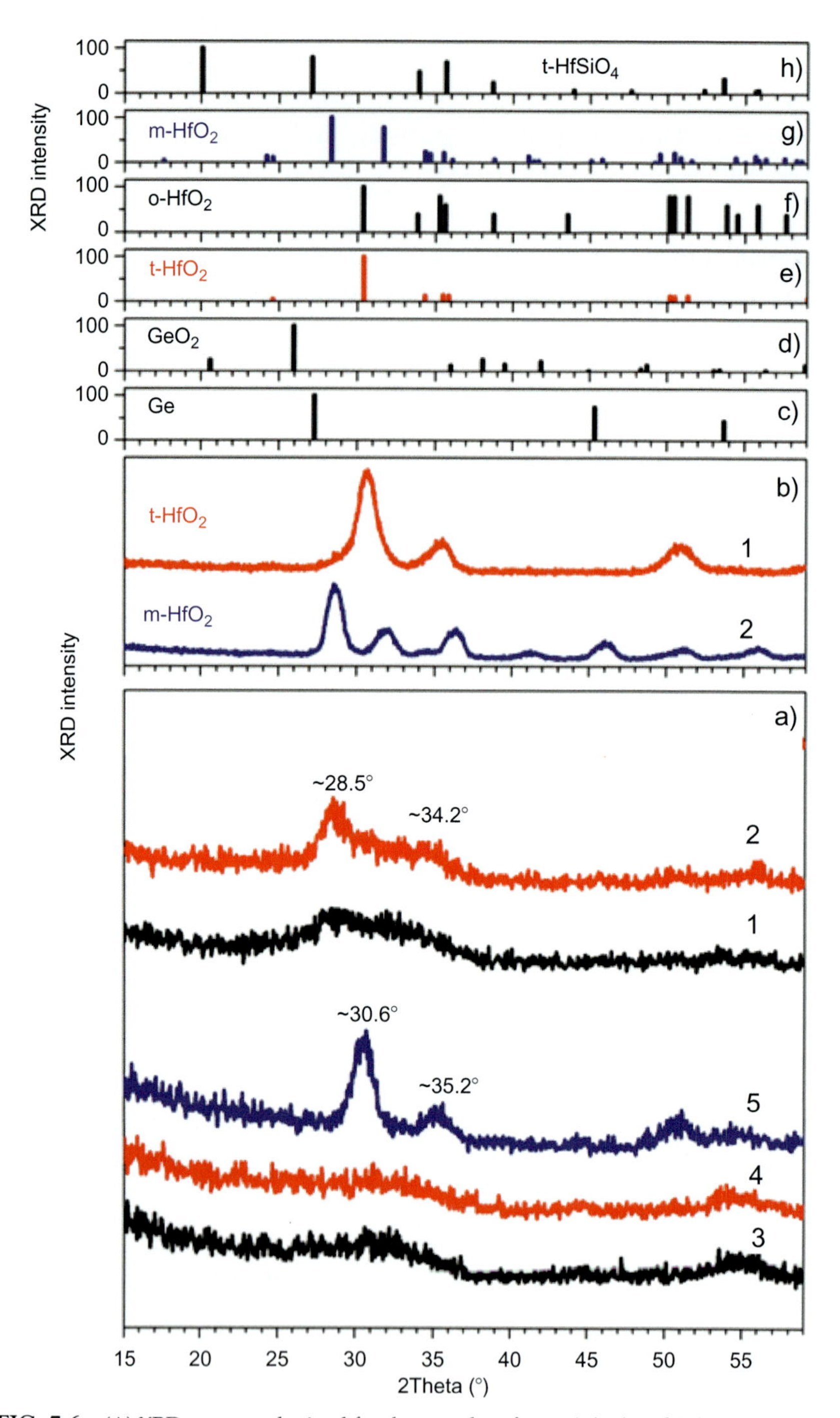

FIG. 7.6 (A) XRD patterns obtained for the samples of type A (1,2) and B (3–5): as deposited (1,3), annealed at 600 °C (2,4) and at 800 °C (5); (B) XRD patterns of monoclinic (1) and tetragonal (2) pure HfO_2 phases; (C–H) standard XRD data for pure Ge (c), GeO_2 (d), tetragonal HfO_2 (E), orthorhombic HfO_2 (F), monoclinic HfO_2 (G), $HfSiO_4$ (H) phases. The XRD patterns in (A) and (B) are shifted for clarity. *From L. Khomenkova, X. Portier, M. Carrada, C. Bonafos, B. S. Sahu, A. Slaoui, F. Gourbilleau, Ge-doped Hafnia-based dielectrics for non-volatile memory applications, ECS Trans. 45(3) (2012) 331–344.*

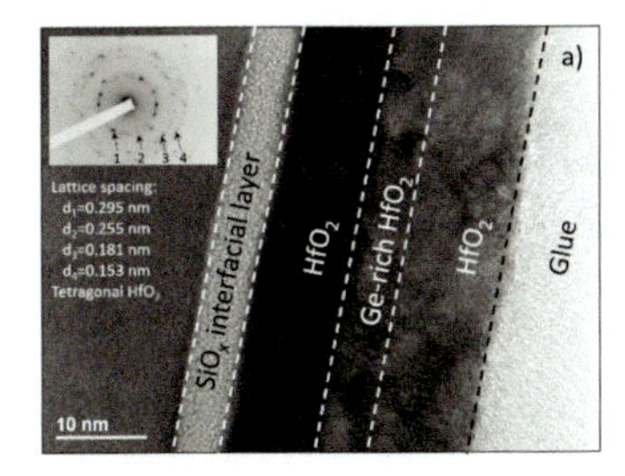

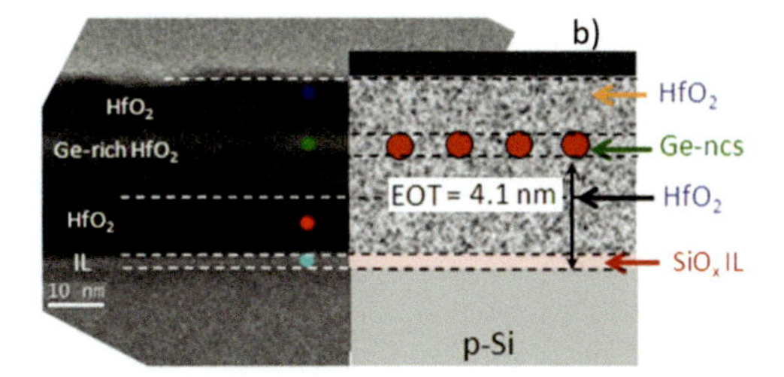

FIG. 7.7 (A) Cross-section TEM image of structure annealed at 615°C during 15min (B); Zero-loss cross-section image and schematic presentation of $HfO_2/HfGeO_x/HfO_2$ (type A sample). *Taken from L. Khomenkova, X. Portier, M. Carrada, C. Bonafos, B. S. Sahu, A. Slaoui, F. Gourbilleau, Ge-doped Hafnia-based dielectrics for non-volatile memory applications, ECS Trans. 45(3) (2012) 331–344.*

layer required for the stabilization of tetragonal phase can also affect electrical properties of the structure.

The EELS analysis was performed for different points of the sample cross-section (Fig. 7.7B). It was confirmed that tunnel and control layers are pure HfO_2. The analysis in the core-loss L23 ionization edge of Ge revealed that the NCs are enriched in germanium and appear in the $HfGeO_x$ layer only [90]. The estimated sizes of Ge-NCs are 4–6nm (Fig. 7.7B) that are in the agreement with Raman scattering data. No Ge diffusion toward the interface between tunnel HfO_2 layer and Si substrate was observed. Furthermore, Ge-NCs situated close to the interface between Ge-rich HfO_2 and control HfO_2 layers that increases effective thickness of tunnel oxide layer. In this regard, the EOT can be considered as mentioned in the Fig. 7.7B.

Electrical properties of the structures

The charge storage properties of all sets have also been investigated versus annealing treatment. Earlier we demonstrated that MIS capacitors fabricated from single pure HfO_2 or $HfSiO_x$ layers did not show any C-V hysteresis loop. Moreover, memory effect was not observed for the MIS capacitors with the as-deposited trilayer structures. The memory window from the samples investigated in the present work appears only upon thermal treatment and this means that all the charge storage properties are due to the $HfGeO_x$ layer evolution.

The $Al/HfO_2/HfGeO_x/HfO_2$/p-Si/Al capacitors (type-A samples) exhibit counter clockwise hysteresis loop (Fig. 7.8A). The memory window (ΔV_{fb}), estimated from flat-band voltage values, shows the largest value for $T_A = 610$°C. As it was shown above, the Ge-NCs were formed in the $HfGeO_x$ layers (Fig. 7.6).

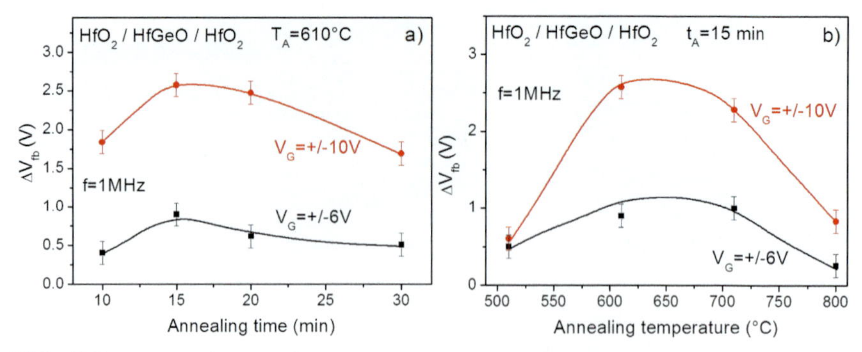

FIG. 7.8 The evolution of $\Delta V_{fb}(V_G)$ versus annealing time (A) and temperature (B) presented for $V_G = \pm 6$ and ± 10 V. The annealing temperature for (A) is $T_A = 610$ °C and the annealing time for (B) is $t_A = 15$ min. The measurement frequency is f = 1 MHz for both cases.

An additional annealing in the forming gas at 400 °C was performed for the set-A samples. This treatment allowed a significant increase of memory window (Fig. 7.9). The C-V hysteresis loop behavior is still counter clockwise, indicating charge trapping in the capacitor through a charge injection mechanism from the substrate (Fig. 7.9). While there is indication of electron charging and trapping under application of positive gate voltages, C-V hysteresis is mainly due to injection of holes from the substrate and trapping into the gate dielectric for negative sweeping voltages that go back to the substrate at positive voltages.

The C-V curves shift toward positive bias with decreasing frequency, and the shift is more obvious in the low frequency range. The effect of the series resistance and variation of the dielectric constant with altering the measurement frequency is very low. Meanwhile, the capacitance shift can be attributed to the presence of fast traps and/or border traps (near-interfacial traps), which can have a rapid connection with the underlying Si-substrate. It is interesting to note that the increase of the sweep voltage value up to $V_G = \pm 14$ V causes the increase of the memory window value to $\Delta V_{fb} = 7$ V (Fig. 7.9B). Further increase of V_G value did not cause the variation of the ΔV_{fb} (not shown here).

Compared to set-A samples, the $HfSiO_x/HfGeO_x/HfSiO_x$ samples (type B samples) annealed at 610 °C followed by additional pre-metallization forming gas annealing exhibit a more efficient charging at low gate voltage (V_G) sweeps which saturates ($\Delta V_{fb} \sim 3.5$ V) at a ± 4 V regime (see Fig. 7.10). Such differences may be attributed to the properties (thickness and permittivity) of the HfO_2 and $HfSiO_x$ bottom/top dielectric layers (see Fig. 7.3). Contrary to A-samples, annealing at $T_A = 710$ °C significantly improves charging efficiency of the B-samples ($\Delta V_{fb} \sim 5$ V for $V_G = \pm 4$ V), which saturates at $V_G = \pm 8$ V ($\Delta V_{fb} \sim 6$ V). This can be due

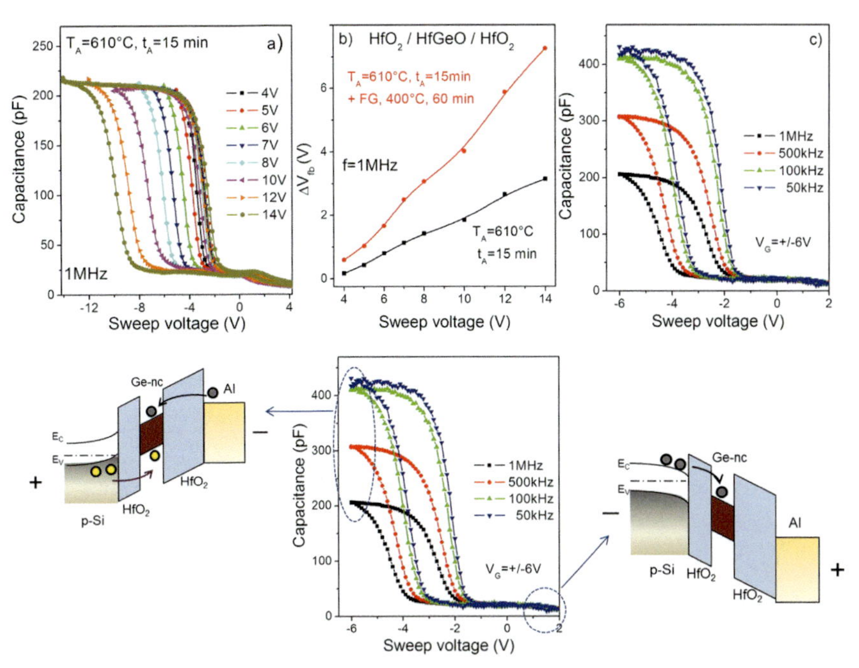

FIG. 7.9 Electrical properties of $HfO_2/HfGeO_x/HfO_2$ structure annealed at $T_A=610\,°C$ for 15 min followed by an additional annealing in forming gas at 400 °C for 60 min: (A) C-V characteristics versus sweep voltage measured at 1 MHz; (B) evolution of $\Delta V_{fb}(V_G)$ stimulated by additional forming gas annealing; (C) frequency dispersion behavior after double annealing treatment measured at $V_G=\pm 6\,V$. *From L. Khomenkova, X. Portier, M. Carrada, C. Bonafos, B. S. Sahu, A. Slaoui, F. Gourbilleau, Ge-doped Hafnia-based dielectrics for non-volatile memory applications, ECS Trans. 45 (3) (2012) 331–344.*

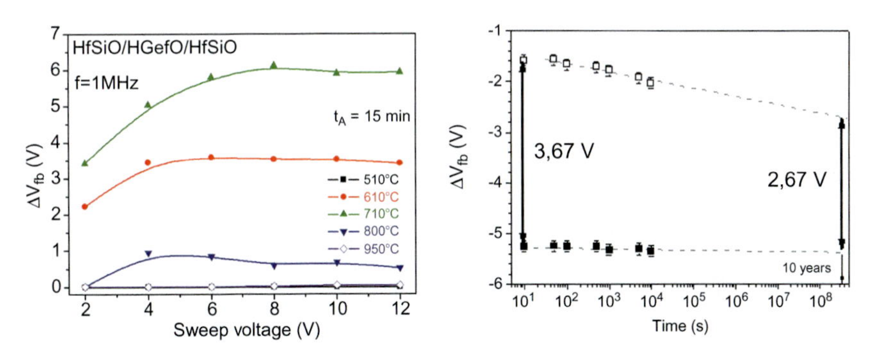

FIG. 7.10 The evolution of $\Delta V_{fb}(V_G)$ versus sweep voltage for different annealing temperatures *(left)* and operation time *(right)*. Annealing was performed at mentioned temperatures in nitrogen flow during $t_A=15$ min and then in forming gas at 400 °C for 60 min. The measurement frequency is f = 1 MHz. For charge retention measurements, the charging time and voltage were 10 s and + 12 V, respectively; data were recorded at ±10 V. *(Left) Taken from L. Khomenkova, X. Portier, M. Carrada, C. Bonafos, B. S. Sahu, A. Slaoui, F. Gourbilleau, Ge-doped Hafnia-based dielectrics for non-volatile memory applications, ECS Trans. 45(3) (2012) 331–344.*

to an improvement of the insulating properties of the control oxide and probably to substantial changes in NC properties (size, density and distribution). Thus, one can conclude that the optimal T_A value is 710 °C for type B—and 610 °C for type A—structures. One can assume that additional effect occurs due to the stresses in the $HfGeO_x$ layer that allowed lower annealing temperature to form Ge-NCs. Moreover, it is observed at lower sweep voltages for set-A and set-B samples. The thinner HfSiOx tunnel layer can explain this observation.

In conclusion, fully high-κ based non-volatile memory structures were fabricated by RF magnetron sputtering technique. Clear memory effect was observed in $HfO_2/HfGeO_x/HfO_2$ and $HfSiO_x/HfGeO_x/HfSiO_x$ stacks and the optimal annealing conditions were found to achieve memory effect, indicating the utility of these stack structures for non-volatile memory devices. The obtained results demonstrate that the RF magnetron sputtering can be a viable technique for fabricating different NCs-based non-volatile memory devices, which can be compatible with the current CMOS technology.

7.2.3.2 *Synthesis of Si-NCs in $HfSiO_x$ films by magnetron sputtering*

As mentioned above, interest in Si-rich HfO_2 materials is mainly driven by their application in microelectronics as alternative "high-κ" dielectric to SiO_2. Their capacity to be used as floating-gate layer in non-volatile memory (NVM) technology has not been firmly demonstrated yet and. However, till present time the formation of Si-NCs in HfO_2 matrix via phase separation process is questionable. There are only few reports on the fabrication of Si-NCs in high-k materials using this approach [33,91,92].

It is worth to point out that in most cases Si-rich HfO_2 composite films are considered as those built from SiO_2 and HfO_2 units, i.e., as $(SiO_2)_x(HfO_2)_{1-x}$ solid solution. In this case, formation of Si-NCs due to thermally stimulated phase separation is hardly possible because of stoichiometric ratio between Si, Hf and O atoms. Earlier we have shown that Si-NCs can be formed in HfO_2 films when such films are ternary compound $HfSiO_x$ built as random bonding of constituent elements [91–93]. When $x=4$, phase separation occurs via formation of SiO_2 and HfO_2 phases. However, when $x<4$, phase separation will have more complex behavior. To get insight on this process in $HfSiO_x$ samples, one can take into account chemical properties of Si and Hf ions, Si—O and Hf—O bonds as well as thermodynamic parameters of related oxides (Table 7.1).

It is known that thermal stability of oxide-based material depends on the coordination number of ions, M-O bond lengths and their nature (ionic or covalent). Materials with higher coordination number, shorter M-O bond length and covalent nature of this bond demonstrate usually thermal stability. The nature of M-O bonding is determined by the difference in the

TABLE 7.1 Chemical properties of Si, Ge, Hf and O and thermodynamic parameters of related oxides.

	Element			
Parameter	**Si**	**Ge**	**Hf**	**O**
Ionic radius, Å	0.40 [Si^{4+}]	0.53 [Ge^{4+}]	0.71 [Hf^{4+}]	1.40 [O^{2-}]
Electronegativity, χ	1.9	2.01	1.3	3.44
Electronegativity difference upon bond formation, χ_M–χ_O	1.54	1.43	2.14	0
Type of M-O bond	Covalent polar	Covalent polar	Ionic	–
Coordination number in the M-O bond	2	4	7	
Length of M-O bond, Å	1.602	1.77	2.2	–
Standard molar enthalpy of the oxide formation at 298.15K, $\Delta_f H^0$ kJ/mol	−910.7 (SiO_2)	−261.9 (GeO) −580.0 (GeO_2)	−1144.7 (HfO_2)	–
Standard molar Gibbs energy of the oxide formation at 298.15K, $\Delta_f G^0$, kJ/mol	−856.3 (SiO_2)	−237.2 (GeO) −521.4 (GeO_2)	−1088.2 (HfO_2)	–

electronegativity of the elements (χ) that composed this bond. When $\chi_M-\chi_O=0$–0.5, the bond is nonpolar covalent, while for $\chi_M-\chi_O=0.5$–1.6 it is polar covalent. For $\chi_M-\chi_O \geq 2.0$, the bond has ionic character. For $\chi_M-\chi_O=1.6$–2.0, if a metal is involved, then the bond is considered ionic, but if only nonmetals are involved, the bond is considered polar covalent. From Table 7.1, one can see that the Hf—O bond is ionic one, whereas Si—O bond is polar covalent. It is worth to note that the ionic strength increases with the increase of the $\chi_M-\chi_O$ difference. For covalent bonding this relation is opposite.

Upon thermal treatment, phase separation in the films is governed by the minimization of the Gibbs energy. Considering the data of Table 7.1, one can assume that upon thermal treatment of Si-rich HfO_2 in inert atmosphere, the formation of HfO_2 phase is most favorable. This means that this phase will form at first during the annealing treatment. Then, after HfO_2 phase formation, depending on the amount of oxygen in the film, either SiO_2 (for $x=4$) or SiO_x (for $x<4$) phase will be formed. In the former, the films will consist of HfO_2 and SiO_2 units, while in the latter, one can expect the formation of Si nuclei and their crystallization (for high temperature annealing) due to further phase separation in SiO_x phase. This statement was confirmed by the investigation of phase segregation in $HfSiO_x$

($x < 4$) [72,82]. In that work, the formation of Si agglomerates (or nanoclusters) in Si-rich HfO_2 films grown by RF-MS was described with emphasis on the homogeneous distribution of Si, Hf and O elements in the as-deposited $HfSiO_x$ film.

The transformation of the film microstructure under annealing was investigated by different techniques such as TEM, X-ray diffraction, Raman scattering and FTIR spectroscopies. However, none of these methods allow for (reliable?) detection of Si nanoclusters in the films. In the case of X-ray diffraction, the close lattice parameters of Si ($a = 5.43$ Å) and HfO_2 ($a = b = 5.14$ Å) hampers to distinguish the Si phase formation due to overlapping of XRD peaks. Similar overlapping of the peaks from Si phase and HfO_2 phase appeared in Raman scattering spectra. Along with the latter, FTIR analysis showed the formation of HfO_2 phase. The presence of Si nanoclusters in the samples annealed at T = 950–1050 °C was revealed by APT.

As detailed in paragraph 2.2.2, APT is a 3D analytical microscopy technique allowing spatial and chemical mapping of atoms in a sample. The APT technique is based on the field evaporation of surface atoms from tip-shaped specimen with a curvature radius smaller than 50 nm. More details about this method can be found in [83,94,95]. Fig. 7.11 shows the three-dimensional atomic maps obtain by APT of $HfSiO_x$ film annealed at 950 °C for 15 min in nitrogen atmosphere. In order to make the maps clear, only 10% of Si and O atoms are represented. The analysis of the atomic distribution of Si and Hf atoms clearly indicates the presence of three different phases in the sample. Hf-rich and Hf-poor zones as well as Si-rich zones (not correlated with Hf-poor) are revealed, indicating that a phase separation between Si-rich and Hf-rich zones occurs during the high-temperature annealing of the sample. This is visible by the non-homogeneous mapping of Si and Hf atoms on Fig. 7.11.

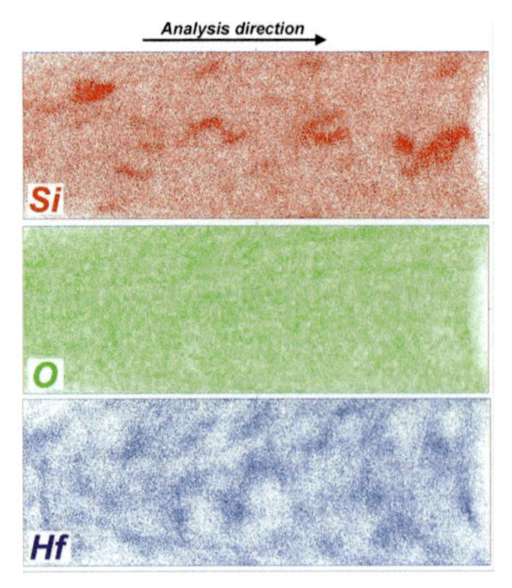

FIG. 7.11 3D atomic maps obtained from APT analysis of the Silicon *(red)*, Oxygen *(green)* and Hafnium *(blue)* in an analyzed volume of 28 × 28 × 80 nm^3. *From E. Talbot, M. Roussel, C. Genevois, P. Pareige, L. Khomenkova, X. Portier, F. Gourbilleau Atomic scale observation of phase separation and formation of silicon clusters in Hf higk-k silicates, J. Appl. Phys. 111 (2012) 103519.*

Isoconcentration maps of Si and Hf atoms were calculated to reveal more insight into phase separation and volume distribution (see Fig. 7.12A and B). Such imaging indicates the existence of a three-phase (α, β and γ) system. The α and β phases relates to Si-rich regions while the γ phase is associated to Hf-rich regions. The resolution of APT allows measuring the local concentrations of each phase that are reported in Table 7.2.

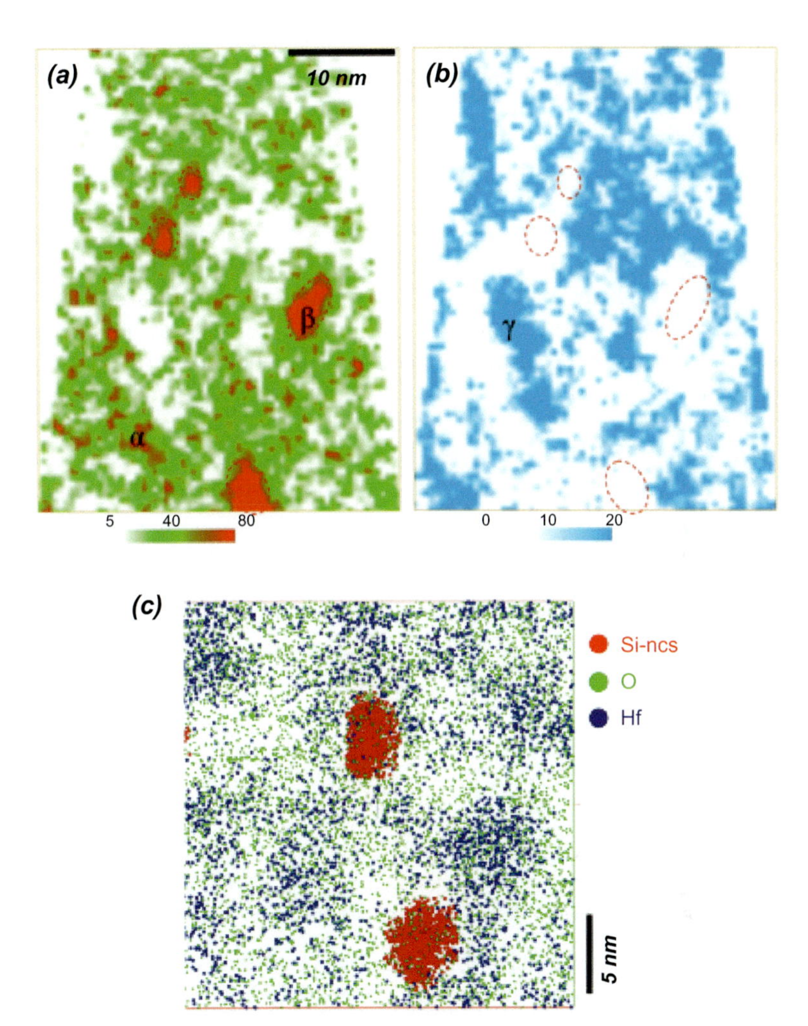

FIG. 7.12 Isoconcentration maps of Silicon atoms (A) and Hafnium atoms (B) in selected volume ($5 \times 40 \times 50\,nm^3$). Pure Si agglomerates are highlighted with dash-circle. Atomic scale mapping of a selected region highlighting Si atoms belonging to Si agglomerates (C). *From E. Talbot, M. Roussel, L. Khomenkova, F. Gourbilleau, P. Pareige, Atomic scale microstructures of high-k HfSiO thin films fabricated by magnetron sputtering, Mater. Sci. Eng. B 177 (2012) 717–720.*

TABLE 7.2 Concentration of Si, Hf and O atoms in different regions on the film.

Label	X_{Si} (at.%)	X_O (at.%)	X_{Hf} (at.%)	Phase
α	32.5	66.2	1.3	SiO_2
β	98.8	1.2	0.0	Si
γ	0.0	68.2	31.8	HfO_2

The presence of pure Si agglomerates in the analyzed volume is clearly demonstrated by the composition of the β phase. Fig. 7.12C shows atomic mapping in a selected region of the analyzed volume evidencing the different chemical regions previously identified. For clarity, only silicon atoms belonging to Si agglomerates are represented. It is worth to point that APT method allows the detection of both amorphous and crystallized Si agglomerates without making difference between them. These agglomerates are homogeneously distributed in the whole volume. Moreover, Si-rich and Hf-rich regions correspond to pure silica and hafnia, confirming a quasi-total phase separation after annealing step at 950 °C during 15 min. The "interface" between the different chemical regions is composed of HfSiO mixture. As pure Si agglomerates do not overlap the Hf-rich regions, one can conclude that phase separation upon annealing occurs via the formation of the HfO_2 phase at first, and then the residual SiO_x phase separates in Si agglomerates and SiO_2 phase. Fig. 7.12C shows that the Si agglomerates have nearly spherical shape with a size distribution between 1 and 6.5 nm and a mean radius of 2.8 nm. The density of agglomerates was estimated to be $2.9 \pm 0.4 \times 10^{17}/cm^3$. One can note that the measured sizes are compatible with quantum confinement effect. Red emission similar to that of Si-NCs in SiO_2 materials was detected from photoluminescence spectra of the films; a finding in favor of the formation of crystalline instead of amorphous Si agglomerates that supports the formation of Si-NCs in HfO_2 from $SiHfO_x$ materials. It is worth to point out that despite such promising materials results, Si-NCs memory devices with attractive performance properties could not be demonstrated yet.

7.2.4 Ultra-low energy Si implantation into HfO_2-based layers

As mentioned previously, few data are available with respect to the synthesis of Si-NCs in a high-κ matrix, especially in hafnium oxide alloys. While promising device results using Si- or Ge-NCs embedded in HfO_2 or Al_2O_3 gate dielectrics have been previously described and can be found in the literature [31,33], the fabrication of semiconducting NCs in high-κ materials is not straightforward, in particular using IBS. Concerning Si-NCs, the main problems, in terms of materials science, related to the use of a matrix other than SiO_2 are identified in Ref. [96], and mainly concern the diffusion and oxidation of silicon as will be discussed hereafter. In addition, as commented previously, the detection of Si- or Ge-NCs in HfO_2 layers is tricky and renders difficult their structural study.

Recently, we extended the ULE-IBS approach to synthesize a plane of Si-NCs within thin high-κ dielectrics (5–10 nm thick HfO_2 and HfSiO layers deposited by MOCVD on Si wafers). After implantation, all samples were annealed at high temperature (>900 °C) for the purpose of NCs formation. Structural and chemical studies were carried out at the atomic scale by HREM (Fig. 7.13) and low-loss EFTEM. The RGB bitmap images constructed with EFTEM maps obtained by filtering around plasmon signatures representative of Si, SiO_2 and HfO_2 reveal that most of the implanted Si in the HfO_2-based layers was oxidized (Fig. 7.14). In addition, the SiO_2 interfacial layer (IL) separating the HfO_2 layer and the Si substrate is larger than 6 nm instead of 1 nm for the as-deposited layer which might be attributed to the diffusion of oxygen toward the Si substrate. The oxidation of Si implanted in very thin SiO_2 matrix has been

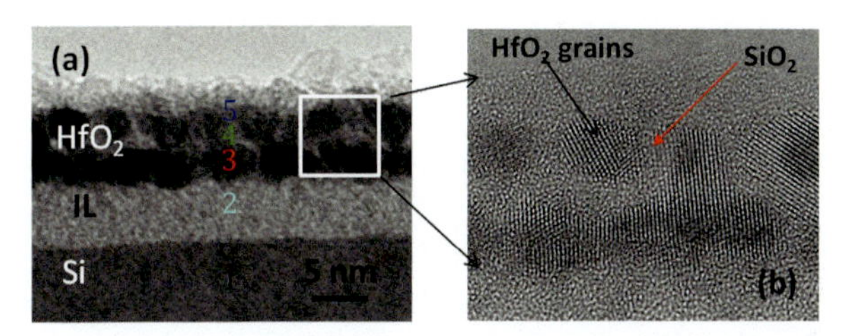

FIG. 7.13 (A) Bright field and (B) HREM images of the HfO_2 layer implanted with Si^+ at 3 keV for a Si excess of 26 at. %, after annealing at 1000 °C for 30 s under N_2 [55]. The HfO_2 grains are separated by amorphous Si rich regions. *From P.E. Coulon, S. Schamm, G. BenAssayag, B. Pecassou, M. Carrada, S. Bhabani, A. Slaoui, S. Lhostis, C. Bonafos, Ultra-low Energy ion implantation of Si into HfO2-based layers for non volatile memory applications, in: Y. Fujisaki, R. Waser, T. Li, C. Bonafos (Eds.), Materials Research Society Symposium Proceedings, vol. 1160 (2009) H01–03.*

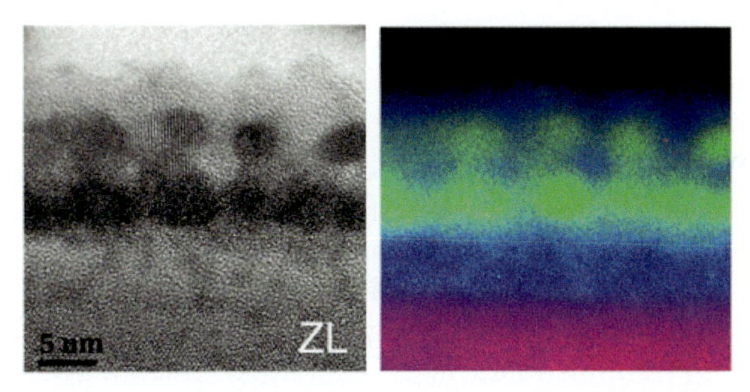

FIG. 7.14 Zero-loss and RGB image of low-loss EFTEM maps of HfO_2, SiO_2 and Si (Si red, SiO_2 green and HfO_2 blue) revealing the oxidized nature of the Si in the high-κ layer.

observed and discussed in detail elsewhere [97]. This so-called "anomalous" oxidation takes place before any annealing, immediately after implantation because the heavily damaged SiO_2 layers absorb humidity. The water molecules are driven in the layers, dissociate and finally react with the implanted Si to form SiO_2. During annealing, further dissociation of OH takes place and finally most of the H atoms diffuse out to the surface. The penetration and final concentration of water molecules do not depend on the relative humidity in the atmosphere but are only limited by the degree of damage, i.e., by the concentration of defects, in the SiO_2 matrix [98]. In the case of Si implanted in SiO_2, we have evaluated that 40% of the implanted Si atoms are oxidized. The remaining non-oxidized Si atoms precipitate to form Si-NCs. In the case of HfO_2, it seems that moisture penetration and implanted oxidation is even more efficient. However, HfO_2 shows as SiO_2 a positive ΔG of the moisture absorption reactions, meaning a small moisture absorption reaction speed [99]. This is consistent with experimental results since there have been few studies showing moisture absorption phenomenon in HfO_2. Nevertheless, even if the reaction speed with moisture is small for bare HfO_2 layers, implanted HfO_2 probably shows a high number of channels for moisture diffusion.

Finally, unexpected structures as HfO_2-NCs embedded within SiO_2 matrix have been obtained (see Fig. 7.13). Capacitance-to-voltage characteristics of MIS capacitors with NCs revealed some hysteresis in the flat-band voltage after application of gate-voltage round-sweeps [100]. These results suggest charge trapping and storage in the implanted/annealed high-κ layers. It has to be noticed that ULE-II of Ge^+ in the same HfO_2 layers leads to the diffusion and pile-up of Ge at the interface with the Si substrate during the implantation process, which prevents Ge-NCs formation.

7.2.5 ULE-II synthesis of Si- and Ge-NCs memories using $SiN/HfO_2/SiO_2$ stacks

To overcome the above limitation, alternative structures have been recently proposed, where HfO_2 is used as tunnel oxide and the NCs are embedded in a SiN overlayer [49,101]. The top SiN layer with a thickness of about 12 nm was deposited with electron cyclotron resonance plasma enhanced chemical vapor deposition method under a flow of SiH_4 and N_2 (instead of NH_3) to minimize the H content in the films. Using the ULE-IBS method formation of Si- and Ge-NCs in a SiN matrix located onto a thin stack of dielectrics comprising HfO_2 layer and an oxide interfacial layer has been demonstrated (see Fig. 7.15A). Such a NCs-$SiN/HfO_2/Si_xO_y$ stack is interesting for the following reasons. First, it exhibits an asymmetric tunnel barrier, which may improve the trade-off between

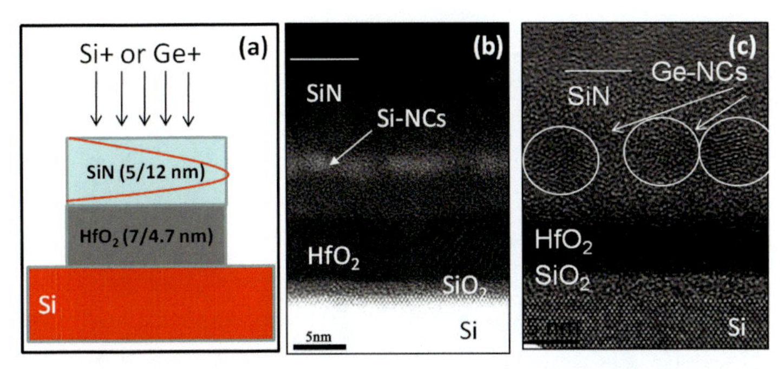

FIG. 7.15 (A) NCM dielectric stack comprising a SiN layer on top of a high-κ tunnel oxide made of HfO_2 materials. This stack is ULE implanted with Si or Ge ions and further annealed for the purpose of NCs formation in the SiN layer; (B) Cross-sectional EFTEM and (C) HREM images respectively showing Si-NCs and Ge-NCs formed in the SiN layer. *From C. Bonafos, M. Carrada, G. Benassayag, S. Schamm-Chardon, J. Groenen, V. Paillard, B. Pecassou, A. Claverie, P. Dimitrakis, E. Kapetanakis, V. Ioannou-Sougleridis, P. Normand, B. Sahu, A. Slaoui, Si and Ge nanocrystals for future memory devices, Mater. Sci. Semicond. Process. 15 (2012) 615–626.*

program/erase efficiency and data retention. Typically, such asymmetric barriers consist of double-stack insulating layers having different bandgap energies. Second, taking advantage of the excellent diffusion barrier properties of SiN materials, well-defined O_2- (and moisture)-protected Si and Ge-NCs are expected to form in the nitride layer. Moreover, SiN contains intrinsic discrete traps (see previous section) that enhance the charge storage capability of the structures.

Fig. 7.15 B and C are cross-sectional TEM images of the SiN/HfO_2 stacks implanted with Si (B) and Ge (C) ions and annealed at high temperature. A plane of Si- (B) and Ge-NCs (C) is formed very close (<2 nm) to the SiN/HfO_2 interface. For the Ge-implanted layers, the presence of Ge-NCs has been also confirmed by Raman scattering. Note that the thermal budget necessary for the phase separation is different than for Si (Rapid Thermal Annealing at 1050°C for 60 s) and Ge (800°C for 30 min). The interfacial layer between HfO_2 and Si remains in both cases smaller than 1 nm after implantation and annealing. This is an evidence of the efficiency of the top SiN layer for blocking O_2 diffusion (compare to Fig. 7.13). Finally, the Equivalent Oxide Thickness (EOT) of the injection oxide separating the Si- and Ge-NCs from the Si substrate (channel) is 4.5 ± 0.5 nm and 3.9 ± 0.5 nm, respectively.

Fig. 7.16 shows typical high-frequency (500 kHz) C-V curves of Al/SiN/HfO_2/Si_xO_y/Si capacitors with Si- and Ge-NCs embedded in the SiN layer. The observed counter-clockwise hysteresis loops reveal significant charge trapping in the dielectric stacks at low applied voltages (especially in the case of the Si-NCs samples) through electron and hole injection from the substrate. In this study, no smearing-out effect was observed in both Si- and Ge-NCs dielectric stacks.

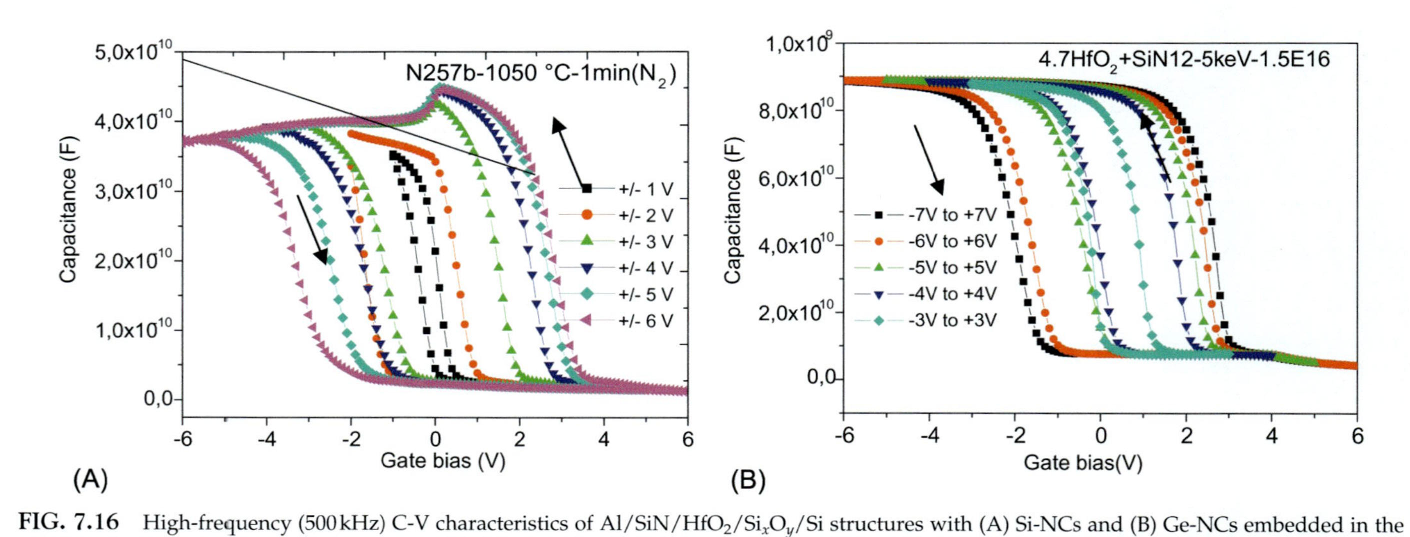

FIG. 7.16 High-frequency (500 kHz) C-V characteristics of Al/SiN/HfO_2/Si_xO_y/Si structures with (A) Si-NCs and (B) Ge-NCs embedded in the SiN layer. *From C. Bonafos, M. Carrada, G. Benassayag, S. Schamm-Chardon, J. Groenen, V. Paillard, B. Pecassou, A. Claverie, P. Dimitrakis, E. Kapetanakis, V. Ioannou-Sougleridis, P. Normand, B. Sahu, A. Slaoui, Si and Ge nanocrystals for future memory devices, Mater. Sci. Semicond. Process. 15 (2012) 615–626.*

For a better understanding of the observed charging effect, frequency dependent C-V and G-V measurements were further carried out in the frequency range of 10–500 kHz (not shown). This is to ascertain that the charging effect originates mainly from the NCs and/or NCs-related traps with minimal influence from interface traps, which typically lead to frequency dispersion in the C-V and G-V characteristics. In both cases, no distortion in the C-V curves due to slow traps and/or large surface density was observed in the samples under study. It is noticed that the full-width-at-half-maximum (FWHM) of the conductance peak is small and almost constant in the frequency range of 10–500 kHz, indicating that the hysteresis and conductance peak are of the same origin. Fig. 7.17 shows the charge retention characteristics of both layers from the application of a stress voltage of ±6 V (positive for electron charging and negative for hole charging) for 3 s. The retention curves exhibit a logarithmic dependence on the waiting time. For the Si-NCs dielectric stacks, the electron loss is very fast during the first seconds but still shows a memory window of 1.1 V after 10^4 s. Charge retention is enhanced in the case of Ge-NCs dielectric stacks, which exhibit a memory window of 2.4 V after a waiting time of 10^4 s and a 10 years extrapolated window of about 1.1 V. As compared to the Si-NCs samples, this enhanced charge retention could be attributed to charge confinement in Ge-NCs [101].

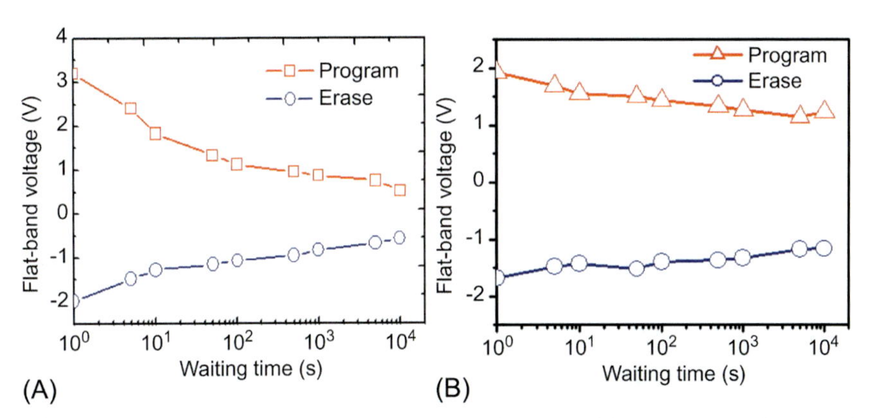

FIG. 7.17 Evolution of the flat-band voltage as a function of the waiting time for the Al/SiN/HfO_2/Si_xO_y/Si structures with (A) Si-NCs and (B) Ge-NCs after application of a stress voltage of +6 V (electron charging) and −6 V (hole charging) for 3 s. *From C. Bonafos, M. Carrada, G. Benassayag, S. Schamm-Chardon, J. Groenen, V. Paillard, B. Pecassou, A. Claverie, P. Dimitrakis, E. Kapetanakis, V. Ioannou-Sougleridis, P. Normand, B. Sahu, A. Slaoui, Si and Ge nanocrystals for future memory devices, Mater. Sci. Semicond. Process. 15 (2012) 615–626.*

7.2.6 ULE-II synthesis of Ge-NCs memories using Al_2O_3 as gate oxide

Al_2O_3 as a gate dielectric is a reasonable choice for NC-NVM cells, since it combines a relatively large dielectric constant (8–10) compared to SiO_2 (4) with barrier heights similar to SiO_2 compared to Si substrate and the majority of common metal electrode materials (e.g., Al). In addition, Al_2O_3 shows a small negative Gibbs free energy change (ΔG) for the moisture absorption reaction characteristics of a low-speed reaction with H_2O- and OH-complexes, even if more hygroscopic than HfO_2 and SiO_2 for which ΔG is positive. As for HfO_2, the implantation of Si at low energy (1 keV) and a dose of $10^{16}\,cm^{-2}$ in thin Al_2O_3 layer (6.5 nm) leads after annealing at high temperature (900 °C under N_2) to the formation of SiO_2 domains instead of Si-NCs. In a second set of samples, ALD grown Al_2O_3 layers of 5 and 7 nm were further implanted with 1 keV Ge ions to doses of 0.5 and $10^{16}\,cm^{-2}$ and subsequently covered with a 10 nm thick ALD Al_2O_3 layer. Next, post-implantation furnace annealing in N_2 ambient for 20 min. at temperatures ranging from 800 °C to 1050 °C were applied for the purpose of Ge-NCs formation. HREM observations confirm the formation of pure Ge-NCs with a mean diameter of 5 nm (see Fig. 7.18). This result is consistent with the highest stability of SiO_2 compared to GeO_2 [102]. STEM-EELS studies based on the low loss region performed by scanning a nanometric probe as a function of the coordinate perpendicular to the film/substrate interface (see Fig. 7.18) confirm the chemical nature of the NCs, i. e., that they are made of pure Ge [103].

According to TEM studies of the control samples, we can conclude that the low-temperature ALD deposition of the Al_2O_3 layers prevents IL development at the Si substrate/Al_2O_3 interface. IL formation after annealing may be due to the increased concentration of residual oxygen atoms in the Al_2O_3 layers [101]. Nevertheless, the IL thickness was found

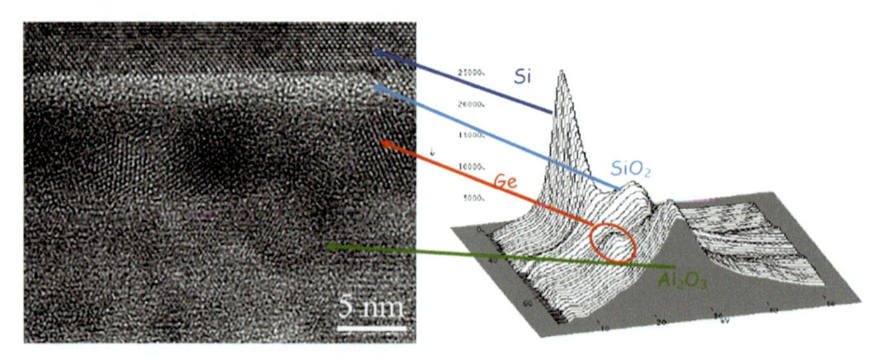

FIG. 7.18 HREM image and STEM-EELS analysis of the nanocomposite layer consisting of a plane of Ge-NCs embedded in a 7 nm-thick alumina matrix obtained after Ge^+ ULE-II.

to be the same (i.e., 1 nm, see Fig. 7.18) after implantation and annealing, indicating that excess oxygen atoms created by ion implantation did not contribute to the IL formation. EELS studies in the low loss region as a function of the coordinate perpendicular to the film/substrate interface (see Fig. 7.18) confirm the chemical nature of the NCs and allow determining the chemical nature of interfacial layer which was found to be SiO_2-rich. At last, HREM images revealed that the non-annealed Al_2O_3 control layers were amorphous while after annealing they were crystallized into the γ-cubic phase.

High-frequency (1 MHz) C-V measurements were used for charge storage evaluation of the implanted and annealed samples. Typical results demonstrating the evolution of the C-V hysteresis as a function of the implantation conditions and annealing temperatures for the 5 and 7 nm-thick implanted layers are presented in Fig. 7.19A and C. It should be emphasized that in any case the hysteresis was clockwise: starting voltage sweep from the inversion regime, holes are injected from the substrate and

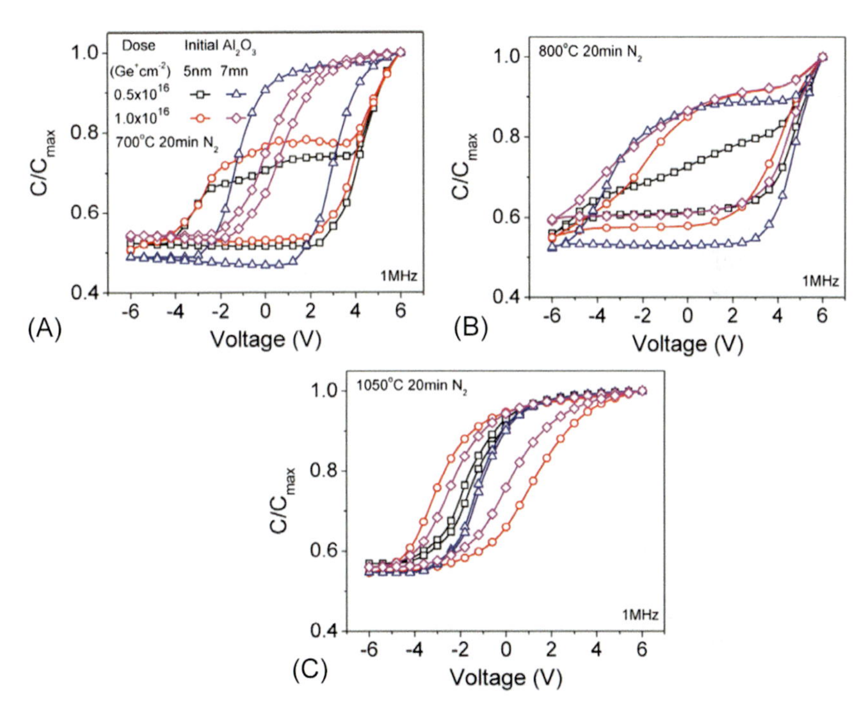

FIG. 7.19 Typical high frequency (1 MHz) C-V characteristics for implanted samples after post-implantation annealing at (A) 700 °C, (B) 800 °C and (C) 1050 °C. *From P. Dimitrakis, A. Mouti, C. Bonafos, S. Schamm, G. BenAssayag, V. Ioannou-Sougleridis, B. Schmidt, J. Becker, P. Normand, Ultra-low-energy ion-beam-synthesis of Ge nanocrystals in thin ALD Al2O3 layers for memory applications, Microelectron. Eng. 86 (2009) 1838–1841.*

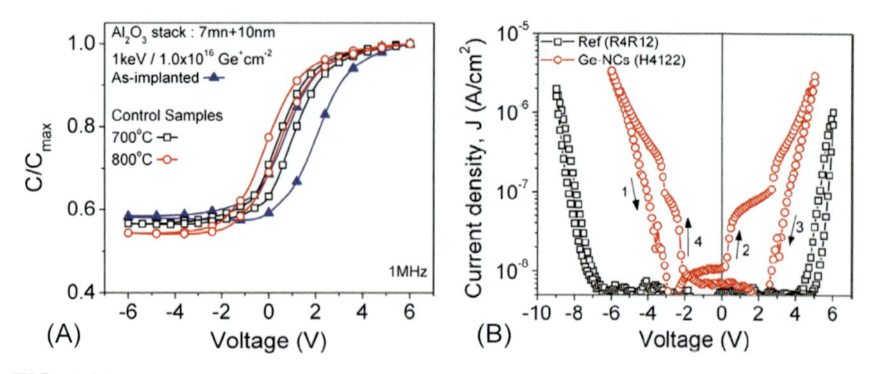

FIG. 7.20 (A) Typical high frequency (1 MHz) C-V characteristics for as-implanted and control samples after post implantation annealing at 700 °C and 800 °C. (B) Typical I-V characteristics (back and forth voltage sweeps) for the as-implanted sample and sample after post-implantation annealing at 800 °C [103].

trapped into the implanted dielectric region, while in the accumulation regime, holes are moved back to the substrate and next electrons are injected into the implanted region. In the case of the 5 nm implanted Al_2O_3 layers, the shape of the C-V characteristics are strongly stretched-out for annealing temperatures of 700 °C and 800 °C. No significant difference between the implanted doses was detected. In the same temperature regime, the samples with 7 nm implanted Al_2O_3 layers exhibit C-V characteristics without any distortion or stretch-out. Annealing at higher temperature improves significantly the shape of the C-V characteristics of the 5 nm samples but reduces the hysteresis in terms of ΔV_{fb} for both 5 and 7 nm-thick samples.

In Fig. 7.20, typical C-V curves of the control and as-implanted 7 nm-thick samples are presented. No distortion of the shape of the C-V characteristics is observed. The C-V curve of the as-implanted sample is similar to those measured after thermal treatment and exhibits higher ΔV_{fb} due to the defects generated by the implantation process. I-V characteristics of the 7 nm-thick sample implanted at the highest dose, before and after post implantation annealing are shown in Fig. 7.20. The as-implanted sample exhibits a very small hysteresis due to the implantation-induced defects and the current leakage is still very low. On the contrary, after annealing I-V curves exhibit strong non-linearity and hysteresis at low electric fields.

The shape of the C-V curves for the samples with 5 nm implanted Al_2O_3 layers annealed in the 700–800 °C regime can be explained assuming that despite a very efficient hole-injection process [104], the Fermi level is pinned due to the large number of Ge nanoclusters and AlO defects at the Al_2O_3/IL interface causing the capacitance to remain constant in the depletion regime. The trapped positive charge screens effectively the surface potential preventing the formation of a strong accumulation layer, i.e., insufficient electron injection conditions. Thus, during the backward sweep only hole rejection is observed. In the case of the 7nm implanted Al_2O_3 layers annealed at the same regime, the interface is less affected by the implantation process. Furthermore, the presence of a thicker crystalline Al_2O_3 region in this case mitigates the leakage current through the implanted layers and thus reduces the C-V curve distortion. Annealing at higher temperature enhances the Ostwald ripening process and a significant amount of implanted Ge is lost due to the formation of GeO volatile species [104]. This improves significantly the C-V characteristics but decreases the hysteresis effect due to the reduction of the implantation defects and dissolution of the Ge nanoclusters.

According to the results presented in Fig. 7.20A, the observed ΔV_{fb} is the same for control samples annealed at 700°C and 800°C, suggesting that the performed thermal treatment does not cause the generation of additional Al_2O_3 defects. However, the higher the annealing temperature, the higher the reduction of the positive fixed charges present into the dielectric stack. Clearly, the as-implanted sample does not exhibit characteristics related to a damaged Al_2O_3/Si substrate interface or suffering from a large concentration of bulk defects. The above experimental evidence leads us to the conclusion that in the case of the annealed 7nm Al_2O_3 implanted samples, the formation of Ge-NCs is responsible for the enhanced charge storage.

I-V characteristics, shown in Fig. 7.20B, reveal that the as-implanted samples can withstand a higher electric field compared to the annealed samples before current begins to flow. This indicates that the implantation-induced-defects do not contribute to the leakage current through the dielectric stack. This suggests also that the implantation process does not substantially affect the Al_2O_3/IL interface quality. After post implantation annealing, the formation of the SiO_2-rich IL facilitates direct tunneling of carriers from Si into the Ge-NCs through the nanocrystalline Al_2O_3 regions. It should be here remembered that according to TEM investigations (see Fig. 7.18) the Ge-NCs are distributed all over the implanted Al_2O_3 layer and the closest NCs are at a direct tunneling distance from the Si substrate. The current flow through the dielectric stack is governed by tunneling of carriers between adjacent NCs.

7.3 Conclusion

In summary, this chapter gave a short overview of 10years efforts dedicated to the development of NCs memories involving semiconducting NCs embedded in metal oxide high-κ matrix and in particular HfO_2 and its compounds (HfSiO), and Al_2O_3 thin layers. Radio-frequency magnetron sputtering has been shown to be a powerful technique for the fabrication of HfSiO layers and HfGeO layers with specific composition. For HfSiO, the phase separation process occurs through the formation of HfO_2, SiO_2, and Si phases. The appearance of the Si phase strongly depends on the film composition and is possible for Hf-silicates with lower Hf content. In the case of $HfGeO_x$, the phase separation process is more thermodynamically favorable and the use of moderate temperatures gives rise to the formation of Ge-NCs embedded in HfO_2 matrix. In the case of ultra-low energy ion beam synthesis, Si and Ge supersaturation are incorporated in HfO_2 or HfSiO matrices by ion implantation. In the case of HfO_2, moisture penetration and further oxidation prevents the formation of pure Si and Ge-NCs. To overcome the above limitation, alternative structures have been proposed, where HfO_2 is used as tunnel oxide and the Si and Ge-NCs are embedded in a SiN overlayer, which is an efficient oxygen blocking layer. Ge-NCs have also been successfully formed in thin alumina matrix by using the same technique.

It has to be pointed out that, whatever the system and elaboration method, the structural characterization of the nanocomposite oxide layer is crucial for optimizing the electrical properties of the produced structures. Nevertheless, evidencing and characterizing the formation of Si- and Ge-NCs embedded in metal oxide matrices (HfO_2 and associated silicates, Al_2O_3) can be tricky and a panel of characterization techniques is often necessary to evidence NCs formation and determine without any ambiguity their composition. Finally STEM-EELS in the core-loss region and APT are the best methods for this delicate analysis, but present the drawback to be difficult to implement, time-consuming and not widespread in the laboratories.

Furthermore, the electrical properties and memory performance of these structures are interesting but the added-value comparing to simpler systems as Si- or Ge-NCs embedded in SiO_2 is not outstanding. This could be due to the systematic formation of an IL between the MO layer and the Si substrate. The growth of this IL cannot be avoided but is now well controlled and its thickness can be limited during the deposition processes. Nevertheless this layer evolves and its thickness increases during the further annealing processes at high temperature necessary for the phase separation and NCs nucleation, impacting the final EOT and memory performances. In addition, the poly-crystalline nature of the MO layers can also be at the origin of additional charge traps (at the grain boundaries for example) that complicates the interpretation of the electrical characteristics. Anyway, the generated results described herein can be beneficial not only for the memory sector but also for a broader scientific community with interests in NC-based-optoelectronics or to more current memories as RRAM.

Acknowledgments

Part of this work has been supported by the ANR Project ANR/PNANO07-0053-NOMAD.

References

[1] ITRS, International Technology Roadmap for Semiconductors, 2009. Executive summary.
[2] P. Dimitrakis, P. Normand, D. Tsoukalas, Silicon nanocrystal memories, in: L. Khriachtchev (Ed.), Silicon Nanophotonics: Basic Principles, Present Status, and Perspectives, Pan Publishing, Singapore, 2008, pp. 211–244.
[3] Y.-C. Lien, J.-M. Shieh, W.-H. Huang, C.-H. Tu, C. Wang, C.-H. Shen, B.-T. Dai, C.-L. Pan, C. Hu, F.-L. Yang, Fast programming metal-gate Si quantum dot nonvolatile memory using green nanosecond laser spike annealing, Appl. Phys. Lett. 100 (2012), 143501-4.
[4] P. Dimitrakis, Charge-Trapping Non-Volatile Memories Volume 1 – Basic and Advanced Devices, ISBN 978–3–319-15289-9, Springer International Publishing Switzerland, Springer Cham, Heidelberg, New York, Dordrecht, London, 2015.
[5] P. Dimitrakis, P. Normand, V. Ioannou-Sougleridis, C. Bonafos, S. Schamm-Chardon, G. Benassayag, E. Iliopoulos, Quantum dots for memory applications, Phys. Status Solidi A 210 (2013) 1490–1504.
[6] J. Yater, Highly optimized nanocrystal-based split gate flash for high performance and low power microcontroller applications, in: Proceedings of the 3rd International Memory Workshop (IMW), 2011.
[7] R. Gusmeroli, C. Monzio Compagnoni, A.-S. Spinelli, Statistical constraints in nanocrystal memory scaling, Microelectron. Eng. 84 (2007) 2869–2874.
[8] P. Dimitrakis, P. Normand, Parasitic memory effects in shallow-trench-isolated nanocrystal memory devices, Solid State Electron. 51 (2007) 147.
[9] P. Dimitrakis, S. Schamm-Chardon, C. Bonafos, P. Normand, Nanoparticle-based memories: concept and operation principles, in: R.S. Chaughule, S.C. Watawe (Eds.), Applications of Nanomaterials, 2012, pp. 1–26. Chap. 2.
[10] L. Wang, C.-H. Yang, J. Wen, Physical principles and current status of emerging nonvolatile solid state memories, Electron. Mater. Lett. 11 (2015) 505–543.
[11] S. Tiwari, F. Rana, H. Hanafi, A. Hartstein, E.F. Crabbe, K. Chan, A silicon nanocrystals based memory, Appl. Phys. Lett. 68 (1996) 1377–1379.
[12] J. De Blauwe, M. Ostraat, M. Green, G. Weber, T. Sorsch, A. Kerber, F. Klemens, R. Cirelli, E. Ferry, J. Grazul, F. Baumann, Y. Kim, W. Mansfield, J. Bude, J. Lee, S. Hillenius, R. Flagan, H. Atwater, Novel, aerosol-nanocrystal floating-gate device for non volatile memory applications, in: International Electron Devices Meeting, 2000, pp. 683–686.
[13] P. Normand, D. Tsoukalas, E. Kapetanakis, J. Van den Berg, D.G. Armour, J. Stoemenos, C. Vieu, Formation of 2-D arrays of silicon nanocrystals in thin SiO_2 films by very-low energy Si+ ion implantation, Electrochem. Solid-State Lett. 1 (1998) 88.
[14] S. Jacob, G. Festes, S. Bodnar, R. Coppard, J. Thiery, T. Pate-Cazal, T. Pedron, B. De Salvo, L. Perniola, E. Jalaguier, F. Boulanger, S. Deleonibus, Integration of CVD silicon nanocrystals in a 32 Mb NOR flash memory, in: Proc. IEEE ESSDERC, 2007, pp. 410–413.
[15] M. Kanoun, C. Busseret, A. Poncet, A. Souifi, T. Baron, E. Gautier, On the saturation mechanism in the Ge nanocrystals-based non-volatile memory, Solid State Electron. 50 (2006) 1310–1314.
[16] P.D. Szkutnik, A. Karmous, F. Bassani, A. Ronda, I. Berbezier, K. Gacem, A. El Hdiy, M. Troyon, Ge nanocrystals formation on SiO_2 by dewetting: application to memory, Eur. Phys. J. Appl. Phys. 41 (2008) 103–106.

[17] S. Duguay, J.J. Grob, A. Slaoui, Y. Le Gall, M. Amann-Liess, Structural and electrical properties of Ge nanocrystals embedded in SiO_2 by ion implantation and annealing, J. Appl. Phys. 97 (2007) 104330.
[18] Z. Liu, C. Lee, V. Narayanan, G. Pei, E.C. Kan, Metal nanocrystal memories—part I: device design and fabrication, IEEE Trans. Electron Devices 49 (2002) 1606–1613.
[19] W. Chen, T. Chang, P. Liu, P. Lin, C. Tu, C. Chang, Formation of stacked Ni silicide nanocrystals for nonvolatile memory application, Appl. Phys. Lett. 90 (2007) 112. 108-1–112 108-3.
[20] J. Dufourcq, S. Bodnar, G. Gay, D. Lafond, P. Mur, G. Molas, J.P. Nieto, L. Vandroux, L. Jodin, F. Gustavo, T. Baron, High density platinum nanocrystals for non-volatile memory applications, Appl. Phys. Lett. 92 (2008). 073-102-1–073 102-3.
[21] T. Ando, Materials, ultimate scaling of high-κ gate dielectrics: higher-κ or interfacial layer scavenging, Materials 5 (2012) 478–500.
[22] S.F. Huang, H.D. Lin, B.H. Liu, F.K. Liu, C.C. Leu, Development of all-solution-processed nanocrystal memory, J. Alloys Compd. 698 (2017) 484–494.
[23] S. Hattori, M. Terai, H. Nishizawa, Changes in the characteristics of nonvolatile memory depending on au nanoparticle size for p and n-type channels, Jpn. J. Appl. Phys. 55 (2016), 025002.
[24] P. Xia, L. Li, P. Wang, Y. Gan, W. Xu, Low-cost fabrication and polar-dependent switching uniformity of memory devices using alumina interfacial layer and Ag nanoparticle monolayer, AIP Adv. 7 (2017) 115010.
[25] T.-K. Kang, Y.-F. Chen, C.-L. Lin, F.-H. Wang, H.-W. Liu, F.-T. Chien, J.-B. Shia, Improved retention characteristics of Pd-nanocrystal-based nonvolatile memories by a simple timing technique, ECS Solid State Lett. 4 (2015) N20–N22.
[26] J.C. Wang, C.H. Chen, C.C.T. Lin, Thickness dependence of Al2O3/HfO2/Al2O3 stacked tunneling layers ongadolinium oxide nanocrystal nonvolatile memory, Microelectron. Eng. 138 (2015) 52–56.
[27] N. El-Atab, A. Nayfeh, 1D versus 3D quantum confinement in 1–5 nm ZnO nanoparticle agglomerations for application in charge-trapping memory devices, Nanotechnology 27 (2016) 275205.
[28] A.M. Lepadatu, C. Palade, A. Slav, A.V. Maraloiu, S. Lazanu, T. Stoica, C. Logofatu, V.S. Teodorescu, M.L. Ciurea, Single layer of Ge quantum dots in HfO_2 for floating gate memory capacitors, Nanotechnology 28 (2017) 175707.
[29] R. Solanki, A. Mahajan, R. Patrikar, Finite-element modeling of retention in nanocrystal flash memories with high-k interpoly dielectric stack, IEEE Trans. Electron Devices 64 (2017) 4897.
[30] J.H. Chen, Y.Q. Wang, W.J. Yoo, Y. Yeo, G. Samudra, D.S. Chan, A.Y. Du, D. Kwong, Nonvolatile flash memory device using Ge nanocrystals embedded in HfAlO high- κ tunneling and control oxides: device fabrication and electrical performance, IEEE Trans. Electron Devices 51 (2004) 1840–1848.
[31] D. Kim, T. Kim, S. Sanerjee, Memory characterization of SiGe quantum dot flash memories with HfO_2 and SiO_2 tunneling dielectrics, IEEE Trans. Electron Devices 50 (2003) 1823–1829.
[32] J. Lu, Y. Kuo, J. Yan, C. Lin, Nanocrystalline silicon embedded zirconium-doped hafnium oxide high-k memory device, Jpn. J. Appl. Phys. 34 (2006) L901.
[33] J.J. Lee, X. Wang, W. Bai, N. Lu, D.-L. Kwong, Theoretical and experimental investigation of Si nanocrystal memory device with HfO2 high-k tunneling dielectric, IEEE Trans. Electron Devices 50 (2003) 2067–2071.
[34] H.K. Kim, I.-H. Yu, J.H. Lee, C.S. Hwang, Interfacial dead-layer effects in Hf-silicate films with Pt or RuO_2 gates, ACS Appl. Mater. Interfaces 5 (2013) 6769–6772.
[35] R. Chau, S. Datta, M. Doczy, B. Doyle, J. Kavalieros, M. Metz, High-k/metal-gate stack and its MOSFET characteristics, IEEE Electron Device Lett. 25 (2004) 408–410.

[36] M.S. Kim, Y.D. Ko, M. Yun, J.H. Hong, M.C. Jeong, J.M. Myoung, I. Yun, Characterization and process effects of HfO_2 thin films grown by metal–organic molecular beam epitaxy, Mater. Sci. Eng. B 123 (2005) 20–30.
[37] A.C. Jones, P.R. Chalker, Some recent developments in the chemical vapour deposition of electroceramic oxides, J. Phys. D Appl. Phys. 36 (2003) R80–R95.
[38] M. Lemberger, A. Paskaleva, S. Zürcher, A.J. Bauer, L. Frey, H. Ryssel, Electrical properties of hafnium silicate films obtained from a single-source MOCVD precursor, Microelectron. Reliab. 45 (2005) 819–822.
[39] T. Nishide, S. Honda, M. Matsuura, M. Ide, Surface, structural and optical properties of sol–gel derived HfO_2 films, Thin Solid Films 371 (2000) 61–65.
[40] M.G. Blanchin, B. Canut, Y. Lambert, V.S. Teodorescu, A. Barau, M. Zaharescu, Structure and dielectric properties of HfO_2 films prepared by a sol–gel route, J. Sol-Gel Sci. Technol. 47 (2008) 165–172.
[41] H. Hu, C.X. Zhu, Y.F. Lu, Y.H. Wu, T. Liew, M.F. Li, B.J. Cho, W.K. Choi, N. Yakovlev, Physical and electrical characterization of HfO_2 metal–insulator–metal capacitors for Si analog circuit applications, J. Appl. Phys. 94 (2003) 551–557.
[42] S.S. Hullavarad, D.E. Pugel, E.B. Jones, R.D. Vispute, T. Venkatesan, Low leakage current transport and high breakdown strength of pulsed laser deposited HfO_2/SiC metal–insulator–semiconductor device structures, J. Electron. Mater. 36 (2007) 648–653.
[43] L. Khomenkova, C. Dufour, P.-E. Coulon, C. Bonafos, F. Gourbilleau, High-k Hf-based layers grown by RF magnetron sputtering, Nanotechnology 21 (2010), 095704.
[44] L. Khomenkova, X. Portier, J. Cardin, F. Gourbilleau, Thermal stability of high-k Si rich HfO_2 layers grown by RF magnetron sputtering, Nanotechnology 21 (2010) 285707.
[45] V. Beyer, J. Von Borany, Ion-beam synthesis of nanocrystals for multidot memory structures, in: C. Whelan, T. Mikolajick (Eds.), Zschech E, Materials for information technology, Springer, 2005, pp. 139–147.
[46] P. Normand, P. Dimitrakis, E. Kapetanakis, D. Skarlatos, K. Beltsios, D. Tsoukalas, C. Bonafos, H. Coffin, G. Benassayag, A. Claverie, V. Soncini, A. Agarwal, C. Sohl, M. Ameen, Processing issues in silicon nanocrystal manufacturing by ultra-low-energy ion-beam-synthesis for non-volatile memory applications, Microelectron. Eng. 73-74 (2004) 730–735.
[47] C. Bonafos, M. Carrada, N. Cherkashin, H. Coffin, D. Chassaing, G. Ben Assayag, A. Claverie, T. Müller, K.H. Heinig, M. Perego, M. Fanciulli, P. Dimitrakis, P. Normand, Manipulation of two-dimensional arrays of Si nanocrystals embedded in thin SiO_2 layers by low energy ion implantation, J. Appl. Phys. 95 (2004) 5696.
[48] C. Bonafos, H. Coffin, S. Schamm, N. Cherkashin, G. Ben Assayag, P. Dimitrakis, P. Normand, M. Carrada, V. Paillard, A. Claverie, Si nanocrystals by ultra-low-energy ion beam-synthesis for nonvolatile memory applications, Solid State Electron. 49 (2005) 1734–1744.
[49] C. Bonafos, M. Carrada, G. Benassayag, S. Schamm-Chardon, J. Groenen, V. Paillard, B. Pecassou, A. Claverie, P. Dimitrakis, E. Kapetanakis, V. Ioannou-Sougleridis, P. Normand, B. Sahu, A. Slaoui, Si and Ge nanocrystals for future memory devices, Mater. Sci. Semicond. Process. 15 (2012) 615–626.
[50] C. Castro, G. BenAssayag, B. Pecassou, A. Andreozzi, G. Seguini, M. Perego, S. Schamm-Chardon, Nanoscale control of Si nanoparticles within a 2D hexagonal array embedded in SiO_2 thin films, Nanotechnology 28 (2017), https://doi.org/10.1088/0957-4484/28/1/014001, 014001.
[51] W. Banerjee, S. Maikap, S.Z. Rahaman, A. Prakash, T.-C. Tien, W.-C. Li, J.-R. Yang, Crystal that remembers: several ways to utilize nanocrystals in resistive switching memory, J. Electrochem. Soc. 159 (2012) H177–H182.
[52] L. Chen, H.-Y. Gou, Q.-Q. Sun, P. Zhou, H.-L. Lu, P.-F. Wang, S.-J. Ding, D.W. Zhang, Enhancement of resistive switching characteristics in Al2O3-based RRAM with embedded ruthenium nanocrystals, IEEE Electron Device Lett. 32 (2011) 794–796.

[53] C.H. Cheng, P.C. Chen, Y.H. Wu, F.S. Yeh, A. Chin, Long-endurance nanocrystal TiO2 resistive memory using a TaON buffer layer, IEEE Electron Device Lett. 32 (2011) 1749–1751.
[54] Y. Wang, Q. Liu, H. Lu, S. Long, W. Wang, Y.T. Li, S. Zhang, W.T. Lian, J.H. Yang, M. Liu, Improving the electrical performance of resistive switching memory using doping technology, Chin. Sci. Bull. 57 (2012) 1235.
[55] Y.C. Ju, S. Kim, T.G. Seong, S. Nahm, H. Chung, K. Hong, W. Kim, Resistance random access memory based on a thin film of CdS nanocrystals prepared via colloidal synthesis, Small 8 (2012) 2849–2855.
[56] C. Bonafos, G. Benassayag, R. Cours, B. Pecassou, P.V. Guenery, N. Baboux, L. Militaru, A. Souifi, E. Cossec, K. Hamga, S. Ecoffey, D. Drouin, Ion beam synthesis of indium-oxide nanocrystals for improvement of oxide resistive random-access memories, Mater. Res. Express 5 (2018) 015027.
[57] D. Lehninger, L. Khomenkova, C. Röder, G. Gärtner, J. Beyer, F. Schneider, V. Klemm, D. Rafaja, J. Heitmann, Ge nanostructures embedded in ZrO_2 dielectrics for non-volatile memory applications, ECS Trans. 68 (2015) 203.
[58] L. Khomenkova, D. Lehninger, O. Kondratenko, S. Ponomaryov, O. Gudymenko, Z. Tsybrii, V. Yukhymchuk, V. Kladko, J. von Borany, J. Heitmann, Effect of Ge content on the formation of Ge nanoclusters in magnetron-sputtered $GeZrO_x$-based structures, Nanoscale Res. Lett. 12 (2017) 967.
[59] T. Müller, C. Bonafos, K.H. Heinig, M. Tencé, H. Coffin, N. Cherkashin, G. BenAssayag, S. Schamm, G. Zanchi, C. Colliex, W. Möller, A. Claverie, Multi-dot floating-gates for nonvolatile semiconductor memories: their ionbeam synthesis and morphology, Appl. Phys. Lett. 82 (2004) 2373.
[60] H. Hosono, Chemical interaction in ion-implanted amorphous SiO2 and application to formation and modification of nanosize colloid particles, J. Non Cryst. Solids 187 (1995) 457–472.
[61] M. Strobel, Modeling and Computer Simulation of Ion Beam Synthesis of Nanostructures, PhD dissertation, University of Dreden (Germany), 1999.
[62] A.L. Stepanov, Synthesis of silver nanoparticles in dielectric matrix by ion implantation: a review, Rev. Adv. Mater. Sci. 26 (2010) 1–29.
[63] R. Carles, C. Farcau, C. Bonafos, G. Benassayag, B. Pecassou, A. Zwick, The synthesis of single layers of Ag nanocrystals by ultra-low-energy ion implantation for large-scale plasmonic structures, Nanotechnology 20 (2009) 355305.
[64] M. Perego, S. Ferrari, M. Fanciulli, G. BenAssayag, C. Bonafos, M. Carrada, A. Claverie, Detection and characterization of silicon nanocrystals embedded in thin oxide layers, J. Appl. Phys. 95 (2004) 257.
[65] S. Schamm, C. Bonafos, H. Coffin, N. Cherkashin, M. Carrada, G. BenAssayag, A. Claverie, M. Tencé, C. Colliex, Imaging Si nanoparticles embedded in SiO2 layers by (S) TEM-EELS, Ultramicroscopy 108 (2008) 346–357.
[66] A. Cerezo, T.J. Godfrey, G.D.W. Smith, Application of a position-sensitive detector to atom probe microanalysis, Rev. Sci. Instrum. 59 (1988) 862–866.
[67] D. Blavette, A. Bostel, J.M. Sarrau, B. Deconihout, A. Menand, An atom probe for three-dimensional tomography, Nature 363 (1993) 432–435.
[68] M.K. Miller, T.F. Kelly, K. Rajan, S.P. Ringer, The future of atom probe tomography, Mater. Today 15 (2012) 158–165.
[69] B. Han, Y. Shimizu, G. Seguini, E. Arduca, C. Castro, G. Ben Assayag, K. Inoue, Y. Nagai, S. Schamm-Chardon, M. Perego, Evolution of shape, size, and areal density of a single plane of Si nanocrystals embedded in SiO_2 matrix studied by atom probe tomography, RSC Adv. 6 (2016) 3617–3622.
[70] F. Gault, F. Vurpillot, A. Vella, M. Gilbert, A. Menand, D. Blavette, B. Deconihout, Design of a femtosecond laser assisted tomographic atom probe, Rev. Sci. Instrum. 77 (2006), 043705.

[71] E. Talbot, R. Larde, F. Gourbilleau, C. Dufour, P. Pareige, Si nanoparticles in SiO_2: an atomic scale observation for optimization of optical devices, Europhys. Lett. 87 (2009) 26004.
[72] E. Talbot, M. Roussel, L. Khomenkova, F. Gourbilleau, P. Pareige, Atomic scale microstructures of high-k HfSiO thin films fabricated by magnetron sputtering, Mater. Sci. Eng. B 177 (2012) 717–720.
[73] M. Roussel, E. Talbot, F. Gourbilleau, P. Pareige, Atomic characterization of Si nanoclusters embedded in SiO 2 by atom probe tomography, Nanoscale Res. Lett. 6 (2011) 164.
[74] S. Duguay, F. Vurpillot, T. Philippe, E. Cadel, R. Larde, B. Deconihout, G. Servanton, R. Pantel, Evidence of atomic-scale arsenic clustering in highly doped silicon, J. Appl. Phys. 106 (2009) 106102.
[75] E. Cadel, F. Vurpillot, R. Larde, S. Duguay, B. Deconihout, Depth resolution function of the laser assisted tomographic atom probe in the investigation of semiconductors, J. Appl. Phys. 106 (2009), 044908.
[76] E. Cadel, N. Barreau, J. Kessler, P. Pareige, Atom probe study of sodium distribution in polycrystalline $Cu(In,Ga)Se_2$ thin film, Acta Mater. 58 (2010) 2634–2637.
[77] R. Larde, E. Talbot, P. Pareige, H. Bieber, G. Schmerber, S. Colis, V. Pierron-Bohnes, A. Dinia, Evidence of superparamagnetic Co clusters in pulsed laser deposition-grown $Zn_{0.9}Co_{0.1}O$ thin films using atom probe tomography, J. Am. Chem. Soc. 133 (2011) 1451–1458.
[78] L. Khomenkova, C. Labbé, X. Portier, M. Carrada, F. Gourbilleau, Undoped and Nd^{3+} doped Si-based single layers and superlattices for photonic applications, Phys. Status Solidi A 210 (2013) 1532–1543.
[79] O. Debieu, R. Pratibha Nalini, J. Cardin, X. Portier, J. Perrière, F. Gourbilleau, Structural and optical characterization of pure Si-rich nitride thin films, Nanoscale Res. Lett. 8 (2013) 31.
[80] L. Khomenkova, P. Normand, F. Gourbilleau, A. Slaoui, C. Bonafos, Optical, structural and electrical characterizations of stacked Hf-based and silicon nitride dielectrics, Thin Solid Films 617 (2016) 143–149.
[81] L. Khomenkova, Y.-T. An, C. Labbé, X. Portier, X.F. Gourbilleau, Hafnia-based luminescent insulator for phosphor applications, ECS Trans. 45 (2012) 1119.
[82] E. Talbot, M. Roussel, C. Genevois, P. Pareige, L. Khomenkova, X. Portier, F. Gourbilleau, Atomic scale observation of phase separation and formation of silicon clusters in Hf higk-k silicates, J. Appl. Phys. 111 (2012) 103519.
[83] P.M. Lambert, Hafnium Germanate from a hydrous hafnium germanium oxide gel, Inorg. Chem. 37 (1998) 1352–1357.
[84] P. Rao, T. Sakuntala, S.N. Achary, A.K. Tyagi, High pressure behavior of $ZrGeO_4$: a Raman spectroscopic and photoluminescence study, J. Appl. Phys. 106 (2009) 123517.
[85] A.M. Hofmeister, J. Horigan, J.M. Campbell, Infrared spectra of GeO_2 with the rutile structure and prediction of inactive modes for isostructural compounds, Am. Mineral. 75 (1990) 1238–1248.
[86] M. Ardyanian, H. Rinnet, M. Vergnat, Influence of hydrogenation on the structure and visible photoluminescence of germanium oxide thin films, JOL 129 (2009) 729.
[87] A. Singh, C.A. Hogarth, An infrared study of the structure of GeO2-CeO2 thin films, J. Mater. Sci. 24 (1989) 307–312.
[88] H. Richter, Z.P. Wang, L. Ley, The one phonon Raman spectrum in microcrystalline silicon, Solid State Commun. 39 (1981) 625–629.
[89] M. Fujii, S. Hayashi, K. Yamamoto, Growth of Ge microcrystals in SiO_2 thin film matrices: a Raman and electron microscopic study, Jpn. J. Appl. Phys. 30 (1991) 687–694.
[90] L. Khomenkova, X. Portier, M. Carrada, C. Bonafos, B.S. Sahu, A. Slaoui, F. Gourbilleau, Ge-doped Hafnia-based dielectrics for non-volatile memory applications, ECS Trans. 45 (3) (2012) 331–344.

[91] S. Das, K. Das, R.K. Singha, A. Dhar, S.K. Ray, Improved charge injection characteristics of Ge nanocrystals embedded in hafnium oxide for floating gate devices, Appl. Phys. Lett. 91 (2007) 233118.
[92] Y.-T. An, C. Labbé, L. Khomenkova, M. Morales, X. Portier, F. Gourbilleau, Microstructure and optical properties of Pr^{3+}-doped hafnium silicate films, Nanoscale Res. Lett. 8 (2013) 43.
[93] L. Khomenkova, Y.-T. An, D. Khomenkov, X. Portier, C. Labbé, F. Gourbilleau, Spectroscopic and structural investigation of undoped and Er^{3+} doped hafnium silicate layers, Phys. B Condens. Matter 453 (2014) 100.
[94] R. Demoulin, G. Beainy, C. Castro, P. Pareige, L. Khomenkova, C. Labbé, F. Gourbilleau, E. Talbot, Origin of Pr^{3+} luminescence in hafnium silicate films: combined atom probe tomography and TEM investigations, Nano Futures 2 (2018), 035005.
[95] W. Lefebvre-Ulrikson, F. Vurpillot, X. Sauvage (Eds.), Atom Probe Tomography, Academic Press, 2016.
[96] J. Lu, Y. Kuo, J. Yan, C. Lin, Zirconium-doped hafnium oxide high-k memory device, Jpn. J. Appl. Phys. 45 (2006) L901.
[97] A. Claverie, C. Bonafos, G. Ben Assayag, S. Schamm, N. Cherkashin, V. Paillard, P. Dimitrakis, E. Kapetenakis, D. Tsoukalas, T. Muller, B. Schmidt, K.H. Heinig, M. Perego, M. Fanciulli, D. Mathiot, M. Carrada, P. Normand, Materials science issues for the fabrication of nanocrystal memory devices by ultra low energy ion implantation, in: Diffusion in Solids and Liquids, 531, 2006, pp. 258–260.
[98] B. Schmidt, D. Grambole, F. Herrmann, Impact of ambient atmosphere on as-implanted amorphous insulating layers, Nucl. Instrum. Methods Phys. Res., Sect. B 191 (2002) 482.
[99] G. He, Z. Sun (Eds.), High-k Gate Dielectrics for CMOS Technology, Wiley He et Sun Wiley-VCH Verlag & Co. KGaA, 2012.
[100] P.E. Coulon, S. Schamm, G. BenAssayag, B. Pecassou, M. Carrada, S. Bhabani, A. Slaoui, S. Lhostis, C. Bonafos, Ultra-low Energy ion implantation of Si into HfO_2-based layers for non volatile memory applications, in: Y. Fujisaki, R. Waser, T. Li, C. Bonafos (Eds.), Materials Research Society Symposium Proceedings, vol. 1160, 2009, pp. H01–H03.
[101] B.S. Sahu, F. Gloux, A. Slaoui, M. Carrada, D. Muller, J. Groenen, C. Bonafos, S. Lhostis, Effect of ion implantation energy for the synthesis of Ge nanocrystals in SiN films with HfO2/SiO2 stack tunnel dielectrics for memory application, Nanoscale Res. Lett. 6 (2011) 177.
[102] B. Thomas, Reed, Free Energy of Formation of Binary Compounds, MIT Press, Cambridge, MA, 1971.
[103] P. Dimitrakis, A. Mouti, C. Bonafos, S. Schamm, G. BenAssayag, V. Ioannou-Sougleridis, B. Schmidt, J. Becker, P. Normand, Ultra-low-energy ion-beam-synthesis of Ge nanocrystals in thin ALD Al_2O_3 layers for memory applications, Microelectron. Eng. 86 (2009) 1838–1841.
[104] V. Beyer, J. von Borany, A transient electrical model of charging for Ge nanocrystal containing gate oxides, J. Appl. Phys. 101 (2007), 094507.

CHAPTER

8

MOx in ferroelectric memories

Stefan Slesazeck[a], *Halid Mulaosmanovic*[a], *Michael Hoffmann*[a], *Uwe Schroeder*[a], *Thomas Mikolajick*[a,b], *and Benjamin Max*[b]

[a]**NaMLab gGmbh, Dresden, Germany,** [b]**TU Dresden, IHM, Dresden, Germany**

8.1 Introduction

Ferroelectric materials feature two non-zero spontaneous polarization states that can be reversed by application of an external electrical field being larger than the coercive field E_C. The two polarization states can be used to store binary information, representing a logical "0" or a "1." Several implementations of ferroelectric memories are under discussion. The ferroelectric capacitor based 1T1C concepts feature a cell structure that is comparable to the DRAM concept. The stored polarization state can be sensed directly as displacement current throughout the reading operation. Ferroelectric transistor implementations (FeFET) adopt the ferroelectric material as gate oxide such that the threshold voltage of the device depends on the polarization state of the material. In that way, the polarization state can be sensed in a non-destructive way. The internal gain of the FeFET device improves the sensing opportunities as compared to the 1T1C concept. Ferroelectric tunnel junctions (FTJ) rely on the modulation of the effective tunneling barrier by the polarization state of the ferroelectric material layer, thus offering non-destructive read out of the information. The polarization reversal is purely field driven and no current flow through the material stack is mandatory to change the polarization state of the ferroelectric. Therefore, compared to other non-volatile memory concepts the ferroelectric memories feature the intrinsic advantage of low-power operation. In the following sections the basics of ferroelectricity will be explained. Further, the current state of the art of the application of ferroelectric materials in the different memory flavors will be discussed.

Metal Oxides for Non-volatile Memory
https://doi.org/10.1016/B978-0-12-814629-3.00008-8

8.2 Ferroelectricity—A material property

Ferroelectricity is—similar to piezoelectricity—a material property that exists only in crystals that are non-centrosymmetric. From the 32 existing crystal point groups only 20 are non-centrosymmetric and might exhibit piezoelectric properties. However, only 10 out for these groups possess a unique polar axis and, therefore, allow the formation of stable electric dipoles. These permanent dipoles are referred to as spontaneous polarization. Upon change of the temperature these crystals change the dipole moment and are therefore referred to as pyroelectrics. Only the crystals that allow reorientation of the electric dipole by application of an electrical field belong to the ferroelectrics. That is, all ferroelectrics are a subgroup of the pyroelectrics and all pyroelectrics are a subgroup of piezoelectrics as is depicted in Fig. 8.1.

In the case where the direction of the polarization may be switched by 180° the material is referred to as uniaxial ferroelectric, whereas in the presence of multiple possible polarization axis the material is denoted a multiaxial ferroelectric. The ferroelectric response is typically characterized by the polarization response in terms of dielectric displacement to the applied electric field. At zero applied field the material will rest in the previously switched polarization state. Thus, the ferroelectric response results in a hysteresis loop as is shown in Fig. 8.2A. The hysteresis loop is characterized by the coercive field Ec that denotes the required electrical field for polarization reversal. Further, the remanent polarization P_r is the portion of ferroelectric polarization that remains upon release of the electrical field. The name ferroelectricity comes from the analogy to the hysteretic response in magnetization of ferromagnetic materials on magnetic fields.

From a thermodynamic perspective the ferroelectricity can be considered as resulting from a phase transition from a parent nonpolar phase to a polar phase [1]. In this theory the free energy is a function of the

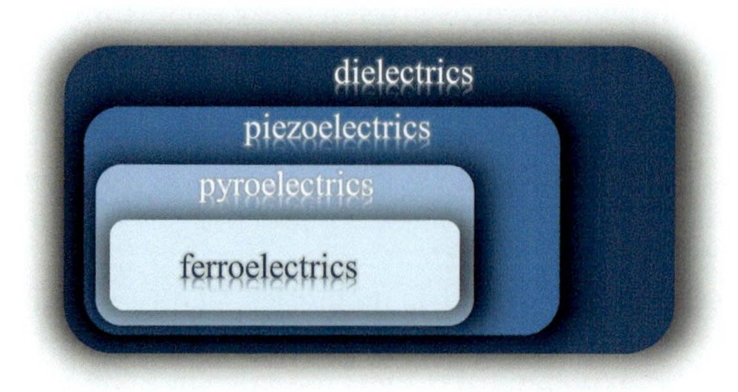

FIG. 8.1 Classification of ferroelectrics.

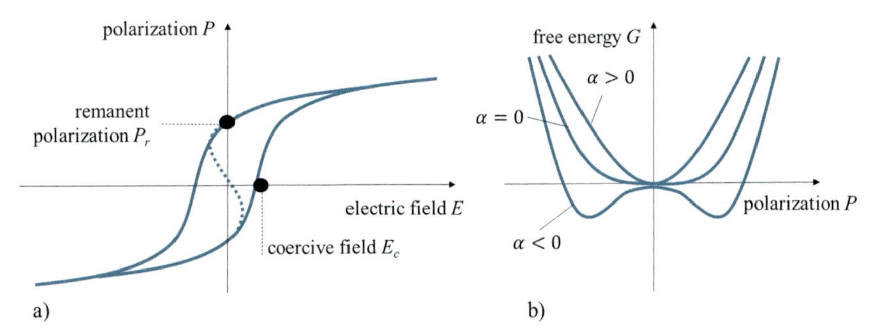

FIG. 8.2 (A) Ferroelectric polarization hysteresis, and (B) free energy G plot versus polarization, indicating two stable minima for $\alpha < 0$.

dielectric susceptibility and the spontaneous polarization that is measured as a dielectric displacement, where other external stimuli such as stress are neglected. The Gibbs free energy is

$$G = \left(\frac{\alpha}{2}\right)P^2 + \left(\frac{\beta}{4}\right)P^4 + \left(\frac{\xi}{6}\right)P^6 - EP, \tag{8.1}$$

where α is a temperature-dependent and β and ξ are temperature-independent ferroelectric anisotropy constants, P is the spontaneous electric polarization, E is the electric field in the ferroelectric. The electric displacement field is then defined as

$$D = \varepsilon_0 \varepsilon_r E + P, \tag{8.2}$$

where ε_0 and ε_r are the vacuum permittivity and relative permittivity of the ferroelectric, respectively. Further, the total polarization in the ferroelectric is given by [2].

$$P_{tot} = P + \varepsilon_0(\varepsilon_r - 1)E \tag{8.3}$$

By minimizing the Gibbs free energy form Eq. (8.1) with respect to the polarization, one obtains an expression for the electric field parallel to the polarization:

$$\frac{\partial G}{\partial P} = 0 \rightarrow E = \alpha P + \beta P^3 + \xi P^5 \tag{8.4}$$

When plotting the relationship between G and P for different α for no applied electric field as depicted in Fig. 8.2B one can observe the occurrence of two minima in the case of $\alpha < 0$. The values of these minima describe the spontaneous polarization as a function of temperature and can be derived by setting $E = 0$. By ignoring the minor contribution of the 5th order term we derive a description of the polarization depending on the ferroelectric anisotropy constants α and β:

$$P^2 = -\frac{\alpha}{\beta} \tag{8.5}$$

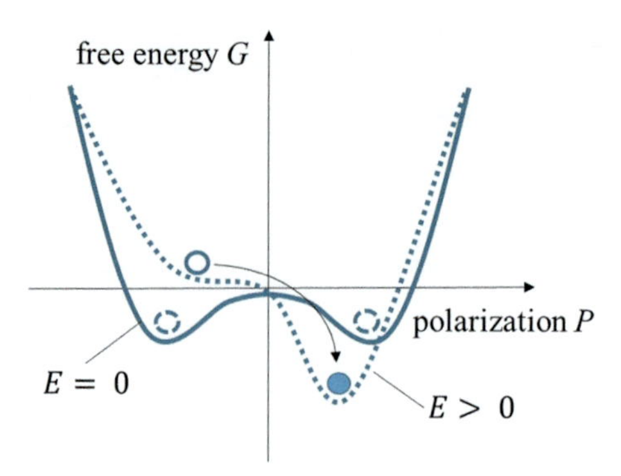

FIG. 8.3 Free energy plot versus polarization for an applied external field larger than the coercive field of the Ferroelectric.

Plotting the relationship between electrical field and polarization as given in Eq. (8.4), one obtains the field dependent polarization hysteresis as is shown schematically in Fig. 8.2A.

Obviously, the spontaneous polarization can rest in two distinct states under no applied electric field that are separated from each other by an energy barrier. That is, the two polarization states can be used to store binary information that can be assessed as a change in the electric displacement according to Eq. (8.2). The switching between the two polarization states, however, is induced by application of an electrical field that is large enough to tilt the energy landscape in a way that results in only one minimum, as is depicted in Fig. 8.3 for an applied electric field $E > 0$.

8.3 Negative capacitance in ferroelectrics

Negative differential capacitance emerges as a consequence of the double-well potential given by Eq. (8.1), that becomes visible in the free energy vs polarization curve as is depicted in Fig. 8.2B. For values of the temperature-dependent ferroelectric anisotropy constant α being smaller than zero the electrical field vs polarization plot exhibits a region with negative slope as indicated by the dashed line in Fig. 8.2A. In this case the ferroelectric capacitor can be used to amplify infinitesimal changes in the applied voltage. Negative capacitance (NC)-based transistors exploiting the ferroelectric instability in a FeFET structure, appear to be a promising solution for ever increasing leakage issues in ultra-scaled electronic devices caused by the Boltzmann limit [3]. However, due to the intrinsically unstable nature of the effect, direct measurements of NC on ferroelectric capacitors are always accompanied by a hysteresis [3–6]. To achieve stable NC operation without hysteresis, it was suggested that

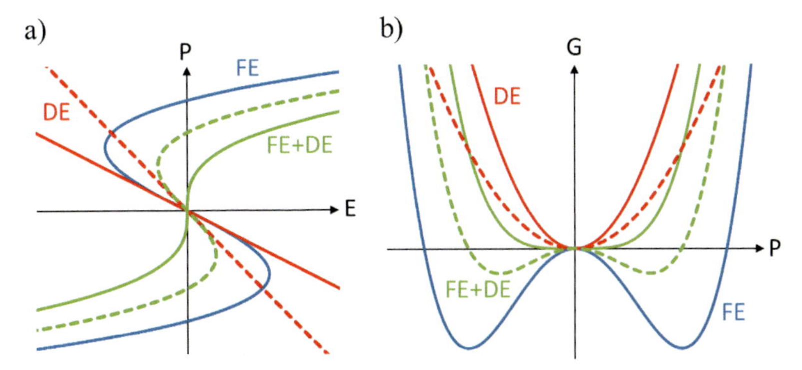

FIG. 8.4 (A) Polarization-electric field curves (*P-E*) and (B) free energy curves for the ferroelectric (FE), dielectric load line (DE) and the total system (FE + DE). Solid red and green lines indicate matched FE and DE capacitances for NC stabilization (no hysteresis). Dashed red and green lines show capacitance mismatch, which leads to hysteresis (NC is unstable).

the positive capacitance of a dielectric or semiconducting material in contact to the ferroelectric has to be added, such that the total capacitance of the system is positive [3,7]. This can be understood from the free energy and P-E curves in Fig. 8.4.

The ferroelectric S-shaped P-E curve is drawn as a blue line. Due to the energy barrier, which separates the two states in a ferroelectric memory, the P-E curve shows a region of NC (negative slope dP/dE). By addition of a dielectric layer with positive capacitance (red lines, parabolic free energy), the total energy of the system (green lines) can be modified such that the global energy minimum is in the NC region. However, this is only possible if the dielectric load line has exactly one intersection with the ferroelectric P-E curve, such that the total P-E curve always shows positive capacitance ($dP/dE > 0$). If there are more intersections of the ferroelectric P-E curve with the dielectric load line (see dashed red line), the NC state will be unstable, since the total system will still have two energy minima at finite polarization.

One has to keep in mind that this is a very simplified picture, where the ferroelectric is in a single-domain state, which does not have to be the case, since ferroelectrics are known to form domains under depolarization fields caused by dielectric "dead" layers [7]. However, Zubko et al. reported that even multi-domain ferroelectric/dielectric superlattices can exhibit stabilized NC effects due to domain wall motion [8]. Nevertheless, it was recently demonstrated that even when the ferroelectric NC state is not stabilized, it is possible to measure NC without hysteresis using pulsed voltage measurements to prevent charge injection to the ferroelectric/dielectric interface [9,10]. This charge injection might also be the reason for the often observed, frequency-dependent hysteresis in many NC devices [11].

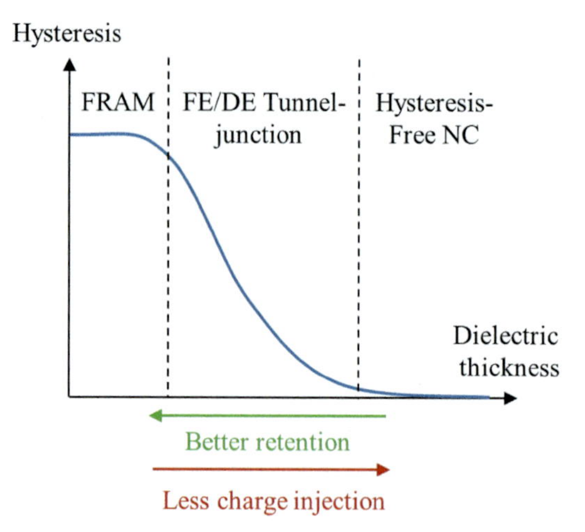

FIG. 8.5 Schematic dependence of the hysteresis as a function of the dielectric thickness in a ferroelectric/dielectric heterostructure capacitor. Depending on the desired application, the thickness can be adjusted accordingly.

When thinking about ferroelectric/dielectric capacitor structures more generally, depending on the thickness of the dielectric layer, there are three main operation regimes which are suited for different memory applications, see Fig. 8.5. For very thin dielectric layers, the device operates similar to a regular FeRAM capacitor (see Section 8.5.1), however, with reduced retention due to the non-zero depolarization field. The FeFET when used as a memory device that is described in more detail in Section 8.5.2, operates in this region as well. Here, the capacitance of the depletion zone in the silicon channel has to be taken into account together with the capacitance of the interfacial layer. For thicker dielectric layers, on the other hand, the device can be operated in the NC regime, where charge injection is negligible and if the capacitances of both layers are matched. In the region in-between, where charge injection still occurs, it is possible to use the device as a ferroelectric/dielectric tunnel junction, as demonstrated recently [12] and will be discussed in more detail in Section 8.5.4.

8.4 Ferroelectricity in hafnium oxide

Throughout CMOS technology development with the goal to further facilitate the Dennard scaling, the hafnium oxide was researched in the late 1990s to be adopted as high-k gate dielectric in its amorphous phase. Hafnium oxide has reasonably high permittivity and bandgap and is stable on silicon and became available in 2007 [13,14]. Moreover, the crystalline phases of HfO_2, that features even higher permittivity than the amorphous phase, was researched to be implemented as dielectric into 3D capacitors of DRAM cells. Different crystalline phases of the hafnium oxide are known and can be formed under different manufacturing conditions. The monoclinic (space group: $P2_1/c$), the tetragonal (space group: $P4_2/nmc$), and the cubic (space group: $Fm\bar{3}m$) phases feature permittivities of 22, 46, and 36, respectively [15]. The thermodynamically most stable phase is the monoclinic phase that in the bulk material can be transformed to the tetragonal and cubic phase at 1973 and 2773 K, respectively. In thin films the crystallization properties can be adjusted by different means, such as introduction of dopant species or applying stress to the film throughout crystallization i.e., by the adoption of stress liners or capping with electrode materials. That is, since the early 2000s the hafnium oxide is a well-established material used in front end of line (FEOL) and back-end of line (BEOL) integration of modern CMOS technologies and has been intensively studied.

Throughout the research on hafnium oxide in 2007 the first indication of ferroelectricity in hafnium oxide was observed and firstly published in 2011 [16], when doping it with silicon in the few % range [16,17]. Already since the first observation of ferroelectricity in HfO, the working theory was that the non-centrosymmetric orthorhombic Pca21 phase, that is not present in the bulk phase diagram [18], is responsible for this behavior. In this phase the polarization of the material is induced by the displacement of four oxygen atoms in the non-centrosymmetric unit cell due to the application of an electrical field. Fig. 8.6 compares the crystal structure of the ferroelectric orthorhombic phase (center) to the monoclinic (left) and tetragonal phase (right). The result of the oxygen displacement are two distinct and stable structural states with corresponding electrical dipoles that in turn induce an electrical field even at zero applied external field [19]. Thus, HfO_2 is the first fluorite-structure oxide material, in which

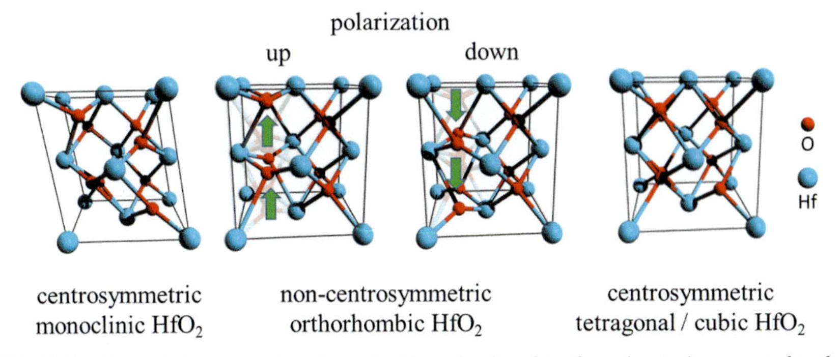

FIG. 8.6 Crystal structure of the ferroelectric orthorhombic phase (center) compared to the monoclinic (left) and tetragonal phase (right).

ferroelectricity was induced. Since then, various dopants such as Zr [20,21], Y [22], Al [23], Gd [24,25], Sr [26], and La [27–29] have been reported to induce ferroelectricity in ALD HfO_2 thin films. Moreover, in [30] it was shown that ferroelectricity can be even observed in undoped HfO_2 thin films.

Besides the electrical characterization of metal-ferroelectric-metal capacitors and FeFETs that were realized by integrating ferroelectric hafnium oxide into existing high-k metal gate (HKMG) technologies the presence of ferroelectric properties was further supported by experimental results of different physical analysis techniques such as piezoresponse force microscopy (PFM) [31] and the confirmation of the presence of the $Pca2_1$ phase by position averaged convergent beam electron diffraction analysis [18]. Moreover, pyroelectricity [32,33] has been demonstrated as well, that should be an inherent property of ferroelectrics, as is depicted in Fig. 8.1. Overall, at the present there is a widely accepted understanding that the ferroelectric hysteresis in hafnium-/zirconium-based ferroelectrics does not originate from artifacts such as leakage [34] or charge trapping.

At present there is still increasing research focus on unveiling all the aspects that lead to the formation of the orthorhombic phase, and the contribution of different factors such as stress, grain size by influencing the surface or interface energy, and doping is still unclear. In [15] it was proposed that surface energy can be an important factor, considering the fact that the gain in free energy due to stress from doping is not sufficient to explain the stabilization of the orthorhombic phase. The finding is supported further by the fact that stabilization is enhanced in thinner films [35]. This theory was recently confirmed in [36] by comparing the proposed model to experimental data in hafnium-zirconium-oxide. Moreover, the observed increasing monoclinic phase fraction with increasing annealing temperature could not be understood. This points to the conclusion that not only thermodynamics, but also kinetic effects need to be considered to explain the physical origin of the formation of the orthorhombic phase [37]. However, based on the results of the research in this field throughout the last decade it is now established that ferroelectricity can be reproducibly achieved in thin hafnium oxide-based films in the thickness range of 5-30 nm, thus being mature enough to be adopted in development of memory devices.

8.4.1 Switching kinetics at nanoscale

(This section is in part a Reprinted with permission from [38]. Copyright 2017 American Chemical Society and from [39]. Copyright 2018 American Chemical Society.)

In ultra-scaled FeFETs only a small number of ferroelectric domains are incorporated in the gate oxide. Thus, the electrical characteristics of such FeFETs are determined by just a single or a few ferroelectric domains. Due to the intrinsic amplification effect of the transistors these unique structures allow for detection of abrupt threshold voltage (V_T) shifts being caused by a discrete polarization reversal of single domains. That fact was adopted in [38] for the analysis of the switching kinetics of single ferroelectric domains in FeFET devices comprising a polysilicon/TiN (8nm)/HfO_2 (10nm)/SiON (1.2nm) gate stack and featuring a channel width and length of $W=80$nm and $L=30$nm, respectively. Fig. 8.7B shows the cross-sectional TEM image of such a device. The top-down view in Fig. 8.7C reveals the presence of only a few grains in the ferroelectric HfO_2 gate oxide after the gate electrode was removed. Details on the deposition, doping and fabrication methods, as well as a demonstration of the ferroelectric properties of the device have been reported elsewhere [42–44].

The polarization state of the FeFET devices can be measured by extracting the threshold voltage from the transfer characteristic as is depicted in Fig. 8.5D. By applying a positive gate voltage, the polarization of the ferroelectric layer can be changed, yielding a downward polarization ($P\downarrow$). The positive polarization surface charge at the HfO_2/SiON interface induces a channel inversion, yielding a low V_T state (red curve). By application of a negative gate pulse, instead, the upwards oriented polarization ($P\uparrow$) results in a high V_T state (blue curve). These two distinct and stable values of V_T are typical for large area FeFETs and can be readily exploited to store binary information, as is elaborated on in more detail in the section "Planar 1T FeFET implementation".

To gain insight into the influence of the domain granularity on the ferroelectric switching, short ($t_{PW}=500$ns) write voltage pulses (of amplitude VP$>$0 or VN$<$0) were applied to the gate electrode (see inset Fig. 8.5E), keeping the other three terminals (source, drain and bulk) grounded. Subsequently, the polarization state was determined. When the VP amplitude is increased progressively form 2 to 5V, a set of ID-VG curves shown in Fig. 8.7E is obtained. Four different and separate current branches can be readily observed. To better visualize this point, the current levels at VG=0.36V and the VT levels corresponding to each VP pulse are extracted from Fig. 8.7E and shown in Figures Fig. 8.77F and G respectively. Interestingly, the switching behavior sensed both as a current level or VT shift is very abrupt. This finding is in a strong contrast to a rather gradual switching observed in large area FeFETs that were reported in [40]. Three distinct VT shifts appear in correspondence to different VP levels, VSW1, VSW2 and VSW3, thus indicating three separate switching events. Similar and completely specular behavior for the transition from low to high VT state is seen when VN is varied and VP kept constant at

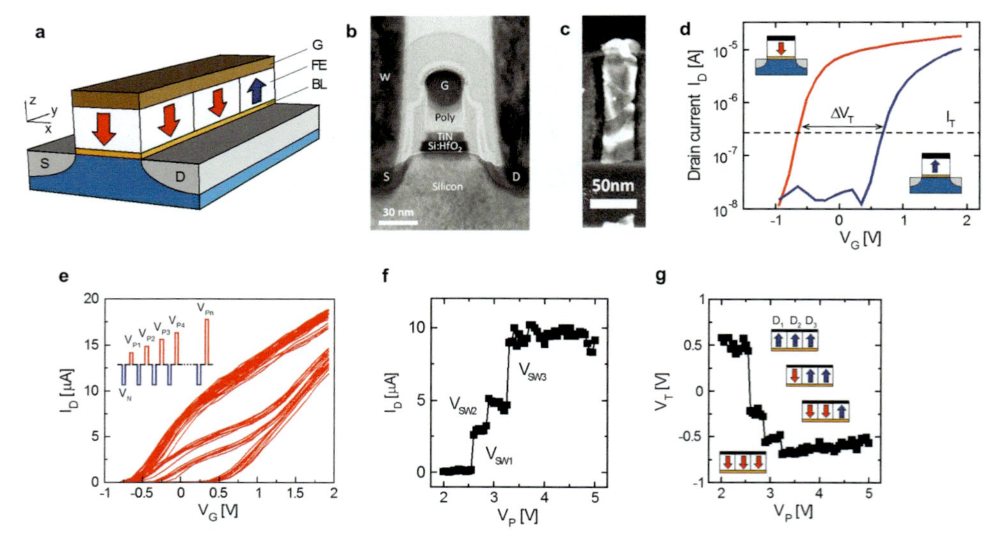

FIG. 8.7 Nanoscale Ferroelectric FET and its abrupt switching events: (A) Schematic illustration of FeFET structure. (B) Cross-sectional TEM image of a Silicon doped HfO_2 FeFET device with L = 30 nm "G," "S," and "D" indicate "gate," "source" and "drain" terminals, respectively. (C) Top-down TEM image showing the ferroelectric HfO_2 after the gate electrode was removed. It reveals the presence of only a few grains. (D) I_D-V_G transfer characteristics of a 30 nm long device after a positive (V_P) and a negative (VN) gate pulse, which correspond to polarization down ($P\downarrow$, *red curve*) and up ($P\uparrow$, *blue curve*), respectively. (E) I_D-V_G curves read-out after each of the progressively increasing V_P pulses in the inset, (F), drain current from (D), extracted at $V_G = 0.36$ V vs the corresponding V_P, showing four different current levels, (G), threshold voltage extracted from (E), vs the corresponding V_P. Three discrete and abrupt V_T shifts correspond to separate switching of three domains within the stack (inset). *Reprinted with permission from H. Mulaosmanovic, et al., Switching kinetics in nanoscale hafnium oxide based ferroelectric field-effect transistors, ACS Appl. Mater. Interfaces 9 (4) (2017) 3792–3798. Copyright 2017 American Chemical Society.*

+4 V as a reference pulse (here not shown). The common understanding is that that these ΔVT shifts result from a discrete switching of three ferroelectric domains (D1, D2 and D3) as schematically depicted in Fig. 8.7G within the gate stack, having three distinct coercive voltages (V_C). The switching voltage VSW is directly proportional to the voltage drop over the ferroelectric at the moment of polarization reversal and hence to the respective coercive voltage VC of the domain. Different VC levels may originate from various grain size [45,46], grain orientation [47] and/or local doping variations. Moreover, once D1 is switched, the electric field distribution in the gate stack changes, which influences the switching of the other two domains. It is, nevertheless, important to underline that the origin of the different VC cannot be uniquely studied using FeFETs, due to their architecture and a single gate electrode covering the whole ferroelectric layer. A possible solution could be to excite single domains on a bare ferroelectric HfO2 film with a tip of a highly accurate PFM, which, however, in practice turned out to be very difficult [47,48].

The polarization reversal at the domain level is known to be a stochastic process [49]. This property can be detected when a domain is excited in the proximity of its VC, where a probabilistic rather than a deterministic switching occurs. Fig. 8.8A shows the resulting VT cycles of a device after performing a sequence of 50 switching pulses for each considered VP, exhibiting only one abrupt ΔVT shift at VSW = 3.2 V (see inset Fig. 8.8A). For VP = 3.15 V an oscillation between the two VT states is clearly visible, indicating the impossibility to switch the domain at every trial. When, instead, VP = 4 V ≫ VSW the domain is always switched into the P↓ state. As a result, each VP amplitude can be associated with a certain probability to switch the polarization state, which leads to a strongly bias-dependent switching probability curve shown in Fig. 8.8B. Interestingly, repeating this experiment on a device with a remarkably large single ΔVT shift (Fig. 8.8C) and adopting a very fine step for increasing VP levels, an oscillatory behavior between few intermediate VT states can be identified. Fig. 8.8D reveals the existence of two intermediate VT levels which are only randomly accessible, thus indicating the presence of at least two domains within the ferroelectric film with very similar VC values.

The kinetics of polarization reversal were analyzed by repeating the experiment in Fig. 8.7E (inset) for different pulse widths (t_q) of VP. As can be seen in Fig. 8.9A and B), the switching voltage VSW decreases with increasing t_{PW}, both for VP and VN sweeps. In general, ferroelectrics do not exhibit a well-defined coercive voltage VC. Instead, it strongly depends upon the length of time over which the electric field is applied [50]. This phenomenon is usually observed in the polarization-voltage (P—V) hysteretic loops, where VC increases with increasing excitation frequency f. Two models are generally accepted to explain this dependence. The Kolmogorov-Avrami-Ishibashi (KAI) model is based on the

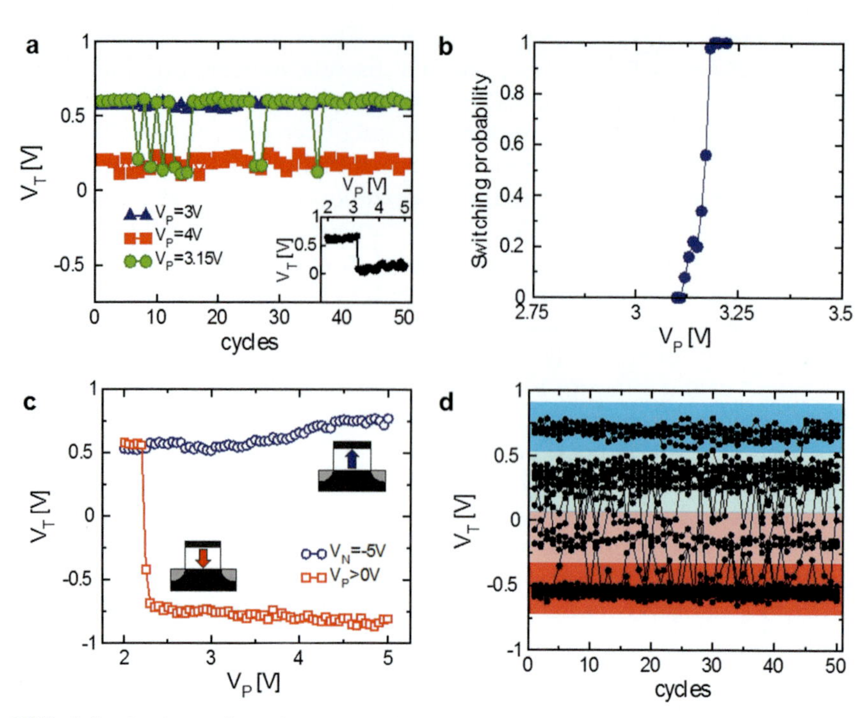

FIG. 8.8 Stochastic ferroelectric switching and intermediate VT states. (A), Oscillatory behavior of the single domain polarization in proximity of the coercive voltage (VP = 3.15 V). (B), Switching probability of the device in function of VP, (C), VT vs VP graph obtained with procedure in Fig. 8.9 (D) (inset) for a device exhibiting a large ΔVT. (D), Oscillatory switching between intermediate VT states for the device in (C), revealing the presence of at least two domains in the ferroelectric layer. VP was varied from 2 to 2.5 V with a step of 10 mV. *Reprinted with permission from H. Mulaosmanovic, et al., Switching kinetics in nanoscale hafnium oxide based ferroelectric field-effect transistors, ACS Appl. Mater. Interfaces 9 (4) (2017) 3792–3798. Copyright 2017 American Chemical Society.*

nucleation and propagation of reversed domains [51]. Here, VC follows the simple power-law relationship VC ∝ f β and has been successfully applied for bulk crystals and clean epitaxial films [52]. On the other hand, polycrystalline disordered ferroelectrics strongly deviate from this relation [53] and several domain nucleation limiting models have been therefore proposed [53,54]. Here, the time of domain wall movement is neglected compared to the nucleation time. The Du-Chen model [54] appears to be particularly suitable in describing field and temperature dependent polarization switching. It assumes the existence of a critical size defined by an attractive potential which nucleating domains must overcome. Following a similar approach of Merz [55] and Landauer [56] for the free energy of domain nucleation within the framework of the classical nucleation theory, a simple expression for the average nucleation waiting time can be found:

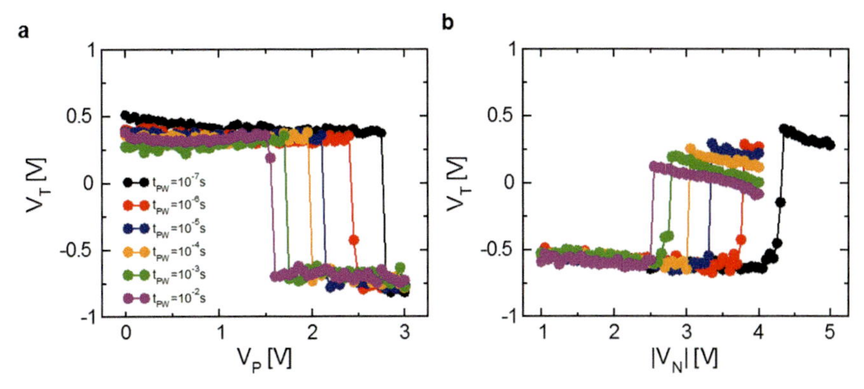

FIG. 8.9 Switching voltage dependence on excitation pulse width. (A), VT vs VP graph and (B), VT vs |VN| graph for pulse widths ranging from 100ns to 10ms. The reduction of ΔVT shift with increasing pulse width tPW in (B) is due to a strong trapping of holes coming from the silicon substrate [40,41]. *Reprinted with permission from H. Mulaosmanovic, et al., Switching kinetics in nanoscale hafnium oxide based ferroelectric field-effect transistors, ACS Appl. Mater. Interfaces 9 (4) (2017) 3792–3798. Copyright 2017 American Chemical Society.*

$$\tau = \tau_0 \cdot \exp\left(\frac{\alpha}{k_B T} \cdot \frac{1}{V_{SW}^2}\right), \tag{8.6}$$

where τ_0 is the shortest nucleation time possible and α is related to intrinsic material properties including domain wall energy and the portion of a switched polarization by the nucleus.

Aiming at a statistical study of the time and field dependence of switching in more detail, the VN-read-VP-read sequence with logarithmically increasing tPW of the VP pulse was adopted (Fig. 8.10A). The procedure was repeated 20 times for each considered VP level. As evident in Fig. 8.10B, the switching time for a fixed VP spreads approximately over one decade and its mean value decreases with increasing VP. Reporting the experimental standard deviation $\sigma_{t_{SW}}$as a function of the average switching time $<t_{SW}>$ in a log–log graph (Fig. 8.10C), the unity slope over several decades can be clearly identified. This strongly hints at the stochastic switching ruled by a Poisson process.

It is known that the domain wall motion in ultra-thin films is extremely slow (e.g., about 1nm/s under electric field of 1MV/cm in PZT [58]) and recently was determined for HfO_2 to be even slower [59]. Therefore, the switching behavior in the thin films was modeled with a purely nucleation-limited scenario. Given the small size of grains, it is expected that they will contain only a single domain [60]. It is supposed that the switching is initiated when a relatively small number of critical nuclei are generated within one grain upon the application of the electric field,

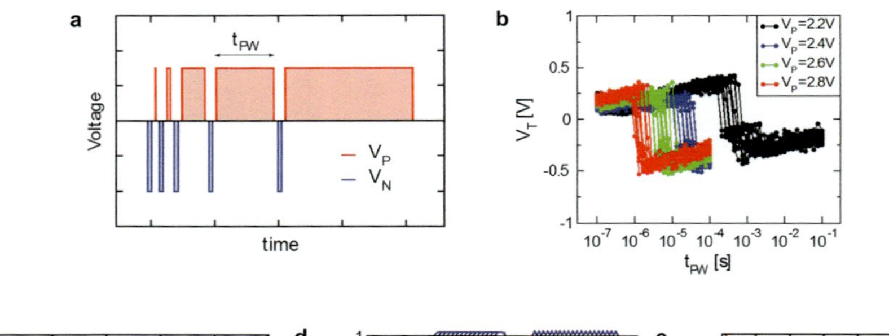

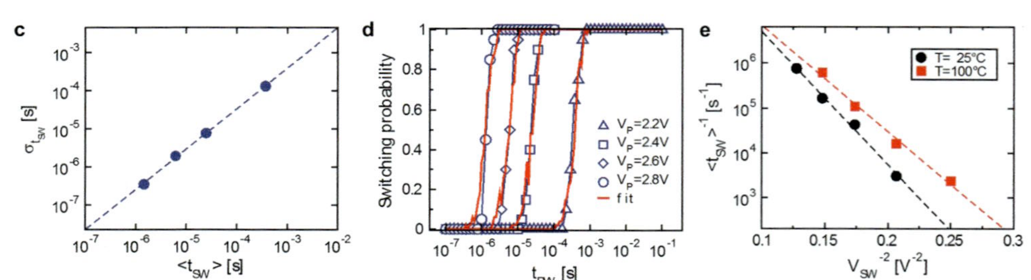

FIG. 8.10 Field, time and temperature dependence of ferroelectric switching in a nucleation-limited scenario. (A), Gate voltage waveform with logarithmically increasing pulse width. After each pulse a fast read out of a transfer curve is carried out. (B), VT vs tPW graph corresponding to 20 repetitions of procedure in a, shown for four different VP levels. For VP = [2.8, 2.6, 2.4] V the measurement was truncated at tPW = 100 µs, in order not to stress the device for longer pulse widths. (C), Standard deviation vs mean value of switching time tSW extracted from (B). (D), Fitting of switching probability curves corresponding to four different VP levels used in b. e, Experimental mean switching time $<tSW>^{-1}$ vs V_{SW}^{-2} for two different temperatures is in agreement with nucleation dominated switching model [57]. *Reprinted with permission from H. Mulaosmanovic, et al., Switching kinetics in nanoscale hafnium oxide based ferroelectric field-effect transistors, ACS Appl. Mater. Interfaces 9 (4) (2017) 3792–3798. Copyright 2017 American Chemical Society.*

with the generation rate given by $\lambda = 1/\tau$ in Eq. (8.6). Within this picture, the polarization of the grain is considered reversed when a certain critical number n of generated nuclei merge together into a single domain occupying the entire grain. Assuming the nucleation to be the Poisson process, which has already been reported for classical ferroelectrics [49] and supported by Fig. 8.10C for the discussed ferroelectric system. The time elapsing between each critical nucleus generation ΔTi will be exponentially distributed as

$$p_{\Delta T_i} = \lambda e^{-\lambda \Delta T_i} \tag{8.7}$$

where $p_{\Delta T_i}$ is the probability density function of ΔTi . Thus the overall grain/domain switching time will be given by the sum of n individual ΔTi intervals corresponding to the number of critical nuclei necessary to form the domain:

$$t_{SW} = \sum_{i=1}^{n} \Delta T_i \tag{8.8}$$

with the mean and variance given by Eqs. (8.9) and (8.10), respectively:

$$< t_{SW} > = \frac{n}{\lambda} \tag{8.9}$$

$$\sigma_{t_{SW}}{}^{2} = \frac{n}{\lambda^2} \tag{8.10}$$

From that it is straightforward to determine n and λ from the experimental data for $<t_{SW}>$ and $\sigma_{t_{SW}}$ using Eqs. (8.9) and (8.10). Generating then n values of exponentially distributed ΔTi as dictated by Eq. (8.7), summing them up according to Eq. (8.8) and repeating this procedure 20 times for each pulse width (tPW) as in the experimental conditions, it is possible to simulate the probabilistic switching. As shown in Fig. 8.10D, the switching probability curves, extracted from Fig. 8.10B, are well fitted over the whole time range, with $n = 5$ nuclei for the considered device. Note that this is a reasonable quantity considering the dimensions of the ferroelectric film. Moreover, plotting a $<t_{SW}>^{-1}$ vs $1/(V_{SW})^2$ graph (Fig. 8.10E), an exponential behavior as in Eq. (8.6) can be deduced. Furthermore, as expected from [57,61] for the discussed model description, the temperature experiments confirm the fact that the switching voltage decreases with increasing T. It is worthwhile to point out, that the analysis presented so far is valid only for the switching of a single domain. When, instead, multi-grain and therefore multi-domain structures are considered, even small fluctuations of α and VSW can lead to an exponentially broad spectrum of τ among different grains. Hence, the Tagantsev's approach [53] might be more appropriate for large area devices [62].

The experiment of Fig. 8.10 clearly reveals that the switching time can be exponentially reduced by increasing the applied bias if high-speed operation is desired. Nevertheless, fast switching and long data retention are highly desirable but often conflicting requirements for non-volatile memory devices. Therefore, the correct tailoring of material parameters τ_0 and α in Eq. (8.6) is of utmost importance for optimizing both requirements.

8.4.2 Accumulative switching

From the relation of mean switching time tSW vs switching voltage as depicted in Fig. 8.10D and E it is clear that the switching under a single pulse having tPW shorter than i.e., 10 μs is precluded for VP = 2.2 V. In [39] it was shown that by applying consecutively several shorter pulses, each of which is insufficient for switching (e.g., VP = 2.2 V, tPW = 1 μs; inset Fig. 8.11A, the switching eventually takes place after a certain critical number of delivered pulses (Fig. 8.11A). This unexpected phenomenon clearly points to a sort of accumulative polarization reversal within the ferroelectric. It is argued that this occurs most probably through a successive formation of ferroelectric nuclei (nanodomains) of the opposite polarization orientation, which gain in their quantity and size with increasing number of electrical pulses, as schematically depicted in Fig. 8.11C. The nuclei formation might be preferentially initiated at the defect sites [63,64], i.e., grain boundaries (due to the polycrystalline film) and ferroelectric/dielectric interface. Following this picture, the succession of gate pulses will generate nanodomains up to a critical point at which the polarization reversal of the entire grain occurs. Consequently, this creates an uniform polarization pathway between the source and drain regions, which turns the device on. This peculiar feature of the accumulation of electrical stimuli through ferroelectric nucleation can be viewed as energy accumulation within the material. To some extent, this represents an analogy to the accumulative crystallization found in phase change materials upon multiple optical [65] or electrical excitations [66,67]. Moreover, this switching mechanism contrasts the traditional binary memory operation mode, where a FeFET cell is deterministically switched using a single pulse.

It should be noted that the transition from OFF to ON state is evident only after all necessary pulses to switch are delivered to the device. This all-or-none behavior can be further appreciated in Fig. 8.11B, which illustrates the evolution of the threshold voltage V_T with the increasing number of excitation pulses. The V_T level remains almost invariant over the first 11 pulses. Then, the 12th pulse abruptly lowers the V_T level to a value which does not change upon additional pulses. The fact that no

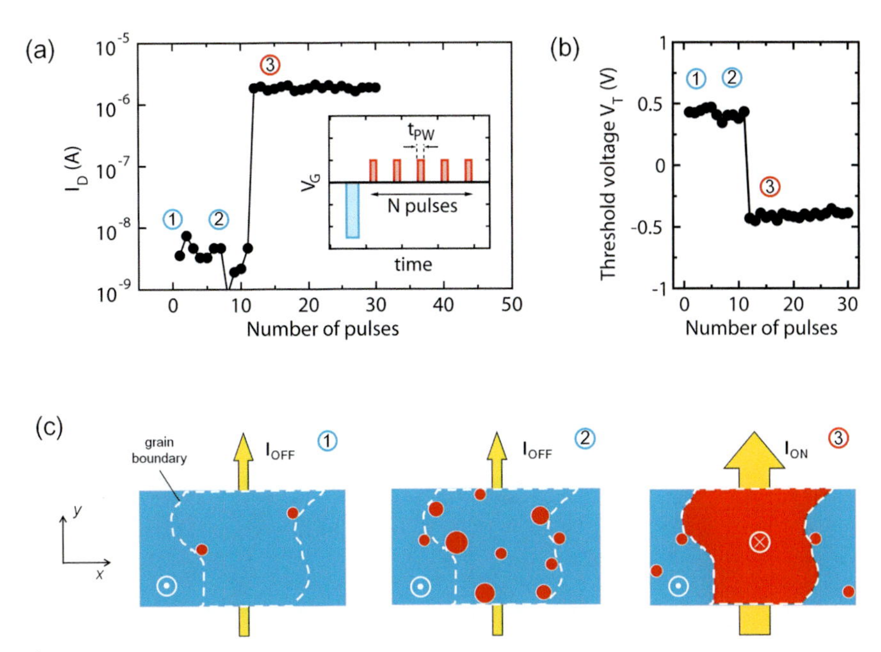

FIG. 8.11 Accumulative polarization reversal in ferroelectric transistors. (A) Sharp switching from OFF to ON after a number of received identical pulses ($V_P = 2.2\,V$, $t_{PW} = 1\,\mu s$) as depicted in the inset. The time distance between single pulses is $\Delta t = 100\,ns$. (B) The threshold voltage V_T was extracted at the drain current level $I_D = 0.1\,\mu A\ W/L$; (C) Schematic illustration of the top-down view of the ferroelectric film (similar to the top-down TEM image of Fig. 8.9B) showing the formation of reverse ferroelectric domains starting from OFF state (polarization up ⊙, *blue coloring*). Three stages illustrate the domain evolution with increasing N as indicated in (A): (1) onset of the nucleation of the domains having polarization down (⊗) (*red coloring*); (2) further nucleation and domain growth upon increasing number of pulses prior to switching; (3) polarization reversal within the entire grain between drain and source regions after a critical number of received pulses, which causes the sharp drain current increase (I_{ON}). *Reprinted with permission from H. Mulaosmanovic, T. Mikolajick, S. Slesazeck, Accumulative polarization reversal in nanoscale ferroelectric transistors, ACS Appl. Mater. Interfaces 10 (28) (2018) 23997–24002. Copyright 2018 American Chemical Society.*

intermediate V_T states are elicited by incoming electrical pulses drastically distinguishes these small-area ferroelectric transistors from the large-area ones. In fact, the latter ones display a gradual polarization switching, which corresponds to a continuum of states between the nominal high-V_T and low-V_T state [68,69], and could be potentially exploited for the multi-level information storage.

As expected from the theory of ferroelectric nucleation [53,55], and confirmed by previous reports for hafnium oxide [62] the switching is both time and field dependent. This implies that longer excitation pulses and/or higher pulse amplitudes will favor the switching. In fact, Fig. 8.12 fully confirms this behavior: when increasing V_P from 1.6V to

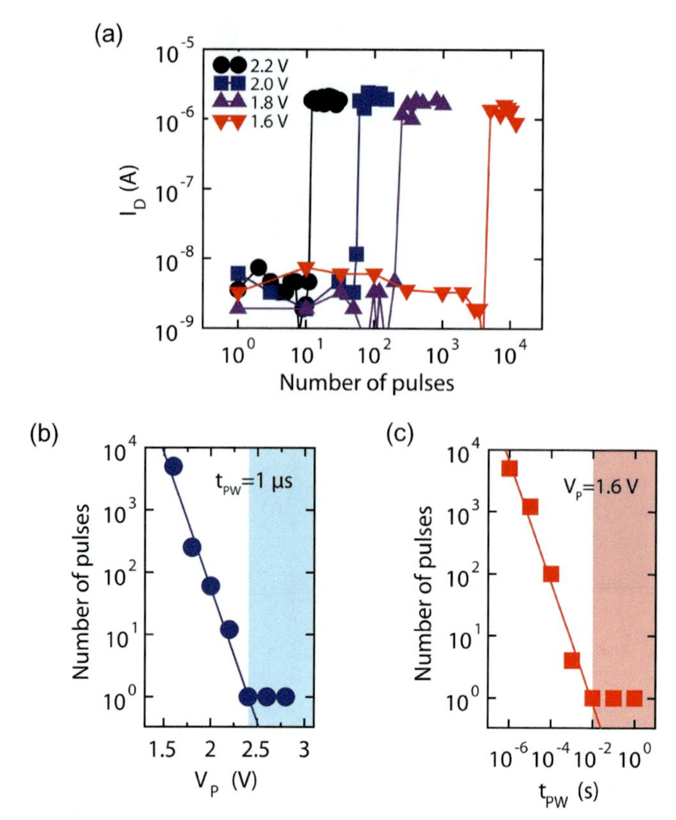

FIG. 8.12 Dynamics of the accumulative switching. (A) Impact of gate pulse amplitude V_P: lower V_P induces a slower domain nucleation and a higher number of pulses is necessary for switching. The abrupt change in drain current upon switching is independent on the strength or duration of individual incoming spikes, but occurs only after all necessary pulses are delivered (all-or-none behavior); Number of pulses to switch increases exponentially with lower V_P (B) and almost linearly with shorter t_{PW} (C), in accordance with ferroelectric switching theory. The shaded area represents the range of V_P (B) and t_{PW} (C) values, where the switching occurs deterministically under a single pulse, i.e., accumulative switching is absent. *Reprinted with permission from H. Mulaosmanovic, T. Mikolajick, S. Slesazeck, Accumulative polarization reversal in nanoscale ferroelectric transistors, ACS Appl. Mater. Interfaces 10 (28) (2018) 23997–24002. Copyright 2018 American Chemical Society.*

2.2 V, while keeping $t_{PW} = 1\,\mu s$, the number of pulses to switch decreases exponentially from more than 5000 to only 12. A further increase in pulse amplitude beyond 2.4 V makes the device to exit the accumulation mode and enter the binary switching mode (shaded area in Fig. 8.12B), where a single pulse is sufficient to reverse the polarization. Fig. 8.12C) shows that by increasing the excitation pulse width while keeping V_P constant, the number of pulses to switch decreases almost linearly. These experiments indicate that one can tailor the number of pulses to switch by appropriate choice of pulse amplitude and duration. Moreover, note that the all-or-none switching behavior is preserved regardless of the number of pulses and pulse parameters applied (Fig. 8.12A).

Fig. 8.13 reveals an additional property of the accumulative switching in FeFETs. When the experiment of Fig. 8.11A) is repeated several times with the unchanged set of parameters, the number of pulses to switch experiences a certain spread of values. This can be understood by considering that the ferroelectric nucleation is inherently a stochastic process. In fact, by studying the nucleation and domain wall growth in $PbTiO_3$ perovskite ferroelectrics by means of molecular dynamics simulations, it was found that nucleation behaves as a Poisson process [49].

Although the accumulative switching property in nano-scaled FeFETs represents an attractive feature, i.e., for its adoption in neuron circuits, it poses constraints when it comes to the memory operation. In fact, in order to obtain high storage densities and compete with existing or emerging memory concepts, FeFETs are usually organized in memory array architectures (e.g., NAND or AND arrays), where many devices share the same gate line. If one device in a row (selected device) is programmed, all other devices sharing the gate line (unselected devices) will experience a certain excitation. Although the impact of such disturb pulses is largely reduced by proper inhibit Schemes [70] (for instance, by applying a certain voltage

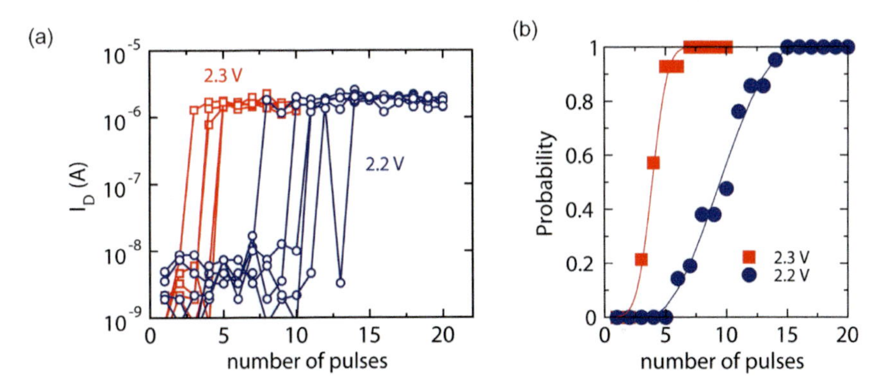

FIG. 8.13 Stochastic accumulative switching. (A) Polarization reversal in nanoscale ferroelectric transistors is a stochastic process due to a probabilistic nucleation mechanism. This leads to a variation of number of pulses to switch, even if the experiment of Fig. 8.11A is repeated with identical set of parameters. Five switching curves for each of the two values of V_P are shown; (B) Probability of switching vs number of received pulses obtained by repeating 20 times the experiment of Fig. 8.11A for each V_P. The solid lines are only guides for the eye. *Reprinted with permission from H. Mulaosmanovic, T. Mikolajick, S. Slesazeck, Accumulative polarization reversal in nanoscale ferroelectric transistors, ACS Appl. Mater. Interfaces 10 (28) (2018) 23997–24002. Copyright 2018 American Chemical Society.*

at source and drain terminals in order to lower the overall voltage drop across the unselected gate stacks), a fraction of the total programming amplitude will however be delivered to unselected devices. Consequently, upon repeated program cycles on the same gate line and due to the accumulation property, these devices could finally undergo an undesired switching. This needs to be taken into account upon analyzing the array disturb effects in dense FeFET memory arrays.

8.5 Ferroelectric memories

Already in 1952 [71,72] first realizations of ferroelectric memories based on barium titanate crystals were investigated. However, already at this time researchers faced the problem of unavailability of a suitable selector devices to enable memory array operation. It took four decades of semiconductor technology development until a first ferroelectric 1T1C memory could be realized in the 1990s [73]. Since then many attempts were made to create a "universal memory" based on ferroelectric materials featuring ultra-low power consumption and high endurance [74,75]. However, due to the incompatibility of the complex perovskite or layered perovskite ferroelectrics with standard front end of line (FEOL) CMOS processing such as hydrogen sensitivity, high thermal budget in the case of FeFET integration or unavailability of suitable deposition techniques that could enable 3D integration of the materials in the case of the 1T1C concept [76,77] the FeRAM technology development stopped at the 130 nm ground rule [78]. Due to the resulting cell-size driven relatively high manufacturing costs per bit the application of the FeRAM stayed limited to niche markets in which low-power write operation and high endurance is of utmost importance. In the FeFET implementation the high permittivity of the perovskite materials (i.e., $SrBi_2Ta_2O_9$ (SBT)) lead to a depolarization field E_{dep} across the ferroelectric in the retention case. It arises from an incomplete polarization surface charge compensation by the confining layers and can be approximated by [78]:

$$E_{dep} = P\left[\varepsilon_0\varepsilon_F\left(\frac{C_{IS}}{C_F}+1\right)\right]^{-1} \tag{8.11}$$

where P, ε_0, ε_F and C_F are the polarization, vacuum permittivity, dielectric constant and capacitance of the ferroelectric, and C_{IS} is the series capacitance of the interface layer and semiconductor. The retention loss due to the depolarization field could be mitigated only by increasing the thickness of the ferroelectric layer into the 100 nm range, thus impeding scalability of the concept. [78,79].

8.5.1 1T1C FeRAM implementation

In the 1T1C FeRAM the data is stored as polarization in a capacitor, which is the most straightforward way to realize a ferroelectric memory. In this case the dielectric of the capacitor is replaced by a ferroelectric. Moreover, the plate line (PL) is pulsed instead of keeping it at half VDD voltage. Pulsing of PL is mandatory, since during destructive readout upon application of a read voltage pulse at the plate line PL the polarization charge is transferred via a selector transistor to the bit line BL where then similar to the DRAM concept a sense amplifier determines the stored logic state. In contrast, in DRAM the information is stored by the charged capacitor. In the first FeRAM devices the information was written complementary into two cells to increase the signal—one written to logical "0" and a second cell written to logical "1." Both cells were combined to realize one bit to increase the signal to noise margin. Later on the read polarization charge was compared to the charge from a reference cell that was used as reference for several bits, thus decreasing the effective cell size. The first products appeared on the market as early as 1993 [73], and today, advanced products using 130-nm technology are on the market [78]. However, further success beyond niche markets is hindered by the fact that a 3D capacitor has not been practically realized using perovskite materials [76,77], despite some success with SBT in low aspect ratio devices [80].

That situation changed with the discovery of ferroelectricity in hafnium oxide that enabled the full 3D integration [81]. However, in the hafnium oxide the high coercive field becomes a limiting factor. Low operation voltages become challenging and the switching of the polarization requires fields close to the breakdown value, resulting in limited cycling endurance [82]. Similar as in the DRAM, after destructive readout the information has to be refreshed. Thus, for continuous operation of the device the trade-off between access time and electric field and thus cycling endurance has to be matched. Both are linked by the intrinsic relation between switching speed and applied electrical field [38]. For example, at a given endurance in the order of 10^{12} switching cycles at 10 year lifetime a 100 μs access time could be realized. Lowering the oxide thickness can result in lower operation voltages at constant electrical field. Oxide thickness as low as 5 nm was shown to exhibit good electrical parameters [83]. Further lowering the thickness gives rise to increased leakage currents that eventually might overwhelm the polarization charge to be sensed during read operation [84,85].

Higher endurance might be attained by application of an antiferroelectric hafnium or zirconium oxide that features smaller coercive fields, as was demonstrated in [23]. Stabilizing ferroelectricity in hafnium oxide by dopants with radius smaller than that of Hf and at comparatively higher doping concentrations than the optimum to achieve ferroelectricity, the anti-ferroelectric hysteresis is observed [16,35,86,87]. Moreover, this behavior can be observed for pure zirconium oxide [21]. Due to the lower coercive field of the anti-ferroelectric and the lower remanent polarization inducing a lower depolarization field and less charge injection the cycling endurance is much higher compared to that of the ferroelectric case [88].

The non-volatility of the anti-ferroelectric material can be attained by shifting the anti-ferroelectric hysteresis curve by work function engineering of the electrodes [88,89] or adding fixed charges [90], thus making use of only one half of the anti-ferroelectric hysteresis loop. Antiferroelectric hysteresis was also observed in dielectric stacks similar to those used in state-of-the-art DRAM capacitors [89]. Therefore, shifting the hysteresis loop by work function engineering could extend state-of-the-art DRAM technologies toward an ultrafast non-volatile memory (NVM) [23].

Effects, such as wakeup, that describes the increase of remanent polarization during first 10 to 100 switching cycles, and fatigue, which is the reduction of remanent polarization for high cycle numbers even before breakdown, still need to be engineered to levels as those found in perovskite ferroelectrics [91,92].

In order to gain sufficient signal to be sensed on the bit line the ferroelectric capacitor has to provide a polarization charge in the order of 15 fC. Hence, for a typical remanent polarization of 10–20 $\mu C/cm^2$ for doped HfO2 a capacitor area of about 0.1 μm^2 is required. Compared to high-density stand-alone DRAM, where the typical charge density of the capacitor is approximately 2 $\mu C/cm^2$ at a k-value of approximately 30 a reduction of the capacitor area by a factor 8 can be attained. For a given DRAM node the potential reduction in the cylinder capacitor area results in a decrease of the aspect ratio from about 50 down to approximately 12 during the capacitor trench etch, as is depicted in Fig. 8.14.

Obviously, the stabilizing support structures that were introduced since the sub-50 nm DRAM nodes [93] could be omitted by reducing the aspect ratio of the capacitor structures. Moreover, the higher charge density of the ferroelectric facilitates the transition from the cup-structure to the pillar structure for the capacitor. In summary, the alleviation of area requirements would lead to a reduction in manufacturing complexity and consequently in the cost per bit. However, until now research has been mainly carried out on single capacitor test structures, but recently integrated 16 and 64 kbit 1T-1C FeRAM array structures based on ferroelectric hafnium oxide have been reported [114,115].

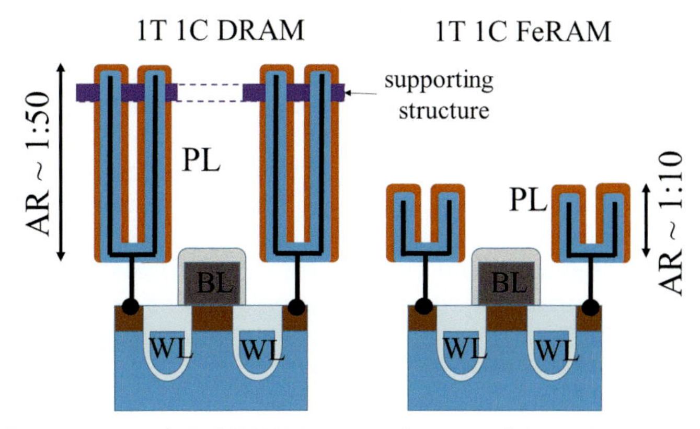

FIG. 8.14 Realization of 1T1C FeRAM compared to state-of-the art DRAM. *Reprinted with permission from S. Slesazeck et al., Embedding hafnium oxide based FeFETs in the memory landscape, in: Proceedings of ICICDT, Otranto, Italy, 2018. Copyright 2018 IEEE.*

8.5.2 Planar 1T FeFET implementation

The high coercive field of the ferroelectric hafnium oxide, that turned out as one of the main issues of a 1T-1C memory cell, translates into an advantage when considering the adoption of the material in the gate stack of a FeFET. Assuming an n-channel transistor and a polarization of the ferroelectric, where the dipole is oriented with the negative charge being close to the channel and the positive charge being close to the gate electrode. In this case the channel will be in accumulation mode. When switching the polarization of the ferroelectric dipole by application of an electrical field between the gate and the channel region, the polarization might be other way around, resulting in a positive charge close to the channel, thus migrating to an inversion condition in the channel. That is, the polarization of the ferroelectric will influence the threshold voltage of the transistor.

A simplified way to understand the creation of the memory window in the FeFET can be derived from considering the flat band voltage case. In an ideal gate stack without any fixed or trapped charge in the dielectrics and no work function difference between channel and gate electrode the flat band case will mean that all capacitor charges in the series connection of the gate stack components have to equal zero. For a normal transistor this is trivial, but in the FeFET case zero surface charge of the ferroelectric layer can only be obtained at the coercive voltages V_{C+} or V_{C-}. That is, the memory window MW can be calculated by the difference in flat band voltage for both polarization states [94].

$$MW = V_{C+} + |V_{C-}| = 2V_C = 2E_C t_{FE} \approx \Delta V_{FB} \approx \Delta V_T.$$

where E_C is the coercive field and t_{FE} denotes the thickness of the ferroelectric. Note that this simple derivation assumes that the polarization charge is high enough to drive the metal-ferroelectric-insulator-semiconductor (MFIS) structure from accumulation to strong inversion when switching. If the polarization is not high enough to achieve this boundary condition, the memory window will be lower and a function of the remanent polarization. For a more realistic assessment one has to take the effect of the depolarization field E_{dep} on the remaining polarization into account. The electric field that is induced by the polarization of the ferroelectric layer also in the resting state under zero applied gate voltage comes already very close to coercive field E_C of the ferroelectric. Under such conditions and for realistic ferroelectric films, where the shape of the polarization hysteresis loop is determined by the distribution of the coercive field for all the individual domains, a significant portion of the domains will already be flipped back yielding a significant decrease of the remanent polarization, thus linking the effect of coercive field and remanent polarization. Nevertheless, with the coercive field E_C being an intrinsic material parameter t_{FE} remains as the main parameter determining the maximum achievable memory window in a given FeFET system. However, the consequence that the maximum attainable memory window is limited by the coercive field holds true and is important. Hence, the high coercive field of hafnium oxide ferroelectrics is actually a benefit rather than a drawback.

Due to the much lower coercive field of the perovskites, much thicker ferroelectric films would be needed to achieve the same memory window based on the classical ferroelectrics, which limits their scalability. In the transistor, a serial capacitor consisting of not only the capacitance of the depletion layer in the semiconductor, but also the unavoidable interface oxide between the channel and the ferroelectric, is an integral part of the structure. This series connection means the voltage that has to be applied to the device for polarization switching is increased compared to the case of a pure ferroelectric capacitor, and in the case of no applied electrical field (e.g., the case of retention), an increased internal depolarization field will be the consequence [78]. Since perovskite-based ferroelectrics have a permittivity in the range of a few hundreds, a large thickness is required to balance this inherent capacitive divider. The much lower permittivity of hafnium oxide in the range of 30 is beneficial and the high coercive field helps to stabilize the polarization in the retention case [44]. Therefore, hafnium-based ferroelectrics seem to be favorable to realize FeFETs.

The first reported demonstrations occurred in 2011 [40] and in 2012. It was verified that use of ferroelectric hafnium oxide can close the scaling gap between FeFETs and conventional FETs [27]. In 2016, the first fully integrated technology where FeFET memory arrays were embedded in a 28-nm CMOS process was demonstrated [95]. Only one year later, scaling to 22-nm fully depleted silicon on insulator (FDSOI) technology together with further advances in the performance of the memory arrays were shown [96]. Fig. 8.15 illustrates the recent gate length scaling of the HfO_2-based FeFETs.

Thus, the vision of making non-volatile memories based on FeFETs has moved to realization much faster than in the 30 years of research before. Comparing to typical Flash devices bares several advantages of the ferroelectric hafnium oxide based FeFET concept. For typical coercive voltages of 1–2 MV/cm a switching voltage in the range of 4 V is mandatory where about half of the applied potential drops over the interfacial silicon oxide layer. Nevertheless, that programming voltage is at least a factor 4 less than typical programming voltages for FLASH devices, resulting in an increased array efficiency. Moreover, low programming voltages in conjunction with negligible programming currents and non-destructive readout makes the FeFET concept the most promising candidate for realization

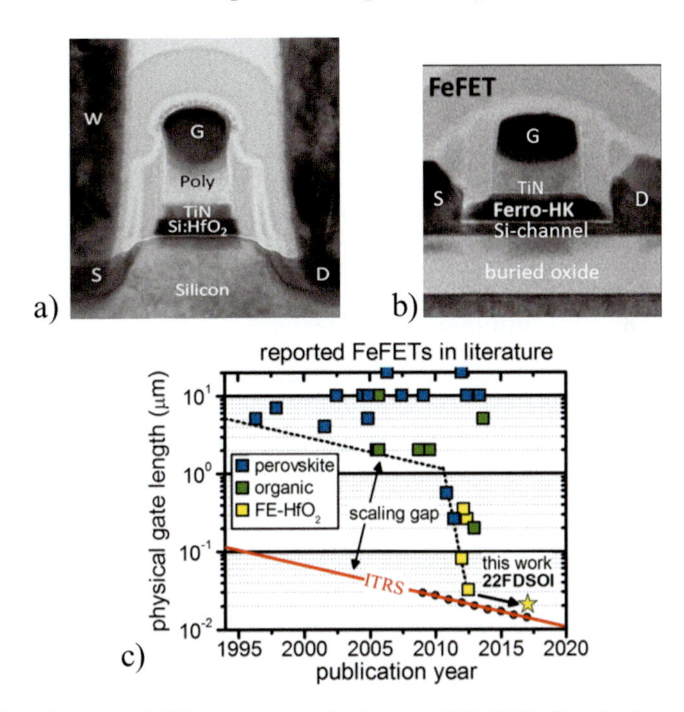

FIG. 8.15 (A) 28 nm FeFET implemented in standard HKMG CMOS technology. (B) 22 nm FeFET implemented in a 22 nm FD-SOI technology and (C) physical gate length scaling of FeFET compared to the embedded NVM logic platforms. *Panel (A) Reprinted (adapted) with permission from H. Mulaosmanovic, et al., Switching kinetics in nanoscale hafnium oxide based ferroelectric field-effect transistors, ACS Appl. Mater. Interfaces 9 (4) (2017) 3792–3798. Copyright 2017 American Chemical Society. Panel (C) Reprinted (adapted) with permission from Dünkel S. et al., A FeFET based super-low-power ultra-fast embedded NVM technology for 22nm FDSOI and beyond, in: IEEE International Electron Device Meeting, San Francisco, CA, 2017 19.7.1–19.7.4. Copyright 2017 IEEE.*

of ultra-low-power memories. The reported endurance of 10^4 for gate first process [95] and up to 10^7 write cycles for gate last process [97] can easily compete with conventional planar NAND devices and makes the FeFET device very suitable as eNVM solution. The endurance of the FeFET devices typically is limited by the endurance of the interfacial oxide layer toward the silicon channel, that is operated close to its break down fields when switching the polarization of the ferroelectric HfO_2 layer.

On the other hand, in terms of cost per bit three detrimental aspects come into play that have to be solved to compete with planar stand-alone NAND. Firstly, from array disturb analysis the realization of AND or NOR array architecture seems to be beneficial [98,99]. That means, a larger number of contacts and metallization lines in the array are mandatory, thus increasing the cell size compared to NAND. Secondly, the polycrystallinity of the ferroelectric material with grain sizes in the range of 10 nm yields to a certain variability in the switching characteristic that can be omitted by increasing transistor size beyond 0.01 μm^2.

Moreover, as was shown in [100], a controllable and reproducible switching of two or few domains within the gate stack can enable to store more than one bit in FeFETs, encoded as possible joint combinations of up and down polarization states of single domains. It was proven that these states are stable both in time and upon cycling, making FeFETs suitable candidates for a multi-level non-volatile storage.

8.5.3 3D FeFET implementation

In order to further scale down the bit size of HfO_2-based FeFET memories one can consider the 3D integration in a similar manner as in state-of-the-art 3D NAND structures, as is illustrated schematically in Fig. 8.16. In such an integration flow one might simply replace the charge trapping layer for realization of high-density 3D FeFET memories. Comparing the electrical performance characteristics of the FeFET to the charge trapping FLASH device the benefit of lower programming voltages comes at the drawback of no multi-level capability. That means, that the effective area per bit would be reduced by a factor of 2 to 3 compared to state-of-the-art 3D NAND devices. A slight counterbalance might be attained by an increased array efficiency due to lower programming voltages for the FeFET that would have to be generated on-chip.

In [101] the first realization of vertically stacked Al-doped HfO_2 devices was demonstrated. However, the reported cycling endurance was shown to be in the range of just 10^2 switching cycles. The degradation of the memory window is mainly attributed to the break-down of the silicon oxide interface between the ferroelectric and the transistor channel. From that data one can deduce that for the integration of the ferroelectric hafnium

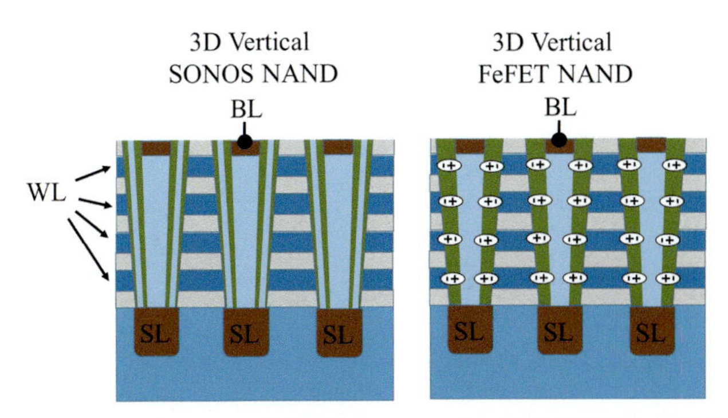

FIG. 8.16 Realization of 3D NAND FeFET compared to 3D NAND FLASH. *Reprinted with permission from S. Slesazeck et al., Embedding hafnium oxide based FeFETs in the memory landscape, in: Proceedings of ICICDT, Otranto, Italy, 2018. Copyright 2018 IEEE.*

oxide in between the two polysilicon structures forming the NAND string channel and the gate planes as depicted in Fig. 8.16, the formation of a reliable silicon oxide interface seems to be the most challenging task. Improvement might be attained by further material development and reduction of the coercive voltage as well as optimization of the interfacial layers to the silicon or the implementation of metallic gate planes. Due to the early stage of research in that field, the concept looks very promising, but many issues such as cell-to-cell disturb and program inhibit schemes for FeFET NAND operation have not been proven yet.

8.5.4 Ferroelectric tunnel junction

In the ferroelectric tunnel junction (FTJ) the polarization reversal of a ferroelectric layer that is sandwiched between two meal electrodes is adopted to alter the tunneling barrier between those electrodes. The polarization state of the device can be detected by measuring the current flow upon application of an electrical field that is smaller than the coercive field of the ferroelectric. The change in the tunneling current is often called the tunneling electro-resistance (TER) effect and is measured as the ratio between the resistance of the device in low-resistance state (LRS) and high-resistance state (HRS). Write operation is performed instead by application of an electrical field larger than the coercive field of the ferroelectric film. In that way the FTJ can be used to store binary information. Since the FTJ is not switched during read operation, it is referred to a non-destructive read operation and in contrast to the 1T1C concept the FTJ requires no refresh operation after read. L. Esaki et al. proposed the concept in the 1970s as a "polar switch." However, because of difficulties in fabrication of an ultra-thin ferroelectric layer that is mandatory to attain

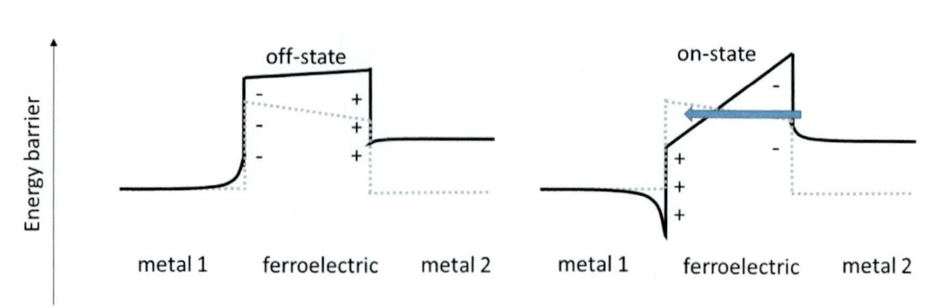

FIG. 8.17 Illustrations of the potential profiles of a FTJ composed of a ferroelectric layer sandwiched between two metal electrodes with different screening length. The dotted lines show the profile without polarization.

reasonable tunneling currents, the concept was demonstrated only in 2009 based on an ultra-thin BaTiO3 ferroelectric layer [102].

Fig. 8.17 depicts the basic working principle of the ferroelectric tunnel junction. Two different metal electrodes with different screening length are used to ensure that the effect of the polarization reversal for both interfaces does not cancel out each other. The TER ratio depends on the barrier height, but also on the conduction mechanism that might change depending on the polarization state [109]. Different mechanisms that might contribute to the resistance change effect in the FTJ have been studied. Among these there is the influence of the asymmetric screening length effect [103,105], strain effects [105], the composite potential barrier effect [105,106], and the reversible metallization effect [107]. Typically, a higher TER ratio can be attained for a thicker ferroelectric layer. However, in case the ferroelectric layer represents the tunneling barrier itself, a higher thickness leads to a reduced current density. Although lower current density would be beneficial for realizing a memory device with larger capacity and lower power consumption, a detectable amount of current in the range of several microampere is mandatory for fast and reliable read operation. Recently it has been shown [12,108,111] that the composite barriers exhibit a larger TER ratio. The operation principle is depicted in Fig. 8.18. The idea of the concept is to separate the tunneling effect and the ferroelectric switching effect from each other. In this way the effect of ferroelectric switching and electron tunneling can be optimized more independently of each other. The results reveal significant enhancement of the TER ratio by introduction of an additional dielectric layer and might eventually lead to larger current density during read operation.

One important parameter that becomes even more pronounced in the composite structure is the depolarization field. According to Eq. (8.11) the depolarization field is inverse proportional to the capacitance ratio of dielectric or interface and ferroelectric layer. It can be formed by finite screening length of metal electrode as well as by capacitance of the dielectric layer. For increasing thickness of the dielectric layer the depolarization

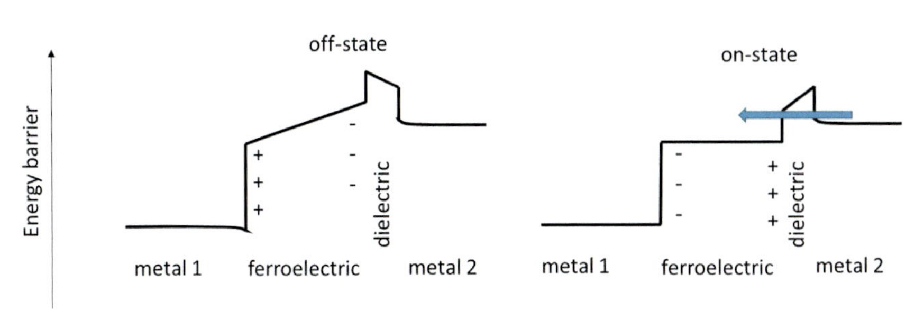

FIG. 8.18 Operation principle of a composite FTJ where a dielectric layer is introduced that separates the ferroelectric polarization layer from the tunneling barrier layer.

field could eventually destabilize or even reverse the polarization [12]. Hence, the precise control of both, thickness of ferroelectric and dielectric layer is of utmost importance for proper FTJ functionality [104].

8.6 Summary and future prospects

Ferroelectrics are an almost ideal candidate for binary non-volatile data storage due to their low power operation that is facilitated by the field-driven switching mechanism. However, traditional ferroelectrics suffered from pure scalability due to the difficulties to integrate the complex materials into CMOS manufacturing process. The recent discovery of ferroelectricity in hafnium oxide (HfO_2) films has changed that situation and has triggered a renewed research activity in order to investigate the origin and properties of this novel ferroelectric system. Since hafnium oxide is a standard material in CMOS processing, the ease of co-integration of FeFETs into state-of-the-art CMOS technologies has opened a broad perspective for their future application as pure memory device, but also for beyond memory applications such as Logic in Memory [110,112] and neuromorphics [69]. The fact that differently optimized FeFETs are also an option to make steep subthreshold devices by stabilizing the negative capacitance region has further inspired the research and development of such devices.

References

[1] A.F. Devonshire, Theory of ferroelectrics, Adv. Phys. 3 (10) (1954) 85–130.
[2] C. Woo, Y. Zheng, Depolarization in modeling nano-scale ferroelectrics using the Landau free energy functional, Appl. Phys. A 91 (2008) 59–63.
[3] S. Salahuddin, S. Datta, Use of negative capacitance to provide voltage amplification for low power nanoscale devices, Nano Lett. 8 (2008) 405–410.

[4] M. Hoffmann, et al., Direct observation of negative capacitance in polycrystalline ferroelectric HfO_2, Adv. Funct. Mater. 26 (47) (2016) 8643–8649.
[5] A.I. Khan, et al., Differential voltage amplification from ferroelectric negative capacitance, Appl. Phys. Lett. 111 (2017) 253501.
[6] M. Hoffmann, et al., Ferroelectric negative capacitance domain dynamics, J. Appl. Phys. 123 (2018) 184101.
[7] M. Hoffmann, et al., On the stabilization of ferroelectric negative capacitance in nanoscale devices, Nanoscale 10 (2018) 10891–10899.
[8] Zubko, et al., Negative capacitance in multidomain ferroelectric superlattices, Nature 534 (2016) 524–528.
[9] Kim, et al., Time-dependent negative capacitance effects in Al2O3/BaTiO3 bilayers, Nano Lett. 16 (7) (2016) 4375–4381.
[10] M. Hoffmann, et al., Unveiling the double well energy landscape in a ferroelectric layer, Nature 565 (2019) 464–467.
[11] Si, et al., Steep-slope hysteresis-free negative capacitance MoS2 transistors, Nat. Nanotechnol. 13 (2018) 24–28.
[12] B. Max, et al., Ferroelectric tunnel junctions based on ferroelectric-dielectric Hf 0.5 Zr 0.5. O 2/A1 2 O 3 capacitor stacks, in: In: 48th European Solid-State Device Research Conference (ESSDERC), 2018, pp. 142–145.
[13] J. Robertson, High dielectric constant oxides, Eur. Phys. J.-Appl. Phys. 28 (2004) 265.
[14] M.T. Bohr, R.S. Chau, T. Ghani, K. Mistry, The high-k solution, IEEE Spectr. 44 (2007) 29.
[15] R. Materlik, C. Künneth, A. Kersch, The origin of ferroelectricity in Hf1−xZrxO2: a computational investigation and a surface energy model, J. Appl. Phys. 117 (2015) 134109.
[16] T.S. Böscke, J. Müller, D. Bräuhaus, U. Schröder, U. Böttger, Ferroelectricity in hafnium oxide thin films, Appl. Phys. Lett. 99 (2011) 102903.
[17] T.S. Böscke, D. St. Teichert, J. Bräuhaus, U. Müller, U. Schröder, T.M. Böttger, Phase transitions in ferroelectric silicon doped hafnium oxide, Appl. Phys. Lett. 99 (2011) 112904.
[18] X. Sang, E.D. Grimley, T. Schenk, U. Schroeder, J.M. LeBeau, On the structural origins of ferroelectricity in HfO2 thin films, Appl. Phys. Lett. 106 (2015) 162905.
[19] P.D. Lomenzo, et al., The effects of layering in ferroelectric Si-doped HfO2 thin films, J. Vac. Sci. Technol. B 32 (2014) 03D123.
[20] J. Müller, T.S. Böscke, D. Bräuhaus, U. Schröder, U. Böttger, J. Sundqvist, P. Kücher, T. Mikolajick, L. Frey, Ferroelectric Zr0.5Hf0.5O2 thin films for nonvolatile memory applications, Appl. Phys. Lett. 99 (2011) 112901.
[21] J. Müller, T.S. Böscke, U. Schröder, S. Mueller, D. Brauhaus, U. Böttger, L. Frey, T. Mikolajick, Ferroelectricity in simple binary ZrO2 and HfO2, Nano Lett. 12 (2012) 4318–4323.
[22] J. Müller, U. Schröder, T.S. Böscke, I. Müller, U. Böttger, L. Wilde, J. Sundqvist, M. Lemberger, P. Kücher, T. Mikolajick, L. Frey, Ferroelectricity in yttrium-doped hafnium oxide, J. Appl. Phys. 110 (2011) 114113.
[23] S. Mueller, J. Mueller, A. Singh, S. Riedel, J. Sundqvist, U. Schroeder, T. Mikolajick, Incipient ferroelectricity in Al-doped HfO2 thin films, Adv. Funct. Mater. 22 (2012) 2412–2417.
[24] S. Mueller, C. Adelmann, A. Singh, S. Van Elshocht, U. Schroeder, T. Mikolajick, Ferroelectricity in Gd-doped HfO2 thin films, ECS J. Solid State Sci. Technol. 1 (2012) N123–N126.
[25] M. Hoffmann, U. Schroeder, T. Schenk, T. Shimizu, H. Funakubo, O. Sakata, D. Pohl, M. Drescher, C. Adelmann, R. Materlik, A. Kersch, T. Mikolajick, Stabilizing the ferroelectric phase in doped hafnium oxide, J. Appl. Phys. 118 (2015), 071006.
[26] T. Schenk, S. Mueller, U. Schroeder, R. Materlik, A. Kersch, M. Popovici, C. Adelmann, S. Van Elshocht, T. Mikolajick, Strontium doped hafnium oxide thin films: wide process window for ferroelectric memories, in: 2013 Proceedings of the European Solid-State Device Research Conference (ESSDERC), 2013, pp. 260–263.

[27] J. Müller, E. Yurchuk, T. Schlösser, J. Paul, R. Hoffmann, S. Mueller, D. Martin, S. Slesazeck, P. Polakowski, J. Sundqvist, M. Czernohorsky, K. Seidel, P. Kücher, R. Boschke, M. Trentzsch, K. Gebauer, U. Schröder, T. Mikolajick, Ferroelectricity in HfO2 enables nonvolatile data storage in 28 nm HKMG, in: 2012 Symposium on VLSI Technology, 2012, pp. 25–26.
[28] A.G. Chernikova, D.S. Kuzmichev, D.V. Negrov, M.G. Kozodaev, S.N. Polyakov, A.M. Markeev, Ferroelectric properties of full plasma-enhanced ALD TiN/La:HfO2/TiN stacks, Appl. Phys. Lett. 108 (2016) 242905.
[29] U. Schroeder, C. Richter, M.H. Park, T. Schenk, M. Pešić, M. Hoffmann, F.P.G. Fengler, D. Pohl, B. Rellinghaus, C. Zhou, C.-C. Chung, J.L. Jones, T. Mikolajick, Lanthanum doped hafnium oxide: a robust ferroelectric material, Inorg. Chem. 57 (2018) 2752–2765.
[30] P. Polakowski, J. Müller, Ferroelectricity in undoped hafnium oxide, Appl. Phys. Lett. 106 (2015) 232905.
[31] D. Martin, J. Müller, T. Schenk, T.M. Arruda, A. Kumar, E. Strelcov, E. Yurchuk, S. Müller, D. Pohl, U. Schröder, S.V. Kalinin, T. Mikolajick, Ferroelectricity in Si-doped HfO2 revealed: a binary lead-free ferroelectric, Adv. Mater. 26 (2014) 8198.
[32] M. Hoffmann, U. Schroeder, C. Künneth, A. Kersch, S. Starschich, U. Böttger, T. Mikolajick, Ferroelectric phase transitions in nanoscale HfO2 films enable giant pyroelectric energy conversion and highly efficient supercapacitors, Nano Energy 18 (2015) 154.
[33] S.W. Smith, A.R. Kitahara, M.A. Rodriguez, M.D. Henry, M.T. Brumbach, J.F. Ihlefeld, Pyroelectric response in crystalline hafnium zirconium oxide (Hf1-xZrxO2) thin films, Appl. Phys. Lett. 110 (2017), 072901.
[34] J.F. Scott, J. Phys, Ferroelectrics go bananas, Condens. Matter 20 (2) (2007) 021001.
[35] E. Yurchuk, J. Müller, S. Knebel, J. Sundqvist, A.P. Graham, T. Melde, U. Schröder, T. Mikolajick, Impact of layer thickness on the ferroelectric behaviour of silicon doped hafnium oxide thin films, Thin Solid Films 533 (2013) 88.
[36] M.H. Park, Y.H. Lee, H.J. Kim, T. Schenk, W. Lee, K.D. Kim, F.P.G. Fengler, T. Mikolajick, U. Schroeder, C.S. Hwang, Surface and grain boundary energy as the key enabler of ferroelectricity in nanoscale hafnia-zirconia: a comparison of model and experiment, Nanoscale 9 (2017) 9973.
[37] M.H. Park, H.J. Kim, T. Moon, K.D. Kim, Y.H. Lee, S.D. Hyun, T. Mikolajick, U. Schroeder, C.S. Hwang, Understanding the formation of the metastable ferroelectric phase in hafnia–zirconia solid solution thin films, Nanoscale 10 (2018) 716.
[38] H. Mulaosmanovic, et al., Switching kinetics in nanoscale hafnium oxide based ferroelectric field-effect transistors, ACS Appl. Mater. Interfaces 9 (4) (2017) 3792–3798.
[39] H. Mulaosmanovic, T. Mikolajick, S. Slesazeck, Accumulative polarization reversal in nanoscale ferroelectric transistors, ACS Appl. Mater. Interfaces 10 (28) (2018) 23997–24002.
[40] J. Müller, T.S. Böscke, U. Schröder, R. Hoffmann, T. Mikolajick, L. Frey, Nanosecond polarization switching and long retention in a novel MFIS-FET based on ferroelectric, IEEE Electron Device Lett. 33 (2) (2012) 185–187.
[41] W.T. Lu, P.C. Lin, T.Y. Huang, C.H. Chien, M.J. Yang, I.J. Huang, P. Lehnen, The characteristics of hole trapping in HfO2/SiO2 gate dielectrics with TiN gate electrode, Appl. Phys. Lett. 85 (16) (2004) 3525–3527.
[42] M.H. Park, Y.H. Lee, H.J. Kim, Y.J. Kim, T. Moon, K.D. Kim, J. Mueller, A. Kersch, U. Schroeder, T. Mikolajick, C.S. Hwang, Ferroelectricity and antiferroelectricity of doped thin HfO2-based films, Adv. Mater. 27 (2015) 1811–1831.
[43] J. Müller, E. Yurchuk, T. Schlösser, J. Paul, R. Hoffmann, S. Müller, D. Martin, S. Slesazeck, P. Polakowski, J. Sundqvist, M. Czernohorsky, K. Seidel, P. Kücher, R. Boschke, M. Trentzsch, K. Gebauer, U. Schröder, T. Mikolajick, Ferroelectricity in

HfO2 enables nonvolatile data storage in 28 nm HKMG, in: IEEE VLSI Tech, 2012, pp. 25–26.

[44] T.S. Böscke, J. Müller, D. Bräuhaus, U. Schröder, U. Böttger, Ferroelectricity in hafnium oxide: CMOS compatible ferroelectric field effect transistors, in: Proc. Intl. Electron Devices Meeting (IEDM), 2011, pp. 24.5.1–24.5.4.

[45] Y. Tan, J. Zhang, Y. Wu, C. Wang, V. Koval, B. Shi, H. Ye, R. McKinnon, G. Viola, H. Yan, Unfolding grain size effects in barium titanate ferroelectric ceramics, Sci. Rep. 5 (2015) 9953.

[46] Y. Jin, X. Lu, J. Zhang, Y. Kan, H. Bo, F. Huang, T. Xu, Y. Du, S. Xiao, J. Zhu, Studying the polarization switching in polycrystalline BiFeO3 films by 2D piezoresponse force microscopy, Sci. Rep. 5 (2015) 12237.

[47] D. Seol, B. Kim, Y. Kim, Non-piezoelectric effects in piezoresponse force microscopy, Curr. Appl. Phys. 17 (5) (2017).

[48] S. Kim, D. Seol, X. Lu, M. Alexe, Y. Kim, Electrostatic-free piezoresponse force microscopy, Sci. Rep. 7 (2017), 41657.

[49] Y. Shin, I. Grinberg, I. Chen, A.M. Rappe, Nucleation and growth mechanism of ferroelectric domain-wall motion, Nature 449 (2007) 881–884.

[50] J.F. Scott, L. Kammerdiner, M. Parris, S. Traynor, V. Ottenbacher, A. Shawabkeh, W. F. Oliver, Switching kinetics of lead zirconate titanate submicron thin-film memories, J. Appl. Phys. 64 (2) (1998) 787–792.

[51] H. Orihara, S. Hashimoto, Y. Ishibashi, A theory of DE hysteresis loop based on the Avrami model, J. Physical Soc. Japan 63 (1994) 1031–1035.

[52] Y.W. So, D.J. Kim, T.W. Noh, J.G. Yoon, T.K. Song, Polarization switching kinetics of epitaxial Pb (Zr0. 4Ti0. 6) O3 thin films, Appl. Phys. Lett. 86 (2005) 092905.

[53] A.K. Tagantsev, I. Stolichnov, N. Setter, J.S. Cross, M. Tsukada, Non-Kolmogorov-Avrami switching kinetics in ferroelectric thin films, Phys. Rev. B 66 (2002) 214109.

[54] X. Du, I.W. Chen, Frequency spectra of fatigue of PZT and other ferroelectric thin films, MRS Proc. 493 (1997) 331.

[55] W.J. Merz, Domain formation and Domain Wall motions in ferroelectric BaTiO3 single crystals, Phys. Rev. 95 (1954) 690–698.

[56] R. Landauer, Electrostatic considerations in BaTiO3 domain formation during polarization reversal, J. Appl. Phys. 28 (1957) 227–234.

[57] D.J. Jung, M. Dawber, J.F. Scott, L.J. Sinnamon, J.M. Gregg, Switching dynamics in ferroelectric thin films: an experimental survey, Integr. Ferroelectr. 48 (2002) 59–68.

[58] T. Tybell, P. Paruch, T. Giamarchi, J.M. Triscone, Domain WALL CREEP IN EPITAXIAL ferroelectric Pb(Zr0.2 Ti0.8)O3 thin films, Phys. Rev. Lett. 89 (9) (2002) 097601.

[59] P. Buragohain, C. Richter, T. Schenk, H. Lu, T. Mikolajick, U. Schroeder, A. Gruverman, Nanoscopic studies of domain structure dynamics in ferroelectric La:HfO2 capacitors, Appl. Phys. Lett. 112 (2018) 222901.

[60] A. Roelofs, T. Schneller, K. Szot, R. Waser, Piezoresponse force microscopy of lead titanate nanograins possibly reaching the limit of ferroelectricity, Appl. Phys. Lett. 81 (2002) 5231–5233.

[61] S.M. Nam, Y.B. Kil, S. Wada, T. Tsurumi, Domain switching kinetics of lead zirconate titinate thin films, Jpn. J. Appl. Phys. 42 (2003) L1519.

[62] S. Mueller, S.R. Summerfelt, J. Müller, U. Schroeder, T. Mikolajick, Ten-nanometer ferroelectric films for next-generation FRAM capacitors, IEEE Electron Device Lett. 33 (2012) 1300–1302.

[63] Y. Kim, H. Han, W. Lee, S. Baik, D. Hesse, M. Alexe, Non-kolmogorov-avrami-ishibashi switching dynamics in nanoscale ferroelectric capacitors, Nano Lett. 10 (2010) 1266–1270.

[64] M.G. Han, J.A. Garlow, M. Bugnet, S. Divilov, M.S. Marshall, L. Wu, M. Dawber, M. Fernandez-Serra, G.A. Botton, S.W. Cheong, F.J. Walker, C.H. Ahn, Y. Zhu, Coupling of bias-induced crystallographic shear planes with charged domain walls in ferroelectric oxide thin films, Phys. Rev. B 94 (2016) 100101.

[65] C.D. Wright, Y. Liu, K.I. Kohary, M.M. Aziz, R.J. Hicken, Arithmetic and biologically-inspired computing using phase-change materials, Adv. Mater. 23 (2011) 3408–3413.
[66] C.D. Wright, P. Hosseini, J.A.V. Diosdado, Beyond von-Neumann computing with nanoscale phase-change memory devices, Adv. Funct. Mater. 23 (18) (2013) 2248–2254.
[67] T. Tuma, A. Pantazi, M. Le Gallo, A. Sebastian, E. Eleftheriou, Stochastic phase-change neurons, Nat. Nanotechnol. 11 (2016) 693–699.
[68] Y. Nishitani, Y. Kaneko, M. Ueda, T. Morie, E. Fujii, Three-terminal ferroelectric synapse device with concurrent learning function for artificial neural networks, J. Appl. Phys. 111 (2012) 124108.
[69] H. Mulaosmanovic, J. Ocker, S. Müller, M. Noack, J. Müller, P. Polakowski, T. Mikolajick, S. Slesazeck, Novel ferroelectric FET based synapse for neuromorphic systems, in: Symp. VLSI Technol., Dig. Technol. Pap, 2017, pp. T176–T177.
[70] S. Mueller, S. Slesazeck, S. Henker, S. Flachowsky, P. Polakowski, J. Paul, E. Smith, J. Müller, T. Mikolajick, Correlation between the macroscopic ferroelectric material properties of Si:HfO2 and the statistics of 28 nm FeFET memory arrays, Ferroelectrics 497 (1) (2016) 42–51.
[71] D.A. Buck, Ferroelectrics for Digital Information Storage and Switching, master's thesis, Massachusetts Institute of Technology Digital Computer Laboratory, 1952.
[72] J.R. Anderson, Ferroelectric materials as storage elements for digital computers and switching systems, Trans. Am. Inst. Electr. Eng. Part 1 71 (1953) 395.
[73] D. Bondurant, Ferroelectronic RAM memory family for critical data storage, Ferroelectrics 112 (1990) 273.
[74] C.-U. Pinnow, T. Mikolajick, Material aspects in emerging nonvolatile memories, J. Electrochem. Soc. 151 (2004) K13.
[75] S.Y. Lee, K. Kim, Future 1T1C FRAM technologies for highly reliable, high density FRAM, Digest. International Electron Devices Meeting, IEEE IEDM, 2002, pp. 547–550.
[76] J.-M. Koo, B.-S. Seo, S. Kim, S. Shin, J.-H. Lee, H. Baik, J.-H. Lee, J.H. Lee, B.-J. Bae, J.-E. Lim, D.-C. Yoo, S.-O. Park, H.-S. Kim, H. Han, S. Baik, J.-Y. Choi, Y.J. Park, Y. Park, Fabrication of 3D trench PZT capacitors for 256Mbit FRAM device application, Digest. International Electron Device Meeting, IEEE IEDM, 2005, pp. 340–343.
[77] C.-P. Yeh, M. Lisker, B. Kalkofen, E.P. Burte, Fabrication and investigation of three-dimensional ferroelectric capacitors for the application of FeRAM, AIP Adv. 6 (3) (2016), 035128.
[78] T.P. Ma, J.-P. Han, Why is nonvolatile ferroelectric memory field-effect transistor still elusive? IEEE Electron Devices Lett. 23 (2002) 386.
[79] P. Polakowski, S. Riedel, W. Weinreich, M. Rudolf, J. Sundqvist, K. Seidel, J. Müller, Ferroelectric deep trench capacitors based on Al:HfO2 for 3D nonvolatile memory applications, Int. Memory Workshop, IEEE, 2014, pp. 1–4.
[80] D.J. Wouters, et al., Integration of SrBi2Ta2O9 thin films for high density ferroelectric random access memory, IEEE J. Solid State Circuits 39 (2004) 667.
[81] E. Yurchuk, S. Mueller, D. Martin, S. Slesazeck, U. Schroeder, T. Mikolajick, J. Müller, J. Paul, R. Hoffmann, J. Sundqvist, T. Schlosser, R. Boschke, R. van Bentum, M. Trentzsch, Origin of the endurance degradation in the novel HfO 2-based 1T ferroelectric non-volatile memories, in: International Reliability Physics Symposium, IEEE IRPS, 2014, pp. 2E.5.1–2E.5.5.
[82] J. Müller, P. Polakowski, S. Mueller, T. Mikolajickb, Ferroelectric hafnium oxide based materials and devices: assessment of current status and future prospects, ECS J. Solid State Sci. Technol. 4 (5) (2015) N30.
[83] A. Chernikova, M. Kozodaev, A. Markeev, D. Negrov, M. Spiridonov, S. Zarubin, O. Bak, P. Buragohain, H. Lu, E. Suvorova, A. Gruverman, A. Zenkevich, Ultrathin Hf0. 5Zr0.5O2 ferroelectric films on Si, ACS Appl. Mater. Interfaces 8 (2016) 7232.

[84] T. Schenk, U. Schroeder, T. Mikolajick, Correspondence - Dynamic leakage current compensation revisited, IEEE Trans. Ultrason. Ferroelectr. Freq. Control 62 (3) (2015) 596.
[85] M. Pešić, M. Hoffmann, C. Richter, S. Slesazeck, U. Schroeder, T. Mikolajick, Anti-ferroelectric-like ZrO 2 non-volatile memory: inducing non-volatility within state-of-the-art DRAM, Proc. Nonvolatile Memory Technol. Symp, IEEE NVMTS, 2017.
[86] U. Schroeder, E. Yurchuk, J. Müller, D. Martin, T. Schenk, P. Polakowski, C. Adelmann, M.I. Popovici, S.V. Kalinin, T. Mikolajick, Impact of different dopants on the switching properties of ferroelectric hafniumoxide, Jpn. J. Appl. Phys. 53 (2014) 08LE02.
[87] M. Pešić, S. Knebel, M. Hoffmann, C. Richter, T. Mikolajick, U. Schroeder, How to make DRAM non-volatile? Anti-ferroelectrics: a new paradigm for universal memories, IEEE Int. Electron Devices Mtg, IEEE IEDM, 2016, pp. 11.6.1–11.6.4.
[88] M. Pešić, M. Hoffmann, C. Richter, T. Mikolajick, U. Schroeder, Nonvolatile random access memory and energy storage based on antiferroelectric like hysteresis in ZrO2, Adv. Funct. Mater. 26 (2016) 7486.
[89] M. Pešić, M. Hoffmann, C. Richter, S. Slesazeck, T. Kämpfe, L.M. Eng, T. Mikolajick, U. Schroeder, Anti-ferroelectric ZrO2, an enabler for low power non-volatile 1T-1C and 1T random access memories, Eur. Solid-State Device Res. Conf., IEEE, 2017.
[90] T. Schenk, M. Hoffmann, J. Ocker, M. Pešić, T. Mikolajick, U. Schroeder, Complex internal bias fields in ferroelectric hafnium oxide, ACS Appl. Mater. Interfaces 7 (2015) 20224.
[91] M. Pešić, F.P.G. Fengler, L. Larcher, A. Padovani, T. Schenk, E.D. Grimley, X. Sang, J.M. LeBeau, S. Slesazeck, U. Schroeder, T. Mikolajick, Physical mechanisms behind the field-cycling behavior of HfO2-based ferroelectric capacitors, Adv. Funct. Mater. 26 (2016) 4601.
[92] S. Slesazeck, et al., Embedding hafnium oxide based FeFETs in the memory landscape, in: Proceedings of ICICDT, Otranto Italy, 2018.
[93] S.L. Miller, P.J. McWhorter, Physics of the ferroelectric nonvolatile memory field effect transistor, J. Appl. Phys. 72 (1992) 5999.
[94] N. Gong, T.P. Ma, Why is FE–HfO2 more suitable than PZT or SBT for scaled nonvolatile 1-T memory cell? A retention perspective, IEEE Electron Device Lett. 37 (2016) 1123.
[95] M. Trentzsch, S. Flachowsky, R. Richter, J. Paul, B. Reimer, D. Utess, S. Jansen, H. Mulaosmanovic, S. Müller, S. Slesazeck, J. Ocker, M. Noack, J. Müller, P. Polakowski, J. Schreiter, S. Beyer, T. Mikolajick, B. Rice, A 28 nm HKMG super low power embedded NVM technology based on ferroelectric FETs, IEEE Int. Electron Devices Mtg, IEEE IEDM, 2016, pp. 11.5.1–11.5.4.
[96] S. Dünkel, et al., A FeFET based super-low-power ultra-fast embedded NVM technology for 22nm FDSOI and beyond, in: 2017 IEEE International Electron Devices Meeting (IEDM), San Francisco, CA, 2017, pp. 19.7.1–19.7.4.
[97] K. Chatterjee, et al., Self-aligned, gate last, FDSOI, ferroelectric gate memory device with 5.5-nm Hf 0.8 Zr 0.2 O2, high endurance and breakdown recovery, IEEE Electron Device Lett. 38 (10) (2017) 1379–1382.
[98] S. Mueller, et al., From MFM capacitors toward ferroelectric transistors: endurance and disturb characteristics of HfO2-based FeFET devices, IEEE Trans. Electron Devices 60 (12) (2013) 4199–4205.
[99] A. Sharma, K. Roy, 1T non-volatile memory design using sub-10nm ferroelectric FETs, IEEE Electron Device Lett. 39 (3) (2018) 359–362.
[100] H. Mulaosmanovic, S. Slesazeck, J. Ocker, M. Pesic, S. Müller, S. Flachowsky, J. Müller, P. Polakowski, J. Paul, S. Jansen, S. Kolodinski, C. Richter, S. Piontek, T. Schenk, A. Kersch, C. Künneth, R. Bentum, U. Schröder, T. Mikolajick, Evidence of single domain switching in hafnium oxide based FeFETs: enabler for multi-level FeFET memory cells, in: Proc. Intl. Electron Devices Meeting (IEDM), 2015, pp. 26.8.1–26.8.3.

[101] K. Florent, et al., First demonstration of vertically stacked ferroelectric Al doped HfO2 devices for NAND applications, in: 2017 Symposium on VLSI Technology, Kyoto, 2017, pp. T158–T159.

[102] V. Garcia, S. Fusil, K. Bouzehouane, S. Enouz-Vedrenne, N.D. Mathur, A. Barthelemy, M. Bibes, Giant tunnel electroresistance for non-destructive readout of ferroelectric states, Nature 460 (2009) 81, https://doi.org/10.1038/nature08128.

[103] M. Ye, R.F.S. Zhuravlev, S.S. Jaswal, E.Y. Tsymbal, Giant electroresistance in ferroelectric tunnel junctions, Phys. Rev. Lett. 94 (2005) 246802, https://doi.org/10.1103/PhysRevLett.94.246802.

[104] Z.J. Ma, T.J. Zhang, R.K. Pan, M.G. Duan, M. He, Optimal dielectric thickness for ferroelectric tunnel junctions with a composite barrier, J. Appl. Phys. 111 (2012) 074311, https://doi.org/10.1063/1.3700245.

[105] H. Kohlstedt, N.A. Pertsev, J. Rodríguez Contreras, R. Waser, Theoretical current-voltage characteristics of ferroelectric tunnel junctions, Phys. Rev. B 72 (2005) 125341, https://doi.org/10.1103/PhysRevB.72.125341.

[106] M. Ye, Y.W. Zhuravlev, S. Maekawa, E.Y. Tsymbal, Tunneling electroresistance in ferroelectric tunnel junctions with a composite barrier, Appl. Phys. Lett. 95 (2009) 052902, https://doi.org/10.1063/1.3195075.

[107] X. Liu, J.D. Burton, E.Y. Tsymbal, Enhanced tunneling electroresistance in ferroelectric tunnel junctions due to the reversible metallization of the barrier, Phys. Rev. Lett. 116 (2016) 197602, https://doi.org/10.1103/PhysRevLett.116.197602.

[108] L. Wang, M.R. Cho, Y.J. Shin, J.R. Kim, S. Das, J.-G. Yoon, J.-S. Chung, T.W. Noh, Overcoming the fundamental barrier thickness limits of ferroelectric tunnel junctions through BaTiO3/SrTiO3 composite barriers, Nano Lett. 16 (2016) 3911, https://doi.org/10.1021/acs.nanolett.6b01418.

[109] D. Pantel, M. Alexe, Electroresistance effects in ferroelectric tunnel barriers, Phys. Rev. B 82 (2010) 134105, https://doi.org/10.1103/PhysRevB.82.134105.

[110] E.T. Breyer, H. Mulaosmanovic, T. Mikolajick, S. Slesazeck, Reconfigurable NAND/NOR Logic Gates in 28 nm HKMG and 22 nm FD-SOI FeFET Technology, IEDM, 2017, pp. 28.5.1–28.5.4.

[111] S. Fujii, Y. Kamimuta, T. Ino, Y. Nakasaki, R. Takaishi, M. Saitoh, First demonstration and performance improvement of ferroelectric HfO2-based resistive switch with low operation current and intrinsic diode property, in: 2016 IEEE Symposium on VLSI Technology, 2016, p. 148, https://doi.org/10.1109/VLSIT.2016.7573413.

[112] E.T. Breyer, H. Mulaosmanovic, S. Slesazeck, T. Mikolajick, Demonstration of Versatile Nonvolatile Logic Gates in 28nm HKMG FeFET Technology, ISCAS, 2018, pp. 1–5.

[114] Okuno J, Kunihiro T, Konishi K, Maemura H, Shute Y, Sugaya F, Materano M, Ali T, Kuehnel K, Seide K, Schroeder U., SoC compatible 1 T1 C FeRAM memory array based on ferroelectric Hf0.5Zr0.5O2, IEEE Symposium on VLSI Technology 2020 Jun 16 (2020) 1–2.

[115] Francois T, Grenouillet L, Coignus J, Blaise P, Carabasse C, Vaxelaire N, Magis T, Aussenac F, Loup V, Pellissier C, Slesazeck S., Demonstration of BEOL-compatible ferroelectric Hf 0.5 Zr 0.5 O 2 scaled FeRAM co-integrated with 130nm CMOS for embedded NVM applications, IEEE International Electron Devices Meeting (IEDM) (2019) 15–17.

CHAPTER

9

"Metal oxides in magnetic memories": Current status and future perspectives

Andreas Kaidatzis[a], *Georgios Giannopoulos*[a,b], *and Dimitris Niarchos*[a]

[a]Institute of Nanoscience and Nanotechnology, NCSR "Demokritos", Athens, Greece, [b]Department of Physics and Astronomy, University College London, London, United Kingdom

9.1 Introduction

Magnetic data recording has been used for information storage since the creation of the first **hard-disk drive (HDD)** [1] in the 1950s and is based on the characteristic feature of ferromagnetic materials, which is the spontaneous appearance of magnetization [2]. A key property of ferromagnetic material data information elements is that they retain their magnetization once initially magnetized, thus allowing for non-volatile data storage. The direction of magnetization of a ferromagnetic material element may be used for binary data storage, as it can be forced to point only toward two directions, representing the "0" or "1" bits of information. Ferromagnets have been also used for **random access memory (RAM)** applications between the 1950s and 1970s, in the technology called "magnetic core memory": a woven web of bulk toroidal magnets and wires allowing writing and reading information [3]. In both of the above-mentioned technologies, information is stored in a ferromagnetic material element, written by means of the Oersted magnetic field produced around a current-currying wire, and read by means of sensing the stray magnetic field emanating around the ferromagnet. The advent of miniaturization and spin electronics (or spintronics) allowed the development of novel magnetic memory

Metal Oxides for Non-volatile Memory
https://doi.org/10.1016/B978-0-12-814629-3.00009-X

technologies, taking advantage of advanced reading and writing schemes and combining high density, non-volatility, and fast access.

9.1.1 Spintronics

"Traditional" semiconductor electronics are based on the electron or hole mass and charge. The pioneering work of Albert Fert [4] and Peter Grünberg [5] (joint Nobel prize in Physics, 2007) in the 1980s, resulted in the discovery of the **giant magnetoresistance (GMR)** effect and the introduction of the electron spin as a new degree of freedom. The spin is a quantum mechanical intrinsic property of an electron and can be either "up" or "down." Within only a few years, technological applications of the GMR effect were invented, and commercially used in the magnetic information storage industry, inaugurating the era of magnetoelectronics [6] or **spintronics** [7].

The GMR effect is explained on the basis of spin-dependent electron transport. Ferromagnetic materials have an imbalance of the electron spin populations at the Fermi level, as the density-of-states available to the spin-up and spin-down electrons is shifted in energy with respect to each other (Fig. 9.1). This results in spin-up and spin-down carriers at the Fermi level having unequal number, character, and mobility. The archetypical spin-polarized device consists of a **ferromagnetic metal/normal metal/ferromagnetic metal (FM/NM/FM)** trilayer; electrons traverse the stack getting "spin-polarized" in the first FM and then "spin-analyzed" in the second FM, after having crossed the NM. When the magnetization of the two ferromagnetic metals is aligned in **parallel (P)** orientation the resistance is low, whereas the resistance is high when it is aligned in **antiparallel (AP) orientation**. The NM decouples the magnetization of the two FM layers, thus allowing the independent control of the magnetization of each layer and the achievement of the P or AP state.

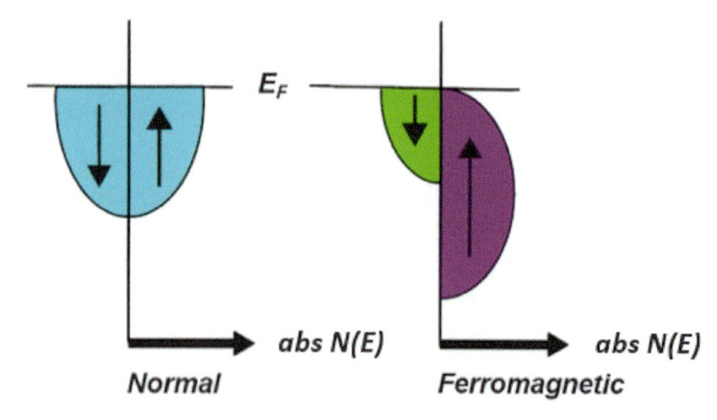

FIG. 9.1 Simplified schemas of the available density of electronic states in a normal metal (left) and in a ferromagnetic metal (right) whose majority spin states are completely filled. E is the electron energy, E_F is the Fermi level and $N(E)$ is the density of states.

9.1.2 Magnetic tunnel junction (MTJ) and tunneling magnetoresistance (TMR)

The building block element of modern magnetic memories is a **metal/insulator/metal (MIM)** junction, called **magnetic tunnel junction (MTJ)**. MTJs are thin-film multilayer stacks in which two thin ferromagnetic layers (typically 1–2.5 nm thick) are separated by a very thin (typically 1–1.5 nm thick) insulating layer. When bias is applied across the junction, electrons cross the insulating layer due to the quantum mechanical tunneling effect; thus, the term "tunnel barrier" is used to denote the insulator. The electrical resistance of the junction in the direction perpendicular to the plane of the layers depends on the magnetic configuration of the two ferromagnetic layers, giving rise to a magnetoresistance effect, in analogy to GMR. When, the two ferromagnetic layers have parallel alignment the resistance is low, whereas when they have anti-parallel alignment the resistance is high (see Fig. 9.2). This phenomenon is called the **tunneling magnetoresistance (TMR)** effect. For memory applications, MTJs are more suitable than all-metallic magnetic multilayers exhibiting GMR, because the former show larger magnetoresistance and much higher electrical resistance (up to several kOhms), which make them more suitable for **complementary metal-oxide-semiconductor (CMOS)** applications.

The first report of the TMR phenomenon actually precedes the discovery of GMR. In 1975, Julliere reports on the conductance of Fe-Ge-Co junctions and its dependence on the parallel or antiparallel configuration of the two ferromagnetic layers, due to the spin polarization of the conduction

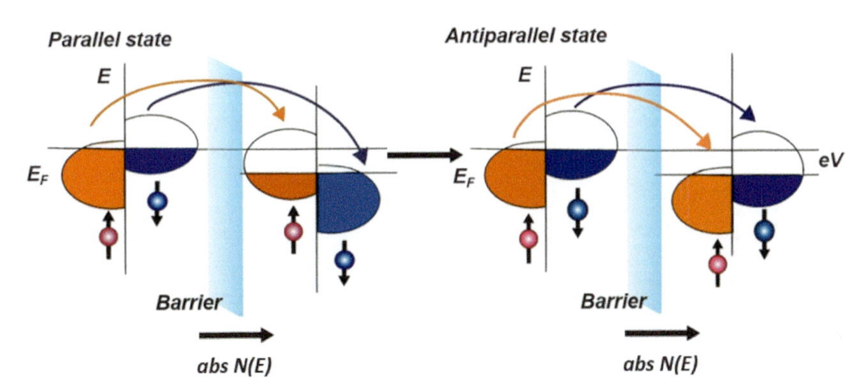

FIG. 9.2 Simplified schemas explaining the tunnel magnetoresistance in the case of two identical ferromagnetic metal layers separated by a non-magnetic insulating tunnel barrier. The simplified density-of-states of the two ferromagnetic layers of the magnetic tunnel junction is shown. Since the tunneling process conserves the spin, when electron states on each side of the barrier are spin-polarized, then electrons will more easily find free states to tunnel to when the magnetizations are parallel (top) than when they are antiparallel (bottom).

electrons [8]. A 14% TMR value is reported at 4.2 K. According to Julliere's model, widely accepted today, the explanation of the TMR phenomenon is due to spin-dependent electron tunneling between the ferromagnetic electrodes (Fig. 9.2); spin is conserved during tunneling and tunnel current depends on the density of states at the Fermi level of the two electrodes, which is spin-polarized in ferromagnets, thus, tunneling probability depends on the relative magnetization orientation of the two layers.

Although spin-dependent tunneling has been demonstrated at low temperature, technological applications of TMR require room-temperature (RT) operation. The first such reports appeared two decades late; in 1995 Miyazaki et al [9] and Moodera et al. [10] almost simultaneously report for the first time on room-temperature TMR of MTJs, using Al_2O_3 as the oxide layer. Fig. 9.3 shows the original RT magnetoresistance variation of a MTJ, as reported by Moodera et al., as a function of applied magnetic field. The ferromagnetic layers used were Co and CoFe and the magnetoresistance curves of single Co and CoFe films is also shown for comparison. Striking differences are observed to the shape of the curves, as well as to their magnetoresistance magnitude and sign. Particularly, TMR is almost two orders of magnitude higher than the "conventional" magnetoresistance of single ferromagnetic layers, manifesting the great potential of MTJs for practical applications.

9.1.3 Spin-transfer torque

In the MTJ schemes described in the previous section, magnetization reversal of the ferromagnetic layers occurs via the application of an external magnetic field. Although this method is effective, requires the

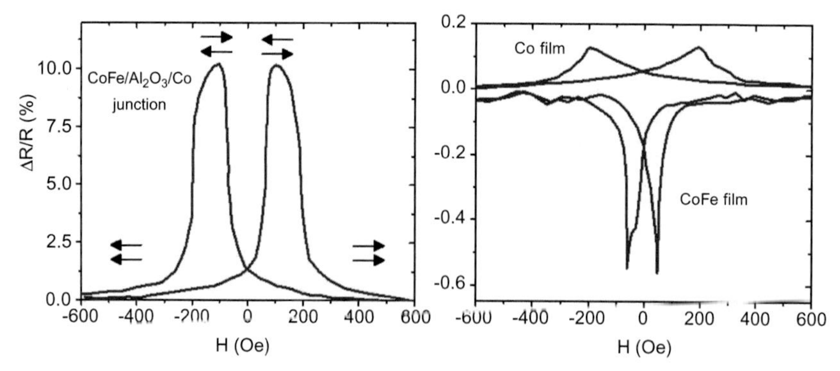

FIG. 9.3 Resistance of a CoFe/Al_2O_3/Co tunnel junction plotted as a function of the applied magnetic field (*H*) in the film plane, at 295 K. The single CoFe and Co film resistance is also shown for comparison. The arrows indicate the direction of magnetization in the two films. *Data obtained from reference J.S. Moodera, L.R. Kinder, T.M. Wong, R. Meservey, Large magnetoresistance at room temperature in ferromagnetic thin film tunnel junctions, Phys. Rev. Lett. 74 (1995) 3273–3276.*

presence of current-currying wires constituting the employed MTJ configurations "bulky." Furthermore, the magnetic field created around the current-currying wire is not localized only at the point of interest; instead, it also affects a wider space. The above render MTJ technology improper for miniaturization and state-of-the-art microelectronics applications.

This issue is tackled by employing the phenomenon of spin-transfer torque (STT), theoretically predicted in 1996 [11] [12] and experimentally demonstrated a few years later [13]. STT allows for reversing the magnetization of a ferromagnet element by using spin-polarized electron currents. This is possible because magnetization in solids arises from the magnetic moments of their atomic electrons, which they have an intrinsic magnetic moment due to their spin [2]. Thus, a spin-polarized current exerts a torque on the free layer magnetization, resulting in reorientation of its spin polarization (Fig. 9.4) [14]. This interaction between the spin-current and magnetization can be precisely directed, as it only occurs locally, in areas in which spin-currents flow. STT allows for all-electronic manipulation of the magnetization of magnetic nanostructures, without the need of external magnetic fields or moving parts. The disadvantage of this method is that current densities between 10^6 and 10^7 A/cm^2 [15] are required for surpassing the "onset" STT-switching current.

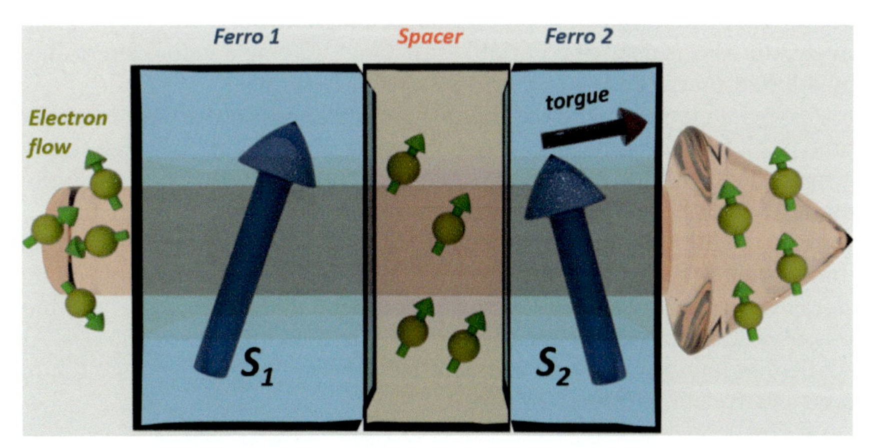

FIG. 9.4 Schematic representation of spin-transfer torques in a ferromagnetic/non-magnetic spacer/ferromagnetic multilayer stack. The current is spin-polarized in the left part of the device with a polarization defined by the left ferromagnetic layer magnetization. When electrons traverse the non-magnetic spacer they conserve their spin-polarization. When the electrons incoming from the left side enter the right ferromagnetic layer, they are polarized along the right layer magnetization, as a torque is exerted on the electrons which rotates their spin angular momentum. However, as angular momentum is conserved, the electrons exert an equal but opposite torque on the right ferromagnetic layer.

9.2 Magnetic random access memory (MRAM)

Magnetic random access memory (MRAM) is a new class of magnetic memory based on spintronic phenomena occurring in MTJs [16]. MRAM, particularly STT-MRAM, is attracting an increasing interest in microelectronics industry, leading to an intense research effort regarding physical concepts, materials used, memory designs, and the combination of magnetics and CMOS electronics [17,18].

9.2.1 MRAM storage principle

MRAM is a ferroic material memory and exploits the spontaneous magnetization of ferromagnets. An MTJ is at the "core" of each MRAM cell. For MRAM applications, MTJs are engineered having two ferromagnetic layers of different magnetic properties: the one has hard-magnetic and the other has soft-magnetic properties; the former has a "fixed" magnetization direction and is called the "reference" layer, whereas the later, called the "storage" layer, may have magnetization pointing toward two predefined directions: parallel or antiparallel to the reference layer. The specific magnetization direction of the storage layer is exploited for information storage and the two possible orientations are used to materialize the binary storage scheme. The action of "writing" information occurs via switching of the storage layer magnetization direction. Effectively, the MTJ is used as a variable resistance, having a value determined by applied magnetic field or spin-polarized current.

A crucial characteristic for memory applications is the data retention time, which is the time that the memory cell is capable of retaining its information without further requirements in energy, that is, in standby mode. The HDD industry requires a minimum data retention time of 10 years. In all magnetic memory media, data retention depends on the unintentional thermally activated reversal of the magnetization of the storage layer (Fig. 9.5). This phenomenon is related to materials parameters, as well as size and shape of the magnetic element, in case of MTJs. An Arrhenius law governs the characteristic thermally activated switching time as a function of temperature T:

$$\tau = \tau_0 \exp\left(\frac{\Delta E}{k_B T}\right)$$

where k_B is the Boltzmann constant. The factor $\Delta = \Delta E / k_B T$ is called the thermal stability factor and defines the data retention time of magnetic memories. Storage duration depends on material properties and the nanostructure size, particularly the volume of the magnetic storage layer. Furthermore, as chip storage capacity increases, data retention duration

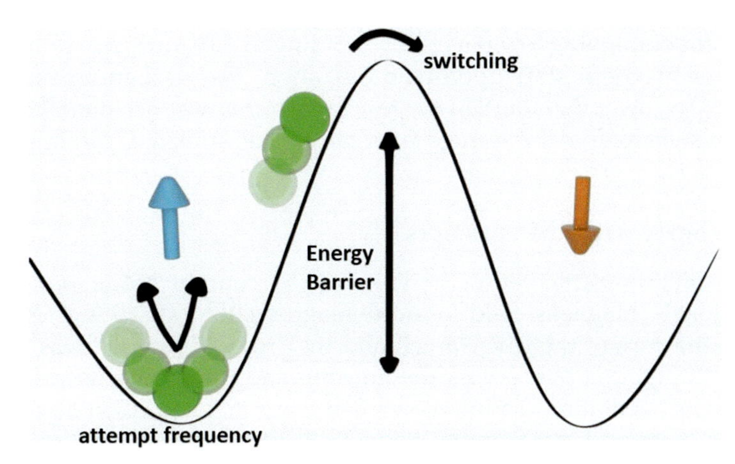

FIG. 9.5 Basic principle of thermally activated magnetization switching: thermal agitation results in surpassing the effective activation energy barrier, resulting in magnetization reversal.

decreases—taking into account same material proprieties—thus, materials engineering is required to achieve sufficiently high thermal stability factor. The thermal stability factor must be greater than 67 for a 32 Mbit chip and greater than 70 for a 1 Gbit chip for sustaining *n* acceptable level of failure probability time during 10 years in standby mode [16]. The higher the memory capacity, the larger the thermal stability factor has to be.

9.2.2 MRAM read principle

All MRAM schemes employ the same data read principle: the measurement of the MTJ cell resistance, which is governed by the TMR effect. The magnetic configuration of each single MTJ cell is determined indirectly exploiting the change of cell resistance between the P and AP states, thus making possible to read each bit of stored information. To perform the read operation, current is flowing through the MTJ cell and its value depends on the magnetization configuration of the MTJ; the output resistance value is compared to a reference resistance value which is the average of the low- and high-resistance states. The read current is chosen so that during the read operation the voltage across the MTJ is in the 0.1–0.2 V range. A low MTJ bias has to be maintained, first of all because the TMR value decreases as the junction bias increases [19]. This occurs since the spin filtering mechanism, giving rise to TMR, becomes less effective as voltage bias increases. Additionally, tunneling electrons injected into the receiving electrode at energies higher than the Fermi level generate magnetic excitations that depolarize the tunneling current. A second reason for employing low bias is for avoiding STT disturbance of the storage layer magnetic state upon read operation. The read current must be sufficiently lower than the STT write current, otherwise accidental magnetization switching of the storage layer may occur upon the read operation.

9.2.3 Early MRAM concepts

The first implementation of the MRAM concept made use of the "traditional" magnetic field-induced magnetization switching. An MTJ was at the core of the MRAM cell and the "read" operation was carried out by applying a bias across the junction and measuring its resistance. The "write" operation requires control over the magnetization direction of the free layer and the Lorentz magnetic field created around a current-carrying wire was exploited for this purpose (Fig. 9.6); current pulses injected through the "write" current line create a magnetic field of appropriate strength and direction for manipulating at will the free layer magnetization.

Implementations of μm-scale junctions ($0.6 \times 1.2\,\mu m^2$), utilizing Al_2O_3 tunnel barrier, achieved TMR values of 40% and RA values of tens kΩ μm^2 [20]; this study also highlighted the importance of shape and aspect ratio of the junction to the magnetization switching process, as well as the microstructure of element boundaries. CMOS compatibility was achieved and a read and program address access time of 14 ns was obtained. The memory cell design was composed of a single transistor and a single MTJ element; the transistor is turned on, and a read current flows through the tunnel junction to read a single bit. This design allowed the fabrication of a 1 Mbit MRAM chip, where MTJs were CMOS integrated using a 0.6 μm process

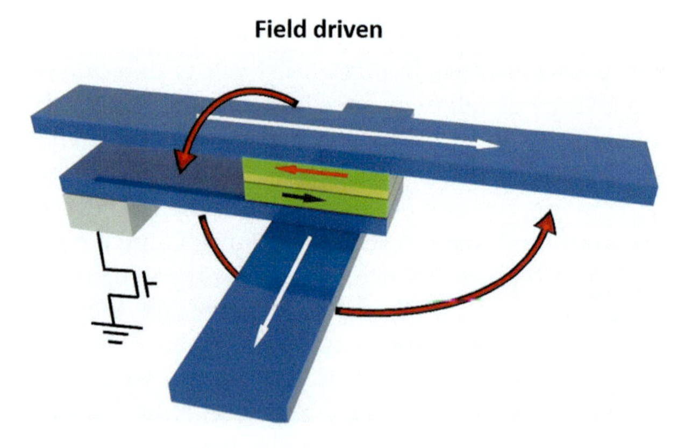

FIG. 9.6 MRAM design utilizing magnetic fields created by current-currying wires for writing magnetic bits. The particular scheme corresponds to the "Toggle MRAM" configuration (see text).

and copper interconnects [21]. At the same time, 1 Gb chips of flash memory—a widely used non-volatile memory technology—were fabricated using 0.13 μm processes [22].

An issue of field-induced MRAM design arises during the write operation; when a current-carrying wire is used for creating the write magnetic field, this field is not spatially confined within the limits of the memory cell to be addressed, but disturbs as well adjacent cells. This selectivity issue is tackled by implementing the "Toggle MRAM" concept [23,24]. In this design, two orthogonal current-carrying wires are used for creating the appropriate "write" magnetic field at the intersection of the Lorentz fields created by each single wire (Fig. 9.6). A single current-carrying wire cannot create sufficient magnetic field to reverse the magnetization of a MTJ cell's free layer. In addition, a multi-step procedure is required for reversing the magnetization of the storage layer. This way, the weak magnetic field created by a single wire does not disturb adjacent memory cells, and the "write" operation is taking place only at the cell to be addressed, thus allowing for a denser MTJ packing. Furthermore, the Toggle MRAM magnetization reversal scheme allows for using "synthetic anti-ferromagnets" as storage layer materials; these are trilayers of ferromagnetic materials separated by a non-magnetic metal, which allows for the antiferromagnetic coupling of the two magnetic layers and results in a higher magnetic anisotropy and thermal stability of the cell. The Toggle MRAM design allowed the development of 4 Mbit MRAM chips, using 0.18 μm CMOS processes.

Another development of the field-induced switching design is Thermally Assisted (TA) MRAM, which reduced power consumption and improved thermal stability and write selectivity of MRAM cells [25,26]. In TA-MRAM, the pinned layer of the MTJ has a much higher blocking temperature than the storage layer, allowing the manipulation of magnetization through the temperature control of the MTJ. In this operating mode, first a current pulse is injected through the junction, producing enough heat to raise the junction's temperature above the blocking temperature of the storage layer. Thereafter, a current-carrying wire crates a magnetic field to switch the storage layer magnetization; however, this magnetic field is significantly lower than in the case of room-temperature operation, thus, much lower current is required. Overall, there is energy saving, as the gain in write energy is higher than the loss due to resistive heating. TA-MRAM has also high selectivity of bit cells, as a combination of magnetic field and heating current is needed to select a particular junction, dramatically decreasing any addressing errors.

To summarize, field-induced MRAM provided a CMOS compatible, low consumption non-volatile magnetic memory technology, first commercialized by Freescale in 2006 in the form of a 4 Mbit chip utilizing MgO tunnel barrier and in-plane magnetization stacks; the first applications include its use in Airbus flight controllers. Although this MRAM technology was effective and reliable, the necessity for additional current lines implied a "bulkier" design, prohibiting further miniaturization and achieving high memory density.

9.2.4 State-of-the-art MRAMs

The first roadblock encountered by the MRAM schemes described above is related to the scalability of MRAM cells to smaller bit sizes. Decreasing size of the in-plane magnetization field-induced switching MTJs below 0.1 μm^2 creates issues related to the control of switching field, device uniformity, long term bit stability, and power consumption. At such small dimensions, the particular shape of the junction, as well as the material's magnetic anisotropy, critically govern the magnitude of the switching field. Furthermore, the width of the switching field distribution is enlarged upon decrease of the size and aspect ratio of the bit.

An important breakthrough on MRAM development was the employment of STTs for switching the magnetization of the nanostructured MTJs using a spin-polarized current [27–29]. This way, the need for the generation of magnetic field using current-carrying wires is eliminated and each MTJ is connected in series to one selection transistor (Fig. 9.7). Furthermore, each bit cell is independently addressed, without any interference occurring to the adjacent cells, as a current-only process is used for the "write" operation. The "0" and "1" bits of information may be written to the cell by applying current pulses of opposite polarity. The fact that the critical current for magnetization switching scales proportionally to the MTJ area—assuming a constant layer thickness and material properties—contributes to the improved downsize scalability of STT-MRAM. Overall, a high bit density and fast read/write operation is

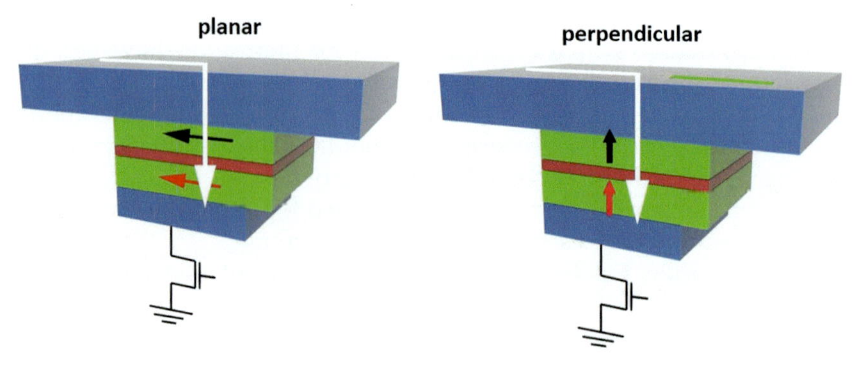

FIG. 9.7 MRAM design employing STTs for writing bits. The MTJs employed have in-plane (left) and perpendicular (right) magnetization.

ensured. This technology reached the market in 2012, used in the 64Mb "Spin Torque MRAM" chip of Everspin.

Although STT-MRAM provided a major breakthrough in the advancement of the technology, the use of in-plane magnetization materials constituted an obstacle to further miniaturization due to their reduced magnetic anisotropy and thus, the reduced thermal stability of the MTJ cells as they decreased in size. This issue was tackled by using storage and reference layer materials with **perpendicular magnetic anisotropy (PMA)**, that is, with easy-axis magnetization perpendicular to the film plane, usually dubbed **perpendicular-MTJs (p-MTJs)**. Such a design allows for denser packing of the MTJs, while simultaneously maintaining an adequate data retention time, due to the higher magnetic anisotropy energy of the materials, which result in higher thermal stability of the nano-patterned magnetic elements.

The first attempt to fabricate p-MTJs was reported in 2008 by Toshiba, demonstrating a continuous MTJ stack having a TMR value of more than 100% at RT [30]. However, this work concerned MTJs comprising free- and reference-layers of chemically ordered $L1_0$-FePt alloy, which has stringent requirements for material growth—including high-temperature deposition—for obtaining the desired crystal structure; furthermore, elaborate interface engineering has to be performed for achieving a high TMR value. In the same year, Toshiba demonstrated STT switching of p-MTJs fabricated using TbCoFe/CoFeB as magnetic electrodes [31], having good device data retention and fast switching. However, the high spin-orbit coupling of the Tb rare-earth metal resulted in high magnetization switching current density (4.7 MA/cm^2); in addition, a low TMR value of 10% has been obtained.

A significant breakthrough came when interfacial PMA at CoFeB/MgO interfaces was discovered [32,33]; the employed materials combination was already used in the in-plane magnetization MRAMs industry, being compatible to CMOS processes. The role of the bcc (001) CoFeB/MgO interface became even more important, as the crystallinity of the metal/oxide interface is critical to the emergence of PMA. The high magnetic anisotropy energy allowed for patterning circular MTJ devices with a diameter as low as 40 nm. Soon afterwards, STT magnetization switching was demonstrated in single CoFeB/MgO-based MTJ cells [34], paving the way for the use of p-MTJs in the STT-MRAM technology.

A crucial disadvantage of STT-MRAM is related to the high current density (up to MA/cm^2) required for STT magnetization switching; this current flows through the ultra-thin oxide tunnel barrier, resulting in its degradation and ultimately, to its failure. This issue is tackled by exploiting a physical concept recently demonstrated: the spin-orbit torque (SOT) magnetization switching effect [35,36]. Using this scheme (Fig. 9.8), a three-terminal device is realized, comprising the MTJ stack grown on

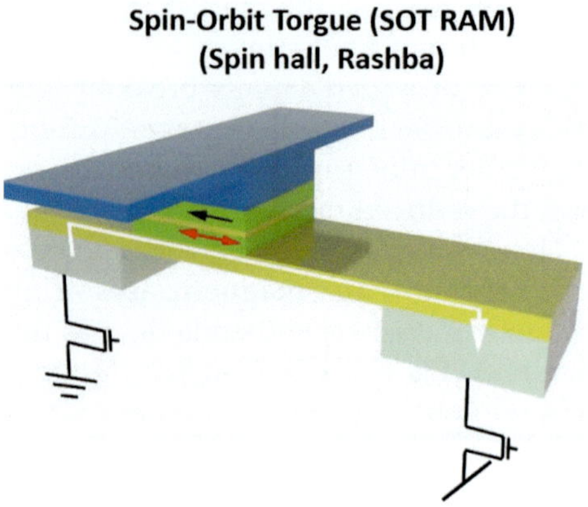

Spin-Orbit Torgue (SOT RAM)
(Spin hall, Rashba)

FIG. 9.8 The three-terminal SOT-MRAM scheme employing spin-orbit torques for magnetization switching.

top of a high spin-orbit coupling metal. In-plane current flowing into the heavy metal/storage layer results in inducing a torque to the later—due to the SOT effect—which eventually switches its magnetization. This **SOT-MRAM** does not require injecting high current densities through the tunnel barrier for magnetization switching. The "read" operation still requires biasing the junction, thus injecting current through the oxide layer; however, the current density required for the "read" operation is several times lower and does not contribute to the degradation of the oxide tunnel barrier. Changing the polarity of the current allows for switching the storage layer magnetization toward the desired direction. Proof-of-concept SOT-MRAM reports used Ta as heavy-metal and CoFeB/MgO/CoFeB MTJs [37], although W is being also studied due to its high spin-orbit coupling coefficient [38,39]. It should be noted that as the SOT effect is an interface phenomenon, it may be exploited only using ultra-thin magnetic storage layers. However, this does not constitutes an issue as p-MTJs also require ultra-thin CoFeB storage layers for exploiting the perpendicular interface magnetic anisotropy.

A recent advancement on magnetic memories is the use of "multiferroic" materials; these are complex single phase ternary or quaternary metallic oxides demonstrating more than one type of ferroic behavior, which is ferromagnetism, ferroelectricity, and ferroelasticity. The most relevant to non-volatile memories multiferroic materials combine ferromagnetic and ferroelectric properties [40]. Of particular importance in multiferroic memories is the electric field control of magnetization; this has been demonstrated to MTJs employing magnetoelectric/ferromagnetic bilayers, where the magnetization state of the ferromagnetic layer is controlled—through exchange coupling—by applying an electric field to the magnetoelectric thin film [41]. Also, electric field pulses applied on ferromagnetic/ferroelectric/ferromagnetic tunnel junctions have been shown to control the TMR of such stacks, due to the effect of bias on the electronic band structure at the ferromagnetic/ferroelectric interface [42]. However, it should be noted that the emergence of multiferroic properties requires thin films having high-quality crystal structure, often of non-equilibrium phases stabilized through strain engineering. The above impose the use of elaborate synthesis equipment, e.g., molecular beam epitaxy or pulsed layer deposition, or stringent experimental conditions, including incompatible to CMOS processes high temperatures and monocrystalline oxide substrates.

9.3 Metal oxides in MRAMs

An MRAM cell employs metal oxides as tunnel barriers in its core component, which is an MTJ. Besides the crucial role that the tunnel barrier plays in the spin-filtering process that results in the measured TMR effect, it is also of great importance to other aspects of the science and engineering of MTJs. The metal oxide barrier is important to the magnitude of the TMR value, as well as its sign. Furthermore, the oxide layer is an integral part of the MTJ structure, affecting the crystal growth of the ferromagnetic layers, in addition to the significant role of the ferromagnetic metal/oxide interface. Finally, the tunnel barrier is of paramount importance to the reliability of the MTJ, as its breakdown characteristics govern the lifespan of the MRAM cell. In the following, the above mentioned topics are discussed. Overall, MTJs are patterned out of highly complex multilayer stacks comprising tens of alternating layer of various materials, each one serving a specific purpose. However, the technology has been mastered and volume manufacturing of MTJ micro/nanometric cells is a reality since several years. Stringernt requirments have to be taken into account during stack growth and the barrier material plays a key role on the crystalline structure of the stack, greatly affecting MTJ performance.

9.3.1 Tunnel barrier and TMR magnitude

The first report of MTJs concerned the use of Ge—a semiconductor—as a tunnel barrier, resulting in low TMR values, obtained only at liquid helium temperature—4.2 K [8]. The employment of alumina as a tunnel barrier constituted a significant increase in TMR, most importantly, obtained at RT, making MTJs relevant to industrial applications [9,10]. Miyazaki et al. employed e-beam evaporated Fe as FM electrodes and RF sputtered AlO_x as tunneling barrier [9]. Moodera et al. employed evaporated CoFe and Co or NiFe FM electrodes and oxygen plasma oxidized metallic Al layers as tunneling barriers [10]. The reported TMR values at RT were 18% and 11.8%, respectively. In the following years, TMR values increased, mainly due to employing magnetron sputtering as thin film deposition method [43] and by devising advanced oxidation schemes for Al [44]; the maximum reported TMR value at RT using alumina tunnel barriers was 80% [45]. Initial industrial applications employed such TMR stacks for use as hard-disk drives read-heads and early design of MRAMs.

In all the reported studies of MTJs based on alumina, the TMR value is always lower than 100%, a value that would allow further adoption of MTJs for industrial applications. This is a result of the amorphous alumina tunnel barrier; due to its lack of any crystallographic symmetry, various Bloch states—conduction electrons wave functions—having density-of-states of different spin polarizations, can all tunnel through the oxide barrier (Fig. 9.9). This results in a spin polarization of the tunnel electrons which is lower than 0.5 at RT, for the usual ferromagnets used in MTJs. A major improvement of this issue was the employment of crystalline MgO (001) tunnel barriers [19,46]. In this case, electrons tunnel coherently through the oxide barrier, retaining their orbital symmetry (Fig. 9.9). Theoretical studies through first-principles calculations demonstrated that Bloch states with spherical symmetry ($\Delta1$ states) predominately tunnel through the MgO (001) barrier. Since the $\Delta1$ states of body-centered-cubic (bcc) Fe(001) have 100% spin-polarization, that is, only the $\Delta1$ band majority spins have available density-of-states at the Fermi level, theoretically, TMR ratios higher than 1000% are expected [47,48].

The first experimental materializations of MTJs with MgO tunnel barriers were reported in 2004. Yuasa et al. [19] fabricated high-quality, fully epitaxial Fe(001)/MgO(001)/Fe(001) MTJs by using molecular beam epitaxy growth and achieved TMR ratios of up to 180% at RT (Fig. 9.10). Almost

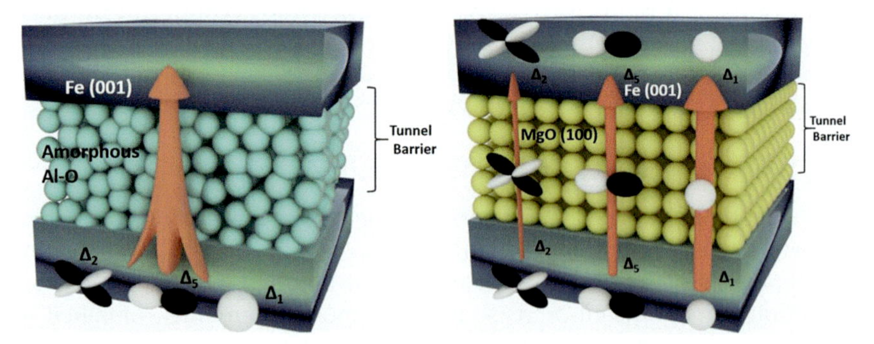

FIG. 9.9 Amorphous alumina tunnel barrier (left) results in incoherent electron tunneling and low TMR. On the contrary, a crystalline MgO tunnel barrier (right) allows for coherent tunneling and high TMR.

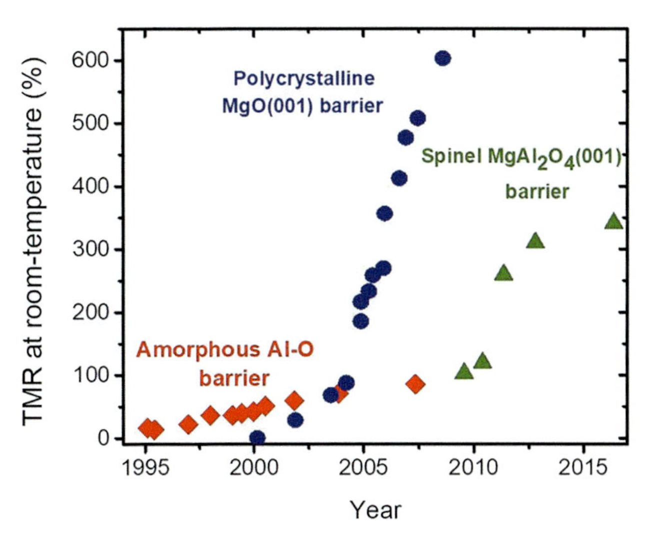

FIG. 9.10 Evolution of the room-temperature TMR ratio of MTJs over the years. Only coherently crystallized ferromagnet/oxide/ferromagnet stacks may result in TMR ratios considerably higher than 100%.

simultaneously, Parkin et al. [46] fabricated MTJs with a highly oriented polycrystalline (or textured) MgO(001) barrier using sputter deposition and obtained even higher MR ratios of up to 220% at RT. The larger TMR values provided by the MgO tunnel barrier allow for a much improved read margin and faster read in memory devices. However, the obtained values are much lower than the theoretically predicted, as a result of non-perfect crystal structure of the MTJ layer and the incoherent Fe/MgO interface due to lattice mismatch and different crystal structure.

An advancement resulting in further increase of the TMR values was using CoFeB as magnetic electrode. The key point of this approach is that the as-deposited CoFeB layers are amorphous, whereas the as-deposited MgO tunnel barrier is polycrystalline with bcc crystal structure. Upon annealing, the CoFeB layers of the MTJ crystallize as well to the bcc structure, since the MgO oxide layer acts as a seed layer for crystallization. This produces a coherent CoFeB/MgO interface that results in higher degree of spin conservation and higher TMR values. This approach initially resulted in 230% TMR at RT [49,50], ultimately surpassing 600% at RT by applying elaborate materials engineering [51]. The use of the CoFeB/MgO approach is also crucial in obtaining p-MTJs, as described in the previous section; coherent crystallization of the CoFeB/MgO interface to the bcc crystal structure results in interface perpendicular anisotropy, which is exploited for high-density MRAM applications.

Recenlty, efforts to minimize the magnetic electrode/tunnel barrier lattice mismatch, which is approximately 3.8% for Fe/MgO, have been reported. A promisng route for doing so is to use novel oxides as tunnel barriers. It has been demonstrated that crystalline $MgAl_2O_4$ (001) spinel layers [52], with smaller in-plane lattice constants, may be used for reducing the lattice misfit with various ferromagnetic metals, including bcc CoFe [52] and Fe [53]. These are suited for making fully epitaxial MTJs with dislocation-free interfaces and a lattice mismatch between Fe and $MgAl_2O_4$(001) lower than 1%. The highest TMR value obtained thus far for $MgAl_2O_4$-based MTJs is 342% at RT (see Fig. 9.10), using Co_2FeAl Heusler alloy electrodes [54], whereas a TMR value of 245% at RT is obtained using Fe electrodes [55]. Alternatively, $MgGa_2O_4$-based MTJs have been also studied [56]; although they show a TMR value of 121% at RT, their low resistance-area product allows for lower power consumption memory cells. In addition to the above, $MgAl_2O_4$-based MTJs have shown to obtain perpendicular magnetic anisotorpy [57], making them very promising for future high-density STT-MRAM applications.

9.3.2 The importance of oxide on MTJ crystal structure

The metal oxide tunnel barrier is an integral part of the MTJ and plays a crucial role during crystal growth process of the multilayer stack. As it was discussed in the previous section, the MgO barrier—having bcc crystal structure—acts as seed for CoFeB crystallization to the bcc phase upon annealing, despite that the face centered cubic crystal structure of CoFeB is more stable [58]. This crystallization process is very delicate and the actual crystal structure of the MgO layer before annealing plays a crucial role; it has been found that process parameters during MgO deposition are critical for obtaining high PMA at the CoFeB/MgO interface [59]. This happens due to the coherent crystallization of the layers, resulting in fully epitaxial CoFeB (001)/MgO (001)/CoFeB (001) stacks. The importance of such process on the TMR of the MTJ has been shown in Lee et al. [60]; the authors demonstrate that, the lattice mismatch between MgO and CoFeB varies with the CoFeB stoichiometry, resulting in partial crystallization of CoFeB and low TMR values for high Co content, whereas high TMR is obtained when CoFeB crystallized to bcc (001) along the entire MgO/CoFeB interface, due to high Fe content (Fig. 9.11).

An important aspect of MTJ growth is its polycrystalline morphology, which results in in homogeneities of the individual MTJ cells. As the MTJ bit size approaches 10 nm, the magnetic and magnetotransport MTJ properties strongly depend on the presence of grains and grain boundaries in polycrystalline MTJs. These cause large bit-to-bit variations in performance. Fully epitaxial MTJs with single-crystal barriers and electrode

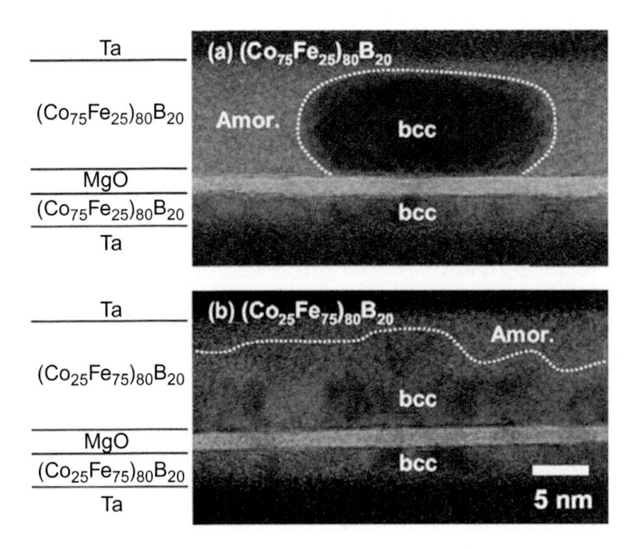

FIG. 9.11 Cross-sectional high-resolution transmission electron microscopy (HRTEM) images of CoFeB/MgO/CoFeB MTJs. (top) A Co-rich sample results in partially crystallized CoFeB/MgO interface and low TMR. (B) A Fe-rich sample results in fully crystallized CoFeB/MgO interface and high TMR. *Reproduced from Y.M. Lee, J. Hayakawa, S. Ikeda, F. Matsukura, H. Ohno, Effect of electrode composition on the tunnel magnetoresistance of pseudo-spin-valve magnetic tunnel junction with a MgO tunnel barrier, Appl. Phys. Lett. 90 2007 212507 with the permission of AIP Publishing.*

layers grown on Si (001) substrates may provide a solution to this issue. However, fully epitaxial MTJs do not ensure crystallographic homogeneity; the large lattice mismatch between MgO and CoFeB results in many misfit dislocations [19]. These result in the degradation of interface PMA, in addition to the large bit-to-bit variations. For significantly reducing misfit dislocations, new crystalline tunnel barrier materials with smaller in-plane lattice constants have been developed, e.g., $MgAl_2O_4$ (001) spinel tunnel barriers [54–56], and are therefore suited for making MTJs with dislocation-free interfaces.

9.3.3 Tunnel barrier breakdown

As with any MIM junction, MTJs have a certain characteristic insulating barrier voltage where breakdown occurs, defined as the maximum voltage difference that can be applied across the junction before the tunnel barrier conducts. This is, in fact, a catastrophic event for the MTJ, as the sudden current surge results in permanent physical changes of the material and conducting paths are created within the barrier oxide. The breakdown voltage of an MTJ defines a form of failure of the junction and it is not a definite value as there is a statistical probability whether the material will fail at a given voltage. The mean breakdown voltage of a large set of MTJs defines the given breakdown voltage value.

The breakdown voltage is a particularly crucial parameter, as it defines the endurance of the memory cell. Although the voltages used during MRAM operation are rather low—in the mV range—compatible to microelectronic circuits, the very low thickness of the tunnel barrier (typically between 1 and 2 nm) results in extremely high electric fields, on the order of MV/m. Another crucial parameter during barrier breakdown is current density. The currents running through MTJs are in the mA range. However, MTJs are spatially confined to the nm-scale dimensions; currently, commercial MRAMs make use of 45-nm wide MTJ cells and state-of-the-art MTJs have been demonstrated in the sub-20 nm range [61]. Thus, this very low junction area results in high current densities. In addition, although reading the cell—that is probing its resistance—requires low current densities (approximately 10^8 A/m^2), writing a cell by means of STT switching requires much higher current density, on the order of 10^{11} A/m^2.

The issue of breakdown voltage in MTJs has been studied since the initial steps of the technology. Early on, current-ramp experiments on MTJs of wide area (greater than 1 μm^2), having ultra-thin amorphous alumina tunnel barriers, determined two types of dielectric breakdown [62,63]. The first was an abrupt breakdown, related to the material intrinsic properties occurring at an electric field of 10 MV/cm; the interaction of the applied electric field with the dipole moment of a bond results in the breaking of the later, leading to electrical stress-induced defect creation. The second was a gradual breakdown mechanism related to defects in the tunnel barrier—a factor extrinsic to the material—that is caused by the particular manufacturing process and can be tailored by reducing the roughness of the bottom electrode. It was demonstrated that the presence of pinholes in the initial material results in this extrinsic breakdown mechanism; the pinholes get wider as the applied current increases, due to the local electric field in the pinhole and/or localized heating of the pinhole. Also, interfacial roughness results in hot spots in the barrier where higher electric fields are created. Another extrinsic factor is related to the MTJ pillar microfabrication, particularly the ion milling process, and to redeposition of metallic species at the edges of the MTJ.

The advent of MgO as a crystalline tunnel barrier, in contrast to amorphous alumina, spurred various studies of its dielectric breakdown characteristics. Initial studies concerned voltage-ramp experiments. The junction area, polarity of the applied voltage, and material processing parameters have been found to affect the breakdown voltage [64]. In all of the above mentioned cases, a permanent alteration of the tunnel barrier occurs upon breakdown, as suggested by the respective electrical characterization. These permanent material modifications have been imaged through HRTEM measurements [65,66], providing insights on the microscopic processes occurring upon breakdown, in addition to the large amounts of samples studied through electrical transport measurements. Initial studies of MgO-based MTJs have shown that multiple pinholes appear upon breakdown of the oxide barrier due to voltage-stressing [65]. However, studies of MTJs for STT-MRAM applications have shown a different breakdown mechanism; the high current densities injected into the MTJ for STT operation and the laterally confined MTJ dimensions result in the coalescence of multiple pinholes and the creation of an area of the junction where the barrier has been completely destroyed due to electromigration of material [66] (Fig. 9.12).

All of the above-mentioned studies concern the DC breakdown voltage, a measure of the maximum voltage an MTJ may withstand. However, MRAM operation requires the use of repeated voltage pulses for reading and writing information, thus, the pulsed voltage breakdown is defined; a series of voltage pulses is applied until electrical breakdown of the MTJ tunnel barrier. This dynamic stressing of the junction reveals breakdown mechanisms distinct of the ones observed during static stressing. The time

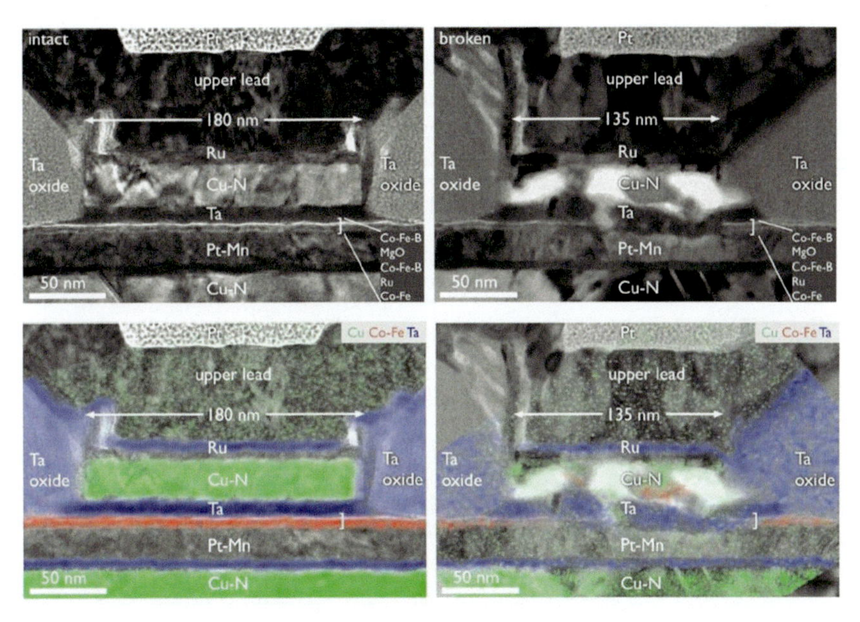

FIG. 9.12 TEM image of an MTJ before (top-left) and after (top-right) barrier breakdown. Energy dispersive X-ray stoichiometric analysis of an MTJ before (bottom-left) and after (bottom-right) barrier breakdown. Cu, Co—Fe, and Ta distributions are mapped in the image. *Reproduced from M. Schäfers, V. Drewello, G. Reiss, A. Thomas, K. Thiel, G. Eilers, M. Münzenberg, H. Schuhmann, M. Seibt, Electric breakdown in ultrathin MgO tunnel barrier junctions for spin-transfer torque switching, Appl. Phys. Lett. 95 2009 232119 with the permission of AIP Publishing.*

interval between two pulses plays a crucial role, in addition to the pulse duration and voltage, and a charge trapping-detrapping breakdown mechanism is revealed [67]. A characteristic time interval of 100 ns was determined; below this value breakdown occurs due to the high average charge trapped in the oxide barrier, whereas above this value the large temporal modulation of trapped charges causes alternating stress of the oxide eventually leading to barrier breakdown (Fig. 9.13). The study highlights the importance of manufacturing conditions, as well as materials choices, for reducing the amount of charge trapping sites by reducing the material defects (e.g., interstitial O or Mg vacancies or local in homogeneities, like the presence of BO), as well as the oxide/magnetic electrode interface defects due to increased lattice mismatch. A high MTJ endurance has been obtained for time interval approximately equal to 100 ns, which corresponds to the characteristic escape time of electrons trapped in the oxide tunnel barrier. For this junction, breakdown occurred after 1.15×10^{11} pulses of +1.3 V amplitude separated by a delay of 100 ns, revealing an endurance better than non-volatile Flash memory (10^4 to 10^6 cycles), but worse than volatile DRAM memory (10^{14} to 10^{16} cycles).

Finally, an interesting correlation has been found between the $1/f$ junction electrical noise before any electrical stress has been applied and the number of write cycles that the junction may withstand before barrier breakdown [68]. Unstressed junctions of large $1/f$ noise have lower write endurance, as they can withstand fewer write pulses. On the contrary, junctions that exhibit low $1/f$ noise before any electrical stress has

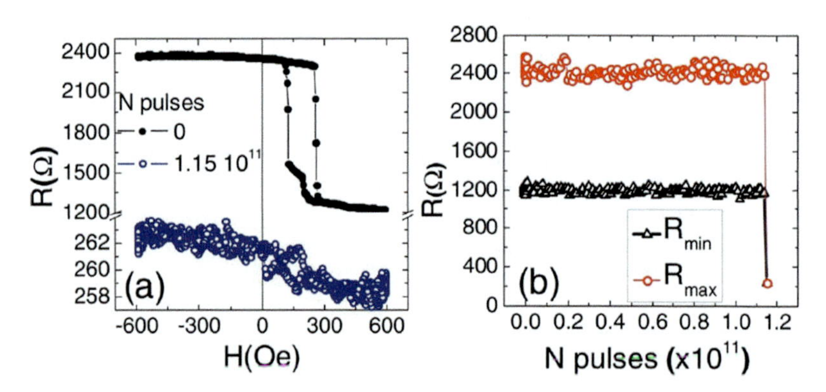

FIG. 9.13 (A) TMR value versus applied magnetic field, before (top) and after barrier breakdown (bottom). (B) Evolution of minimum cell resistance (parallel magnetic configuration) and maximum cell resistance (antiparallel magnetic configuration) as a function of number of pulses. A sharp drop is observed after 1.15×10^{11} pulses when the breakdown abruptly occurs. *Reproduced from S. Amara-Dababi, R.C. Sousa, M. Chshiev, H. Bea, J. Alvarez-He Rault, L. Lombard, I.L. Prejbeanu, K. Mackay, B. Dieny, Charge trapping-detrapping mechanism of barrier breakdown in MgO, Appl. Phys. Lett. 99 2011 083501 with the permission of AIP Publishing.*

been applied, withstand more applied pulses. As $1/f$ noise is related to the amount of electron trapping sites present in the barrier, it was suggested that junction endurance is also sensitive to the amount of barrier defects. Thus, the measurement of $1/f$ noise of unstressed MTJs may be used for non-destructively predicting the endurance of single memory cells.

9.4 Perspectives

Non-volatile magnetic memories have the potential to serve as universal computer memory, fulfilling the cache, main memory, and mass storage requirements. The first MRAM commercial use—employing magnetic field-induced magnetization switching—concerned its use as an embedded memory. During the past years the technology has matured and the current champion—STT-MRAM—is commercially exploited in discrete memory chips (Everspin already ships 256 Mb chips) and in Solid State Drives as write cache memory (according to IBM's recent announcement). Furthermore, Everspin has already announced the production of 1 Gb STT-MRAM chips employing 28 nm CMOS technology.

A crucial materials limitation at such low dimension scales is related to magnetic anisotropy energy of p-MTJs and the thermal stability of recorded data. The use of CoFeB as ferromagnetic material and the perpendicular magnetic anisotropy of the CoFeB/MgO interface—although has greatly driven MRAM development the past 10 years—is now approaching its limits. Recently, the use of Mn-based magnetic materials has been introduced for MTJ fabrication; particularly, it has been shown that MnGa-based MTJs—using MgO tunnel barriers—exhibit sufficiently high perpendicular magnetic anisotropy energy for producing MTJs down to the 10–20 nm scale, with sufficient thermal stability [69]. In addition, SOT-induced magnetization switching has been also demonstrated in MnGa/Pt stacks with perpendicular magnetic anisotropy [70], allowing the development of SOT-MRAM cells of high thermal stability.

Another critical aspect of current MRAM technology—as described in the previous sections—concerns the high write power and long latency associated to current-induced switching. A potential solution to this issue has been recently emerged due to the discovery of the **voltage-controlled magnetic anisotropy (VCMA)** effect [71]. In VCMA, biasing an MTJ stack results in the creation of an electric field through the tunnel barrier. When considering a CoFeB-based MTJ, upon applying voltage across the junction, electron charges accumulate at the CoFeB/MgO interface, inducing a change in the occupation of atomic orbitals; this enhances or reduces the interfacial PMA energy, depending on the applied voltage polarity. In effect, this is equivalent to decreasing or increasing the energy barrier between the two stable magnetization configuration states, thus allowing voltage-assisted or voltage-induced magnetization switching of the MTJ. Much lower energy dissipation may be achieved employing the VCMA effect for writing data into an MTJ, since Joule heating and Ohmic losses are greatly reduced, compared to standard STT-MRAM technology. In addition, faster precessional switching of an MTJ is enabled (down to hundreds of ps) by lowering the energy barrier between the two stable magnetization states.

References

[1] C.J. Bashe, L.R. Johnson, J.H. Palmer, E.W. Pugh, IBM's Early Computers, MIT Press, Cambridge, Massachusetts, 1986, ISBN: 0-262-02225-7.
[2] J.M.D. Coey, Magnetism and Magnetic Materials, Cambridge University Press, Cambridge, UK, 2010, ISBN: 978-0-521-81614-4.
[3] J.W. Forrester, Digital information storage in three dimensions using magnetic cores, J. Appl. Phys. 22 (1951) 44.
[4] M.N. Baibich, J.M. Broto, A. Fert, F. Nguyen Van Dau, F. Petroff, P. Etienne, G. Creuzet, A. Friederich, J. Chazelas, Giant magnetoresistance of (001)Fe/(001)Cr magnetic superlattices, Phys. Rev. Lett. 61 (1988) 2472.
[5] G. Binasch, P. Grünberg, F. Saurenbach, W. Zinn, Enhanced magnetoresistance in layered magnetic structures with antiferromagnetic interlayer exchange, Phys. Rev. B 39 (1989). 4828(R).
[6] G.A. Prinz, Magnetoelectronics, Science 282 (1988) 1660.
[7] S.A. Wolf, D.D. Awschalom, R.A. Buhrman, J.M. Daughton, S. von Molnar, M.L. Roukes, A.Y. Chtchelkanova, D.M. Treger, Spintronics: a spin-based electronics vision for the future, Science 294 (2001) 1488.
[8] M. Julliere, Tunneling between ferromagnetic films, Phys. Lett. 54A (1975) 225–226.
[9] T. Miyazaki, N. Tezuka, Giant magnetic tunneling effect in Fe/Al2O3/Fe junction, J. Magn. Magn. Mater. 139 (1995) L231–L234.
[10] J.S. Moodera, L.R. Kinder, T.M. Wong, R. Meservey, Large magnetoresistance at room temperature in ferromagnetic thin film tunnel junctions, Phys. Rev. Lett. 74 (1995) 3273–3276.
[11] J.C. Slonczewski, Current-driven excitation of magnetic multilayers, J. Magn. Magn. Mater. 159 (1996) L1–L7.
[12] L. Berger, Emission of spin waves by a magnetic multilayer traversed by a current, Phys. Rev. B 54 (1996) 9353.
[13] E.B. Myers, D.C. Ralph, J.A. Katine, R.N. Louie, R.A. Buhrman, Current-induced switching of domains in magnetic multilayer devices, Science 285 (1999) 867–870.
[14] A. Brataas, A.D. Kent, H. Ohno, Current-induced torques in magnetic materials, Nat. Mater. 11 (2012) 372–381.
[15] A. Hirohata, K. Yamada, Y. Nakatani, I.-L. Prejbeanu, B. Diény, P. Pirroe, B. Hillebrands, Review on spintronics: principles and device applications, J. Magn. Magn. Mater. 509 (2020) 166711.
[16] B. Dieny, I.L. Prejbeanu, Magnetic random-access memory, [book auth.] R. B. Goldfarb, K.-J. Lee, in: B. Dieny (Ed.), Introduction to Magnetic Random-Accesss Memory, first ed., John Wiley&Sons, Inc., S.l., 2017.
[17] S. Yuasa, K. Hono, G. Hu, D.C. Worledge, Materials for spin-transfer-torque magnetoresistive random-access memory, MRS Bull. 43 (2018) 352–357.

[18] M.C. Gaidis, Magnetic back-end technology, [book auth.] R. B. Goldfarb, K.-J. Lee, in: B. Dieny (Ed.), Introduction to Magnetic Random-Access Memory, first ed., John Wiley&-Sons, Inc., S.l., 2017.
[19] S. Yuasa, T. Nagahama, A. Fukushima, Y. Suzuki, K. Ando, Giant room-temperature magnetoresistance in single-crystal Fe/MgO/Fe magnetic tunnel junctions, Nat. Mater. 3 (2004) 868–871.
[20] S. Tehrani, B. Engel, J.M. Slaughter, E. Chen, M. DeHerrera, M. Durlam, P. Naji, R. Whig, J. Janesky, J. Calder, Recent developments in magnetic tunnel junction MRAM, IEEE Trans. Magn. 36 (2000) 2752–2757.
[21] M. Durlam, P.J. Naji, A. Omair, M. DeHerrera, J. Calder, J.M. Slaughter, B.N. Engel, N.D. Rizzo, G. Grynkewich, B. Butcher, C. Tracy, K. Smith, K.W. Kyler, J.J. Ren, J.A. Molla, W.-A. Feil, R.G. Williams, S. Tehrani, A 1-Mbit MRAM based on 1T1MTJ bit cell integrated with copper interconnects, IEEE J. Solid State Circuits 38 (2003) 769–773.
[22] K. Yoshida, O. Tsuchiya, Y. Yamaguchi, J. Kishimoto, Y. Ikeda, S. Narumi, Y. Takase, K. Furusawa, K. Izawa, T. Yoshitake, T. Kobayashi, H. Kurata, M. Kanemitsu, A 1Gb multi-level AG-AND-type flash memory with 10MB/s programming througput for mass storage application, in: ISSCC 2003 Proceedings, 2003. Paper 16.6.
[23] B.N. Engel, J. Åkerman, B. Butcher, R.W. Dave, M. DeHerrera, M. Durlam, G. Grynkewich, J. Janesky, S.V. Pietambaram, N.D. Rizzo, J.M. Slaughter, K. Smith, J.J. Sun, S. Tehrani, A 4-Mb toggle MRAM based on a novel bit and switching method, IEEE Trans. Magn. 41 (2005) 132–136.
[24] L. Savtchenko, B.N. Engel, N.D. Rizzo, M. DeHerrera, J. Janesky, Method Of Writing to Scalable Magnetoresistance Random Access Memory Element, 6,545,906 B1 US Patent Office, 2003.
[25] I.L. Prejbeanu, W. Kula, K. Ounadjela, R.C. Sousa, O. Redon, B. Dieny, J.-P. Nozières, Thermally assisted switching in exchange-biased storage layer magnetic tunnel junctions, IEEE Trans. Magn. 40 (2004) 2625–2627.
[26] I.L. Prejbeanu, M. Kerekes, R.C. Sousa, H. Sibuet, O. Redon, B. Dieny, J.P. Nozieres, Thermally assisted MRAM, J. Phys. Condens. Matter 19 (2007) 165218.
[27] Y. Huai, F. Albert, P. Nguyen, M. Pakala, T. Valet, Observation of spin-transfer switching in deep submicron-sized and low-resistance magnetic tunnel junctions, Appl. Phys. Lett. 84 (2004) 3118.
[28] G.D. Fuchs, N.C. Emley, I.N. Krivorotov, P.M. Braganca, E.M. Ryan, S.I. Kiselev, J.C. Sankey, D.C. Ralph, R.A. Buhrman, J.A. Katine, Spin-transfer effects in nanoscale magnetic tunnel junctions, Appl. Phys. Lett. 85 (2004) 1205.
[29] H. Kubota, A. Fukushima, Y. Ootani, S. Yuasa, K. Ando, H. Maehara, K. Tsunekawa, D.-D. Djayaprawira, N. Watanabe, Y. Suzuki, Magnetization switching by spin-polarized current in low-resistance magnetic tunnel junction with MgO (001) barrier, IEEE Trans. Magn. 41 (2005) 2633–2635.
[30] M. Yoshikawa, E. Kitagawa, T. Nagase, T. Daibou, M. Nagamine, K. Nishiyama, T. Kishi, H. Yoda, Tunnel magnetoresistance over 100% in MgO-based magnetic tunnel junction films with perpendicular magnetic L10-FePt electrodes, IEEE Trans. Magn. 44 (2008) 2573–2576.
[31] M. Nakayama, T. Kai, N. Shimomura, M. Amano, E. Kitagawa, T. Nagase, M. Yoshikawa, T. Kishi, S. Ikegawa, H. Yoda, Spin transfer switching in TbCoFe/CoFeB/MgO/CoFeB/TbCoFe magnetic tunnel junctions with perpendicular magnetic anisotropy, J. Appl. Phys. 103 (2008) 07A710.
[32] S. Ikeda, K. Miura, H. Yamamoto, K. Mizunuma, H.D. Gan, M. Endo, S. Kanai, J. Hayakawa, F. Matsukura, H. Ohno, A perpendicular-anisotropy CoFeB–MgO magnetic tunnel junction, Nat. Mater. 9 (2010) 721–724.
[33] D.C. Worledge, G. Hu, D.W. Abraham, J.Z. Sun, P.L. Trouilloud, J. Nowak, S. Brown, M.-C. Gaidis, E.J. O'Sullivan, R.P. Robertazzi, Spin torque switching of perpendicular ta/CoFeB/MgO-based magnetic tunnel junctions, Appl. Phys. Lett. 98 (2011), 022501.

[34] H. Sato, M. Yamanouchi, K. Miura, S. Ikeda, H.D. Gan, K. Mizunuma, R. Koizumi, F. Matsukura, H. Ohno, Junction size effect on switching current and thermal stability in CoFeB/MgO perpendicular magnetic tunnel junctions, Appl. Phys. Lett. 99 (2011), 042501.

[35] I.M. Miron, G. Gaudin, S. Auffret, B. Rodmacq, A. Schuhl, S. Pizzini, J. Vogel, P. Gambardella, Current-driven spin torque induced by the Rashba effect in a ferromagnetic metal layer, Nat. Mater. 9 (2010) 230–234.

[36] I.M. Miron, K. Garello, G. Gaudin, P.-J. Zermatten, M.V. Costache, S. Auffret, S. Bandiera, B. Rodmacq, A. Schuhl, P. Gambardella, Perpendicular switching of a single ferromagnetic layer induced by in-plane current injection, Nature 476 (2011) 189–194.

[37] M. Cubukcu, O. Boulle, M. Drouard, K. Garello, C.O. Avci, I.M. Miron, J. Langer, B. Ocker, P. Gambardella, G. Gaudin, Spin-orbit torque magnetization switching of a three-terminal perpendicular magnetic tunnel junction, Appl. Phys. Lett. 104 (2014), 042406.

[38] C.-F. Pai, L. Liu, Y. Li, H.W. Tseng, D.C. Ralph, R.A. Buhrman, Spin transfer torque devices utilizing the giant spin hall effect of tungsten, Appl. Phys. Lett. 122404 (2012) 101.

[39] A. Kaidatzis, C. Bran, V. Psycharis, M. Vazquez, J.M. Garcia-Martın, D. Niarchos, Tailoring the magnetic anisotropy of CoFeB/MgO stacks onto W with a ta buffer layer, Appl. Phys. Lett. 106 (2015) 262401.

[40] L.W. Martin, Y.-H. Chu, R. Ramesh, Emerging multiferroic memories, [book auth.] Auciello O., Wouters D. in: S. Hong (Ed.), Emerging Non-Volatile Memories, Springer, Boston, 2014.

[41] C. Binek, B. Doudin, Magnetoelectronics with magnetoelectrics, J. Phys. Condens. Matter 17 (2005) 39–44.

[42] M. Hambe, A. Petraru, N.A. Pertsev, P. Munroe, V. Nagarajan, H. Kohlstedt, Crossing an interface: ferroelectric control of tunnel currents in magnetic complex oxide heterostructures, Adv. Funct. Mater. 20 (2010) 2436–2441.

[43] D. Wang, C. Nordman, J.M. Daughton, Z. Qian, J. Fink, 70% TMR at room temperature for SDT sandwich junctions with CoFeB as free and reference layers, IEEE Trans. Magn. 40 (2004) 2269–2271.

[44] M. Tsunoda, K. Nishikawa, S. Ogata, M. Takahashi, 60% magnetoresistance at room temperature in co–Fe/Al–O/co–Fe tunnel junctions oxidized with Kr–O2 plasma, Appl. Phys. Lett. 80 (2002) 3135–3137.

[45] H.X. Wei, Q.H. Qin, M. Ma, R. Sharif, X.F. Hana, 80% tunneling magnetoresistance at room temperature for thin Al–O barrier magnetic tunnel junction with CoFeB as free and reference layers, J. Appl. Phys. 101 (2007) 09B501.

[46] S.S.P. Parkin, C. Kaiser, A. Panchula, P.M. Rice, B. Hughes, M. Samant, S.-H. Yang, Giant tunnelling magnetoresistance at room temperature with MgO (100) tunnel barriers, Nat. Mater. 3 (2004) 862–867.

[47] W.H. Butler, X.-G. Zhang, T.C. Schulthess, J.M. MacLaren, Spin-dependent tunneling conductance of FezMgOzFe sandwiches, Phys. Rev. B 63 (2001), 054416.

[48] J. Mathon, A. Umerski, Theory of tunneling magnetoresistance of an epitaxial FeÕMgOÕFe001… junction, Phys. Rev. B 63 (2001). 220403(R).

[49] D.D. Djayaprawira, K. Tsunekawa, M. Nagai, H. Maehara, S. Yuasa, Y. Suzuki, K. Ando, 230% room-temperature magnetoresistance in CoFeB/MgO/CoFeB magnetic tunnel junctions, Appl. Phys. Lett. 86 (2005), 092502.

[50] S. Yuasa, D.D. Djayaprawira, Giant tunnel magnetoresistance in magnetic tunnel junctions with a crystalline MgO(001) barrier, J. Phys. D. Appl. Phys. 40 (2007) R337–R354.

[51] S. Ikeda, J. Hayakawa, Y. Ashizawa, Y.M. Lee, K. Miura, H. Hasegawa, M. Tsunoda, F. Matsukura, H. Ohno, Tunnel magnetoresistance of 604% at 300 K by suppression of ta diffusion in CoFeB/MgO/CoFeB pseudo-spin-valves annealed at high temperature, Appl. Phys. Lett. 93 (2008), 082508.

[52] R. Shan, H. Sukegawa, W.H. Wang, M. Kodzuka, T. Furubayashi, T. Ohkubo, S. Mitani, K. Inomata, K. Hono, Demonstration of half-metallicity in fermi-level-tuned heusler alloy Co2FeAl0.5Si0.5 at room temperature, Phys. Rev. Lett. 102 (2009) 246601.
[53] H. Sukegawa, H. Xiu, T. Ohkubo, T. Furubayashi, T. Niizeki, W. Wang, S. Kasai, S. Mitani, K. Inomata, K. Hono, Tunnel magnetoresistance with improved bias voltage dependence in lattice-matched Fe/spinel MgAl2O4/Fe(001), Appl. Phys. Lett. 212505 (2010) 96.
[54] T. Scheike, H. Sukegawa, K. Inomata, T. Ohkubo, K. Hono, S. Mitani, Chemical ordering and large tunnel magnetoresistance in Co2FeAl/MgAl2O4/Co2FeAl(001) junctions, Appl. Phys. Express 9 (2016).
[55] M. Belmoubarik, H. Sukegawa, T. Ohkubo, S. Mitani, K. Hono, MgAl2O4(001) based magnetic tunnel junctions made by direct sputtering of a sintered spinel target, Appl. Phys. Lett. 108 (2016) 132404.
[56] H.I. Sukegawa, Y. Kato, M. Belmoubarik, P.-H. Cheng, T. Daibou, N. Shimomura, Y. Kamiguchi, J. Ito, H. Yoda, T. Ohkubo, S. Mitani, K. Hono, MgGa2O4 spinel barrier for magnetic tunnel junctions: coherent tunneling and low barrier height, Appl. Phys. Lett. 110 (2017) 122404.
[57] H. Sukegawa, J.P. Hadorn, Z. Wen, T. Ohkubo, S. Mitani, K. Hono, Perpendicular magnetic anisotropy at lattice-matched Co2FeAl/MgAl2O4(001) epitaxial interfaces, Appl. Phys. Lett. 110 (2017) 112403.
[58] S. Yuasa, Y. Suzuki, T. Katayama, K. Ando, Characterization of growth and crystallization processes in CoFeB / MgO / CoFeB magnetic tunnel junction structure by reflective high-energy electron diffraction, Appl. Phys. Lett. 87 (2005) 242503.
[59] A. Kaidatzis, C. Serletis, D. Niarchos, Dependence of magnetic anisotropy on MgO sputtering pressure in Co20Fe60B20/MgO stacks, J. Phys. Conf. Ser. 903 (2017), 012019.
[60] Y.M. Lee, J. Hayakawa, S. Ikeda, F. Matsukura, H. Ohno, Effect of electrode composition on the tunnel magnetoresistance of pseudo-spin-valve magnetic tunnel junction with a MgO tunnel barrier, Appl. Phys. Lett. 90 (2007) 212507.
[61] N. Perrissin, S. Lequeux, N. Strelkov, A. Chavent, L. Vila, L.D. Buda-Prejbeanu, S. Auffret, R.C. Sousa, I.L. Prejbeanua, B. Dieny, A highly thermally stable sub-20 nm magnetic random-access memory based on perpendicular shape anisotropy, Nanoscale 10 (2018) 12187–12195.
[62] B. Oliver, Q. He, X. Tang, J. Nowak, Dielectric breakdown in magnetic tunnel junctions having an ultrathin barrier, J. Appl. Phys. 91 (2002) 4348.
[63] B. Oliver, G. Tuttle, Q. He, X.I. Tang, J. Nowak, Two breakdown mechanisms in ultrathin alumina barrier magnetic tunnel junctions, J. Appl. Phys. 95 (2004) 1315.
[64] A.A. Khan, J. Schmalhorst, A. Thomas, O. Schebaum, G. Reiss, Dielectric breakdown in Co–Fe–B/MgO/Co–Fe–B magnetic tunnel junction, J. Appl. Phys. 103 (2008) 123705.
[65] A. Thomas, V. Drewello, M. Schäfers, A. Weddemann, G. Reiss, G. Eilers, M. Münzenberg, K. Thiel, M. Seibt, Direct imaging of the structural change generated by dielectric breakdown in MgO based magnetic tunnel junctions, Appl. Phys. Lett. 93 (2008) 152508.
[66] M. Schäfers, V. Drewello, G. Reiss, A. Thomas, K. Thiel, G. Eilers, M. Münzenberg, H. Schuhmann, M. Seibt, Electric breakdown in ultrathin MgO tunnel barrier junctions for spin-transfer torque switching, Appl. Phys. Lett. 95 (2009) 232119.
[67] S. Amara-Dababi, R.C. Sousa, M. Chshiev, H. Bea, J.A.-H. Rault, L. Lombard, I.L. Prejbeanu, K. Mackay, B. Dieny, Charge trapping-detrapping mechanism of barrier breakdown in MgO, Appl. Phys. Lett. 99 (2011), 083501.
[68] S. Amara-Dababi, H. Béa, R.C. Sousa, C. Baraduc, B. Dieny, Correlation between write endurance and electrical low frequency noise in MgO based magnetic tunnel junctions, Appl. Phys. Lett. 052404 (2013) 102.

[69] K.Z. Suzuki, R. Ranjbar, J. Okabayashi, Y. Miura, A. Sugihara, H. Tsuchiura, S. Mizukami, Perpendicular magnetic tunnel junction with a strained Mn-based nanolayer, Sci. Rep. 6 (2016) 30249.
[70] R. Ranjbar, K.Z. Suzuki, Y. Sasaki, L. Bainsla, S. Mizukami, Current-induced spin–orbit torque magnetization switching in a MnGa/Pt film with a perpendicular magnetic anisotropy, Jpn. J. Appl. Phys. 55 (2016) 120302.
[71] W.-G. Wang, M. Li, S. Hageman, C.L. Chien, Electric-field-assisted switching in magnetic tunnel junctions, Nat. Mater. 11 (2012) 64–68.

CHAPTER

10

Correlated transition metal oxides and chalcogenides for Mott memories and neuromorphic applications

Laurent Cario, Julien Tranchant, Benoit Corraze, and Etienne Janod

Institut des Matériaux Jean Rouxel (IMN), Nantes, France

10.1 Introduction

Mott insulators are a large class of materials that can undergo various kinds of insulator to metal transitions (IMT) also called Mott transitions in response to different external perturbations like pressure, heating, and doping [1]. These IMT were extensively studied in transition metal oxides during the last decades and led to the discovery of astonishing properties like high Tc superconductivity and colossal magnetoresistance. But despite these efforts Mott insulators hardly find applications in the field of microelectronics. This is mainly due to the difficulty to use perturbations like pressure, heating and doping in real devices. But the recent discovery of insulator to metal transitions driven by voltage pulses in Mott insulators, i.e., Electric Mott Transitions (EMT), will enable the creation of simple devices easily controllable [2–5]. This review aims to describe the current understanding of Electric Mott Transition and its use to develop a new electronic based on Mott insulators: the Mottronics [6].

The paper is organized as follows. Section 10.2 describes briefly the theoretical background of Mott insulators, the different ways to break the Mott insulating state under equilibrium conditions and the most famous Mott insulator compounds $(V_{1-x}Cr_x)_2O_3$, $NiS_{2-x}Se_x$ and $GaTa_4Se_8$. Section 10.3 gives an overview of Electric Mott transitions. Three different

Metal Oxides for Non-volatile Memory
https://doi.org/10.1016/B978-0-12-814629-3.00010-6

mechanisms of EMT are described that are closely related to the filling, bandwidth and temperature controlled IMT observed at equilibrium. This review focuses particularly in Section 10.4 on the electric Mott transition mechanism based on bandwidth controlled IMT which is triggered by an avalanche breakdown. Finally, Section 10.5 shows how Electric Mott transitions may be used to build up new microelectronic devices based on Mott insulators such as Mott memories or neuromorphic components including artificial synapses and neurons.

10.2 Mott insulators and Mott transitions

10.2.1 Mott insulators: Definition

According to conventional band theories, compounds with an integer number of unpaired electrons per site should have their Fermi level crossing a band and therefore should behave as normal metals (see Fig. 10.1A). However, early on it was realized that many of such materials, like NiO for example, behave as insulators [7]. Sir Nevil Mott was the first to realize that this discrepancy was related to the on-site electron–electron repulsion which is not properly taken into account in conventional band theories [8,9]. For this reason, this type of materials are nowadays called Mott insulators. In fact, when the Coulomb repulsion (Hubbard) energy U becomes higher than the bandwidth W, it splits the band into two sub-bands, the Lower (LHB) and Upper (UHB) Hubbard Bands. This situation is shown in Fig. 10.1A in the case of a single band system with one unpaired electron per site (half-filled system). When U is larger than the bandwidth W, a Mott-Hubbard gap E_G opens up between the LHB and the UHB, roughly equal to $E_G \approx U$ - W. Fig. 10.1B shows also the case of a valence band mainly built up from anionic state that lies higher that the LHB. This type of correlated materials called charge transfer insulators have a reduced gap compared to the Mott Hubbard gap.

Understanding the Mott insulating state has been a long-standing problem [10,11], but recently the dynamical mean field theory (DMFT) has succeeded to give a proper theoretical description of Mott insulators and predict its phase diagram. Numerous review papers [12–15] describe how changing correlation strength (i.e., the U/W ratio) destroys the Mott insulating state and stabilizes a correlated metal that displays as main feature a quasi-particle peak (see Fig. 10.2A). The typical phase diagram of Mott insulators as predicted from DMFT is sketched in Fig. 10.2B. A first order (Mott) transition line around $U/W \approx 1.2$ separates the paramagnetic metallic (PM) domain at low U/W from the Mott Paramagnetic Insulator (PI) domain at high U/W. This transition is driven by

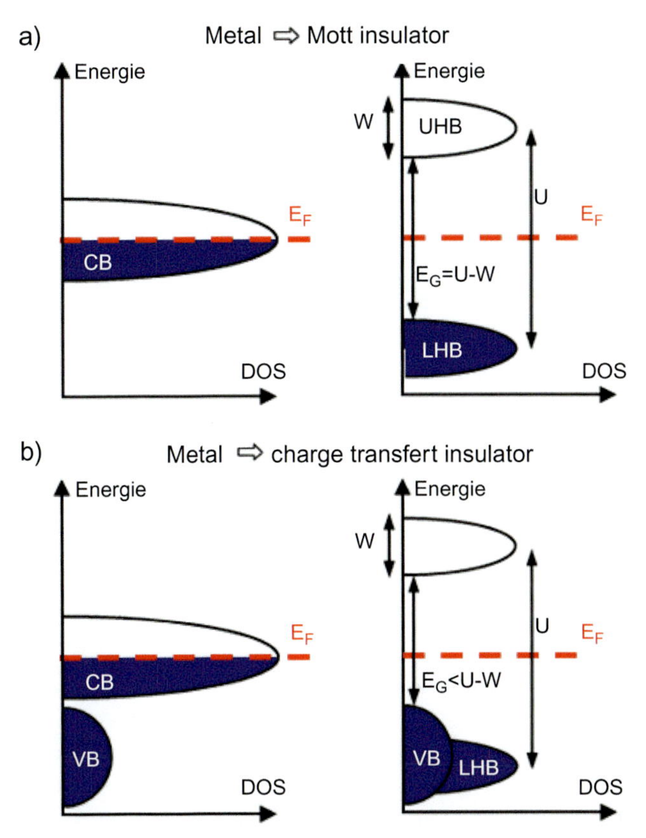

FIG. 10.1 Density of states represented for a half-filled system in the case of a normal metal ($U = 0$) and a correlated insulators ($U > W$). Panel (A) illustrate the case of Mott insulator for which the gap is open between the Upper (UHB) and Lower Hubbard Band (LHB) and panel (B) illustrate the case of charge transfer insulator for which the gap opens between the anionic Valence Band (VB) and the UHB. W, U, CB, E_G stand respectively for bandwidth, Hubbard energy, Conduction Band and gap.

the electronic (charge) degree of freedom and both insulating and metallic phases keep the same crystallographic symmetry. Only a volume contraction occurs on crossing the Mott transition line toward the metallic phase. Whereas the Mott line and the high temperature part of the phase diagram are universal, the low temperature part is non-universal and material-dependent and can present various kinds of long-range (for example magnetic or orbital) orders. As sketched in Fig. 10.2C this theoretical phase diagram is related to the experimental temperature—pressure (T–P) phase diagram as the physical pressure directly affects the bandwidth and therefore the U/W term.

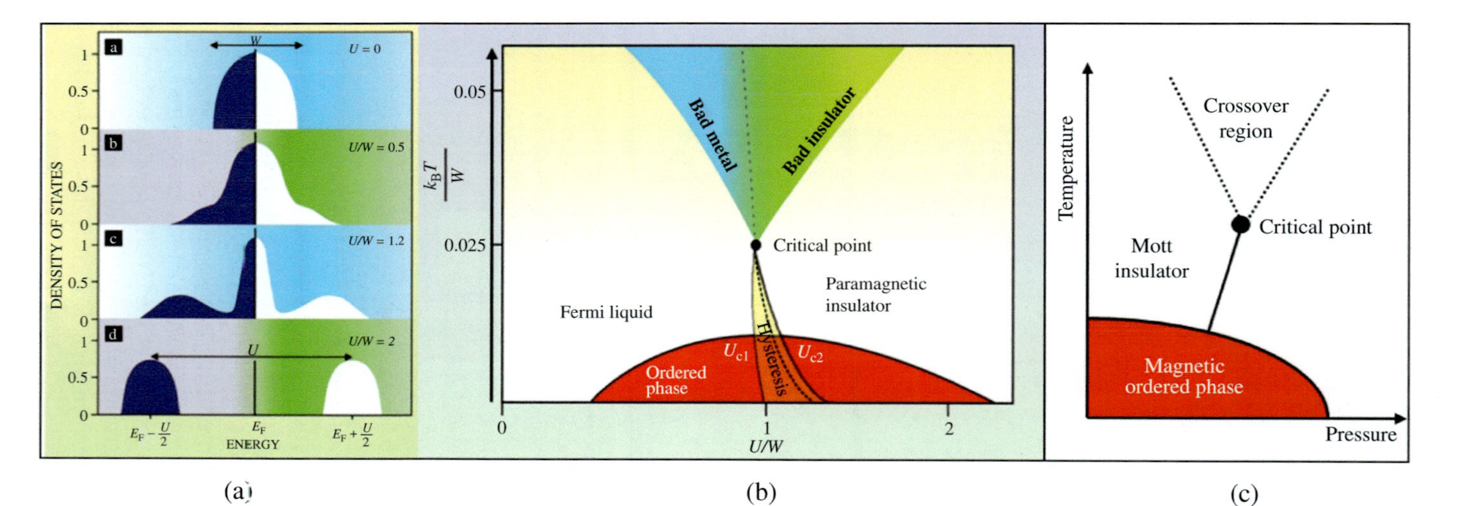

FIG. 10.2 (A) Evolution of the electronic Density of States (DOS) of a half-filled system as predicted from DMFT when the correlation strength U/W is decreased from 2 to 0. A Mott insulator to metal transition occurs slightly above $(U/W)_c \approx 1.2$. (B) Schematic phase diagram displayed as k_BT/W *vs.* U/W predicted from DMFT for half-filled compounds. T, U and W are the temperature, the Hubbard electron–electron repulsion term and the bandwidth, respectively. (C) Schematic experimental temperature—pressure phase diagram often observed in Mott insulator systems. *Panel A and B Reproduced with permission from Ref. G. Kotliar, D. Vollhardt, Strongly correlated materials: insights from dynamical mean-field theory, Phys. Today 57 (2004) 53. Copyright 2004, American Institute of Physics.*

10.2.2 Insulator to metal transition in Mott insulators

According to numerous literature there exists three main ways to break the Mott insulating state and induce an Insulator to Metal Transition (IMT) [1].

(1) Bandwidth controlled IMT. This IMT can be achieved by applying an external pressure. Under pressure the orbitals overlaps are enhanced which increases the bandwidth *W* and therefore reduces the correlation strength *U/W* [1,16]. This corresponds to crossing the Mott transition line along the horizontal axis of the *T–P* phase diagram (see Fig. 10.2C).

(2) Temperature controlled IMT. This IMT, is driven by the temperature and corresponds to crossing the Mott transition line along the vertical axis in the *T–P* phase diagram (see Fig. 10.2C). This IMT occurs therefore between the low temperature paramagnetic metal and the high temperature Mott paramagnetic insulator [1,17]. This transition strongly contrasts with the more usual transitions from a low-*T* insulator to a high-*T* metal.

(3) Filling controlled IMT. This IMT, occurs when the band filling deviates from an exactly integer number of unpaired electron per site (Fig. 10.3A) [1]. This may be achieved by tuning the electronic filling thanks to e.g. chemical doping. Fig. 10.3B shows the filling-controlled Mott transition observed in titanates $Sm_{1-x}Ca_xTiO_3$ [18]. It shows the existence of a critical hole doping $p_c \neq 0$ necessary to induce the

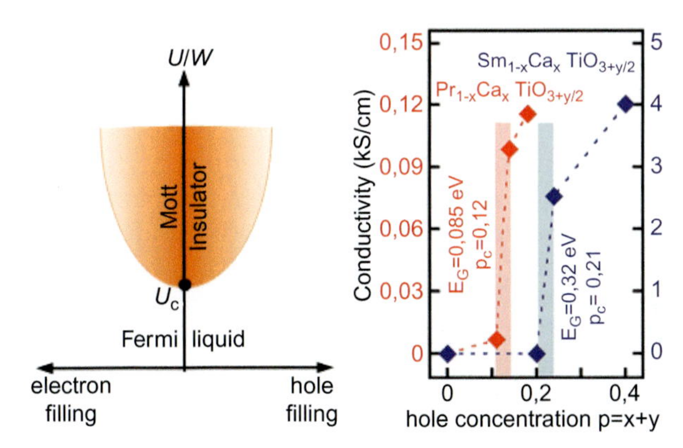

FIG. 10.3 (A) Diagram *U/W* versus filling for the doped Mott insulators, (B) a critical hole doping $p_c = 0.12$ and $p_c = 0.21$ is necessary to induce the transition from the Mott insulator to the correlated metal in the $Pr_{1-x}Ca_xTiO_3$ and $Sm_{1-x}Ca_xTiO_3$ compounds, respectively. *Data extracted from Ref. T. Katsufuji, Y. Taguchi, Y. Tokura, Transport and magnetic properties of a Mott-Hubbard system whose bandwidth and band filling are both controllable: R1-XCaxTiO3 +y/2, Phys. Rev. B 56 (1997) 10145.*

transition from the Mott insulator to the correlated metal (i.e., Fermi Liquid). This critical doping increases with the correlation strength U/W and thus with the Mott-Hubbard gap as shown on the phase diagram U/W versus doping in Fig. 10.3A. Experimentally, this trend is well illustrated by the $RTiO_3$ (R=La, Pr, Nd, Sm, Y) Mott insulators (see Fig. 10.3B), with a critical hole doping p_c increasing from $p_c \approx 0.12$ for $PrTiO_3$ (Mott-Hubbard gap $E_G \approx 0.085$ eV) to $p_c \approx 0.21$ for $SmTiO_3$ ($E_G \approx 0.21$ eV) and even to $p_c \approx 0.35$ for $YTiO_3$ ($E_G \approx 0.45$ eV) [18,19]. Theoretically, this issue of a finite doping necessary to achieve an insulator to metal transition both in band and Mott insulators has been first discussed on the basis of the "Mott criterion." [20] A more recent theoretical development specific to Mott insulators proposes the existence of a finite value of doping which scales with $\sqrt{U/W-(U/W)_C}$, in good agreement with the behavior observed in titanates [21].

10.2.3 Renowned examples of Mott and charge transfer insulators

Numerous Mott insulators have been identified, among inorganic or organic compounds. This section will describe the main characteristic of inorganic oxide $(V_{1-x}Cr_x)_2O_3$, or chalcogenide $NiS_{2-x}Se_x$, and AM_4Q_8 (A=Ga, Ge; M=V, Nb, Ta, Mo; Q=S, Se) Mott insulators.

10.2.3.1 *The vanadium oxide $(V_{1-x}Cr_x)_2O_3$*

The chromium doped Vanadium sesquioxide $(V_{1-x}Cr_x)_2O_3$ compounds represents the archetypical Mott insulator system [1,22–24]. At room temperature V_2O_3 adopts the corundum structure with a space group $R\overline{3}c$ (see Fig. 10.4A) and is a correlated metal [25]. A proper isovalent substitution of vanadium per chromium turns this system into a Mott insulating phase. According to the early studies of this system, this substitution does not affect the band filling nor the structure type but induces a tensile strain (i.e., a negative chemical pressure effect). Fig. 10.4B present the whole phase diagram of the Mott insulator system $(V_{1-x}Cr_x)_2O_3$ that was unveiled in the 70s by McWhan et al. [22,23,26,27]. In this work they established an equivalence between the application of an external physical pressure and the internal chemical pressure related to the V /Cr and V/Ti substitutions (see Fig. 10.4B). The empirical rule proposed then was a 1% change of composition x in $(V_{1-x}M_x)_2O_3$ is equivalent to applying a positive pressure (M=Ti) or negative pressure (M=Cr) of 4 kbar. In broad terms, this equivalence remains relevant, but recent work has come to refine this vision [28]. It shows that the equivalence pressure - substitution is not complete if we look in detail at the electronic structure as will be discussed below. The

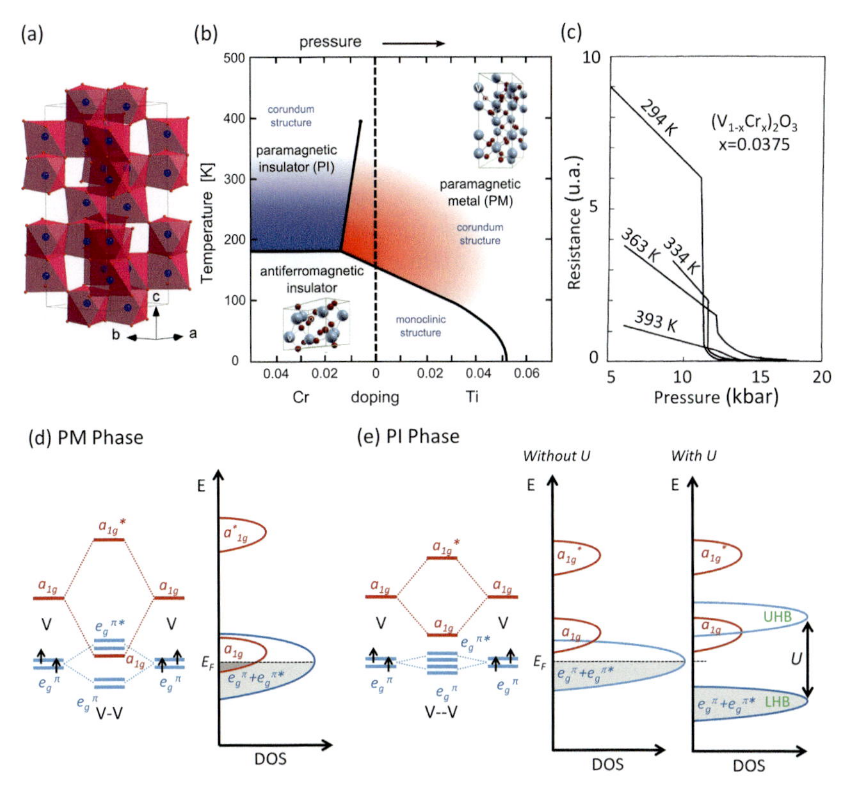

FIG. 10.4 (A) Crystallographic structure of V_2O_3 highlighting the presence of V—V dimers along the c direction (B) Phase diagram of $(V_{1-x}M_x)_2O_3$, with M = Cr and Ti. In this system, changing the V/M ratio by 1% is equivalent to applying an external pressure of ≈4 kbar [23]. (C) Resistance versus pressure across the Mott insulator to metal transition in $(V_{0.9625}Cr_{0.0375})_2O_3$. (D) Schematic representation of the electronic structure of the paramagnetic metal (PM) phase of V_2O_3. (E) Schematic representation of the electronic structure of the paramagnetic insulating (PI) phase of $(V_{1-x}Cr_x)_2O_3$ with or without taking into account the Hubbard energy *U*. *Panel B Reproduced with permission from Ref. P. Hansmann, A. Toschi, G. Sangiovanni, T. Saha-Dasgupta, S. Lupi, M. Marsi, K. Held, Mott–Hubbard transition in V2O3 revisited, Phys. Status Solidi B 250 (2013) 1251. Copyright © 2013, Wiley-VCH. Panel C Adapted with permission from Ref. A. Jayaraman, D.B. McWhan, J.P. Remeika, P.D. Dernier, Critical behavior of the Mott transition in Cr-doped V2O3, Phys. Rev. B 2 (1970) 3751. Copyright 1970, American Physical Society.*

phase diagram of the $(V_{1-x}Cr_x)_2O_3$ system is presented in Fig. 10.4B. It exhibits a Mott IMT line ending around 450 K separating a Mott insulating phase from a correlated metallic phase. Applying a pressure of about 12 kbar at room temperature on $(V_{0.9625}Cr_{0.0375})_2O_3$ induces a bandwidth-controlled IMT as shown in Fig. 10.4C [26]. As observed experimentally, the IMT occurs without any crystallographic symmetry breaking and the space group remains $R\overline{3}c$ for the metallic and insulating phases. A strong decrease of the unit cell volume (~1%) occurs at this IMT indicating the first order nature of the transition [23]. For a narrow range of composition $(V_{1-x}Cr_x)_2O_3$ with $0.7\% < x < 1.77\%$ this Mott first order transition line can be also crossed by tuning the temperature. This thermally controlled IMT occurs at a temperatures between 190 and 400 K depending on the composition as shown in Fig. 10.4B. At low temperature, pure and Cr-substituted V_2O_3 display a phase transition toward a monoclinic (space group *I2/a*) antiferromagnetic insulating (AFI) phase. In $(V_{1-x}Cr_x)_2O_3$ with $x > 1.77\%$, the transition occurs between two insulating phases while in pure V_2O_3 the transition occurs at ≈165 K between the insulating AFI phase and the correlated metallic phase [27]. This insulator to metal transition does not correspond to a Mott transition as it involves a crystallographic symmetry breaking and an additional magnetic ordering.

The electronic structure of vanadium sesquioxide has long been controversial [1]. As can be seen in Fig. 10.4D, the energy levels responsible for the electronic properties are derived from the vanadium 3d orbitals. In the octahedral environment, the five degenerated 3d levels of the vanadium atoms are split by the crystal field into the triple degenerated t_{2g} and doubly degenerated e_g levels. However, the deformation of the octahedron in V_2O_3, related to rhombohedral symmetry, split again these t_{2g} levels into a_{1g} and double degenerate e_g^π levels. In addition, the existence of V—V vanadium pairs, oriented along the *c*-axis and located in two face-sharing octahedra, further separates these levels, essentially to form a bonding level a_{1g} and an anti-bonding level a_{1g}*. The main debate regarding the electronic structure of V_2O_3 is whether the bonding level a_{1g} is above or below the e_g^π levels. In 1978, Castellani et al. have proposed that the a_{1g} level is the lowest in energy, and that this level is populated by two paired electrons from the two vanadium of the dimer [29,30]. In this model, each vanadium has only one unpaired electron and therefore has a spin $S=1/2$. However, this vision was questioned in 2000 by polarized X-ray absorption measurements showing that vanadium actually has a spin $S=1$, with an a_{1g} bonding level very close to the e_g^π (Fig. 10.4D) [31]. Although the debate is not closed, LDA (local density approximation) + DMFT calculations have reached the same conclusions [32,33].

Recent works have shown that the substitution of vanadium per chromium modifies slightly the electronic structure (Fig. 10.4E) [28]. Indeed, the bonding orbitals a_{1g}, partially populated in V_2O_3 (PM phase) are rising in energy, and become empty, going toward the insulating PI phase of Mott $(V_{1-x}Cr_x)_2O_3$ system. This effect is related to the increase of the vanadium distance in the V—V pair. The mechanism of the insulator–metal transition would then be related to a decrease of the separation between bonding a_{1g} and anti-bonding a_{1g}* levels concomitant with the increase of the distance between vanadium in the V—V pair. The metal–insulator transition induced by the chromium substitution thus differs somewhat from the simple view of the bandwidth-controlled Mott transition. In contrast, the pressure-induced Mott insulator to metal transition in Cr-substituted V_2O_3 is much closer to the theoretical situation considered by the DMFT for a canonical Mott insulator, since pressure application modifies only the bandwidth without noticeable change of the a_{1g} band filling [34].

10.2.3.2 $NiS_{2-x}Se_x$

If $(V_{1-x}Cr_x)_2O_3$ is considered to be the canonical Mott-Hubbard insulator, NiS_2 is generally recognized as an archetypical type of charge transfer insulator [1,35,36[. It belongs to the isostructural group of pyrites of formula MS_2 with M = Fe, Co, Ni, Cu and Zn. This structure can be described as a NaCl type arrangement of M atoms and strongly covalent bonded S_2^{2-} pairs (see Fig. 10.5A) [37,38[. In this structure, the transition metal at oxidation state M^{2+} occupies an octahedral environment which is highlighted in Fig. 10.5A. The resulting crystal field leads to the splitting of its five 3d orbitals into three degenerate orbitals of symmetry t_{2g} and two degenerate orbitals of symmetry e_g electronically active. The electronic count leads to the prediction of an insulating band character for FeS_2 (filled t_{2g} and empty e_g levels) and ZnS_2 (filled t_{2g} and e_g levels), which they prove to be experimentally. Likewise, the incomplete electronic filling of CoS_2 and CuS_2 leads to a metallic behavior of the materials. NiS_2 has its t_{2g} levels completely filled and two electrons on the two e_g levels. It represents a half-filled two degenerated bands system. In first approximation this

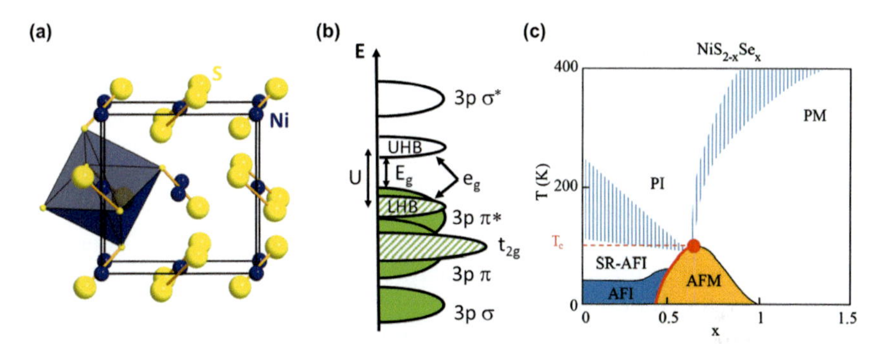

FIG. 10.5 (A) Crystallographic structure of the NiS_2 compound (pyrite structure). The octahedral environment of Ni is highlighted. (B) Schematic density of state of NiS_2 considered to be the best to describe the electronic structure of the compound. This electronic structure corresponds to a charge transfer Mott insulator. (C) Temperature phase diagram T versus x the amount of Selenium in the $NiS_{2-x}Se_x$ solid solution. "P" means paramagnetic, "AF" antiferromagnetic, "SR" short range. "I" indicates an insulating state and "M" a metallic state. The *red line* indicates the first-order Mott insulator–metal transition line. *Adapted with permission from Ref. Y. Sekine, H. Takahashi, N. Môri, T. Matsumoto, T. Kosaka, Effect of pressure on transport properties of Ni(S1 − xSex)2, Phys. B Condens. Matter 237–238 (1997) 148. Copyright 2001, Elesevier.*

material should be expected as a metal, but it exhibits a Mott insulating behavior due to strong electronic correlations. Sulfur orbitals, however, are also involved at the Fermi level. The $(S_2)^{2-}$ molecular dimers present bonding and anti-bonding levels of the molecular orbitals σ and π resulting from the covalent S—S interaction. These sulfur orbitals play an important role in the electronic properties of the material. It is thus generally accepted that the gap is created between the sulfur p-bands, strongly hybridized with LHB, and the UHB, as sketched in Fig. 10.5B. It would therefore be a charge-transfer gap as supported by photoemission measurements [36]. From these measurements, and cluster configuration interaction calculations, the values $E_g = 1.8\,\text{eV}$ and $U = 3.3\,\text{eV}$ were extracted. The exact nature of Mott's gap in NiS_2 is however regularly questioned and is still an open question for some authors [36,39].

As expected for a Mott insulator, NiS_2 undergoes a Mott insulator to metal transition under pressure (see Fig. 10.6A) [40]. In the $NiS_{2-x}Se_x$ system, the substitution of sulfur by selenium induces also a Mott insulator to metal transition (first-order character, no structural nor electronic filling modifications) (see Fig. 10.6B) [41,42]. The phase diagram of this system is represented in Fig. 10.5C and was established mainly from transport and magnetic measurements [36,41]. In addition X-ray structural studies evidenced a clear jump of cell parameter at the Mott transition and the two phases coexistence associated with the first order transition (see Fig. 10.6C and D) [43].

The phase diagram for the $NiS_{2-x}Se_x$ solid solution is reminiscent of the diagram predicted by DMFT (compare Fig. 10.2 and Fig. 10.5). The substitution of S for Se, is often discussed as a chemical pressure effect equivalent to the application of an external physical pressure, which also induces the Mott transition in NiS_2 as seen in transport measurements (see Fig. 10.6A) [42]. However, this equivalence is questioned in the literature. Whereas the physical pressure tends to widen all the bands, the substitution of S by Se would increase only the width of the p-bands of the chalcogens. Moreover, it would decrease essentially the anti-bonding bonding gap of the p-levels, thus closing the gap in a completely different manner. Many modeling of the electronic structure and experimental studies were reported in the literature and a consensus is obtained as to the non-equivalence between chemical substitution and external pressure [39].

10.2.3.3 AM_4Q_8 compounds

The AM_4Q_8 compounds (A = Ga, Ge; M = V, Nb, Ta, Mo; Q = S, Se, Te) have emerged recently as a new example of canonical Mott insulator. These compounds exhibit a lacunar spinel structure, in which the electronic sites correspond to the tetrahedral transition metal clusters M_4 shown in the inset of Fig. 10.7A [44]. Fig. 10.7B represents the molecular orbital scheme diagram of the clusters. The t_{2g} levels of the transition

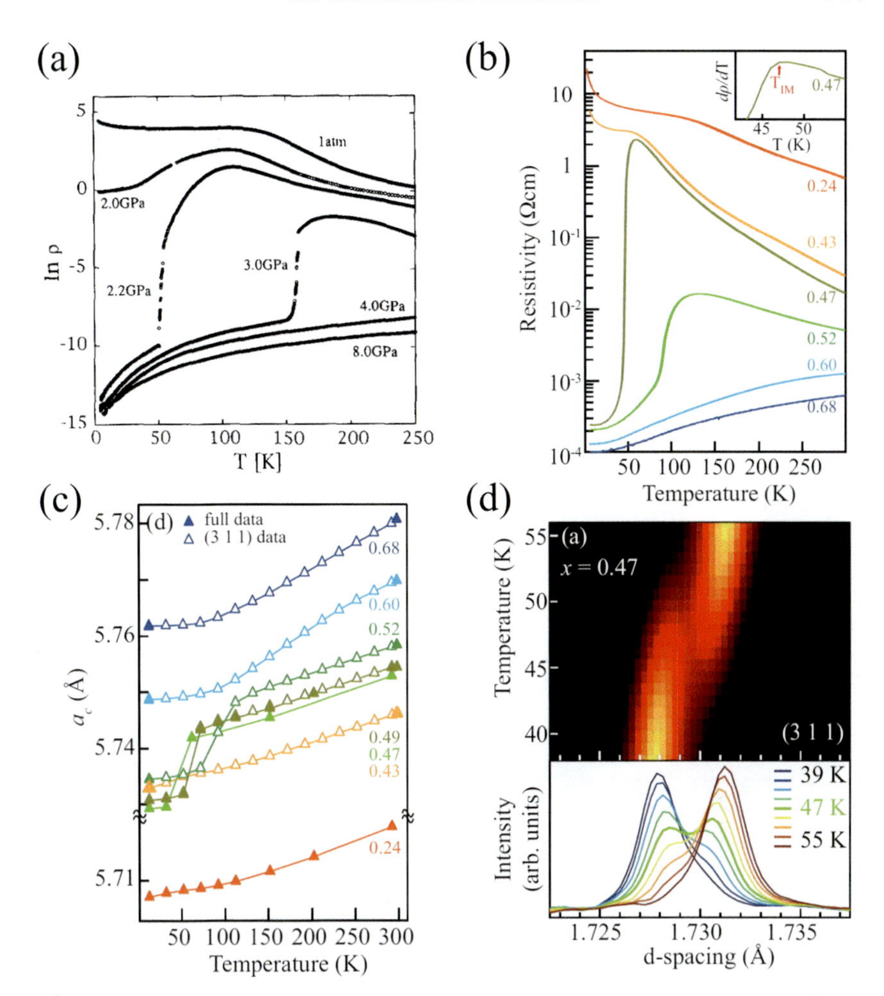

FIG. 10.6 (A) Evolution of the resistivity as a function of temperature under different pressures for the NiS_2 compound. (B) Evolution of the resistivity as a function of temperature for different values of selenium content x in the compounds $NiS_{2-x}Se_x$. (C) Variation of the unit cell parameter a_c with temperature for different values of selenium content x values in $NiS_{2-x}Se_x$. A jump of unit cell volume occurs for composition allowing the crossing of the first-order Mott transition line. (D) In the vicinity of the Mott line, a splitting of the Bragg peaks is observed, demonstrating the coexistence of the two phases, metallic and insulating, expected for a first-order transition. *Panel A Reproduced with permission from Ref. Y. Sekine, H. Takahashi, N. Môri, T. Matsumoto, T. Kosaka, Effect of pressure on transport properties of Ni(S1 − xSex)2, Phys. B Condens. Matter 237–238 (1997) 148. Copyright 1997, Elsevier; Panels B, C and D Reproduced with permission from Ref. G. Han, S. Choi, H. Cho, B. Sohn, J.-G. Park, C. Kim, Structural investigation of the insulator-metal transition in NiS2-xSex compounds, Phys. Rev. B 98 (2018) 125114. Copyright 2018, American Physical Society.*

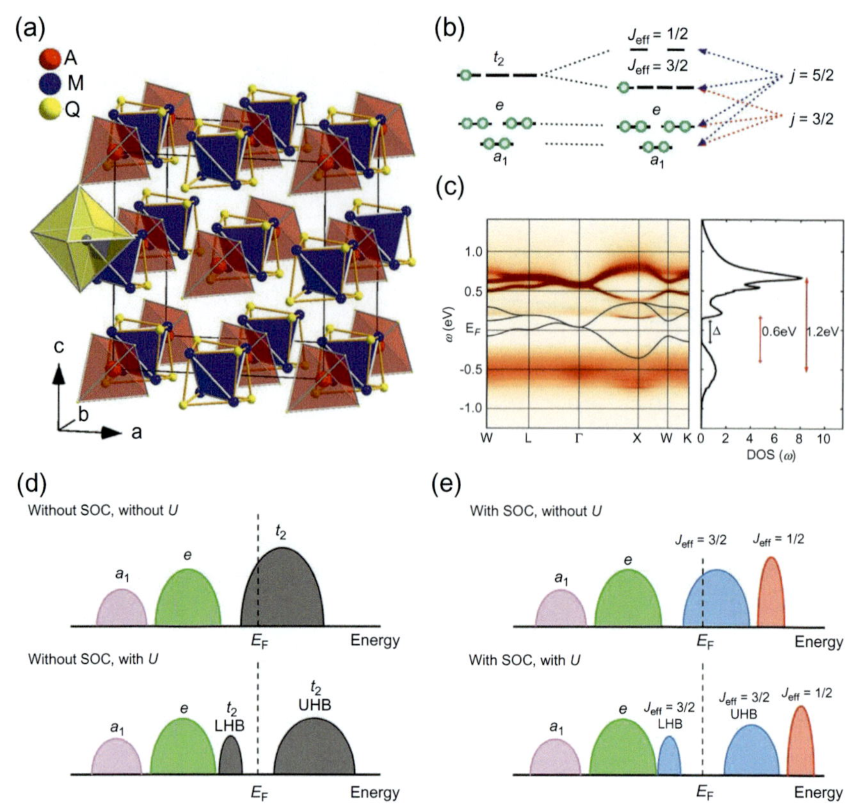

FIG. 10.7 (A) Crystallographic structure of AM_4Q_8 (A = Ga, Ge; M = V, Nb, Ta, Mo; Q = S, Se) compounds, highlighting the M_4 tetrahedral clusters in blue. (B) Representation of the molecular orbital diagram of the M_4 clusters. (C) Result of the LDA + DMFT calculation showing the formation of the gap between the Lower (LHB) and Upper (UHB) Hubbard bands due to the coulomb energy U. The formation of the gap is explained in (D) that sketches the density of state with or without taking into account the effect of *U*. (E) sketch of the density of states explaining the formation of the gap in AM_4Q_8 by taking into account the effect of the Spin Orbit Coupling (SOC). *Panel B Reproduced from Ref. M.Y. Jeong, S.H. Chang, B.H. Kim, J.-H. Sim, A. Said, D. Casa, T. Gog, E. Janod, L. Cario, S. Yunoki, M. J. Han, J. Kim, Direct experimental observation of the molecular JEff = 3/2 ground state in the lacunar spinel GaTa4Se8, Nat. Commun. 8 (2017) 782; Panel C Reproduced with permission from Ref. A. Camjayi, C. Acha, R. Weht, M.G. Rodríguez, B. Corraze, E. Janod, L. Cario, M.J. Rozenberg, First-order insulator-to-metal Mott transition in the paramagnetic 3D system GaTa4Se8, Phys. Rev. Lett. 113 086404 (2014), Copyright 2018, American Physical Society; Panel E Reproduced from Ref. M.Y. Jeong, S.H. Chang, B.H. Kim, J.-H. Sim, A. Said, D. Casa, T. Gog, E. Janod, L. Cario, S. Yunoki, M.J. Han, J. Kim, Direct experimental observation of the molecular JEff = 3/2 ground state in the lacunar spinel GaTa4Se8, Nat. Commun. 8 (2017) 782.*

metals interact to form a_1 level, doubly degenerated e level and triply degenerated t_2 level [45]. As represented Fig. 10.7B each M_4 cluster contain one unpaired electron among seven when M = V, Nb, Ta. In the case M = Mo the clusters hold 11 d electrons. These compounds have a narrow gap of 0.1–0.3 eV, which can be tuned by chemical substitution on the chalcogenide site [46]. LDA + DMFT calculations on $GaTa_4Se_8$ confirms that the gap opens thanks to on-site correlation [47]. The band structure and density of states calculated by LDA + DMFT are represented around the Fermi level in Fig. 10.7C, while a scheme of the density of state is shown in Fig. 10.7D. Resonant inelastic X-ray scattering measurements and LDA +U calculations have moreover revealed that the Spin Orbit Coupling (SOC) is an important ingredient to take into account in this family of compounds [48,49]. $GaTa_4Se_8$ is indeed a rare example of SOC + U Mott insulator (see Fig. 10.7B and E) [48,49].

Beside their electronic structure, the AM_4Q_8 compounds display two important characteristics of canonical Mott insulators: they are paramagnetic insulators above 55 K [45,50,51] and do *not* exhibit any temperature-controlled IMT up to 800 K at ambient pressure [2]. Moreover, as represented in Fig. 10.8A, $GaTa_4Se_8$ and $GaNb_4Q_8$ (Q = S, Se) undergo a bandwidth-controlled insulator to metal transition under pressure, with superconductivity at $T_C \approx 2$–7 K in the pressurized metallic state above 11 GPa [50,52]. Transport property measurements under pressure in $GaTa_4Se_8$ have proven that this pressure-induced bandwidth-controlled IMT is of first order with an hysteresis (see Fig. 10.7B) [47]. Moreover, optical conductivity measurements displayed in Fig. 10.8B confirms the formation of a quasi-particle peak and therefore of a correlated metal under pressure [53]. Numerous studies have also demonstrated that the AM_4Q_8 undergo filling-controlled IMT when doped on the A site or on the M site [54]. The AM_4Q_8 compounds display therefore the filling and bandwidth control IMT expected in canonical Mott insulators. This family of compound exhibit also astonishing electronic properties such as negative colossal magnetoresistance [54], ferromagnetic half metallicity [54], multiferroïcity [55], and Néel type skyrmion lattice [56,57], that are encountered in other family of Mott insulators.

10.2.4 The famous example of VO_2 that exhibit an IMT but not a Mott transition

Beyond the examples of canonical Mott and charge transfer insulators many other oxide insulators display IMT potentially interesting for memory applications. A representative example is VO_2, whose temperature–pressure phase diagram is shown in Fig. 10.9A. While the VO_2 phase diagram shares some similarities with the low temperature

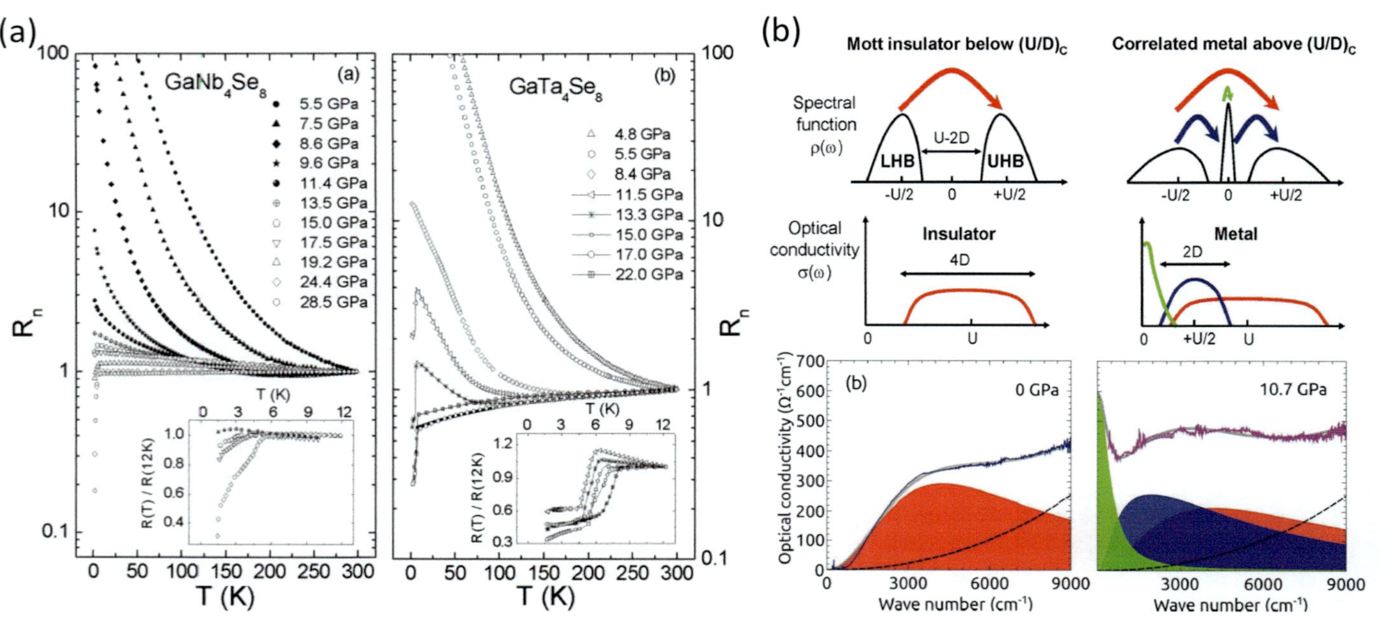

FIG. 10.8 (A) Resistivity *vs.* temperature at different pressures in $GaNb_4Se_8$ and $GaTa_4Se_8$ showing the bandwidth control insulator to metal transition. These compounds become also superconducting under pressure at low temperature. (B) Optical conductivity *vs.* wave number measured for $GaTa_4Se_8$ in the insulating and in the metallic state appearing beyond the Mott line under pressure (10.7 GPa). The sketches explain the low energy excitations (red, blue, and green arrows) expected in the Mott insulator and in the correlated metal. In the correlated metal the low energy contribution (in green) corresponds to the quasiparticle peak, a typical signature of electronic correlation. *Panel A Reproduced with permission from Ref. M.M. Abd-Elmeguid, B. Ni, D.I. Khomskii, R. Pocha, D. Johrendt, X. Wang, K. Syassen, Transition from Mott insulator to superconductor in GaNb4Se8 and GaTa4Se8 under high pressure, Phys. Rev. Lett. 93 (2004). Copyright 2004, American Physical Society, Panel B Reproduced with permission from Ref. (a) Dorolti, L. Cario, B. Corraze, E. Janod, C. Vaju, H.-J. Koo, E. Kan, M.-H. Whangbo, Half-metallic ferromagnetism and large negative magnetoresistance in the new lacunar spinel GaTi3VS8, J. Am. Chem. Soc. 132 (2010) 5704; (b) C. Vaju, J. Martial, E. Janod, B. Corraze, V. Fernandez, L. Cario, Metal – metal bonding and correlated metallic behavior in the new deficient spinel Ga0.87Ti4S8, Chem. Mater. 20 (2008) 2382; (c) E. Janod, E. Dorolti, B. Corraze, V. Guiot, S. Salmon, V. Pop, F. Christien, L. Cario, Negative colossal magnetoresistance driven by carrier type in the ferromagnetic Mott insulator GaV4S8, Chem. Mater. 27 (2015) 4398, Copyright 2013, American Physical Society.*

2.4. The famous example of VO_2that exhibit an IMT but not a Mott transition.

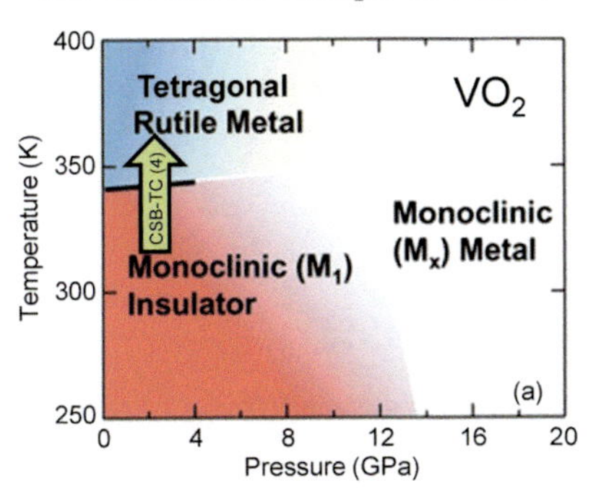

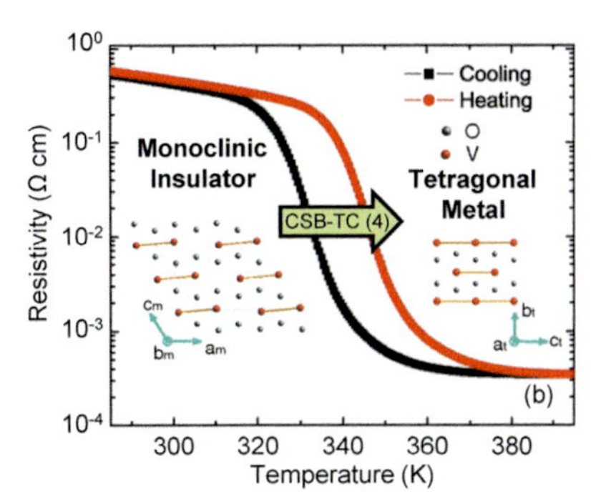

FIG. 10.9 (A) Temperature–pressure phase diagram of VO_2. (B) Resistivity *vs.* temperature at the thermally driven insulator to metal transition in VO_2. This IMT is associated with a Crystallographic Symmetry Breaking between the monoclinic low-T and the tetragonal high-T phases. *Panel A Adapted with permission from Ref. H. Wen, L. Guo, E. Barnes, J.H. Lee, D.A. Walko, R.D. Schaller, J.A. Moyer, R. Misra, Y. Li, E.M. Dufresne, D.G. Schlom, V. Gopalan, J.W. Freeland, Structural and electronic recovery pathways of a photoexcited ultrathin VO2 film, Phys. Rev. B 88 (2013) 165424. Copyright 2014, American Institute of Physics, Panel B Reproduced with permission from Ref. T.F. Qi, O.B. Korneta, S. Parkin, L.E. De Long, P. Schlottmann, G. Cao, Negative volume thermal expansion via orbital and magnetic orders in Ca2Ru1-xCrxO4 (0<x<0.13), Phys. Rev. Lett. 105 (2010) 177203, Copyright 2013, American Physical Society.*

and non-universal part of the phase diagram of Mott insulators (see Fig. 10.2C and Fig. 10.4B for a comparison), its high temperature part above 340 K is radically different, without any almost vertical line separating a metallic from an insulating phase. In this context and surprisingly, the thermally driven IMT observed at $T_{IMT}=340$ K in VO_2 is often referred to as a Mott transition. But clearly this should not be the case as the transition is associated with a crystallographic symmetry breaking [58,59] which attests that the charge degree of freedom is not the only one involved in this transition as it should be the case for a Mott transition. In order to explain the wide range of phenomena observed in VO_2 many recent studies point to a combined effect of lattice distortion and Coulomb correlation [4]. Similar thermally driven insulator to metal transitions that cannot be classified as Mott type are observed in other compounds like Ca_2RuO_4 ($T_{IMT}=357$ K) [60] [61], ,NbO_2 ($T_{IMT}=1070$ K) [62] and $ANiO_3$ perovskites [63,64].

10.3 Electric Mott transitions

Beyond the three types of Mott transitions described above there has been an increasing interest in the last years to induce a Mott transition thanks to an electric field. Numerous studies have demonstrated non-linear electrical behaviors and volatile or non-volatile resistive switching in Mott insulators. Depending on the materials, different mechanisms are involved to explain this out of equilibrium behaviors. As described below these mechanisms are intimately connected to the three types of Mott transitions encountered at equilibrium.

10.3.1 Joule heating and thermally driven IMT

As discussed in Part 2.2, temperature can be used as a tuning parameter to trigger the Mott IMT. The application of an electric field at $T < T_{IMT}$ can also lead to the IMT if the Joule self-heating makes the sample temperature to exceed T_{IMT} and cross the Mott line. Such a thermal mechanism is at play in correlated metal in the close vicinity of the Mott line, as recently confirmed by a DMFT theoretical study [65]. The main signature of this field induced IMT is a sudden increase of the resistance which is volatile as it is maintained only under electric field. This was observed in compounds such as $(V_{1-x}Cr_x)_2O_3$ ($x \approx 0.01$) [66] and $NiS_{2-x}Se_x$ ($x \approx 0.45$) [67,68], where it involves the transition from the low T metallic phase to the high T paramagnetic Mott insulator phase (see Fig. 10.4B and Fig. 10.5C). A similar behavior was also observed on $GaTa_4Se_8$ under pressure in the vicinity of the Mott IMT line [47].

A thermally driven mechanism convincingly explains the sudden change of resistance observed under voltage in numerous correlated compounds that exhibit a phase transition involving a symmetry breaking. This is the case in VO_2 [69,70], Ca_2RuO_4 [71], in pure V_2O_3 below the AFI—metal transition temperature [67,72,73] and in magnetite Fe_3O_4 [74,75]. In all these compounds, Joule heating-driven IMT also called Resistive switching were observed when the temperature chosen for the application of the voltage was just below the temperature of the IMT. This resistive switching occurs above a threshold voltage corresponding to a Joule heating threshold. This phenomenon is essentially volatile (i.e., low resistance state is maintained only under electric field) but a non-volatile resistive switching (i.e., the low resistance state remains even after the end of electric pulses) can be also achieved by choosing the working temperature within the hysteresis domain of the first order IMT, as demonstrated in VO_2 [76,77]. Correlated materials showing a Joule heating induced thermally driven IMT were barely studied in the context of ReRAM applications but were investigated as selectors in ReRAM crossbar arrays [78], in order to suppress the undesired sneak path currents (see Ref. [79]).

Finally, it is worth noting that a resistive switching was also observed under voltage pulse in NbO_2 [80], which might not be related to the crossing of the IMT observed at high temperature ($T_{IMT} = 1070\,K$). In this case a thermal runaway process may explain the transition [81,82].

10.3.2 Electromigration and filling controlled IMT

In oxide Mott insulators filling controlled insulator to metal transition IMT is easily achieved by tuning the oxygen content [1]. Alternatively, the oxygen content may also be changed at the local scale by applying a voltage pulse. In non-stoichiometric transition metal oxides, the migration of oxygen vacancies observed under electric field (i.e., electro-migration process), along grain boundaries or dislocations, induces a modification of the metal/oxygen ratio and a valence change of the cations in the vicinity of theses defects. At the local scale, a filling controlled IMT may occur and lead to a non-volatile reversible bipolar resistive switching[a] by the formation/destruction of a metallic filamentary path between the electrodes [83,84]. Alternatively, the electro-migration phenomenon can also occur close to the metallic electrode/insulator oxide interface and lead to a bipolar resistive switching by modification of a Schottky barrier. This type of resistive switching mechanism called Valence Change Mechanism (VCM) was intensively studied in band insulators and led to the realization of a new type of Resistive Random Access Memory [83,85]. But this mechanism was also observed in various Mott or correlated transition metal oxides. For example, interfacial VCM type resistive switching was reported in systems like La_2CuO_4 [85], $Pr_{0.7}Ca_{0.3}MnO_3$ [86,87], and $YBa_2Cu_3O_{7-x}$ [88]. Alternatively, filamentary VCM type resistive switching was reported in many transition metal oxide Mott insulators [85],[b] such as NiO [89,90], CuO [91,92], and CoO [93], Fe_2O_3 [94], and MnO_x [95]. The VCM type of resistive switching may involve the formation of oxidized or reduced phases. For example, resistive switching in CoO films was proposed to be related to the formation of an oxidized phase Co_3O_4 [95]. Conversely, in CuO films the resistive switching was associated to the formation and destruction of conducting filaments made of a reduced phase, namely Cu_2O and possibly metallic Cu [93]. In NiO the resistive switching is also related to the creation of reduced Ni metallic phase. In this case the Ni filament is formed by a thermally assisted ionic migration process while its destruction occurs due to Joule heating. Consequently, unipolar resistive switching was mainly reported for NiO [85,90,96].

Regardless the nature of electromigration, interfacial or filamentary, this mechanism imposes a severe constraint for application. It indeed requires a forming step, i.e., the application of a specific electrical pulse, usually of high power/energy, to generate the initial filament or interface.

[a]A bipolar resistive switching depends on the polarity of the applied pulse: some filament or interfacial states are created with one polarity and destroyed by the opposite one. Conversely, both polarities have similar effects for a unipolar resistive switching.

[b]The Term "Mott insulator" is Used Here in its Broad Sense, Including both Mott-Hubbard and Charge-Transfer Insulators, see Ref. [11].

10.3.3 Dielectric breakdown and bandwidth controlled IMT

During the last few years, numerous experiments have demonstrated that a Mott insulator subjected to an electric field exceeding a threshold value undergoes an insulator to metal transition that cannot be explained by the two previous mechanisms. This case of volatile resistive switching was first reported in the quasi-one-dimensional Mott insulators Sr_2CuO_3 and $SrCuO_2$ by Taguchi et al. [97] and in the insulating charge-ordered state of $La_{2-x}Sr_xNiO_4$ by Yamanouchi et al. [98]. The so-called dielectric breakdown occurs in these compounds above a threshold field of the order of 0.1–10 kV/cm. The same phenomenon was observed afterwards for the canonical Mott insulators $(V_{1-x}Cr_x)_2O_3$ [99,100], $NiS_{2-x}Se_x$ [99], and AM_4Q_8 (A = Ga, Ge; M = V, Nb, Ta, Mo; Q = S, Se) [101–103], and for molecular Mott insulators K-TCNQ [104], and κ-(BEDT-TTF)$_2$Cu[N(CN)$_2$]Br [105], and [Au(Et-thiazdt)$_2$] [106].

To explain the dielectric breakdown of Mott insulators under electric fields many theoretical works were first devoted to the Zener breakdown [107–110]. For instance, calculations were performed in 1D Hubbard chains using exact diagonalization [107], and time-dependent density matrix renormalization group [108], or in the limit of large dimensions using dynamical mean field theory [109]. All these theoretical studies have predicted nonlinear behavior in the current–voltage characteristics, and the existence of a threshold field (E_{th}) beyond which a field induced metal appears. Zener type breakdown should occur when the electric field is such that it bends the Hubbard bands by the gap energy E_G within the length ξ of the order of the unit cell. Hence, the Zener breakdown is predicted to occur for strength of the electric field $E_{th} \sim E_G/\xi$ of the order of 10^6–10^7 V/cm [107]. It cannot explain the experimental values of the threshold field that are at least two orders of magnitude lower [97,98,102,105]. As a consequence, volatile resistive switching in Mott insulators cannot be explained by a Zener breakdown scenario. Alternatively, recent studies on the AM_4Q_8 Mott insulators support that the dielectric breakdown originates from an electric field induced electronic avalanche phenomenon [111,112]. This breakdown may induce a nonvolatile resistive switching resulting from the creation of conducting filamentary paths made of compressed metallic zone. The mechanism involves therefore a Bandwidth Controlled IMT at the local scale. A complete description of this mechanism will be given in the next section.

10.4 Electric Mott transition by dielectric breakdown: Detailed mechanism

10.4.1 Phenomenology of the dielectric breakdown

The most prominent signature of the electric Mott transition observed in canonical Mott insulators is the volatile resistance drop, also called resistive switching, that occurs under an electric field. Fig. 10.10 displays

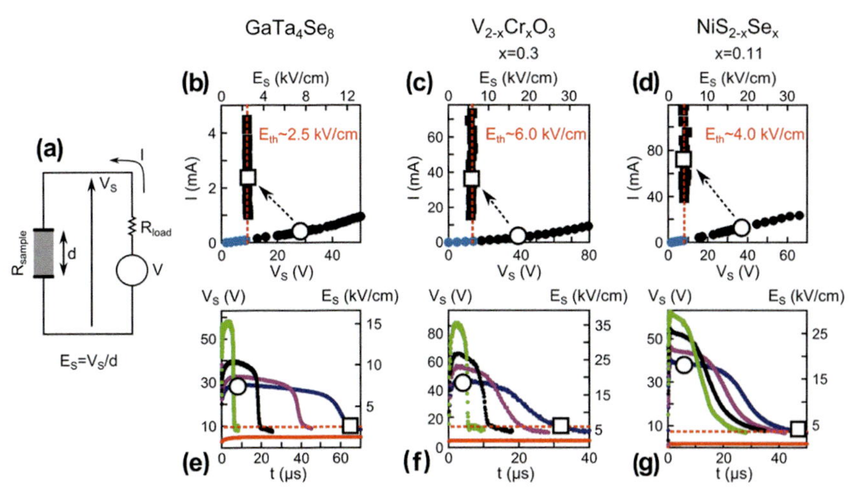

FIG. 10.10 (A) Shows the schematics of the experimental setup. Universal dielectric breakdown I—V characteristics (top panels B, C, D) and time dependence of the sample voltage $V_S(t)$ (bottom panels E, F, G) are displayed for three different types of narrow gap Mott insulators. Blue dots correspond to the region below E_{th}, where no breakdown is observed. Black symbols correspond to the I—V characteristic in the resistive switching region, above E_{th}. The black dots show the initial *I—V*, before the breakdown, and the black squares indicate the final state. The open symbols highlight a particular breakdown transition for easier visualization. Measurements on $GaTa_4Se_8$ were performed at 77K, on $V_{2\text{-}x}Cr_xO_3$ (x=0.3) at 164K and on $NiS_{2\text{-}x}Se_x$ (x=0.11) at 4K. *Reproduced with permission from P. Stoliar, L. Cario, E. Janod, B. Corraze, C. Guillot-Deudon, S. Salmon-Bourmand, V. Guiot, J. Tranchant, M. Rozenberg, Universal electric-field-driven resistive transition in narrow-gap Mott insulators, Adv. Mater. 25 (2013) 3222, Copyright 2013,* WILEY-VCH Verlag GmbH & Co.

typical resistive switching observed on canonical Mott insulators $GaTa_4Se_8$, $(V_{1-x}Cr_x)_2O_3$ (x=0.15) and $NiS_{2\text{-}x}Se_x$ (x=0.11) when voltage pulses are applied to the circuit made of the Mott insulator crystal in series with a load resistance (Fig. 10.10A). Each voltage pulse exceeding the threshold value (of the order of a few kV/cm) induces a volatile resistive switching after a delay time. Fig. 10.10E, F and G show that the resistive switching occurs only above a threshold electric field E_{th} of a few kV/cm (see red dotted line) and after a time t_{delay} which decreases as the voltage across the sample increases. The sample field E_s after the resistive switching event always lies on the same value E_{th} that also corresponds to the lower voltage that can induce a resistive switch in DC measurements. As a consequence the Mott insulator compounds exhibit a very specific current–voltage characteristic with two branches. The first one corresponds to the non-transited state which deviates only slightly from the Ohm's law. The second branch, which is almost vertical and lies at the threshold field, corresponds to the "transited" state (see Fig. 10.10B, C and D).

10.4.2 The initial spark: Creation of hot electrons and electronic avalanche

All Mott insulator compounds exhibit the same type of *I(V)* characteristic with threshold electric field in the 1–10 kV/cm range [99]. The magnitude of the threshold field in AM_4Q_8 Mott insulators as well as their *I(V)* characteristics shown in Fig. 10.11A–E compare well with the threshold field values and *I(V)* characteristics observed for avalanche breakdowns in narrow gap semiconductors [113]. For this reason, it was proposed that the resistive switching observed in the Mott Insulators AM_4Q_8 originates from an avalanche breakdown phenomenon [111]. In semiconductors, the avalanche threshold field varies as a power law of the band gap and follows the universal law $E_{th} \propto E_G^{2.5}$ [114,115]. Fig. 10.11F reveals that AM_4Q_8 compounds have a similar variation of the threshold field as a function of the Mott-Hubbard gap [111]. This power law behavior provides a first evidence that supports the avalanche breakdown scenario in these Mott insulators.

The pioneering works of Fröhlich and Seitz [116–120] have established two different regimes of dielectric breakdown. A first regime, corresponding to a "clean" limit, occurs preferentially at low temperature in ultrapure semiconductors with long mean free paths. In this case, the avalanche process is initiated by the tiny number of electrons available, which gain energy independently to each other's. In classical semiconductors such as Si, Ge or GaAs, avalanche breakdown is the consequence of an impact ionization: electrons in the conduction band, accelerated by an electric field, can gain enough energy to induce an impact ionization. This process generates electron–holes pairs and promotes new electrons in the conduction band. The repetition of this process leads to a free-carriers multiplication if the average ionization impact rate exceeds the electron–hole recombination rate. In this regime, the electric-field-induced energy gain in the conduction band is only limited by the electron–phonon (e^- - ph) scattering. As e^- - ph scattering raises with temperature, the threshold electric field E_{th} also increases with *T* in this regime. Alternatively, a second regime may appear when e^- -e^- scattering is not negligible anymore. This may occur in the realistic situation where defects induce discrete energy levels in the gap distributed on a typical energy width $\Delta\varepsilon$. This specific density of states is depicted in Fig. 10.12A and may be found in realistic Mott insulators as shown in Fig. 10.12B and C. Unlike the electrons in the conduction band, the trapped electrons cannot be accelerated by an electric field because these levels are not continuously distributed. However, the trapped electron give rise to an efficient scattering with the conduction electrons and the phonon. As a consequence, the thermal balance of the electronic system consists of the energy transfer rates from the electric field to the conduction electrons (gain P_{in}) and from the electrons localized in shallow levels to the lattice (loss P_{out}),

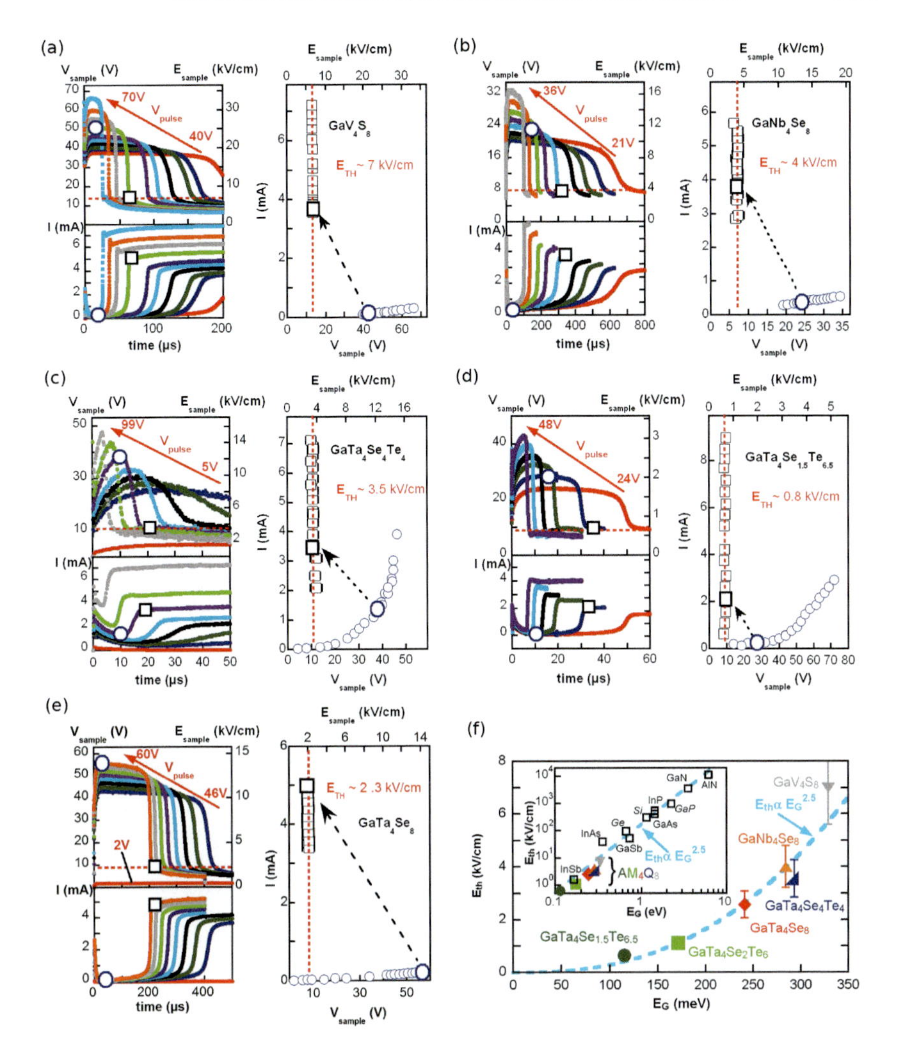

FIG. 10.11 (A)–(E) Time dependence of the voltage and intensity measured at 77 K across different AM_4Q_8 single crystals when they are subjected to voltages pulses. All the resistive switching observed during the pulses are volatile, i.e., the resistance is the same before and after the electric pulse. The current–voltage characteristics are plotted from the values of I and V measured during the pulses, before (blue circles) and after (open squares) the volatile transition (see corresponding symbols in I—V plot). (F) Dependence of E_{th} in Mott insulators and semiconductors. Threshold electric field as a function of the Mott gap E_G for various AM_4Q_8 compounds. The solid blue curve corresponds to a power law dependence $E_{th} \propto E_G^{2.5}$. Inset: comparison of the threshold fields versus gap dependence for the AM_4Q_8 compounds and for classical semiconductors. The solid blue line displays the universal law $E_{th}[kV/cm] = 173\ (E_G[eV])^{2.5}$ observed for semiconductors. *Reproduced with permission from V. Guiot, L. Cario, E. Janod, B. Corraze, V. T. Phuoc, M. Rozenberg, P. Stoliar, T. Cren, D. Roditchev, Avalanche breakdown in GaTa4Se8 – xTex narrow-gap Mott insulators, Nat. Commun. 4 (2013) 1. Copyright 2013, Macmillan Publishers Limited.*

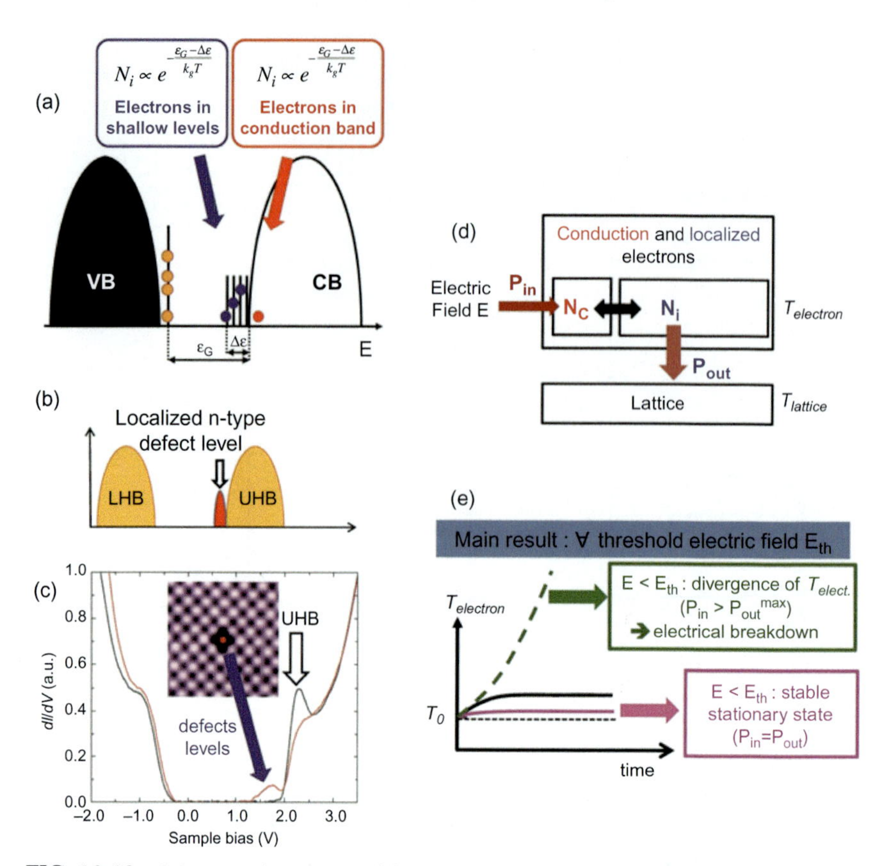

FIG. 10.12 Schematic description of the main ingredients involved in the dielectric breakdown in the dirty limit according to the Fröhlich model. (A) The density of states displays a gap ε_G between the fundamental impurity level and the conduction band, with discrete impurity levels spread on the energy width $\Delta\varepsilon$. Some electrons in the impurity levels (N_i) and in the conduction band (N_c) are indicated by the blue and red points. (B) typical DOS in a slightly *n*-doped Mott-Hubbard insulator, according to Ref. [12,13]. An additional peak related to localized *n*-type defects level appears contiguous to the Upper Hubbard Band. (C) Experimental evidence of such defects level unraveled by STM/STS experiments performed close (red curve) and far (black curve) from a missing Cl defect in $Ca_2CuO_2Cl_2$. (D) Thermal balance of the electronic system. Electrons have a temperature distinct from the lattice temperature. P_{in} and P_{out} are the heating and cooling powers sustained by the electrons. The thermalization between electrons, the heating by the electric field and the cooling through the lattice are described. (E) Time evolution of the electronic temperature under several values of the electric field. The dielectric breakdown regime appears above a threshold electric field E_{th}. *Panel C From Ref. C. Ye, P. Cai, R. Yu, X. Zhou, W. Ruan, Q. Liu, C. Jin, Y. Wang, Visualizing the atomic-scale electronic structure of the Ca2CuO2Cl2 Mott insulator, Nat. Commun. 4 (2013) 1365.*

as shown Fig. 10.12D. Under electric field, the electronic temperature T_e rises above the lattice-bath temperature T_0 until the two energy transfer rates (gain and loss) equilibrate (see Fig. 10.12E). But above a threshold field E_{th} the system becomes unstable as the energy rates gained and lost by the electrons can no longer be equilibrated. Consequently, the electronic temperature raises drastically which induces the electrical breakdown. The model leads to two main predictions [116,117]. First at low field the resistance of the sample should deviates from the Ohm's law in a predictable manner without external tunable parameter. Second the threshold electric field should exhibit a thermally activated behavior at high temperature and a crossover to the low temperature regime, as depicted in Fig. 10.13A. These predictions were tested in the series of AM_4Q_8 compounds [112]. A nonlinear electrical behavior was indeed found at low field. Moreover, Fig. 10.13B shows the temperature dependence of the threshold field E_{th} observed for several AM_4Q_8 single crystals. A remarkable agreement is found with the theory: existence of a threshold electric field, of two temperature regimes and of an activated dependence of E_{th} at high temperature [112]. These experimental observations support therefore that the resistive switching in canonical Mott insulators originates from the creation of hot electrons under electric field.

10.4.3 Runaway process and creation of a conducting filamentary path

To capture the main features of the Electric Mott Transition a phenomenological model was developed based on a very simple assumption [99,121]. Under electric field, some hot carriers are produced that break

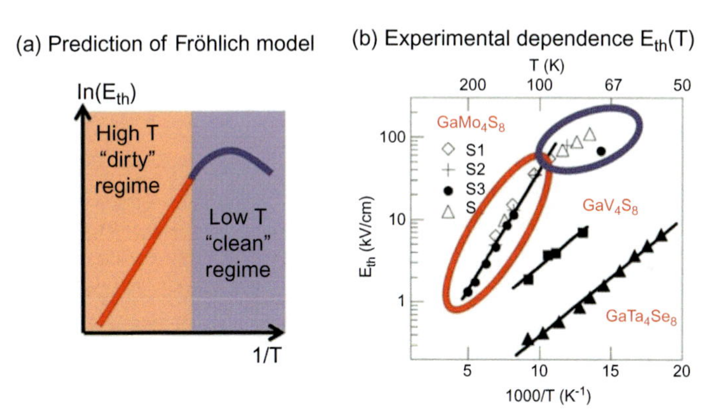

FIG. 10.13 Evolution of the threshold electric field E_{th} with temperature. (A) theoretical predictions of Fröhlich in the dirty (high temperature) and clean (low temperature) regimes. (B) Temperature dependence of the threshold field E_{th} for three compounds of the AM_4Q_8 family. In the case of $GaMo_4S_8$, four samples (S1-S4) have been measured. The black lines are the fit with an activation law.

locally the Mott insulating state and generate a correlated metal phase. To account for the competition between the two phases under electric field, a two-minima energy landscape, where the Mott insulator (MI) state is at the lowest energy and the correlated metal (CM) state at higher energy, was considered and implemented in a resistor network. The probability of transition from the Mott insulating phase to the correlated metallic phase depends on the applied voltage while the probability of the transition back is a thermal activation process over a fixed energy barrier. This modeling work shows that a volatile resistance drop is obtained thanks to the apparition of metallic filamentary path between the electrodes. Fig. 10.14A and Fig. 10.14B display the dynamics of the creation (and relaxation) of metallic sites and of the filamentary path. At the beginning of the pulse, all sites are in the insulating state but under electric field some insulating sites switch into metallic sites. During the pulse, the production rate of metallic sites largely overcome their relaxation rate. The fraction of metallic sites hence increases up to a critical threshold above which a runaway process starts resulting in the formation of a metallic filamentary path across the sample, bridging the two electrodes [99,121]. This process is illustrated in the Fig. 10.14C. Calculations combining the energy landscape model with a thermal model support that the onset of the resistive transition is solely driven by a purely electronic transition, while Joule heating occurs once the metallic filament is created and the current starts to raise in the circuit [121].

Finally, the electric Mott transition and concomitant resistive switching during an electric pulse can be summarized as schematized in Fig. 10.15. Above a threshold electric field, free carriers are accelerated within the Mott insulators, which enhances the electronic temperature and promotes the massive creation of new carriers [99,112]. This phenomenon leads to an electronic avalanche breakdown which induces the formation of a metallic filamentary path within the material after a time delay (t_{delay}) which varies as the inverse of the applied voltage (Fig. 10.15A and B). The resistive switching occurs when the filamentary path bridges the opposite electrodes (Fig. 10.15C, D, E and F). Once this filamentary path is created, two situations may be encountered after the end of the voltage pulse as depicted in Fig. 10.15H and G. The filamentary path can either fully or only partially dissolve. In the last situation, the resistance does not relax back to its initial value, if the remaining filamentary path percolates between electrodes. This type of resistive switching is called non-volatile and is usually obtained when the applied voltage is large compared to the threshold voltage. Conversely when the filament self dissolves, the resistance relaxes back to its initial value and for this reason this type of resistive switching is called volatile. This volatile resistive switching is usually observed for voltages slightly higher than the threshold field. Recently, it was proposed that the stability of the filamentary path could be related to

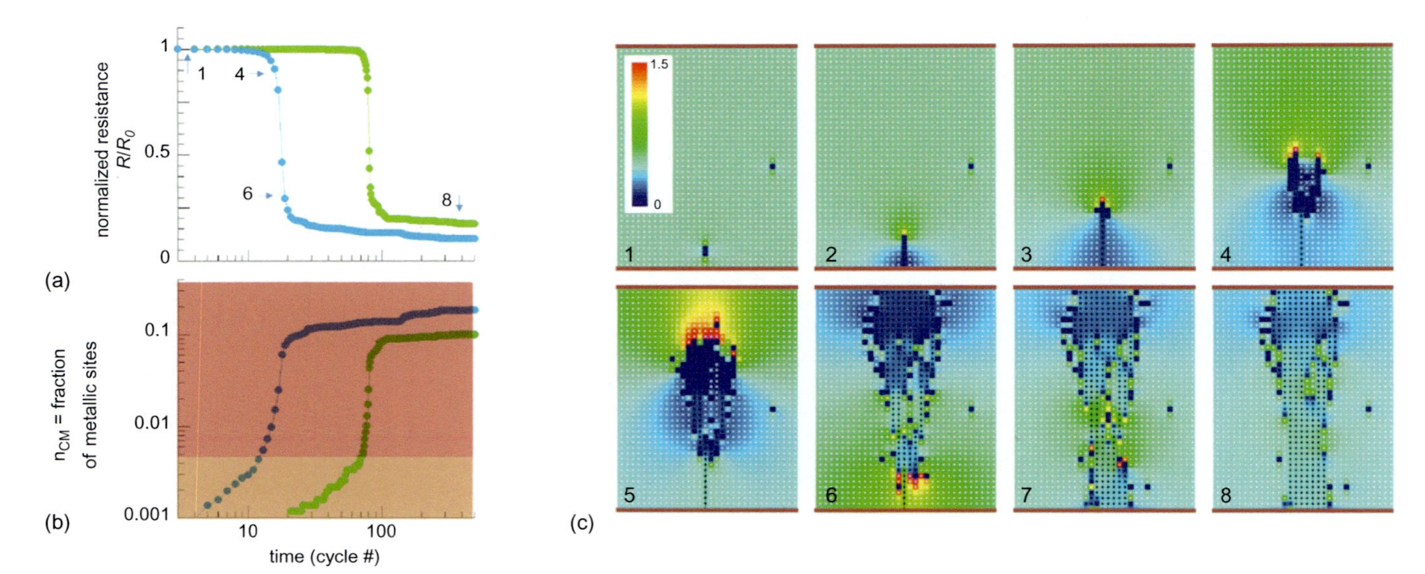

FIG. 10.14 Modeling of the behavior of a Mott insulator under electric field (A) Resistive switching observed after a delay time decreasing with the applied voltage. This applied voltage is higher for the blue curve than for the green one [31]. (B) Concomitant increase of the fraction of metallic sites in the resistor network. The orange area corresponds to the increase of metallic cells before the creation of the filamentary percolating path (threshold value is ~0.00375). The red area corresponds to the increase of metallic cell number during and after the creation of this filament. (C) Resistor network evolution as a function of time showing the filament formation. The red lines at the top and the bottom stand for the electrodes where the voltage is applied. The local electric field is represented by the color scale in snapshot 1: from zero in dark blue color to red in the higher. Time of snapshot 1, 4, 6, and 8 are indicated in (A). *Reproduced with permission from Ref. C. Adda, L. Cario, J. Tranchant, E. Janod, M.-P. Besland, M. Rozenberg, P. Stoliar, B. Corraze, First demonstration of "leaky integrate and fire" artificial neuron behavior on* $(V_{0.95}Cr_{0.05})_2O_3$ *thin film, MRS Commun. 8 (2018) 835, Copyright 2018, Material Research Society.*

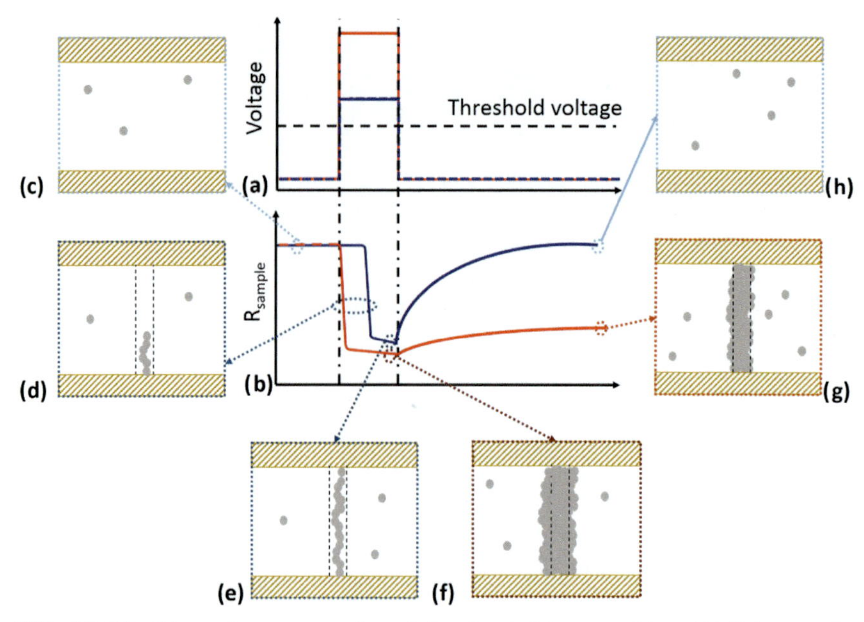

FIG. 10.15 Schematic representation of the electric Mott transition. (A) Voltage pulse applied on the Mott insulator. The blue and red pulse correspond to different voltages above the threshold voltage, (B) Variation of the sample resistance before, during and after the voltage pulses. Before the voltage pulse, the sample resistance is in the high resistive state. During the voltage pulse, resistive switching to low resistive states are observed after a delay time which decrease with increasing voltage. After the pulse, the sample resistance either returns to its initial value, (blue curve) meaning that the resistive switching is volatile, or the sample resistance remains in the low resistance state, leading hence to a non-volatile resistive switching. (C, D, E, F, G, H) snapshots correspond to the time evolution of the formation of the metallic filamentary path within the sample associated to the resistive switching. The brown hatched rectangles represent the electrodes. *Reproduced with permission from Ref. C. Adda, B. Corraze, P. Stoliar, P. Diener, J. Tranchant, A. Filatre-Furcate, M. Fourmigué, D. Lorcy, M.-P. Besland, E. Janod, L. Cario, Mott insulators: a large class of materials for leaky integrate and fire (LIF) artificial neuron, J. Appl. Phys. 124 (2018) 152124. Copyright 2018, Elsevier.*

elastic constraints and that below a critical size the filamentary path could not be stable by itself and would completely dissolve whereas larger filaments would be metastable 122.

10.4.4 Non-volatile electric Mott transition: A consequence of lattice compression specific to Mott physics

As discussed above, narrow gap Mott Insulators exhibit a non-volatile resistive switching for electric fields well above the avalanche threshold field. Fig. 10.16 shows the resistance *vs.* temperature curves measured

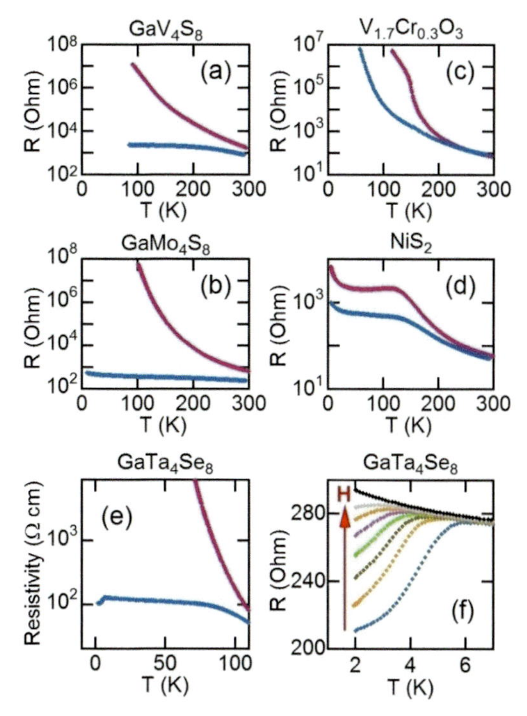

FIG. 10.16 Variation of resistance as a function of temperature for various narrow gap Mott insulators in pristine state (pink curves) and after the application of an electric pulse inducing a non-volatile resistive switch (blue curves), in (A) GaV_4S_8, (B) $GaMo_4S_8$, (C) $(V_{0.85}Cr_{0.15})_2O_3$, (D) NiS_2 and (E) $GaTa_4Se_8$. (F) Resistance *vs.* temperature for various magnetic fields (from 0 to 5 Tesla) in a transited $GaTa_4Se_8$ single crystal [2].

at low bias level in the pristine and transited states of several crystals belonging to the AM_4Q_8 compounds, $(V_{0.85}Cr_{0.15})_2O_3$ and NiS_2. In all cases, the transited states present a significant resistance drop after non-volatile transition.

Several in-depth studies have unveiled important features of this non-volatile low resistance state. In $GaTa_4Se_8$, Scanning Tunneling Microscopy (STM) measurements have revealed that the non-volatile resistive switching is related to an electronic phase separation at the nanoscale 101,123. While the surface topography of pristine $GaTa_4Se_8$ crystals is structure less, granular filamentary structures qualitatively oriented along the direction of the electric pulses are observed after resistive switching inside the crystal after cleaving. Spectroscopic (STS) experiments have confirmed the existence of electric-field-induced metallic nano-domains of about 30–70 nm within these granular filamentary paths. These nanodomains were carefully investigated by Energy Dispersive X-Ray spectroscopy, and by Transmission Electron Microscopy. No chemical composition change nor any crystallographic symmetry breaking or amorphization between the electrodes were detected at the nanometric scale [2]. This excludes the formation of conducting bridge-like filaments [124], amorphous-crystalline transition as observed in phase change materials [125] or phase transition similar to the monoclinic-tetragonal phase change observed in VO_2 (see discussion in Section 10.2.4). However, as shown in Fig. 10.16F the presence of granular superconductivity was observed at low temperature in a transited crystal of $GaTa_4Se_8$. This is reminiscent of the superconducting transition observed under pressure in this compound (see Fig. 10.8 and Section 10.2.3.3), and directly suggests the presence of compressed metallic nano-domains in transited crystals of $GaTa_4Se_8$ [11].

Similar conclusions can be drawn from the non-volatile resistive switching observed in $(V_{1-x}Cr_x)_2O_3$ [126]. The phase diagram of this compound reproduced in Fig. 10.17A (see also discussion in Section 10.2.3.1) shows indeed that the transition temperature from the Mott insulating phase (PI) to the AFI phase at low temperature does not change much with chromium content or pressure while a huge variation is observed for the transition temperature from the metallic (PM) to the AFI phase. This was used to characterize the possible formation under electric pulse of compressed metallic domains. Fig. 10.1B shows the resistance versus temperature (blue curve) measured on a $(V_{0.975}Cr_{0.025})_2O_3$ single crystal. Above 190K, the crystal shows an insulating behavior consistent with the expected PI state. The transition from the PI to the AFI phase occurs at around 180K which is in good agreement with the value expected from the phase diagram (see Fig. 10.17A). After non-volatile resistive switching the resistance $R(T)$ of the $(V_{0.975}Cr_{0.025})_2O_3$ crystal is decreased by several orders of magnitude at low temperature. Fig. 10.17B compares pristine state (blue curve) and transited state (red curve). Both curves show the signature of the PI-AFI transition around 180K. In the switched crystal, it corresponds to the transition of the $(V_{0.975}Cr_{0.025})_2O_3$ matrix embedding the filamentary conducting path (light blue regions in Fig. 10.17C). However, a new transition is clearly observable around 40K in the $R(T)$ curve of the switched $(V_{0.975}Cr_{0.025})_2O_3$ crystal (in red). According to the phase diagram of this system, this may be viewed as the PM$\rightarrow$AFI transition of $(V_{0.975}Cr_{0.025})_2O_3$ under a pressure of 30kbar [26]. This observation supports the creation of compressed metallic domains under electric field. After resistive switching the metallic domains correspond to $(V_{0.975}Cr_{0.025})_2O_3$ domains under pressure, as indicated by the red dotted line in Fig. 10.17A. From these experiments, it was proposed that the transited crystal of $(V_{1-x}Cr_x)_2O_3$ contains some conducting granular filamentary paths embedded in a pristine like matrix (see Fig. 10.17C). Moreover, these granular filamentary paths include some compressed regions that have undergone a bandwidth control insulator to metal

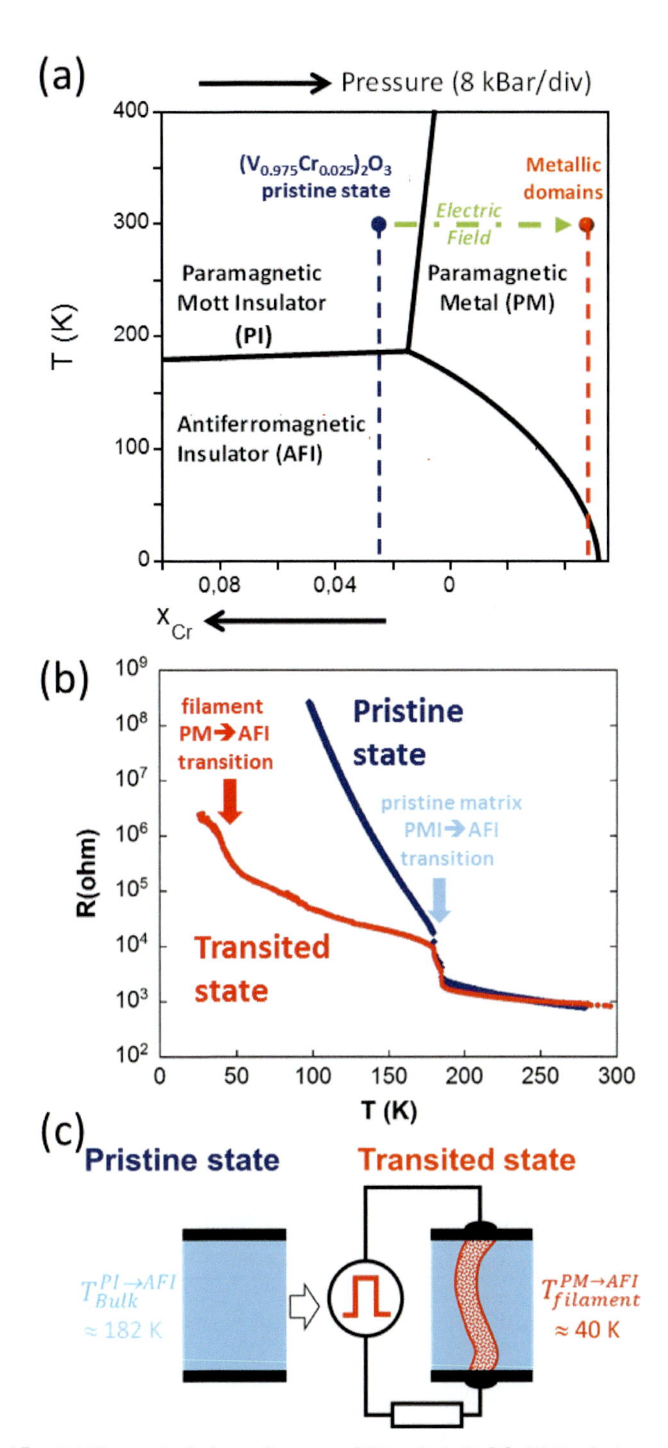

FIG. 10.17 (A) Theoretical phase diagram of $(V_{1-x}Cr_x)_2O_3$ [5]. (B) Evolution of resistance vs temperature of a $(V_{0.975}Cr_{0.025})_2O_3$ crystal in its pristine state and in a transited state. As expected, the PMI-AFI phase transition occurs around 180 K in the pristine state (see blue dotted line in (A) and blue arrow in (B)). In the transited state a new transition occurs at 40 K suggesting the formation of compressed metallic domains (see red dotted line in (A) and red arrow in (B)). (C) Schematical drawing of the formation of a granular filamentary path in the transited state of a $(V_{0.95}Cr_{0.05})_2O_3$ crystal containing some compressed domains with a PM-AFI phase transition around 40 K as indicated by the red dotted line in (A).

transition. This was recently confirmed by a Raman and micro-X-Ray Diffraction studies showing a lattice compression in the transited region of $(V_{1-x}Cr_x)_2O_3$ [127].

All these results suggest that the electronic avalanche breakdown induces the collapse at the local scale of the Mott insulating state into a correlated metallic state and leads to the formation of a conductive filament formed of compressed domains. Such a strong response of the lattice (e.g. a volume contraction) is also observed at the bandwidth controlled IMT achieved under pressure [22]. Moreover the Dynamical Mean Field Theory predicts that this lattice response follows from a dramatic change in the electronic wavefunction across the IMT, which has a direct effect on the compressibility of the lattice [16]. The compression of the lattice observed at the Electric Mott transition appears therefore as a strong signature of the Mott physics.

Finally the detailed study of the non-volatile Electric Mott Transition has shown that the successive application of unipolar electric pulses to Mott insulators makes the resistance switch back and forth between high and low resistance states [102]. As discussed above, the SET transition corresponds to the formation of a metastable filamentary path thanks to an avalanche breakdown. On the other hand, the RESET transition from the low resistance state to the high resistance state corresponds to the partial or total dissolution of the filamentary path. The mechanism of the RESET transition is likely related to a thermal relaxation of the metallic domains back to the insulating state [128]. Fig. 10.18 summarizes this scenario and provides schematic representations of the evolution of the filament during the volatile, the "SET" and thermally assisted "RESET" transitions. This reversibility of the non-volatile transition enables to build memories based on these materials [129]. Noteworthy intermediate levels between high and low resistance states can be reached [128,130], which could be of interest for multi-level data storage or memristive applications (artificial synapses). This will be discussed in detail in the next section.

10.5 Microelectronic applications of Mott insulators: Toward Mottronics

10.5.1 Mott memories

The huge non-volatile memory market is currently led by the Flash technology, used e.g. in Flash SD cards and Solid State Drives. However, the limit of this technology in downscaling will hinder its development in a near future [131]. Several emerging Random Access Memories (RAM) [132], i.e., Phase-Change RAM (PCRAM) [133], Magnetic RAMs (MRAM) [134], and Resistive RAM (ReRAM) [3], are currently considered as interesting candidates to overcome the shortcomings of Flash

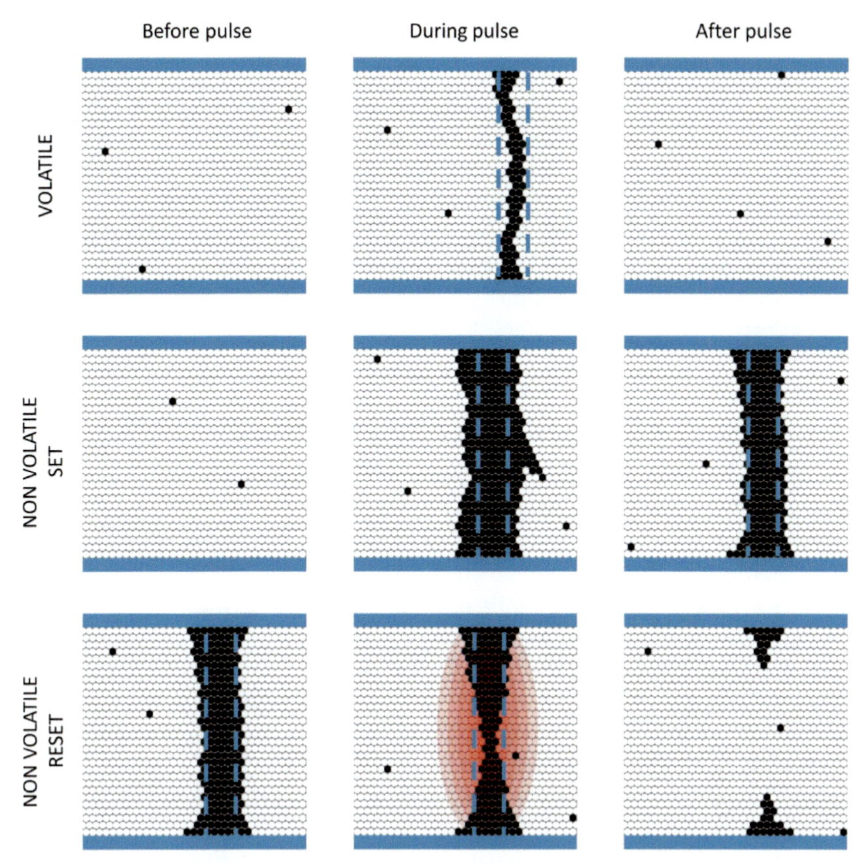

FIG. 10.18 Schematic illustration of the filament evolution before, during and after the application of a pulse inducing a volatile transition, a "SET" non-volatile transition and a "RESET" non-volatile transition. White and black domains represent respectively Mott Insulating and Correlated Metal regions of the material. Top and bottom electrodes are depicted in blue, and dashed blue lines represent a critical radius of stability for the filament. *Reproduced with permission from Ref. E. Janod, J. Tranchant, B. Corraze, M. Querré, P. Stoliar, M. Rozenberg, T. Cren, D. Roditchev, V. T. Phuoc, M.-P. Besland, L. Cario, Resistive switching in Mott insulators and correlated systems, Adv. Funct. Mater. 25 (2015) 6287, Copyright 2015, WILEY-VCH Verlag GmbH & Co.*

memories. ReRAMs appear as a very appealing solution among these potential candidates, thanks to a very simple architecture and promising memory performances [131]. In that context the EMT's observed in Mott insulators were investigated to build a Mott based ReRAM called the Mott-memory. In that purpose, the transition needs to be non-volatile which in practice is only achieved with the EMTs based on filling and bandwidth control insulator to metal transitions. The next paragraphs will therefore mainly concentrate on these two types of Mott memories.

10.5.1.1 *Mott memories based on a thermally driven mechanism*

Thermally driven IMT generally leads to reversible resistive states but it is noteworthy that non-volatile transitions have been mastered in the correlated insulator VO_2 by a fine control of temperature in the vicinity of the first order phase transition between monoclinic insulating and tetragonal metallic phases. Control is obtained either by external temperature regulation around 68 °C [76], either through sample self-heating by Joule effect [77][77,79]. But temperature variations of the device environment exclude realistic memory applications based on this type of transition.

10.5.1.2 *Mott memories based on filling-controlled mechanism*

As described in Section 10.4.2., oxygen electromigration accounting for Valence Change Mechanism (VCM) in transition metal oxides is also at the origin of the filling controlled IMT observed under electric field in oxide Mott insulators. One of the most studied Mott system for building ReRAM devices is found in the perovskite manganites $A_{0.7}B_{0.3}MnO_3$ (where A = Pr, La, Sm; B = Ca, Sr). These manganites are generally sandwiched between two metallic electrodes such as Pt or $SrRuO_3$ as a bottom electrode (BE) [87] and Ti, Al, W, Ag, Au, Pt, or Ta as a top electrode (TE). In these devices, TE material plays a crucial role on the occurrence of a redox reaction or not at the TE/$Pr_{0.7}Ca_{0.3}MnO_3$ (PCMO)-like interface, since it determines if an oxide layer can form at the interface to generate a Schottky-like barrier, as shown in Fig. 10.19A. Indeed, materials such as Ti, Al or Ta are active toward injection/release of O ions, whereas Au, Pt or Ag are not [5]. Transport properties are then dominated by the TE/PCMO interface.

Performances of manganites-based devices were evidenced for BE diameters down to 42 nm [135] and film thicknesses as low as 3 nm [136]. Resistance switching generally occurs after a forming step, even if forming-free device was recently mentioned [137]. As a consequence of interfacial phenomenon, the I—V characteristic of those devices is area dependent, and the High Resistance State (HRS)/Low Resistance State (LRS) ratio remains constant as device size decreases [87]. Nevertheless, this ratio is strongly affected by pulse duration, and sub-μs pulse programming, down to 50 ns, often results in HRS/LRS <3 [138,139], whereas highest HRS/LRS ($>10^3$) are obtained with pulses in the 0.1–10 ms range [140]. Oxygen vacancy diffusion also results in bipolar voltage pulsing protocols with asymmetry between SET and RESET pulse voltages, mostly in the 3-6 V range [88,137,138,140–142]. Cycling endurance reaches 10^4.

cycles in most studies, but 10^7 cycles could be obtained in the case of 50 ns pulses leading to HRS/LRS ratio <2 [141] as a result. The optimization tradeoff is generally referred to the time-voltage dilemma. Finally, a

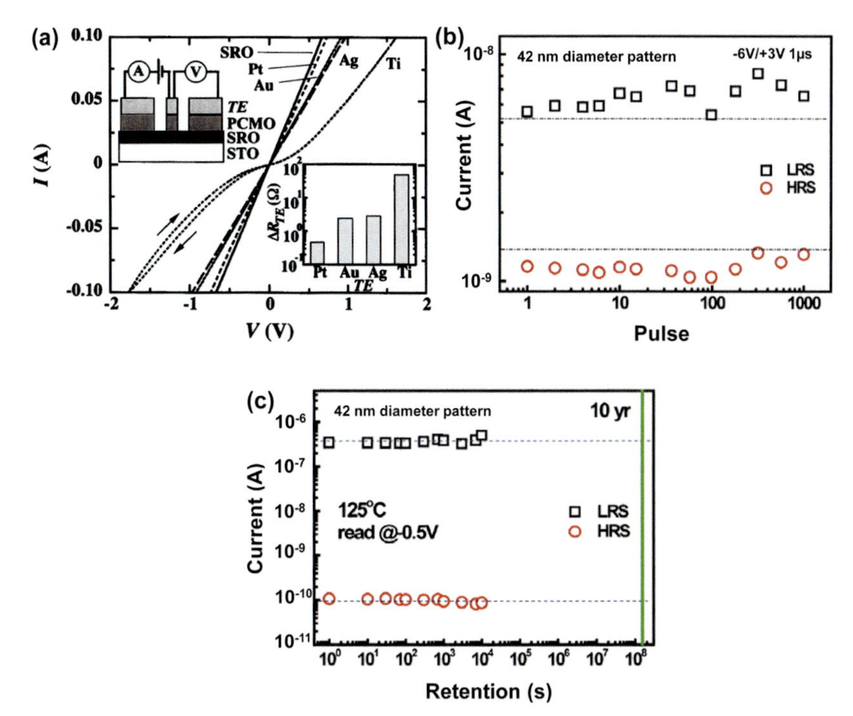

FIG. 10.19 (A) I—V characteristics of TE/PCMO/SRO layered structures with five different TEs. Here, TE, PCMO, and SRO stand for top electrode, $Pr_{0.7}Ca_{0.3}MnO_3$, and $SrRuO_3$, respectively. The upper panel of the insets shows a schematic of the samples. The lower one shows ΔR_{TE} ($=R_{TE} - R_{SRO}$) for TE=Pt, Au, Ag, and Ti. (B) AC endurance properties of Al/$La_{0.7}Sr_{0.3}MnO_3$ stack. Pulse width is 1 μs. The programming voltages for the LRS and the HRS were −6 V, and +3 V. The inset shows schematic diagram of nanoscale device. (C) Excellent retention properties of LRS and HRS at 125 °C. The result predicts 10 years lifetime. *Panel A Reproduced with permission and adapted from S.Q. Liu, N.J. Wu, A. Ignatiev, Electric-pulse-induced reversible resistance change effect in magnetoresistive films, Appl. Phys. Lett.76 (2000) 2749, Copyright 2004, AIP Publishing LLC; Panel B and C reproduced with permission from J. Lee, H. Choi, D. Seong, J. Yoon, J. Park, S. Jung, W. Lee, M. Chang, C. Cho, H. Hwang, The impact of Al interfacial layer on resistive switching of La0.7Sr0.3MnO3 for reliable ReRAM applications, Microelectron. Eng. 86 (2009) 1933, Copyright 2004, AIP Publishing LLC.*

good data retention is generally observed, up to 10 years at 125 °C (extrapolated), as displayed in Fig. 10.19B and C [135].

10.5.1.3 Mott memories based on bandwidth-controlled mechanism

The electric Mott transitions based on bandwidth controlled IMT discovered on Mott insulator compounds like AM_4Q_8, $(V_{0.95}Cr_{0.05})_2O_3$ or NiS_2 leads to non-volatile resistive switching which makes them potential candidates for ReRAM applications [2,102,104,130,143–145]. Studies on this type of ReRAM have mainly focused on devices made of $(V_{0.95}Cr_{0.05})_2O_3$ and

GaV_4S_8. The following sections present therefore the realization of Metal/ Insulator/Metal (MIM) devices using these narrow gap Mott insulators and describe the performances obtained in the context of ReRAM applications.

10.5.1.3.1 Performances of Mott insulator based ReRAM devices

The bandwidth controlled IMT observed in single crystals of the AM_4Q_8 family of compounds (see Section 10.4.3.) could be transposed to thin films of GaV_4S_8 [146]. Polycrystalline stoichiometric layers of this AM_4Q_8 compound were obtained either by non-reactive sputtering [147] or by Ar/H_2S reactive sputtering [148]. As presented in Fig. 10.20, the application of electric pulses to $Au/GaV_4S_8/Au$ symmetric MIM structures with 2×2 μm^2 bottom electrode size enabled to obtain resistive switching cycles with $4 < R_{OFF}/R_{ON} < 6$, as displayed in Fig. 10.20- R_{OFF} and R_{ON} are the resistance values of high (*OFF*) and low (*ON*) resistance states, respectively. In this case, the pulse protocol consists of an alternation of 3.2V /7 × 500 ns multi-pulses for SET transitions with long pulses of lower voltage for RESET (1.6 V/500 μs) [149]. Importantly the RS cycles

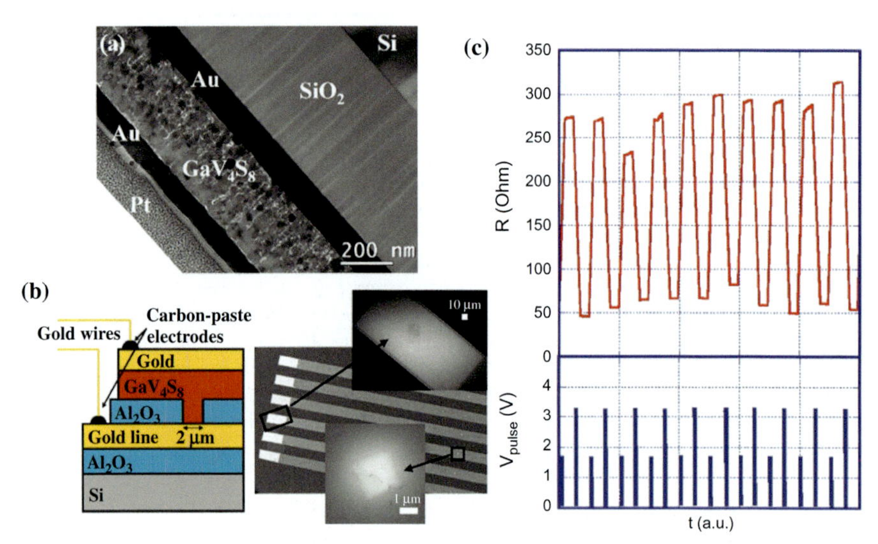

FIG. 10.20 (A) Top-view TEM picture of the 50 × 50 μm² $Au/GaV_4S_8/Au$ MIM structure. The platinum layer on gold top electrode is due to the sample preparation by FIB (B) Schematic drawing of a 2 × 2 μm² $Au/GaV_4S_8/Au$ MIM structure cross-section and SEM pictures of the corresponding substrate surface, before deposition of the GaV_4S_8 thin film and the top gold electrode. (C) Switching performance of the 150 nm thick GaV_4S_8 layer by applying successive 3.2 V/7 × 500 ns pulses separated by 3.5 μs off period, and 1.6 V/500 μs pulses. *Adapted with permission from J. Tranchant, E. Janod, L. Cario, B. Corraze, E. Souchier, J.-L. Leclercq, P. Cremillieu, P. Moreau, M.-P. Besland, Electrical characterizations of resistive random access memory devices based on GaV4S8 thin layers, Thin Solid Films 533 (2013) 61, Elsevier.*

could be obtained without forming step and with pulse duration as low as 100 ns. In terms of performances, more than 65,000 cycles were obtained with a 0.01% error rate, and data retention could be extrapolated to 10 years at room temperature without resistance change [150].

The property of bandwidth controlled IMT was also transferred from single crystals to polycrystalline stoichiometric thin films of $(V_{0.95}Cr_{0.05})_2O_3$ deposited by Ar/O_2 reactive magnetron sputtering of V and Cr targets [100].

Electrical performances were evaluated on symmetric Metal/Insulator/Metal (MIM) TiN / $(V_{0.95}Cr_{0.05})_2O_3$ / TiN devices with via diameters in the 330–1600 nm range—see Fig. 10.21 [145]. The main results such as programming current, endurance, scalability, and retention times are displayed in Fig. 10.22 [145]. The application of unipolar 3.5 V/100 ns pulses to a device with 330 nm vias and a 100 nm thick $(V_{0.95}Cr_{0.05})_2O_3$ film triggers resistive switching cycles with an average R_{OFF}/R_{ON} ratio of 50, without any

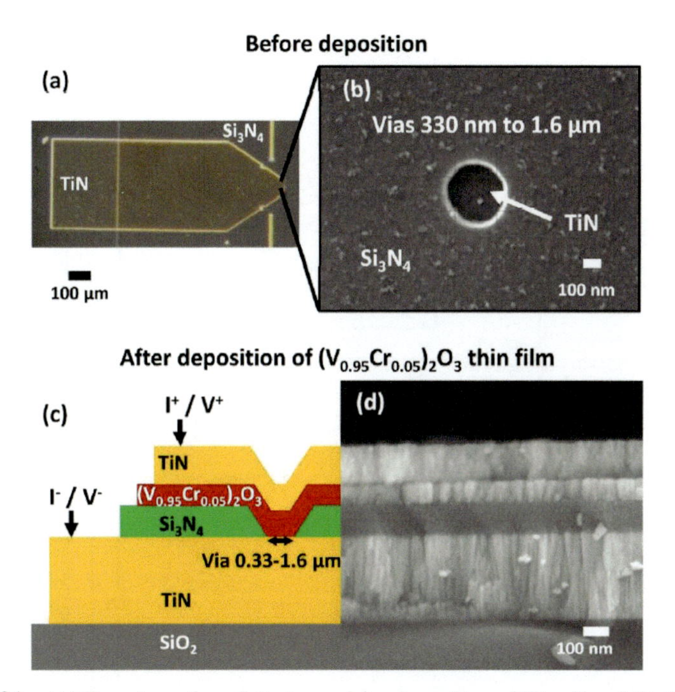

FIG. 10.21 (A) Top view of a substrate used for deposition of $(V_{0.95}Cr_{0.05})_2O_3$ thin films and (B) SEM picture of a bottom electrode with via of 330 nm in diameter. (C) Cross-sectional schematic view of the device based on $(V_{0.95}Cr_{0.05})_2O_3$ and (D) corresponding back scattered electron SEM picture of the multi-layer. *Reproduced with permission from M. Querré,- E. Janod, L. Cario, J. Tranchant, B. Corraze, V. Bouquet, S. Deputier, S. Cordier, M. Guilloux-Viry,- M.-P. Besland, Metal–insulator transitions in (V1-XCrx)2O3 thin films deposited by reactive direct current magnetron Co-sputtering, Thin Solid Films 617 (Part B) (2016) 56, Copyright 2018 IEEE.*

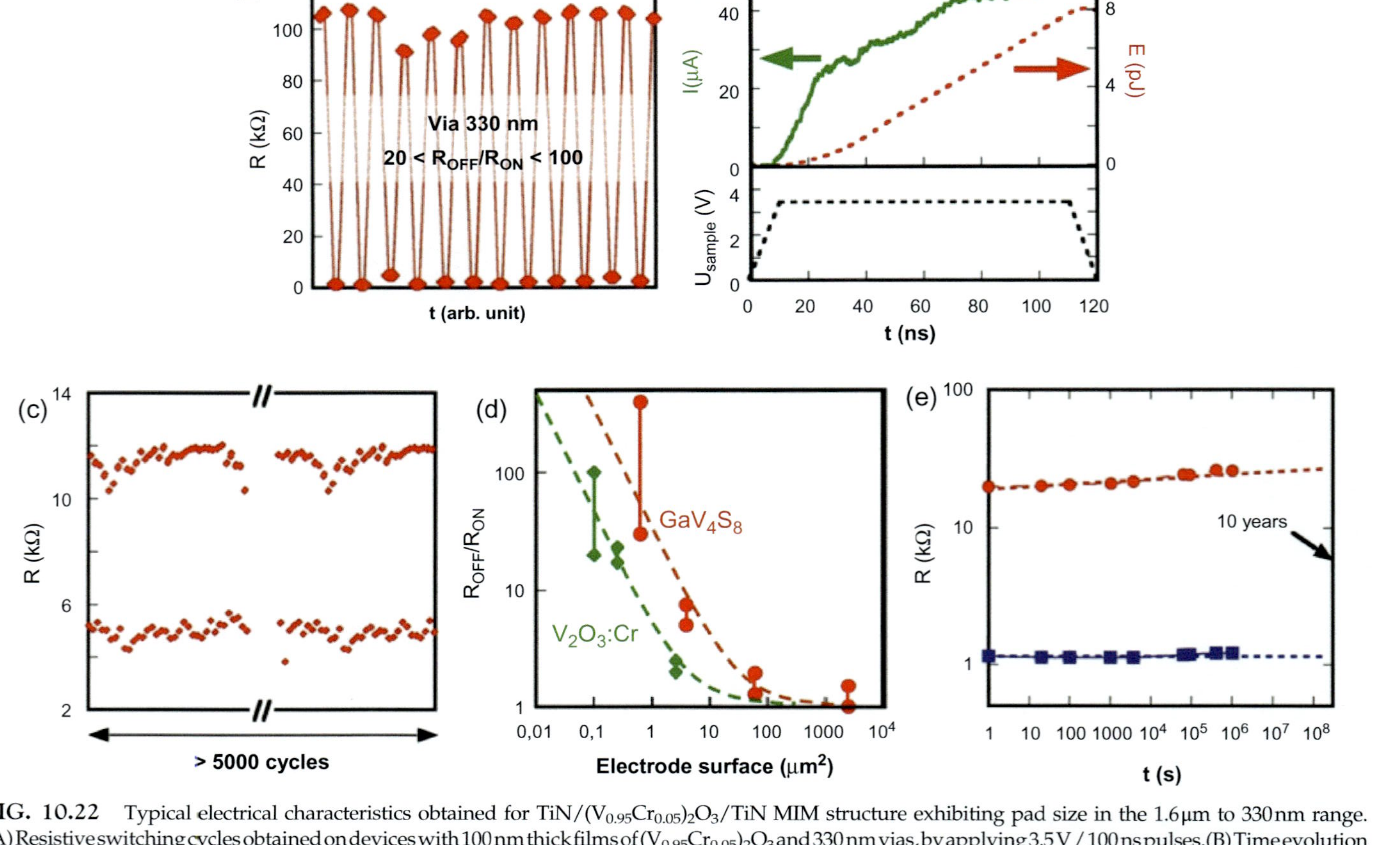

FIG. 10.22 Typical electrical characteristics obtained for TiN/$(V_{0.95}Cr_{0.05})_2O_3$/TiN MIM structure exhibiting pad size in the 1.6μm to 330nm range. (A) Resistive switching cycles obtained on devices with 100nm thick films of $(V_{0.95}Cr_{0.05})_2O_3$ and 330nm vias, by applying 3.5V / 100ns pulses. (B) Time evolution of current (I_{max} = 48μA) and energy (E = 8pJ) during a 3.5V / 100ns pulse inducing one of the OFF→ON transitions in Fig. 10.4A. (C) Cycling endurance obtained on a device with a 1.6μm via and a 880nm thick film demonstrates more than 5000 RS cycles. After this endurance test, RS cycles with a similar R_{OFF}/R_{ON} ratio than Fig. 10.4C were obtained on this device. (D) Evolution of R_{OFF}/R_{ON} ratio with the electrode surface for thin films of the Mott insulators GaV_4S_8 and $(V_{0.95}Cr_{0.05})_2O_3$. (E) Data retention at room temperature, extrapolated to 10 years, for OFF (red circles) and ON (blue squares) resistance states on a 880nm thick film of $(V_{0.95}Cr_{0.05})_2O_3$. *Adapted with permission from J. Tranchant, E. Janod, L. Cario, B. Corraze, E. Souchier, J.-L. Leclercq, P. Cremillieu, P. Moreau, M.-P. Besland, Electrical characterizations of resistive random access memory devices based on GaV4S8 thin layers, Thin Solid Films 533 (2013) 61, Copyright 2018 IEEE.*

forming step before cycling (see Fig. 10.22A). In this case a programming current of 48 μA with an energy per pulse of 8 pJ was sufficient to induce the RS (Fig. 10.22B). Endurance measured on another $(V_{0.95}Cr_{0.05})_2O_3$ based device exceeds 5000 successive cycles, as shown in Fig. 10.22C.

The downscaling properties were also investigated on MIM structures by reducing the electrode via diameter from 1.6 μm down to 330 nm. As displayed in Fig. 10.22D the R_{OFF}/R_{ON} ratio strongly increases with decreasing pad area and reaches values up to 100 for via diameter of 330 nm. This scaling of R_{OFF}/R_{ON} ratio with $1/S$ is consistent with previous observations on GaV_4S_8 thin films [150]. This result can be easily explained considering the filamentary model depicted in Section 10.5. As long as the cycling involves the creation/full dissolution of a single filament, R_{OFF} is indeed expected to scale with the inverse of the pad area $1/S$ while R_{ON} is expected to depend only on the resistance of the filamentary conducting path expected to be independent of the pad area. As a consequence, R_{OFF}/R_{ON} should increase as $1/S$ for small pads area, which is observed experimentally for areas below 2 μm^2 (Fig. 10.22D) and should keep increasing as long as pad sizes remain larger than the filamentary conducting paths. R_{OFF}/R_{ON} ratios larger than the 10^3–10^4 current values can thus be expected with further pad size downscaling.

The stability of high and low resistive states obtained on MIM structures has been investigated at room temperature. Extrapolation of R_{OFF} and R_{ON} to 10 years shows respectively slight increase and stagnation of these resistance levels (Fig. 10.22E). Both states exhibit therefore good retention times, which is promising for data storage.

An interesting feature of the resistive switching based on the Bandwidth control mechanism is that it results from an avalanche breakdown which is triggered above a threshold electric field (a few kV/cm). Consequently, the SET voltages measured on single crystals (typically 30–50 V for 10–30 μm inter-electrode distance) are much larger than those observed on thin films (down to 3.5 V for 100 nm). This enables a very simple way to tune the SET voltage by changing the thickness of the device. A SET voltage lower than 1 V is expected for sub-30 nm thin films targeted in future devices.

To summarize, the endurance of Mott-RAM devices is very promising compared to values ranging from 10^3 to 10^7 cycles currently obtained in Flash technology [131]. The writing time and the erasing time of 100 ns are favorable compared to characteristics achieved in Flash technology, i.-e., writing time of 1 μs and even much better than the typical erasing times of 10 ms. In addition, the writing/erasing voltage envisioned in future devices in the 1 V range stands as a huge advantage when compared to the 12 V reported for Flash memories [131]. Among other ReRAM emerging technologies, Mott-insulator based ReRAM devices could be thus considered as really promising candidates to take over the Flash technology.

10.5.2 New components based on Mott insulators for neurocomputing

During the last half century, the tremendous development of computers has led to the revolution of the information technology. Nevertheless, the way computers store and process the information has scarcely changed since their inception and relies on the concepts proposed by von Neumann in the 40's. This von Neumann architecture, based on a clear separation between the memory and the processing unit, is extremely powerful in many cases such as high-speed processing of large data streams. However, von Neumann computers are outperformed by the mammal brain in numerous data-processing applications such as pattern recognition and data mining [144,151–154]. In fact, the brain is organized with a very different architecture, based on a network of closely connected neurons and synapses. Neuromorphic engineering aims to mimic brain-like behavior through the implementation of artificial neural networks based on the combination of a large number of artificial neurons massively interconnected by an even larger number of artificial synapses [144,152]. In most cases, artificial neural networks are software-implemented in conventional hardware; they are programmed in computers with standard architectures, i.e., on von Neumann architectures. In order to effectively implement artificial neural networks directly in hardware it is mandatory to develop artificial neurons and synapses as downscalable components [153,154]. In that context, Mott insulators were recently investigated to build up this new types of components.

10.5.2.1 Artificial synapses

In biological neuron networks, an action potential is generated in the form of an electrical spike by a pre-synaptic neuron. This signal propagates via the axon through a synapse to a post-synaptic neuron, as schematically displayed in Fig. 10.23A [155]. The synaptic conductance (or weight) is dynamically tuned by the delay between pre-synaptic and post-synaptic signals. The ability of synapses to modify their weight depending on the history of electrical spikes they have received is called Spike-Timing-Dependent Plasticity (STDP). The works of Strukov et al. [84] have demonstrated that non-volatile multi-level resistive switching memories (ReRAM) can have an analogue behavior as synapses and that such components, called memristors, can implement STDP.

Interestingly, non-volatile multi-level resistive switching has been reported in several Mott insulator systems [130,156]. Hence, the bandwidth-controlled IMT based filamentary RS observed in GaV_4S_8 can lead to multiple resistance states between HRS and LRS (Fig. 10.23B) [130]. As displayed in Fig. 10.23C, the formation/reduction of an oxide layer at the interface between the Al top electrode and a manganite

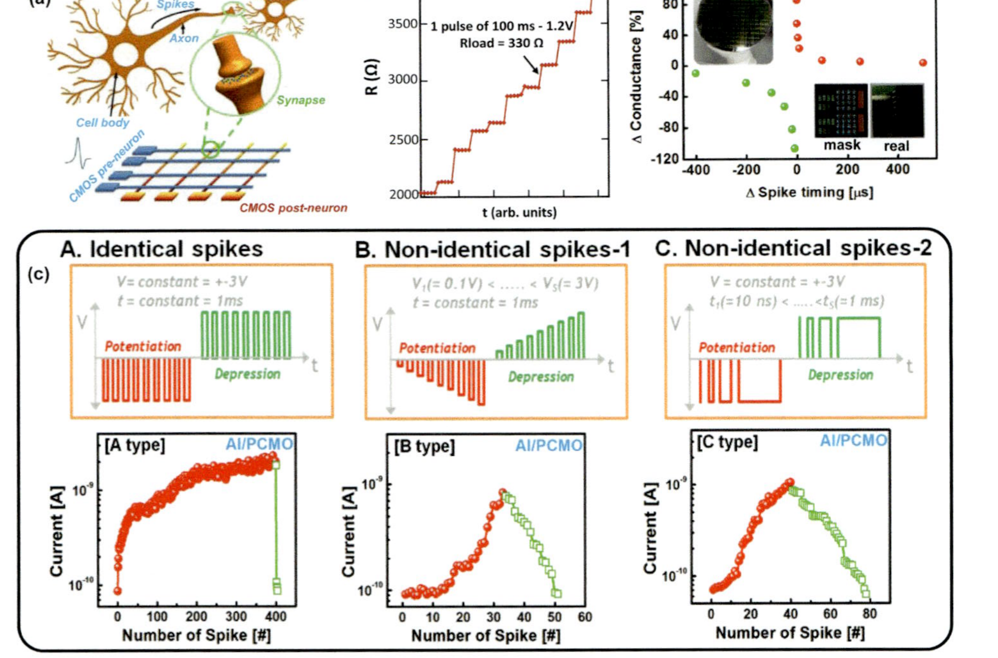

FIG. 10.23 (A) Schematic illustration of a neuromorphic device to emulate the biological neuron system. (B) Control of the intermediate resistance levels achievable from successive reset pulses in a TiN/GaV_4S_8/TiN MIM device. (C) Responding to programming spike (with potentiation and depression) by using various Schemes. A shows the abrupt nature of the reset transition in Al/PCMO devices. Precise control of the resistance was not possible during the pulsed reset process in this device due to large nonlinearity between positive and negative region and different speed of SET/RESET operations. To overcome these issues, various programming schemes are proposed (B and C). (D) STDP characteristic of 1 k-bit ReRAM array. Inset shows the 8-in. wafer (left, top) and photo images of mask and real 1 k-bit ReRAM array. *Panel A Reproduced with permission and adapted from Ref. S. Park, M. Chu, J. Kim, J. Noh, M. Jeon, B. Hun Lee, H. Hwang, B. Lee, B. Lee, Electronic system with memristive synapses for pattern recognition, Sci. Rep. 5 (2015) 10123, Copyright 2018 IEEE; Panel B Reproduced with permission from Ref. B. Das, A. Lele, P. Kumbhare, J. Schulze, U. Ganguly, PrxCa1–XMnO3-based memory and Si time-keeping selector for area and energy efficient synapse, IEEE Electron Device Lett. 40 (2019) 850, Copyright 2017 ECS—The Electrochemical Society; Panel C Adapted with permission from Ref. S. Park, M. Chu, J. Kim, J. Noh, M. Jeon, B. Hun Lee, H. Hwang, B. Lee, B. Lee, Electronic system with memristive synapses for pattern recognition, Sci. Rep. 5 (2015) 10123, Copyright 2018 IEEE; Panel D Reproduced with permission from Ref. Jo, M., Seong, D., Kim, S., Lee, J., Lee, W., Park, J.-B., Park, S., Jung, S., Shin, J., Lee, D., Hwang, H., 2010 Symposium on VLSI Technology. 53 (2010), Copyright 2012 IEEE.*

($Pr_{0.7}Ca_{0.3}MnO_3$, PCMO) layer can also result in many gradual resistance states, that can be controlled via various pulsing strategies to generate potentiation (synaptic weight increase) or depression (synaptic weight decrease) in a PCMO-based 1 kbit crosspoint array [155]. Such programming strategies can be applied to mimic STDP in PCMO, as shown in Fig. 10.23D.

Finally, PCMO-based artificial synapses were implemented in more complex architectures of Hardware Neural Networks (HNN) and have demonstrated capabilities in many brain-inspired activities, such as speech recognition [155], learning and recognition of human thought patterns using electroencephalography signals generated while a subject imagines speaking vowels [157] or hand-written digit recognition [158]. Besides, STDP implementation through such neuromorphic systems result in high energy efficiency [158] and can lead to 1000 times acceleration over biological systems [143]. Due to their potential for ultra-high-integration density, Mott insulator based synapses can then be considered as very promising candidates for neural network computation systems in the era of "Big Data" and "Internet of things".

10.5.2.2 Artificial neurons

Neurons in the mammal brain fulfill three main functions called Leaky, Integrate and Fire. The neurons receive electric pulses coming from other neurons, which contribute to enhance their membrane potential (integrate function). The membrane potential relaxes between incoming pulses, a feature called "Leaky." Above a threshold value of the membrane, the neuron emits an action potential toward other neurons, an event called "Fire." Lapicque already realized in 1907 that the neuron behaves like a parallel RC circuit before the Fire event. His model, called Leaky Integrate and Fire (LIF), states that incoming spikes will charge the capacitor with electrical charges, which will then "leak" through the resistor in parallel [159,160]. When the membrane potential reaches a given threshold the neuron fires an output electric spike. Interestingly, artificial neural networks (ANN) using discrete spikes to compute and transmit information (Spiking Neural Networks) are technologically appealing thanks to their high energy efficiency. The building block of such ANNs is usually a Leaky Integrate and Fire artificial neuron [34] but more complex neuron model may be also used such as the Hodgkin Huxley model that is based on a more realistic description of the neuronal behavior.

Due to their unconventional behavior under electric pulse Mott insulators were early on considered to realize artificial neurons either of the Hodgkin Huxley or of the LIF types.

10.5.2.2.1 Neuristor based on NbO_2

The Hodgkin–Huxley model is a mathematical model describing the action potential generated thanks to the diffusion of two types of ions, Na^+ and K^+, across the membrane of a biological neuron. In 2012, Pickett and co-workers proposed to use the Resistive Switching property of NbO_2 to emulate an electronic analog of the Hodgkin-Huxley model [161]. The proposed device includes two nanoscale NbO_2 samples (one mimicking a part of the Na^+ channel, the other the K^+ channel) associated with several other capacitors and resistors. This device is called neuristor, in reference to the seminal work of H. Crane in the early 60s [162]. Importantly it was shown that the spiking behavior of the NbO_2-based neuristor, i.e., the emission of a spike-like output signal when an input signal exceeds a threshold voltage, can be used to form the components necessary for a Turing machine [163]. These results are a proof-of-principle evidencing that a network of resistive switching entities can be used to implement universal computing.

Interestingly the neuristor proposed in Ref. [161] is based on the volatile resistive switching of the strongly correlated material NbO_2. The mechanism initially proposed to explain this switching involved a thermally induced "electronic *phase* transition," i.e., a local Joule heating sufficiently strong to reach and exceed the insulator-to-metal and structural phase transition temperature at $T_{IMT}=1070\,K$. However, subsequent studies have definitively proved that the local temperature remains lower than T_{IMT} during the resistive switching process [81,82]. Therefore the operation of the NbO_2-based neuristor does not rest on a structural and electronic phase transition as initially proposed, but on a thermal runaway phenomenon.

Finally, compared to traditional but more complex Complementary-Metal-Oxide-Semiconductor (CMOS) transistor-based implementation of a neuristor [164], the NbO_2-based neuristor has an interesting potential for miniaturization. However, it still contains at least seven components including two capacitors [165], the latter remaining hardly downscalable. The NbO_2-based neuristor hence represents an important step toward beyond CMOS neurocomputing, but remains quite distant from the ideal artificial neuron which must be both single component and downscalable to dimensions lower than 20–30 nm.

10.5.2.2.2 Leaky integrate and fire neuron based on Mott insulators

Recently it was demonstrated that narrow gap Mott insulator behaves, under a train of electric pulses, as leaky integrator system [166,167]. This is illustrated in Fig. 10.24 that shows the response of the Mott insulator $GaTa_4Se_8$ to trains of short pulses that varies in duration t_{ON} or separation t_{OFF}. The Mott insulator appears as an imperfect integrator of the signal. Indeed for a given V_{PULSE}, raising t_{ON} or t_{OFF} induces opposite effects on the leaky integration and lead respectively to a decrease or an increase of N_{FIRE} (Fig. 10.24B and C).

Based on a modeling work it was proposed that for Mott insulators the accumulation of correlated metal sites plays a similar role as the

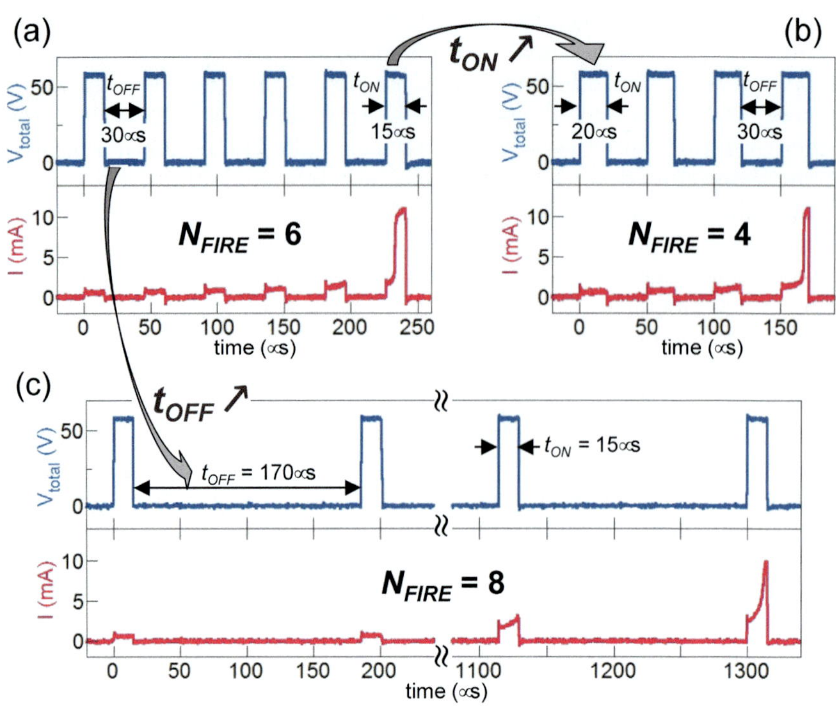

FIG. 10.24 Experimental resistive switching obtained by applying trains of short pulses of various t_{ON} and t_{OFF} on a $GaTa_4Se_8$ sample at 78 K. Using $t_{ON}=15\,\mu s$ and $t_{OFF}=30\,\mu s$ leads to $N_{FIRE}=6$ (A). Increasing t_{ON} to $20\,\mu s$ leads to a decrease of N_{FIRE} to 4 (B). Increasing t_{OFF} to $170\,\mu s$ leads to an increase of N_{FIRE} to 8 (C). *Reproduced with permission from C. Adda, J. Tranchant, P. Stoliar, B. Corraze, E. Janod, E. Gay; E. Llopis, M.P. Besland, L.E. Hueso, L. Cario, An artificial neuron founded on resistive switching of Mott insulators, in:* 2017 IEEE International Memory Workshop (IMW), *2017, pp. 1–4, Copyright 2017 IEEE.*

accumulation of charges in the cellular membrane of the neuron [[166,167]]. Fig. 10.25 compares schematically the behavior of a biological neuron and of the LIF artificial neuron made with a Mott insulator. The Electric Mott transition opens therefore a unique opportunity to implement the Leaky integrate and Fire (LIF) functionalities and build up a LIF artificial neuron in a very simple manner. This is of great interest in the context of neurocomputing and first miniaturized devices were therefore realized using either the Mott insulators GaV_4S_8 (polished crystals) or $(V_{0.95}Cr_{0.05})_2O_3$ (thin films) [168]. Fig. 10.26 shows the device prepared from the $(V_{0.95}Cr_{0.05})_2O_3$ thin films. The electrical measurements performed on this sample with train of pulses are similar to those obtained on crystals and demonstrates that miniaturized LIF artificial neurons may be downscalable and used in artificial neural networks.

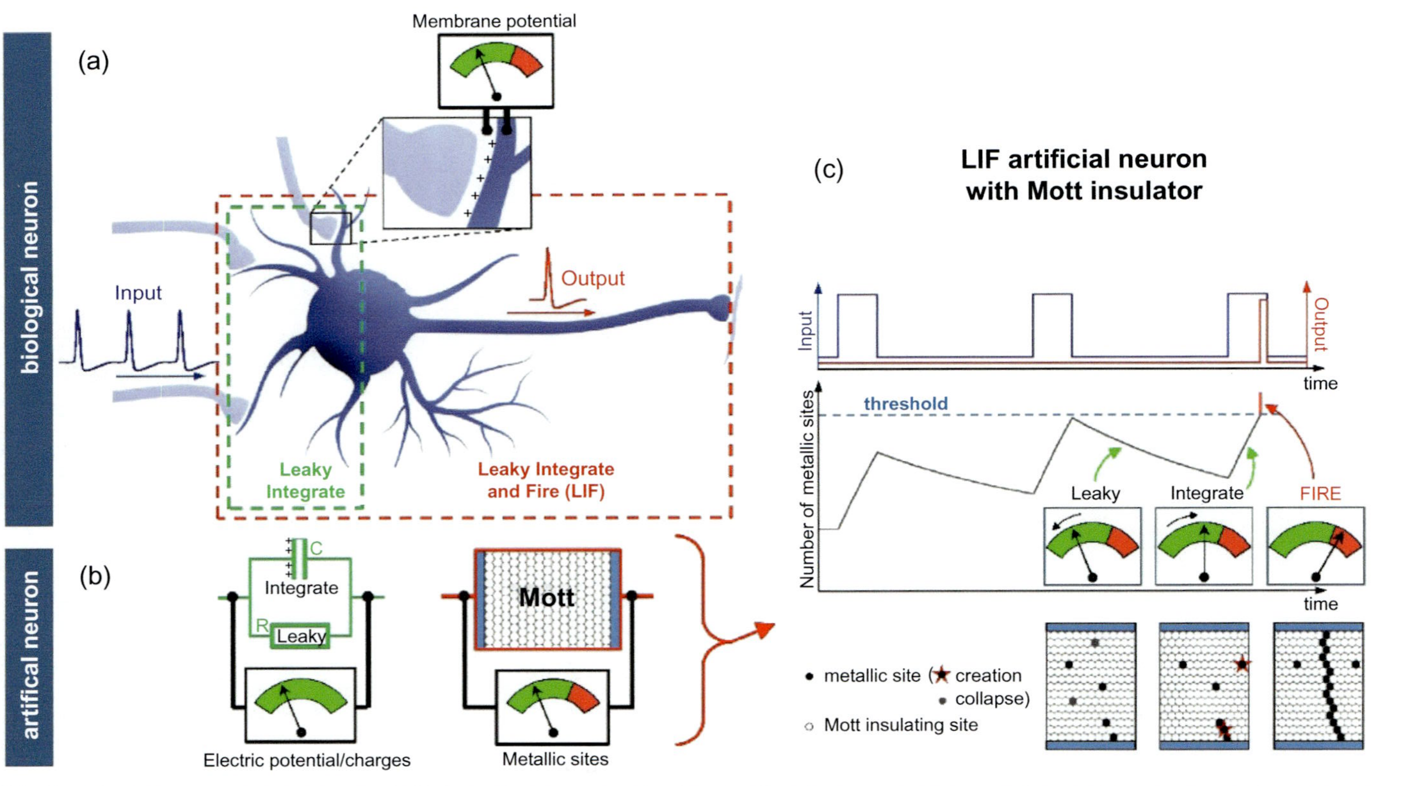

FIG. 10.25 Schematic representation of a biological neuron receiving input spikes from other neurons and triggering an ouput action potential when the membrane potential reaches the threshold value (A). The LIF artificial neurons based on Lapicke's model reproduces the evolution of the membrane potential thanks to an RC circuit accumulating electrical charges (B). The Mott artificial neuron sketched in (C) reproduces the LIF behavior thanks to the accumulation of correlated metallic sites and triggers and output signal when a filamentary path is created within the sample. *Adapted from P. Stoliar, J. Tranchant, B. Corraze, E. Janod, M.-P. Besland, F. Tesler, M. Rozenberg, L. Cario, A leaky-integrate-and-fire neuron analog realized with a Mott insulator, Adv. Funct. Mater. 27 (2017) 1604740, Copyright 2015, WILEY-VCH Verlag GmbH & Co.*

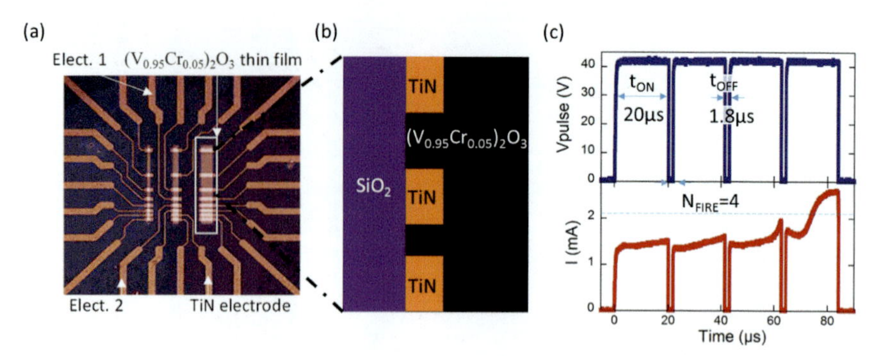

FIG. 10.26 (A) Picture of the miniaturized neuron made of a $(V_{0.95}Cr_{0.05})_2O_3$. Three bands of thin film with various width (80, 40 and 20 μm) were deposited on Si/SiO_2 substrates holding TiN electrodes with various inter-electrodes distances (2, 3, 5, 20, 40, 80, and 160 μm). (B) Schematic cross-sectional view of the stacking of the device. (C) LIF behavior observed on the $(V_{0.95}Cr_{0.05})_2O_3$ planar device. *Reproduced with permission from C. Adda,- L. Cario, J. Tranchant, E. Janod, M.-P. Besland, M. Rozenberg, P. Stoliar, B. Corraze, First demonstration of "leaky integrate and fire" artificial neuron behavior on (V0.95Cr0.05)2O3 thin film, MRS Commun. 8 (2018) 835, Copyright 2018, Material Research Society.*

10.6 Conclusion

Mott insulators are systems with an integer number of unpaired electrons that should be metallic according to standard band structure calculations but are insulating owing to strong electronic correlations. The most prominent examples of these type of systems are $(V_{1-x}Cr_x)_2O_3$, $NiS_{2-x}Se_x$ and AM_4Q_8 compounds. According to the literature there are three main ways to destabilize the Mott insulating state which consist in either applying pressure (Bandwidth-Controlled IMT), changing the temperature in the vicinity of the Mott transition line (Temperature-controlled IMT) or doping the system away from half filling (Filling-controlled IMT). But recently another way to destabilize the Mott insulating state was reported consisting in applying an electric field. Under this tuning parameter, Mott insulators may experience an IMT also called Electric Mott Transition whose main signature is a volatile or non-volatile resistive switching. This way to induce an IMT in Mott insulator is very promising in terms of applications as the electric field is the main control parameter in microelectronic devices. It may therefore open the door to new electronics based on Mott insulators, the long sought after Mottronics.

Three types of EMT's mechanisms may be encountered depending on the type of IMT involved in the change of resistance. A first type of EMT can be explained by a Joule heating induced Temperature-Controlled IMT. This thermal mechanism of resistive switching is encountered in Mott insulator systems like $(V_{1-x}Cr_x)_2O_3$ ($x \approx 0.01$) and $NiS_{2-x}Se_x$ ($x \approx 0.45$) and in many correlated insulators like VO_2, Ca_2RuO_4,or Fe_3O_4. A second type of EMT observed in correlated and Mott Insulators is based on an ionic migration process. In transition metal oxides migration of oxygen under electric field can indeed induce a filling-controlled IMT either along filamentary paths or at the oxide-metal electrode interface. This type of resistive switching first described in band insulators like $SrTiO_3$ is called VCM for Valence Change Mechanism. Filamentary VCM type resistive switching occurs in Mott insulators like NiO while interfacial VCM type resistive switching occurs for various metal- insulator junctions made of correlated materials like $Pr_{0.7}Ca_{0.3}MnO_3$ or $YBa_2Cu_3O_{7-x}$. Finally, the last type of EMT mechanism is related to the creation of hot carriers under electric field leading to an electronic avalanche breakdown. This avalanche breakdown induces the collapse of the Mott insulating state at the local scale and triggers the formation of filamentary conducting paths with compressed metallic domains. For this reason this type of EMT is closely related to the Bandwidth-Controlled IMT observed under physical pressure. This type of EMT was found in many inorganic Mott insulators like $(V_{1-x}Cr_x)_2O_3$, $NiS_{2-x}Se_x$ and the AM_4Q_8 family and also in molecular Mott insulators like κ-$(BEDT\text{-}TTF)_2Cu[N(CN)_2]Br$, and $[Au(Et\text{-}thiazdt)_2]$.

EMT are interesting in the context of Resistive RAM. In the case of a non-volatile resistive switching the Mott insulators may indeed be used to develop a new type of ReRAM called the Mott memory. Two mechanisms of EMT are currently investigated to build up Mott memory devices. They use either the electromigration process inducing Filling Controlled IMT or the avalanche breakdown inducing Bandwidth-Controlled IMT. In both cases promising performances are observed compared to the Flash technology used nowadays with writing and erasing times reduced by more than three orders of magnitude, and switching voltages reduced by a factor of 5. Without need for a forming step, this type of memories display other promising features compared to existing emerging memory technologies such as low writing current, and promising endurance or reliable data retention up to 10 years. On the other hand, EMT is also very interesting in the context of neuromorphic components. Downscalable artificial synapses were realized based on resistive switching involving a filling controlled IMT in Mott insulators. On the other hand the volatile resistive switching observed in Mott insulator due to Temperature-Controlled IMT or the avalanche breakdown mechanisms were used to build up artificial neurons. Mott insulators shows therefore great potential to build up efficient memory and neuromorphic components and more generally new electronic devices based on the electric Mott transition.

References

[1] M. Imada, A. Fujimori, Y. Tokura, Metal-insulator transitions, Rev. Mod. Phys. 70 (1998) 1039.

[2] E. Janod, J. Tranchant, B. Corraze, M. Querré, P. Stoliar, M. Rozenberg, T. Cren, D. Roditchev, V.T. Phuoc, M.-P. Besland, L. Cario, Resistive switching in Mott insulators and correlated systems, Adv. Funct. Mater. 25 (2015) 6287.

[3] F. Pan, S. Gao, C. Chen, C. Song, F. Zeng, Recent Progress in resistive random access memories: materials, switching mechanisms, and performance, Mater. Sci. Eng. R. Rep. 83 (2014) 1.
[4] Y. Zhou, S. Ramanathan, Mott memory and neuromorphic devices, Proc. IEEE 103 (2015) 1289.
[5] Y. Wang, K.-M. Kang, M. Kim, H.-S. Lee, R. Waser, D. Wouters, R. Dittmann, J.J. Yang, H.-H. Park, Mott-transition-based RRAM, Mater. Today 28 (2019) 63.
[6] Y. Tokura, M. Kawasaki, N. Nagaosa, Emergent functions of quantum materials, Nat. Phys. 13 (2017) 11.
[7] J.H. de Boer, E.J.W. Verwey, Semi-conductors with partially and with completely filled 3d-lattice bands, Proc. Phys. Soc. 49 (1937) 59.
[8] N.F. Mott, R. Peierls, Discussion of the paper by de Boer and Verwey, Proc. Phys. Soc. 49 (1937) 72.
[9] N.F. Mott, The basis of the electron theory of metals, with special reference to the transition metals, Proc. Phys. Soc. A 62 (1949) 416.
[10] J. Hubbard, Electron correlations in narrow energy bands. II. The degenerate band case, Proc. R. Soc. Lond. A 277 (1964) 237.
[11] J. Zaanen, G.A. Sawatzky, J.W. Allen, Band gaps and electronic structure of transition-metal compounds, Phys. Rev. Lett. 55 (1985) 418.
[12] A. Georges, G. Kotliar, W. Krauth, M.J. Rozenberg, Dynamical mean-field theory of strongly correlated fermion systems and the limit of infinite dimensions, Rev. Mod. Phys. 68 (1996) 13.
[13] G. Kotliar, S.Y. Savrasov, K. Haule, V.S. Oudovenko, O. Parcollet, C.A. Marianetti, Electronic structure calculations with dynamical mean-field theory, Rev. Mod. Phys. 78 (2006) 865.
[14] G. Kotliar, D. Vollhardt, Strongly correlated materials: insights from dynamical mean-field theory, Phys. Today 57 (2004) 53.
[15] R. Bulla, T.A. Costi, D. Vollhardt, Finite-temperature numerical renormalization group study of the Mott transition, Phys. Rev. B 64 (2001), 045103.
[16] S.R. Hassan, A. Georges, H.R. Krishnamurthy, Sound velocity anomaly at the Mott transition: application to organic conductors and V_2O_3, Phys. Rev. Lett. 94 (2005), 036402.
[17] H. Kuwamoto, J.M. Honig, J. Appel, Electrical properties of the $(V_{1-x}Cr_x)_2O_3$ system, Phys. Rev. B 22 (1980) 2626.
[18] T. Katsufuji, Y. Taguchi, Y. Tokura, Transport and magnetic properties of a Mott-Hubbard system whose bandwidth and band filling are both controllable: $R_{1-x}Ca_xTiO_{3+y/2}$, Phys. Rev. B 56 (1997) 10145.
[19] Y. Taguchi, Y. Tokura, T. Arima, F. Inaba, Change of electronic structures with carrier doping in the highly correlated electron system $Y_{1-x}Ca_xTiO_3$, Phys. Rev. B 48 (1993) 511.
[20] P.P. Edwards, T.V. Ramakrishnan, C.N.R. Rao, The metal-nonmetal transition: a global perspective, J. Phys. Chem. 99 (1995) 5228.
[21] C.-H. Yee, L. Balents, Phase separation in doped Mott insulators, Phys. Rev. X 5 (2015), 021007.
[22] D.B. McWhan, J.P. Remeika, T.M. Rice, W.F. Brinkman, J.P. Maita, A. Menth, Electronic specific heat of metallic Ti-doped V2O3, Phys. Rev. Lett. 27 (1971) 941.
[23] D.B. McWhan, T.M. Rice, J.P. Remeika, Mott transition in Cr-doped V2O3, Phys. Rev. Lett. 23 (1969) 1384.
[24] P. Limelette, A. Georges, D. Jérome, P. Wzietek, P. Metcalf, J.M. Honig, Universality and critical behavior at the Mott transition, Science 302 (2003) 89.
[25] F. Rodolakis, B. Mansart, E. Papalazarou, S. Gorovikov, P. Vilmercati, L. Petaccia, A. Goldoni, J.P. Rueff, S. Lupi, P. Metcalf, M. Marsi, Quasiparticles at the Mott transition in V2O3: wave vector dependence and surface attenuation, Phys. Rev. Lett. 102 (2009), 066805.

[26] A. Jayaraman, D.B. McWhan, J.P. Remeika, P.D. Dernier, Critical behavior of the Mott transition in Cr-doped V_2O_3, Phys. Rev. B 2 (1970) 3751.
[27] D.B. McWhan, J.P. Remeika, Metal-insulator transition in $(V_{1-x}Cr_x)_2O_3$, Phys. Rev. B 2 (1970) 3734.
[28] P. Hansmann, A. Toschi, G. Sangiovanni, T. Saha-Dasgupta, S. Lupi, M. Marsi, K. Held, Mott–Hubbard transition in V_2O_3 revisited, Phys. Status Solidi B 250 (2013) 1251.
[29] C. Castellani, C.R. Natoli, J. Ranninger, Magnetic structure of V_2O_3 in the insulating phase, Phys. Rev. B 18 (1978) 4945.
[30] C. Castellani, C.R. Natoli, J. Ranninger, Metal-insulator transition in pure and Cr-doped V_2O_3, Phys. Rev. B 18 (1978) 5001.
[31] J.-H. Park, L.H. Tjeng, A. Tanaka, J.W. Allen, C.T. Chen, P. Metcalf, J.M. Honig, F.M.F. de Groot, G.A. Sawatzky, Spin and orbital occupation and phase transitions in V2O3, Phys. Rev. B 61 (2000) 11506.
[32] K. Held, G. Keller, V. Eyert, D. Vollhardt, V. Anisimov, Mott-Hubbard metal-insulator transition in paramagnetic V2O3: an LDA+DMFT(QMC) study, Phys. Rev. Lett. 86 (2001) 5345.
[33] G.G. Keller, K. Held, V. Eyert, D. Vollhardt, V.I. Anisimov, Electronic structure of paramagnetic V2O3: strongly correlated metallic and Mott insulating phase, Phys. Rev. B 70 (2004) 205116.
[34] F. Rodolakis, P. Hansmann, J.-P. Rueff, A. Toschi, M.W. Haverkort, G. Sangiovanni, A. Tanaka, T. Saha-Dasgupta, O.K. Andersen, K. Held, M. Sikora, I. Alliot, J.-P. Itié, F. Baudelet, P. Wzietek, P. Metcalf, M. Marsi, Inequivalent routes across the Mott transition in V_{2}O_{3} explored by X-ray absorption, Phys. Rev. Lett. 104 (2010), 047401.
[35] A. Fujimori, K. Mamiya, T. Mizokawa, T. Miyadai, T. Sekiguchi, H. Takahashi, N. Môri, S. Suga, Resonant photoemission study of pyrite-type NiS_2, CoS_2 and FeS_2, Phys. Rev. B 54 (1996) 16329.
[36] J.M. Honig, J. Spałek, Metal–insulator phase transitions and related critical phenomena in $NiS_{2-x}Se_x$, Curr. Opinion Solid State Mater. Sci. 5 (2001) 269.
[37] J.M. Honig, Correlation effects in the $NiS_{2-x}Se_x$ system, J. Solid State Chem. 147 (1999) 68.
[38] J.M. Honig, J. Spałek, Electronic properties of $NiS_{2\text{-}x}Se_x$ single crystals: from magnetic Mott–Hubbard insulators to normal metals, Chem. Mater. 10 (1998) 2910.
[39] J. Kuneš, L. Baldassarre, B. Schächner, K. Rabia, C. Kuntscher, D. Korotin, V. Anisimov, J. McLeod, E. Kurmaev, A. Moewes, Metal-insulator transition in $NiS_{2-x}Se_x$, Phys. Rev. B 81 (2010), 035122.
[40] Y. Sekine, H. Takahashi, N. Môri, T. Matsumoto, T. Kosaka, Effect of pressure on transport properties of $Ni(S_{1-x}Se_x)_2$, Phys. B Condens. Matter 237–238 (1997) 148.
[41] M. Matsuura, H. Hiraka, K. Yamada, Y. Endoh, Magnetic phase diagram and metal-insulator transition of $NiS_{2\text{-}x}Se_x$, J. Phys. Soc. Jpn. 69 (2000) 1503.
[42] T. Miyadai, M. Saitoh, Y. Tazuke, Metal-insulator transition in NiS2 − xSex system—volume effect, J. Magn. Magn. Mater. 104–107 (1992) 1953.
[43] G. Han, S. Choi, H. Cho, B. Sohn, J.-G. Park, C. Kim, Structural investigation of the insulator-metal transition in $NiS_{2\text{-}x}Se_x$ compounds, Phys. Rev. B 98 (2018) 125114.
[44] H. Benyaich, J. Jegaden, M. Potel, M. Sergent, A. Rastogi, R. Tournier, New chalcogenides and chalcohalogenides with tetrahedral Nb_4 or Ta_4 clusters, J. Less-Common Met. 102 (1984) 9.
[45] Pocha, D. Johrendt, R. Pöttgen, Electronic and structural instabilities in GaV4S8 and GaMo4S8, Chem. Mater. 12 (2000) 2882.
[46] V. Guiot, E. Janod, B. Corraze, L. Cario, Control of the electronic properties and resistive switching in the new series of Mott insulators $GaTa_4Se_{8\text{-}Y}Te_y$ ($0 \le y \le 6.5$), Chem. Mater. 23 (2011) 2611.

[47] A. Camjayi, C. Acha, R. Weht, M.G. Rodríguez, B. Corraze, E. Janod, L. Cario, M.J. Rozenberg, First-order insulator-to-metal Mott transition in the paramagnetic 3D system $GaTa_4Se_8$, Phys. Rev. Lett. 113 (2014), 086404.
[48] M.Y. Jeong, S.H. Chang, B.H. Kim, J.-H. Sim, A. Said, D. Casa, T. Gog, E. Janod, L. Cario, S. Yunoki, M.J. Han, J. Kim, Direct experimental observation of the molecular $J_{Eff} = 3/2$ ground state in the lacunar spinel $GaTa_4Se_8$, Nat. Commun. 8 (2017) 782.
[49] H. Lee, M.Y. Jeong, J.-H. Sim, H. Yoon, S. Ryee, M.J. Han, Charge density functional plus U calculation of lacunar spinel GaM_4Se_8 (M = Nb, Mo, Ta, and W), EPL 125 (2019) 47005.
[50] R. Pocha, D. Johrendt, B. Ni, M.M. Abd-Elmeguid, Crystal structures, electronic properties, and pressure-induced superconductivity of the tetrahedral cluster compounds $GaNb_4S_8$, $GaNb_4Se_8$, and $GaTa_4Se_8$, J. Am. Chem. Soc. 127 (2005) 8732.
[51] H. Müller, W. Kockelmann, D. Johrendt, The magnetic structure and electronic ground states of Mott insulators GeV_4S_8 and GaV_4S_8, Chem. Mater. 18 (2006) 2174.
[52] M.M. Abd-Elmeguid, B. Ni, D.I. Khomskii, R. Pocha, D. Johrendt, X. Wang, K. Syassen, Transition from Mott insulator to superconductor in $GaNb_4Se_8$ and $GaTa_4Se_8$ under high pressure, Phys. Rev. Lett. 93 (2004).
[53] V. Ta Phuoc, C. Vaju, B. Corraze, R. Sopracase, A. Perucchi, C. Marini, P. Postorino, M. Chligui, S. Lupi, E. Janod, L. Cario, Optical conductivity measurements of $GaTa_4Se_8$ under high pressure: evidence of a bandwidth-controlled insulator-to-metal Mott transition, Phys. Rev. Lett. 110 (2013), 037401.
[54] a) L.C. Dorolti, B. Corraze, E. Janod, C. Vaju, H.-J. Koo, E. Kan, M.-H. Whangbo, Half-metallic ferromagnetism and large negative magnetoresistance in the new lacunar spinel GaTi3VS8, J. Am. Chem. Soc. 132 (2010) 5704. b) C. Vaju, J. Martial, E. Janod, B. Corraze, V. Fernandez, L. Cario, Metal – metal bonding and correlated metallic behavior in the new deficient spinel $Ga_{0.87}Ti_4S_8$, Chem. Mater. 20 (2008) 2382. c) E. Janod, E. Dorolti, B. Corraze, V. Guiot, S. Salmon, V. Pop, F. Christien, L. Cario, Negative colossal magnetoresistance driven by carrier type in the ferromagnetic Mott insulator GaV4S8, Chem. Mater. 27 (2015) 4398.
[55] K. Singh, C. Simon, E. Cannuccia, M.-B. Lepetit, B. Corraze, E. Janod, L. Cario, Orbital-ordering-driven multiferroicity and magnetoelectric coupling in GeV4S8, Phys. Rev. Lett. 113 (2014) 137602.
[56] I. Kézsmárki, S. Bordács, P. Milde, E. Neuber, L.M. Eng, J.S. White, H.M. Rønnow, C.D. Dewhurst, M. Mochizuki, K. Yanai, H. Nakamura, D. Ehlers, V. Tsurkan, A. Loidl, Néel-type skyrmion lattice with confined orientation in the polar magnetic semiconductor GaV_4S_8, Nat. Mater. 14 (2015) 1116.
[57] Á. Butykai, S. Bordács, I. Kézsmárki, V. Tsurkan, A. Loidl, J. Döring, E. Neuber, P. Milde, S.C. Kehr, L.M. Eng, Characteristics of ferroelectric-ferroelastic domains in Néel-type Skyrmion host GaV 4 S 8, Sci. Rep. 7 (2017) 1.
[58] W.-P. Hsieh, M. Trigo, D.A. Reis, G.A. Artioli, L. Malavasi, W.L. Mao, Evidence for photo-induced monoclinic metallic VO_2 under high pressure, Appl. Phys. Lett. 104 (2014), 021917.
[59] H. Wen, L. Guo, E. Barnes, J.H. Lee, D.A. Walko, R.D. Schaller, J.A. Moyer, R. Misra, Y. Li, E.M. Dufresne, D.G. Schlom, V. Gopalan, J.W. Freeland, Structural and electronic recovery pathways of a photoexcited ultrathin VO_2 film, Phys. Rev. B 88 (2013) 165424.
[60] T.F. Qi, O.B. Korneta, S. Parkin, L.E. De Long, P. Schlottmann, G. Cao, Negative volume thermal expansion via orbital and magnetic orders in $Ca_2Ru_{1-x}Cr_xO_4$ ($0 < x < 0.13$), Phys. Rev. Lett. 105 (2010) 177203.
[61] C. Weber, D.D. O'Regan, N.D.M. Hine, M.C. Payne, G. Kotliar, P.B. Littlewood, Vanadium dioxide: a Peierls-Mott insulator stable against disorder, Phys. Rev. Lett. 108 (2012) 256402.
[62] V. Eyert, The metal-insulator transition of NbO_2: an embedded Peierls instability, EPL 58 (2002) 851.

[63] J.B. Torrance, P. Lacorre, A.I. Nazzal, E.J. Ansaldo, C. Niedermayer, Systematic study of insulator-metal transitions in perovskites $RNiO_3$ (R=Pr,Nd,Sm,Eu) due to closing of charge-transfer gap, Phys. Rev. B 45 (1992) 8209.
[64] H. Park, A.J. Millis, C.A. Marianetti, Site-selective Mott transition in rare-earth-element nickelates, Phys. Rev. Lett. 109 (2012) 156402.
[65] J. Li, C. Aron, G. Kotliar, J.E. Han, Electric-field-driven resistive switching in the dissipative Hubbard model, Phys. Rev. Lett. 114 (2015) 226403.
[66] F.A. Chudnovskii, A.L. Pergament, G.B. Stefanovich, P.A. Metcalf, J.M. Honig, Switching phenomena in chromium-doped vanadium sesquioxide, J. Appl. Phys. 84 (1998) 2643.
[67] F.A. Chudnovskii, A.L. Pergament, P. Somasundaram, J.M. Honig, Delay time measurements of $NiS_{2-x}Se_x$-based switches, Phys. Status Solidi A 172 (1999) 131.
[68] F.A. Chudnovskii, A.L. Pergament, G.B. Stefanovich, P. Somasundaram, J.M. Honig, N-type negative resistance in $M/NiS_{2-x}Se_x/M$ structures, Phys. Status Solidi A 161 (1997) 577.
[69] J. Kim, C. Ko, A. Frenzel, S. Ramanathan, J.E. Hoffman, Nanoscale imaging and control of resistance switching in VO_2 at room temperature, Appl. Phys. Lett. 96 (2010) 213106.
[70] A. Zimmers, L. Aigouy, M. Mortier, A. Sharoni, S. Wang, K.G. West, J.G. Ramirez, I.K. Schuller, Role of thermal heating on the voltage induced insulator-metal transition in VO_2, Phys. Rev. Lett. 110 (2013), 056601.
[71] F. Nakamura, M. Sakaki, Y. Yamanaka, S. Tamaru, T. Suzuki, Y. Maeno, Electric-field-induced metal maintained by current of the Mott insulator Ca_2RuO_4, Sci. Rep. 3 (2013) 2536.
[72] J.S. Brockman, L. Gao, B. Hughes, C.T. Rettner, M.G. Samant, K.P. Roche, S.S.P. Parkin, Subnanosecond incubation times for electric-field-induced metallization of a correlated electron oxide, Nat. Nanotechnol. 9 (2014) 453.
[73] S. Guénon, S. Scharinger, S. Wang, J.G. Ramírez, D. Koelle, R. Kleiner, I.K. Schuller, Electrical breakdown in a V_2O_3 device at the insulator-to-metal transition, EPL 101 (2013) 57003.
[74] T. Burch, P.P. Craig, C. Hedrick, T.A. Kitchens, J.I. Budnick, J.A. Cannon, M. Lipsicas, D. Mattis, Switching in magnetite: a thermally driven magnetic phase transition, Phys. Rev. Lett. 23 (1969) 1444.
[75] A.A. Fursina, R.G.S. Sofin, I.V. Shvets, D. Natelson, The origin of hysteresis in resistive switching in magnetite is joule heating, Phys. Rev. B 79 (2009) 245131.
[76] T. Driscoll, H.-T. Kim, B.-G. Chae, M. Di Ventra, D.N. Basov, Phase-transition driven memristive system, Appl. Phys. Lett. 95 (2009), 043503.
[77] S.-H. Bae, S. Lee, H. Koo, L. Lin, B.H. Jo, C. Park, Z.L. Wang, The Memristive properties of a single VO_2 nanowire with switching controlled by self-heating, Adv. Mater. 25 (2013) 5098.
[78] M.D. Pickett, R. StanleyWilliams, Sub-100 fJ and sub-nanosecond thermally driven threshold switching in niobium oxide crosspoint nanodevices, Nanotechnology 23 (2012) 215202.
[79] M.-J. Lee, Y. Park, D.-S. Suh, E.-H. Lee, S. Seo, D.-C. Kim, R. Jung, B.-S. Kang, S.-E. Ahn, C.B. Lee, D.H. Seo, Y.-K. Cha, I.-K. Yoo, J.-S. Kim, B.H. Park, Two series oxide resistors applicable to high speed and high density nonvolatile memory, Adv. Mater. 19 (2007) 3919.
[80] S. Kim, J. Park, J. Woo, C. Cho, W. Lee, J. Shin, G. Choi, S. Park, D. Lee, B.H. Lee, H. Hwang, Threshold-switching characteristics of a nanothin-NbO_2-layer-based Pt/NbO_2/Pt stack for use in cross-point-type resistive memories, Microelectron. Eng. 107 (2013) 33.
[81] S. Kumar, Z. Wang, N. Davila, N. Kumari, K.J. Norris, X. Huang, J.P. Strachan, D. Vine, A.L.D. Kilcoyne, Y. Nishi, R.S. Williams, Physical origins of current and temperature controlled negative differential resistances in NbO_2, Nat. Commun. 8 (2017) 1.

[82] G.A. Gibson, Designing negative differential resistance devices based on self-heating, Adv. Funct. Mater. 28 (2018) 1704175.
[83] W.R. Waser, M. Aono, Nanoionics-based resistive switching memories, Nat. Mater. 6 (2007) 833.
[84] D.B. Strukov, G.S. Snider, D.R. Stewart, R.S. Williams, The missing memristor found, Nature 453 (2008) 80.
[85] A. Sawa, Resistive switching in transition metal oxides, Mater. Today 11 (2008) 28.
[86] H.S. Lee, S.G. Choi, H.-H. Park, M.J. Rozenberg, A new route to the Mott-Hubbard metal-insulator transition: strong correlations effects in $Pr_{0.7}Ca_{0.3}MnO_3$, Sci. Rep. 3 (2013) 1704.
[87] a S.Q. Liu, N.J. Wu, A. Ignatiev, Electric-pulse-induced reversible resistance change effect in magnetoresistive films, Appl. Phys. Lett. 76 (2000) 2749. b A. Sawa, T. Fujii, M. Kawasaki, Y. Tokura, Hysteretic current–voltage characteristics and resistance switching at a rectifying Ti / Pr0.7Ca0.3MnO3 Interface, Appl. Phys. Lett. 85 (2004) 4073.
[88] a) M.J. Rozenberg, M.J. Sánchez, R. Weht, C. Acha, F. Gomez-Marlasca, P. Levy, Mechanism for bipolar resistive switching in transition-metal oxides, Phys. Rev. B 81 (2010) 115101. b) C. Acha, M.J. Rozenberg, Non-volatile resistive switching in the dielectric superconductor $YBa_2Cu_3O_{7-\delta}$, J. Phys. Condens. Matter 21 (2009), 045702.
[89] D.C. Kim, S. Seo, S.E. Ahn, D.-S. Suh, M.J. Lee, B.-H. Park, I.K. Yoo, I.G. Baek, H.-J. Kim, E.K. Yim, J.E. Lee, S.O. Park, H.S. Kim, U.-I. Chung, J.T. Moon, B.I. Ryu, Electrical observations of filamentary conductions for the resistive memory switching in NiO films, Appl. Phys. Lett. 88 (2006) 202102.
[90] K. Kinoshtia, T. Okutani, H. Tanaka, T. Hinoki, K. Yazawa, K. Ohmi, S. Kishida, Opposite bias polarity dependence of resistive switching in N-type Ga-doped-ZnO and p-type NiO thin films, Appl. Phys. Lett. 96 (2010) 143505.
[91] T. Yajima, K. Fujiwara, A. Nakao, T. Kobayashi, T. Tanaka, K. Sunouchi, Y. Suzuki, M. Takeda, K. Kojima, Y. Nakamura, K. Taniguchi, H. Takagi, Spatial redistribution of oxygen ions in oxide resistance switching device after forming process, Jpn. J. Appl. Phys. 49 (2010), 060215.
[92] K. Fujiwara, T. Nemoto, M.J. Rozenberg, Y. Nakamura, H. Takagi, Resistance switching and formation of a conductive bridge in metal/binary oxide/metal structure for memory devices, Jpn. J. Appl. Phys. 47 (2008) 6266.
[93] H. Shima, F. Takano, Y. Tamai, H. Akinaga, I.H. Inoue, Synthesis and characterization of Pt/Co–O/Pt trilayer exhibiting large reproducible resistance switching, Jpn. J. Appl. Phys. 46 (2007) L57.
[94] S B Lee, Chae S C, Chang S H, Liu C, Jung C U, Seo S, Kim D-W, J. Korean Phys. Soc. 51 (2007) S96.
[95] S. Zhang, S. Long, W. Guan, Q. Liu, Q. Wang, M. Liu, Resistive switching characteristics of MnO x -based ReRAM, J. Phys. D. Appl. Phys. 42 (2009), 055112.
[96] K.M. Kim, D.S. Jeong, C.S. Hwang, Nanofilamentary resistive switching in binary oxide system: a review on the present status and outlook, Nanotechnology 22 (2011) 254002.
[97] Y. Taguchi, T. Matsumoto, Y. Tokura, Dielectric breakdown of one-dimensional Mott insulators Sr_2CuO_3 and $SrCuO_2$, Phys. Rev. B 62 (2000) 7015.
[98] S. Yamanouchi, Y. Taguchi, Y. Tokura, Dielectric breakdown of the insulating charge-ordered state in $La_{2-x}Sr_xNiO_4$, Phys. Rev. Lett. 83 (1999) 5555.
[99] P. Stoliar, L. Cario, E. Janod, B. Corraze, C. Guillot-Deudon, S. Salmon-Bourmand, V. Guiot, J. Tranchant, M. Rozenberg, Universal electric-field-driven resistive transition in narrow-gap Mott insulators, Adv. Mater. 25 (2013) 3222.
[100] M. Querré, E. Janod, L. Cario, J. Tranchant, B. Corraze, V. Bouquet, S. Deputier, S. Cordier, M. Guilloux-Viry, M.-P. Besland, Metal–insulator transitions in (V1-XCrx) 2O3 thin films deposited by reactive direct current magnetron Co-sputtering, Thin Solid Films 617 (Part B) (2016) 56.

[101] C. Vaju, L. Cario, B. Corraze, E. Janod, V. Dubost, T. Cren, D. Roditchev, D. Braithwaite, O. Chauvet, Electric-pulse-driven electronic phase separation, insulator–metal transition, and possible superconductivity in a Mott insulator, Adv. Mater. 20 (2008) 2760.

[102] L. Cario, C. Vaju, B. Corraze, V. Guiot, E. Janod, Electric-field-induced resistive switching in a family of Mott insulators: towards a new class of RRAM memories, Adv. Mater. 22 (2010) 5193.

[103] I.H. Inoue, M.J. Rozenberg, Taming the Mott transition for a novel Mott transistor, Adv. Funct. Mater. 18 (2008) 2289.

[104] R. Kumai, Y. Okimoto, Y. Tokura, Current-induced insulator-metal transition and pattern formation in an organic charge-transfer complex, Science 284 (1999) 1645.

[105] F. Sabeth, T. Iimori, N. Ohta, Insulator–metal transitions induced by electric field and photoirradiation in organic Mott insulator deuterated κ-$(BEDT\text{-}TTF)_2Cu[N(CN)_2]Br$, J. Am. Chem. Soc. 134 (2012) 6984.

[106] P. Stoliar, P. Diener, J. Tranchant, B. Corraze, B. Brière, V. Ta-Phuoc, N. Bellec, M. Fourmigué, D. Lorcy, E. Janod, L. Cario, Resistive switching induced by electric pulses in a single-component molecular Mott insulator, J. Phys. Chem. C 119 (2015) 2983.

[107] a T. Oka, R. Arita, H. Aoki, Breakdown of a Mott insulator: a nonadiabatic tunneling mechanism, Phys. Rev. Lett. 91 (2003), 066406. b T. Oka, H. Aoki, Dielectric breakdown in a Mott insulator: many-body Schwinger-Landau-Zener mechanism studied with a generalized Bethe ansatz, Phys. Rev. B 81 (2010), 033103. c T. Oka, H. Aoki, Ground-state decay rate for the Zener breakdown in band and Mott insulators, Phys. Rev. Lett. 95 (2005) 137601.

[108] F. Heidrich-Meisner, I. González, K.A. Al-Hassanieh, A.E. Feiguin, M.J. Rozenberg, E. Dagotto, Nonequilibrium electronic transport in a one-dimensional Mott insulator, Phys. Rev. B 82 (2010) 205110.

[109] M. Eckstein, T. Oka, P. Werner, Dielectric breakdown of Mott insulators in dynamical mean-field theory, Phys. Rev. Lett. 105 (2010) 146404.

[110] H. Aoki, N. Tsuji, M. Eckstein, M. Kollar, T. Oka, P. Werner, Nonequilibrium dynamical mean-field theory and its applications, Rev. Mod. Phys. 86 (2014) 779.

[111] V. Guiot, L. Cario, E. Janod, B. Corraze, V.T. Phuoc, M. Rozenberg, P. Stoliar, T. Cren, D. Roditchev, Avalanche breakdown in GaTa4Se8 − xTex narrow-gap Mott insulators, Nat. Commun. 4 (2013) 1.

[112] P. Diener, E. Janod, B. Corraze, M. Querré, C. Adda, M. Guilloux-Viry, S. Cordier, A. Camjayi, M. Rozenberg, M.P. Besland, L. Cario, How a Dc electric field drives Mott insulators out of equilibrium, Phys. Rev. Lett. 121 (2018), 016601.

[113] M.E. Levinshteĭn, J. Kostamovaara, S. Vainshtein, Breakdown Phenomena in Semiconductors and Semiconductor Devices, World Scientific, 2005.

[114] J.L. Hudgins, G.S. Simin, E. Santi, M.A. Khan, An assessment of wide bandgap semiconductors for power devices, IEEE Trans. Power Electron. 18 (2003) 907.

[115] J.L. Hudgins, Wide and narrow bandgap semiconductors for power electronics: a new valuation, J. Electron. Mater. 32 (2003) 471.

[116] H. Fröhlich, N.F. Mott, On the theory of dielectric breakdown in solids, Proc. R. Soc. Lond. A Math. Phys. Sci. 188 (1947) 521.

[117] S. Whitehead, Dielectric Breakdown of Solids, Clarendon Press, 1951.

[118] H. Frohlich, Theory of electrical breakdown in ionic crystals, Proc. R. Soc. Lond. A 160 (1937) 230.

[119] F. Seitz, On the theory of electron multiplication in crystals, Phys. Rev. 76 (1949) 1376.

[120] H. Fröhlich, F. Seitz, Notes on the theory of dielectric breakdown in ionic crystals, Phys. Rev. 79 (1950) 526.

[121] P. Stoliar, M. Rozenberg, E. Janod, B. Corraze, J. Tranchant, L. Cario, Nonthermal and purely electronic resistive switching in a Mott memory, Phys. Rev. B 90 (2014), 045146.

[122] F. Tesler, C. Adda, J. Tranchant, B. Corraze, E. Janod, L. Cario, P. Stoliar, M. Rozenberg, Relaxation of a spiking Mott artificial neuron, Phys. Rev. Appl. 10 (2018), 054001.
[123] V. Dubost, T. Cren, C. Vaju, L. Cario, B. Corraze, E. Janod, F. Debontridder, D. Roditchev, Resistive switching at the nanoscale in the Mott insulator compound GaTa4Se8, Nano Lett. 13 (2013) 3648.
[124] Schindler, S.C.P. Thermadam, R. Waser, M.N. Kozicki, Bipolar and unipolar resistive switching in Cu-doped SiO2, IEEE Trans. Electron Devices 54 (2007) 2762.
[125] A. Pirovano, A.L. Lacaita, A. Benvenuti, F. Pellizzer, R. Bez, Electronic switching in phase-change memories, IEEE Trans. Electron Devices 51 (2004) 452.
[126] M. Querré, J. Tranchant, B. Corraze, S. Cordier, V. Bouquet, S. Députier, M. Guilloux-Viry, M.-P. Besland, E. Janod, L. Cario, Non-volatile resistive switching in the Mott insulator (V1 − xCrx)2O3, Phys. B Condens. Matter 536 (2018) 327.
[127] D. Babich, J. Tranchant, C. Adda, B. Corraze, M.-P. Besland, P. Warnike, D. Bedau, P. Bertoncini, J.-Y. Mevellec, B. Humbert, J. Rupp, T. Hennen, D. Wouters, R. Llopis, L. Cario, E. Janod, Lattice contraction induced by resistive switching in chromium-doped V_2O_3: a hallmark of Mott physics, hal-03224199v2, Nat. Commun., 2021. Submitted for publication.
[128] J. Tranchant, E. Janod, B. Corraze, P. Stoliar, M. Rozenberg, M.-P. Besland, L. Cario, Control of resistive switching in AM_4Q_8 narrow gap Mott insulators: a first step toward neuromorphic applications, Phys. Status Solidi A 212 (2015) 239.
[129] International technology roadmap for semiconductors 2013. www.itrs.net, in Emerging Research Devices and in Emerging Research Materials.
[130] J. Tranchant, J. Sandrini, E. Janod, D. Sacchetto, B. Corraze, M.-P. Besland, J. Ghanbaja, G.D. Micheli, P.-E. Gaillardon, L. Cario, Control of resistive switching in Mott memories based on TiN/AM4Q8/TiN MIM devices, ECS Trans. 75 (2017) 3.
[131] International Technology Roadmap for Semiconductors, Emerging Research Devices, 2011. http://www.itrs.net/.
[132] Y. Fujisaki, Review of emerging new solid-state non-volatile memories, Jpn. J. Appl. Phys. 52 (2013), 040001.
[133] H.-S.P. Wong, S. Raoux, S. Kim, J. Liang, J.P. Reifenberg, B. Rajendran, M. Asheghi, K.E. Goodson, Phase change memory, Proc. IEEE 98 (2010) 2201.
[134] J.-G. Zhu, Magnetoresistive random access memory: the path to competitiveness and scalability, Proc. IEEE 96 (2008) 1786.
[135] J. Lee, H. Choi, D. Seong, J. Yoon, J. Park, S. Jung, W. Lee, M. Chang, C. Cho, H. Hwang, The impact of Al interfacial layer on resistive switching of La0.7Sr0.3MnO3 for reliable ReRAM applications, Microelectron. Eng. 86 (2009) 1933.
[136] M. Hasan, R. Dong, H.J. Choi, D.S. Lee, D.-J. Seong, M.B. Pyun, H. Hwang, Uniform resistive switching with a thin reactive metal interface layer in metal-La0.7-Ca0.3MnO3-metal heterostructures, Appl. Phys. Lett. 92 (2008) 202102.
[137] P. Kumbhare, I. Chakraborty, A. Khanna, U. Ganguly, Memory performance of a simple Pr0.7Ca0.3MnO3-based selectorless RRAM, IEEE Trans. Electron Devices 64 (2017) 3967.
[138] Z.L. Liao, Z.Z. Wang, Y. Meng, Z.Y. Liu, P. Gao, J.L. Gang, H.W. Zhao, X.J. Liang, X.D. Bai, D.M. Chen, Categorization of resistive switching of metal-Pr0.7Ca0.3MnO3-metal devices, Appl. Phys. Lett. 94 (2009) 253503.
[139] C.J. Kim, I.-W. Chen, Effect of top electrode on resistance switching of (Pr, Ca)MnO3 thin films, Thin Solid Films 515 (2006) 2726.
[140] S. Park, H. Kim, M. Choo, J. Noh, A. Sheri, S. Jung, K. Seo, J. Park, S. Kim, W. Lee, J. Shin, D. Lee, G. Choi, J. Woo, E. Cha, J. Jang, C. Park, M. Jeon, B. Lee, B.H. Lee, H. Hwang, RRAM-based synapse for neuromorphic system with pattern recognition function, in: 2012 International Electron Devices Meeting, 2012, pp. 10.2.1–10.2.4.

[141] S.-L. Li, D.S. Shang, J. Li, J.L. Gang, D.N. Zheng, Resistive switching properties in oxygen-deficient $Pr_{0.7}Ca_{0.3}MnO_3$ junctions with active Al top electrodes, J. Appl. Phys. 105 (2009), 033710.
[142] M. Jo, D. Seong, S. Kim, J. Lee, W. Lee, J.-B. Park, S. Park, S. Jung, J. Shin, D. Lee, H. Hwang, 2010 Symposium on VLSI Technology, 2010, p. 53.
[143] B. Das, A. Lele, P. Kumbhare, J. Schulze, U. Ganguly, PrxCa1–XMnO3-based memory and Si time-keeping selector for area and energy efficient synapse, IEEE Electron Device Lett. 40 (2019) 850.
[144] A.K. Jain, J. Mao, K.M. Mohiuddin, Artificial neural networks: a tutorial, Computer 29 (1996) 31.
[145] J. Tranchant, M. Querre, E. Janod, M. Besland, B. Corraze, L. Cario, Mott memory devices based on the mott insulator $(V_{1-x}Cr_x)_2O_3$, in: 2018 IEEE International Memory Workshop (IMW), IEEE, 2018, pp. 1–4.
[146] E. Souchier, L. Cario, B. Corraze, P. Moreau, P. Mazoyer, C. Estournès, R. Retoux, E. Janod, M.-P. Besland, First evidence of resistive switching in polycrystalline GaV4S8 thin layers, Phys. Status Solidi RRL 5 (2011) 53.
[147] E. Souchier, M.-P. Besland, J. Tranchant, B. Corraze, P. Moreau, R. Retoux, C. Estournès, P. Mazoyer, L. Cario, E. Janod, Deposition by radio frequency magnetron sputtering of GaV4S8 thin films for resistive random access memory application, Thin Solid Films 533 (2013) 54.
[148] J. Tranchant, A. Pellaroque, E. Janod, B. Angleraud, B. Corraze, L. Cario, M.-P. Besland, Deposition of GaV_4S_8 thin films by H_2S/Ar reactive sputtering for ReRAM applications, J. Phys. D. Appl. Phys. 47 (2014), 065309.
[149] J. Tranchant, E. Janod, B. Corraze, M.-P. Besland, L. Cario, From resistive switching mechanisms in AM4Q8 Mott insulators to Mott memories, in: Memory Workshop (IMW), 2015 IEEE International, IEEE, 2015, pp. 1–4.
[150] J. Tranchant, E. Janod, L. Cario, B. Corraze, E. Souchier, J.-L. Leclercq, P. Cremillieu, P. Moreau, M.-P. Besland, Electrical characterizations of resistive random access memory devices based on GaV4S8 thin layers, Thin Solid Films 533 (2013) 61.
[151] G. Indiveri, T.K. Horiuchi, Frontiers in neuromorphic engineering, Front. Neurosci. 5 (2011) 118.
[152] C. Mead, Neuromorphic electronic systems, Proc. IEEE 78 (1990) 1629.
[153] K. Boahen, Neuromorphic microchips, Sci. Am. 292 (2005) 56.
[154] J. Misra, I. Saha, Artificial neural networks in hardware: a survey of two decades of progress, Neurocomputing 74 (2010) 239.
[155] S. Park, A. Sheri, J. Kim, J. Noh, J. Jang, M. Jeon, B. Lee, B. Lee, B. Lee, H. Hwang, Neuromorphic speech systems using advanced ReRAM-based synapse, in: 2013 IEEE International Electron Devices Meeting, 2013.
[156] A.V. Babu, S. Lashkare, U. Ganguly, B. Rajendran, Stochastic learning in deep neural networks based on nanoscale PCMO device characteristics, Neurocomputing 321 (2018) 227.
[157] S. Park, M. Chu, J. Kim, J. Noh, M. Jeon, B. Hun Lee, H. Hwang, B. Lee, B. Lee, Electronic system with memristive synapses for pattern recognition, Sci. Rep. 5 (2015) 10123.
[158] S. Lashkare, N. Panwar, P. Kumbhare, B. Das, U. Ganguly, PCMO-based RRAM and NPN bipolar selector as synapse for energy efficient STDP, IEEE Electron Device Lett. 38 (2017) 1212.
[159] L. Lapicque, Recherches quantitatives sur l'excitation électrique des nerfs traitée comme une polarisation, J. Physiol. Pathol. Gen. 9 (1907) 567–578.
[160] N. Brunel, M.C.W. van Rossum, Lapicque's 1907 paper: from frogs to integrate-and-fire, Biol. Cybern. 97 (2007) 337.
[161] M.D. Pickett, G. Medeiros-Ribeiro, R.S. Williams, A scalable neuristor built with Mott memristors, Nat. Mater. 12 (2012) 114.

[162] H.D. Crane, The neuristor, IEEE Trans. Electron. Comput. EC-9 (1960) 370.
[163] M.D. Pickett, R.S. Williams, Phase transitions enable computational universality in neuristor-based cellular automata, Nanotechnology 24 (2013) 384002.
[164] G. Rachmuth, C.-S. Poon, Transistor analogs of emergent iono-neuronal dynamics, HFSP J. 2 (2008) 156.
[165] H. Lim, V. Kornijcuk, J.Y. Seok, S.K. Kim, I. Kim, C.S. Hwang, D.S. Jeong, Reliability of neuronal information conveyed by unreliable neuristor-based leaky integrate-and-fire neurons: a model study, Sci. Rep. 5 (2015).
[166] P. Stoliar, J. Tranchant, B. Corraze, E. Janod, M.-P. Besland, F. Tesler, M. Rozenberg, L. Cario, A leaky-integrate-and-fire neuron analog realized with a Mott insulator, Adv. Funct. Mater. 27 (2017) 1604740.
[167] C. Adda, B. Corraze, P. Stoliar, P. Diener, J. Tranchant, A. Filatre-Furcate, M. Fourmigué, D. Lorcy, M.-P. Besland, E. Janod, L. Cario, Mott insulators: a large class of materials for leaky integrate and fire (LIF) artificial neuron, J. Appl. Phys. 124 (2018) 152124.
[168] C. Adda, L. Cario, J. Tranchant, E. Janod, M.-P. Besland, M. Rozenberg, P. Stoliar, B. Corraze, First demonstration of "leaky integrate and fire" artificial neuron behavior on (V0.95Cr0.05)2O3 thin film, MRS Commun. 8 (2018) 835.

CHAPTER

11

The effect of external stimuli on the performance of memristive oxides

Yang Li, Dennis Valbjørn Christensen, Simone Sanna, Vincenzo Esposito, and Nini Pryds

Department of Energy Conversion and Storage, Technical University of Denmark (DTU), Fysikvej, Kongens Lyngby, Denmark

Abbreviations

AMR anisotropic magnetoresistance
BE bottom electrode
BMP bound magnetic polarons
CAFM conductive atomic force microscopy
CC compliance current
DMS diluted magnetic semiconductor
ECM electrochemical metallization memory
EDX energy dispersive X-ray spectroscopy
FeRAM ferroelectric random access memory
HRS/R_{OFF} high resistance state
IRS initial resistance state
LRS/R_{ON} low resistance state
MIM metal–insulator–metal
MRAM magnetoresistive random access memory
PCM phase change memory
PDMS polydimethylsiloxane
PEN poly-ethylene-naphthalate
PET poly-ethylene-terephthalate
RRAM resistive (switching) random access memory
SEE single-event effect
SEM scanning electron microscopy
STEM scanning transmission electron microscopy
STM scanning tunneling microscopy
TE top electrode

Metal Oxides for Non-volatile Memory
https://doi.org/10.1016/B978-0-12-814629-3.00011-8

TEM	transmission electron microscopy
TID	total ionizing dose
UV	ultraviolet light
VCM	valance change memory
XPS	X-ray photoelectron spectroscopy

Symbols list

$I(t)$	current as a function of time
ϕ	generalized flux
$V(t)$	voltage as a function of time
$q(t)$	charge as a function of time
$M(q(t))$	memristance
R_{OFF}	high resistance state
R_{ON}	low resistance state
I_{CC}	compliance current
M	electrochemical active metal
M^{z+}	metal cations
e^-	electrons
$V_O^{\cdot\cdot}$	oxygen vacancy
H	magnetic field
$\triangle R$	resistance difference
$R(H)$	resistance under external magnetic field
$R(H_0)$	resistance at zero field
M_S	saturation magnetization
H_C	coercive field
T_{crit}	critical temperature
R	bending radius
θ	bending angle
ε	strain
t	thickness of the substrate
θ_c	critical incident value
P_r	ferroelectric polarization
P_s	saturated ferroelectric polarization

11.1 Introduction

The memristor concept was first introduced and predicted by Leon Chua [1,2] as the fourth basic circuit element after resistor, capacitor, and inductor. It is a two-terminal passive device linking the relationship between the charge q as time integral of current $I(t)$ and the generalized flux ϕ as the time integral of voltage $V(t)$. If the resistance representing the ratio of V/I depends on the charge $q(t)$ passed through the device, it becomes a memristance $M(q(t))$ according to

$$V(t) = M(q(t))/t.$$

This theoretical concept was first verified experimentally by R. Stanley Williams and his colleagues by fabricating nanoscale memristor based on TiO_{2-x} thin films sandwiched between two electrodes [3]. The memristance, i.e., the property of a material that enables it to remember the last resistance it had before being shut off, shows hysteretic behavior, which can be exploited as a nonvolatile memory device. Memristors are also known as resistive switching memories, and they can be grouped in different categories depending on the underlying physical mechanism responsible for the nonvolatile behavior: ferroelectric random access memory (FeRAM) [4,5], magnetoresistive random access memory (MRAM) [6,7], phase-change memory (PCM) [8,9], and resistive (switching) random access memory (RRAM) [10–13].

We here focus on RRAMs and provide an overview of how devices respond to various external stimuli (electrical field, magnetic field, temperature, strain, and radiation). We also explain the physical mechanisms behind these observations. For the other types of novel nonvolatile memory technologies mentioned above, the reader is referred to many relevant reports, including review papers and books [4–9], which will not be included in this book chapter.

A typical RRAM device exhibits of three-layers consisting of a metal–insulator–metal (MIM) structure. Fig. 11.1A shows a schematic illustration of the MIM device structure, which consists of two metal electrodes (top electrode (TE) and bottom electrode (BE)) and an insulating, resistive switching layer (oxide) sandwiched in between. During the switching operation, the top electrode is connected to an external electrical stimulus, while the bottom electrode is grounded. Fig. 11.1B shows a typical *I- V* curve of the RRAM devices switching process. Usually, the initial resistance state (IRS) of the device is high, and a so-called forming process

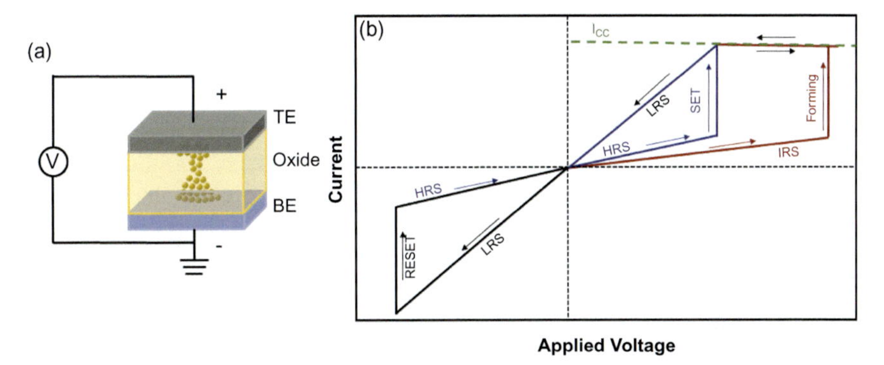

FIG. 11.1 (A) A schematic illustration of a metal–insulator–metal (MIM) structured RRAM device under an electric field. The device consists of two metal electrodes (top electrode (TE) and bottom electrode (BE)) and an insulating resistive switching layer (oxide) sandwiched in between. During the switching operation, the top electrode is connected to an external electrical stimulus, while the bottom electrode is grounded. (B) Typical IV curves for the Forming (red line), SET (blue line), and RESET (black line) of an RRAM device. The compliance current (green dash line) is labeled as I_{CC}. IRS, HRS, and LRS stands for initial resistance state, high resistance state, and low resistance state, respectively.

is required to trigger the resistive switching behavior. The red line in Fig. 11.1B illustrates the *I- V* curve of a "forming process." Initially the current is low and the device resistance is at IRS. As the voltage increases, the current increases and changes abruptly at the point called the "forming process." However, the voltage is limited by the compliance current (I_{CC}). The I_{CC} is usually achieved by externally connecting a transistor in series with the RRAM device or alternatively, it is an integrated part of the measurement equipment to protect the RRAM device from break-down, which will destroy the switching ability of the device. During the forming process of most resistive switching devices, conductive filaments are formed, resulting in switching the device to a low resistance state (LRS). The black line in Fig. 11.1B shows the *I-V* curve of the RESET process where the voltage increases in the opposite direction and the current drops at the RESET point. The device switches from LRS to a high resistance state (HRS) during the RESET process. The opposite process from HRS to LRS is called SET and presented by the blue line in Fig. 11.1B. The device switches under the electrical stimuli between two different resistance states, i.e., high resistance state (HRS or R_{OFF}) and low resistance state (LRS or R_{ON}) with a typical value of resistance ratio (R_{OFF}/R_{ON}), e.g., for some materials exceeding 10 [13].

Redox-based ion-related RRAM devices include two types of memory devices: electrochemical metallization memories (ECM) and valance change memories (VCM) [13–16]. More detailed explanation of RRAM devices and the corresponding switching mechanisms can be found in Refs [10,11,17–24]. In the remaining chapter, emphasis will be put on the effect of external stimuli on device behavior.

11.2 Electrical field

An external applied electrical field is the most basic and common method to tune the resistance states of RRAM devices. The ECM based device is a memory cell with a metal electrode supplying electrochemically active elements, such as Ag and Cu, and an inert counter electrode, such as Pt and W. The distance between the electrodes can be minimal, e.g., a few nanometers to a few tens of nanometers. A conductive filament formed during the switching process consists of active electrodeposition of metal bridging the two electrodes. The SET process of ECM involves three steps: (i) anodic dissolution of electrochemical active metal M ($M \rightarrow M^{z+} + ze^-$, with M^{z+} representing the metal cations in the resistive oxide thin film and e^- representing electrons), (ii) migration of the metal cations under the external electrical field, and (iii) reduction and electrocrystallization of M on the surface of the inert electrode ($M^{z+} + ze^- \rightarrow M$). In the RESET process, the formed conductive filaments are dissolved partially under reverse external stimuli [25]. VCM device consists of electrodes that do not inject metal cations but rather form conductive filaments consisting of oxygen vacancies $V_{\ddot{O}}$. The conductive filaments are driven by point defects, such as oxygen vacancies $V_{\ddot{O}}$ [26–28].

Direct experimental observations of conductive filaments were reported using different experimental methods, including energy dispersive X-ray spectroscopy (EDX) [29], scanning electron microscopy (SEM) [29], transmission electron microscopy (TEM) [26,27,30–34], scanning TEM (STEM) [35], conductive atomic force microscopy (CAFM) [36,37], X-ray absorption spectroscopy, [28,38] and scanning tunneling microscopy (STM) [39].

Pt/SiO_2/Ag structured RRAM is an ECM device, and its formation and dissolution dynamics of conductive filaments were studied by TEM, as shown in Fig. 11.2A–C. Fig. 11.2A shows the initial device state, i.e., without any conductive filaments inside the SiO_2 layer (the bright area between the Ag layer and the Pt layer). During the forming process, conductive Ag filaments were formed (Fig. 11.2B), and later dissolved during the RESET process (Fig. 11.2C). A schematic illustration of the Ag ions transportation and reduction process, which results in dendrite-shaped filaments, is given in Fig. 11.2D [40].

As explained earlier, in the voltage sweep switching mode, the device resistance states can be controlled by the SET process' compliance current [41–49] and the voltage in the RESET process [41,50–57]. By limiting the compliance current (I_{CC}) during the SET process, the diameter of the conductive filament and the LRS can be controlled [42,50,58]. Fig. 11.3A shows the *I-V* curves of the W/Ta/TaO_x/Pt RRAM device [42] in the voltage sweep switching mode with different I_{CC} varying from 30 to 300 μA while the LRS after the SET process is modulated from 23.7 to 1.3 kΩ. Fig. 11.3B shows the measured LRS as a function of I_{CC} for different materials, i.e., TaO_x [41–44], HfO_x [44–46], TiO_x [47], CuO [48], and NiO [49]. It was found that the LRS after the SET process is proportional to the reciprocal of I_{CC} ($LRS \propto I_{CC}^{-1}$), irrespective of the type of material used.

In the voltage sweep RESET process, the resistance can be modulated by the RESET stop voltage V_{stop}. Fig. 11.4A shows a TiN/HfO_x/AlO_x/Pt RRAM device [51] in the voltage sweep switching mode. The HRS after the RESET process increases from 299.5 kΩ to 13.6 MΩ as the absolute value of RESET stop voltage $|V_{stop}|$ increases from 2.1 to 3.3 V. Fig. 11.4B shows the measured HRS as a function of the absolute value of RESET stop voltage $|V_{stop}|$, for the following materials: Ta_2O_5 [41], AlO_x [52,53], HfO_x/AlO_x [51,59], HfO_x [50], HfO_x/TiO_x [55], ZnO [56], $ZnSnO_x$/ZrO_x [57], and Ag:SiO_2 [54]. Here, it is clearly shown that the higher the RESET stop voltage leads to a higher HRS, and the HRS is proportional to the RESET stop voltage ($\log(HRS) \propto V_{stop}$).

The examples in Figs. 11.3 and 11.4 illustrate the use of voltage sweep to switch the RRAM devices. Besides voltage sweep, the external voltage can also be applied in the form of a voltage pulse [59–66]. Compared to the voltage sweep switching method, the voltage pulse switching method has several advantages, such as fast switching speed and low switching

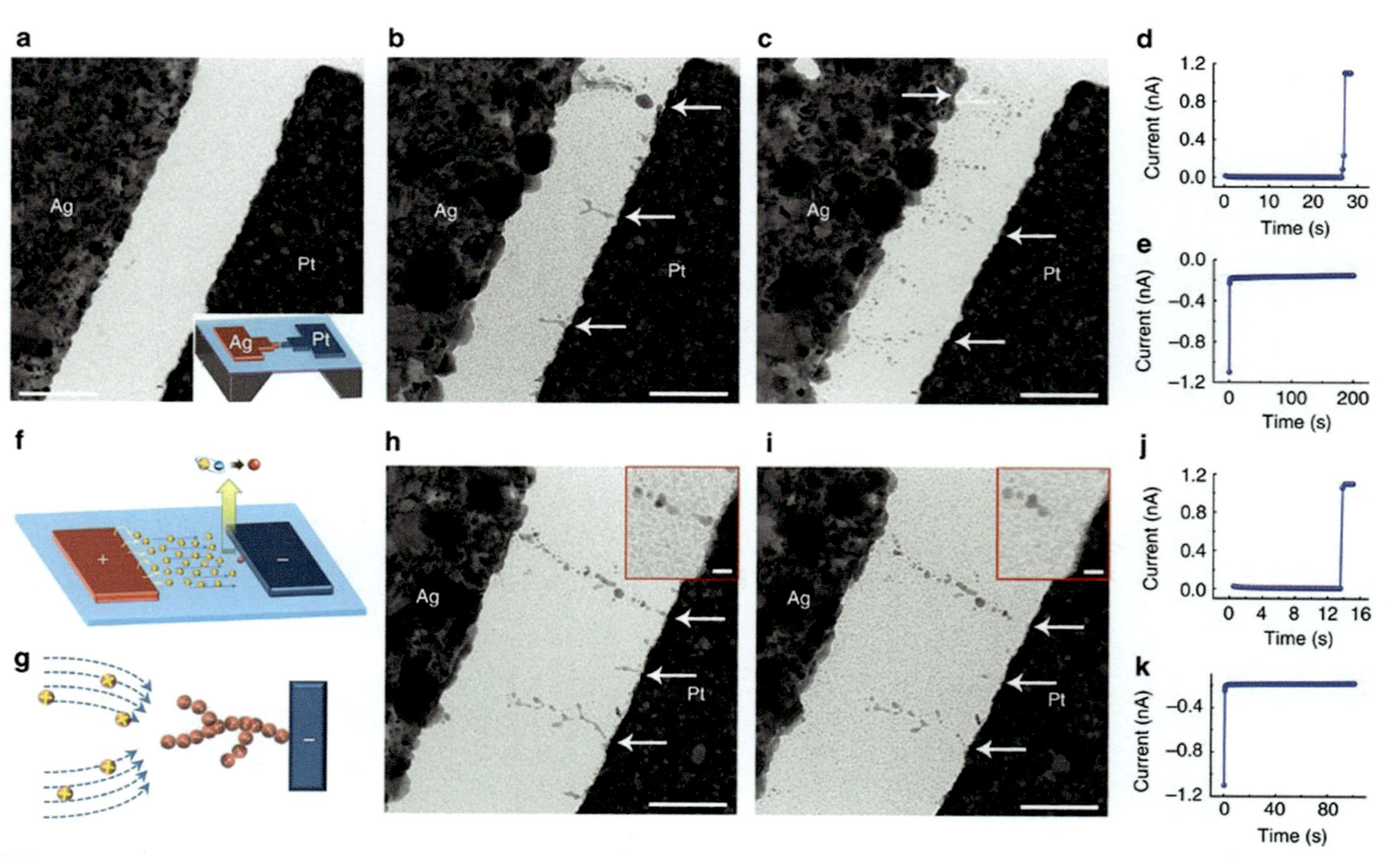

FIG. 11.2 Observation of Ag conductive filament dynamics in SiO_2-based RRAM. Scale bar, 200 nm. (A) TEM image of an as-fabricated Ag/SiO_2/Pt structured RRAM device with an inset showing the device structure schematic illustration. (B) TEM image of the same device after the forming process. The arrows highlight several representative filaments. (C) TEM image of the same device after RESET. (D) A schematic illustration of the conductive filament formation process under external electrical stimuli [40].

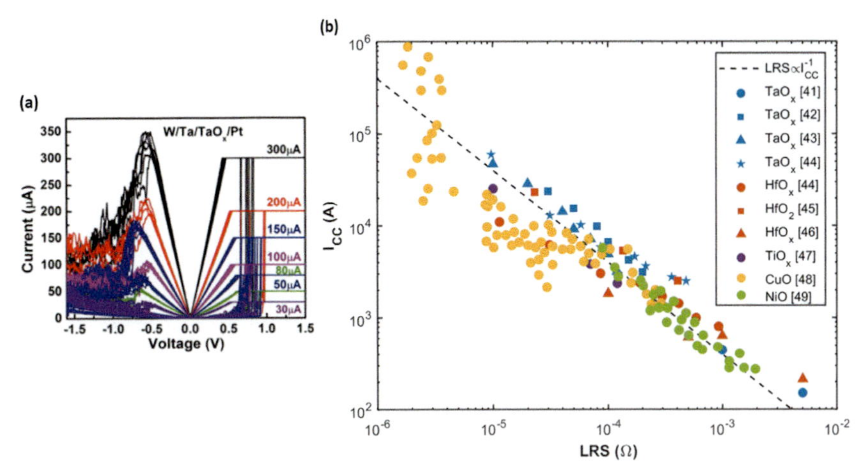

FIG. 11.3 (A) IV curves of the W/Ta/TaO$_x$/Pt RRAM device [42] in the voltage sweep switching mode using different compliance currents I_{CC} ranging from 30 to 300 μA, (B) The low state resistance (LRS) as a function of the compliance current I_{CC}. Data were plotted from different materials including TaO$_x$ [41–44], HfO$_x$ [44–46], TiO$_x$ [47], CuO [48], and NiO [49]. Irrespective of the type of material, the LRS is proportional to the reciprocal of I_{CC}.

energy. For example, the fastest switching time by voltage pulse is about 100 ps for the SET process and 120 ps for the RESET process [65], whereas the time scale for the voltage sweep switching method is a few seconds. The switching energy consumption by voltage pulse is as low as less than 0.1 pJ per bit [66]. The voltage pulse amplitude and width can also

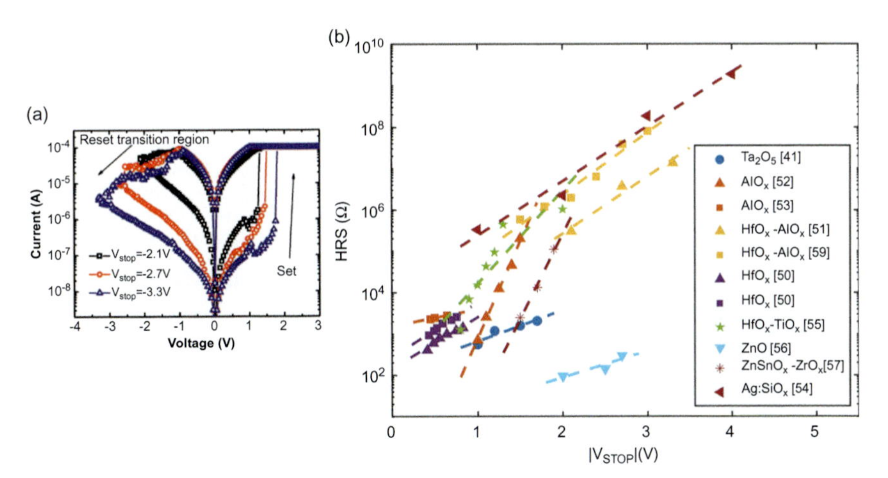

FIG. 11.4 (A) TiN/HfO$_x$/AlO$_x$/Pt RRAM device [51] in the voltage sweep switching mode with RESET stop voltage $|V_{stop}|$ increases from 2.1 to 3.3 V, (B) measured HRS as a function of the absolute value of the RESET stop voltage $|V_{stop}|$ including different oxide materials of Ta$_2$O$_5$ [41], AlO$_x$ [52,53], HfO$_x$/AlO$_x$ [51,59], *HfO$_x$* [50], *HfO$_x$/TiO$_x$* [55], *ZnO* [56], *ZnSnO$_x$/ZrO$_x$* [57], *and Ag:SiO$_2$* [54].

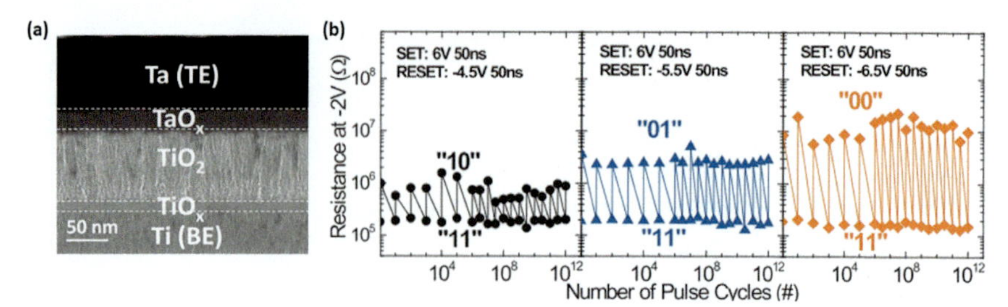

FIG. 11.5 (A) A cross-sectional TEM picture of the $Ti/TiO_x/TiO_2/TaO_x/Ta$ RRAM, (B) three different HRS labeled by "10," "01," "00," were achieved by using different voltage pulse amplitudes of −4.4, −5.5, and −6.5V in the RESET process, respectively. An excellent endurance property of 10^{12} cycles without obvious resistance degradation was also verified using pulse switching mode [67].

influence the resistance states of RRAM devices. In the $Ti/TiO_x/TiO_2/TaO_x/Ta$ RRAM devices [67], the HRS can be modulated by different voltage pulse amplitudes. Fig. 11.5A is a cross-sectional TEM picture of the device showing the device microstructure. The Tantalum (Ta) layer acts as TE, and the Titanium (Ti) layer acts as BE. By using 50ns wide RESET voltage pulses with different amplitudes of −4.5, −5.5, and −6.5V, the HRS of the device can be modulated into three different HRS labeled as "10," "01," and "00" (Fig. 11.5B) with up to 10^{12} cycles endurance for each state. The LRS is labeled as "11" in all the examples.

In Pt/PCMO/Al/W RRAM devices [68], the HRS can be modulated by different RESET voltage pulse widths. Fig. 11.6A show a cross-sectional TEM picture of the RRAM microstructure. The device is switched by using

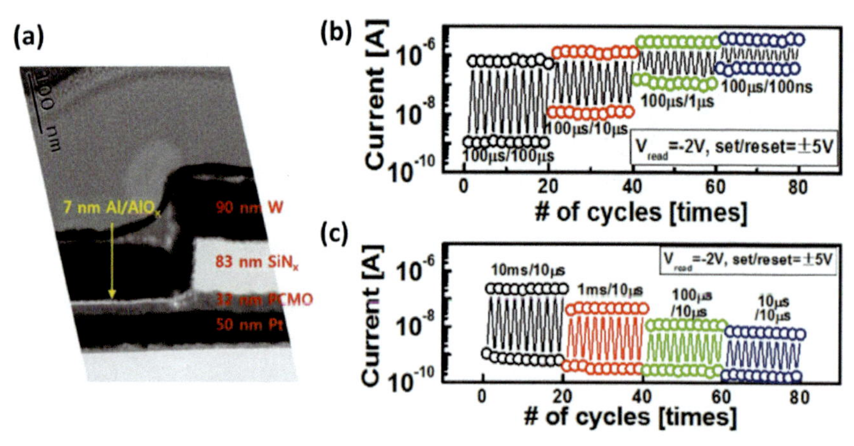

FIG. 11.6 (A) A cross-sectional TEM picture of Pt/PCMO/Al/W RRAM device, (B) SET pulse width fixed at 100μs and RESET pulse width varied from 100μs to 100ns (C) RESET pulse width fixed at 10μs and SET pulse width varied from 10ms to 10μs. The SET and RESET voltage pulses are 5 and −5V in amplitude. The currents are read at −2V [68].

SET and RESET voltage pulses of ±5 V. In Fig. 11.6B, the SET pulse width is set to be 100 μs, and the RESET voltage pulse width varied from 100 μs to 100 ns. The lower currents read at −2 V in Fig. 11.6B increase as the RESET voltage pulse widths decrease, illustrating that the HRS increases as the RESET voltage width decreases. In Fig. 11.6C, the RESET pulse width is set to be 10 μs and the SET voltage pulse widths varied from 10 ms to 10 μs. The higher currents read at −2 V Fig. 11.6C decrease as the SET voltage pulse widths decrease, which means that the LRS decreases as the SET voltage width decreases.

11.3 Magnetic field

So far, only a few works have been focused on a device's responses to the external magnetic field [69–74]. The research question is whether the magnetic field can influence resistive switching—If the answer is positive, this can provide a unique opportunity to encode information in resistive switching with the help of magnetic modulation [69].

For ECM devices, such as $Pt/TiO_2/Ni$ [70], $Pt/HfO_2/Ni$ [71], and $Pt/HfO_2/Co$ RRAM [72], the conductive filaments consist of magnetic metal ions such as Ni and Co. The anisotropic magnetoresistance (AMR) behavior in $Pt/HfO_2/Ni$ device is shown in Fig. 11.7 [70]. Fig. 11.7D shows the directions of the external magnetic field. Magnetic field H_x and H_y are in plane of the substrate and magnetic field H_z is perpendicular to the substrate. Magnetoresistance is defined as $\Delta R/R = [R(H) - R(H_0)]/R(H_0) \times 100(\%)$, where $R(H)$ is the resistance under the external magnetic field and $R(H_0)$ is the resistance at zero field. Fig. 11.7A–C show the AMR of device at LRS under different magnetic fields. Low AMR value was observed with a relative resistance change of less than 0.5%. The observed AMR is induced by ferromagnetic conductive filaments formed during the LRS. Whereas no AMR was observed for devices at HRS (data not shown in original references) [70,71].

For VCM devices, such as Pt/Co:ZnO/Pt [69], Pt/Mn-substituted ZnO/Pt [74], and $Ag/BaTiO_3/FeMn/BaTiO_3/Ag$ [73], the magnetization (or magnetic moment) and coercive field of the RRAM devices at both HRS and LRS were tested [69,74]. The resistive switching behavior was found to be controlled by magnetic fields [73]. A model of bound magnetic polarons (BMP), stemming from a coupling between the magnetic moments of the electrons and those of the magnetic ions, was used to explain the resistive switching as a function of magnetic modulation of the VCM [69,74,75]. Fig. 11.8A shows the $Pt/(Zn_{0.95}Co_{0.05})O/Pt$ RRAM device structure and the resistive switching layer $(Zn_{0.95}Co_{0.05})O$, is made of a diluted magnetic semiconductor (DMS). Fig. 11.8B and C show typical magnetization hysteresis loops of devices measured at HRS and LRS. The most eminent feature is that both the saturation magnetization (M_S)

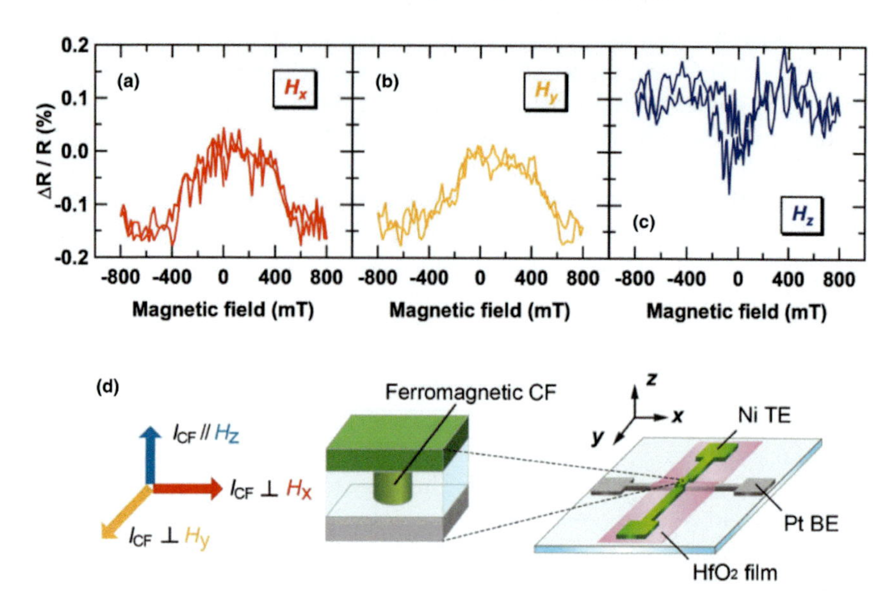

FIG. 11.7 Anisotropic magnetoresistance (AMR) observed in Pt/TiO_2/Ni RRAM devices at LRS. The directions of the applied magnetic field are (A) H_x, (B) H_y, and (C) H_z as indicated in (D) [70].

and coercive field (H_C) changed as the device switched between HRS and LRS. The M_S changed from 3.2 to 5.8 μemu and the H_C changed from 65 to 90 kOe as the RRAM switched from HRS to LRS, respectively. Reversible control of M_S and H_C in a sequence of HRS/LRS cycles is shown in Fig. 11.8D and E. The BMP mechanism is believed to be accounted for the

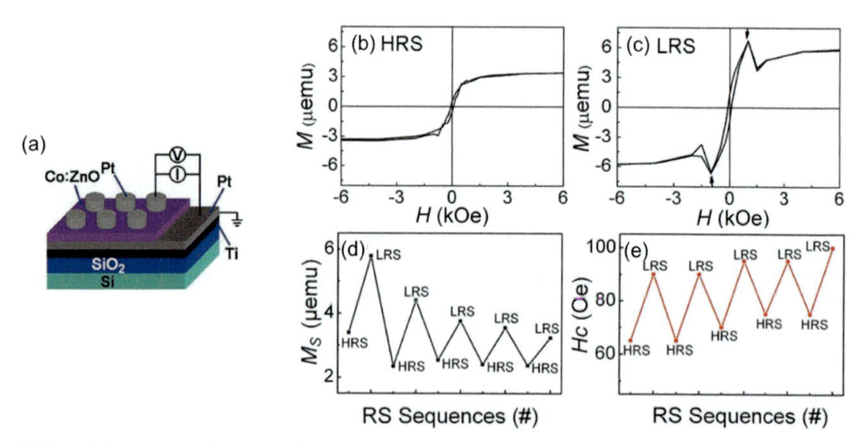

FIG. 11.8 (A) A schematic illustration of a Pt/($Zn_{0.95}Co_{0.05}$)O/Pt RRAM device and measurement setup. Hysteresis loop measured at room temperature of RRAM devices at (B) HRS and (C) LRS. The magnetization M_S (D) and coercive field H_C (E) are altered reversibly when the resistance of RRAM switched between HRS and LRS [69].

observed ferromagnetism. The higher density of well-ordered $V_{\ddot{O}}$ yields a larger volume occupied by BMPs, thus increasing the probability of transferring more Co^{2+} ions into the ferromagnetic domains and enhanced the ferromagnetism. While the model claimed remains phenomenological, the description appears reasonable and efforts to pursue these processes further are highly warranted.

The desire is to achieve multi-level or multi-state storage by combining electrical and magnetic properties. However, a real uncoupled multi-level storage device has not yet been fully realized since the magnetic properties are always coupled to the device resistance states. More detailed physical mechanism models are desirable to better understand the device behavior and its relation to magnetic stimuli.

11.4 Thermochemical treatments

In this part, we would like to discuss the thermochemical treatment, which refers to the annealing of post-deposited RRAM devices under specific gas atmospheres at elevated temperatures. Depending on the gas composition, temperature and time can influence the crystallinity [76–78], defects concentration [61,79–83], and stoichiometry of the materials [82]. In turn, these factors affect the performances of the devices.

Thermochemical treatments can profoundly influence the device crystallinity, i.e., the degree of structural order. The $Pt/TaO_x/W$ system after annealing at 964 °C for 30 min shows a change of the TaO_x layer structure from amorphous to polycrystalline β phase while the resistive switching showed negligible degradation [76]. For the $TiN/HfO_x/TiN$ based system, only the cubic HfO_x layer deposited at 350 °C showed resistive switching behavior while the monoclinic HfO_x deposited at 400 °C did not show resistive switching behavior [84]. Interestingly, in the $Pt/HfO_2/Pt$ system, the monoclinic HfO_x also showed resistive switching [78]. For the $Pt/La_{0.7}Ca_{0.3}MnO_3$ (LCMO)/Ag system, the resistance switching took place only at the crystalline state of the LCMO and not in the amorphous state [85].

During a thermochemical treatment, defects such as hydrogen ions or $V_{\ddot{O}}$, can be introduced into metal oxide. Fig. 11.9A shows that introducing hydrogen impurity into Ta_2O_5 by annealing the Ta_2O_5 layer in NH_3 at 400 °C for 5 min. The RRAM device has a structure of $TiN/Ta_2O_5/Ta$. The presence of hydrogen inside Ta_2O_5 decreases the forming voltage as well as the LRS of the device. Besides, the drift of both HRS and LRS to higher resistances is reduced after treatment with NH_3. Fig. 11.9B and C show the degradation of the resistance after 15 days for samples treated at 250 °C. However, devices that annealed with NH_3 showed less pronounced degradation of the resistance. The authors explained this behavior in the following way: annealing introduces H species, which led to the spontaneous formation of O—H

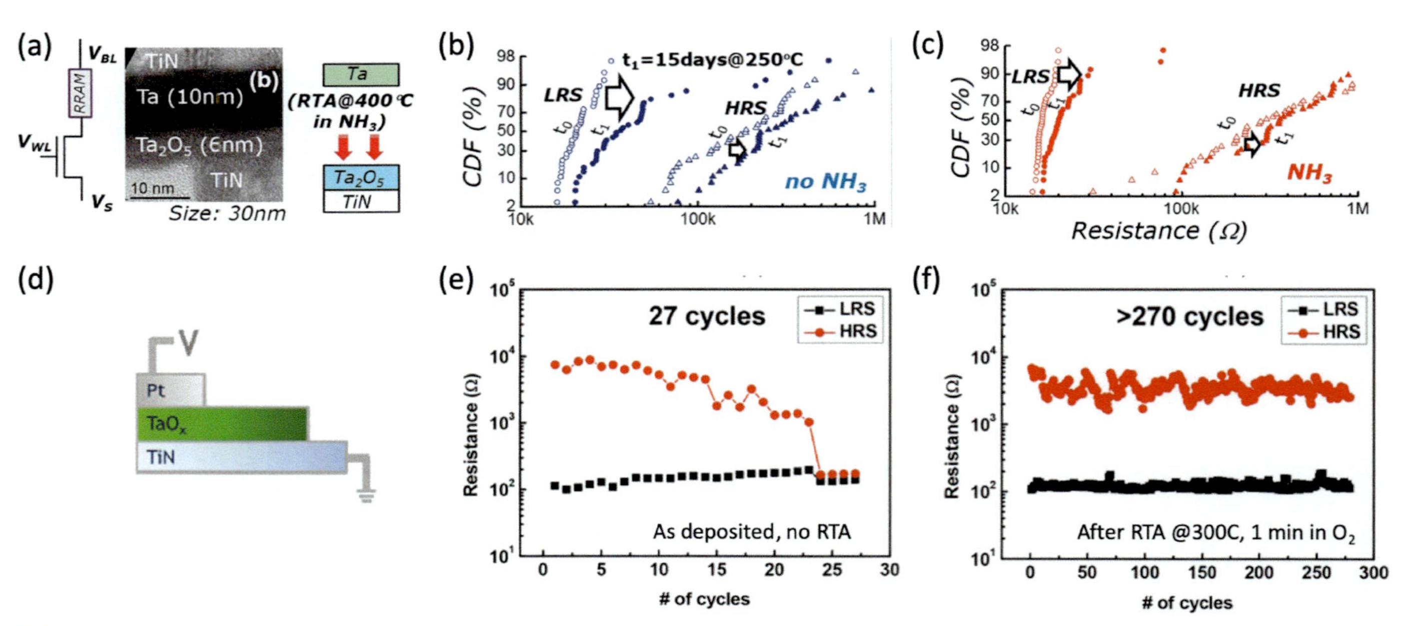

FIG. 11.9 Rapid thermal annealing (RTA) effect in TaO_x based RRAM devices. (A) A schematic illustration of the TiN/Ta_2O_5/Ta RRAM device (left-hand side of the figure) and a TEM cross-sectional view of the RRAM device (right-hand side of the figure). (B) Resistance retention behavior in TiN/Ta_2O_5/Ta RRAM devices without NH_3 annealing, (C) resistance retention behavior in TiN/Ta_2O_5/Ta RRAM devices NH_3 annealing [81]. (D) A schematic illustration of Pt/TaO_x/TiN RRAM, (E) endurance behavior of the Pt/TaO_x/TiN RRAM without RTA, (F) endurance behavior of the Pt/TaO_x/TiN RRAM after RTA [82].

bonds in Ta_2O_5. This process can be expressed using the Kröger–Vink notation by $\frac{1}{2}H_2(g) + O_O^x \rightleftharpoons HO_O^{\cdot} + e^-$, which shows that the formation of O—H act as donors, i.e., leads to the generation of electrons (e^-). After heat treating the devices at 250 °C for 15 days, the introduction of H species led to the better thermal stability of both the HRS and the LRS. At an elevated temperature (250 °C) and over a long time (15 days), the O—H bonds stabilized the $V_O^{\cdot\cdot}$ defects and improved the stability of the conductive filaments. In another example, it was shown that annealing of Al-doped ZnO/ZnO/Au in H_2 at 350 °C for 15 min introduced a layer with a high concentration of $V_O^{\cdot\cdot}$, which acted as oxygen reservoir [80]. This resulted in an improvement of the R_{OFF}/R_{ON} ratio from ~10 to 10^4. In both Pt/Al_2O_3/Cu based ECM type and Pt/HfO_2/Zr based VCM type RRAM, annealing the devices in 1 MPa H_2 at 400 °C for 30 min lowered the forming voltage due to the generation of oxygen vacancies $V_O^{\cdot\cdot}$ and OH^- bounds, which assist the conductive filament formation [79].

Thermochemical treatment can also result in stoichiometry changes. An example of such an effect can be seen in the Pt/TaO_x/TiN system annealed in O_2 at 300 °C for 1 min. Here, the stoichiometry of TaO_x changed from TaO_{2-y} to Ta_2O_{5-z}, which has a profound effect on the device endurance, i.e., changes from 23 cycles to more than 270 cycles, as shown in Fig. 11.9D and E.

11.5 Strain

Due to the potential application of RRAM devices in electronics, roll-up displays, and wearable devices, the impact of stress and flexible (foldable or stretchable) RRAM devices have been widely studied and reported. Various substrates including paper [86,87], polyethersulfone (PES) [88–93], plastic [94–97], polyimide [98–100], poly-ethylene-naphthalate (PEN) [101], poly-ethylene terephthalate (PET) [102–109], parylene [110], polydimethylsiloxane (PDMS) [111], mica [112], and water-soluble substrates [113] have been used in flexible applications.

Exploring the feasibility of printable, foldable and stretchable devices and the effect of strain on the device is of particular interest. It is a key parameter for tuning the functionality of the devices [114]. Three sources of stress can be applied to a device, i.e., stress caused by external mechanical deformation [91], stress introduced by interfaces mismatch, i.e., heterostructure interfaces [114,115] and weak points at grain boundaries [116], and stress-induced by doping [117]. Fig. 11.10 shows schematic illustrations of two types of stress sources, with (a) showing the mechanical bending stress [91] and (b) showing the stress introduced at heterostructure interfaces, either in a multilayer (left) or vertical aligned (right) structure [114]. Doping is a widespread way to influence the performance of the devices. Currently, only a few reports explain or relate the stress caused

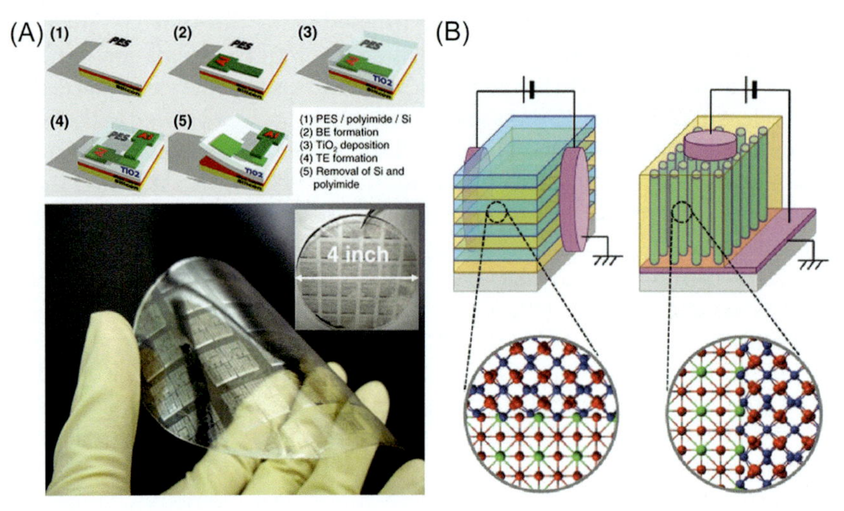

FIG. 11.10 Two sources of stress applied to RRAM device: (A) the fabrication process flow of Al/TiO_2/Al structured RRAM on polyethersulfone (PES) and a photograph of the mechanical bending stress applied to RRAM devices [91]; (B) illustration of stress introduced by heterostructure interfaces inside RRAM device [114].

by doping to the device performance, and therefore, this not be discussed here [43,118–130].

The most-reported applied stress is bending, which is often characterized by the so-called mechanical flexibility and endurance. The flexibility of the device is commonly characterized by the bending radius R and bending angle θ, as illustrated in Fig. 11.11. Mechanical endurance is characterized by the repeatability of the bending process, i.e., the number of bending cycles.

Before reaching a critical crack initiation and propagation which lead to the failure of the device, most devices exposed to mechanical stress showed stable resistive switching, including single-layered Al_2O_3 [88,96], HfO_2

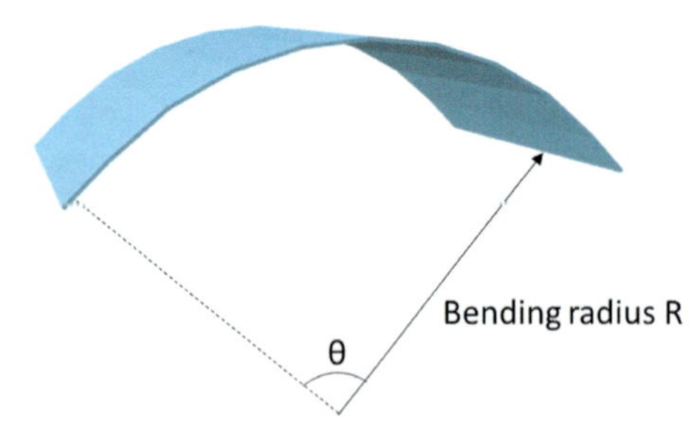

FIG. 11.11 Common mechanical stress applied to RRAM devices with an illustration of the bending radius R and the bending angle θ.

[107,110,113], TiO_2 [87,91,93,95,98,106], ZnO [89,90,108,131], NiO [103], Cu_xO [97], $WO_3{\cdot}H_2O$ nanosheets [105], and α-IGZO [94], bilayered Al_2O_3/HfO_2 [102], $Zr_{0.5}Hf_{0.5}O_2$/InGaZnO [112], Al-In-O/InO_x [132], GeO_x/HfON [99], and TaO_x/TiO_2 [100]. Due to the different ductility of the individual film, detachment of the films or formation of crack often happened under repetitive bending or varying strain conditions [90,105,131]. As the bending radius decreases (i.e., increasing the bending angle), cracks are initiated and propagated leading to deterioration of the resistive switching behavior. The crack density increases as the strain increases [107]. An example using the ITO/HfO_x/ITO device show that the relation between the bending radius and the strain as well as crack density are shown in Fig. 11.12A and C. The strain was estimated by the relation $\varepsilon = t/2R$, where t is the thickness of the substrate and R is the bending radius [107]. As the bending radius decreased from flat state (∞) to 2 mm, the strain increased from 0% to 3.18%. The structure of the HfO_x layer is a mixed amorphous-nanocrystalline and the tensile stresses in HfO_x layer is found to be released by the presence of the amorphous components. As the crack density in the ITO bottom electrode increased from 0 to 35 mm^{-1}, the resistance of the ITO layer increased and the corresponding voltage required for RESET increased as well. At the

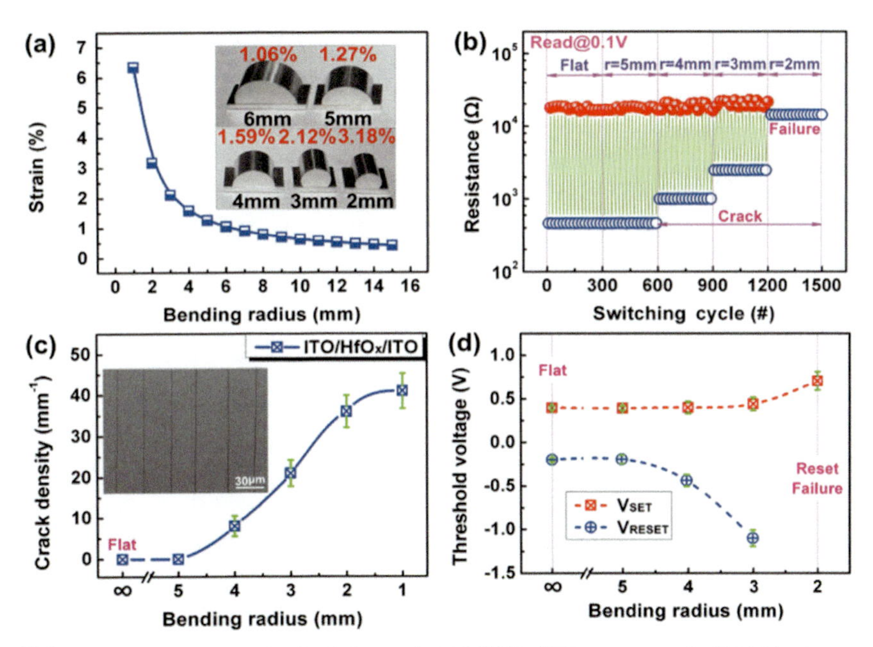

FIG. 11.12 Mechanical flexibility of ITO/HfO_x/ITO structured RRAM device: (A) mechanical strain vs bending radius, (B) resistance evolution as a function of the bending radius, (C) crack density vs bending radius, and (D) switching voltages evolution as a function of the bending radius [107].

critical bending radius of 2mm, the applied voltage mainly dropped on the fractured bottom electrode and the voltage remained on the HfO_x layer is no longer sufficient to rupture the conductive filament. The devices failed to RESET and the R_{OFF}/R_{ON} ratio decreased to 0, as shown in Fig. 11.12B and D [107].

The experimental setup of strained $Ti/TiO_2/TaO_x/Ta$ device deposited on polyimide is shown in Fig. 11.13A and B. In these experiments, both the LRS and HRS increased as the bending radius increased from a flat state to 10mm (Fig. 11.13C) [100]. The physical mechanism behind the behavior of the device resistance under bending is still unclear.

Experiments with $ITO/HfO_x/ITO$ deposited on poly-ethylene-terephthalate (PET) showed no obvious degradation of the resistance after 2.4×10^4 cycles of bending, i.e., at a bending radius of 6mm [107]. Whereas ITO/ZnO/ITO devices deposited on polyethersulfone (PES) began to show a degradation of the resistance already after 10 cycles of bending, i.e., at a bending radius of 20mm. The degradation of the resistance is caused by cracks formation in the ITO bottom electrode [90]. The number of

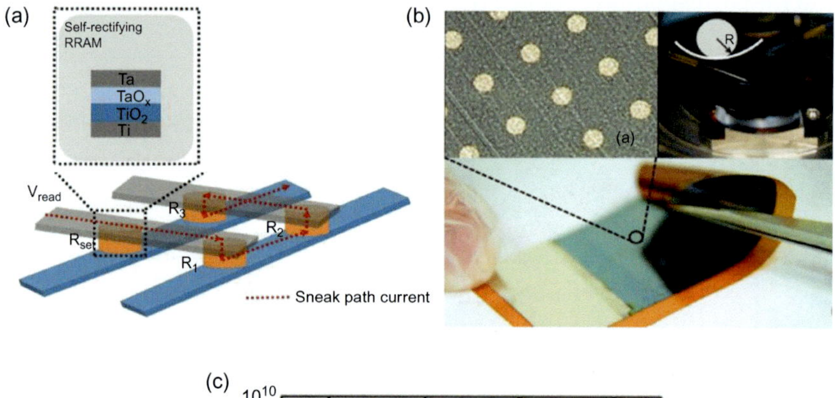

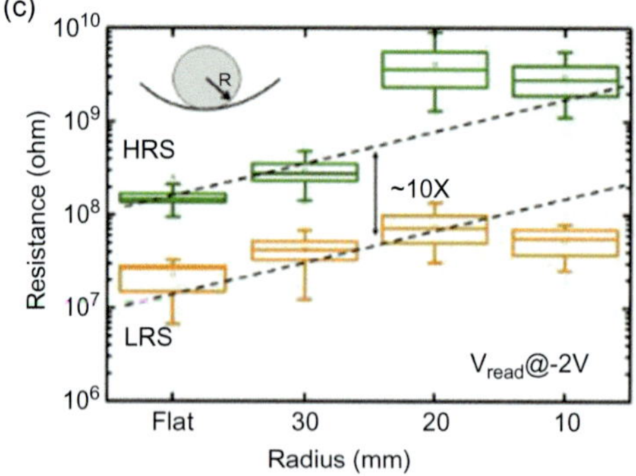

FIG. 11.13 (A) Schematic illustration of the device structure. (B) Image of fabricated RRAM devices, electrical measurement setup and flexible RRAM device. (C) Resistance evolution as the bending radius decreased [100].

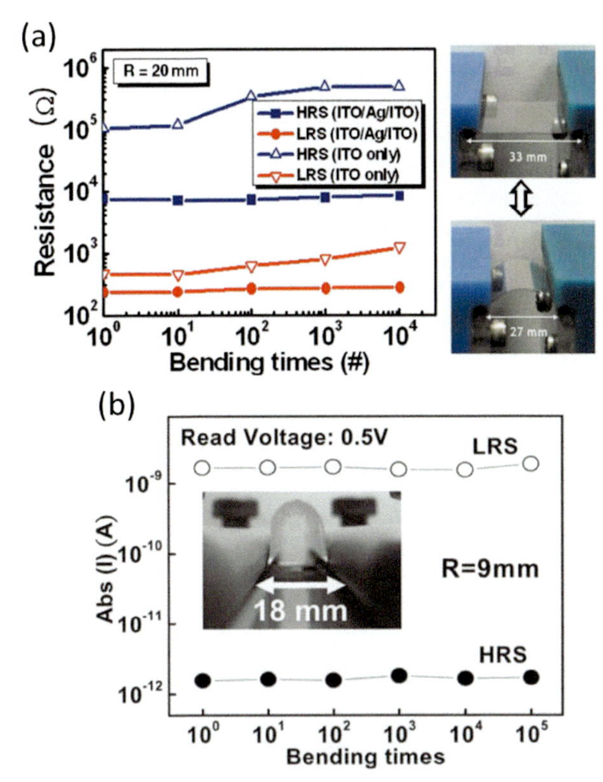

FIG. 11.14 Mechanical endurance of RRAM devices. (A) ITO/ZnO/ITO structured RRAM devices showed severe resistance degradation after 10 times of bending due to the bottom electrode cracking. Improved mechanical endurance was achieved by inserting Ag into the bottom electrode [90]. (B) TaN/HfON/GeO/Ni structured RRAM device showed robust LRS and HRS after 10^5 times of bending [99].

bending cycles was improved from 10 to 10^4 cycles for the PES/ITO/ZnO/ITO by inserting a ductile Ag layer between the ITO bottom electrode and the ZnO layer, as shown in Fig. 11.14A. A record of 10^5 bending cycles was measured for the Ni/GeO/HfON/TaN system, as shown in Fig. 11.14B [99]. Both LRS and HRS remained stable after 10^5 bending cycles. However, the reason for such improvement in the bending resistant was not explicitly explained. Other works [87,103,105,107,112, 131–133] also reported the mechanical robustness of the device by showing stable resistance states after many cycles of bending, yet currently, the maximum bending cycles is still 10^5.

Since a low adhesion between the device and the substrates usually results in the formation of cracks and detachment under repetitive bending, flexible materials rather than metal oxides are often used, e.g., organic materials [86,111,133,134] and graphene oxide [92,104,109] based devices.

Since this chapter deals mainly with metal-oxides, we will not discuss these types of materials by referring the readers interested in these materials to the references above.

Besides bending, stress can also be introduced at the interfaces, such as multilayer heterostructures, by the mismatch of the different layers [135,136]. Many studies in metal oxide heterostructures with a variation of interfaces focused on tuning the ionic conductivity through lattice compressive or tensile strain caused by the mismatch between the layers [137–140]. For example, tuning the $V_{\ddot{O}}$ mobility in $Gd_{0.1}Ce_{0.9}O_{2-\delta}/Er_2O_3$ heterostructures was reported to enhance the mobility as the number of interfaces increases [115]. The strain was systematically varied by increasing the number of individual layers of $Gd_{0.1}Ce_{0.9}O_{2-\delta}$ and Er_2O_3 from 1 to 60 while keeping the overall heterostructure at a constant thickness. By changing the compressive strain in these structures by more than 1.16%, the activation energy for oxygen vacancies migration was altered by 0.31 eV while the ionic conductivity increased. The same authors used the same system $Gd_{0.1}Ce_{0.9}O_{2-\delta}/Er_2O_3$ to fabricate an RRAM device with side electrodes to study the stress effect on the performance of the device along the direction of the interfaces, Fig. 11.15A [115,136,141]. The lattice mismatch between the $Gd_{0.1}Ce_{0.9}O_{2-\delta}$ and the Er_2O_3 interface led to an in-plane compressive strain in $Gd_{0.1}Ce_{0.9}O_{2-\delta}$, as shown in Fig. 11.15B and in the TEM image in Fig. 11.15C and D. By increasing the number of interfaces from 6 to 60 while keeping the ceria-to-erbia ratio and the total film thickness constant, the device transformed from a volatile to a nonvolatile behavior. Hysteresis of the resistance during cycling was also observed, and it was explained by the formation of filaments of ceria at the phase boundary with erbia. The authors speculate that the low mobility of oxygen ions is a result of compressive strain. The low mobility of oxygen vacancies does not allow fully oxidation Ce, which hindered the formation of conductive filaments. This effect retained the device kinetically frozen at LRS.

11.6 Radiation

Radiation is another form of external stimuli which can affect the performance of the resistive switching device. External radiations often cause defects generation in the materials, such as electron–hole pairs [142,143] and Frenkel defects [144], or ion displacement [145–147], which thereby affect the performance of the RRAM devices. A wide range of electromagnetic and particle radiations can affect the RS behavior of the materials, e.g., X-ray [147–150], γ-ray [84,145,147,151–161], ultraviolet light (UV) [148,162], visible light [163–165], protons [147,149,150], α particles [166–168], heavy ions [169–172], and neutrons [143,146]. Interestingly, in space application,

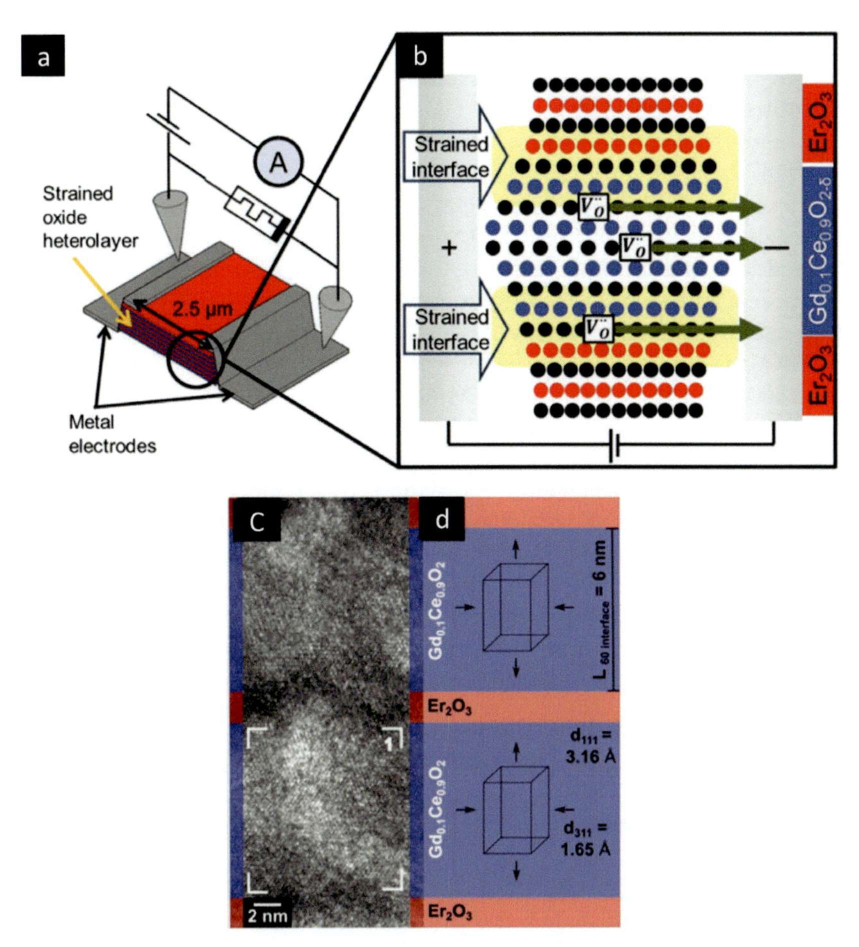

FIG. 11.15 (A) a schematic illustration of a device composed of $Gd_{0.1}Ce_{0.9}O_{2-\delta}/Er_2O_3$ heterostructures and the testing set up, (B) schematic view of one of the $Gd_{0.1}Ce_{0.9}O_{2-\delta}/Er_2O_3$ interface showing the lattice mismatch induced strain and the conduction channels of oxygen vacancies $V_{\ddot{O}}$, (C) bright field high resolution TEM image of a 60 interface sample, and (D) diagram illustration of the Miller plane spacing measurements and the in-plane compressive and out-of-plane tensile strain [141].

the failure of the RRAM devices can be caused by cosmic rays and radiations from the Van Allen radiation belts around the earth [166,173].

The effect of radiation can be categorized into two main types: (1) a cumulative effect and (2) a single-event effect (SEE). Cumulative effects are the device responses to total ionizing dose (TID) integrated over time. While SEEs are caused by a single, energetic particle and is characterized by an instantaneous failure mechanism [166,171,173–176].

Total ionizing dose (TID) is characterized by absorbed dose, i.e., the energy absorbed by matter, with a unit called rad (an acronym for radiation absorbed dose). The effect of TID on RRAM devices were tested using γ-ray generated by ^{60}Co source [84,145,147,151–161], X-ray [147–150], protons [147,149,150], and EUV [148] as radiation sources. The responses of the device to TID depend on metal oxide materials, radiation dose and device dimensions. Different TID tolerance values were reported in different oxide-based RRAM [145,151,160,161]. A threshold TID to change device resistances or switching voltages was reported [148] and a higher dose usually caused more severe device degradation [148,156,159,160]. In Pt/HfO_2/Ti/Pt devices, the TID threshold for HRS change was reported to be 100 Mrad for EUV and 600 Mrad for X-ray [148]. The effect of γ-ray irradiation on the Mo/Ti/HfO_2/Au system is shown in Fig. 11.16. Both HRS and LRS decreased with the irradiation dose increase, i.e., ^{60}Co γ-ray, Fig. 11.16B. In particular, more than 20% change in LRS, 20% change in forming voltage and 16% change in SET voltage were observed as the radiation dose exceeded 3 Mrad, see Fig. 11.16A and D [159]. Here, the decrease in the forming voltage and SET voltage was explained by increased oxygen vacancies inside the HfO_2 layer after radiation. In other experiments, the performance of the device as a response to radiation was explained by radiation-induced charge trapping inside the RRAM devices

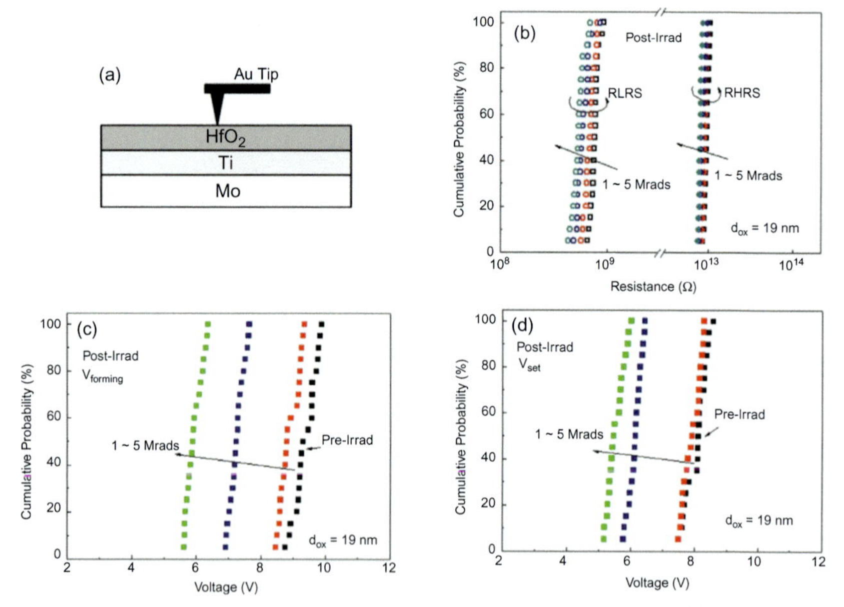

FIG. 11.16 (A) Schematic illustration of a Ti/HfO_2/Au RRAM device. ^{60}Co γ-ray irradiation effect on (B) resistance states, (C) forming voltage and (D) SET voltage in a Ti/HfO_2/Au RRAM [159].

[147,155,156], radiation-induced defects generation [151,159], and metal oxide phase transition [160,161].

The thinner oxide layer and smaller device area showed improved resilience to radiation, e.g., Pt/TaO_x/TiN RRAM devices, as shown in Fig. 11.17A. These devices with a thickness of 25nm were exposed to 180krad (Si) of γ-ray radiation [155]. Devices with the same area of 20μm × 20μm, but higher thickness, i.e., 50nm, show a more extensive HRS degradation after radiation than thinner samples, i.e., 25nm (Fig. 11.17B). An electron–hole pair generation explains this under γ-ray radiation where holes believed to be trapped. Holes are more likely to be trapped in a thick oxide layer or near the oxide/metal electrode interface in this device. This effect was explained by the drift or hopping of holes under different built-in electric field caused by different oxide thickness. Here, the trapped holes in thick oxide may form a path for electron transportation, which resulted in a decreased HRS. Whereas holes tapped near the interface in thin oxide are more easily to escape and tunnel through the electrode, which impeded the formation of an electron conduction path in the thin oxide. In similar devices, with a different area of 2μm × 2μm, 20μm × 20μm, and 80μm × 80μm, higher HRS degradation is observed in larger area devices, as shown in Fig. 11.17C, indicating an improved tolerance to irradiation as devices scale down [155].

RRAM devices were also reported to be resilient to single-event effects (SEE). The TiN/HfO_x/Ti device was irradiated by heavy ions and showed no single event upset [171]. The effects of SEE on the devices were also simulated based on TiN/TaO_x/Ta/TiN RRAM's experimental results. The results indicated that the system is highly resistant to SEEs for over 30years of operation [175].

Radiation can also affect the device resistances through radiation-induced metal-oxygen bond breaking [163,177], photon-excited oxygen ion migration [164], and illumination generated electron–hole pairs [165]. Fig. 11.18A shows the effect of photo-radiation on the performance of RRAM devices for the ITO/ZnO nanorods/Au structured RRAM [163]. After treating the sample with a low surface-tension perfluorinated liquid, the hydrothermally grown ZnO nanorods changed from hydrophilic to superhydrophobic. When submerged in water, a double layer containing a water-air interface structure is formed due to the hydrophobic property of ZnO nanorods, as shown in Fig. 11.18B. The air layer prevented the water from penetrating the device, which caused an operational failure or, even more seriously, an electric shock. The water/air interface completely reflected the light when the incident angle is larger than the critical incident value of $\theta_c = 48.6°$. When the angle of the light incident is larger than θ_c, the light is totally reflected and the device showed resistor-type behavior, as shown in Fig. 11.18C. When the light incident angle is smaller than the critical angle θ_c, the light penetrates the

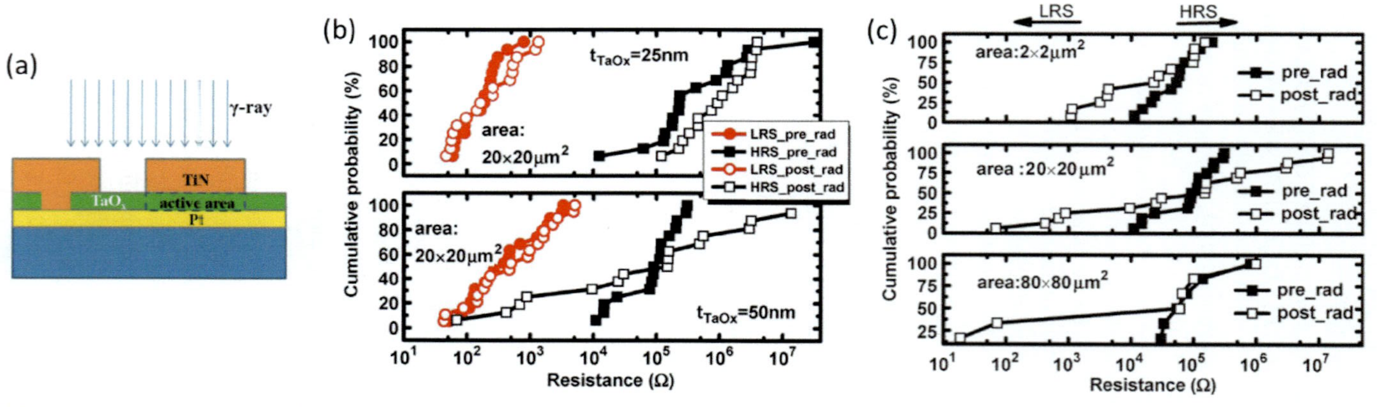

FIG. 11.17 (A) A schematic illustration of 20 μm × 20 μm Pt/TaO_x/TiN RRAM device, (B) resistance degradation after radiation in devices of the same area of 20 μm × 20 μm with 25-nm thick TaO_x and 50-nm thick TaO_x, respectively, (C) HRS degradation after radiation in devices of a different area of 2 × 2 μm, 20 × 20 μm, and 80 × 80 μm [155].

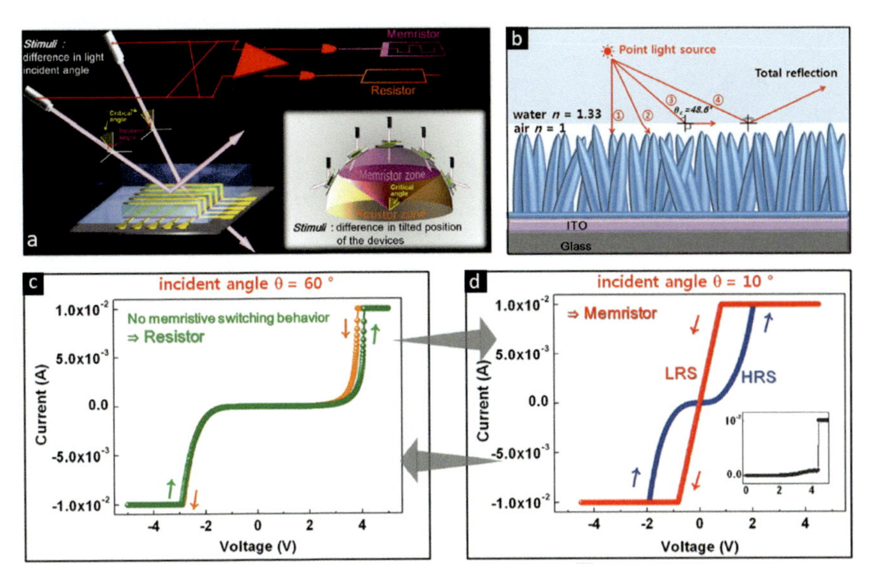

FIG. 11.18 Light radiation (wavelength of 200–2500nm) induced RS behavior inside the ITO/ZnO nanorods/Au structured RRAM devices. (A) Schematic diagram representing the concept of light-controlled reversible device function between memristor-type behavior and resistor-type behavior, (B) schematic side view of RRAM device submerged in water with a total reflection at 48.6 degrees, (C) when the incident angle is 60 degrees, the light is reflected. The device showed resistor-type behavior, (D) when the incident angle is 10 degree, the light can assist the oxygen inside ZnO nanorods, and the device showed memristor-type behavior [163].

water-air interface and device showed memristor-type behavior, as shown in Fig. 11.18D. The presence of light, triggered the resistive switching behavior due to the radiation induced desorption of cheimisorbed oxygen ions from ZnO nanorods. This gave rise to selective illumination direction as a tuning parameter to trigger the RS.

Light illumination can also influence the oxygen ion migration of the devices and lead to the accelerated dissolution of the filament. In TiN/HfO_2/diamond-coated Si tip RRAM device [164] (Fig. 11.19A), light radiation caused the current drop/RESET process inside the RRAM device. Under light illumination, the device current drops and the resistance increases, which indicates a conductive filament rupture process under illumination, as shown in Fig. 11.19B. As the illumination intensity increases, the time it takes to RESET decreases. At an illumination intensity of 1 mW/cm^2, the device resistance can be modulated by controlling the exposure time, as shown in Fig. 11.19C. The illumination induced the drop in the resistance, and it is due to the photons induced excitation of oxygen ions migration. During the formation of conductive

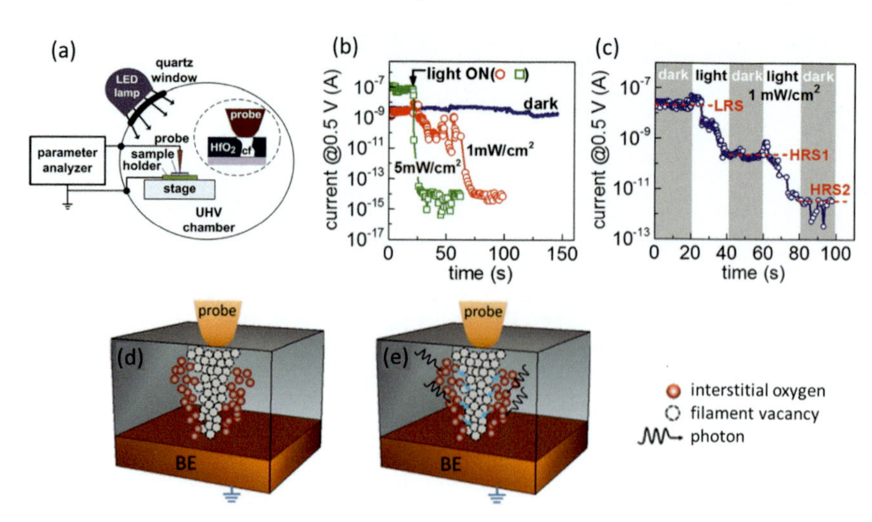

FIG. 11.19 Light-induced resistance modulation due to the light-excited oxygen ion movement. (A) Schematic illustration of the device structure and experimental setup used in this study; (B) Current drop (RESET) process speed dependence on illumination intensity; (C) different intermediate resistance states can be achieved by limiting exposure time duration; (D) Conductive filament comprising oxygen vacancies under darkness; (E) photon excited oxygen migration and recombination with oxygen ions, leading to conductive filament rupture under illumination [164].

filament (comprising oxygen vacancies) in the dark, Fig. 11.19D, oxygen ions populate the interstitial sites in the filament vicinity. Under light illumination, the interstitial oxygen ions migration is excited by photons, migrating and recombining with the filament vacancies. This effect leads to a conductive filament rupture, as shown in Fig. 11.19E.

Fig. 11.20A and B show the Ti/$BiFeO_3$/ITO RRAM device's behavior under different light intensities [165]. As the light intensity increases, both SET voltage (V_{SET}) and RESET voltage (V_{RESET}) decrease (Fig. 11.20C). The HRS drops from 550 MΩ in the dark to about 1 MΩ under a light with an intensity of 20 mW/cm^2 (Fig. 11.20D). The HRS change under light illumination was verified both in the dark and under light illumination repeatedly (Fig. 11.20E). Besides the switching properties, the device also showed a significant photo-ferroelectric effect where both ferroelectric polarization (P_r) and saturated ferroelectric polarization (P_s) increase as the light intensity increases, not shown here. The photo-generated polarized charges explained light-induced modulation of resistive switching in the $BiFeO_3$ layer, which causes an increase in the electric field in the opposite direction leading to a decrease in the HRS and the LRS.

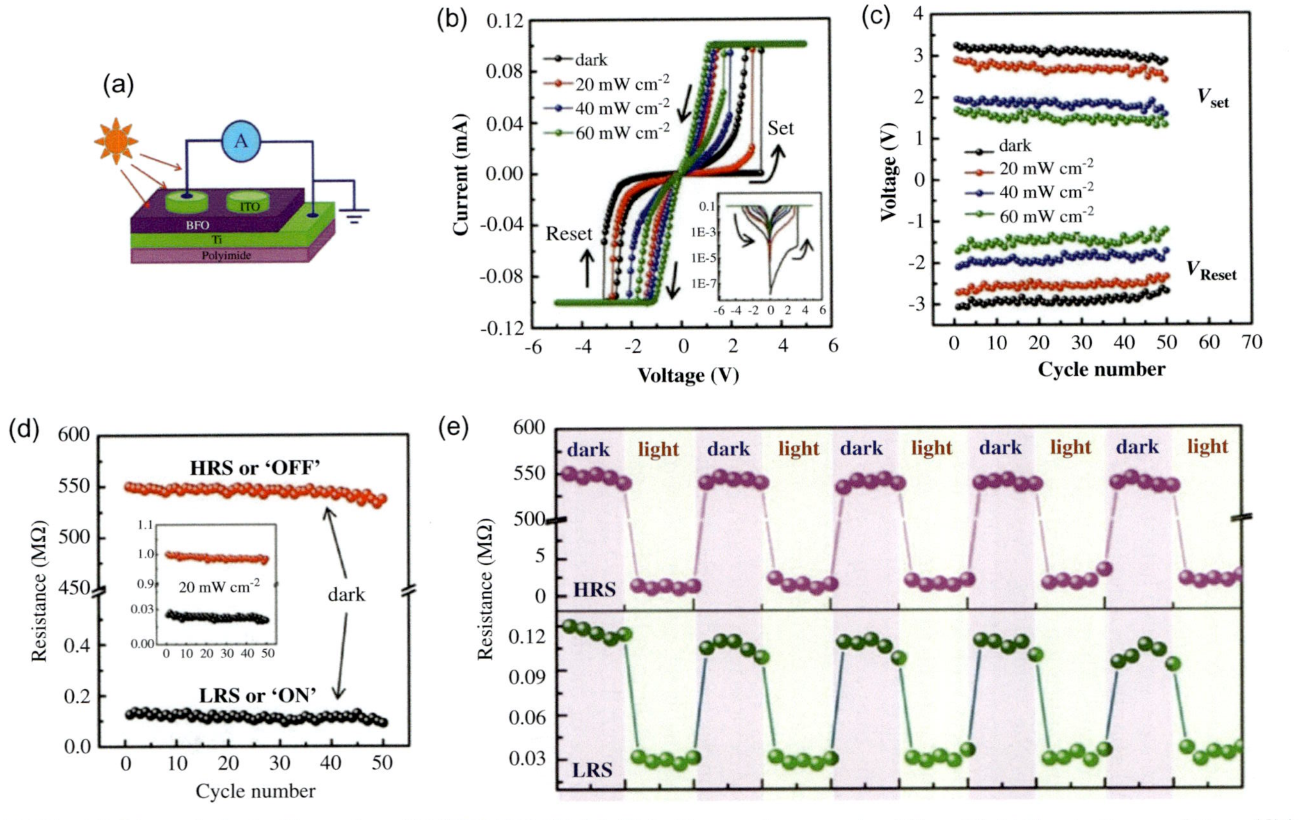

FIG. 11.20 (A) Schematic device illustration of Ti/BFO/ITO RRAM, (B) I—V curve change under different light illumination conditions, (C) light radiation caused decrease of switching voltages, (D) light illumination decreased both HRS and LRS, (E) repeatable verification of resistance change in the dark and under light illumination condition which is explained by photo-generated polarized charges in $BiFeO_3$ layer [165].

11.7 Outlook

Memristors materials capable of responses to external stimuli represent one of the most exciting and emerging areas of scientific interest and new commercial applications. Different compliance currents, sweep stop voltages, and voltage pulse heights/widths are effective regulations on the electrical stimuli that can switch the RRAM devices into more than two resistance states. These methods are extremely useful and open opportunities for the application of RRAMs in higher density information storage and neuromorphic computing. Magnetic field can induce magnetic resistance change whereas electrical stimuli can change the magnetic states, for RRAM devices with ferroelectric conductive filaments. Heat stimuli, such as thermochemical treatment, can improve the stability of the RRAM devices, which is beneficial for devices operating at higher temperature conditions than room temperature. Strain can cause mechanical cracks inside the RRAM material layers or induce films detachment. The study of strain effect on devices behavior and the use of polymer substrates can broaden the applications of RRAMs, such as in wearable electronics. Strain in RRAM devices with heterostructures can change the mobility of the oxygen vacancies thus induce the resistive switching of the device. Radiations including visible light and high-energy radiations can also act as stimuli for resistive switching in RRAM devices. While this field is facing many exciting challenges, there are several opportunities in the design, synthesis, and engineering of memristor systems. However, this research topic is, to a large extent, still unexplored both experimentally and theoretically. This book chapter highlights the recent progress in stimuli-responsiveness of the metal oxide-based memristors. The demand for widespread research on the development of stimuli-responsive memristors is requisite to understand and resolve the future challenges relating to the memristive switching dynamics, which require higher resolutions in experiments and more accuracy in modeling. This strategy can lead to new and yet unexplored properties for future memory and computing technologies. Navigating this multi-dimensional space is an enormous task without additional guidance. Therefore, combining in situ techniques and computational materials science to uncover the detailed switching mechanism is critical for understanding the response of materials to multiple stimuli.

References

[1] L.O. Chua, Memristor—the missing circuit element, IEEE Trans. Circuit Theor. 18 (5) (1971) 507–519, https://doi.org/10.1109/TCT.1971.1083337.

[2] L.O. Chua, S.M. Kang, Memristive devices and systems, Proc. IEEE 64 (2) (1976) 209–223, https://doi.org/10.1109/PROC.1976.10092.

[3] D.B. Strukov, G.S. Snider, D.R. Stewart, R.S. Williams, The missing memristor found, Nature 453 (7191) (2008) 80–83, https://doi.org/10.1038/nature08166.

[4] Y. Arimoto, H. Ishiwara, Current status of ferroelectric random-access memory, MRS Bull. 29 (11) (2004) 823–828, https://doi.org/10.1557/mrs2004.235.

[5] H. Ishiwara, M. Okuyama, Y. Arimoto, Ferroelectric Random Access Memories: Fundamentals and Applications, vol. 93, Springer Science & Business Media, 2004.

[6] J.-G. Zhu, Magnetoresistive random access memory: the path to competitiveness and scalability, Proc. IEEE 96 (11) (2008) 1786–1798, https://doi.org/10.1109/JPROC.2008.2004313.

[7] D. Apalkov, B. Dieny, J.M. Slaughter, Magnetoresistive random access memory, Proc. IEEE 104 (10) (2016) 1796–1830, https://doi.org/10.1109/JPROC.2016.2590142.

[8] M. Salinga, B. Kersting, I. Ronneberger, V.P. Jonnalagadda, X.T. Vu, M. Le Gallo, I. Giannopoulos, O. Cojocaru-Mirédin, R. Mazzarello, A. Sebastian, Monatomic phase change memory, Nat. Mater. 17 (8) (2018) 681–685, https://doi.org/10.1038/s41563-018-0110-9.

[9] G.W. Burr, M.J. BrightSky, A. Sebastian, H.Y. Cheng, J.Y. Wu, S. Kim, N.E. Sosa, N. Papandreou, H.L. Lung, H. Pozidis, E. Eleftheriou, C.H. Lam, Recent progress in phase-change memory technology, IEEE J. Emerging Sel. Top. Circuits Syst. 6 (2) (2016) 146–162, https://doi.org/10.1109/JETCAS.2016.2547718.

[10] B. Mohammad, M.A. Jaoude, V. Kumar, D.M. Al Homouz, H.A. Nahla, M. Al-Qutayri, N. Christoforou, State of the art of metal oxide memristor devices, Nanotechnol. Rev. 5 (3) (2016) 311–329, https://doi.org/10.1515/ntrev-2015-0029.

[11] H.S.P. Wong, H.Y. Lee, S. Yu, Y.S. Chen, Y. Wu, P.S. Chen, B. Lee, F.T. Chen, M.J. Tsai, Metal-oxide RRAM, Proc. IEEE 100 (6) (2012) 1951–1970, https://doi.org/10.1109/JPROC.2012.2190369.

[12] Y. Li, S. Long, Y. Liu, C. Hu, J. Teng, Q. Liu, H. Lv, J. Suñé, M. Liu, Conductance quantization in resistive random access memory, Nanoscale Res. Lett. 10 (1) (2015) 420, https://doi.org/10.1186/s11671-015-1118-6.

[13] R. Waser, R. Dittmann, C. Staikov, K. Szot, Redox-based resistive switching memories nanoionic mechanisms, prospects, and challenges, Adv. Mater. 21 (25–26) (2009) 2632–2663, https://doi.org/10.1002/adma.200900375.

[14] I. Valov, R. Waser, J.R. Jameson, M.N. Kozicki, Electrochemical metallization memories—fundamentals, applications, prospects, Nanotechnology 22 (25) (2011) 254003, https://doi.org/10.1088/0957-4484/22/28/289502.

[15] I. Valov, M.N. Kozicki, Cation-based resistance change memory, J. Phys. D Appl. Phys. 46 (8) (2013), https://doi.org/10.1088/0022-3727/46/7/074005.

[16] J.J. Yang, I.H. Inoue, T. Mikolajick, C.S. Hwang, Metal oxide memories based on thermochemical and valence change mechanisms, MRS Bull. 37 (2) (2012) 131–137, https://doi.org/10.1557/mrs.2011.356.

[17] M.A. Zidan, J.P. Strachan, W.D. Lu, The future of electronics based on memristive systems, Nat. Electron. 1 (1) (2018) 22–29, https://doi.org/10.1038/s41928-017-0006-8.

[18] M. Yi, Y. Cao, H. Ling, Z. Du, L. Wang, T. Yang, Q. Fan, L. Xie, W. Huang, Temperature dependence of resistive switching behaviors in resistive random access memory based on graphene oxide film, Nanotechnology 25 (18) (2014), https://doi.org/10.1088/0957-4484/25/18/185202.

[19] B. Hudec, C.W. Hsu, I.T. Wang, W.L. Lai, C.C. Chang, T. Wang, K. Frohlich, C.H. Ho, C.-H. Lin, T.H. Hou, 3D resistive RAM cell design for high-density storage class memory—a review, Sci. China Inf. Sci. 59 (6) (2016) 1–21, https://doi.org/10.1007/s11432-016-5566-0.

[20] R. Waser, R. Dittmann, M. Salinga, M. Wuttig, Function by defects at the atomic scale—new concepts for nonvolatile memories, Solid State Electron. 54 (9) (2010) 830–840, https://doi.org/10.1016/j.sse.2010.04.043.

[21] J.J. Yang, D.B. Strukov, D.R. Stewart, Memristive devices for computing, Nat. Nanotechnol. 8 (1) (2012) 13–24, https://doi.org/10.1038/nnano.2012.240.
[22] S.G. Hu, S.Y. Wu, W.W. Jia, Q. Yu, L.J. Deng, Y.Q. Fu, Y. Liu, T.P. Chen, Review of nanostructured resistive switching memristor and its applications, Nanosci. Nanotechnol. Lett. 6 (9) (2014) 729–757, https://doi.org/10.1166/nnl.2014.1888.
[23] S. Kim, J. Zhou, W.D. Lu, Crossbar RRAM arrays: selector device requirements during write operation, IEEE Trans. Electron Devices 61 (8) (2014) 2820–2826, https://doi.org/10.1109/TED.2014.2327514.
[24] I. Valov, T. Tsuruoka, Effects of moisture and redox reactions in VCM and ECM resistive switching memories, J. Phys. D Appl. Phys. 51 (2018), https://doi.org/10.1088/1361-6463/aad581, 413001.
[25] S. Larentis, F. Nardi, S. Balatti, D.C. Gilmer, D. Ielmini, Resistive switching by voltage-driven ion migration in bipolar RRAM-part II: modeling, IEEE Trans. Electron Devices 59 (9) (2012) 2468–2475, https://doi.org/10.1109/TED.2012.2202320.
[26] D.-H. Kwon, K.M. Kim, J.H. Jang, J.M. Jeon, M.H. Lee, G.H. Kim, X.-S. Li, G.-S. Park, B. Lee, S. Han, M. Kim, C.S. Hwang, Atomic structure of conducting nanofilaments in TiO_2 resistive switching memory, Nat. Nanotechnol. 5 (2) (2010) 148–153, https://doi.org/10.1038/nnano.2009.456.
[27] G.-S. Park, Y.B. Kim, S.Y. Park, X.S. Li, S. Heo, M.-J. Lee, M. Chang, J.H. Kwon, M. Kim, U.-I. Chung, R. Dittmann, R. Waser, K. Kim, In situ observation of filamentary conducting channels in an asymmetric Ta_2O_{5-x}/TaO_{2-x} bilayer structure, Nat. Commun. 4 (2013) 1–9, https://doi.org/10.1038/ncomms3382.
[28] S. Kumar, C.E. Graves, J.P. Strachan, E.M. Grafals, A.L.D. Kilcoyne, T. Tyliszczak, J.N. Weker, Y. Nishi, R.S. Williams, Direct observation of localized radial oxygen migration in functioning tantalum oxide memristors, Adv. Mater. 28 (14) (2016) 2772–2776, https://doi.org/10.1002/adma.201505435.
[29] Y. Yang, X. Zhang, M. Gao, F. Zeng, W. Zhou, S. Xie, F. Pan, Nonvolatile resistive switching in single crystalline ZnO nanowires, Nanoscale 3 (4) (2011) 1917–1921, https://doi.org/10.1039/c1nr10096c.
[30] J. Yao, L. Zhong, D. Natelson, J.M. Tour, In situ imaging of the conducting filament in a silicon oxide resistive switch, Sci. Rep. 2 (2012) 1–5, https://doi.org/10.1038/srep00242.
[31] Q. Liu, J. Sun, H. Lv, S. Long, K. Yin, N. Wan, Y. Li, L. Sun, M. Liu, Real-time observation on dynamic growth/dissolution of conductive filaments in oxide-electrolyte-based ReRAM, Adv. Mater. 24 (14) (2012) 1844–1849.
[32] X. Tian, L. Wang, J. Wei, S. Yang, W. Wang, Z. Xu, X. Bai, Filament growth dynamics in solid electrolyte-based resistive memories revealed by in situ TEM, Nano Res. 7 (7) (2014) 1065–1072, https://doi.org/10.1007/s12274-014-0469-0.
[33] G.S. Park, X.S. Li, D.C. Kim, R.J. Jung, M.J. Lee, S. Seo, Observation of electric-field induced Ni filament channels in polycrystalline Ni Ox film, Appl. Phys. Lett. 91 (22) (2007) 1–4, https://doi.org/10.1063/1.2813617.
[34] Q. Liu, S. Long, H. Lv, W. Wang, J. Niu, Z. Huo, J. Chen, M. Liu, Controllable growth of nanoscale conductive filaments in solid-electrolyte-based ReRAM by using a metal nanocrystal covered bottom electrode, ACS Nano 4 (10) (2010) 6162–6168, https://doi.org/10.1021/nn1017582.
[35] J. Lee, C. Du, K. Sun, E. Kioupakis, W.D. Lu, Tuning ionic transport in memristive devices by graphene with engineered nanopores, ACS Nano 10 (3) (2016) 3571–3579, https://doi.org/10.1021/acsnano.5b07943.
[36] U. Celano, L. Goux, A. Belmonte, A. Schulze, K. Opsomer, C. Detavernier, O. Richard, H. Bender, M. Jurczak, W. Vandervorst, Conductive-AFM tomography for 3D filament observation in resistive switching devices, in: Tech. Dig.—Int. Electron Devices Meet. IEDM, 2013, pp. 574–577, https://doi.org/10.1109/IEDM.2013.6724679.

[37] U. Celano, Y. Yin Chen, D.J. Wouters, G. Groeseneken, M. Jurczak, W. Vandervorst, Filament observation in metal-oxide resistive switching devices, Appl. Phys. Lett. 102 (12) (2013) 3–6, https://doi.org/10.1063/1.4798525.
[38] J.P. Strachan, G. Medeiros-Ribeiro, J.J. Yang, M.X. Zhang, F. Miao, I. Goldfarb, M. Holt, V. Rose, R.S. Williams, Spectromicroscopy of tantalum oxide memristors, Appl. Phys. Lett. 98 (24) (2011), https://doi.org/10.1063/1.3599589.
[39] A. Wedig, M. Luebben, D.Y. Cho, M. Moors, K. Skaja, V. Rana, T. Hasegawa, K.K. Adepalli, B. Yildiz, R. Waser, I. Valov, Nanoscale cation motion in TaO_x, HfO_x and TiO_x memristive systems, Nat. Nanotechnol. 11 (1) (2016) 67–74, https://doi.org/10.1038/nnano.2015.221.
[40] Y. Yang, P. Gao, S. Gaba, T. Chang, X. Pan, W. Lu, Observation of conducting filament growth in nanoscale resistive memories, Nat. Commun. 3 (2012) 732–738, https://doi.org/10.1038/ncomms1737.
[41] W. Hu, L. Zou, C. Gao, Y. Guo, D. Bao, High speed and multi-level resistive switching capability of Ta_2O_5 thin films for nonvolatile memory application, J. Alloys Compd. 676 (2016) 356–360, https://doi.org/10.1016/j.jallcom.2016.03.211.
[42] A. Prakash, J. Park, J. Song, J. Woo, E.J. Cha, H. Hwang, Demonstration of low power 3-bit multilevel cell characteristics in a TaO_x-based RRAM by stack engineering, IEEE Electron Device Lett. 36 (1) (2015) 32–34, https://doi.org/10.1109/LED.2014.2375200.
[43] S.H. Misha, N. Tamanna, J. Woo, S. Lee, J. Song, J.J. Park, S. Lim, J.J. Park, H. Hwang, Effect of nitrogen doping on variability of TaO_x-RRAM for low-power 3-bit MLC applications, ECS Solid State Lett. 4 (3) (2015) P25, https://doi.org/10.1149/2.0011504ssl.
[44] M. Azzaz, E. Vianello, B. Sklenard, P. Blaise, A. Roule, C. Sabbione, S. Bernasconi, C. Charpin, C. Cagli, E. Jalaguier, S. Jeannot, S. Denorme, P. Candelier, M. Yu, L. Nistor, C. Fenouillet-Beranger, L. Perniola, Endurance/retention trade off in HfO_x and TaO_x based RRAM, in: *2016 IEEE 8th International Memory Workshop (IMW)*, 2016, pp. 1–4, https://doi.org/10.1109/IMW.2016.7495268.
[45] H.Y. Lee, P.S. Chen, T.Y. Wu, Y.S. Chen, C.C. Wang, P.J. Tzeng, C.H. Lin, F. Chen, C.H. Lien, M.J. Tsai, Low power and high speed bipolar switching with a thin reactive Ti buffer layer in robust HfO_2 based RRAM, in: *2008 IEEE International Electron Devices Meeting (IEDM*, 2008, pp. 1–4, https://doi.org/10.1109/IEDM.2008.4796677.
[46] D. Ielmini, Filamentary-switching model in RRAM for time, energy and scaling projections, in: 2011 International Electron Devices Meeting, 2011, pp. 17.2.1–17.2.4, https://doi.org/10.1109/IEDM.2011.6131571.
[47] J. Park, S. Jung, J. Lee, W. Lee, S. Kim, J. Shin, H. Hwang, Resistive switching characteristics of ultra-thin TiO_x, Microelectron. Eng. 88 (7) (2011) 1136–1139, https://doi.org/10.1016/j.mee.2011.03.050.
[48] T.-N. Fang, S. Kaza, S. Haddad, A. Chen, Y.C. Wu, Z. Lan, S. Avanzino, D. Liao, C. Gopalan, S. Choi, S. Mahdavi, M. Buynoski, Y. Lin, C. Marrian, C. Bill, M. Vanbuskirk, M. Taguchi, Erase mechanism for copper oxide resistive switching memory cells with nickel electrode, in: 2006 International Electron Devices Meeting (IEDM, 2006, pp. 1–4, https://doi.org/10.1109/IEDM.2006.346731.
[49] F. Nardi, D. Ielmini, C. Cagli, S. Spiga, M. Fanciulli, L. Goux, D.J. Wouters, Control of filament size and reduction of reset current below 10 μA in NiO resistance switching memories, Solid State Electron. 58 (1) (2011) 42–47, https://doi.org/10.1016/j.sse.2010.11.031.
[50] F. Nardi, S. Larentis, S. Balatti, D.C. Gilmer, D. Ielmini, Resistive switching by voltage-driven ion migration in bipolar RRAM-part I: experimental study, IEEE Trans. Electron Devices 59 (9) (2012) 2461–2467, https://doi.org/10.1109/TED.2012.2202320.
[51] S. Yu, Y. Wu, H.S.P. Wong, Investigating the switching dynamics and multilevel capability of bipolar metal oxide resistive switching memory, Appl. Phys. Lett. 98 (10) (2011) 98–101, https://doi.org/10.1063/1.3564883.

[52] Y. Wu, S. Yu, H.S.P. Wong, Y.S. Chen, H.Y. Lee, S.M. Wang, P.Y. Gu, F. Chen, M.J. Tsai, AlOx-based resistive switching device with gradual resistance modulation for neuromorphic device application, in: *2012 4th IEEE International Memory Workshop (IMW)*, 2012, pp. 12–15, https://doi.org/10.1109/IMW.2012.6213663.

[53] B. Sarkar, B. Lee, V. Misra, Understanding the gradual reset in Pt/Al_2O_3/Ni RRAM for synaptic applications, Semicond. Sci. Technol. 30 (10) (2015) 105014, https://doi.org/10.1088/0268-1242/30/10/105014.

[54] F. Wu, S. Si, T. Shi, X. Zhao, Q. Liu, L. Liao, H. Lv, S. Long, M. Liu, Negative differential resistance effect induced by metal ion implantation in SiO_2 film for multilevel RRAM application, Nanotechnology 29 (5) (2018) 054001, https://doi.org/10.1088/1361-6528/aaa065.

[55] W. Ma, L. Liu, Y. Wang, Z. Chen, B. Chen, B. Gao, X. Liu, J. Kang, Multilevel resistive switching in HfO_x/TiO_x/HfO_x/TiO_x multilayer-based RRAM with high reliability, in: *2014 12th IEEE International Conference on Solid-State and Integrated Circuit Technology (ICSICT)*, 2014, pp. 5–7, https://doi.org/10.1109/ICSICT.2014.7021236.

[56] C.H. Huang, J.S. Huang, C.C. Lai, H.W. Huang, S.J. Lin, Y.L. Chueh, Manipulated transformation of filamentary and homogeneous resistive switching on ZnO thin film memristor with controllable multistate, ACS Appl. Mater. Interfaces 5 (13) (2013) 6017–6023, https://doi.org/10.1021/am4007287.

[57] M. Ismail, H. Abbas, C. Choi, S. Kim, Stabilized and RESET-voltage controlled multi-level switching characteristics in ZrO_2-based memristors by inserting a-ZTO interface layer, J. Alloys Compd. 835 (2020) 155256, https://doi.org/10.1016/j.jallcom.2020.155256.

[58] D. Ielmini, Modeling the universal set/reset characteristics of bipolar RRAM by field- and temperature-driven filament growth, IEEE Trans. Electron Devices 58 (12) (2011) 4309–4317, https://doi.org/10.1109/TED.2011.2167513.

[59] S. Yu, Y. Wu, Y. Chai, J. Provine, H.-S.P.S.P. Wong, Characterization of switching parameters and multilevel capability in HfO_x/AlO_x bi-layer RRAM devices, in: *Proceedings of 2011 International Symposium on VLSI Technology, Systems and Applications*, 2011, pp. 1–2, https://doi.org/10.1109/VTSA.2011.5872251.

[60] G. Wang, S. Long, Z. Yu, M. Zhang, Y. Li, D. Xu, H. Lv, Q. Liu, X. Yan, M. Wang, H. Liu, B. Yang, M. Liu, Impact of program/erase operation on the performances of oxide-based resistive switching memory, Nanoscale Res. Lett. 10 (1) (2015) 39, https://doi.org/10.1186/s11671-014-0721-2.

[61] S. Lee, J.B. Park, M.J. Lee, J.J. Boland, Multilevel resistance in ZnO nanowire memristors enabled by hydrogen annealing treatment, AIP Adv. 6 (12) (2016) 125010, https://doi.org/10.1063/1.4971820.

[62] W. Wang, G. Pedretti, V. Milo, R. Carboni, A. Calderoni, N. Ramaswamy, A.S. Spinelli, D. Ielmini, Learning of spatiotemporal patterns in a spiking neural network with resistive switching synapses, Sci. Adv. 4 (9) (2018) 1–9, https://doi.org/10.1126/sciadv.aat4752.

[63] D. Bhattacharjee, W. Kim, A. Chattopadhyay, R. Waser, V. Rana, Multi-valued and fuzzy logic realization using TaO_x memristive devices, Sci. Rep. 8 (1) (2018) 8, https://doi.org/10.1038/s41598-017-18329-3.

[64] K.-P. Chang, Y.-C. Chen, W.-C. Chien, E.-K. Lai, K.-P. Chang, Y.-C. Chen, W.-C. Chien, E.-K. Lai, Operating Method of Electrical Pulse Voltage for RRAM Application, 2012. U. S. Patent No. 8,134,865.

[65] A.C. Torrezan, J.P. Strachan, G. Medeiros-Ribeiro, R.S. Williams, Sub-nanosecond switching of a tantalum oxide memristor, Nanotechnology 22 (48) (2011) 485203, https://doi.org/10.1088/0957-4484/22/48/485203.

[66] B. Govoreanu, G.S. Kar, Y.Y. Chen, V. Paraschiv, S. Kubicek, A. Fantini, I.P. Radu, L. Goux, S. Clima, R. Degraeve, N. Jossart, O. Richard, T. Vandeweyer, K. Seo, P. Hendrickx, G. Pourtois, H. Bender, M. Jurczak, $10 \times 10 nm^2$ Hf/HfO_x crossbar resistive

RAM with excellent performance, reliability and low-energy operation, in: *2011 International Electron Devices Meeting*, 2011, pp. 729–732, https://doi.org/10.1109/IEDM.2011.6131652.

[67] C.-W. Hsu, I.-T. Wang, C.-L. Lo, M. Chiang, W. Jang, C.-H. Lin, T.-H. Hou, Self-rectifying bipolar TaO_x/TiO_2 RRAM with superior endurance over 10^{12} cycles for 3D high-density storage-class memory, in: *2013 Symposium on VLSI Circuits (VLSI)*, 2013, pp. T166–T167.

[68] S. Park, H. Kim, M. Choo, J. Noh, A. Sheri, S. Jung, K. Seo, J. Park, S. Kim, W. Lee, J. Shin, D. Lee, G. Choi, J. Woo, E. Cha, J. Jang, C. Park, H. Hwang, RRAM-based synapse for neuromorphic system with pattern recognition function, in: *2012 International Electron Devices Meeting (IEMD)*, 2012, pp. 10–12, https://doi.org/10.1109/IEDM.2012.6479016.

[69] G. Chen, C. Song, C. Chen, S. Gao, F. Zeng, F. Pan, Resistive switching and magnetic modulation in cobalt-doped ZnO, Adv. Mater. 24 (26) (2012) 3515–3520, https://doi.org/10.1002/adma.201201595.

[70] S. Otsuka, Y. Hamada, T. Shimizu, S. Shingubara, Ferromagnetic nano-conductive filament formed in $Ni/TiO_2/Pt$ resistive-switching memory, Appl. Phys. A Mater. Sci. Process. 118 (2) (2014) 613–619, https://doi.org/10.1007/s00339-014-8769-5.

[71] S. Otsuka, Y. Hamada, D. Ito, T. Shimizu, S. Shingubara, Magnetoresistance of conductive filament in $Ni/HfO_2/Pt$ resistive switching memory, Jpn. J. Appl. Phys. 54 (5) (2015), https://doi.org/10.7567/JJAP.54.05ED02.

[72] L. Li, Y. Liu, J. Teng, S. Long, Q. Guo, M. Zhang, Y. Wu, G. Yu, Q. Liu, H. Lv, M. Liu, Anisotropic magnetoresistance of nano-conductive filament in $co/HfO_2/Pt$ resistive switching memory, Nanoscale Res. Lett. 12 (1) (2017) 10–15, https://doi.org/10.1186/s11671-017-1983-2.

[73] H. Li, S. Wu, P. Hu, D. Li, G. Wang, S. Li, Light and magnetic field double modulation on the resistive switching behavior in $BaTiO_3/FeMn/BaTiO_3$ trilayer films, Phys. Lett. A 381 (25–26) (2017) 2127–2130, https://doi.org/10.1016/j.physleta.2017.04.039.

[74] X.L. Wang, Q. Shao, C.W. Leung, R. Lortz, A. Ruotolo, Nonvolatile, electric control of magnetism in Mn-substituted ZnO, Appl. Phys. Lett. 104 (6) (2014) 062409, https://doi.org/10.1063/1.4865428.

[75] D.R. Yakovlev, W. Ossau, Magnetic polarons, in: Introduction to the Physics of Diluted Magnetic Semiconductors, Springer, Berlin, Heidelberg, 2010, pp. 221–262, https://doi.org/10.1007/978-3-642-15856-8.

[76] C. Chen, C. Song, J. Yang, F. Zeng, F. Pan, Oxygen migration induced resistive switching effect and its thermal stability in W/TaOx/Pt structure, Appl. Phys. Lett. 100 (25) (2012) 253509, https://doi.org/10.1063/1.4730601.

[77] N.S. Kamarozaman, M.F.M. Soder, M.Z. Musa, R.A. Bakar, S.H. Herman, M. Rusop, Effect of post-deposition annealing process on the resistive switching behavior of TiO_2 thin films by sol-gel method, Appl. Mech. Mater. 393 (2013) 63–67, https://doi.org/10.4028/www.scientific.net/AMR.925.125.

[78] Y.M. Kim, J.S. Lee, Reproducible resistance switching characteristics of hafnium oxide-based nonvolatile memory devices, J. Appl. Phys. 104 (11) (2008), https://doi.org/10.1063/1.3041475.

[79] D. Lee, J. Woo, E. Cha, S. Lee, H. Hwang, Effects of high-pressure hydrogen annealing on the formation of conducting filaments in filament-type resistive random-access memory, J. Electron. Mater. 43 (9) (2014) 3635–3639, https://doi.org/10.1007/s11664-014-3265-4.

[80] Y. Sun, X. Yan, X. Zheng, Y. Liu, Y. Zhao, Y. Shen, Q. Liao, Y. Zhang, High on-off ratio improvement of ZnO-based forming-free memristor by surface hydrogen annealing,

ACS Appl. Mater. Interfaces 7 (13) (2015) 7382–7388, https://doi.org/10.1021/acsami.5b01080.
[81] L. Goux, J.Y. Kim, B. Magyari-Kope, Y. Nishi, A. Redolfi, M. Jurczak, H-treatment impact on conductive-filament formation and stability in Ta_2O_5-based resistive-switching memory cells, J. Appl. Phys. 117 (12) (2015) 124501, https://doi.org/10.1063/1.4915946.
[82] J. Hong, W. Jang, H. Song, C. Kang, H. Jeon, Endurance improvement due to rapid thermal annealing (RTA) of a TaOx thin film in an oxygen ambient, J. Korean Phys. Soc. 66 (5) (2015) 721–725, https://doi.org/10.3938/jkps.66.721.
[83] K. Miller, K.S. Nalwa, A. Bergerud, N.M. Neihart, S. Chaudhary, Memristive behavior in thin anodic Titania, IEEE Electron Device Lett. 31 (7) (2010) 737–739, https://doi.org/10.1109/LED.2010.2049092.
[84] K.A. Morgan, R. Huang, K. Potter, C. Shaw, W. Redman-White, C.H. De Groot, Total dose hardness of $TiN/HfO_x/TiN$ resistive random access memory, IEEE Trans. Nucl. Sci. 61 (6) (2014) 2991–2996, https://doi.org/10.1109/tns.2014.2365058.
[85] D.S. Shang, L.D. Chen, Q. Wang, W.D. Yu, X.M. Li, J.R. Sun, B.G. Shen, Crystallinity dependence of resistance switching in $La_{0.7}Ca_{0.3}MnO_3$ films grown by pulsed laser deposition, J. Appl. Phys. 105 (6) (2009), https://doi.org/10.1063/1.3082762.
[86] B.H. Lee, D. Il Lee, H. Bae, H. Seong, S.B. Jeon, M.L. Seol, J.W. Han, M. Meyyappan, S.G. Im, Y.K. Choi, Foldable and disposable memory on paper, Sci. Rep. 6 (December) (2016) 1–8, https://doi.org/10.1038/srep38389.
[87] D.H. Lien, Z.K. Kao, T.H. Huang, Y.C. Liao, S.C. Lee, J.H. He, All-printed paper memory, ACS Nano 8 (8) (2014) 7613–7619, https://doi.org/10.1021/nn501231z.
[88] S. Kim, Y.K. Choi, Resistive switching of aluminum oxide for flexible memory, Appl. Phys. Lett. 92 (22) (2008) 1–4, https://doi.org/10.1063/1.2939555.
[89] S. Kim, H. Moon, D. Gupta, S. Yoo, Y.K. Choi, Resistive switching characteristics of sol-gel zinc oxide films for flexible memory applications, IEEE Trans. Electron Devices 56 (4) (2009) 696–699, https://doi.org/10.1109/TED.2009.2012522.
[90] J. Won Seo, J.W. Park, K.S. Lim, S.J. Kang, Y.H. Hong, J.H. Yang, L. Fang, G.Y. Sung, H.-K. Kim, Transparent flexible resistive random access memory fabricated at room temperature, Appl. Phys. Lett. 95 (13) (2009) 1–4, https://doi.org/10.1063/1.3242381.
[91] S. Kim, O. Yarimaga, S.J. Choi, Y.K. Choi, Highly durable and flexible memory based on resistance switching, Solid State Electron. 54 (4) (2010) 392–396, https://doi.org/10.1016/j.sse.2009.10.021.
[92] H.Y. Jeong, J.Y. Kim, J.W. Kim, J.O. Hwang, J.E. Kim, J.Y. Lee, T.H. Yoon, B.J. Cho, S.O. Kim, R.S. Ruoff, S.Y. Choi, Graphene oxide thin films for flexible nonvolatile memory applications, Nano Lett. 10 (11) (2010) 4381–4386, https://doi.org/10.1021/nl101902k.
[93] H.Y. Jeong, Y.I. Kim, J.Y. Lee, S.Y. Choi, A low-temperature-grown TiO_2 based device for the flexible stacked RRAM application, Nanotechnology 21 (11) (2010) 115203, https://doi.org/10.1088/0957-4484/21/11/115203.
[94] Y.C. Liu, Z.Q. Wang, Y.X. Liu, X.H. Li, H.Y. Xu, X.T. Zhang, Flexible resistive switching memory device based on amorphous InGaZnO film with excellent mechanical endurance, IEEE Electron Device Lett. 32 (10) (2011) 1442–1444, https://doi.org/10.1109/led.2011.2162311.
[95] S. Kim, H.Y. Jeong, S.K. Kim, S.Y. Choi, K.J. Lee, Flexible memristive memory array on plastic substrates, Nano Lett. 11 (12) (2011) 5438–5442, https://doi.org/10.1021/nl203206h.
[96] Y.W. Dai, L. Chen, W. Yang, Q.Q. Sun, P. Zhou, P.F. Wang, S.J. Ding, D.W. Zhang, F. Xiao, Complementary resistive switching in flexible RRAM devices, IEEE Electron Device Lett. 35 (9) (2014) 915–917, https://doi.org/10.1109/LED.2014.2334609.
[97] H.G. Yoo, S. Kim, K.J. Lee, Flexible one diode-one resistor resistive switching memory arrays on plastic substrates, RSC Adv. 4 (38) (2014) 20017–20023, https://doi.org/10.1039/c4ra02536a.

[98] J.J. Huang, Y.M. Tseng, W.C. Luo, C.W. Hsu, T.H. Hou, One selector-one resistor (1S1R) crossbar array for high-density flexible memory applications, in: 2011 *Int. Electron Devices Meet*, 2011, pp. 31–37, https://doi.org/10.1109/IEDM.2011.6131653.

[99] C.H. Cheng, F.S. Yeh, A. Chin, Low-power high-performance nonvolatile memory on a flexible substrate with excellent endurance, Adv. Mater. 23 (7) (2011) 902–905, https://doi.org/10.1002/adma.201002946.

[100] C. Chou, C. Hsu, C. Chang, T. Hou, Self-rectifying Ta/TaO_x/TiO_2/Ti cell for high-density flexible RRAM, in: *2014 International Conference on Solid State Devices and Materials*, 2014, pp. 422–423.

[101] J. Jang, F. Pan, K. Braam, V. Subramanian, Resistance switching characteristics of solid electrolyte chalcogenide Ag_2Se nanoparticles for flexible nonvolatile memory applications, Adv. Mater. 24 (26) (2012) 3573–3576, https://doi.org/10.1002/adma.201200671.

[102] R.-C. Fang, Q.-Q. Sun, P. Zhou, W. Yang, P.-F. Wang, D.W. Zhang, High-performance bilayer flexible resistive random access memory based on low-temperature thermal atomic layer deposition, Nanoscale Res. Lett. 8 (1) (2013) 92, https://doi.org/10.1186/1556-276x-8-92.

[103] S. Kim, J.H. Son, S.H. Lee, B.K. You, K.-I. Park, H.K. Lee, M. Byun, K.J. Lee, Flexible crossbar-structured resistive memory arrays on plastic substrates via inorganic-based laser lift-off, Adv. Mater. 26 (44) (2014) 7418, https://doi.org/10.1002/adma.201470300.

[104] A. Midya, N. Gogurla, S.K. Ray, Flexible and transparent resistive switching devices using Au nanoparticles decorated reduced graphene oxide in polyvinyl alcohol matrix, Curr. Appl. Phys. 15 (6) (2015) 706–710, https://doi.org/10.1016/j.cap.2015.03.008.

[105] L. Liang, K. Li, C. Xiao, S. Fan, J. Liu, W. Zhang, W. Xu, W. Tong, J. Liao, Y. Zhou, B. Ye, Y. Xie, Vacancy associates-rich ultrathin nanosheets for high performance and flexible nonvolatile memory device, J. Am. Chem. Soc. 137 (8) (2015) 3102–3108, https://doi.org/10.1021/jacs.5b00021.

[106] K.N. Pham, V.D. Hoang, C.V. Tran, B.T. Phan, TiO_2 thin film based transparent flexible resistive switching random access memory, Adv. Nat. Sci. Nanosci. Nanotechnol. 7 (1) (2016) 015017, https://doi.org/10.1088/2043-6262/7/1/015017.

[107] J. Shang, W. Xue, Z. Ji, G. Liu, X. Niu, X. Yi, L. Pan, Q. Zhan, X.H. Xu, R.W. Li, Highly flexible resistive switching memory based on amorphous-nanocrystalline hafnium oxide films, Nanoscale 9 (21) (2017) 7037–7046, https://doi.org/10.1039/c6nr08687j.

[108] B. Sun, X. Zhang, G. Zhou, T. Yu, S. Mao, S. Zhu, Y. Zhao, Y. Xia, A flexible nonvolatile resistive switching memory device based on ZnO film fabricated on a foldable PET substrate, J. Colloid Interface Sci. 520 (2018) 19–24, https://doi.org/10.1016/j.jcis.2018.03.001.

[109] L. Wu, J. Guo, W. Zhong, W. Zhang, X. Kang, W. Chen, Y. Du, Flexible, multilevel, and low-operating-voltage resistive memory based on MoS_2–rGO hybrid, Appl. Surf. Sci. 463 (July 2018) (2019) 947–952, https://doi.org/10.1016/j.apsusc.2018.09.022.

[110] X. Zhao, R. Wang, X. Xiao, C. Lu, F. Wu, R. Cao, C. Jiang, Q. Liu, Flexible cation-based threshold selector for resistive switching memory integration, Sci. China Inf. Sci. 61 (6) (2018) 0–8, https://doi.org/10.1007/s11432-017-9352-0.

[111] M. Yang, X. Zhao, Q. Tang, N. Cui, Z. Wang, Y. Tong, Y. Liu, Stretchable and conformable synapse memristors for wearable and implantable electronics, Nanoscale 10 (38) (2018) 18135–18144, https://doi.org/10.1039/c8nr05336g.

[112] X. Yan, Z. Zhou, J. Zhao, Q. Liu, H. Wang, G. Yuan, J. Chen, Flexible memristors as electronic synapses for neuro-inspired computation based on scotch tape-exfoliated mica substrates, Nano Res. 11 (3) (2018) 1183–1192, https://doi.org/10.1007/s12274-017-1781-2.

[113] H. Bae, B.H. Lee, D. Lee, M.L. Seol, D. Kim, J.W. Han, C.K. Kim, S.B. Jeon, D. Ahn, S.J. Park, J.Y. Park, Y.K. Choi, Physically transient memory on a rapidly dissoluble paper

for security application, Sci. Rep. 6 (December) (2016) 1–7, https://doi.org/10.1038/srep38324.
[114] S. Lee, A. Sangle, P. Lu, A. Chen, W. Zhang, J.S. Lee, H. Wang, Q. Jia, J.L. MacManus-Driscoll, Novel electroforming-free nanoscaffold memristor with very high uniformity, tunability, and density, Adv. Mater. 26 (36) (2014) 6284–6289, https://doi.org/10.1002/adma.201401917.
[115] S. Schweiger, M. Kubicek, F. Messerschmitt, C. Murer, J.L.M. Rupp, A microdot multilayer oxide device: let us tune the strain-ionic transport interaction, ACS Nano 8 (5) (2014) 5032–5048, https://doi.org/10.1021/nn501128y.
[116] Y. Shi, Y. Ji, F. Hui, M. Nafria, M. Porti, G. Bersuker, M. Lanza, In situ demonstration of the link between mechanical strength and resistive switching in resistive random-access memories, Adv. Electron. Mater. 1 (4) (2015) 1400058, https://doi.org/10.1002/aelm.201400058.
[117] S. Ambrogio, S. Balatti, S. Choi, D. Ielmini, Impact of the mechanical stress on switching characteristics of electrochemical resistive memory, Adv. Mater. 26 (23) (2014) 3885–3892, https://doi.org/10.1002/adma.201306250.
[118] Y.Y. Chen, R. Roelofs, A. Redolfi, R. Degraeve, D. Crotti, A. Fantini, S. Clima, B. Govoreanu, M. Komura, L. Goux, L. Zhang, A. Belmonte, Q. Xie, J. Maes, G. Pourtois, M. Jurczak, Tailoring switching and endurance/retention reliability characteristics of HfO_2/Hf RRAM with Ti, Al, Si dopants, in: 2014 *Symp. VLSI Technol. Dig. Tech. Pap*, 2014, pp. 1–2, https://doi.org/10.1109/VLSIT.2014.6894403. no. 59.
[119] H. Zhang, L. Liu, B. Gao, Y. Qiu, X. Liu, J. Lu, R. Han, J. Kang, B. Yu, Gd-doping effect on performance of HfO_2 based resistive switching memory devices using implantation approach, Appl. Phys. Lett. 98 (4) (2011) 1–4, https://doi.org/10.1063/1.3543837.
[120] C.S. Peng, W.Y. Chang, Y.H. Lee, M.H. Lin, F. Chen, M.J. Tsai, Improvement of resistive switching stability of HfO_2 films with Al doping by atomic layer deposition, Electrochem. Solid St. 15 (4) (2012) 88–90, https://doi.org/10.1149/2.011204esl.
[121] K.C. Chang, T.M. Tsai, T.C. Chang, Y.E. Syu, S.L. Chuang, C.H. Li, D.S. Gan, S.M. Sze, The effect of silicon oxide based RRAM with tin doping, Electrochem. Solid St. 15 (3) (2012) 65–68, https://doi.org/10.1149/2.013203esl.
[122] N. Raghavan, R. Degraeve, A. Fantini, L. Goux, D.J. Wouters, G. Groeseneken, M. Jurczak, Stochastic variability of vacancy filament configuration in ultra-thin dielectric RRAM and its impact on OFF-state reliability, in: Tech. Dig.—Int. Electron Devices Meet. IEDM, 2013, pp. 21.1.1–21.1.4, https://doi.org/10.1109/IEDM.2013.6724674.
[123] B. Traore, P. Blaise, E. Vianello, H. Grampeix, S. Jeannot, L. Perniola, B. De Salvo, Y. Nishi, On the origin of low-resistance state retention failure in HfO_2-based RRAM and impact of doping/alloying, IEEE Trans. Electron Devices 62 (12) (2015) 4029–4036, https://doi.org/10.1109/TED.2015.2490545.
[124] H. Xu, D.H. Kim, Z. Xiahou, Y. Li, M. Zhu, B. Lee, C. Liu, Effect of co doping on unipolar resistance switching in Pt/Co:ZnO/Pt structures, J. Alloys Compd. 658 (2016) 806–812, https://doi.org/10.1016/j.jallcom.2015.11.018.
[125] N. Sedghi, H. Li, I.F. Brunell, K. Dawson, R.J. Potter, Y. Guo, J.T. Gibbon, V.R. Dhanak, W.D. Zhang, J.F. Zhang, J. Robertson, S. Hall, P.R. Chalker, The role of nitrogen doping in ALD Ta_2O_5 and its influence on multilevel cell switching in RRAM, Appl. Phys. Lett. 110 (10) (2017) 1–6, https://doi.org/10.1063/1.4978033.
[126] L.F. Liu, J.F. Kang, N. Xu, X. Sun, C. Chen, B. Sun, Y. Wang, X.Y. Liu, X. Zhang, R.Q. Han, Gd doping improved resistive switching characteristics of TiO_2-based resistive memory devices, Jpn. J. Appl. Phys. 47 (4 PART 2) (2008) 2701–2703, https://doi.org/10.1143/JJAP.47.2701.
[127] B. Gao, H.W. Zhang, S. Yu, B. Sun, L.F. Liu, X.Y. Liu, Y. Wang, R.Q. Han, J.F. Kang, B. Yu, Y.Y. Wang, Oxide-based RRAM: uniformity improvement using a new material-oriented methodology, in: Dig. Tech. Pap. Symp. VLSI Technol, 2009, pp. 30–31.

[128] H. Zhang, B. Gao, B. Sun, G. Chen, L. Zeng, L. Liu, X. Liu, J. Lu, R. Han, J. Kang, B. Yu, Ionic doping effect in ZrO_2 resistive switching memory, Appl. Phys. Lett. 96 (12) (2010) 98–101, https://doi.org/10.1063/1.3364130.
[129] C.Y. Liu, X.J. Lin, H.Y. Wang, C.H. Lai, Improved resistive switching dispersion of NiO_x thin film by Cu-doping method, Jpn. J. Appl. Phys. 49 (5 PART 1) (2010) 0565071–0565074, https://doi.org/10.1143/JJAP.49.056507.
[130] B. Gao, J.F. Kang, H.W. Zhang, B. Sun, B. Chen, L.F. Liu, X.Y. Liu, R.Q. Han, Y.Y. Wang, B. Yu, Z. Fang, H.Y. Yu, D.L. Kwong, Oxide-based RRAM: physical based retention projection, in: 2010 Proc. Eur. Solid State Device Res. Conf. ESSDERC 2010, 2010, pp. 392–395, https://doi.org/10.1109/ESSDERC.2010.5618200.
[131] F.M. Simanjuntak, D. Panda, K.H. Wei, T.Y. Tseng, Status and prospects of ZnO-based resistive switching memory devices, Nanoscale Res. Lett. 11 (1) (2016), https://doi.org/10.1186/s11671-016-1570-y.
[132] W. Duan, Y. Tang, X. Liang, C. Rao, J. Chu, G. Wang, Y. Pei, Solution processed flexible resistive switching memory based on Al-In-O self-mixing layer, J. Appl. Phys. 124 (10) (2018) 1–10, https://doi.org/10.1063/1.5041469.
[133] Y. Cai, J. Tan, L. Yefan, M. Lin, R. Huang, A flexible organic resistance memory device for wearable biomedical applications, Nanotechnology 27 (27) (2016) 275206, https://doi.org/10.1088/0957-4484/27/27/275206.
[134] T.Y. Wang, Z.Y. He, L. Chen, H. Zhu, Q.Q. Sun, S.J. Ding, P. Zhou, D.W. Zhang, An organic flexible artificial bio-synapses with long-term plasticity for neuromorphic computing, Micromachines 9 (5) (2018), https://doi.org/10.3390/mi9050239.
[135] C. Korte, J. Keppner, A. Peters, N. Schichtel, H. Aydin, J. Janek, Coherency strain and its effect on ionic conductivity and diffusion in solid electrolytes—an improved model for nanocrystalline thin films and a review of experimental data, Phys. Chem. Chem. Phys. 16 (44) (2014) 24575–24591, https://doi.org/10.1039/c4cp03055a.
[136] S. Schweiger, J.L.M. Rupp, Strain and Interfaces for Metal Oxide-Based Memristive Devices, Elsevier Inc, 2018, https://doi.org/10.1016/b978-0-12-811166-6.00014-5.
[137] J.L.M. Rupp, S. Schweiger, F. Messerschmitt, Strained Multilayer Resistive-Switching Memory Elements, Washington, DC: U.S. Patent and Trademark Office, 2017.
[138] S. Azad, O.A. Marina, C.M. Wang, L. Saraf, V. Shutthanandan, D.E. McCready, A. El-Azab, J.E. Jaffe, M.H. Engelhard, C.H.F. Peden, S. Thevuthasan, Nanoscale effects on ion conductance of layer-by-layer structures of gadolinia-doped ceria and zirconia, Appl. Phys. Lett. 86 (13) (2005) 131906, https://doi.org/10.1063/1.1894615.
[139] J. Garcia-Barriocanal, A. Rivera-Calzada, M. Varela, Z. Sefrioui, E. Iborra, C. Leon, S.J. Pennycook, J. Santamaria, Colossal ionic conductivity at interfaces of epitaxial ZrO_2: $Y_2O_3/SrTiO_3$ heterostructures, Science 321 (5889) (2008) 676–680, https://doi.org/10.1126/science.1156393.
[140] S. Sanna, V. Esposito, A. Tebano, S. Licoccia, E. Traversa, G. Balestrino, Enhancement of ionic conductivity in sm-doped ceria/yttria-stabilized zirconia heteroepitaxial structures, Small 6 (17) (2010) 1863–1867, https://doi.org/10.1002/smll.200902348.
[141] S. Schweiger, R. Pfenninger, W.J. Bowman, U. Aschauer, J.L.M. Rupp, Designing strained interface heterostructures for memristive devices, Adv. Mater. 29 (15) (2017), https://doi.org/10.1002/adma.201605049.
[142] K.H. Chen, K.C. Chang, T.C. Chang, T.M. Tsai, S.P. Liang, T.F. Young, Y.E. Syu, S.M. Sze, Illumination effect on bipolar switching properties of Gd:SiO_2 RRAM devices using transparent indium tin oxide electrode, Nanoscale Res. Lett. 11 (1) (2016) 2–6, https://doi.org/10.1186/s11671-016-1431-8.
[143] K.W. Hsu, H. Ren, R.J. Agasie, S. Bian, Y. Nishi, J.L. Shohet, Effects of neutron irradiation of ultra-thin HfO_2 films, Appl. Phys. Lett. 104 (3) (2014) 1–4, https://doi.org/10.1063/1.4863222.

[144] A. Mehonic, T. Gerard, A.J. Kenyon, Light-activated resistance switching in SiO_x RRAM devices, Appl. Phys. Lett. 111 (23) (2017) 233502, https://doi.org/10.1063/1.5009069.
[145] J. Holt, N. Cady, J. Yang-Scharlotta, Radiation testing of tantalum oxide-based resistive memory, in: *2015 IEEE International Integrated Reliability Workshop (IIRW)*, 2015, pp. 155–158, https://doi.org/10.1109/IIRW.2015.7437091.
[146] A.N. Mikhaylov, E.G. Gryaznov, A.I. Belov, D.S. Korolev, A.N. Sharapov, D.V. Guseinov, D.I. Tetelbaum, S.V. Tikhov, N.V. Malekhonova, A.I. Bobrov, D.A. Pavlov, S.A. Gerasimova, V.B. Kazantsev, N.V. Agudov, A.A. Dubkov, C.M.M. Rosário, N.A. Sobolev, B. Spagnolo, Field- and irradiation-induced phenomena in memristive nanomaterials, Phys. Status Solidi 13 (10–12) (2016) 870–881, https://doi.org/10.1002/pssc.201600083.
[147] M.J. Marinella, S.M. Dalton, P.R. Mickel, P.E.D. Dodd, M.R. Shaneyfelt, E. Bielejec, G. Vizkelethy, P.G. Kortula, Initial assessment of the effects of radiation on the electrical characteristics of TaO_x memristive memories, IEEE Trans. Nucl. Sci. 59 (6) (2012) 2987–2994, https://doi.org/10.1109/tns.2012.2224377.
[148] B.Y. Tsui, K.C. Chang, B.Y. Shew, H.Y. Lee, M.J. Tsai, Investigation of radiation hardness of HfO_2 resistive random access memory, in: *Proceedings of Technical Program—2014 International Symposium on VLSI Technology, Systems and Application, VLSI-TSA* 2014, 2014, pp. 1–2, https://doi.org/10.1109/VLSI-TSA.2014.6839675.
[149] J.S. Bi, Z.S. Han, E.X. Zhang, M.W. McCurdy, R.A. Reed, R.D. Schrimpf, D.M. Fleetwood, M.L. Alles, R.A. Weller, D. Linten, M. Jurczak, A. Fantini, The impact of X-ray and proton irradiation on HfO_2/Hf-based bipolar resistive memories, IEEE Trans. Nucl. Sci. 60 (6) (2013) 4540–4546, https://doi.org/10.1109/TNS.2013.2289369.
[150] S.L. Weeden-Wright, W.G. Bennett, N.C. Hooten, E.X. Zhang, M.W. McCurdy, M.P. King, R.A. Weller, M.H. Mendenhall, M.L. Alles, D. Linten, M. Jurczak, R. Degraeve, A. Fantini, R.A. Reed, D.M. Fleetwood, R.D. Schrimpf, TID and displacement damage resilience of 1T1R HfO_2/Hf resistive memories, IEEE Trans. Nucl. Sci. 61 (6) (2014) 2972–2978, https://doi.org/10.1109/tns.2014.2362538.
[151] H. García, M.B. González, M.M. Mallol, H. Castán, S. Dueñas, F. Campabadal, M.C. Acero, L. Sambuco Salomone, A. Faigón, Electrical characterization of defects created by γ-radiation in HfO_2-based MIS structures for RRAM applications, J. Electron. Mater. 47 (9) (2018) 5013–5018, https://doi.org/10.1007/s11664-018-6257-y.
[152] N. Arun, K.V. Kumar, A.P. Pathak, D.K. Avasthi, S.V.S.N. Rao, Hafnia-based resistive switching devices for nonvolatile memory applications and effects of gamma irradiation on device performance, Radiat. Eff. Defects Solids 173 (3–4) (2018) 239–249, https://doi.org/10.1080/10420150.2018.1425863.
[153] N. Arun, K. Vinod Kumar, A. Mangababu, S.V.S. Nageswara Rao, A.P. Pathak, Influence of the bottom metal electrode and gamma irradiation effects on the performance of HfO_2-based RRAM devices, Radiat. Eff. Defects Solids 174 (1–2) (2019) 66–75, https://doi.org/10.1080/10420150.2019.1579213.
[154] J. Bi, Y. Duan, F. Zhang, M. Liu, Total ionizing dose effects of 1 Mb HfO_2-based resistive-random-access-memory, in: *Proc. Int. Symp. Phys. Fail. Anal. Integr. Circuits, IPFA*, 2018-July, 2018, pp. 1–4, https://doi.org/10.1109/IPFA.2018.8452173.
[155] L. Zhang, R. Huang, D. Gao, P. Yue, P. Tang, F. Tan, Y. Cai, Y. Wang, Total ionizing dose (TID) effects on TaO_x-based resistance change memory, IEEE Trans. Electron Devices 58 (8) (2011) 2800–2804, https://doi.org/10.1109/TED.2011.2148121.
[156] Y.L. Wu, C.Y. Huang, J.J. Lin, Effects of radiation on the bipolar resistive switching characteristics of Al/HfO_2/ITO structure, in: *2012 IEEE 11th International Conference on Solid-State and Integrated Circuit Technology*, 2012, pp. 1–3, https://doi.org/10.1109/ICSICT.2012.6467646.
[157] R. Fang, Y. Gonzalez Velo, W. Chen, K.E. Holbert, M.N. Kozicki, H. Barnaby, S. Yu, Total ionizing dose effect of γ-ray radiation on the switching characteristics and

filament stability of HfO_x resistive random access memory, Appl. Phys. Lett. 104 (18) (2014), https://doi.org/10.1063/1.4875748.

[158] F. Yuan, S. Shen, Z. Zhang, γ-ray irradiation effects on $TiN/HfO_x/Pt$ resistive random access memory devices, in: *IEEE Aerospace Conference Proceedings*, 2015, pp. 1–7, https://doi.org/10.1109/AERO.2015.7119258.

[159] S.H. Lin, Y.L. Wu, Y.H. Hwang, J.J. Lin, Study of radiation hardness of HfO_2-based resistive switching memory at nanoscale by conductive atomic force microscopy, Microelectron. Reliab. 55 (11) (2015) 2224–2228, https://doi.org/10.1016/j.microrel.2015.03.009.

[160] K. Agashe, N. Sarwade, S. Joshi, M. Thakurdesai, S. Surwase, P. Tirmali, K. Asokan, Effect of gamma irradiation on resistive switching of $Al/TiO_2/n^+$ Si ReRAM, Nucl. Instrum. Methods Phys. Res., Sect. B 403 (2017) 38–44, https://doi.org/10.1016/j.nimb.2017.04.091.

[161] W. Duan, J. Wang, X. Zhong, The effect of γ-rays irradiation on the electrical properties of WO_x film-based memory cells, Europhys. Lett. 119 (2) (2017) 27003, https://doi.org/10.1209/0295-5075/119/27003.

[162] K.C. Liu, W.H. Tzeng, K.M. Chang, Y.C. Chan, C.C. Kuo, Effect of ultraviolet light exposure on a HfO_x RRAM device, Thin Solid Films 518 (24) (2010) 7460–7463, https://doi.org/10.1016/j.tsf.2010.05.024.

[163] J. Park, S. Lee, J. Lee, K. Yong, A light incident angle switchable ZnO nanorod memristor: reversible switching behavior between two nonvolatile memory devices, Adv. Mater. 25 (44) (2013) 6423–6429, https://doi.org/10.1002/adma.201303017.

[164] Y. Zhou, K.S. Yew, D.S. Ang, T. Kawashima, M.K. Bera, H.Z. Zhang, G. Bersuker, White-light-induced disruption of nanoscale conducting filament in hafnia, Appl. Phys. Lett. 107 (7) (2015) 0–5, https://doi.org/10.1063/1.4929324.

[165] B. Sun, M. Tang, J. Gao, C.M. Li, Light-controlled simultaneous resistive and ferroelectricity switching effects of $BiFeO_3$ film for a flexible multistate high-storage memory device, ChemElectroChem 3 (6) (2016) 896–901, https://doi.org/10.1002/celc.201600002.

[166] R.C. Baumann, Radiation-induced soft errors in advanced semiconductor technologies, IEEE Trans. Device Mater. Reliab. 5 (3) (2005) 305–315, https://doi.org/10.1109/TDMR.2005.853449.

[167] H.J. Barnaby, S. Malley, M. Land, S. Charnicki, A. Kathuria, B. Wilkens, E. Deionno, W.-M. Tong, Impact of alpha particles on the electrical characteristics of TiO_2 memristors, IEEE Trans. Nucl. Sci. 58 (6 PART 1) (2011) 2838–2844, https://doi.org/10.1109/TNS.2011.2168827.

[168] W.M. Tong, J.J. Yang, P.J. Kuekes, D.R. Stewart, R.S. Williams, E. DeIonno, E.E. King, S.-C. Witczak, M.D. Looper, J.V. Osborn, Radiation hardness of TiO_2 memristive junctions, IEEE Trans. Nucl. Sci. 57 (3 PART 3) (2010) 1640–1643, https://doi.org/10.1109/TNS.2010.2045768.

[169] B.V. Mistry, S.J. Trivedi, U.V. Chhaya, S.A. Khan, D.K. Avasthi, U.S. Joshi, RRAM properties of swift heavy ion irradiated $Ag/In_2O_3/Pt/Si$ heterostructures, Radiat. Eff. Defects Solids 168 (7–8) (Aug. 2013) 625–629, https://doi.org/10.1080/10420150.2013.792815.

[170] M. Alayan, M. Bagatin, S. Gerardin, A. Paccagnella, L. Larcher, E. Vianello, E. Nowak, B. De Salvo, L. Perniola, Experimental and simulation studies of the effects of heavy-ion irradiation on HfO_2-based RRAM cells, IEEE Trans. Nucl. Sci. 64 (8) (2017) 2038–2045, https://doi.org/10.1109/TNS.2017.2721980.

[171] M. Alayan, M. Bagatin, S. Gerardin, A. Paccagnella, E. Vianello, E. Nowak, B. De Salvo, L. Larcher, L. Perniola, Heavy-ion upset immunity of RRAM cells based on thin HfO_2 layers, in: *16th European Conference on Radiation and Its Effects on Components and Systems (RADECS)*, vol. 2016-Septe, 2016, pp. 1–5, https://doi.org/10.1109/RADECS.2016.8093174.

[172] F. Tan, R. Huang, X. An, Y. Cai, Y. Pan, W. Wu, H. Feng, X. Zhang, Y. Wang, Investigation on the response of TaO_x-based resistive random-access memories to heavy-ion irradiation, IEEE Trans. Nucl. Sci. 60 (6) (2013) 4520–4525, https://doi.org/10.1109/TNS.2013.2287615.
[173] P.E. Dodd, L.W. Massengill, Basic mechanisms and modeling of single-event upset in digital microelectronics, IEEE Trans. Nucl. Sci. 50 III (3) (2003) 583–602, https://doi.org/10.1109/TNS.2003.813129.
[174] Y. Gonzalez-Velo, H.J. Barnaby, M.N. Kozicki, Review of radiation effects on ReRAM devices and technology, Semicond. Sci. Technol. 32 (8) (2017) 083002, https://doi.org/10.1088/1361-6641/aa6124.
[175] Z. Ye, R. Liu, J.L. Taggart, H.J. Barnaby, S. Yu, Evaluation of radiation effects in RRAM-based neuromorphic computing system for inference, IEEE Trans. Nucl. Sci. 66 (1) (2019) 97–103, https://doi.org/10.1109/TNS.2018.2886793.
[176] W.G. Bennett, N.C. Hooten, R.D. Schrimpf, R.A. Reed, M.H. Mendenhall, M.L. Alles, J. Bi, E.X. Zhang, D. Linten, M. Jurzak, A. Fantini, Single- and multiple-event induced upsets in HfO_2/Hf 1T1R RRAM, IEEE Trans. Nucl. Sci. 61 (4) (2014) 1717–1725, https://doi.org/10.1109/TNS.2014.2321833.
[177] C.C. Shih, K.C. Chang, T.C. Chang, T.M. Tsai, R. Zhang, J.H. Chen, K.H. Chen, T.F. Young, H.L. Chen, J.C. Lou, T.J. Chu, S.Y. Huang, D.H. Bao, S.M. Sze, Resistive switching modification by ultraviolet illumination in transparent electrode resistive random access memory, IEEE Electron Device Lett. 35 (6) (2014) 633–635, https://doi.org/10.1109/LED.2014.2316673.

CHAPTER

12

Nonvolatile MO_X RRAM assisted by graphene and 2D materials

Qi Liu[a] *and Xiaolong Zhao*[b]

[a]Frontier Institute of Chip and System, Fudan University, Shanghai, China,
[b]School of Microelectronics, University of Science and Technology of China, Hefei, China

Nonvolatile MO_X resistive random access memory (RRAM) devices with high switching speed, low power dissipation, good endurance, long retention time, high I_{ON}/I_{OFF} ratio and high reliability, has been developed, while none of them possesses all these excellent performance at once. Therefore, improvement from the view of new materials and device structures is necessary and this has been explored on the way. 2D materials have shown excellent structural stability, mechanical strength (strong covalent bonds), thermal performance, optical and electrical properties [1]. For example, graphene possesses unparalleled breaking strength, high conductivity, ultra-high chemical and thermal stabilities [2,3]; MoS_2 has shown good flexibility, large Young's modulus [4], excellent thermal stability up to 1100 °C (melting point ~1185 °C) [1,5], and two different phases with different conductivity (conductive Tetragonal phase and less conductive Hexagonal phase, bandgap 1.8 eV) [1,6]; h-BN exhibits fascinating chemical inertness, inherently flexibility, high mechanical strength (breaking strength 15.7 Nm^{-1}), thermal stability (melting point ~2600 °C, stability in air up to 1000 °C, in Ar up to 2200 °C and in N_2 2400 °C), and excellent dielectric properties (4.69–5.9 eV band gap, 2–4 dielectric constant) [1,7–10]; Black phosphorus (BP) shows thickness-depended and tunable direct bandgap (0.3–2 eV), in-plane anisotropy and broadband nonlinear optical response [11,12], but owing to its poor environmental stability in the presence of oxygen and moisture, BP were generally

Metal Oxides for Non-volatile Memory
https://doi.org/10.1016/B978-0-12-814629-3.00012-X

functionalized before it is utilized for RRAM application [12,13]. The 2D materials have been successfully introduced into the fields of field-effect transistors (FETs), sensors, energy, storage and optoelectronics and bring fascinating results [14–18].

Introduction of 2D materials is also considered as a promising strategy in both academy and industry to improve the robustness of MO_X RRAM devices by combing the advantages of 2D materials and conventional MO_X. Various intrinsic and functionalized 2D material layers, such as graphene, graphene oxide, MoS_2, $MoSe_2$, h-BN and even black phosphorous have been studied in MO_X RRAM devices for the potential of additional functionalities. Since both the thickness and roughness of 2D layered materials can be controlled accurately at the atomic scale, the reliability and uniformity of the electronic devices based on such materials could also be optimized. Especially, various MO_X RRAM devices based on 2D materials have been demonstrated to possess excellent transparency, flexibility, printability, or thermal stability even under harsh conditions. In addition, the large-area synthetization of 2D materials (e.g., graphene, MoS_2, BN and so on) by state-of-art chemical vapor deposition (CVD) method illuminates a clear future of MO_X RRAM device based on 2D materials [19–23].

12.1 MO_X RRAM with graphene-based electrodes

Sheet resistance of graphene is $\sim 125\,\Omega/\text{square}$ at a single-layer thickness of 3 Å, which is significantly lower than that of any metal [20], making it an excellent choice as a flexible and transparent electrode, especially for wearable electronics. Development of large scale and high quality graphene has been realized by state-of-art CVD method [19,20,24]. Application of graphene electrode is agile, because it can be transferred to other arbitrary substrates with large area after removal of the supporting substrate and provides strong interfacial adhesion, namely van der Waals interaction (adhesive energy of $0.45\,J/m^2$ on SiO_2) [25]. Furthermore, the low-temperature ($\leq 400\,°C$) growth of graphene on Cu, Si et al. have been developed to make it a CMOS compatible technology for integration without transferring and high-temperature processes [26,27]. Graphene electrodes have been demonstrated to benefit the excellent transparency, flexibility, low power dissipation, and even the high-temperature stability characteristics of the MO_X RRAM devices [28–35].

12.1.1 Transparent and flexible graphene electrode

MO_X RRAM generally consists of an Electrode-Insulator-Electrode-Substrate structure. The MOx RRAM is transparent if all individual parts are transparent. To make the MO_X RRAM device with graphene electrode transparent, another transparent counter electrode is needed, and the commercially available glass substrate coated with a transparent ITO seems an extraordinary choice [31–34,36,37]. CVD graphene electrode devices not only show considerable RS performances, but also show good transparent characteristics. Shown in Fig. 12.1, all the given graphene/MO_X/ITO stacks show >70% optical transmittance in the 400–800nm wavelength region, while the graphene/SiO_x/graphene stack shows the

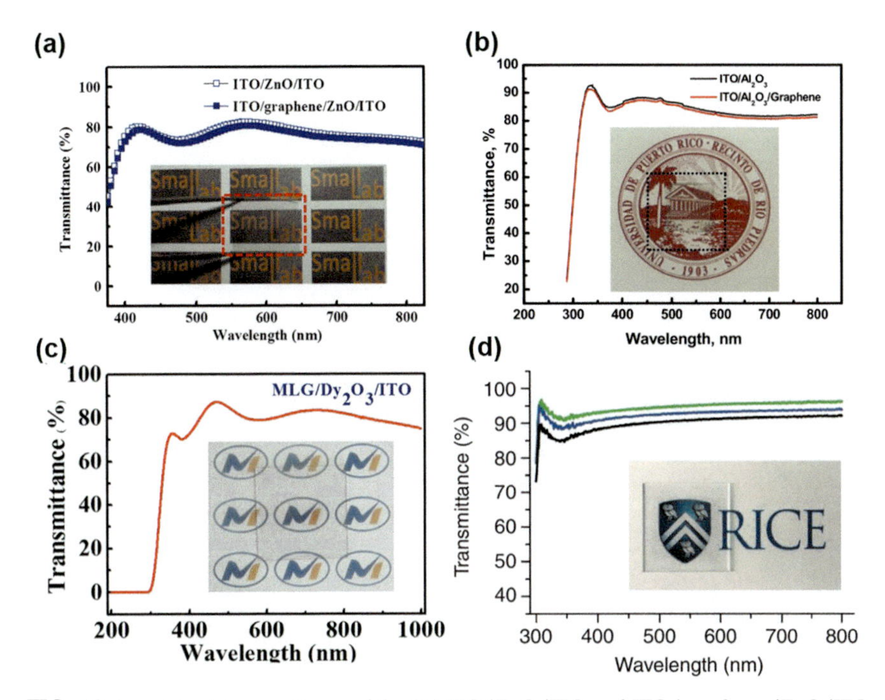

FIG. 12.1 Optical transmittance of the (A) ITO/ZnO/ITO and ITO/graphene/ZnO/ITO devices, (B) ITO/Al_2O_3 and ITO/Al_2O_3/graphene stacks, (C) MLG/Dy_2O_3/ITO device and (D) G/SiO_x/G-layered structure with different layer thicknesses of graphene. The green, blue and black curves correspond to the transmittance in SLG/SiO_x/SLG, SLG/SiO_x/BLG and BLG/SiO_x/BLG structures. Here SLG, BLG, and MLG denote single-layer graphene, bilayer graphene, and multilayer graphene, respectively. The insets show the photograph of the corresponding structures. *Panel (A) Reproduced with permission P.-K. Yang, W.-Y. Chang, P.-Y. Teng, S.-F. Jeng, S.-J. Lin, P.-W. Chiu, J.-H. He, Fully transparent resistive memory employing graphene electrodes for eliminating undesired surface effects, Proc. IEEE 101 (7) (2013) 1732–1739. Copyright 2013, IEEE. Panel (B) Reproduced with permission S. Dugu, S.P. Pavunny, T.B. Limbu, B.-R. Weiner, G. Morell, R.S. Katiyar, A graphene integrated highly transparent resistive switching memory device, APL Mater., 6 (5) (2018) 058503. Copyright 2018, American Institute of Physics. Panel (C) Reproduced with permission H. Zhao, H. Tu, F. Wei, J. Du, Highly transparent dysprosium oxide-based RRAM with multilayer graphene electrode for low-power nonvolatile memory application, IEEE Trans. Electron Devices 61 (5) (2014) 1388–1393. Copyright 2014, IEEE. Panel (D) Reproduced with permission J. Yao, J. Lin, Y. Dai, G. Ruan, Z. Yan, L. Li, L. Zhong, D. Natelson, J.M. Tour, Highly transparent nonvolatile resistive memory devices from silicon oxide and graphene, Nat. Commun. 3 (2012) 1101. Copyright 2012, Macmillan Publishers Limited.*

best transparent over 90%. With both single-layer graphene (SLG) as the electrodes, the optical transmittance can be as high as 95%.

Ji et al. made a comparison about flexibility between the MLG and oxide ITO electrodes [38]. Optical image (see Fig. 12.2A) of ITO film on PET substrate after bending 100 times displays many cracks and the sheet resistance of ITO film after bending increased dramatically, indicating bending instability of ITO film. While no obvious cracks were found on MLG on PET after bending 1000 times (see Fig. 12.2B) and the change of sheet resistance of MLG film before and after bending was negligible (within a few tens of Ω/square). Therefore, the graphene electrode shows

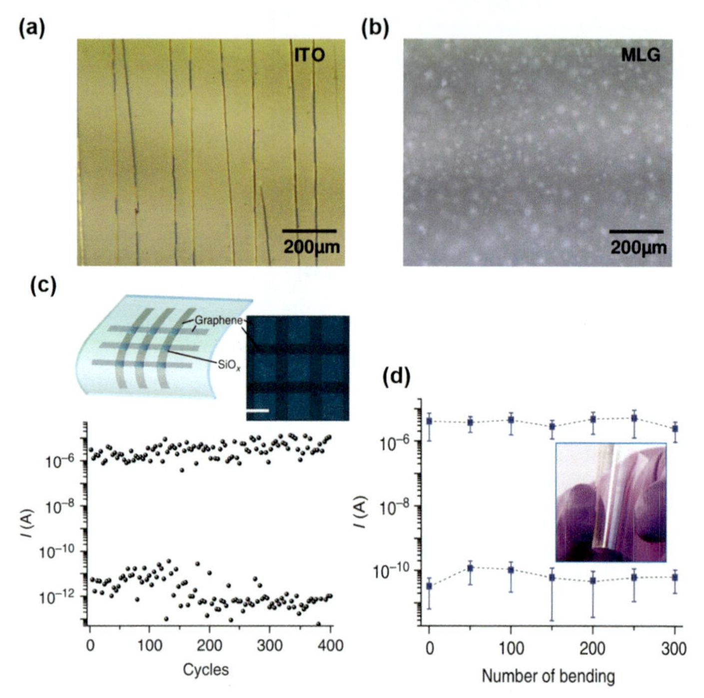

FIG. 12.2 (A) Optical image of ITO/PET film after 100 times bending. (B) Optical image of MLG/PET film after 1000 times bending. (C) Endurance of one crossbar device using +5 and +14 V as SET and RESET voltages. The memory states (current) were recorded at +1 V. The inset shows the schematic of $BLG/SiO_x/BLG$ crossbar structures on a plastic (fluoropolymer) substrate and corresponding optical image. Scale bar, 100 μm. (D) Evolution of both ON and OFF currents (read at +1 V) of the crossbar device during repeated bending test around a ~1.2 cm diameter curvature. The inset shows the actual transparent memory devices using the pillar structures on the plastic substrate. *Panels (A and B) Reproduced with permission Y. Ji, S. Lee, B. Cho, S. Song, T. Lee, Flexible organic memory devices with multilayer graphene electrodes, ACS Nano, 5 (7) (2011) 5995–6000. Copyright 2011, American Chemical Society. Panels (C and D) Reproduced with permission J. Yao, J. Lin, Y. Dai, G. Ruan, Z. Yan, L. Li, L. Zhong, D. Natelson, J. M. Tour, Highly transparent nonvolatile resistive memory devices from silicon oxide and graphene, Nat. Commun. 3 (2012) 1101. Copyright 2012, Macmillan Publishers Limited.*

much better flexibility to resist the destruction from bending stress, this might be ascribed the high mechanical strength of graphene. In order to make the graphene/MO_X-based RRAM device applicable for flexible electronics, this stack should be transferred onto a flexible substrate, which is the most popular way to fabricate a flexible RRAM device. Just like the situation in Ref. [31], after transplanted to a plastic substrate (fluoropolymer, melting point >280°C), the BLG/SiO_x/BLG cell with both graphene electrodes still shows stable memory switching characteristics (see Fig. 12.2C). And no obvious RS degradation was observed even after bending >300 times at a bending radius ≈1.2cm (see Fig. 12.2D).

12.1.2 Low power dissipation graphene electrode

High out-of-plane resistance [39–41], poor van der Waals interfaces with metal oxide, formation of graphene oxide [34] and thermal barrier property [16,42] of graphene help to decrease the RESET current. Compared with the TiN/Ti/HfO_2/Pt control device (C-RRAM) without graphene, the G-RRAM in Ref. [30] requires a higher forming voltage due to the built-in series resistance in G-RRAM (see Fig. 12.3A). Because the inserted SLG increases the overall low resistance state (LRS) of G-RRAM, a significant reduction of RESET current is observed under the same current compliance (not shown here). While, Fig. 12.3B shows the reduction of RESET current under the optimum SET I_{CC} condition of both G-RRAM (10 μA) and C-RRAM (100 μA). As shown in Fig. 12.3C, the G-RRAM shows 22 times lower averaged RESET current compared to C-RRAM. And even reported in Ref. [41], the reduction of RESET current can be as much as 330 times lower, companied with the improvement of switching uniformity. Fig. 12.3D shows the resistance distribution of the two samples. G-RRAM device offers higher LRS resistance and better high resistance state (HRS) uniformity. In Fig. 12.3E, the programming power consumption (calculated as the product of RESET voltage and RESET current) shows ~47 times reduction for G-RRAM as compared to C-RRAM. As shown in Fig. 12.3F–I, even relative higher LRS/HRS were obtained, introduction of graphene didn't degrade the endurance (120 DC cycles), retention (100° C, 10^5 s), read immunity (0.3 V, over 1000 s) and tunable LRS performances by I_{CC} (1–50 μA) of the device.

Similar low power situation is also reported by Qian et al. based on a Pt/Ti/TiO_x/G structure with a quality-tunable graphene electrode [40]. The device shows a reduction of nearly three orders of magnitude in the switching power dissipation owing to the fact that low quality graphene leads to higher LRS and HRS. Even for phase change memory, graphene can also act as a thermal barrier to decrease the programming current, as reported by Ahn et al. based on a Pt/$Ge_2Sb_2Te_5$/graphene/W

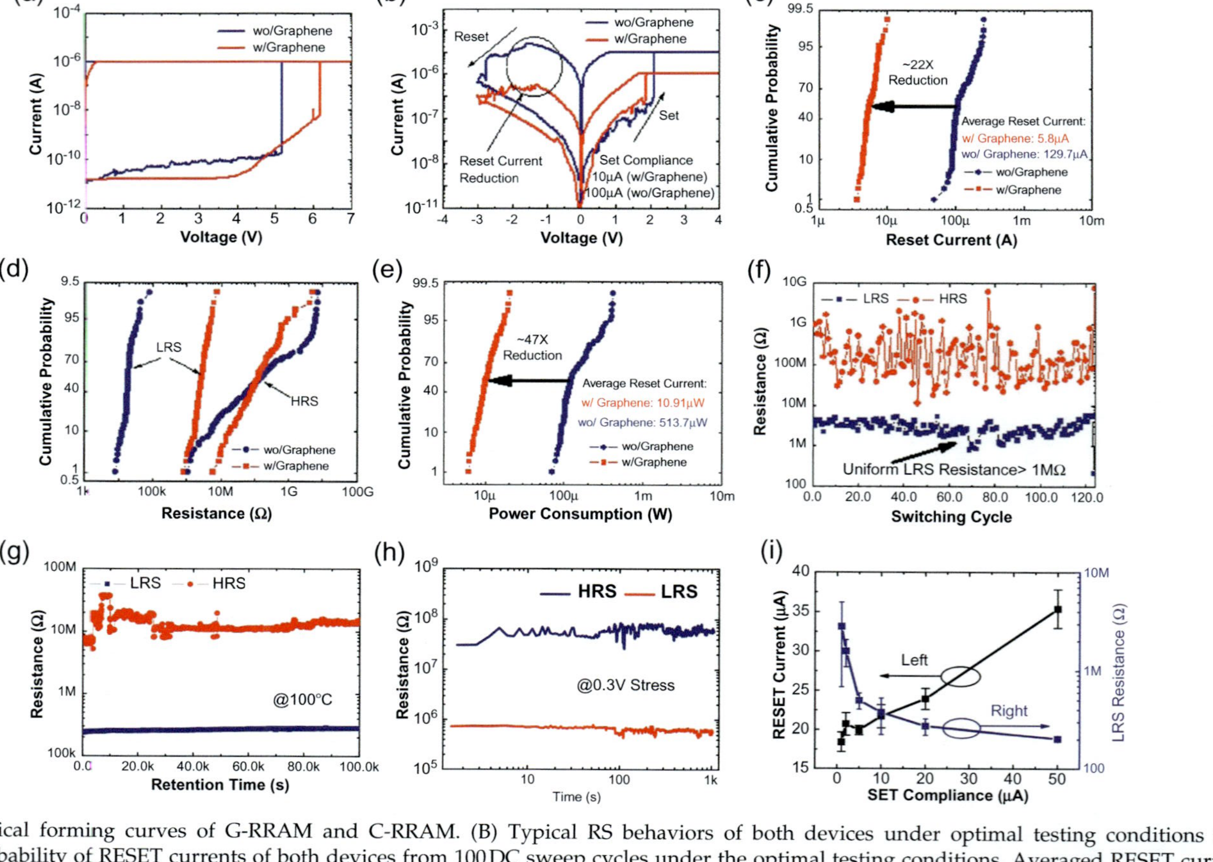

FIG. 12.3 (A) Typical forming curves of G-RRAM and C-RRAM. (B) Typical RS behaviors of both devices under optimal testing conditions respectively. (C) Accumulative probability of RESET currents of both devices from 100 DC sweep cycles under the optimal testing conditions. Averaged RESET current reduced up to 22 times. (D) Accumulative probability of HRS/LRS resistances obtained by DC sweep. The LRS resistance of G-RRAM is ~100 times higher due to the insertion of SLG. (E) Accumulative probability of RESET power consumption of both devices. Approximately 47 times reduction of power consumption is observed for G-RRAM. (F) DC switching endurance of G-RRAM. (G) Retention measurement of G-RRAM for 10^5 s at 100 °C is achieved. (H) Read immunity test of G-RRAM, where good Read-disturb immunity is achieved under a constant voltage stress of 0.3 V. (I) The RESET current and LRS resistance as a function of the SET I_{CC} of G-RRAM. *Reproduced with permission H. Tian, H.-Y. Chen, B. Gao, S. Yu, J. Liang, Y. Yang, D. Xie, J. Kang, T.-L. Ren, Y. Zhang, H. S. P. Wong, Monitoring oxygen movement by Raman spectroscopy of resistive random access memory with a graphene-inserted electrode. Nano Lett. 13 (2) (2013) 651–657. Copyright 2013, American Chemical Society.*

structure [42]. Owing to the introduction of graphene electrode, much more Joule heating generated at the graphene/oxide interface because of the high out-of-plane resistance and the high interfacial resistance of the graphene/oxide interface, this finally contributes to the reduced RESET current and device power dissipation. Above all, graphene is an ideal thermal barrier electrode for thermal interface engineering to decrease the device power dissipation.

12.1.3 Scaling down of MO_X RRAM by graphene edge electrode

As the most promising candidate for next generation nonvolatile memories, extreme scalability of MO_X RRAM down to sub-5nm makes it relatively competitive. The two-terminal metal–insulator–metal (MIM) structure is beneficial for the high-density 3D layered integration of RRAM to enable ultra-high-density storage. Many researches have been done to find out what is the scaling limit of the MO_X RRAM by scaling the size of the electrode [43] or the RS layer [44]. And the extreme lateral size ($6 \times 6\,nm^2$) of MO_X RRAM has been proved with considerable performance utilizing carbon nanotube (CNT) electrodes [45]. The key of device scaling is keeping the conductive filament (CF) stable when shrinking the device size. During the device shrinking process, fewer CFs with smaller sizes form in the effective device region. Reversible formation/rupture of stable CF renders the reliable operation of MO_X RRAM. Therefore, exploration of the limit CF size of the MO_X RRAM is critical to help understand the nature of CF inside the RRAM device.

Wong's group first exploited the atomically thin nature of the graphene edge to assemble RRAM devices (G-RRAM) in a vertical 3D cross-point architecture, as shown in Fig. 12.4A, where the yellow, green and black represents the TiN electrode, HfO_x RS layer and the graphene electrode, respectively. The RRAM cells are formed at the intersections of the TiN pillar electrode and the graphene plane electrode. The graphene plane electrode is really the thinnest (~3 Å thick) electrode with excellent conductivity among all electrodes, leading to the thinnest effective device region ever reported. HfO_x RS layer surrounds the TiN pillar electrode and is also in contact with the graphene plane electrode. The enlarged schematic illustration of the device cross-section is shown in Fig. 12.4B. TEM images of the device's cross-section is presented in Fig. 12.4C–E. The region where the graphene edge contacts the RS layer (HfO_x) is highlighted in red. In this experiment, RRAMs based on platinum electrodes (Pt-RRAM) were fabricated as control devices. Fig. 12.4F and G are the 3D cross-point architecture of the Pt-RRAM and the enlarged device details.

RS performances of both G-RRAM and Pt-RRAM are studied systematically. A comparison of the typical SET/RESET switching cycle of the G-RRAM with the Pt-RRAM is shown in Fig. 12.5A, while the inset shows

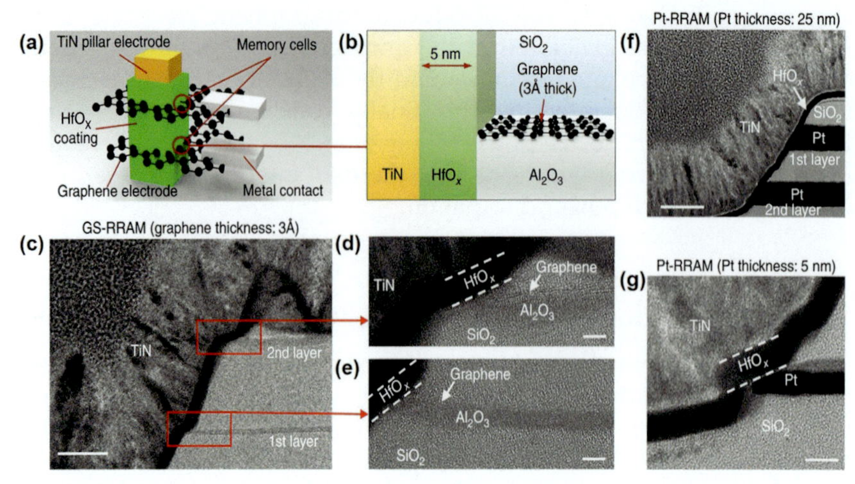

FIG. 12.4 Structure of graphene-based and Pt-based RRAM in vertical 3D cross-point architecture. (A) An illustration of G-RRAM in a vertical cross-point architecture. (B) A schematic cross-section of the graphene-based RRAM. (C) High-resolution TEM image of the two-stack G-RRAM structure. The RRAM memory elements are highlighted in red. Scale bar, 40nm. (D,E) First and second layer of G-RRAM with graphene on top of the Al_2O_3 layer. Scale bars, 5nm. (F,G) TEM image of the two-stack Pt-RRAM. Scale bars, 40nm (F) and 5nm (G). *Reproduced with permission S. Lee, J. Sohn, Z. Jiang, H.Y. Chen, H.S. Philip Wong, Metal oxide-resistive memory using graphene-edge electrodes, Nat. Commun. (6) (2015) 8407. Copyright 2015, Macmillan Publishers Limited.*

the magnified view of G-RRAM plot. The SET operation is achieved by applying a positive voltage to the TiN electrode in the Pt-RRAM and a negative voltage to the TiN electrode in the G-RRAM. The SET/RESET voltage, the RESET current and HRS/LRS (at 0.1 V bias) distribution of G-RRAM and Pt-RRAM after 50 switching cycles are shown in Fig. 12.5B and C. Importantly, the SET/RESET voltages and the RESET currents of G-RRAM are considerably lower than those of Pt-RRAM. Even with such low programming voltages and currents, the memory window is larger for G-RRAM compared with Pt-RRAM (right panel of Fig. 12.5C). Owing to such low SET/RESET voltages and currents, the power consumption (product of the programming voltages and the currents) of the G-RRAM is 300 times lower than that of the Pt-RRAM (see Fig. 12.5D). From the pulse-mode endurance test with 500ns width pulse, the switching energy ($P = V \times I \times t = 0.2\,V \times 2.3\,mA \times 500\,ns$) was found to be around 230 fJ. The temperature-accelerated LRS retention-time measurement indicates the excellent stability of the oxygen vacancy CF in the G-RRAM for information storage. Over 1600 endurance switching cycles without obvious write/read disturbances were observed under pulse mode with 500ns write/erase width and produced a >70 switching window. The CF can be limited to atomic scale by the ultra-thin graphene edge electrode, but the G-RRAM devices still show perfect RS performances, especially including the

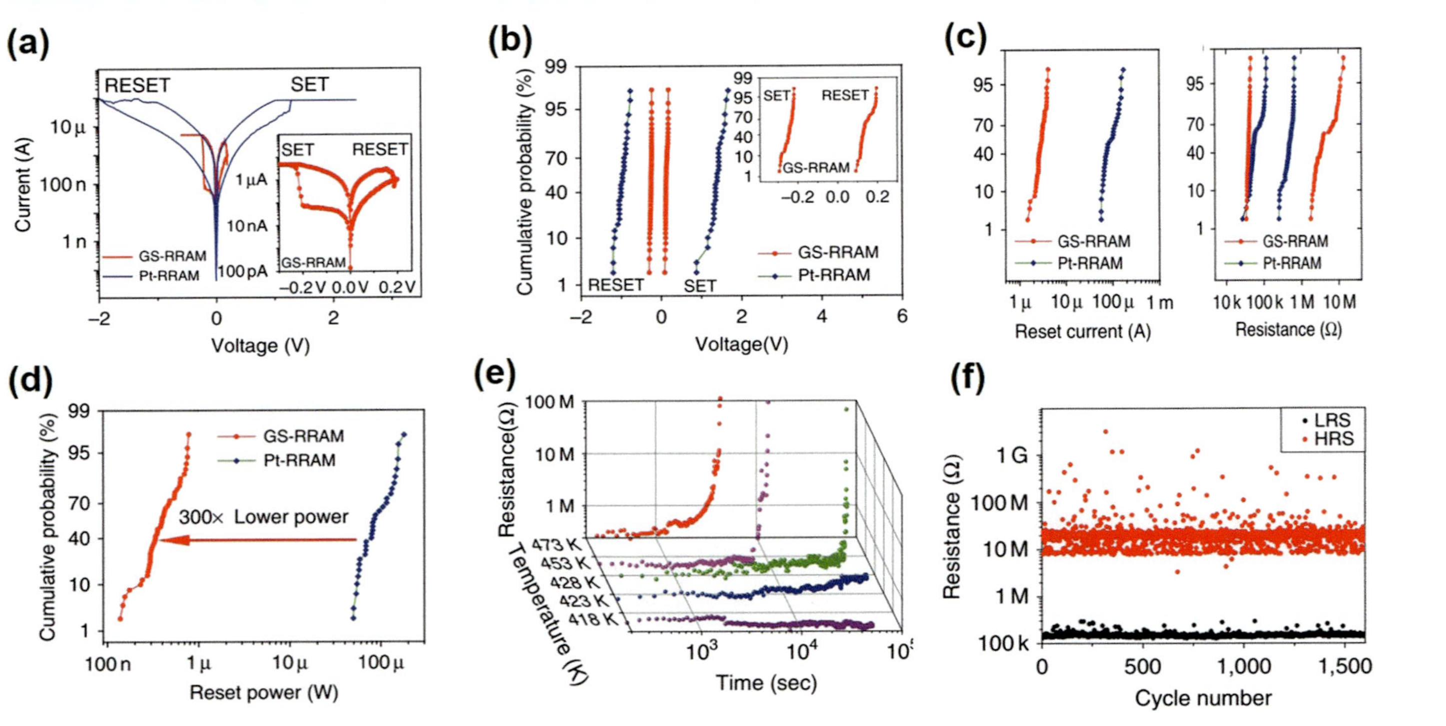

FIG. 12.5 The device characteristics of G-RRAM and Pt-RRAM. (A) Typical DC I–V switching characteristics of G-RRAM and Pt-RRAM. For Pt-RRAM, SET process is observed when positive voltage is applied to TiN. For G-RRAM, SET process is observed when positive voltage is applied to graphene. The SET compliances for G-RRAM and Pt-RRAM are 5 and 80 mA, respectively, for optimum conditions. A magnified plot of G-RRAM is shown as inset. (B) The SET and RESET voltage distribution of G-RRAM and Pt-RRAM after 50 cycles of switching. The SET/RESET voltages of G-RRAM are noticeably lower (inset). (C) RESET current distribution (left panel) and resistance distribution after 50 cycles for both device at 0.1 V. G-RRAM exhibits much lower RESET current and larger memory windows compared with Pt-RRAM. (D) RESET power distribution of G-RRAM and Pt-RRAM. The power consumption of G-RRAM is 300 times lower than that of Pt-RRAM. This is from the combined effect of lower programming voltages and currents. (E) Temporal evolution of G-RRAM LRS resistance at temperatures ranging from 418 to 473 K near 0.1 V bias. (F) Pulse endurance test of G-RRAM. Device switched with over 70× difference in HRS and LRS, and suffered no read/write disturbance after 1600 cycles. *Reproduced with permission S. Lee, J. Sohn, Z. Jiang, H.Y. Chen, H.S. Philip Wong, Metal oxide-resistive memory using graphene-edge electrodes, Nat. Commun. 6 (2015) 8407. Copyright 2015, Macmillan Publishers Limited.*

stability under high temperature (418–473 K), indicating the extremely shrinkable potential of the MO_X RRAM for high-density storage application.

12.1.4 Heat-resistant graphene electrode for high-temperature device

Graphene is a native heat-resistant material [3], as generally grown under high temperature by CVD method. Annealing under high temperature progressively helps to cure the lattice defects even under 2800 °C [46], which can be easily identified by Raman spectra. Romanenko et al. found that the electrical conductivity of MLG and graphite flake-like structures increases with the increment of temperature, demonstrating the negative temperature coefficient of their resistance [47]. And the graphene paper has also been proved that its electronic conductivity enhances with increased temperature in Ref. [48]. All these results are coincident with the calculation model of the graphene conductivity based on Boltzmann transport equation and 2D electron gas theory [49]. The calculated conductivity of few layer graphene (FLG) and relative thicker graphene nanosheet (GN) structures as a function of thickness and temperature is shown in Fig. 12.6, where the conductivity decreases with increasing thickness, while the conductivity increases with increment of temperature (inset).

Owing to the excellent heat-resistant property of graphene as mentioned above, graphene electrode can be utilized to improve the thermal stability of electronic device, which is of great importance, especially in the field of in military, aerospace, automobile and energy industries. Wang et al. [35] designed the MLG/$MoS_{2-x}O_x$/MLG (GMG device, layer thicknesses of 8/40/8 nm) MO_X RRAM device, where the exfoliated and oxidized MoS_2 ($MoS_{2-x}O_x$) is utilized as insulator sandwiched by exfoliated MLG top and bottom electrode (BE). As shown in Fig. 12.7, this tailored MO_X RRAM structure shows stable I–V curves from 20 to 340 °C with small switching voltages, nice endurance at 100 200 and 300 °C, and robust retention property at 340 °C for more than 10^5 s. This GMG RRAM device shows the most excellent heat-resistant performance among various reported RRAM devices, where the MO_X and organic RRAM devices were reported to operate under no more than 200 °C [50,51], and even some of them show increased leakage current and decreased R_{OFF}/R_{ON} ratio under elevated temperature.

12.1.5 MO_X RRAM with reduced graphene oxide (rGO) electrode

As the derivates of graphene, graphene oxide (GO) shows insulator property and has been explored for RRAM application [52–55], while the reduced graphene oxide (rGO) can act as both insulator and conductor with

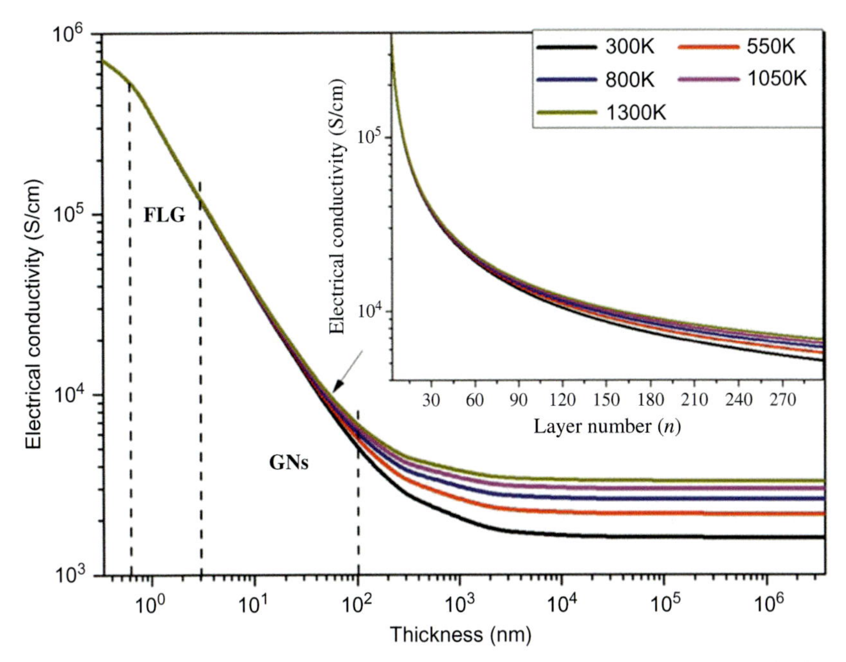

FIG. 12.6 Thickness and temperature dependent of FLG/GN conductivity calculated based on Boltzmann transport equation and 2D electron gas theory. *Reproduced with permission X.-Y. Fang, X.-X. Yu, H.-M. Zheng, H.-B. Jin, L. Wang, M.-S. Cao, Temperature- and thickness-dependent electrical conductivity of few-layer graphene and graphene nanosheets, Phys. Lett. A 379 (37) (2015) 2245–2251 Copyright 2015, ELSEVIER.*

excellent electronic conductivity as graphene, according to the degree of reduction [56]. rGO has been studied as the electrode in solar cells [57], FETs [58,59], and RRAM device [60–62]. And even, an all rGO based RRAM has been reported for WORM (write-once-read-many) electronics application with a ON/OFF ratio of 10^2 and nice flexibility, with the rGO obtained by the mostly used solution and annealing process of GO film [56].

Xu et al. reported the utilization of rGO electrode in MO_X RRAM device with a rGO/MoS_2 nanosphere/ITO structure [62]. To fabricate the rGO electrode onto the MoS_2 nanosphere film, the GO film was prepared first via spin-coating method on a SiO_2/Si substrate. Afterward, the GO film was reduced at 1000 °C to obtain the rGO film, which was later transferred to the as-grown MoS_2/ITO stack and patterned as the top electrode. As shown in Fig. 12.8, the device shows low SET/RESET voltage (~2 V) and high ON/OFF resistance ratio (~10^4). The RS behavior here may ascribe to the tunable polarization potential barriers between the MoS_2 nanospheres by external electric field.

Tian et al. fabricated a cost-effective, transfer-free and flexible MO_X RRAM using laser-scribed rGO patterning technology, where the as-grown GO film was reduced by laser to form the bottom electrode [61]. The resultant rGO lines could be directly used as the electrode for

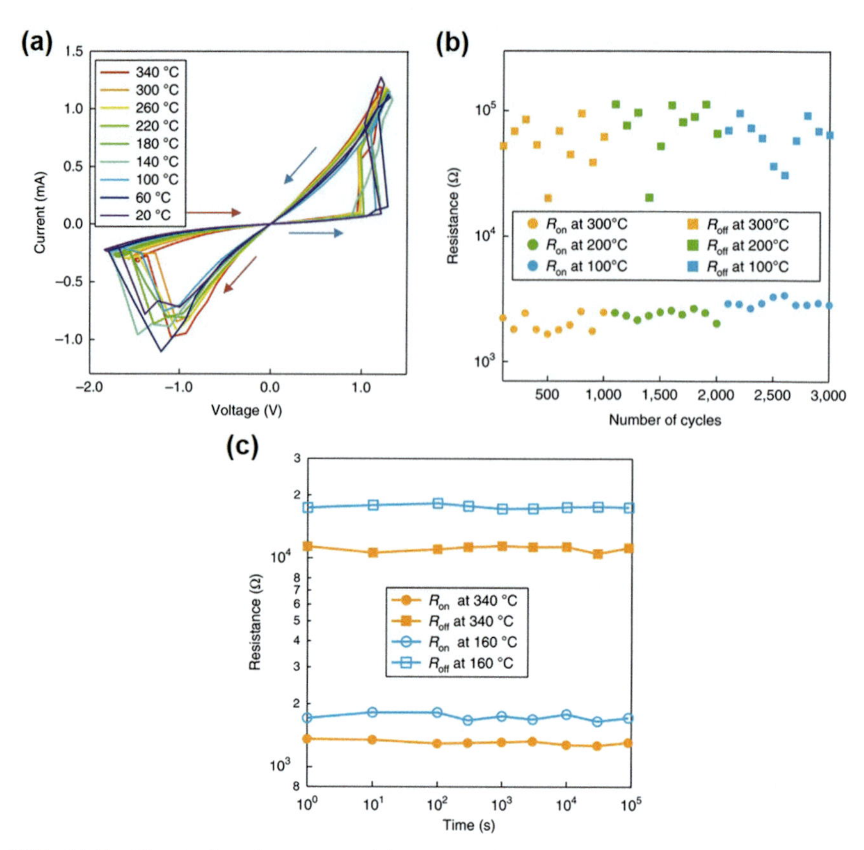

FIG. 12.7 Electrical performances of the MLG/$MoS_{2-x}O_x$/MLG devices at elevate temperatures. (A) RS curves of a GMG device at different temperatures. The arrows indicate the switching directions. The GMG device can operate at a high temperature up to 340 °C, demonstrating high thermal stability. (B) 1000 RS cycle endurance under 1 μs pulse at 300 °C (orange), 200 °C (green) and 100 °C (blue), respectively. (C) Both HRS and LRS can maintain 10^5 s at 340 °C (orange) and 160 °C (blue) without obvious exacerbation. The resistance values were read at 0.1 V. The good endurance and retention time further demonstrate the high thermal stability of the GMG devices. *Reproduced with permission M. Wang, S. Cai, C. Pan, C. Wang, X. Lian, Y. Zhuo, K. Xu, T. Cao, X. Pan, B. Wang, S.-J. Liang, J.J. Yang, P. Wang, F. Miao, Robust memristors based on layered two-dimensional materials, Nat. Electron. 1 (2) (2018) 130–136. Copyright 2018, Macmillan Publishers Limited.*

integration of electronic devices. Then, 10 nm HfO_x thin film was deposited onto the patterned rGO electrode as the active layer followed by the deposition of curable silver paste as the top electrode. It is attractive that the laser-scribing technology is very time-efficient and the patterning of a memory array with an area up to 100 cm^2 could be finished within 25 min. Compared to conventional lithography, no photoresist is required, thus the patterned rGO could remain clean for further device fabrication.

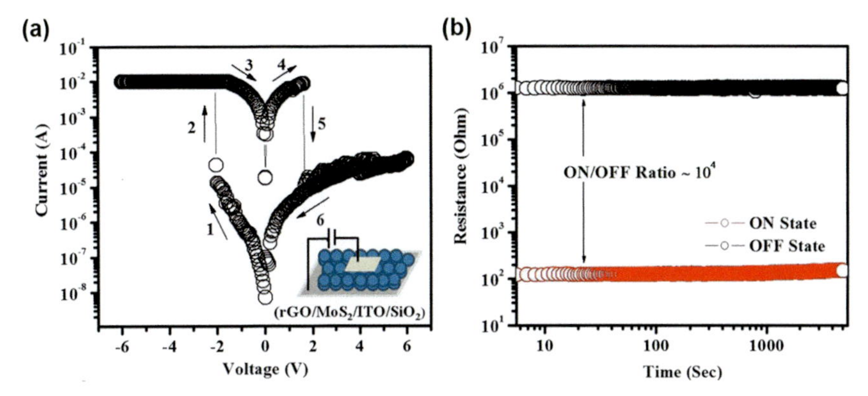

FIG. 12.8 (A) Typical I–V curve and schematic illustration (inset) of the $rGO/MoS_2/ITO$ RRAM device. (B) Retention tests of the ON and OFF resistance states with the a ratio of around 10^4 at -1 V. *Reproduced with permission X.-Y. Xu, Z.-Y. Yin, C.-X. Xu, J. Dai, J.-G. Hu, Resistive switching memories in MoS2 nanosphere assemblies, Appl. Phys. Lett. 104 (3) (2014) 033504. Copyright 2014, American Institute of Physics.*

The fabricated MO_X RRAM device with a configuration of $Ag/HfO_x/rGO$ exhibited a reliable RS memory behavior (see Fig. 12.9), including uniform I–V endurance for more than 100 cycles with excellent reproducibility, multiple LRS obtained by modulating the I_{CC} and stable LRS/HRS retention for more than 10^4 s at room temperature without obvious degradation. The excellent performance of the device demonstrated that the laser-scribing technology is potentially feasible for the construction of more cost- and time-effective electronic devices based on rGO electrodes. Similar to graphene, the rGO has been proved a proper flexible electrode. It's obvious that almost all these devices are with large sizes (tens of micrometers), this might be ascribed to CMOS-incompatible methods during the device fabrication, especially the preparation process of rGO electrode. Therefore, the fabrication of tiny RRAM devices with rGO electrodes is stills a big challenge.

12.2 Modulating ion migration in MO_X RRAM by 2D materials

Generally, ion migration and the redox in the MIM structure happen randomly under external electric field both in electrochemical mechanism (ECM) and valence change mechanism (VCM) RRAM devices. The resulted random formation/rupture of CFs contributes to the dispersive switching parameters [63]. Accumulation of cations in the RS layer of an ECM device after long-time repeating switching cycles leads to the deterioration of resistance states and device reliability [64–66]. The continuous loss of anions (e.g., oxygen ions) in the RS layer of a VCM device after

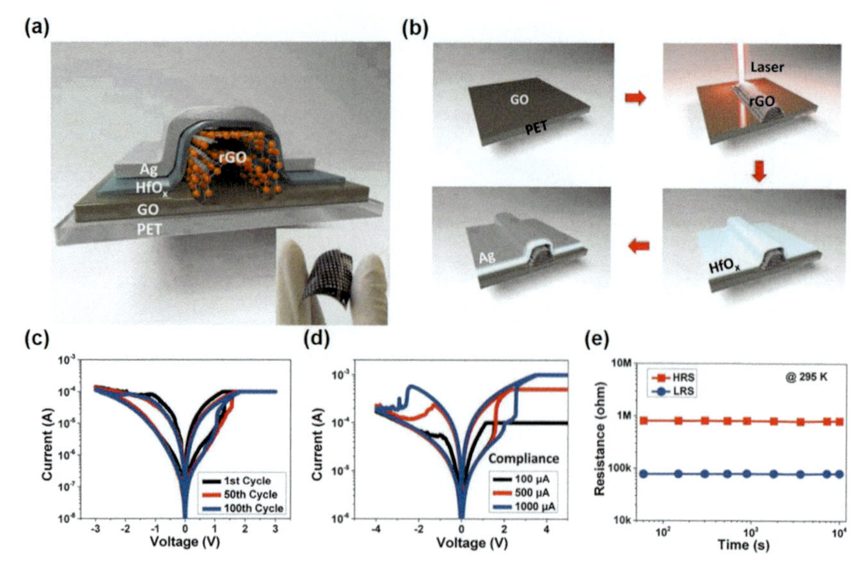

FIG. 12.9 (A) Schematic illustrations of the Ag/HfO_x/rGO/PET device. The inset shows the photograph of devices. (B) Schematic illustrations of the fabrication processes of the device. (C) RS I–V curves of the device at the 1st, 50th and 100th cycle, respectively. (D) RS I–V curves of the device under different I_{CC}. (E) Retention measurement of the device at room temperature. *Reproduced with permission H. Tian, H.Y. Chen, T.L. Ren, C. Li, Q.T. Xue, M.A. Mohammad, C. Wu, Y. Yang, H.S. Wong, Cost-effective, transfer-free, flexible resistive random access memory using laser-scribed reduced graphene oxide patterning technology, Nano Lett. 14 (6) (2014) 3214–9.Copyright 2014, American Chemical Society.*

long-time repeating switching cycles may also lead to the RESET failure and even hard breakdown of the RS layer [67–69]. Therefore, the random formation of CFs and over injection/loss of cations/anions are the basic scientific issues of the RRAM devices. Modulation of ion migration by 2D materials provides potential approaches to solve these problems.

12.2.1 Inducement of ion migration by sporadic 2D material fragments

Even though state-of-art CVD technology have been developed to synthetize high-quality 2D materials, such as graphene, MoS_2, BN and so on, mass production of 2D materials are generally based on solution-assisted-exfoliated 2D material fragments. Sporadic 2D material fragments can also play an important role in the ion transmission and CF formation in the RRAM devices, thus modulating the RS performance. For example, graphene quantum dots (GQDs) help to better control of the RS behaviors for analog switching [70,71]; SiO_2 film doped by network-type graphene oxide has been used as an intercalation to suppress the overshoot issue and enhance the switching stability [72]; MoS_2, WS_2 and black phosphorus nanosheets or quantum dots were generally dispersed into organic solvent for organic RRAM devices or even ink for printable electronics (organic devices are beyond the discussion of this chapter) [11,60,73,74] .

Graphene quantum dots (GQDs) are edge-functionalized graphene nanosheets bonded with abundant and easily detached oxygen functional groups, which can be validated by photoluminescence spectrometry and Fourier transform infrared spectrometry [70,71]. Bonding with various functional groups, GQDs with ample material availability possess fascinating electronic properties because of quantum confinement and edge effects [75,76]. The poor repeatable analog resistance states of memristive devices require numerous training epochs, resulting in low-efficient operation and energy waste of controllable synaptic behavior for neuromorphic computing [77]. Basically, the poor repeatability of analog resistance states is closely associated with the random nature of CF formation. Highly repeatable analog switching of MO_X RS device is urgently needed for high-density artificial neural networks. Wang et al. proposed to improve the analog switching behavior of Pt/FeO_x/Pt (Control device) by introducing GQDs at the FeO_x/TE interface (GQDMem) [71]. The GQDs were synthesized by hydrothermal methods and identified to possess oxygen functional groups, such as –COOH, –OH and epoxides (see Fig. 12.10A). As results, the GQDMem devices show larger initial OFF current, smaller forming voltage, tightly distributed HRS/LRS, lower and uniformed operating voltages under DC measurement (see Fig. 12.10B–F). Furthermore, compared with the control devices, the highly repeatable analog resistance states of GQDMem devices can be validated under different RESET pulse magnitudes with more uniform four-level resistance distribution (see Fig. 12.10G–J). With these samples, analog resistance states of GQDMem devices with highly tight distribution were achieved with nearly 85% reduction in variations. Besides, switching voltages are reduced by 40% and their distribution is also narrowed by 84%.

Two possible reasons may contribute to the improvement of the analog switching of GQDMem device: (i) modified energy band by functional groups of GQDs; (ii) weak bond of oxygen functional groups. The embedded GQDs may lower the conduction band offset for electrons and facilitate the tunneling at low bias region, contributing to the lower forming voltage and SET/RESET voltages (see Fig. 12.11A and B). Abundant oxygen functional groups (e.g., oxygen anions) can release from GQDs, making GQDs as a role of oxygen-reservoirs. Oxygen-based atomic level oxygen functional groups detach from the GQD framework under Joule heating, because of the weak chemical bond and migrate into the RS layer acting as trapping centers to suppress the stochastic nature of CF. These trapping centers localize the formation/rupture of CF along the GQDs and contribute to the uniformed and reduced operating voltages (see

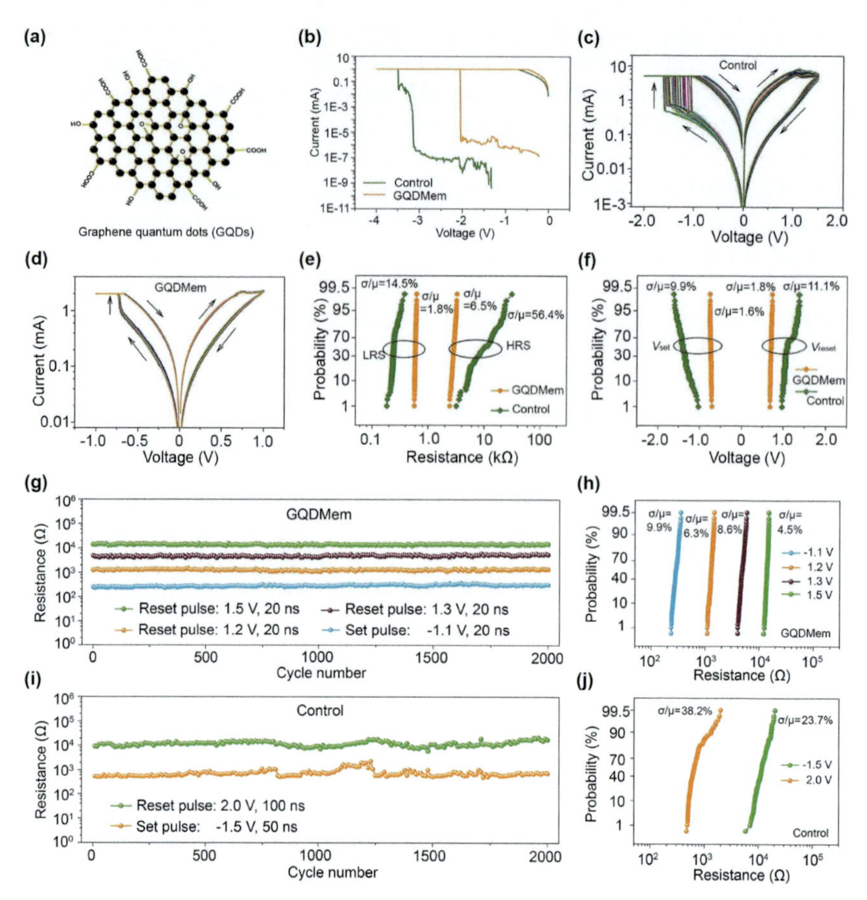

FIG. 12.10 (A) Schematic diagram of GQD with various oxygen functional groups. (B) Electroforming processes of GQDMem and control device. (C,D) DC voltage sweeping of the control and GQDMem devices for 50 consecutive cycles. (E,F) Distribution of HRS/LRS and SET/RESET voltages of both devices for 100 cycles. (G) Endurance of GQDMem device with various voltage amplitudes, the resistances were read every 10 pulses. (H) Distribution of multilevel resistances in GQDMem. (I) Endurance of the control devices, the resistances were read every 10 pulses. (J) Distribution of HRS and LRS in the control device. All the data were read at 0.2 V. *Reproduced with permission C. Wang, W. He, Y. Tong, Y. Zhang, K. Huang, L. Song, S. Zhong, R. Ganeshkumar, R. Zhao, Memristive devices with highly repeatable analog states boosted by graphene quantum dots, Small 2017. Copyright 2017, Wiley-VCH.*

Fig. 12.11C–F). This work in Ref [71] may pave the way to engineer highly repeatable analog switching for controllable artificial synapses, which may be used to build neural networks with accurate and efficient learning capability.

Tsai et al. [78] demonstrated that the SiO_2 film with h-BN inside it (fabricated by RF magnetron co-sputtering) can suppress overshoot issue

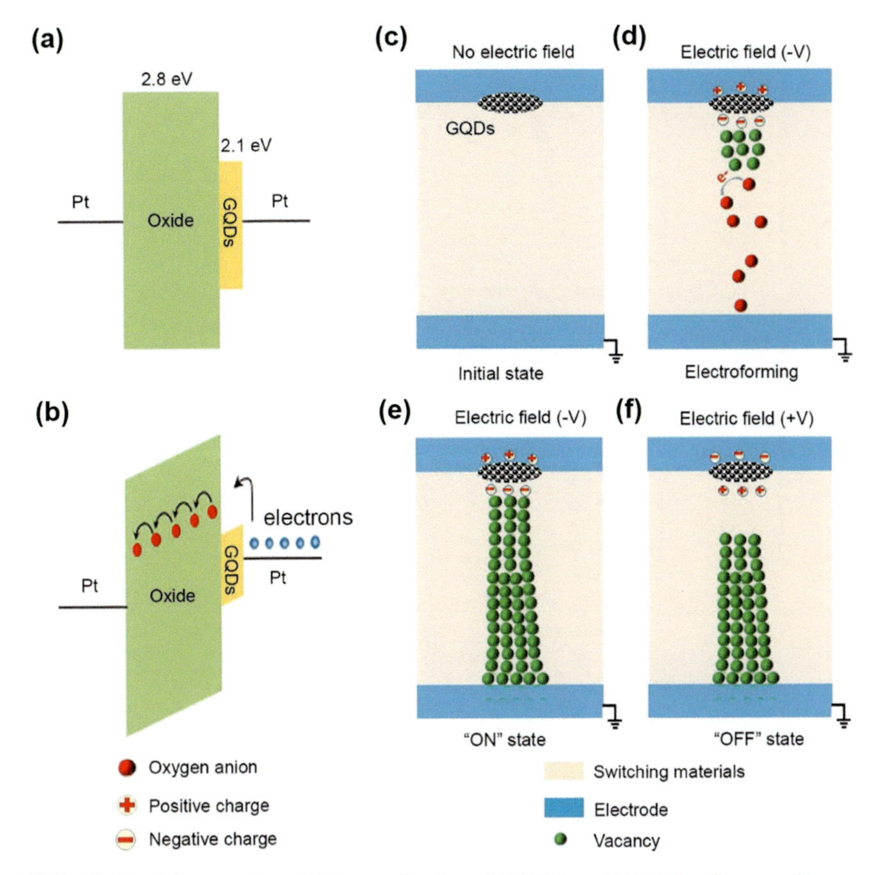

FIG. 12.11 Microscopic switching mechanism of GQDMem. (A) Flat band energy diagram of GQDMem device. Representative bandgaps of FeO_x and GQDs are 2.8 and 2.1 eV, respectively. (B) Energy band diagram of biased GQDMem device. (C) Pristine device. (D) Electroforming process. Oxygen anions released from GQDs facilitate the growth of oxygen vacancy CF. (E) SET process with CF connecting the bottom and top electrodes. (F) RESET process. CF rupture around GQDs. *Reproduced with permission C. Wang, W. He, Y. Tong, Y. Zhang, K. Huang, L. Song, S. Zhong, R. Ganeshkumar, R. Zhao, Memristive devices with highly repeatable analog states boosted by graphene quantum dots, Small 2017. Copyright 2017, Wiley-VCH.*

of RRAM device. They proposed a resistive memory with outstanding comprehensive performance by inserting buffer layers of SiO_2 film doped with h-BN (BN:SiO_2) into HfO_2-based RRAM. BN:SiO_2 can be designed to fit one or two surfaces of the HfO_2 film (see Fig. 12.12). As results, the Pt/BN:SiO_2/HfO_2/BN:SiO_2/TiN structure, where the BN:SiO_2 fits both surfaces of the HfO_2 film, was observed to have higher reliability and superior switching endurance (>10^{12}cycles). Similar endurance improvement behaviors are also observed in TiN/SLG/HfO_x/Pt [30] and Al/SLG/WO_3/Al [79] devices. Confirmed by X-ray photoelectron

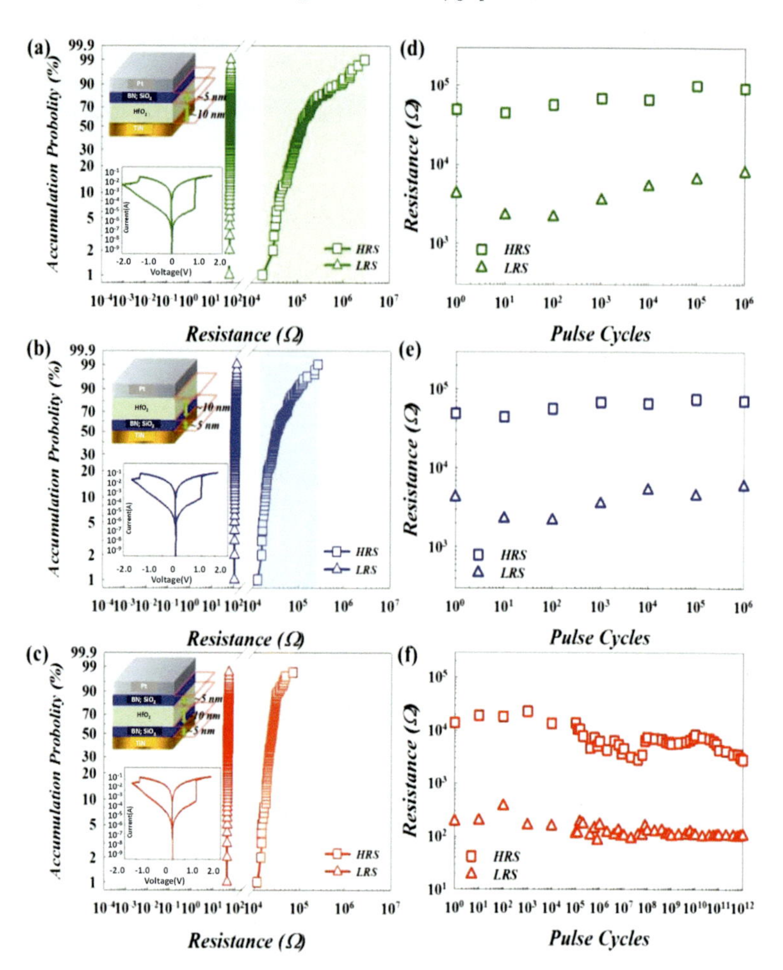

FIG. 12.12 Schematic structure, HRS/LRS distributions and RS characteristics of (A) Pt/ $BN:SiO_2/HfO_2/TiN$ RRAM devices, (B) $Pt/HfO_2/BN:SiO_2/TiN$ devices and (C) Pt/BN: $SiO_2/HfO_2/BN:SiO_2/TiN$ structure devices. (D–F) are the endurance properties of the corresponding devices mentioned in (A), (B) and (C), respectively. *Reproduced with permission T.-M. Tsai, C.-H. Wu, K.-C. Chang, C.-H. Pan, P.-H. Chen, N.-K. Lin, J.-C. Lin, Y.-S. Lin, W.-C. Chen, H. Wu, N. Deng, H. Qian, Controlling the degree of forming soft-breakdown and producing superior endurance performance by inserting BN-based layers in resistive random access memory, IEEE Electron Device Lett. 38 (4) (2017) 445–448. Copyright 2017, IEEE.*

spectroscopy (XPS) spectra, h-BN exists in the $BN:SiO_2$ layer, which are formed during the sputter process. The enhanced RS performance of the device can be attributed to the localization of adsorption/desorption of oxygen ions by h-BN flakes, just like the situation of graphene intercalation [30]. Therefore, the degree of soft-breakdown or damage in the insulator (caused by oxygen ions during the forming process) is better controllable in the BN-based RRAMs than in the HfO_2-based RRAMs.

12.2.2 Suppressing the over-growth of CF by 2D material barrier

Owing to excellent impermeability of graphene, it has been studied as a barrier materials to suppress the interpenetration of atoms/molecules/ion from both sides [34,80–85]. Introduction of graphene can also eliminate the undesired surface effects by insulating the RS layer and the surrounding atmosphere [33]. The moving ion under electric field can also be suppressed as demonstrated by Liu et al. based on HRTEM observation and first principles calculation [81]. Having a similar lattice structure to graphene, h-BN has also been reported to be able to localize the oxygen ion transmission inside the MO_X RRAM device [78]. Liu et al. first demonstrated the impermeability of graphene in ECM MO_X RRAM device by suppressing CF overgrowth in memory [81]. Negative-SET phenomena are generally found in various ECM MO_X RRAM [86–89], which damages the reliability of RRAM device and leads to the storage malfunction. Taking the $Ag/ZrO_2/Pt$ device as an example (Pt grounded and Ag biased), the device can suffer thousands of SET ($0 \rightarrow 1.2$ V)/RESET ($0 \rightarrow -3$ V) switching cycles. The typical negative-SET behavior may happen occasionally and randomly during the frequently SET/RESET operation as shown in Fig. 12.13A. During the negative RESET process, the device switched to HRS at -1 V, but returned back to LRS again at -2.5 V before the bias reaching to -3 V. Characterized by HRTEM and EDS mapping, the overgrowth of the Ag CF into the inert electrode is the primary reason (see Fig. 12.13B and C), which might provide enough active Ag resources to connect the ruptured CF for this unintentional back-switching. When inserting a graphene layer to form an $Ag/ZrO_2/G/Pt$ device, the negative-SET phenomenon was eliminated fundamentally, even the RESET process was enlarged to -4 V. HRTEM and EDS characterization (see Fig. 12.13E and F) of Ag CF inside the $Ag/ZrO_2/G/Pt$ demonstrated the impeded CF growth toward the inert Pt electrode, which cannot provide active Ag resources to induce the negative-SET behavior. Schematic illustrations in Fig. 12.13G elucidate the basic mechanism of negative-SET behavior and how graphene eliminated it. In addition, the introduction of graphene doesn't degrade the overall performances of the devices, the $Ag/ZrO_2/G/Pt$ devices still show fast switching speed within 30 ns under ± 4 V pulse and good endurance for more than 10^6 cycles.

12.2.3 Monitoring ion migration by 2D material barrier

Formation and rupture of oxygen vacancy CF are considered to dominate the RS behavior of VCM RRAM [90–92], therefore immediate observation of ion movement process is important to understand the switching mechanism. The adsorption of external ions may doping/decorating the graphene which can be detected immediately by Raman spectroscopy. Tian et al. integrated SLG into RRAM as an electrode to investigate the

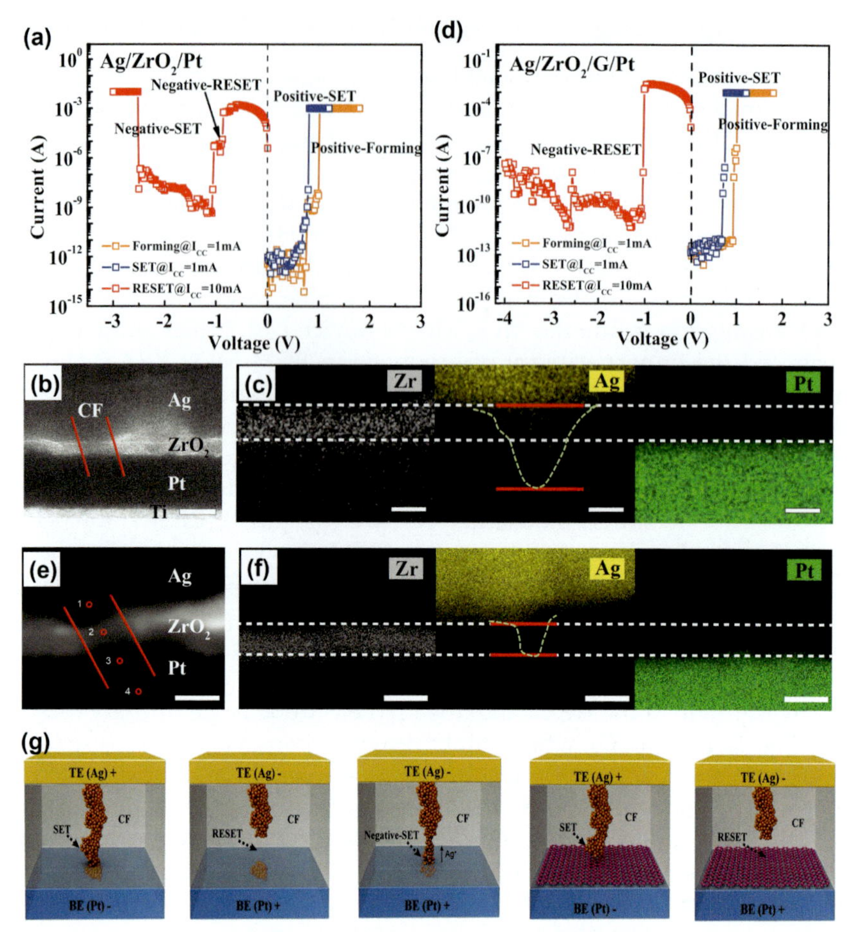

FIG. 12.13 Eliminating negative-SET behavior by suppressing nanofilament overgrowth in cation-based memory utilizing a graphene barrier. (A) Negative-SET behavior in the $Ag/ZrO_2/Pt$ device under DC voltage sweep mode. (B) HRTEM image of $Ag/ZrO_2/Pt$ device in ON-state. A possible CF region is highlighted by the two red lines. The scale bar is 30 nm. (C) EDS elemental mapping images of Zr, Ag, and Pt of the CF region in (B). The interfaces of Ag/ZrO_2 and ZrO_2/Pt are highlighted by the white dashed lines. The boundaries of Ag CF are marked by two red lines in Ag elemental mapping image. Scale bar is 10 nm. (D) Typical I–V curves of $Ag/ZrO_2/G/Pt$ device under DC voltage sweep mode. Negative-SET behavior has not been observed even the voltage increasing to −4 V during the negative-RESET process. (E) HRTEM image of $Ag/ZrO_2/G/Pt$ device in ON-state. A possible CF region is highlighted. (F) EDS elemental mapping images of Zr, Ag and Pt of the CF region. The boundaries of Ag CF are highlighted by two red lines in Ag elemental mapping image. Scale bar is 15 nm. (G) Mechanism of Negative-SET behavior in $Ag/ZrO_2/Pt$ device and how it is eliminated. Precipitation of Ag into the inert electrode contributes to the Negative-SET behavior, after inserting graphene at the ZrO_2/Pt interface, precipitation of Ag won't happen again, leading to the elimination of Negative-SET behavior. *Reproduced with permission S. Liu, N. Lu, X. Zhao, H. Xu, W. Banerjee, H. Lv, S. Long, Q. Li, Q. Liu, M. Liu, Eliminating negative-SET behavior by suppressing nanofilament overgrowth in cation-based memory, Adv. Mater. 28 (48) (2016) 10623–10629. Copyright 2016, WILEY-VCH.*

migration process of oxygen ions combing electrical measurement and Raman spectroscopy [30]. Fig. 12.14A shows the schematic illustrations of the TiN/Ti/SLG/HfO_2/Pt device (G-RRAM) and the Raman characterization method, while the optional and SEM image of the crossbar device are shown in Fig. 12.14B and C, respectively. SLG was used as an "oxygen barrier" to prevent oxygen ions from migrating deeply into the Ti electrode of the Ti/SLG/HfO_2/Pt device (G-RRAM device). During the SET/RESET switching cycling, noticeable changes in the D-band, G-band and 2D-band signals were observed. Fig. 12.14D shows single-point Raman measurements over nine cycles of SET/RESET programming at the same location, which is 1 μm from the cross point. The changes of the D, G and 2D peaks are shown in Fig. 12.14E–G. The D peak intensity decreases gradually during the SET/RESET programming operations. During SET, the G peak moves toward higher wavenumbers and moves backward during RESET, respectively. While the 2D peak intensity after SET is always lower than that after RESET in each cycle, a reversible change of the 2D peak intensity is consistently observed. Thermal annealing during the SET/RESET operation benefits the restoration of the damaged lattice, contributing to the decreased D peak intensity. G peak shift and 2D peak intensity change reflect the change of doping concentrations in graphene. It is possible that the G peak shift and the change of 2D peak intensity might be resulted from the doping changes due to oxygen ion migration: during SET process, oxygen ions escape from the oxide layer and diffuse laterally until they form covalent bonds with dangling bonds of defects on SLG surface. This process would dope the graphene into p-type, which will shift the G-peak to larger wavenumber and decrease the intensity of the 2D-peak. During the RESET process, oxygen ions move back to oxide layer due to the reversed applied electric field. This process would change the graphene toward intrinsic doping, which will shift the G-peak to a smaller wavenumber and increase the intensity of the 2D-peak. Monitoring the oxygen ion migration dynamics during the switching process assisted by Raman spectroscopy of 2D graphene electrodes opens up an important analysis method for investigation of the switching mechanism of MO_X RRAM.

12.2.4 Confining ion migration in MO_X RRAM by 2D material defect engineering

Since graphene can depress the electromigration of ions inside the MO_X RRAM, Zhao et al. go again this way to fabricate defective graphene (DG) at the interface of the active electrode (AE) and RS layer to allow certain cations to be injected into RS layer of based on the ECM RRAM cell [93]. Therefore, the CF formation can be confined at the tailored defective region of the graphene and optimize the device performance. Graphene

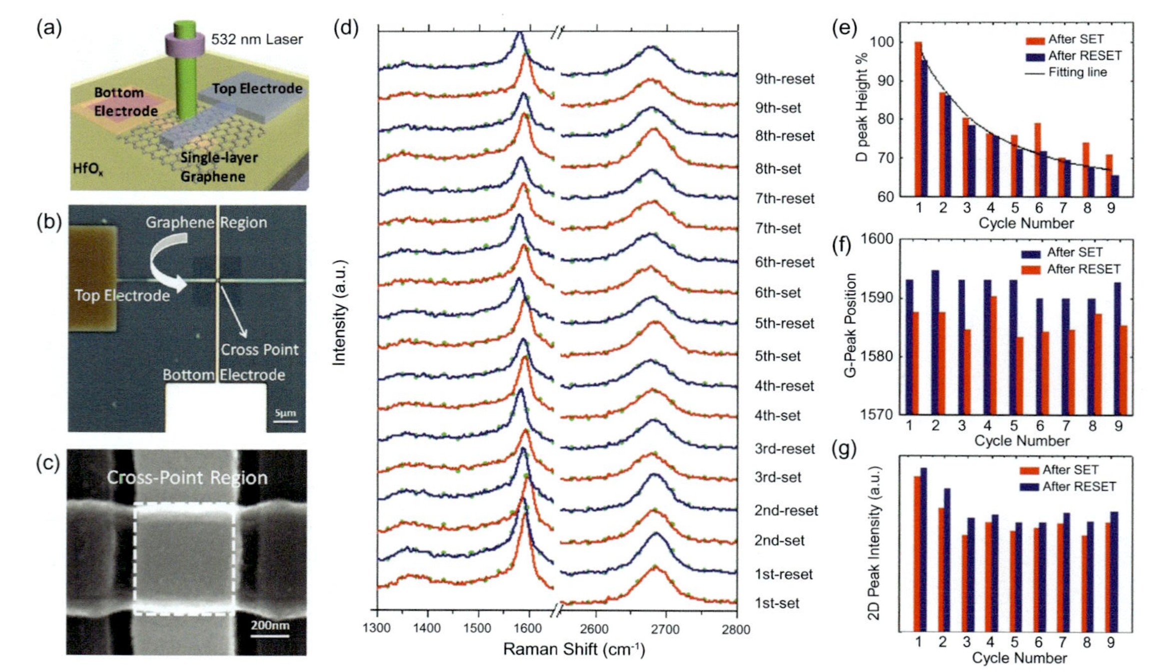

FIG. 12.14 (A) Schematic illustrations of the TiN/Ti/G/HfO_2/Pt device and the Raman characterization method. (B) An optical image of $0.5 \times 0.5\,\mu m^2$ cross-point region inside the $10 \times 10\,\mu m^2$ SLG. (C) SEM image of $0.5 \times 0.5\,\mu m^2$ cross-point region under high magnification. (D) Single-point Raman data measured after SET and RESET cycles at the same location, which is 1 μm from the cross-point. (E) Statistical analysis of D peak drop as a function of switching cycle. It is observed that the D peak gradually decreases. (F) The position of G peak as a function of switching cycle. The G-peak position of the SET is always higher than the RESET in each cycle, and reversible shifts of G peak position with RRAM programming actions are observed. (G) The intensity of 2D peak as a function of switching cycle. The 2D-peak intensity of the SET is always lower than the RESET in each cycle, and reversible changes of 2D peak intensity with RRAM programming events are observed. *Reproduced with permission H. Tian, H.-Y. Chen, B. Gao, S. Yu, J. Liang, Y. Yang, D. Xie, J. Kang, T.-L. Ren, Y. Zhang, H. S. P. Wong, Monitoring oxygen movement by Raman spectroscopy of resistive random access memory with a graphene-inserted electrode, Nano Lett. 13 (2) (2013) 651–657. Copyright 2013, American Chemical Society.*

with fabricated nanopores has shown promising results as size-selective barriers for water and gas purification [83,84,94], DNA sequencing [95,96], and single-molecule detectors [85], where the desired particles can transport only through pores in the graphene layer. The site control of cation migration and CF formation by defective graphene at the designed locations was validated by conductive atomic force microscope (C-AFM) with high currents emerging at the defective region. RS is actually a localized behavior, however, cations are injected from the whole area of the active electrode into the RS layer supplying excessive cations beyond what is required for CF formation, leading to deterioration of device uniformity and reliability. Two kinds of graphene defects were fabricated. One was nanoscale concentrated defects (single nanohole with a ~45 nm diameter) fabricated by electron beam lithography (EBL) and oxygen plasma etching with the photoresist acting as a mask. The other was atomic discrete defects fabricated by bombardment of accelerated Si^+ ion isolated from SiH_4 source by an ion implanter. Correspondingly, three devices were fabricated (see Fig. 12.15A), namely $Ag/SiO_2/Pt$ control device without graphene, $Ag/DG/SiO_2/Pt$ device with concentrated-defect graphene (CDG device) and $Ag/DG/SiO_2/Pt$ device with discrete-defect graphene (DDG device). Typical RS characteristics of the control, DDG (10^{13} Si^+/cm^2 irradiation fluence) and CDG devices are compared under various I_{CC} (0.1–500 μA) (see Fig. 12.15A). Regular differences are observed based on their typical I–V curves. The CDG device shows a 0.1 μA maximum volatile ON-state current, which is much lower than that (100 μA) of the control device, while the DDG device shows an elevated 500 μA maximum volatile ON-state current compared with the control device. When the I_{CC} is higher than the maximum volatile ON-state current, all devices transform from volatile threshold switching (TS) behavior into nonvolatile memory switching (MS) behavior. The difference of these two behaviors is that, volatile TS behavior loses its LRS and returns back to HRS immediately after removal of external stimulus, while nonvolatile MS behavior maintains its LRS quite well [97]. The former one is broadly studied for selector in 1S1R scheme of RRAM applications [98], while the latter one takes the responsibility for information storage.

The different RS characteristics in the three devices can be explained by the tremendous difference in Ag CF stability controlled by the DG-modulated cation migration (see Fig. 12.15B). With the external stimulus applied onto the Ag AE, Ag atoms in the control device randomly migrate toward the counter electrode from the whole device area leading to the multiple nucleation/formation of CFs and sometimes the narrowing of the RS layer [64,99]. CF atoms can diffuse into RS layer from the surfaces of the multiple CFs, and smallest CFs rupture and transform into nanoparticles first especially when giving a small compliance current (<100 μA). While for the CDG device, graphene nanohole as a

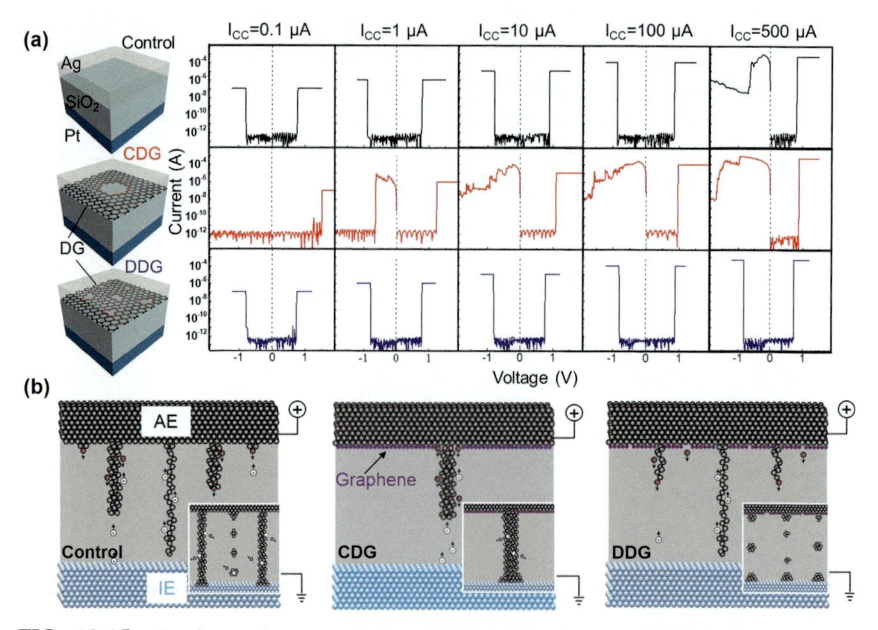

FIG. 12.15 Confining the cation injection toward the RS layer by defective graphene. (A) Schematic illustrations of the Ag/SiO_2/Pt control device (first line), the CDG device (second line) and the DDG device (third line), and their typical RS I–V characteristic comparison under various I_{CC} (0.1–500 μA). The maximum volatile ON-state currents are 100, 0.1, and 500 μA, respectively, indicating the difference of CF stability modulated by the defective graphene. (B) Schematic illustrations of the cation migration, CF morphology, and CF stability modulated by defective graphene. AE and IE denote the active Ag electrode and the inert Pt electrode. In the control device, active cations migrate randomly toward the counter electrode from the whole device area and lead to formation of multiple CFs. After removal of external bias, atoms from CFs can diffuse into the RS layer from the large surface of the CFs, and the smallest CFs rupture and transform into nanoparticles first (inset). In the CDG device, concentrated cation migration at the nanohole leads to the localized formation of single or a few centralized robust CFs along the graphene nanohole. The relatively larger atomic density and smaller effective total surface area contribute to relatively weaker atom diffusion and much better CF stability (inset). In the DDG device, cation migration is restrained along the discrete atomic graphene defects, thus leading to the formation of discrete fragile tiny CFs, which can evolve from continuous bridge to discrete nanoparticles spontaneously (inset). *Reproduced with permission X. Zhao, J. Ma, X. Xiao, Q. Liu, L. Shao, D. Chen, S. Liu, J. Niu, X. Zhang, Y. Wang, R. Cao, W. Wang, Z. Di, H. Lv, S. Long, M. Liu, Breaking the current-retention dilemma in cation-based resistive switching devices utilizing graphene with controlled defects, Adv. Mater. 30 (14) (2018) 1705193. Copyright 2018, WILEY-VCH.*

concentrated graphene defects, induces the cation migration confined at the nanohole regions. This leads to the formation of fewer, but more robust CFs with a relatively larger atomic density and smaller effective total surface area. Relative smaller surface/volume ratio of the robust CFs facilitates relatively weaker atom diffusion and much better CF stability for nonvolatile RS memory application. On the contrary, the discrete atomic scale graphene defects in the DDG device facilitate the decentralization of much tinier CFs. Fragile tiny CFs with poor stability evolve spontaneously from bridges to discontinuous nanoparticles after the removal of external stimulus owing to nano/quantum size effects and interfacial-energy-related Gibbs–Thomson effect [100], making the DDG device advantageous for volatile TS selector application even under a higher I_{CC}. Therefore, by centralizing/decentralizing the CF distribution to modulate the cation transmission and CF formation, both robust/fragile CF were obtained for memory/selector application. The optimized CDG nonvolatile memory shows considerable multiple-level storage, low power dissipation ($<1.5\,\mu W$), fast switching speed (SET/RESET switching within 100/200 ns) and good endurance ($>10^7$) properties. While the DDG volatile selector shows excellent endurance ($>10^6$), sharp turn-on slope ($<1\,mV\,dec^{-1}$), nice bias stress and fast ON/OFF speed (100 ns/1 μs).

These findings mainly uncovered the underlying regulation between CF morphology and CF stability, assisted by the modulation of cation injection utilizing defective graphene. Ag electrode seems a nice choice for volatile selector while not a suitable choice for nonvolatile memory, because of the relative larger solubility or diffusion coefficient in a certain MO_X dielectric compared to the promising Cu electrode [101,102]. A detailed study about Cu/nanohole-graphene/HfO_2/Pt (NG-based device) and Cu/HfO_2/Pt (control) RRAM device was also performed to study the effects of confined cation injection at a nanohole region on the RS performances (see Fig. 12.16) [103]. Confined at the nanohole region, the formation/rupture of fewer or just single robust CF contributes to the enhanced RS performances, including low power dissipation, uniform switching parameter, nice switching endurance ($>10^7$) and excellent retention property (2×10^5 s, 125 °C). Basically, spatial limitation of cation injection can be implemented by limiting the cation source or the RS layer. Wong et al. also observed significant improvement of cycling endurance for Cu-based RRAM by scaling the RS layer down to 30 nm in diameter [44].

Different from Zhao's works, Lu et al. confine the anion transmission in the VCM RRAM by nanopore graphene [69]. First, the comparison between Ta/single-layer-graphene/Ta_2O_5/Pd (w/G device) and Ta/Ta_2O_5/Pd (w/o G device) was performed (see Fig. 12.17A–C). Fig. 12.17A shows the typical I–V curves of both devices. The w/G device shows more reliable RS operations, while the w/o G device shows much higher RESET-fail rate under no compliance. The relative higher initial resistance of the w/G device leads to smaller initial current and larger forming voltage (see Fig. 12.17B). In addition, the w/G devices show lower ON/OFF currents and improved device-to-device uniformity. Limited by the impermeable graphene, O^{2-} anion migration and redox reactions at the Ta/Ta_2O_5 interface were depressed, leading to less oxygen vacancies and higher initial resistance. The native defects in the form of nanopores were expected to induce the formation of fewer and tinier CFs, which contribute to the lower ON/OFF current in the w/G device. To test this hypothesis, Ta/multi-layer-graphene/

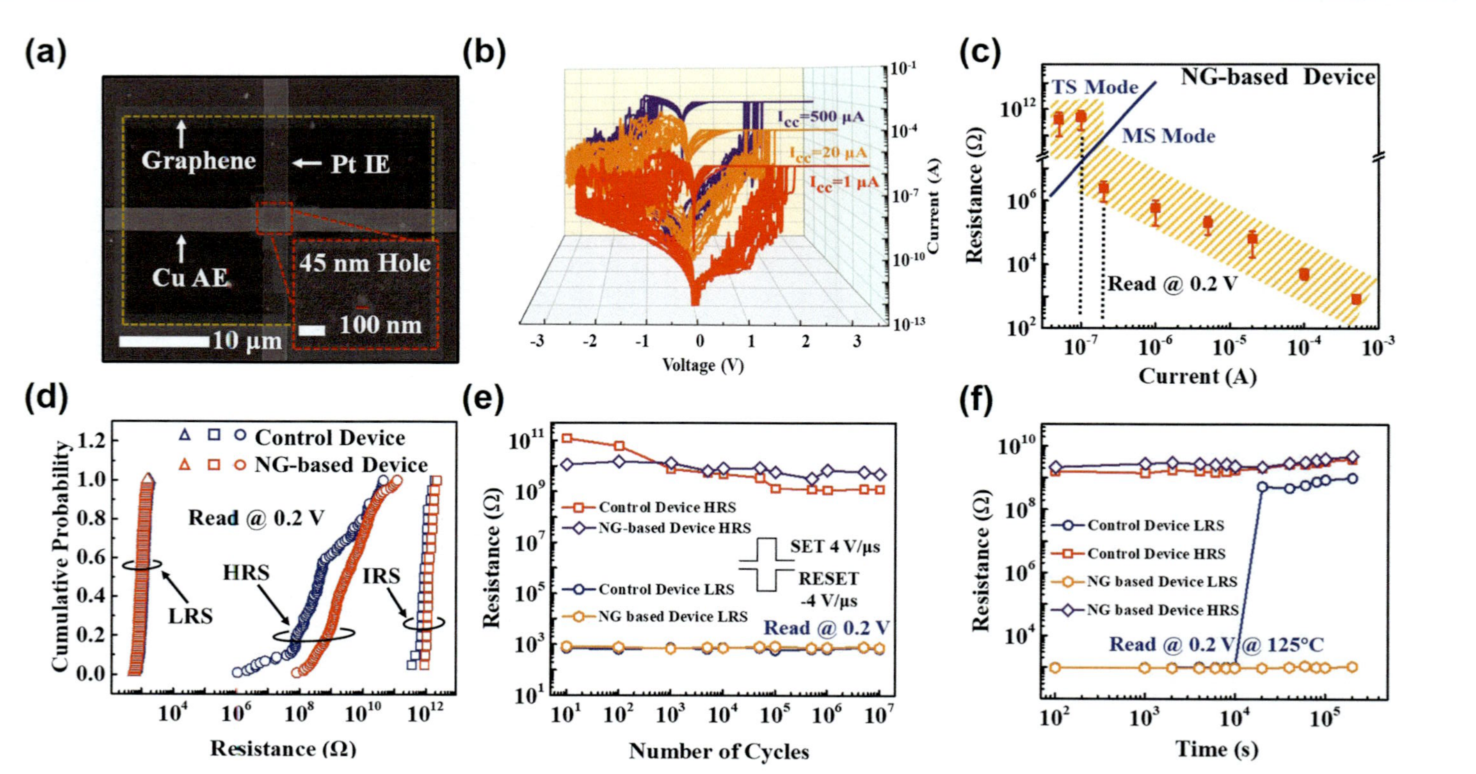

FIG. 12.16 Improvement of the RS performance by confining the cation injection utilizing a nanohole graphene base on Cu/HfO_2/Pt material system. (A) SEM image of the NG-based device. The inset SEM image, captured before Cu AE was deposited, shows the single graphene nanohole marked by the red line. (B) Ten typical I–V curves of the device under 500, 20, and 1 μA I_{CC}, respectively. (C) The resistances (read at 0.2 V) distribution of the NG-based devices after SET process with various I_{CC}. The TS and MS modes are separated by the blue line and the transition intervals of both modes are marked by the black dashed lines. (D) The device-to-device accumulative probability of resistances (IRS, LRS, and HRS), based on 100 switching cycles from 20 randomly selected cells for each of the control and NG-based devices. The NG-based device shows better HRS uniformity. (E) The pulse endurance results of both the devices with 10^7 switching cycles, with 500 ns/4 V SET and 500 ns/−4 V RESET pulse. Deterioration of HRS is only observed in the control device. (F) Resistance retention property of both the devices for 2×10^5 s at 125 °C. LRS of the control device failed much earlier. *Reproduced with permission X. Zhao, S. Liu, J. Niu, L. Liao, Q. Liu, X. Xiao, H. Lv, S. Long, W. Banerjee, W. Li, S. Si, M. Liu, Confining cation injection to enhance cbram performance by nanopore graphene layer, Small 13 (35) (2017) 1603948. Copyright 2018, WILEY-VCH.*

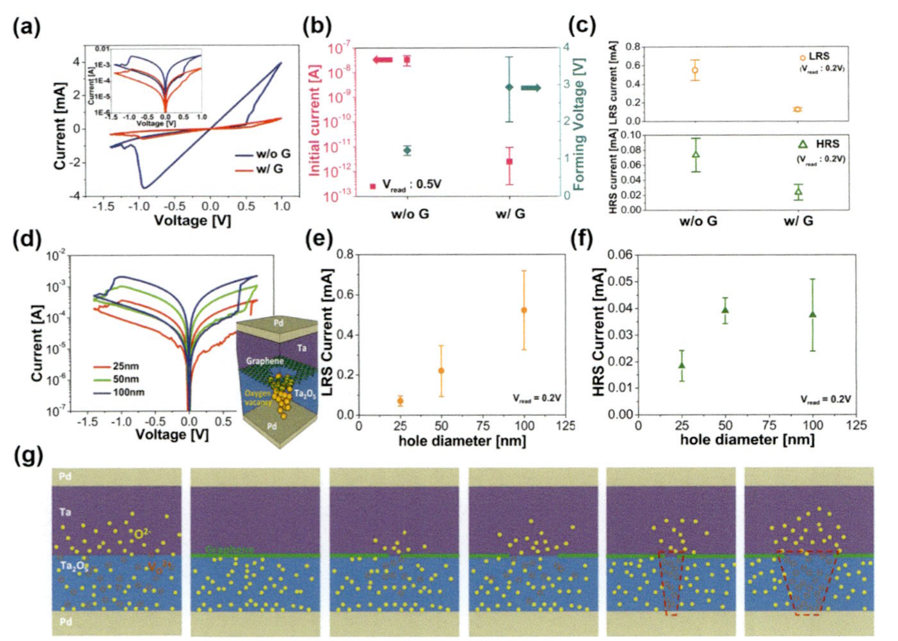

FIG. 12.17 (A) Resistive switching I – V characteristic of the Ta/Ta_2O_5 and $Ta/G/Ta_2O_5$ devices (inset: log-scale). (B) Distribution of the initial virgin state current and the forming voltage for the Ta/Ta_2O_5 and $Ta/G/Ta_2O_5$ devices. The $Ta/G/Ta_2O_5$ devices are ~4 orders of magnitude more resistive than the Ta/Ta_2O_5 devices in the virgin state. The read voltage is 0.5 V. (C) Current levels in LRS (upper panel) and HRS (lower panel) of the Ta/Ta_2O_5 and $Ta/G/Ta_2O_5$ devices, showing the reduced operating current in the graphene-inserted devices and tighter variability control. The read voltage is 0.2 V. (D) RS I – V characteristic of the $Ta/MLG/Ta_2O_5$ devices with different sized nanopores fabricated in the graphene layer. The inset shows the schematic of the graphene-inserted memristor structure where oxygen ions transport only through the nanopore in the graphene layer, forming a CF with oxygen vacancies. (E) Current levels in LRS and (F) HRS of the graphene-inserted devices with different nanopore sizes. Read voltage is 0.2 V. (G) Schematic illustration of the ion intermixing situations at the Ta/Ta_2O_5 interface and CF formation in the Ta/Ta_2O_5, $Ta/MLG/Ta_2O_5$ and $Ta/MLG/Ta_2O_5$ devices with small/large nanopores. *Reproduced with permission J. Lee, C. Du, K. Sun, E. Kioupakis, W.D. Lu, Tuning ionic transport in memristive devices by graphene with engineered nanopores, ACS Nano 10 (3) (2016) 3571–9. Copyright 2016, American Chemical Society.*

Ta_2O_5/Pd devices were fabricated and measured with engineered nanopores on MLG (see Fig. 12.17D–F). MLG was used to eliminate the effect of native pores of graphene. Fig. 12.17D shows the RS I – V characteristic of the Ta/MLG/Ta_2O_5 devices with different sized nanopores fabricated in the MLG layer. The inset shows the schematic of the graphene-inserted RRAM structure where oxygen ions migrate only through the nanopore in the graphene layer, forming a CF with oxygen vacancies. From Fig. 12.17E and F, devices with small nanopores show smaller ON/OFF current, indicating the tunable switching characteristics of the Ta/Ta_2O_5 devices by nanopore MLG. Shown in Fig. 12.17G, the injection of the vacancies and CF formation/rupture occur only at the engineered nanopores. The CF sizes can be modulated by the sizes of the nanopores, leading to controllable and optimized RS performance. However, in the situation of MLG without nanopore on it, Lübben et al. reported that the device may change from VCM mechanism to ECM with a Ta/MLG/Ta_2O_5/Pt structure [104]. Since the oxygen ion migration through the MLG/Ta_2O_5 may be fully inhibited and the participation of Ta-based redox from the graphene boundaries may be activated to dominate the RS behavior of device with a large device area ranges from 25×25 to $1000 \times 1000\,\mu m^2$.

12.3 MO_X RRAM assisted by additional 2D intercalation layer

The nature of the interface between the MO_X layer and the electrode plays an important role in determining the RS characteristics, and clear understanding of the interface is critical to undertake advanced research. 2D materials provide a feasible way to modify the interface at the atomic level.

12.3.1 MO_X RRAM with intrinsic 2D materials

Kim et al. [79] took both CVD-grown graphene and h-BN to modify the Al/WO_3 interface at the atomic level and made a comparison among the Al/WO_3/Al, Al/graphene/WO_3/Al and Al/h-BN/WO_3/Al RRAM devices (see Fig. 12.18A–C). All devices were forming-free owing to the sufficient oxygen deficiencies in the WO_3 films. The graphene modified RRAM device exhibited stabilized RS I–V curves during the cycling measurement, enhanced endurance and retention properties, while the device modified by h-BN showed degraded performance, especially the large vibration of ON/OFF currents during endurance measurement. During the SET process, the top electrode was negatively biased, oxygen ions were pushed away from the electrode/oxide interface and oxygen vacancy CF formed across the WO_3 film. While during the RESET process, the top electrode (TE) is positively biased, CF ruptured and oxygen ions migrated toward the electrode/oxide interface (see Fig. 12.18D). As demonstrated in Ref [30], graphene acting as an "oxygen barrier" can prevent the oxygen ions

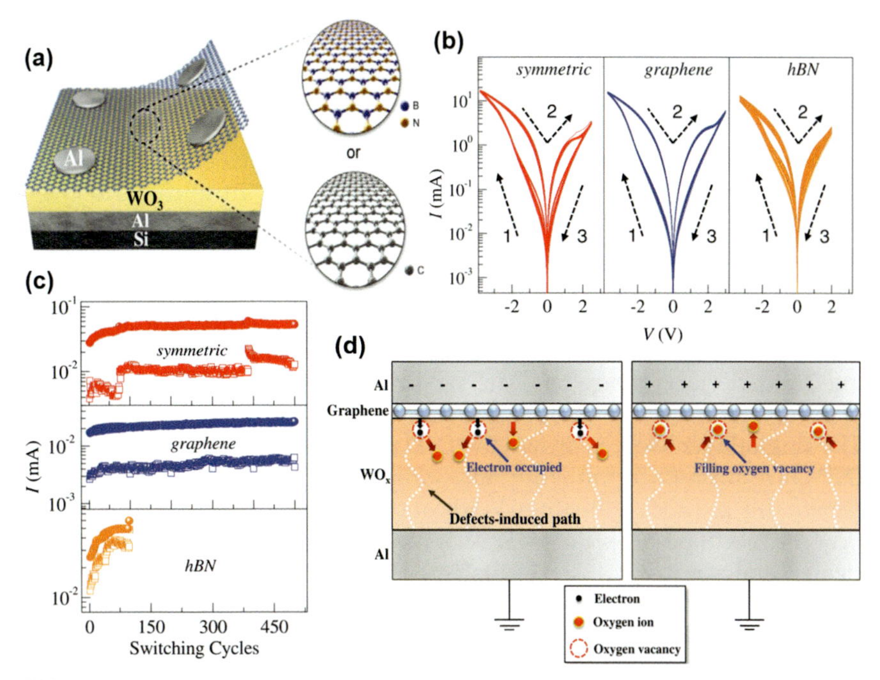

FIG. 12.18 (A) Schematic illustrations of an $Al/WO_3/Al$ resistive switching memory cell. Both CVD-grown single-layer graphene and h-BN were inserted at the Al/WO_3 interface to study the role of the interface on RS performances. (B) Typical I–V cures of the $Al/WO_3/Al$ (500 cycles), $Al/graphene/WO_3/Al$ (500 cycles) and $Al/h\text{-}BN/WO_3/Al$ (500 cycles) RRAM devices. (C) The endurance of the devices for ON/OFF current. The $Al/graphene/WO_3/Al$ device is stable up to 500 cycles with a stable ON/OFF ratio of ~10, while other samples are unstable showing a dramatic change in the current level. In particular, the h-BN inserted sample survived only up to ~100 switching cycles. (D) The "oxygen barrier" ability of the graphene contributes to the stabilized redox inside the MO_X RRAM device. *Reproduced with permission J. Kim, D. Kim, Y. Jo, J. Han, H. Woo, H. Kim, K. K. Kim, J. P. Hong, H. Im, Impact of graphene and single-layer BN insertion on bipolar resistive switching characteristics in tungsten oxide resistive memory, Thin Solid Films 589 (2015) 188–193. Copyright 2015, ELSEVIER.*

from diffusing into the Al electrode and contribute to stabilizing the chemical process under consecutive switching operations. For the $Al/h\text{-}BN/WO_3/Al$ device, the same redox-reaction for the RESET process is not fully recovered due to its electrical insulative nature and low permittivity [7–10], leading to the increase in current with increasing RS cycle. From above, atomic configuration of the electrode/oxide interface plays a key role in determining the RS performances of the RRAM devices. Based on different MIM structures, the insertion of certain 2D materials does not always contribute to enhancing the RS behavior. Therefore, more in-depth and systematic researches about 2D-assisted RRAM in the same MO_x material system are needed to explore the influence of various emerging 2D materials.

There is always a dilemma between the leakage current and the operating voltage in MO_X RRAM devices. A thick MO_X layer generally results in relatively high program voltage while a thin one causes large leakage current and a small window. Introducing of metal nanoparticles/nanotips [105–107] and substituting TE materials [101,108] have been explored to reduce the operating voltages, but without obvious contributions to lower leakage current. On the other hand, interface engineering by introducing 2D materials (such as MoS_2, graphene or BN) might enable a decreased OFF current, it doesn't reduce the operating voltages immediately [30,79,81,109]. Wang et al. [109] proposed a novel 2D–0D interface engineering idea to reduce the programming voltage on the base of introducing 2D materials, where the MoS_2–Pd nanoparticles hybrid structure is used to engineer the oxide/electrode interface of HfO_x-based RRAM. The single-layer MoS_2 is grown by CVD method and the Pd nanoparticles (NPs) are formed by utilizing an anodic aluminum oxide template. As shown in Fig. 12.19A, the three kinds of devices, namely the conventional, Pd NPs modified and MoS_2–Pd NPs modified RRAM devices, were fabricated and compared. As results (see Fig. 12.19B–E), the MoS_2–Pd NPs modified RRAM exhibits good transparency, lowered SET voltage (from -3.5 to -0.8 V), more uniform ON/OFF current and an enlarged resistive switching window (30 times larger than the conventional one). In the conventional device with thick MO_X film, the CFs grow irregular which needs more energy to form conductive paths. By introducing a less conductive MoS_2 layer to increase the HRS, and the SET voltage can be lowered and uniformed by embedding Pd NPs. The large resistance at MoS_2/HfO_x interface reduces the OFF current more than one order of magnitude (from $\sim 10^{-5}$ to $\sim 10^{-6}$) and greatly improve resistive switching window. Besides, the application of MoS_2–Pd NPs plays a significant role in lowering the SET voltage with better uniformity, which is hard to be achieved by inserting only MoS_2 layer. Conventional RRAMs have SET voltages in the range of -3.5 to -6 V, while the Pd-modified ones are -0.8 to -2 V. This is mainly related to the enhanced electric field effect by the Pd NPs [105], which can be used as the seeds to facilitate the formation of CFs with smaller switching voltage. Above all, 2D–0D interface-engineered RRAMs are potential candidates for data storage in transparent circuits and wearable electronics with relatively low supply voltage.

12.3.2 Oxidized film of 2D materials for MO_X RRAM

To combine the advantages of both conventional MO_X and emerging 2D materials into RRAM devices, the most straightforward way is taking the oxidized films of the layer materials as RS layer. Oxidized films from h-BN [110], WS_2 [111], MoS_2 [21,35,111], ZnS [112] and black phosphorous

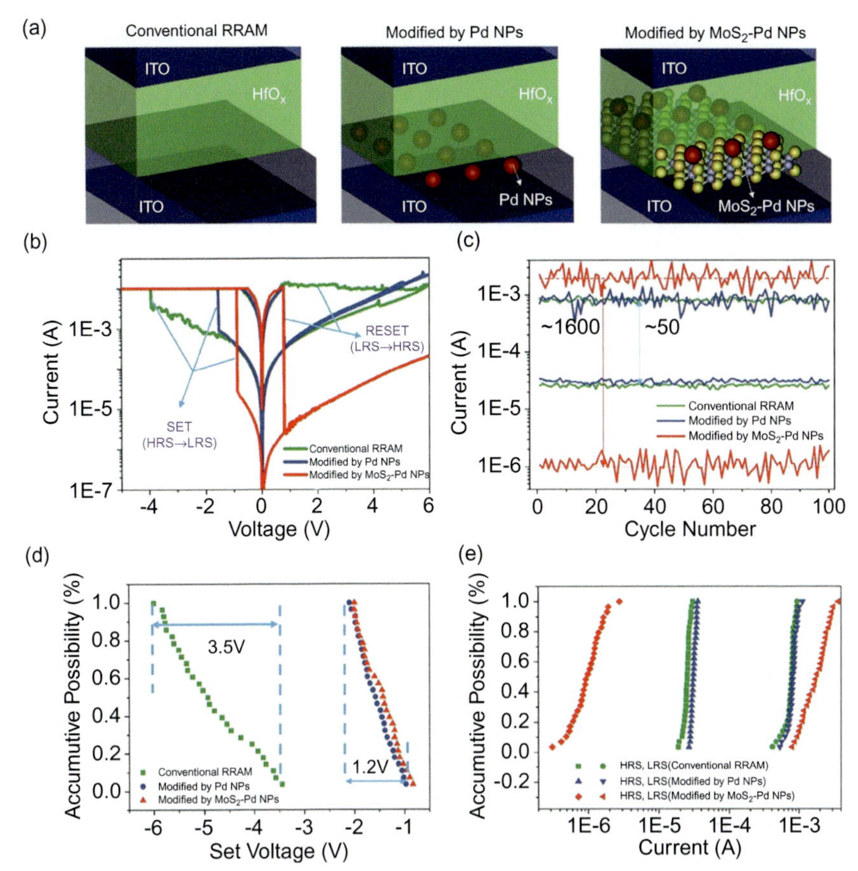

FIG. 12.19 (A) Schematic illustrations of conventional, Pd NPs modified and MoS_2–Pd NPs modified RRAM devices. (B) Typical I–V curves of the RRAM devices. (C) Endurance of the devices, with the ON/OFF current of the three kinds of RRAM read at −0.1 V bias voltage. (D) The SET voltage accumulative possibility of three kinds of RRAM. (E) ON/OFF current accumulative possibility of the devices. *Reproduced with permission X.-F. Wang, H. Tian, H.-M. Zhao, T.-Y. Zhang, W.-Q. Mao, Y.-C. Qiao, Y. Pang, Y.-X. Li, Y. Yang, T.-L. Ren, Interface engineering with MoS2-Pd nanoparticles hybrid structure for a low voltage resistive switching memory, Small 14 (2) (2018). Copyright 2017, Wiley-VCH.*

[13] have been reported with excellent RS performances. These devices have been studied for various applications, including ultrasensitive memristor application, ultralow power dissipation device with ultrathin (<1 nm) switching layer, flexible and printable RRAM devices, large scale flexible storage and display systems and even heat-resistant memory applications under extremely harsh condition [13,21,35,110–112].

The MO_X RS layer thickness plays an important role in dimensions and morphology of the resulting filament. Aggressive scaling of this active layer thickness is critical toward reducing the operating current, voltage, and energy consumption in filamentary-type RRAM. Oxidized 2D materials have shown their potential to shrink the vertical size of the RS layer to less than 3nm and even to sub-1nm, as reported in Refs [110, 111]. Even though BNO_x film fabricated by RF magnetron co-sputtering have been used as RS layer, no systematic RS performances were reported except its ability to suppress the overshoot issue in MO_X RRAM device owing to the existing of h-BN [78]. Zhao et al. [110] demonstrated the sub-pA operation current and femtojoule per bit energy consumption of the Ag/BNO_x/graphene structure (see Fig. 12.20). The BNO_x with ultrathin 0.9nm thickness was obtained by oxidization of the exfoliated BN film under reactive ion etching and sandwiched by exfoliate MLG and evaporated Ag electrode. Furthermore, this RRAM device with extremely thin oxide RS layer showed low leakage current of 0.1pA at 0.4V bias and excellent RS reliability for 100 switching cycles under 5pA I_{CC}. While under the optimal 2–3nm ALD MO_X (e.g., HfO_2, Al_2O_3 and SiO_2), poor dielectric property may lead to high leakage current or even the breakdown of the film [113,114]. For the thinner thickness less than 2nm, device design based on emerging 2D materials can further reduce RS layer thickness, which facilitates the vertical CF growth after the first cation hops into the RS layer by enhanced electric field effect as simulated by Monte Carlo method (see Fig. 12.20I). The Ag/BNO_x/graphene RRAM device in Ref [110] decreased the filament length to sub-nanometer or atomic scale within an ultrathin 2D layered RS insulator and suppressed the lateral expansion of CF growth. Formation and rupture of such an atomic filament to dominate the RS behavior led to the minimization of the device power dissipation. This experiment shows a way to fabricate RRAM devices with sub-5nm total thickness when companied with a pair of 2D electrodes, such as graphene electrodes [35,115]. These attractive ultra-low energy devices based on oxidized 2D materials pave the way toward realizing femtojoule and sub-femtojoule electronic device applications.

Explosive growth of the printed electronics industry has expedited the search for advanced memory materials suitable for manufacturing flexible devices [74,116]. Bessonov et al. [111] fabricated Ag/MoO_x/MoS_2/Ag RRAM devices on a flexible poly(ethylene naphthalate) (PEN) plastic substrate (see Fig. 12.21A). The MoO_x/MoS_2 stack was solution-processed from MoS_2 flakes (1–8 layers with lateral size of 100–400nm) by heat-assisted air oxidization (200°C, 3h) and then sandwiched between two printed silver electrodes. The annealing oxidization process can be alternatively substituted by xenon flash light exposure technology. The bottom electrode (BE) and TE are fabricated by inkjet printing and screen printing of silver nanoparticle ink, respectively. A forming or initialization process is not required owing to the oxygen-deficient area of MoO_x RS layer with the thickness of less than 3nm. The devices exhibit an unprecedentedly large and tunable electrical resistance range from 10^2 to $10^8\,\Omega$ combined

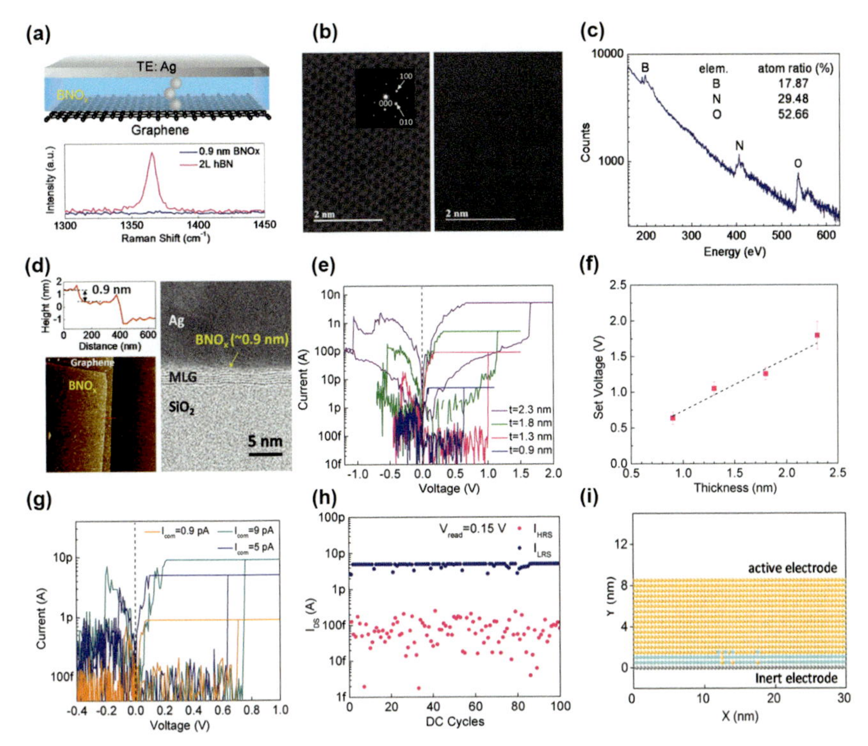

FIG. 12.20 BNO_x characterization and device performance. (A) Upper panel: The schematic of an atomically thin RRAM device. Bottom panel: Raman spectra of a bilayer h-BN flake, before (red) and after (blue) oxygen plasma treatment. (B) High-resolution STEM images show the crystalline lattice of h-BN before the oxygen plasma treatment (left panel), and after the sample becomes amorphous (right panel). The scale bar is 2 nm. The insets of the left panel are the electron diffraction pattern of the crystalline h-BN, where crystal planes and interplanar spacing are indicated by Miller indices. (C) EELS spectrum of a BNO_x flake. The element atom ratio confirmed the abundance of oxygen in the sample after the oxygen plasma treatment. (D) The lower left panel is the AFM image of a BNO_x stacked on a five-layer graphene, of which the height profile is shown at the upper left panel. The right panel is the cross-sectional STEM image of a device made from the sample in the left panel. The layered structure of five-layer graphene and the amorphous morphology of BNO_x can be clearly observed. (E) The SET–RESET I–V characteristics of devices with various BNO_x layer thicknesses measured under current compliances of 5, 90, 500, and 2000 pA, respectively. (F) The linear dependence of the device SET voltage on the BNO_x layer thickness. For each device, several (>5) switching cycles were measured to obtain a statistical distribution, indicated by the error bar. (G) SET–RESET I–V characteristics of a device with 0.9 nm BNO_x switching layer subject to different current compliances of 0.9, 5, and 9 pA. (H) The current read at 0.15 V over 100 continuous switching cycles. (I) Monte Carlo simulation of the filament formation process in mediums with 0.9 nm oxide thickness, showing the possibility of forming a single atomic chain filament. *Reproduced with permission H. Zhao, Z. Dong, H. Tian, D. DiMarzi, M.G. Han, L. Zhang, X. Yan, F. Liu, L. Shen, S.J. Han, S. Cronin, W. Wu, J. Tice, J. Guo, H. Wang, Atomically thin femtojoule memristive device, Adv. Mater. (2017). Copyright 2017, Wiley-CVH.*

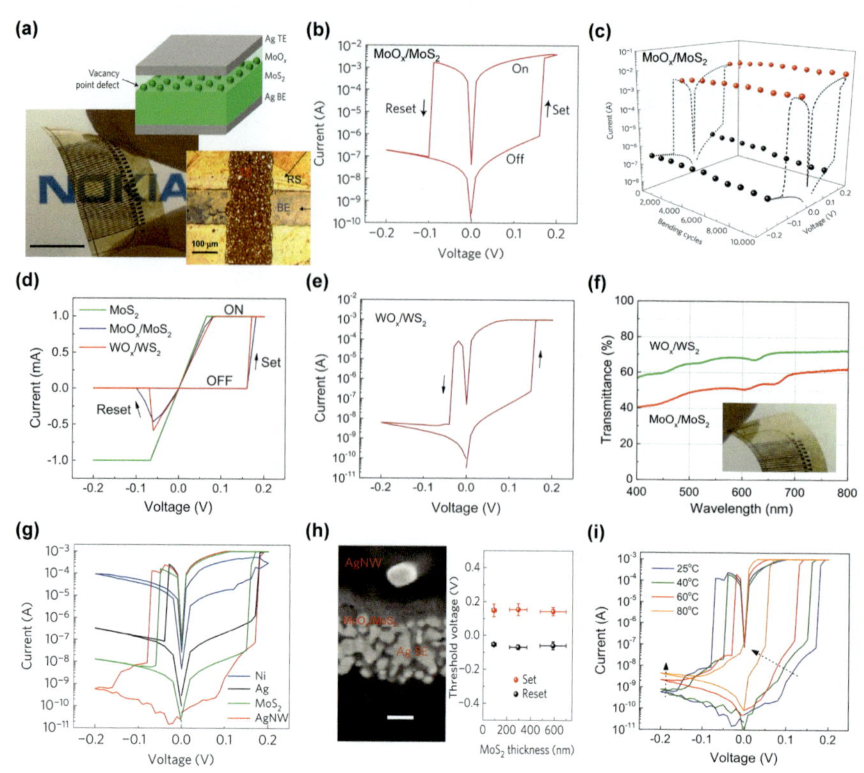

FIG. 12.21 (A) Photograph of a cross-point RRAM array on PEN plastic substrate showing relatively high visual transparency. Scale bar, 10mm. The upper inset is the schematic of the $Ag/MoO_x/MoS_2/Ag$ RRAM device. The lower inset shows the optical image of a cross-point area ($\sim 130 \times 170\,\mu m^2$) with inkjet-printed bottom electrode (BE) and screen-printed top electrode (TE). (B) Typical I–V curve of the $Ag/MoO_x/MoS_2/Ag$ RRAM device with printed electrodes. A forward bias is applied on TE. (C) Sustainability of I_{off} (0.1V, −0.2V) and I_{on} (0.2V, −0.05V) with repeated bending up to a radius of curvature of 10mm. (D) The I–V curves of MoO_x/MoS_2, WO_x/WS_2 heterostructures and non-oxidized MoS_2 sandwiched between two screen-printed Ag electrodes. (E) I–V curve of the WO_x/WS_2 heterostructures with printed electrodes in log scale. (F) Transmittance spectra of MoO_x/MoS_2 and WO_x/WS_2 films on the 50μm thick PEN substrate. (G) Tailoring the I–V characteristics under 1mA I_{CC} using various TE materials while the Ag BE is fabricated by screen printing. (H) Left: cross-section SEM image of the $AgNW/MoO_x/MoS_2/Ag$ device. Scale bar, 100nm. Right: switching voltages of the $AgNW/MoO_x/MoS_2/Ag$ devices vs the MoS_2 thickness. This indicates that MoO_x is fully responsible for the resistance switching. (I) I–V curves of the $AgNW/MoO_x/MoS_2/Ag$ devices under various temperatures. *Reproduced with permission A.A. Bessonov, M.N. Kirikova, D.I. Petukhov, M. Allen, T. Ryhanen, M.J. Bailey, Layered memristive and memcapacitive switches for printable electronics, Nat. Mater. 14 (2) (2015) 199–204. Copyright 2015, Macmillan Publishers Limited.*

with ultralow programming voltages of 0.1–0.2 V, even under repeated bending up to 10^4 cycles at a curvature radius of 10 mm (see Fig. 12.21B and C), indicating its excellent mechanical durability. Fig. 12.21D shows the I–V curves of MoO_x/MoS_2, WO_x/WS_2 heterostructures (manufactured by the same method) and non-oxidized MoS_2 sandwiched between two screen-printed Ag electrodes, while Fig. 12.21E shows the log scale I–V curve of the WO_x/WS_2 heterostructures. Having similar structural and electronic properties, MoO_x/MoS_2 and WO_x/WS_2 heterostructures exhibit almost identical bipolar RS I–V behavior, while the non-oxidized MoS_2 device show linear I–V curve, indicating that the MoS_2 layer used in this experiment is relatively conductive. Therefore, the relatively thick MoS_2 film in the $Ag/MoO_x/MoS_2/Ag$ structure serves as a mechanical support smoothing the rough silver surface. Furthermore, both the MoO_x/MoS_2 and WO_x/WS_2 films show considerable transmittance on the 50 μm thick PEN substrate as shown in Fig. 12.21F. As shown in Fig. 12.21G, I_{OFF} depends on various TEs (Ni, Ag, MoS_2 and AgNW) which might be caused by different Schottky barrier heights, but the basic switching type remains unchanged, indicating that MoS_2/MoO_x is responsible for the resistance switching. As an alternative TE, Ag nanowires in organic media provides a tunable effective electrode area, which is comparable to the dimension of conducting channels, thus significantly reducing the I_{OFF} current and increasing the ON/OFF ratio. The left panel of Fig. 12.21H is the cross-section SEM image of the formed $AgNW/MoO_x/MoS_2/Ag$ RRAM device. Shown in the right panel of Fig. 12.21H, the MoS_2 thickness shows insignificant effect on the switching voltages of the $AgNW/MoO_x/MoS_2/Ag$ device, indicating that the high conductive MoS_2 film does not influence the voltage drop of ultrathin MoO_x (<3 nm), which fully dominates the RS behavior of the MoO_x/MoS_2 based RRAM. Moreover, the $AgNW/MoO_x/MoS_2/Ag$ device can also work under elevated temperature up to 80 °C but with clear temperature dependence of the SET voltage (see Fig. 12.21I). Above all, excellent electrical and mechanical reliability of the $Ag/MoO_x/MoS_2/Ag$ RRAM device fulfill the requirements imposed by system-in-foil technology. Low cost and easier availability are the critical factors for printed electronics. The work in Ref. [111] provides a viable way for active materials to enable printable and flexible MO_X RRAM devices.

Generally, it is a big challenge for electronic devices to exhibit both good flexibility and high thermal stability: inorganic material based devices usually are deficient in mechanical flexibility while organic material based devices are lacking of good thermal stability. In Ref [35], van der Waals heterostructures, full 2D graphene/$MoS_{2-x}O_x$/graphene (GMG structure, 8/40/8 nm), are formed with atomic-level precision by stacking layers of different 2D materials (see Fig. 12.22A–D) for flexible and thermally stabile device application. The $MoS_{2-x}O_x$ film were obtained by

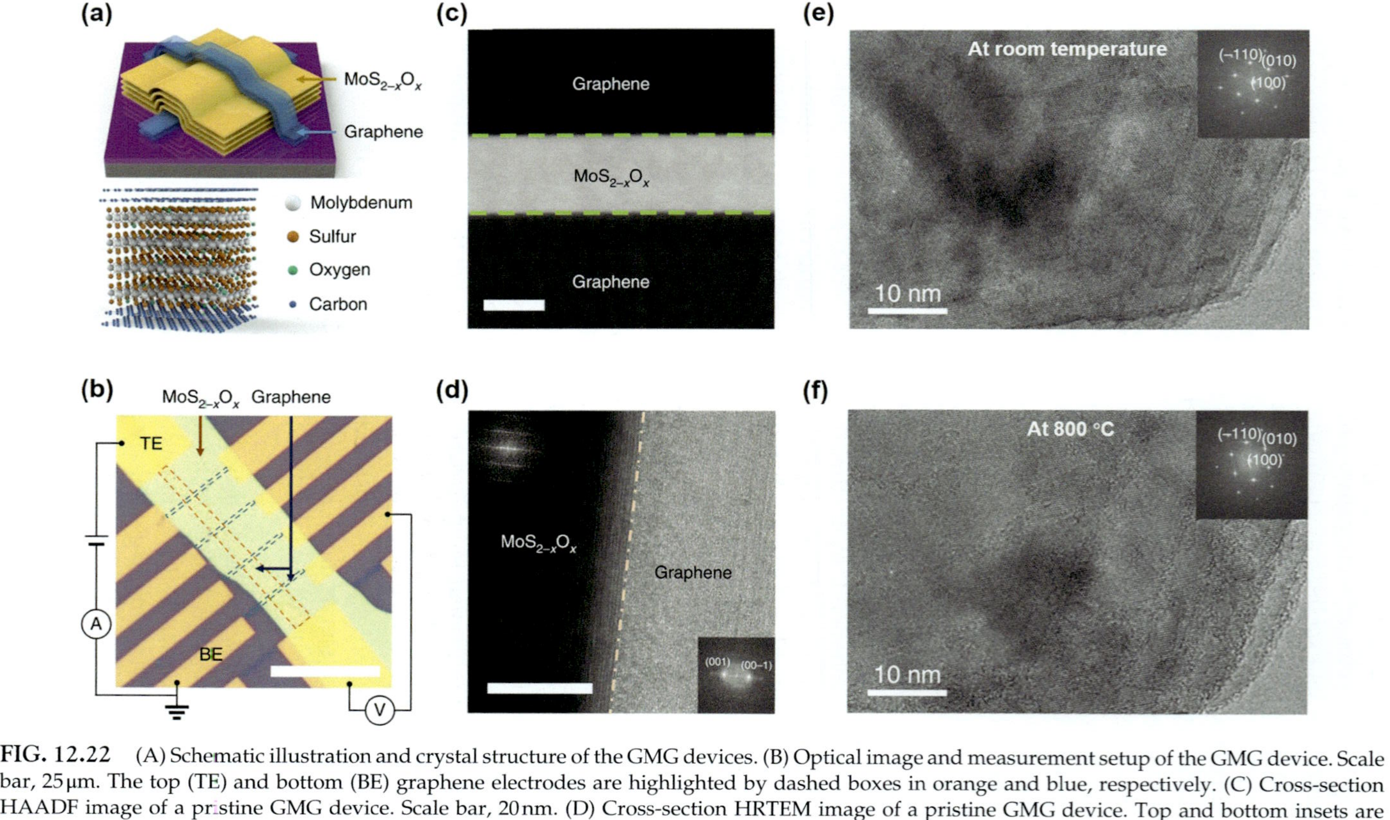

FIG. 12.22 (A) Schematic illustration and crystal structure of the GMG devices. (B) Optical image and measurement setup of the GMG device. Scale bar, 25 μm. The top (TE) and bottom (BE) graphene electrodes are highlighted by dashed boxes in orange and blue, respectively. (C) Cross-section HAADF image of a pristine GMG device. Scale bar, 20 nm. (D) Cross-section HRTEM image of a pristine GMG device. Top and bottom insets are the power spectra of the $MoS_{2-x}O_x$ and graphene films in the HRTEM image, respectively. $MoS_{2-x}O_x$ maintains layered crystal structure after the oxidization and shaping excellent contact with graphene. Scale bar, 20 nm. (E,F) HRTEM images of the same $MoS_{2-x}O_x$ membrane at room temperature and 800 °C, respectively. $MoS_{2-x}O_x$ maintains excellent layered crystal structure at 800 °C. Insets show the power spectra of the corresponding HRTEM images of the $MoS_{2-x}O_x$ membrane at different temperatures. *Reproduced with permission M. Wang, S. Cai, C. Pan, C. Wang, X. Lian, Y. Zhuo, K. Xu, T. Cao, X. Pan, B. Wang, S.-J. Liang, J.J. Yang, P. Wang, F. Miao, Robust memristors based on layered two-dimensional materials, Nat. Electron. 1 (2) (2018) 130–136. Copyright 2018, Macmillan Publishers Limited.*

oxidization of mechanically exfoliated MoS_2 in ambient air on a hot plate under 160 °C for 1.5 h. Remarkably, confirmed by in situ HRTEM observation, $MoS_{2-x}O_x$ maintains excellent crystal structure even at temperatures up to 800 °C, unchanged with the same one at room temperature (see Fig. 12.22E and F). Actually, pristine MoS_2 can still be used as a lubricant at 1300 °C, indicating its high stability at very high temperatures [5]. Therefore, the superior high thermal stability of the $MoS_{2-x}O_x$ layer should be attributed to the structural stability of MoS_2. Such high thermal stability has not been reported in the general binary transition metal oxides used in conventional MO_X RRAM. The top and bottom electrodes are exfoliated MLG connected with the Ti/Au metal conductive pad for measurement. The GMG structure was stacked by standard polyvinyl alcohol (PVA) transfer method.

As shown in Fig. 12.23, the GMG RRAM devices exhibit excellent RS performance, including reproducible I–V curves, high endurance of up to 10^7, fast ON/OFF speed <100 ns and uniform switching parameters (compared to the Au/$MoS_{2-x}O_x$/Au control device). Obviously, the GMG RRAM devices exhibit excellent flexibility as a combination of flexible graphene and MoS_2 when transferred onto a flexible polyimide substrate. The devices can maintain critical HRS/LRS states and survive over 1000 bending cycles at a strain of ~0.6%. The designed GMG RRAM device shows robust RS behaviors and good thermal stability under high operating temperature of up to 340 °C (see Fig. 12.7), which is lacking in conventional memristors. Different from the graphene/BN/graphene full 2D stack, where graphene boundary and B vacancies facilitate the electrochemical redox of Ti from Ti/Au protective coat [115,117], this GMG RRAM device is based on oxygen vacancy CF. The atomically sharp interface between the MLG electrodes and the layered structure of $MoS_{2-x}O_x$ film enables a well-defined conduction channel. In addition, the thermal stability of $MoS_{2-x}O_x$ layer may also enable the original state of non-switching regions in the devices even at elevated temperatures, contributing to the robust RS behaviors of the GMG RRAM devices. This work offers the way to design new devices with atomic-level precision, flexibility and thermal stability based on 2D materials.

12.4 Conclusion

This chapter introduces the research and development status of nonvolatile MO_X RRAM devices based on graphene and other 2D materials. With excellent electronic conductivity, flexibility, transparency, and excellent impermeability, 2D graphene has attracted predominant attention for MO_X RRAM applications. With the MO_X acting as the resistive switching layer, graphene as electrode or assistant intercalation layer has been

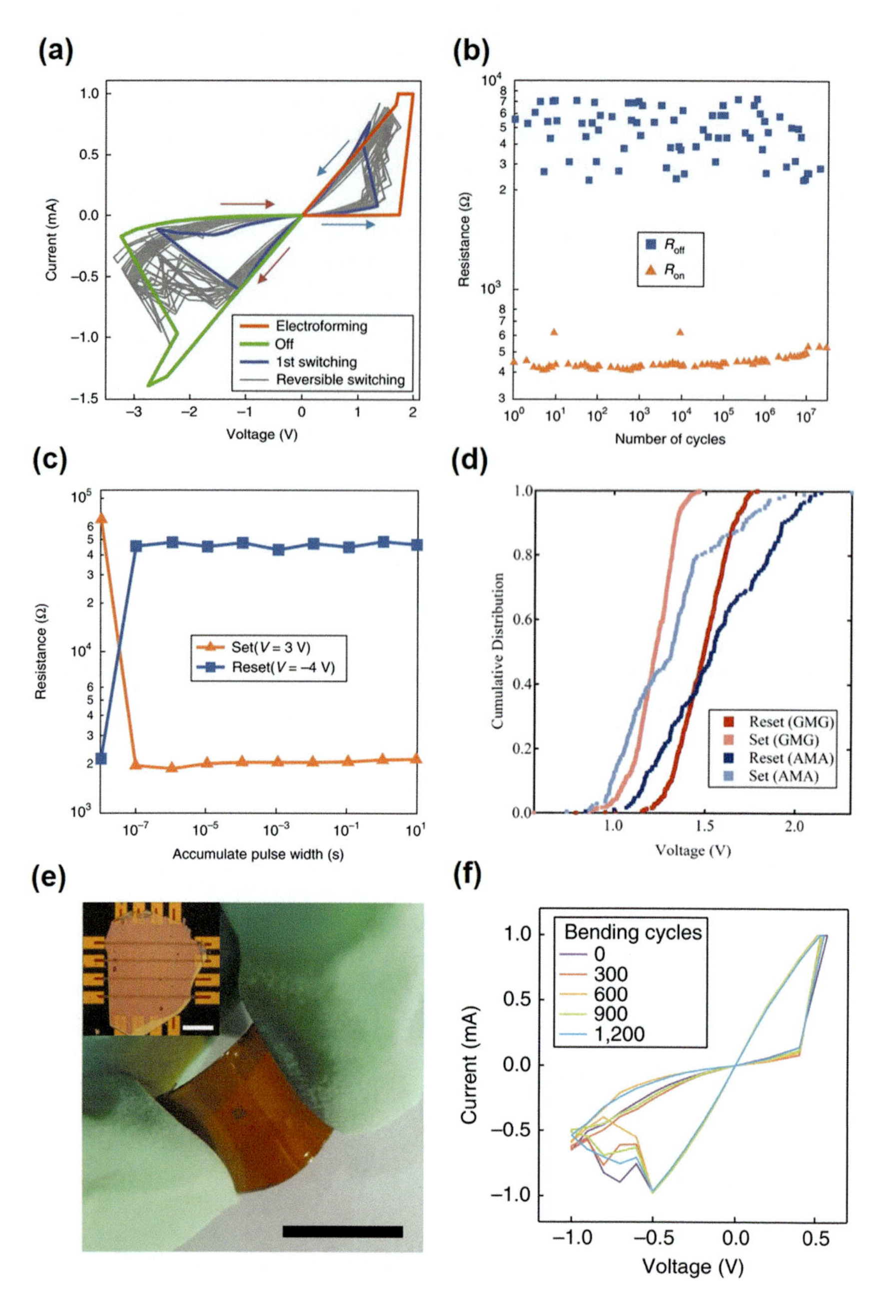

FIG. 12.23 (A) DC RS curves of a GMG device with 1 mA electroforming I_{CC}. (B) Over 2×10^7 switching cycles obtained with 1-μs pulse width. (C) Switching speed test of the GMG device. The device can be switched in less than 100 ns with +3 V and −4 V bias for SET and RESET operation. (D) The cumulative distribution of RESET (absolute value) and SET voltages of the GMG device and Au/$MoS_{2-x}O_x$/Au control device. (E) Photograph of a crossbar array on a polyimide substrate under bending. Scale bar, 10 mm. Inset: optical image of a GMG crossbar device. Scale bar, 20 μm. (F) RS curves of a GMG device against repeated mechanical bending. *Reproduced with permission M. Wang, S. Cai, C. Pan, C. Wang, X. Lian, Y. Zhuo, K. Xu, T. Cao, X. Pan, B. Wang, S.-J. Liang, J. J. Yang, P. Wang, F. Miao, Robust memristors based on layered two-dimensional materials, Nat. Electron. Nature Electronics 1 (2) (2018) 130–136. Copyright 2018, Macmillan Publishers Limited.*

extensively studied to optimize the device structure. MO_X RRAMs enhanced by graphene show excellent resistive switching performance in reliability, power dissipation, retention, endurance and so on. In addition, extremely scaling size (sub-10 nm) of MO_X RRAM has been demonstrated by the integration of graphene. Other emerging nitride or transition metal dichalcogenide 2D materials such as BN, MoS_2 et al. have also been introduced to improve the performance of the conventional MO_X RRAM. Especially, various MO_X RRAM devices based on 2D materials have been demonstrated to possess excellent transparency, flexibility or printability, even under harsh conditions (e.g., high temperature condition). There may be still many challenges to overcome, but the massive explorations of the application of 2D materials in MO_X RRAM provide feasible approaches doubtlessly to improve the storage performance for next generation NVM applications.

References

[1] P. Miro, M. Audiffred, T. Heine, An atlas of two-dimensional materials, Chem. Soc. Rev. 43 (18) (2014) 6537–6554.

[2] A.K. Geim, Graphene: status and prospects, Science 324 (5934) (2009) 1530–1534.

[3] A.S. Barnard, I.K. Snook, Thermal stability of graphene edge structure and graphene nanoflakes, J. Chem. Phys. 128 (9) (2008) 094707.

[4] S. Bertolazzi, J. Brivio, A. Kis, Stretching and breaking of ultrathin MoS2, ACS Nano 5 (12) (2011) 9703–9709.

[5] W.O. Winer, Molybdenum disulfide as a lubricant: a review of the fundamental knowledge, Wear 10 (6) (1967) 422–452.

[6] P. Zhang, C.X. Gao, B.H. Xu, L. Qi, C.J. Jiang, M.Z. Gao, D.S. Xue, Structural phase transition effect on resistive switching behavior of MoS2-polyvinylpyrrolidone nanocomposites films for flexible memory devices, Small 12 (15) (2016) 2077–2084.

[7] A. Lipp, K.A. Schwetz, K. Hunold, Hexagonal boron nitride: Fabrication, properties and applications, J. Eur. Ceram. Soc. 5 (1) (1989) 3–9.

[8] L. Song, L. Ci, H. Lu, P.B. Sorokin, C. Jin, J. Ni, A.G. Kvashnin, D.G. Kvashnin, J. Lou, B.-I. Yakobson, P.M. Ajayan, Large scale growth and characterization of atomic hexagonal boron nitride layers, Nano Lett. 10 (8) (2010), 3209–15.

[9] K. Watanabe, T. Taniguchi, H. Kanda, Direct-bandgap properties and evidence for ultraviolet lasing of hexagonal boron nitride single crystal, Nat. Mater. 3 (6) (2004) 404–409.

[10] K.K. Kim, A. Hsu, X. Jia, S.M. Kim, Y. Shi, M. Dresselhaus, T. Palacios, J. Kong, Synthesis and characterization of hexagonal boron nitride film as a dielectric layer for graphene devices, ACS Nano 6 (10) (2012) 8583–8590.

[11] S.-T. Han, L. Hu, X. Wang, Y. Zhou, Y.-J. Zeng, S. Ruan, C. Pan, Z. Peng, Black phosphorus quantum dots with tunable memory properties and multilevel resistive switching characteristics, Adv. Sci. (2017) 1600435.

[12] Y. Cao, X. Tian, J. Gu, B. Liu, B. Zhang, S. Song, F. Fan, Y. Chen, Covalent functionalization of black phosphorus with conjugated polymer for information storage, Angew. Chem. Int. Ed. 57 (17) (2018) 4543–4548.

[13] C. Hao, F. Wen, J. Xiang, S. Yuan, B. Yang, L. Li, W. Wang, Z. Zeng, L. Wang, Z. Liu, Y. Tian, Liquid-exfoliated black phosphorous nanosheet thin films for flexible resistive random access memory applications, Adv. Funct. Mater. (2016) n/a-n/a.

[14] T. Yang, B. Zheng, Z. Wang, T. Xu, C. Pan, J. Zou, X. Zhang, Z. Qi, H. Liu, Y. Feng, W. Hu, F. Miao, L. Sun, X. Duan, A. Pan, Van der Waals epitaxial growth and optoelectronics of large-scale WSe2/SnS2 vertical bilayer p-n junctions, Nat. Commun. 8 (1) (2017) 1906.
[15] C.R. Dean, A.F. Young, I. Meric, C. Lee, L. Wang, S. Sorgenfrei, K. Watanabe, T. Taniguchi, P. Kim, K.L. Shepard, J. Hone, Boron nitride substrates for high-quality graphene electronics, Nat. Nanotechnol. 5 (10) (2010) 722–726.
[16] G. Fiori, F. Bonaccorso, G. Iannaccone, T. Palacios, D. Neumaier, A. Seabaugh, S.K. Banerjee, L. Colombo, Electronics based on two-dimensional materials, Nat. Nanotechnol. 9 (10) (2014) 768–779.
[17] C. Liu, X. Yan, X. Song, S. Ding, D.W. Zhang, P. Zhou, A semi-floating gate memory based on van der Waals heterostructures for quasi-non-volatile applications, Nat. Nanotechnol. 13 (2018) 404–410.
[18] V.K. Sangwan, H.-S. Lee, H. Bergeron, I. Balla, M.E. Beck, K.-S. Chen, M.C. Hersam, Multi-terminal memtransistors from polycrystalline monolayer molybdenum disulfide, Nature 554 (7693) (2018) 500.
[19] Y. Lee, S. Bae, H. Jang, S. Jang, S.E. Zhu, S.H. Sim, Y.I. Song, B.H. Hong, J.H. Ahn, Wafer-scale synthesis and transfer of graphene films, Nano Lett. 10 (2) (2010) 490–493.
[20] S. Bae, H. Kim, Y. Lee, X. Xu, J.-S. Park, Y. Zheng, J. Balakrishnan, T. Lei, H. Ri Kim, Y.I. Song, Y.-J. Kim, K.S. Kim, B. Özyilmaz, J.-H. Ahn, B.H. Hong, S. Iijima, Roll-to-roll production of 30-inch graphene films for transparent electrodes, Nat. Nanotechnol. 5 (2010) 574.
[21] D. Son, S.I. Chae, M. Kim, M.K. Choi, J. Yang, K. Park, V.S. Kale, J.H. Koo, C. Choi, M. Lee, J.H. Kim, T. Hyeon, D.-H. Kim, Colloidal synthesis of uniform-sized molybdenum disulfide nanosheets for wafer-scale flexible nonvolatile memory, Adv. Mater. 28 (42) (2016) 9326.
[22] T. Kobayashi, M. Bando, N. Kimura, K. Shimizu, K. Kadono, N. Umezu, K. Miyahara, S. Hayazaki, S. Nagai, Y. Mizuguchi, Production of a 100-m-long high-quality graphene transparent conductive film by roll-to-roll chemical vapor deposition and transfer process, Appl. Phys. Lett. 102 (2) (2013) 023112.
[23] J. Han, J.-Y. Lee, H. Kwon, J.-S. Yeo, Synthesis of wafer-scale hexagonal boron nitride monolayers free of aminoborane nanoparticles by chemical vapor deposition, Nanotechnology 25 (14) (2014) 145604.
[24] X. Xu, Z. Zhang, L. Qiu, J. Zhuang, L. Zhang, H. Wang, C. Liao, H. Song, R. Qiao, P. Gao, Z. Hu, L. Liao, Z. Liao, D. Yu, E. Wang, F. Ding, H. Peng, K. Liu, Ultrafast growth of single-crystal graphene assisted by a continuous oxygen supply, Nat. Nanotechnol. 11 (2016) 930–935.
[25] S.P. Koenig, N.G. Boddeti, M.L. Dunn, J.S. Bunch, Ultrastrong adhesion of graphene membranes, Nat. Nanotechnol. 6 (9) (2011) 543.
[26] Z. Li, P. Wu, C. Wang, X. Fan, W. Zhang, X. Zhai, C. Zeng, Z. Li, J. Yang, J. Hou, Low-temperature growth of graphene by chemical vapor deposition using solid and liquid carbon sources, ACS Nano 5 (4) (2011) 3385–3390.
[27] D. Wei, Y. Lu, C. Han, T. Niu, W. Chen, A.T. Wee, Critical crystal growth of graphene on dielectric substrates at low temperature for electronic devices, Angew. Chem. Int. Ed. 52 (52) (2013), 14121–6.
[28] J. Sohn, S. Lee, Z. Jiang, H.-Y. Chen, and H. S. P. Wong, n.d. "Atomically Thin Graphene Plane Electrode for 3D RRAM." pp. 5.3. 1–5.3. 4.
[29] S. Lee, J. Sohn, Z. Jiang, H.Y. Chen, H.S. Philip Wong, Metal oxide-resistive memory using graphene-edge electrodes, Nat. Commun. 6 (2015) 8407.
[30] H. Tian, H.-Y. Chen, B. Gao, S. Yu, J. Liang, Y. Yang, D. Xie, J. Kang, T.-L. Ren, Y. Zhang, H.S.P. Wong, Monitoring oxygen movement by Raman spectroscopy of resistive random access memory with a graphene-inserted electrode, Nano Lett. 13 (2) (2013) 651–657.
[31] J. Yao, J. Lin, Y. Dai, G. Ruan, Z. Yan, L. Li, L. Zhong, D. Natelson, J.M. Tour, Highly transparent nonvolatile resistive memory devices from silicon oxide and graphene, Nat. Commun. 3 (2012) 1101.

[32] S. Dugu, S.P. Pavunny, T.B. Limbu, B.R. Weiner, G. Morell, R.S. Katiyar, A graphene integrated highly transparent resistive switching memory device, APL Mater. 6 (5) (2018) 058503.

[33] P.-K. Yang, W.-Y. Chang, P.-Y. Teng, S.-F. Jeng, S.-J. Lin, P.-W. Chiu, J.-H. He, Fully transparent resistive memory employing graphene electrodes for eliminating undesired surface effects, Proc. IEEE 101 (7) (2013) 1732–1739.

[34] H. Zhao, H. Tu, F. Wei, J. Du, Highly transparent dysprosium oxide-based RRAM with multilayer graphene electrode for low-power nonvolatile memory application, IEEE Trans. Electron Devices 61 (5) (2014) 1388–1393.

[35] M. Wang, S. Cai, C. Pan, C. Wang, X. Lian, Y. Zhuo, K. Xu, T. Cao, X. Pan, B. Wang, S.-J. Liang, J.J. Yang, P. Wang, F. Miao, Robust memristors based on layered two-dimensional materials, Nat. Electron. 1 (2) (2018) 130–136.

[36] J. Shang, G. Liu, H. Yang, X. Zhu, X. Chen, H. Tan, B. Hu, L. Pan, W. Xue, R.-W. Li, Thermally stable transparent resistive random access memory based on all-oxide heterostructures, Adv. Funct. Mater. 24 (15) (2014) 2171–2179.

[37] G. Khurana, P. Misra, R.S. Katiyar, Forming free resistive switching in graphene oxide thin film for thermally stable nonvolatile memory applications, J. Appl. Phys. 114 (12) (2013) 124508.

[38] Y. Ji, S. Lee, B. Cho, S. Song, T. Lee, Flexible organic memory devices with multilayer graphene electrodes, ACS Nano 5 (7) (2011) 5995–6000.

[39] H.-Y. Chen, H. Tian, B. Gao, S. Yu, J. Liang, J. Kang, Y. Zhang, T.-L. Ren, H.S.P. Wong, Ieee, Electrode/oxide interface engineering by inserting single-layer graphene: application for HfOx-based resistive random access memory, in: 2012 *Ieee International Electron Devices Meeting (Iedm)*, 2012.

[40] M. Qian, Y. Pan, F. Liu, M. Wang, H. Shen, D. He, B. Wang, Y. Shi, F. Miao, X. Wang, Tunable, ultralow-power switching in memristive devices enabled by a heterogeneous graphene-oxide interface, Adv. Mater. 26 (20) (2014) 3275–3281.

[41] B.C. Jang, H. Seong, J.Y. Kim, B.J. Koo, S.K. Kim, S.Y. Yang, S.G. Im, S.-Y. Choi, Ultra-low power, highly uniform polymer memory by inserted multilayer graphene electrode, 2D Materials 2 (4) (2015) 044013.

[42] C. Ahn, S.W. Fong, Y. Kim, S. Lee, A. Sood, C.M. Neumann, M. Asheghi, K.E. Goodson, E. Pop, H.P. Wong, Energy-efficient phase-change memory with graphene as a thermal barrier, Nano Lett. 15 (10) (2015) 6809–6814.

[43] F. Yuan, Z. Zhang, C. Liu, F. Zhou, H.M. Yau, W. Lu, X. Qiu, H.S.P. Wong, J. Dai, Y. Chai, Real-time observation of the electrode-size-dependent evolution dynamics of the conducting filaments in a SiO2 layer, ACS Nano 11 (4) (2017) 4097–4104.

[44] S. Fujii, J.A.C. Incorvia, F. Yuan, S. Qin, F. Hui, Y. Shi, Y. Chai, M. Lanza, H.S.P. Wong, Scaling the CBRAM switching layer diameter to 30 nm improves cycling endurance, IEEE Electron Device Lett. 39 (1) (2018) 23–26.

[45] Y. Wu, Y. Chai, H.-Y. Chen, S. Yu, H.-S.P. Wong, Resistive switching AlOx-based memory with CNT electrode for ultra-low switching current and high density memory application, in: IEEE Symposium on VLSI Technology, 2011, pp. 26–27.

[46] J. Campos-Delgado, Y.A. Kim, T. Hayashi, A. Morelos-Gómez, M. Hofmann, H. Muramatsu, M. Endo, H. Terrones, R.D. Shull, M.S. Dresselhaus, M. Terrones, Thermal stability studies of CVD-grown graphene nanoribbons: defect annealing and loop formation, Chem. Phys. Lett. 469 (1–3) (2009) 177–182.

[47] A.I. Romanenko, O.B. Anikeeva, V.L. Kuznetsov, A.N. Obrastsov, A.P. Volkov, A.V. Garshev, Quasi-two-dimensional conductivity and magnetoconductivity of graphite-like nanosize crystallites, Solid State Commun. 137 (11) (2006) 625–629.

[48] H. Chen, M.B. Müller, K.J. Gilmore, G.G. Wallace, D. Li, Mechanically strong, electrically conductive, and biocompatible graphene paper, Adv. Mater. 20 (18) (2008) 3557–3561.

[49] X.-Y. Fang, X.-X. Yu, H.-M. Zheng, H.-B. Jin, L. Wang, M.-S. Cao, Temperature- and thickness-dependent electrical conductivity of few-layer graphene and graphene nanosheets, Phys. Lett. A 379 (37) (2015) 2245–2251.

[50] H.Y. Lee, P.S. Chen, T.Y. Wu, Y.S. Chen, C.C. Wang, P.J. Tzeng, C.H. Lin, F. Chen, C.H. Lien, M.J. Tsai, Ieee, Low Power and High Speed Bipolar Switching with A Thin Reactive Ti Buffer Layer in Robust HfO(2) Based RRAM, 2008.
[51] C. Chen, C. Song, J. Yang, F. Zeng, F. Pan, Oxygen migration induced resistive switching effect and its thermal stability in W/TaO x/Pt structure, Appl. Phys. Lett. 100 (25) (2012) 253509.
[52] M. Rogala, P.J. Kowalczyk, P. Dabrowski, I. Wlasny, W. Kozlowski, A. Busiakiewicz, S. Pawlowski, G. Dobinski, M. Smolny, I. Karaduman, L. Lipinska, R. Kozinski, K. Librant, J. Jagiello, K. Grodecki, J.M. Baranowski, K. Szot, Z. Klusek, The role of water in resistive switching in graphene oxide, Appl. Phys. Lett. 106 (26) (2015) 263104.
[53] L.-H. Wang, W. Yang, Q.-Q. Sun, P. Zhou, H.-L. Lu, S.-J. Ding, D.W. Zhang, The mechanism of the asymmetric SET and RESET speed of graphene oxide based flexible resistive switching memories, Appl. Phys. Lett. 100 (6) (2012).
[54] D.Y. Yun, T.W. Kim, Nonvolatile memory devices based on Au/graphene oxide nanocomposites with bilateral multilevel characteristics, Carbon 88 (2015) 26–32.
[55] S.K. Pradhan, B. Xiao, S. Mishra, A. Killam, A.K. Pradhan, Resistive switching behavior of reduced graphene oxide memory cells for low power nonvolatile device application, Sci. Rep. 6 (2016) 26763.
[56] J. Liu, Z. Yin, X. Cao, F. Zhao, L. Wang, W. Huang, H. Zhang, Fabrication of flexible, all-reduced graphene oxide non-volatile memory devices, Adv. Mater. 25 (2) (2013) 233–238.
[57] Q. Liu, Z. Liu, X. Zhang, N. Zhang, L. Yang, S. Yin, Y. Chen, Organic photovoltaic cells based on an acceptor of soluble graphene, Appl. Phys. Lett. 92 (22) (2008) 223303.
[58] Y. Gim, B. Kang, B. Kim, S.-G. Kim, J.-H. Lee, K. Cho, B.-C. Ku, J.H. Cho, Atomically-thin molecular layers for electrode modification of organic transistors, Nanoscale 7 (33) (2015) 14100–14108.
[59] Y.J. Choi, J.S. Kim, J.Y. Cho, H.J. Woo, J. Yang, Y.J. Song, M.S. Kang, J.T. Han, J.H. Cho, Tunable charge injection via solution-processed reduced graphene oxide electrode for vertical schottky barrier transistors, Chem. Mater. 30 (3) (2018) 636–643.
[60] J. Liu, Z. Zeng, X. Cao, G. Lu, L.H. Wang, Q.L. Fan, W. Huang, H. Zhang, Preparation of MoS(2)-polyvinylpyrrolidone nanocomposites for flexible nonvolatile rewritable memory devices with reduced graphene oxide electrodes, Small 8 (22) (2012) 3517–3522.
[61] H. Tian, H.Y. Chen, T.L. Ren, C. Li, Q.T. Xue, M.A. Mohammad, C. Wu, Y. Yang, H.S. Wong, Cost-effective, transfer-free, flexible resistive random access memory using laser-scribed reduced graphene oxide patterning technology, Nano Lett. 14 (6) (2014), 3214–9.
[62] X.-Y. Xu, Z.-Y. Yin, C.-X. Xu, J. Dai, J.-G. Hu, Resistive switching memories in MoS2 nanosphere assemblies, Appl. Phys. Lett. 104 (3) (2014) 033504.
[63] H.S.P. Wong, H.-Y. Lee, S. Yu, Y.-S. Chen, Y. Wu, P.-S. Chen, B. Lee, F.T. Chen, M.-J. Tsai, Metal-oxide RRAM, Proc. IEEE 100 (6) (2012) 1951–1970.
[64] U. Celano, G. Giammaria, L. Goux, A. Belmonte, M. Jurczak, W. Vandervorst, Nanoscopic structural rearrangements of the Cu-filament in conductive-bridge memories, Nanoscale 8 (29) (2016) 13915–13923.
[65] H. Lv, X. Xu, H. Liu, R. Liu, Q. Liu, W. Banerjee, H. Sun, S. Long, L. Li, M. Liu, Evolution of conductive filament and its impact on reliability issues in oxide-electrolyte based resistive random access memory, Sci. Rep. 5 (2015) 7764.
[66] K. Rajan, A. Chiappone, D. Perrone, S. Bocchini, I. Roppolo, K. Bejtka, M. Castellino, C.-F. Pirri, C. Ricciardi, A. Chiolerio, Ionic liquid-enhanced soft resistive switching devices, RSC Adv. 6 (96) (2016) 94128–94138.
[67] X.Y. Li, X.L. Shao, Y.C. Wang, H. Jiang, C.S. Hwang, J.S. Zhao, Thin TiOx layer as a voltage divider layer located at the quasi-Ohmic junction in the Pt/Ta2O5/Ta resistance switching memory, Nanoscale 9 (6) (2017) 2358–2368.

[68] J.J. Yang, F. Miao, M.D. Pickett, D.A. Ohlberg, D.R. Stewart, C.N. Lau, R.S. Williams, The mechanism of electroforming of metal oxide memristive switches, Nanotechnology 20 (21) (2009) 215201.
[69] J. Lee, C. Du, K. Sun, E. Kioupakis, W.D. Lu, Tuning ionic transport in memristive devices by graphene with engineered nanopores, ACS Nano 10 (3) (2016) 3571–3579.
[70] A.C. Obreja, D. Cristea, I. Mihalache, A. Radoi, R. Gavrila, F. Comanescu, C. Kusko, Charge transport and memristive properties of graphene quantum dots embedded in poly(3-hexylthiophene) matrix, Appl. Phys. Lett. 105 (8) (2014).
[71] C. Wang, W. He, Y. Tong, Y. Zhang, K. Huang, L. Song, S. Zhong, R. Ganeshkumar, R. Zhao, Memristive devices with highly repeatable analog states boosted by graphene quantum dots, Small 13 (20) (2017) 1603435.
[72] K.-C. Chang, R. Zhang, T.-C. Chang, T.-M. Tsai, J.C. Lou, J.-H. Chen, T.-F. Young, M.-C. Chen, Y.-L. Yang, Y.-C. Pan, G.-W. Chang, T.-J. Chu, C.-C. Shih, J.-Y. Chen, C.-H. Pan, Y.-T. Su, Y.-E. Syu, Y.-H. Tai, S.M. Sze, Origin of hopping conduction in graphene-oxide-doped silicon oxide resistance random access memory devices, IEEE Electron Device Lett. 34 (5) (2013) 677–679.
[73] M.M. Rehman, G.U. Siddiqui, Y.H. Doh, K.H. Choi, Highly flexible and electroforming free resistive switching behavior of tungsten disulfide flakes fabricated through advanced printing technology, Semicond. Sci. Technol. 32 (9) (2017).
[74] M.M. Rehman, G.U. Siddiqui, J.Z. Gul, S.-W. Kim, J.H. Lim, K.H. Choi, Resistive switching in all-printed, flexible and hybrid MoS2-PVA nanocomposite based memristive device fabricated by reverse offset, Sci. Rep. 6 (2016).
[75] S.K. Hämäläinen, Z. Sun, M.P. Boneschanscher, A. Uppstu, M. Ijäs, A. Harju, D. Vanmaekelbergh, P. Liljeroth, Quantum-confined electronic states in atomically well-defined graphene nanostructures, Phys. Rev. Lett. 107 (23) (2011) 236803.
[76] K.A. Ritter, J.W. Lyding, The influence of edge structure on the electronic properties of graphene quantum dots and nanoribbons, Nat. Mater. 8 (3) (2009) 235.
[77] S. Yu, B. Gao, Z. Fang, H. Yu, J. Kang, H.S. Wong, A low energy oxide-based electronic synaptic device for neuromorphic visual systems with tolerance to device variation, Adv. Mater. 25 (12) (2013) 1774–1779.
[78] T.-M. Tsai, C.-H. Wu, K.-C. Chang, C.-H. Pan, P.-H. Chen, N.-K. Lin, J.-C. Lin, Y.-S. Lin, W.-C. Chen, H. Wu, N. Deng, H. Qian, Controlling the degree of forming soft-breakdown and producing superior endurance performance by inserting BN-based layers in resistive random access memory, IEEE Electron Device Lett. 38 (4) (2017) 445–448.
[79] J. Kim, D. Kim, Y. Jo, J. Han, H. Woo, H. Kim, K.K. Kim, J.P. Hong, H. Im, Impact of graphene and single-layer BN insertion on bipolar resistive switching characteristics in tungsten oxide resistive memory, Thin Solid Films 589 (2015) 188–193.
[80] J. Hong, S. Lee, S. Lee, H. Han, C. Mahata, H.-W. Yeon, B. Koo, S.-I. Kim, T. Nam, K. Byun, B.-W. Min, Y.-W. Kim, H. Kim, Y.-C. Joo, T. Lee, Graphene as an atomically thin barrier to cu diffusion into Si, Nanoscale 6 (13) (2014) 7503–7511.
[81] S. Liu, N. Lu, X. Zhao, H. Xu, W. Banerjee, H. Lv, S. Long, Q. Li, Q. Liu, M. Liu, Eliminating negative-SET behavior by suppressing nanofilament overgrowth in cation-based memory, Adv. Mater. 28 (48) (2016) 10623–10629.
[82] J.S. Bunch, S.S. Verbridge, J.S. Alden, A.M. van der Zande, J.M. Parpia, H.G. Craighead, P.L. McEuen, Impermeable atomic membranes from graphene sheets, Nano Lett. 8 (8) (2008) 2458–2462.
[83] P.R. Kidambi, M.S. Boutilier, L. Wang, D. Jang, J. Kim, R. Karnik, Selective nanoscale mass transport across atomically thin single crystalline graphene membranes, Adv. Mater. 29 (19) (2017) 1605896.
[84] S. Si, W. Li, X. Zhao, M. Han, Y. Yue, W. Wu, S. Guo, X. Zhang, Z. Dai, X. Wang, X. Xiao, C. Jiang, Significant radiation tolerance and moderate reduction in thermal

transport of a tungsten nanofilm by inserting monolayer graphene, Adv. Mater. 29 (3) (2017) 1604623.
[85] S.P. Koenig, L. Wang, J. Pellegrino, J.S. Bunch, Selective molecular sieving through porous graphene, Nat. Nanotechnol. 7 (11) (2012) 728–732.
[86] T. Tsuruoka, K. Terabe, T. Hasegawa, M. Aono, Forming and switching mechanisms of a cation-migration-based oxide resistive memory, Nanotechnology 21 (42) (2010) 425205.
[87] W. Guan, S. Long, Q. Liu, M. Liu, W. Wang, Nonpolar nonvolatile resistive switching in cu doped $\hbox {ZrO} _ {2} $, IEEE Electron Device Lett. 29 (5) (2008) 434–437.
[88] L. Zhong, L. Jiang, R. Huang, C. De Groot, Nonpolar resistive switching in Cu/SiC/Au non-volatile resistive memory devices, Appl. Phys. Lett. 104 (9) (2014) 093507.
[89] C. Chen, S. Gao, G. Tang, C. Song, F. Zeng, F. Pan, Cu-embedded AlN-based nonpolar nonvolatile resistive switching memory, IEEE Electron Device Lett. 33 (12) (2012) 1711–1713.
[90] D.-H. Kwon, K.M. Kim, J.H. Jang, J.M. Jeon, M.H. Lee, G.H. Kim, X.-S. Li, G.-S. Park, B. Lee, S. Han, M. Kim, C.S. Hwang, Atomic structure of conducting nanofilaments in TiO2 resistive switching memory, Nat. Nanotechnol. 5 (2) (2010) 148–153.
[91] J.-Y. Chen, C.-L. Hsin, C.-W. Huang, C.-H. Chiu, Y.-T. Huang, S.-J. Lin, W.-W. Wu, L.-J. Chen, Dynamic evolution of conducting nanofilament in resistive switching memories, Nano Lett. 13 (8) (Aug, 2013) 3671–3677.
[92] Y. Yang, X. Zhang, L. Qin, Q. Zeng, X. Qiu, R. Huang, Probing nanoscale oxygen ion motion in memristive systems, Nat. Commun. 8 (2017) 15173.
[93] X. Zhao, J. Ma, X. Xiao, Q. Liu, L. Shao, D. Chen, S. Liu, J. Niu, X. Zhang, Y. Wang, R. Cao, W. Wang, Z. Di, H. Lv, S. Long, M. Liu, Breaking the current-retention dilemma in cation-based resistive switching devices utilizing graphene with controlled defects, Adv. Mater. 30 (14) (2018) 1705193.
[94] K. Celebi, J. Buchheim, R.M. Wyss, A. Droudian, P. Gasser, I. Shorubalko, J.I. Kye, C. Lee, H.G. Park, Ultimate permeation across atomically thin porous graphene, Science 344 (6181) (2014) 289–292.
[95] S. Garaj, W. Hubbard, A. Reina, J. Kong, D. Branton, J.A. Golovchenko, Graphene as a subnanometre trans-electrode membrane, Nature 467 (7312) (2010) 190–193.
[96] S.J. Heerema, C. Dekker, Graphene nanodevices for DNA sequencing, Nat. Nanotechnol. 11 (2) (2016) 127–136.
[97] H. Sun, Q. Liu, C. Li, S. Long, H. Lv, C. Bi, Z. Huo, L. Li, M. Liu, Direct observation of conversion between threshold switching and memory switching induced by conductive filament morphology, Adv. Funct. Mater. 24 (36) (2014) 5679–5686.
[98] J. Song, J. Woo, A. Prakash, D. Lee, H. Hwang, Threshold selector with high selectivity and steep slope for cross-point memory Array, IEEE Electron Device Lett. 36 (7) (2015) 681–683.
[99] R. Midya, Z. Wang, J. Zhang, S.E. Savel'ev, C. Li, M. Rao, M.H. Jang, S. Joshi, H. Jiang, P. Lin, K. Norris, N. Ge, Q. Wu, M. Barnell, Z. Li, H.L. Xin, R.S. Williams, Q. Xia, J.J. Yang, Anatomy of Ag/Hafnia-based selectors with 1010 nonlinearity, Adv. Mater. 29 (12) (2017) 1604457.
[100] Z. Wang, S. Joshi, S.E. Savel'ev, H. Jiang, R. Midya, P. Lin, M. Hu, N. Ge, J.P. Strachan, Z. Li, Q. Wu, M. Barnell, G.-L. Li, H.L. Xin, R.S. Williams, Q. Xia, J.J. Yang, Memristors with diffusive dynamics as synaptic emulators for neuromorphic computing, Nat. Mater. 16 (1) (2017) 101–108.
[101] A. Bricalli, E. Ambrosi, M. Laudato, M. Maestro, R. Rodriguez, D. Ielmini, SiOx-based resistive switching memory (RRAM) for crossbar storage/select elements with high on/off ratio, in: IEEE International Electron Devices Meeting, 2016, pp. 4.3.1–4.3.4.
[102] D.-Y. Cho, S. Tappertzhofen, R. Waser, I. Valov, Bond nature of active metal ions in SiO2-based electrochemical metallization memory cells, Nanoscale 5 (5) (2013) 1781–1784.

[103] X. Zhao, S. Liu, J. Niu, L. Liao, Q. Liu, X. Xiao, H. Lv, S. Long, W. Banerjee, W. Li, S. Si, M. Liu, Confining cation injection to enhance cbram performance by nanopore graphene layer, Small 13 (35) (2017) 1603948.

[104] M. Lübben, P. Karakolis, V. Ioannou-Sougleridis, P. Normand, P. Dimitrakis, I. Valov, Graphene-modified Interface controls transition from VCM to ECM switching modes in ta/TaOx based memristive devices, Adv. Mater. 27 (40) (2015) 6202–6207.

[105] Q. Liu, S. Long, H. Lv, W. Wang, J. Niu, Z. Huo, J. Chen, M. Liu, Controllable growth of nanoscale conductive filaments in solid-electrolyte-based ReRAM by using a metal nanocrystal covered bottom electrode, ACS Nano 4 (10) (2010) 6162–6168.

[106] B.K. You, J.M. Kim, D.J. Joe, K. Yang, Y. Shin, Y.S. Jung, K.J. Lee, Reliable memristive switching memory devices enabled by densely packed silver nanocone arrays as electric-field concentrators, ACS Nano 10 (10) (2016) 9478–9488.

[107] K.-Y. Shin, Y. Kim, F.V. Antolinez, J.S. Ha, S.-S. Lee, J.H. Park, Controllable formation of nanofilaments in resistive memories via tip-enhanced electric fields, Adv. Electron. Mater. 2 (10) (2016) 1600233.

[108] U. Celano, L. Mirabelli, L. Goux, K. Opsomer, W. Devulder, F. Crupi, C. Detavernier, M. Jurczak, W. Vandervorst, Tuning the switching behavior of conductive-bridge resistive memory by the modulation of the cation-supplier alloys, Microelectron. Eng. 167 (2017) 47–51.

[109] X.-F. Wang, H. Tian, H.-M. Zhao, T.-Y. Zhang, W.-Q. Mao, Y.-C. Qiao, Y. Pang, Y.-X. Li, Y. Yang, T.-L. Ren, Interface engineering with MoS2-Pd nanoparticles hybrid structure for a low voltage resistive switching memory, Small 14 (2) (2018).

[110] H. Zhao, Z. Dong, H. Tian, D. DiMarzi, M.G. Han, L. Zhang, X. Yan, F. Liu, L. Shen, S.J. Han, S. Cronin, W. Wu, J. Tice, J. Guo, H. Wang, Atomically thin femtojoule memristive device, Adv. Mater. 29 (47) (2017) 1703232.

[111] A.A. Bessonov, M.N. Kirikova, D.I. Petukhov, M. Allen, T. Ryhanen, M.J. Bailey, Layered memristive and memcapacitive switches for printable electronics, Nat. Mater. 14 (2) (2015) 199–204.

[112] L. Hu, S. Fu, Y. Chen, H. Cao, L. Liang, H. Zhang, J. Gao, J. Wang, F. Zhuge, Ultrasensitive memristive synapses based on lightly oxidized sulfide films, Adv. Mater. 29 (24) (2017) 1606927.

[113] W. Banerjee, X. Xu, H. Liu, H. Lv, Q. Liu, H. Sun, S. Long, M. Liu, Occurrence of resistive switching and threshold switching in atomic layer deposited ultrathin (2 nm) aluminium oxide crossbar resistive random access memory, IEEE Electron Device Lett. 36 (4) (2015) 333–335.

[114] L. Zhao, Z. Jiang, H.-Y. Chen, J. Sohn, K. Okabe, B. Magyari-Köpe, H.-S.P. Wong, Y. Nishi, Ultrathin (~ 2nm) HfO x as the fundamental resistive switching element: thickness scaling limit, stack engineering and 3D integration, in: IEEE International Electron Devices Meeting, 2014, pp. 6.6. 1–6.6. 4.

[115] C. Pan, Y. Ji, N. Xiao, F. Hui, K. Tang, Y. Guo, X. Xie, F.M. Puglisi, L. Larcher, E. Miranda, L. Jiang, Y. Shi, I. Valov, P.C. McIntyre, R. Waser, M. Lanza, Coexistence of grain-boundaries-assisted bipolar and threshold resistive switching in multilayer hexagonal boron nitride, Adv. Funct. Mater. 27 (10) (2017) 1604811.

[116] D.-H. Lien, Z.-K. Kao, T.-H. Huang, Y.-C. Liao, S.-C. Lee, J.-H. He, All-printed paper memory, ACS Nano 8 (8) (2014) 7613–7619.

[117] C. Pan, E. Miranda, M.A. Villena, N. Xiao, X. Jing, X. Xie, T. Wu, F. Hui, Y. Shi, M. Lanza, Model for multi-filamentary conduction in graphene/hexagonal-boron-nitride/graphene based resistive switching devices, 2D Materials 4 (2) (2017).

CHAPTER

13

Ubiquitous memristors on-chip in multi-level memory, in-memory computing, data converters, clock generation and signal transmission

Ioannis Vourkas[a], *Manuel Escudero*[b], *Georgios Ch. Sirakoulis*[c], *and Antonio Rubio*[b]

[a]Department of Electronic Engineering, Universidad Técnica Federico Santa María, Valparaíso, Chile, [b]Department of Electronic Engineering, Universitat Politècnica de Catalunya, Barcelona, Spain, [c]Department of Electrical and Computer Engineering, Democritus University of Thrace, Xanthi, Greece

13.1 Introduction

The existence of the "memristor", nowadays known as the fourth fundamental circuit element, was postulated by Chua in 1971 [1]. Chua theoretically explored the properties of such a nonlinear circuit element and observed it was essentially a resistor with memory, so he called it a MEMoRy resISTOR, thus a "memristor." It is a two-terminal circuit element characterized by a nonlinear relation between the time integrals of the current and the voltage applied to its terminals [2]. Unprecedented attention on this device technology has been drawn ever since 2008 when a team of Hewlett Packard linked the term "memristor" to resistive switching (RS) devices [3,4], as ubiquitous circuit components enabling a novel generation of electronic systems. It is worth mentioning though that RS behavior had been observed much earlier [5]. Currently, the term

Metal Oxides for Non-volatile Memory
https://doi.org/10.1016/B978-0-12-814629-3.00013-1

memristor may refer to any RS device that complies with a set of certain properties known as "fingerprints" [6].

Several material compounds can be the basis of memristor devices [7]. A wide variety of such materials have been studied, aiming to predict the switching profiles and manipulate them appropriately to achieve better on–off ratio, endurance, and state retention. RS devices are usually fabricated in metal–insulator–metal (MIM) structures with the insulating material being a metal oxide [8] such as NbO_x [5,9], HfO_x [10], TiO_x [11], AlO_x [12], TaO_x [13], or NiO [14], a chalcogenide such as GeS_x [15], a selenide (e.g., GeSe [16]), a nitride (e.g., SiN_x [17]), a perovskite [18] or a graphene oxide [19]. It is worth noting though that using such materials as the functional RS layer might not always be compatible with current integrated circuit (IC) industry infrastructure, and this has also led to a focused interest in amorphous silicon and silicon oxide [20], even though HfO_x, NbO_x, as well as TaO_x are known to be CMOS-compatible.

Of course, every material can behave in a different way when exposed to the electrical field induced by the applied voltage. On one hand, RS has been identified as originating from the local formation/disruption of conductive filaments (CFs) between the metallic electrodes [21]. The filamentary type switching includes thermochemical (TCM), electrochemical (ECM), and valance change (VCM) mechanisms. On the other hand, RS can be caused by an interfacial mechanism [22] with distributed migration of traps at the metal–insulator interface (e.g., oxygen vacancies distributed along an interface that controls the conductivity of the MIM structure). The interface-type switching includes VCM, purely electronic and electrostatic effects, and ferroelectric polarization mechanism. Filamentary devices generally show sharp conductance transitions whereas interfacial (or homogeneous) switching devices change their resistance by a uniform interface effect, thus tend to respond more slowly to the applied input signals, modifying their state in a more incremental manner.

For a proper analysis of the main differences resulting in filamentary versus interfacial RS types, from a device performance viewpoint, see [23], whereas for a proper classification of RS devices in terms of functional characteristics and material selection, and a review of modeling approaches, read further in [24]. Finally, [25] reviews the recent advances in memristive materials for the development of artificial synaptic devices. Most importantly, while this technology is continuously maturing, the first commercially available discrete RS devices were released recently by *Knowm Inc.* [26] who provided affordable access to real memristors, bringing this technology closer to scientists, researchers, and enthusiasts that might not have access to fabrication facilities [27]. This notwithstanding, Panasonic was the first company into mass production with resistive random-access memory (ReRAM) technology, as it has been shipping a microcontroller with ReRAM on-chip since 2012 [28].

Memristors demonstrate many promising features, such as plasticity, analog nature, non-volatility, along with a low power consumption, high density, and excellent scalability, converting them to an emerging trend in modern electronics with an ever growing variety of potential applications. Applicability of memristors is as simple as it sounds: *a memristor could essentially replace any resistor that has a particular role in a given circuit* (e.g., a feedback resistor in an amplifier or a reference resistor in a comparator, to name a few), thus improving the overall functionality and/or upgrading the circuit from static to programmable [29,30]. In this chapter we present a brief overview of selected related applications concerning a broad spectrum of *on-chip* memristor workability, aiming to reveal the potential impact of this emerging device technology in literally any dimension of future electronic circuits and systems.

13.2 Multi-level memory and in-memory arithmetic structures

In the era of *big data*, memristors embedded in novel nanoelectronic platforms constitute perhaps the most promising key-enabling technology for the storage and processing of massive amounts of data, enabling true *memcomputing* [31], i.e., using memory to process information. The memristor provides an unconventional computing framework, ideally combining resistance-based information storage and processing in a single device [32]. Multi-digit arithmetic using the multi-level memory characteristics of memristors further emphasize their versatility and potential for future on-chip information processing and mass storage architectures. For instance, recently Wust et al. [33] proposed the implementation of a general-purpose CPU using signed-digit arithmetic, exploiting memristors and their multi-level tuning property to implement multi-valued registers in dense array configurations. Moreover, several works have been also published especially concerning the design, fabrication, modeling and implementation of computation in spatially extended discrete media with many memristors [34,35].

In this context, Cellular Automata (CA), originally postulated in the 1940s by Ulam and von Neumann [36], constitute a well-studied inherently parallel computing paradigm of high efficiency and robustness [37,38]. A Cellular Automaton consists of a regular and uniform *d*-dimensional lattice (or array). At each site of the lattice there is a cell whose state is updated at discrete time steps, based on its current state and the states of its immediate neighborhood. Likewise in other multi-agent system models, CA usually treat time as discrete and state updates are occurring synchronously. The synchronous approach assumes the presence of a global clock to ensure all cells are updated together. In contrast, *asynchronous CA* are able to update individual cells independently, in such a way that the new state of a cell affects the calculation of states in neighboring cells.

Owing to their potential to capture globally emerging behavior from collective interaction of simple and local components, CA have found successful application in several computational problems. Additionally, when CA-based models are implemented in hardware (HW), the circuit design reduces to the design of a single cell and the overall layout results regular with exclusively local interconnections [39,40]. Moreover, the computations are executed fast by exploiting the parallelism of the CA structure, thus meeting the necessary information processing requirements in modern computationally-demanding applications. The CA approach is consistent with the modern notion of unified space–time; memory (CA cell state) and processing unit (local evolution rule) are inseparably related to a CA cell. Therefore, it is quite justifiable to develop computing paradigms which combine the capabilities and the structural simplicity of CA with the unique properties of memristors for *in-memory information processing*. To this end, Itoh and Chua [41] first discussed simulation of CA in networks of memristors, whereas later several other works [42,43] were published exploiting the gradual resistance switching behavior of memristors in novel designs of memristive CA units, using memristors as storage and/or computing elements.

Within the general framework of asynchronous CA, we could embody the work of Pershin et al. [44] which demonstrated experimentally the *memcomputing* version of the mechanical *Pascaline*. The original Pascaline was a mechanical calculator built by the mathematician Blaise Pascal [45], which encoded information in the angles of its mechanical components and was able to sum and subtract integers. More specifically, it encoded 10 digits in each wheel/cylinder (one wheel for each power of the base-10 system), so it was essentially a multi-state machine, in the sense that every memory element encoded more than two values of information. What is more, the results of the computation were stored directly in the states of the processing units themselves, enabling the user to read the result of the computation without any additional storage device, thus demonstrating *multi-state* and *computing-in-memory* features, tempting to be employed in modern computing systems and architectures.

The memcomputing version of the Pascaline is storing the numerical results in the multiple levels of resistance of each memristive device, being also capable of addition, subtraction and even multi-base operations, all realized directly in memory. As shown in Fig. 13.1, the system *resembles a one-dimensional asynchronous CA*; it consists of a number of identical computing blocks (cells), each one responsible for a different digit of a number. The blocks are connected with each other in series so as to propagate the carry.

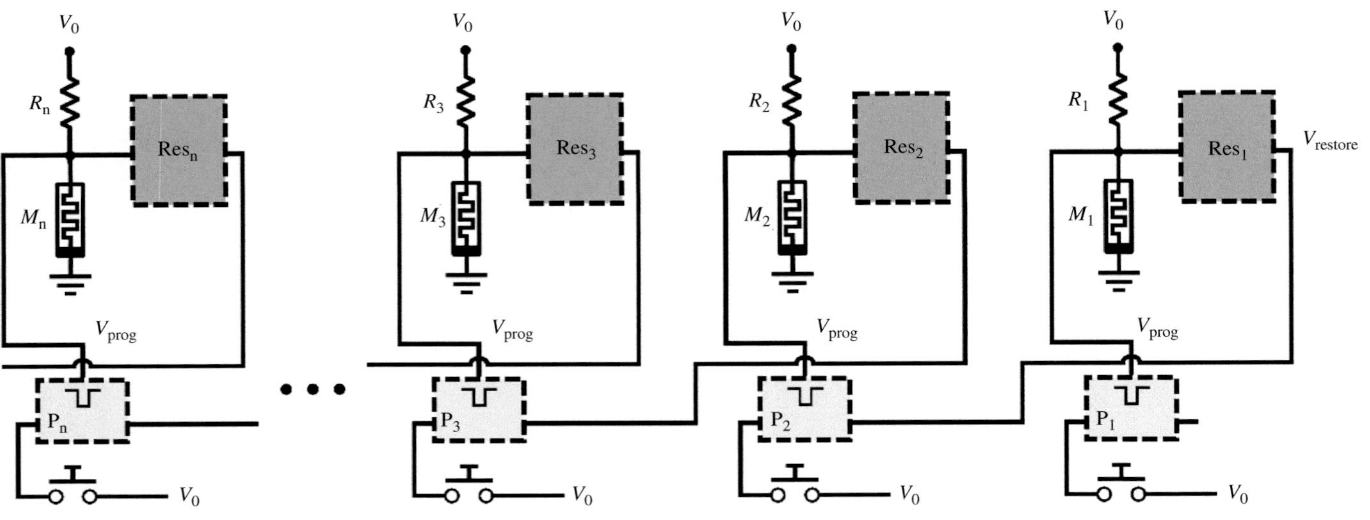

FIG. 13.1 Memristive Pascaline. Pulse generators P_1–P_n (at the bottom) are triggered either by push buttons or carry pulses created by the restore circuits Res_1–Res_n. Restore circuits are used to SET the memristors when their memristance exceeds a specific threshold, at the same time generating the carry pulse that activates the next pulse generator. *Adapted from Y.V. Pershin, L.K. Castelano, F. Hartmann, V. Lopez-Richard, M. Di Ventra, A Memristive Pascaline, IEEE Trans. Circuits Syst. Express Briefs 63 (6) 2016 558–562.*

Each block consists of a memristor M_i which is connected to a power source V_0 through a series resistor R_i. Moreover, there is a restore circuit Res_i and a pulse generator P_i. The digit value (the current state of the cell) is stored in the resistance (memristance) of the memristor and it is represented by the voltage drop on it, as a result of the voltage divider, when no SET ($R_{OFF} \rightarrow R_{ON}$) or RESET ($R_{ON} \rightarrow R_{OFF}$) voltage pulses are applied to it. Every pulse generator P_i (shown at the bottom of Fig. 13.1) is triggered either by a push button or by a carry pulse received from the restore circuit Res_i on its right. The negative programming (RESET) pulses from P_i gradually increase the resistance of M_i and, consequently, the voltage drop on them. Once the pulse is finished, the output node of P_i (V_{prog}) is in high impedance state (floating). This way, such programming pulses do not interfere with the restore circuits Res_i, which in turn serve to restore/initialize memristors when their resistance exceeds a maximum value, at the same time generating the carry pulse (the output of Res_i is connected to the input of P_{i+1}, i.e., the P_i of the immediate more significant digit found on its left side, for carry over).

More specifically, the restore circuit *Res* compares internally the voltage across the memristor V_M with a reset threshold voltage $V_{Res,t}$ and generates a positive restoring pulse as soon as the voltage on the memristor exceeds $V_{Res,t}$. The restore pulse-width is selected sufficiently long to SET the memristor into its minimum resistance (usually referred as R_{MIN} or R_{ON}). Such memristive Pascaline *supports different number bases*, which are defined by the length of the programming reset pulses (from pulse generators P_i) and the configurable threshold voltage $V_{Res,t}$ of the restore circuits. Fig. 13.2 shows an example of individual operation of each cell assuming a base-5 system. The red curve shows the voltage on the

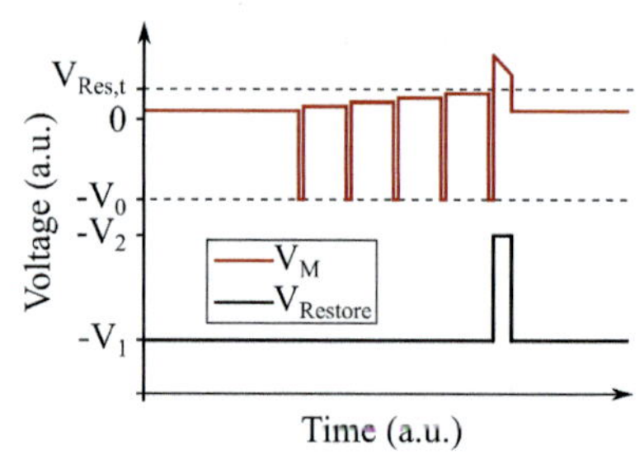

FIG. 13.2 Cartoon graph with the expected response of the circuit to a train of negative pulses (original pulses in [44] had $V_0=2.5V$ amplitude and were 10-ms-wide) applied to the top terminal of M_1. This example concerns a base-5 arithmetic (note how the memristor resets by the fifth pulse). V_M is the voltage across the memristor, whereas $V_{Res,t}$ is the reset threshold voltage of every restore circuit *Res*, and $V_{Restore}$ is the output voltage of the restore circuit. *Adapted from Y.V. Pershin, L.K. Castelano, F. Hartmann, V. Lopez-Richard, M. Di Ventra, A Memristive Pascaline, IEEE Trans. Circuits Syst. Express Briefs, 63 (6) (2016) 558–562.*

memristor while pushing the button 5 times; the last pulse activates the restoring process which reinitializes the memristor state. Further information can be found in [44].

It is worth mentioning that such memristive Pascaline was realized experimentally by the authors of [44] using four memristor emulators, thus being *the first experimental demonstration of multi-digit arithmetic with asynchronous multilevel memory devices* (memristors) and arithmetic with mixed-base numbers (not shown here). Moreover, an interesting concept to cause incremental memristance modulation via the application of consecutive programming pulses, was presented in [46] using composite memristive systems. More specifically, instead of modifying the state of a single memristor, composite memristive switches were employed, consisting of several interconnected memristors, and the equivalent memristance was gradually modified each time by a constant amount, resulting in a behavior similar to that demonstrated in Fig. 13.2.

13.3 ADC and DAC in-memory data converters

Analog-to-digital converters (ADCs) are widely used in modern electronics to represent values of analog signals in digital form for subsequent storage and/or processing. Digital-to-analog converters (DACs) perform the inverse operation, being ubiquitous components existing in every data-driven and mixed-signal circuit. DACs are in fact the link between the digital domain of signal processing and the real-world where analog transducers are found. Highly energy-efficient data converters are very much desired for future computing platforms, hence searching new ways of implementing compact and ultra-low-energy data converters is of significant relevance. It is worth mentioning though that achieving both high resolution and speed of conversion is not an easy task. In fact, this *trade-off* between *performance* and *reliability* is a major bottleneck in data converter design, often leading to special purpose designs and/or sophisticated techniques [47].

Some memristor-based electronic circuits performing analog-to-digital and digital-to-analog conversions were published recently [48–50]. Such circuits may find widespread use, e.g., as interfaces between analog circuits and logic/computing circuits based also on memristors. Among the first such works we distinguish that of Pershin et al. [48], which presented an approach to memristive ADC and DAC conversion in which the digital value of the signal was stored in the resistive states of memristors. Hence, *the same memristive devices* are used for two purposes: as elements *performing conversion* and also as elements *storing the code*. This approach is thus directly compatible with the memristive computing architectures such as those commented in the previous section, as well as with several more memristive logic design schemes that were demonstrated in [32,51–53].

As in the case of the memristive Pascaline, the memristive ADC and DAC also require bipolar memristors with threshold-type switching behavior [54]; i.e., devices whose memristance R_M changes only when the voltage across the memristor V_M is higher than its threshold voltage V_{th} ($|V_M| > V_{th}$, here assuming a symmetric threshold voltage value for both SET and RESET of memristors.) Another desired property is a large resistance window, i.e., $R_{OFF} >> R_{ON}$. Fig. 13.3A shows the main idea behind the ADC implementation. The analog-to-digital conversion is performed when synchronized double-step pulses are simultaneously applied to a set of N memristors that are previously initialized in the low resistive state (LRS, R_{MIN} or R_{ON}), also corresponding here to logic value "1." The duration T of every single pulse is selected in such a way that if the voltage drop on any memristor exceeds the reset threshold V_{th}, then such device is completely RESET to the high resistive state (HRS, R_{MAX} or R_{OFF}), which here corresponds to logic value "0." Furthermore, the amplitude of the pulses applied from the right side (see Fig. 13.3A)

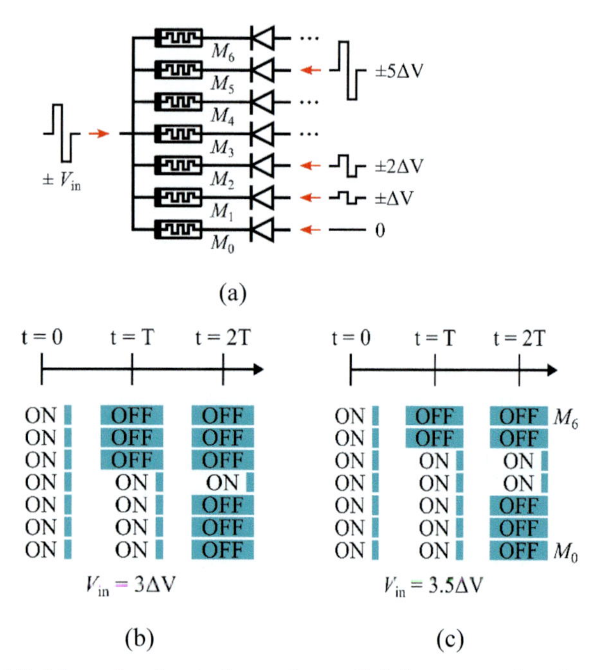

FIG. 13.3 (A) Schematic of unipolar analog-to-digital converter based on memristors. Operation is based on the simultaneous application of double-step square pulses of specific amplitude. (B) and (C) show the states of memristors at different moments of time for two different values of input voltage V_{in}. *Adapted from Y.V. Pershin, E. Sazonov, M. Di Ventra, Analog-to-digital and digital-to-analog conversion with memristive devices, IET Electron. Lett. 48 (2) (2012) 73–74.*

is purposely selected as multiple of a basic quantity ΔV, i.e., $i\Delta V$, where $i=0, \ldots N\text{-}1$ is the index of the memristor, whereas the value suggested in [48] for ΔV is $(4/3) \times V_{th}$. The double pulse applied from the left side is defined by the input voltage amplitude $V_{in} \geq 0$, whereas the diodes are necessary to avoid unwanted posterior SET of memristors.

As far as operation is concerned, the first (positive) input pulse forces all memristors with indexes $i > (V_{in} + V_{th})/\Delta V$ to switch to the HRS state. The consecutive (negative) input pulse does the same with all memristors with $i < (V_{in} - V_{th})/\Delta V$. Eventually, such applied set of pulses leaves only one or two memristors at most in the LRS state; the latter depends on whether V_{in} is closer to $k\Delta V$ or to $(k+0.5)\Delta V$, with k being an integer number. Figs. 13.3B and C present an example of the main switching dynamics in both of the abovementioned cases, respectively. Table 13.I provides the correspondence between the input voltage intervals and the ADC output code stored in the state of the memristors.

Next, Fig. 13.4A presents the corresponding DAC scheme, where it is assumed that the *output code is pre-programmed into the states of memristors.* For the correct operation of the DAC, it is important to apply small amplitude input signals during the conversion process (so that $|V_M| < V_{th}$) to avoid disturbing the stored code in the memristor states. Therefore, the maximum allowable δV in the figure is determined from the relation $(N-1)\delta V < V_{th}$. The circuit utilizes an op-amp to form the weighted sum of all input pulses applied to memristors [48]. A very similar circuit was later experimentally implemented in [49] by L. Gao et al. with a discrete integrated circuit (IC) op-amp (STTL074, which is quite slow but this does not tell anything about the performance of an integrated on-chip solution) and a packaged TiO_2-based memristor chip wired manually

TABLE 13.I ADC code for different ranges of the applied input voltage, for the specific setup configuration and the parameter values as proposed in [48].

Input voltage range	Corresponding code (7 bits)
$0 \leq V_{in} < 0.25\Delta V$	000000**1**
$0.25\Delta V \leq V_{in} < 0.75\Delta V$	00000**11**
$0.75\Delta V \leq V_{in} < 1.25\Delta V$	00000**1**0
$1.25\Delta V \leq V_{in} < 1.75\Delta V$	0000**11**0
$1.75\Delta V \leq V_{in} < 2.25\Delta V$	0000**1**00
$2.25\Delta V \leq V_{in} < 2.75\Delta V$	000**11**00
$2.75\Delta V \leq V_{in} < 3.25\Delta V$	000**1**000
...	...

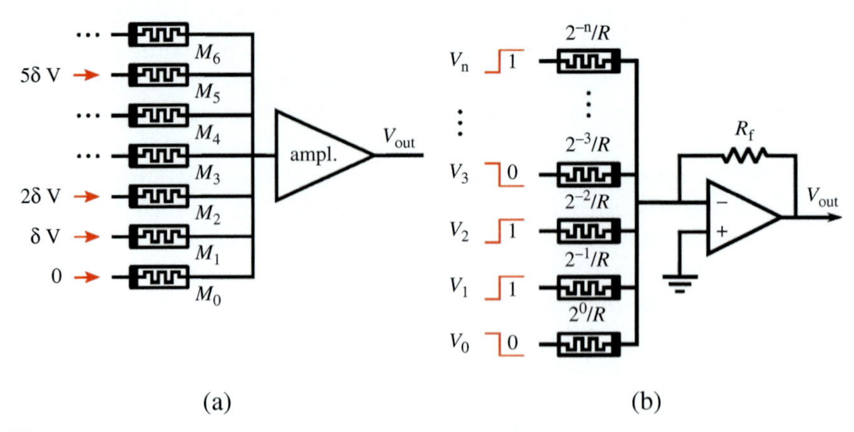

FIG. 13.4 (A) Schematic of a 7-bit digital-to-analog converter. (B) Schematic of a generalized n-bit binary-weighted DAC where memristors, instead of holding the digital code, are rather programmed to a fixed conductance. In both cases, an opamp is used to form the weighted sum of all input pulses. *Panel A, Adapted from Y.V. Pershin, E. Sazonov, and M. Di Ventra, Analog-to-digital and digital-to-analog conversion with memristive devices, IET Electron. Lett., 48 (2) (2012) 73–74. Panel B, Adapted from L. Gao, F. Merrikh-Bayat, F. Alibart, X. Guo, B.D. Hoskins, K.-T. Cheng, D.B. Strukov, Digital-to-analog and analog-to-digital conversion with metal oxide memristors for ultra-low power computing, 2013 IEEE/ACM Int. Symp. on Nanoscale Architectures (NANOARCH), NY, USA, July 15–17 2013, pp. 19–22.*

on a breadboard, according to the topology shown in Fig. 13.4B. In this case, for a set of digital input voltages V_i corresponding to the digital code (where i is from 0 to n) the analog output voltage V_{out} can be expressed as $V_{out} = -(R_f/R)\sum_i 2^{-i}V_i$, when the ith memristor is tuned with precision to a state with a conductance of $2^{-i}/R$ (for a precise tuning algorithm based on a simple circuit setup, check out the *voltage divider* approach proposed in [55]). In the same work [49], a 4-bit ADC was also implemented experimentally with a *Hopfield neural network*, consisting of four inverting amplifiers (neurons), and a 4×6 memristor crossbar array.

We note here that the conversion time of memristive ADCs in [48] is of the order of twice the switching time of the memristors. Therefore, considering that such time could be in the *sub-nanosecond range* as seen in [56], the proposed system could result in conversion frequencies higher than 1 GHz. On the other hand, small threshold voltages ~0.3 V have been observed in certain memristive devices [26,27]. Therefore, for a particular input range, such memristive ADCs could be able to distinguish a large number of voltage levels. Therefore, memristive solutions for data converters seem promising to offer simultaneously both high resolution and high speed of conversion.

It is worth noting that a more recent work by Danial et al. [50] explored innovative approaches for generic, high-precision, high-speed and energy-efficient memristive DACs, applying artificial intelligence (AI) techniques. Parallelism, simplicity, as well as fault tolerance and energy-efficiency are just a few example properties that would significantly enhance conventional DAC circuits. According to results reported in [50], online predictions and cognitive decisions enable DACs that are *self-reconfiguring*, *self-calibrating*, and also *noise tolerant*, utilizing a massive amount of correlated data to *adapt to real-time variations* and the requirements of the running application.

13.4 Memristor-based clock signal generators

In the literature, there are several known methods of using memristors in oscillating circuits [57,58]. For instance, a programmable frequency-relaxation oscillator was proposed in [59] where a memristor was used to set the switching thresholds of a Schmitt trigger. Relaxation oscillators are generally used to produce low frequency signals for applications such as clock signals in digital circuits. In modern electronics, the clock signal is most frequently produced by quartz generators but sometimes just by *RC*-based circuits or alternative approaches. The quartz generators offer a high precision but at the cost of a considerable size, whereas less precise *RC*-circuits usually result being more compact.

In this context, a compact clock signal generator based on memristors was proposed in [60], operating in a *Sisyphus-like* cycle. According to Greek mythology, king Sisyphus was once doomed to endlessly push a huge rock up a steep hill, only to see it to roll down again when reaching the top, repeating this action forever. The authors in [60] called their design a "*memristive Sisyphus*", owing to its cyclic behavior resembling a Sisyphus process. It is a reactance-less oscillator combining only digital logic gates, a single-supply voltage and a threshold-type switching bipolar memristor, likewise in the rest of applications commented in the previous sections. The *frequency and duty cycle are defined by the switching characteristics of the memristor* and two external resistors (each one participating in the SET ($R_{HIGH} \rightarrow R_{LOW}$) and RESET ($R_{LOW} \rightarrow R_{HIGH}$) process, respectively) which can be variable to enable programmability of the circuit.

There are generally two phases/stages of operation; in phase 1 a positive voltage applied to the memristor causes its memristance to increase (RESET). On the other hand, in phase 2 the device polarity is changed and the memristance decreases (SET). So the memristor oscillates between two resistive states, as shown in Fig. 13.5. It is characteristic that the RESET phase lasts quite longer than the SET phase, transitions which resemble the oscillations of the instantaneous rock-height (measured above the ground level) in the aforementioned Sisyphus myth (pushing the rock up the hill takes longer than when it freely rolls down.)

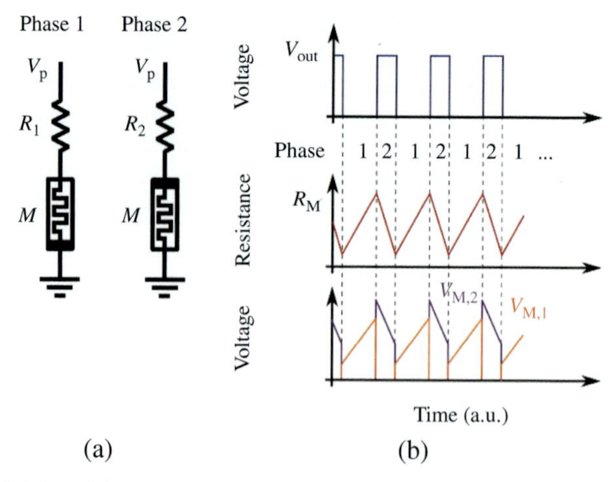

FIG. 13.5 (A) Simplified schematic concerning the equivalent circuits realizing the two-phase **memristive Sisyphus**: the phase 1 when memristance increases (left circuit) and phase 2 when memristance decreases (right circuit). (B) Cartoon graph showing the sequence of clock pulses (V_{out}), memristance R_M evolution with time (oscillations) and voltages V_M at the top and bottom electrode of the memristor, as shown in Fig. 13.6. *Adapted from Y.V. Pershin, S.N. Shevchenko, F. Nori, Memristive Sisyphus circuit for clock signal generation, Sci. Rep. 6 (2016) 26155.*

The circuit schematic of the corresponding clock signal generator is shown in Fig. 13.6. Apart from the memristor, the system includes a logic OR gate, whose inputs pass through a Schmitt-trigger, two *open-drain logic gates*, i.e., an identity gate (buffer) and a NOT gate, resistors R_1 and R_2 which enable configuring the SET and RESET switching times of the memristor, an additional resistor R_3 and a capacitor. The output of the OR gate defines the operation phase: logic 0 corresponds to phase 1 whereas logic 1 corresponds to phase 2. The hysteretic levels of the Schmitt-trigger inputs are employed as voltage thresholds triggering the stage changes, denoted by V_+ for the logic 1, and V_- for the logic 0,

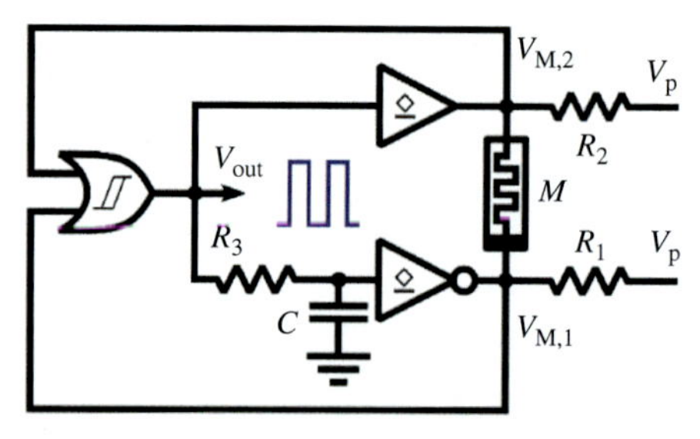

FIG. 13.6 Circuit schematic for to a specific implementation of a clock signal generator based on memristors. V_p is the power supply voltage. *Adapted from Y.V. Pershin, S.N. Shevchenko, F. Nori, Memristive Sisyphus circuit for clock signal generation, Sci. Rep. 6 (2016) 26155.*

where $V_- < V_+$. The circuit operation switches from phase 1 to phase 2 as soon as the voltage drop on the memristor V_M reaches the V_+ threshold. When this happens, the output of the OR gate changes to logic 1, thus driving the bottom terminal of the memristor (denoted by the thick black line) to the ground, at the same time setting the open drain identity gate to the high impedance state. Once the switching is complete and the voltage at the top electrode of the memristor is $V_{M,2} > V_-$, the circuit will remain in phase 2 as soon as the condition $V_{M,2} > V_-$ holds and will change back to phase 1 in a similar way once $V_{M,2} < V_-$. Consequently, both inputs of the OR gate turn to logic 0.

The expected evolution of the voltage levels at the output of the OR gate V_{out}, where the produced clock signal is read, and at the top and bottom electrodes of the memristor, are roughly shown in the cartoon graphs of Fig. 13.5B. It can be observed that while the $V_{M,1}$ and $V_{M,2}$ curves clearly show the two stages of the circuit operation, the output V_{out} corresponds to *a stable clock signal*. In [60] the authors discuss the particular details of their experimental implementation, guiding the reader to reproduce the results using specific electronic components and following the proposed design specifications. It is worth mentioning that cyclic behavior of output signals was also discussed in [61] implementing composite memristive configurations consisting of pairs of anti-serially connected memristors. Such theoretical considerations together with the experimental emulation shown in [60], clearly highlight the potential of such memristor-based circuits for very compact clock generators. The nanoscale realization of memristor devices, supporting switching frequencies in the MHz-GHz range [56], could potentially offer really *small-size alternatives to conventional quartz-based oscillators* for future electronic systems. For a rigorous analysis of the operation of the clock generating circuit, read further in [60].

13.5 Metastable memristive transmission lines

Memristors store information as resistance levels in an analog fashion and enable memory and processing in the same devices, thus eliminating the data transfer bottleneck experienced by conventional computer architectures where memory and processing units are strictly separated. Nevertheless, information transfer remains a very important aspect of modern information technologies. In this context, conventional well-known transmission line models employ capacitors and inductors, while line losses are taken into consideration by including resistive elements in the line. Therefore, the transmission of signals is generally considered almost always in the framework of transmission line models. Recently, *reconfigurable transmission lines* were proposed in [62]. The transmission characteristics of such lines could be *pre/re-programmed on demand* owing to the existence

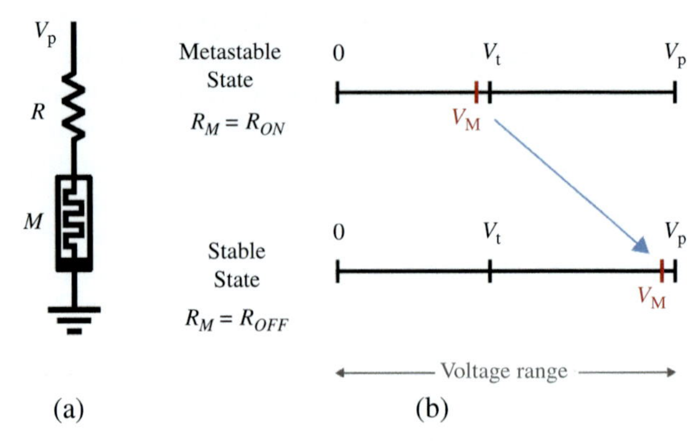

FIG. 13.7 (A) Metastable memristor-based circuit (voltage divider). (B) Metastable circuit state when the voltage across M is slightly below its threshold voltage V_t while $R_M = R_{ON}$ and the transition to the stable state when $R_M = R_{OFF}$. *Adapted from V. Slipko, Y. Pershin, Metastable memristive lines for signal transmission and information processing applications," Phys. Rev. E 95 (2017) 042213 1–6.*

of *memcapacitors* instead of simple capacitors. However, according to a different approach presented in [63] by Slipko et al., signal transmission (more correctly in this case described as "edge propagation") could be possible even using uniquely resistive elements.

More specifically, such idea is based upon the simple metastable circuit shown in Fig. 13.7A, concerning a voltage divider between a resistor R and a memristor M. Likewise in the previous sections, this application assumes bipolar and threshold-type switching devices. Depending on the polarity of M, the resistance R_M increases with a positive voltage across M only when the voltage across the memristor is $|V_M| > V_t$, where V_t is the threshold voltage. We assume the resistance change-rate is very fast above (and negligibly slow - or zero - below) the voltage threshold V_t. R_M is initially SET to a low resistance near R_{ON}. The value of the resistor R is selected such that when $R_M = R_{ON}$, V_M is found slightly below V_t (see Fig. 13.7B). Under these conditions, the state of the memristor is not affected and the circuit could remain in this state for an extended period of time. However, such configuration is also called as a "*metastable state*" since any small perturbation resulting instantly in $V_M > V_t$ could trigger a *self-accelerated RESET*, i.e., *OFF switching* (accelerated because as R_M increases, so does V_M and we assume that dR_M/dt is proportionally increasing with increasing V_M) towards a finally stable state near R_{OFF}.

Building upon such a basic component, a *metastable memristive line* is thus composed of a set of in-series connected metastable memristive circuits, i.e., it is *built of only resistive components*. This is why operation generally requires a power source and also a periodic refresh of the memristor states, as we explain next. Therefore, in this sense such metastable

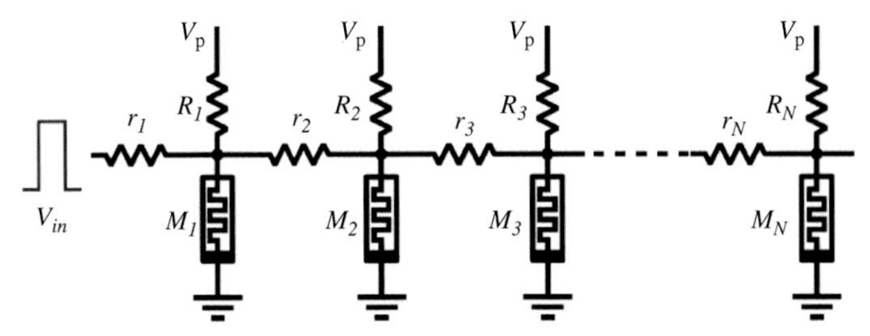

FIG. 13.8 General circuit schematic of a homogeneous metastable memristive line consisting of N metastable resistor-memristor circuits connected via coupling resistors r_i. *Adapted from V. Slipko, Y. Pershin, Metastable memristive lines for signal transmission and information processing applications, Phys. Rev. E 95 (2017) 042213 1–6.*

memristive lines are different from the traditional transmission lines. In fact, it is an architecture capable of simply transferring a signal edge from one space location to another, taking advantage of consecutive switching events of successive memristors. The authors in [63] investigated numerically the dynamics of pulse edge propagation along such types of lines and developed a simple and useful theory of this phenomenon.

Let us consider a chain of metastable memristive circuits connected in series, as shown in Fig. 13.8. Under certain conditions, such line could transfer a signal edge from one point of the line to another. Indeed, the OFF switching of M_1 triggers the OFF switching of M_2, and this in turn triggers the OFF switching of M_3, etc. Therefore, a short input pulse sets off a sequence of OFF switching events permitting the *propagation of a signal front along the line*. Correct operation assumes all the memristors along the line are initially SET to enable metastability, something that has to be done every time a signal needs to be propagated. This is potentially a disadvantage of such an approach. However, as the authors of [63] suggest, metastable memristive lines could find applications in information processing since the time delays introduced could be of use in the development of *race logic architectures* [63,64]; hence processing and transmission could take place simultaneously!

In such an homogeneous metastable memristive line, it is $r_i = r$, $R_i = R$, and $R_{M,i}\,(t=0) = R_{ON}$ for $i = 1, \ldots N$. Figure 13.9 shows roughly the memristive line dynamics when the latter is triggered by a rectangular input voltage pulse. In particular, Slipko et al. in [63] observed that the switching of memristors occurs sequentially with almost the same time interval τ between adjacent switching events. The dynamics in the central part of the lines is determined solely by the line properties and not by the boundary conditions. The reader can go deeper in details reading in [63] about the dynamics of pulse edge propagation in the limit of an infinite line, where we can safely assume that V_i and V_{i+1} are simply time-shifted with respect to each other; i.e., $V_{i+1}(t) = V_i(t-\tau)$, where τ is the pulse edge propagation time per metastable circuit component.

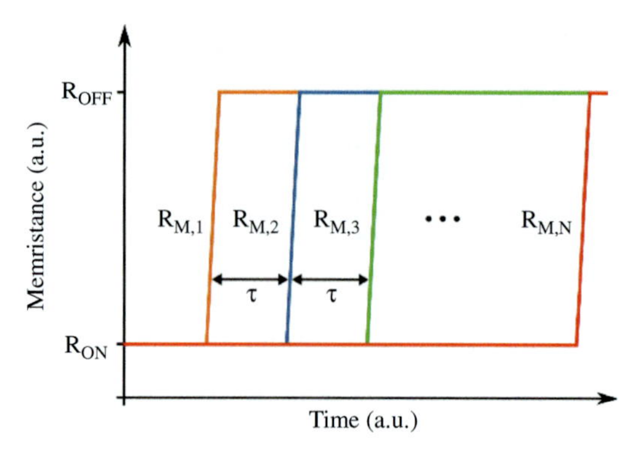

FIG. 13.9 Cartoon graph showing the memristance dynamics during pulse edge propagation in the memristive metastable line of Fig. 13.8. *Adapted from V. Slipko, Y. Pershin, Metastable memristive lines for signal transmission and information processing applications, Phys. Rev. E 95 (2017) 042213 1–6.*

13.6 Conclusions

Memristors are ubiquitous and this chapter reviewed just a small set of possible applications of this emerging and promising device technology. We particularly focused on *on-chip applications and tasks*, such as memory and in-memory processing with multi-base arithmetic units, signal transmission, as well as in the periphery where data conversion and clock signal generation blocks are found. Several more references were included to other relevant applications, highlighting the potential impact of this device technology that extends to almost every dimension of future electronic circuits and systems, including unconventional approaches to computing such as *neuromorphic*, *stochastic*, and *reservoir*-based computing, where memristors are getting increasingly involved.

Acknowledgment

This work was supported in part by the Chilean research grants CONICYT REDES ETAPA INICIAL 2017 No. REDI170604, CONICYT BASAL FB0008, and in part by the Spanish MINECO and ERDF (TEC2016-75151-C3-2-R).

References

[1] L.O. Chua, Memristor—the missing circuit element, IEEE Trans. Circuit Theory 18 (5) (1971) 507–519.
[2] L.O. Chua, Everything you wish to know about Memristors but are afraid to ask, Radioengineering 24 (2) (2015) 319–368.
[3] D.B. Strukov, G.S. Snider, D.R. Stewart, R.S. Williams, The missing memristor found, Nature 453 (2008) 80–83.

[4] R. Williams, How we found the missing memristor, IEEE Spectr. 45 (12) (Dec. 2008) 28–35.
[5] W. Hiatt, T. Hickmott, Bistable switching in niobium oxide diodes, Appl. Phys. Lett. 6 (106) (1965).
[6] L.O. Chua, If it's pinched it's a memristor, Semicond. Sci. Technol. 29 (10) (2014) 104001.
[7] Y.V. Pershin, M. Di Ventra, Memory effects in complex materials and nanoscale systems, Adv. Phys. 60 (2) (2011) 145–227.
[8] S.D. Ha, S. Ramanathan, et al., J. Appl. Phys. 110 (7) (2011) 071101.
[9] T. Mikolajick, H. Wylezich, H. Maehne, S. Slesazeck, Versatile resistive switching in niobium oxide, in: 2016 IEEE International Symposium on Circuits and Systems (ISCAS), Montreal, Canada, May 22–25, 2016.
[10] H. Zhang, et al., Gd-doping effect on performance of HfO2 based resistive switching memory devices using implantation approach, Appl. Phys. Lett. 98 (4) (2011), 042105.
[11] S.J. Song, et al., Real-time identification of the evolution of conducting nano-filaments in TiO2 thin film ReRAM, Sci. Rep. 3 (2013) 3443.
[12] C. Xie, et al., High-performance nonvolatile Al/AlOx/CdTe: Sb nanowire memory device, Nanotechnology 24 (35) (2013) 355203.
[13] S. Menzel, P. Kaupmann, R. Waser, Understanding filamentary growth in electrochemical metallization memory cells using kinetic Monte Carlo simulations, Nanoscale 7 (29) (2015) 12673–12681.
[14] Y. Li, P. Fang, X. Fan, Y. Pei, NiO-based memristor with three resistive switching modes, Semicond. Sci. Technol. 35 (2020), 055004.
[15] F. d'Acapito, E. Souchier, P. Noé, P. Blaise, M. Bernard, V. Jousseaume, Role of Sb dopant in ag: GeSx-based conducting bridge random access memories, Phys. Status Solidi (a) 213 (2) (2016) 311–315.
[16] M. Kozicki, M. Park, M. Mitkova, Nanoscale memory elements based on solid-state electrolytes, IEEE Trans. Nanotechnol. 4 (3) (2005) 331–338.
[17] H.D. Kim, H.M. An, S.M. Hong, T.G. Kim, Unipolar resistive switching phenomena in fully transparent SiN-based memory cells, Semicond. Sci. Technol. 27 (12) (2012) 125020.
[18] C.H. Yang, et al., Electric modulation of conduction in multiferroic ca-doped BiFeO3 films, Nat. Mater. 8 (2009) 485–493.
[19] C.L. He, et al., Nonvolatile resistive switching in graphene oxide thin films, Appl. Phys. Lett. 95 (23) (2009) 232101.
[20] J. Yao, Z.Z. Sun, L. Zhong, D. Natelson, J.M. Tour, Resistive switches and memories from silicon oxide, Nano Lett. 10 (10) (2010) 4105–4110.
[21] Y.C. Yang, et al., Electrochemical dynamics of nanoscale metallic inclusions in dielectrics, Nat. Commun. 5 (2014) 4232.
[22] S. Bagdzevicius, K. Maas, M. Boudard, M. Burriel, Interface-type resistive switching in perovskite materials, J. Electroceram. 39 (2017) 157.
[23] G. Sassine, et al., Interfacial versus filamentary resistive switching in TiO2 and HfO2 devices, J. Vac. Sci. Technol. A B34 (1) (2016). Art (no. 012202).
[24] M.A. Villena, J.B. Roldán, F. Jiménez-Molinos, E. Miranda, J. Suñé, M. Lanza, SIM2RRAM: a physical model for RRAM devices simulation, J. Comput. Electron. 16 (4) (2017) 1095–1120.
[25] S.G. Kim, J.I.S. Han, H. Kim, S.Y. Kim, H.W.N. Jang, Recent advances in memristive materials for artificial synapses, Adv. Mater. Technol. 3 (12) (Dec. 2018), 1800457.
[26] Knowm Inc., Neuromemristive Artificial Intelligence (Online]. Available: [Accessed January 2021) https://knowm.org.
[27] J. Gomez, I. Vourkas, A. Abusleme, Exploring memristor multi-level tuning dependencies on the applied pulse properties via a low-cost instrumentation setup, IEEE Access 7 (2019) 59413–59421. pp. 1.
[28] Panasonic, MN101LSeries MCU (Online]. Available: [Accessed January 2021) https://industrial.panasonic.com/ww/products/semiconductors/microcomputers/mn101l.

[29] Y.V. Pershin, M. Di Ventra, Practical approach to programmable analog circuits with memristors, IEEE Trans. Circuits Syst. I Reg. Papers 57 (8) (Aug. 2010) 1857–1864.
[30] V. Ntinas, I. Vourkas, G.C. Sirakoulis, LC filters with enhanced memristive damping, in: 2015 IEEE International Symposium on Circuits and Systems (ISCAS), IEEE, 2015, pp. 2664–2667. 10.1109/ISCAS.2015.7169234.
[31] MemComputing Inc., Powerful Co-Processors with Speed like no other (Online]. Available: [Accessed November 2018) http://memcpu.com.
[32] I. Vourkas, G.C. Sirakoulis, Emerging memristor-based logic circuit design approaches: a review, IEEE Circ. Syst. Mag 16 (3) (2016) 15–30 (3rd quarter).
[33] D. Wust, D. Fey, J. Knödtel, A programmable ternary CPU using hybrid CMOS/memristor circuits, Int. J. Parallel Emergent Distrib. Syst. 33 (4) (2018) 387–407.
[34] I. Vourkas, G.C. Sirakoulis, Networks of memristors and memristive components, in: Memristor-Based Nanoelectronic Computing Circuits and Architectures, first ed., Springer Int. Publishing, Switzerland, 2016, pp. 173–198.
[35] L. Chua, G.C. Sirakoulis, A. Adamatzky, Handbook of Memristor Networks, Springer Int. Publishing, Switzerland, 2019, https://doi.org/10.1007/978-3-319-76375-0.
[36] J. von Neumann, Theory of Self-Reproducing Automata, University of Illinois, Urbana, IL, 1966.
[37] B. Chopard, Cellular automata modeling of physical systems, in: R.A. Meyers (Ed.), Computational Complexity, Springer Int. Publishing, New York, NY, 2012, pp. 407–433.
[38] V. Ntinas, B. Moutafis, G.A. Trunfio, G.C. Sirakoulis, Parallel fuzzy cellular automata for data-driven simulation of wildfire spreading, J. Comput. Sci. 21 (2017) 469–485.
[39] M.-A. Tsompanas, G.C. Sirakoulis, Modeling and hardware implementation of an amoeba-like cellular automaton, Bioinspir. Biomim. 7 (2012), 036013.
[40] P. Progias, G.C. Sirakoulis, An FPGA processor for modelling wildfire spread, Math. Comput. Model. 57 (5–6) (2013) 1436–1452.
[41] M. Itoh, L.O. Chua, Memristor cellular automata and memristor discrete-time cellular neural networks, Int. J. Bifurcation Chaos 19 (11) (2009) 3605–3656.
[42] I. Vourkas, D. Stathis, G.C. Sirakoulis, Memristor-based parallel sorting approach using one-dimensional cellular automata, Electron. Lett. 50 (24) (2014) 1819–1821.
[43] R.-E. Karamani, V. Ntinas, I. Vourkas, G.C. Sirakoulis, 1-D Memristor-based cellular automaton for pseudo-random number generation, in: Int. Symp. on Power & Timing Modeling, Optimization & Simulation (PATMOS), Thessaloniki, Greece, Sept., 2017, pp. 25–27.
[44] Y.V. Pershin, L.K. Castelano, F. Hartmann, V. Lopez-Richard, M. Di Ventra, A Memristive Pascaline, IEEE Trans. Circuits Syst. Express Briefs 63 (6) (2016) 558–562.
[45] B. Pascal, Lettre Dédicatoire à Monseigneur le Chancelier sur le Sujet de la Machine Nouvellement Inventée par le Sieur B. P. Pour Faire Toutes Sortes D'opérations D'arithmétique par un Mouvement Réglé Sans Plume ni Jetons, Avec un Avis Nécessaire à Ceux qui Auront Curiosité de Voir Ladite Machine et s'en Servir. Suivi du Privilège du Roy, 1645.
[46] G. Papandroulidakis, I. Vourkas, N. Vasileiadis, G.C. Sirakoulis, Boolean logic operations and computing circuits based on Memristors, IEEE Trans. Circuits Syst. Express Briefs 61 (12) (Dec. 2014) 972–976.
[47] R.H. Walden, Analog-to-digital converter survey and analysis, IEEE J. Sel. Areas Commun. 17 (4) (April 1999) 539–550.
[48] Y.V. Pershin, E. Sazonov, M. Di Ventra, Analog-to-digital and digital-to-analog conversion with memristive devices, IET Electron. Lett. 48 (2) (2012) 73–74.
[49] L. Gao, F. Merrikh-Bayat, F. Alibart, X. Guo, B.D. Hoskins, K.-T. Cheng, D.B. Strukov, Digital-to-analog and analog-to-digital conversion with metal oxide memristors for ultra-low power computing, in: 2013 IEEE/ACM Int. Symp. on Nanoscale Architectures (NANOARCH), NY, USA, July 15–17, 2013, pp. 19–22.

[50] L. Danial, N. Wainstein, S. Kraus, S. Kvatinsky, DIDACTIC: a data-intelligent digital-to-analog converter with a trainable integrated circuit using Memristors, IEEE J. Emerging Sel. Top. Circuits Syst. 8 (1) (2018) 146–158.
[51] G. Papandroulidakis, I. Vourkas, A. Abusleme, G.C. Sirakoulis, A. Rubio, Crossbar-based Memristive logic-in-memory architecture, IEEE Trans. Nanotechnol. 16 (3) (2017) 491–501.
[52] M. Escudero, I. Vourkas, A. Rubio, F. Moll, Memristive logic in crossbar memory arrays: variability-aware design for higher reliability, IEEE Trans. Nanotechnol. 18 (Dec. 2019) 635–646.
[53] C. Fernandez, and I. Vourkas, "ReRAM-based ratioed combinational circuit design: a solution for in-memory computing," 2020 Int. Conf. on Modern Circuits and Syst. Technol. (MOCAST), Bremen, Germany, Sept. 07–09.
[54] U.-B. Han, D. Lee, J.-S. Lee, Reliable current changes with selectivity ratio above 10^9 observed in lightly doped zinc oxide films, NPG Asia Mater. 9 (2017), e351.
[55] J. Gomez, I. Vourkas, A. Abusleme, G.C. Sirakoulis, A. Rubio, Voltage divider for self-limited analog state programing of Memristors, IEEE Trans. Circuits Syst. Express Briefs 67 (4) (Apr. 2020) 620–624.
[56] A. Torrezan, J.P. Strachan, G. Medeiros-Ribeiro, R.S. Williams, Sub-nanosecond switching of a tantalum oxide memristor, Nanotechnology 22 (485203) (2011) 1–7.
[57] V. Ntinas, I. Vourkas, G.C. Sirakoulis, A. Adamatzky, Oscillation-based slime Mould electronic circuit model for maze-solving computations, IEEE Trans. Circuits Syst. Regul. Pap. 64 (6) (2017) 1552–1563.
[58] E. Kyriakides, J. Georgiou, A compact, low-frequency, memristor-based oscillator, Int. J. Circuit Theory Appl. 43 (11) (2015) 1801–1806.
[59] A. Mosad, M. Fouda, M. Khatib, K. Salama, A. Radwan, Improved memristor-based relaxation oscillator, Microelectron. J. 44 (9) (2013) 814–820.
[60] Y.V. Pershin, S.N. Shevchenko, F. Nori, Memristive Sisyphus circuit for clock signal generation, Sci. Rep. 6 (2016) 26155.
[61] I. Vourkas, G.C. Sirakoulis, On the generalization of composite Memristive network structures for computational analog/digital circuits and systems, Microelectron. J. 45 (11) (2014) 1380–1391.
[62] Y.V. Pershin, V.A. Slipko, M. Di Ventra, Reconfigurable transmission lines with memcapacitive materials, Appl. Phys. Lett. 107 (253101) (2015) 1–5.
[63] V. Slipko, Y. Pershin, Metastable memristive lines for signal transmission and information processing applications, Phys. Rev. E 95 (042213) (2017) 1–6.
[64] A. Madhavan, T. Sherwood, D. Strukov, Race logic: abusing hardware race conditions to perform useful computation, IEEE Micro 35 (3) (2015) 48–57.

CHAPTER

14

Neuromorphic applications using MO_x-based memristors

S. Brivio[a] and E. Vianello[b]

[a]CNR—IMM, Unit of Agrate Brianza, Agrate Brianza, Italy, [b]CEA-LETI, Université Grenoble Alpes, Grenoble, France

14.1 Introduction on neuromorphic computing

Neuromorphic computing has been recently come back to the interest of scientists and industries as a way to face the tremendous raise of data production in the era of social networks and Internet-of-Things [1]. Fig. 14.1, in fact, reports the recent exponential increase of data production break out into structured and unstructured contributions. The amount of energy required to elaborate such big data wall is forecast to increase by 2040 to the 10^{27} J, which is far beyond what humans will be able to produce [2]. This issue is due to the dominating contribution of unstructured data (see Fig. 14.1). Indeed, power hungry computing machines in big Data Processing Centers usually take advantage of machine learning algorithms, including neural networks and deep neural networks (DNN), to provide a structure and extract useful information from unstructured data.

In a simplistic view, a neural network can be thought as a model function with a large number of free parameters, which have to be fit to a correspondingly large amount of, let us say, experimental data. The continuous update of such free parameters requires a continuous transfer of data between memory and processing units that kills the computing speed and blows up the power consumption. In fact, the physical separation of memory and processing units as originally proposed by von Neumann lead to such so-called von Neumann's bottleneck.

In this scenario, neuromorphic engineering aims at providing computing machines that are efficient in elaborating unstructured data through the application of machine learning algorithms. Currently these software

Metal Oxides for Non-volatile Memory
https://doi.org/10.1016/B978-0-12-814629-3.00014-3

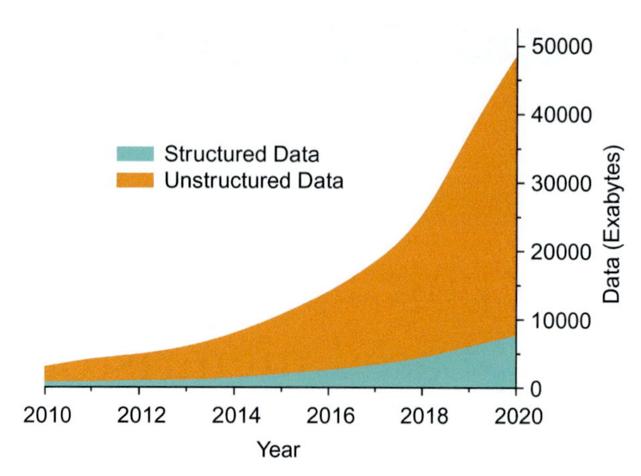

FIG. 14.1 Growth of structured and unstructured data from 2010 to 2020. *Source: IDC The Digital Universe. Dec. 2012.*

run on Graphical Processing Units (GPUs) and Neural Processing Units (NPUs). However, GPUs require external memory units and are still affected by the von Neumann bottleneck. Conversely, NPUs lower the influence of von Neumann bottleneck through the close proximity of volatile memories like Static Random Access Memories (SRAMs) or Dynamic Random Access Memories (DRAMs) to the processing units. However, the NPU architecture requires access to external non-volatile storage [3]. Emerging non-volatile memory devices, or memristor, and the MO_x-based ones in particular (see Chapters 1–2 and the brief recap in Section 14.2), can solve such memory issues and can bring additional acceleration compared to GPU and NPU performances through the possibility of in-memory parallel computing, as we will discuss in the Section 14.4. It is worth mentioning that other memory technologies, like Phase Change Memories (PCM) and Spin Transfer Torque Magnetic Memories (STT-MRAM), are being investigated for the realization of neuromorphic systems. These latter technologies have already achieved a significant technology readiness level and developed systems exploiting them are usually larger and with a higher level of complexity than those comprising memristive devices. On the other hand, memristive technology have some specific advantages over the PCMs and STT-MRAMs that make their investigation necessary. A critical assessment and comparison of all memory technologies in the neuromorphic field can be found in Refs. [4,5].

For the sake of historical fairness, the expression *neuromorphic engineering* was, in some sense, coined by Carver Mead, who already in the 80s foresaw the advantages of analog electronics, inspired by the biological information processing, in the elaboration of ill-conditioned data [6]. Mead's view of brain-inspired analog electronics, in which the physical phenomena are the computational primitives, is still a matter of investigation. In this stream of the neuromorphic research, memristive technology can be barely used as a non-volatile memory technology, as well as, can be an enabler of novel local computational primitives compatible with the neuromorphic framework. This research field will be discussed in Sections 14.3–14.4.

14.2 Recap of MO_x-based memristor technology

The various class of memristive technology based on MO_x have already been introduced in the previous Chapters 1–2. For the sake of completeness, a concise recap of the main properties of the memristive devices that are useful to neuromorphic computing is presented here. Memristive devices are metal/oxide/metal structures that undergo resistance change upon suitably high voltage application between their two metallic terminals. Various classifications can be drawn on the basis of the physical mechanism responsible for the switching or on the constitutive materials, rather than on the long/short retention characteristic of the resistance changes. Fig. 14.2 is an attempt to summarize the different classes of memristive devices.

14.2.1 Switching mechanisms

Memristive devices can be classified on the basis of the extent of the oxide region that hosts the microscopic changes responsible for the overall resistance change. In particular, the resistance change can be due to formation and dissolution of nanometric filamentary regions that short and

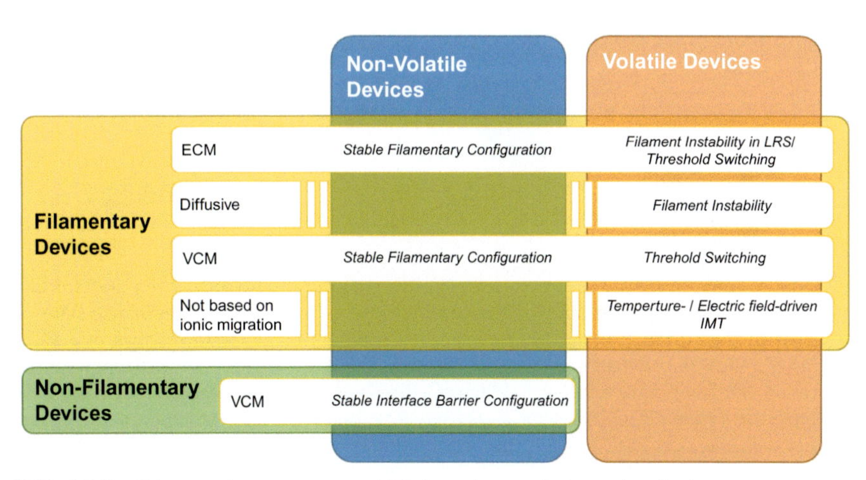

FIG. 14.2 Scheme of most common MO_x-based memristive technologies.

disconnect the two electrodes, giving rise to low and high resistance states (LRS and HRS), respectively [7–9]. These devices are identified as ***filamentary***. Any filamentary device typically requires current limitation by a transistor connected in series during the SET operation [8]. Indeed, as shown in Fig. 14.3A, the transistor voltage gate controls the maximum allowed current flowing during the SET operation, also granting a control over the value of the low resistance state. Pictorially, a low resistance after SET is associated to a big conductive filament (CF). We can distinguish among devices whose CFs are constituted by metal cations or by anion species. Furthermore, devices exist whose switching operation is not due to ion migration.

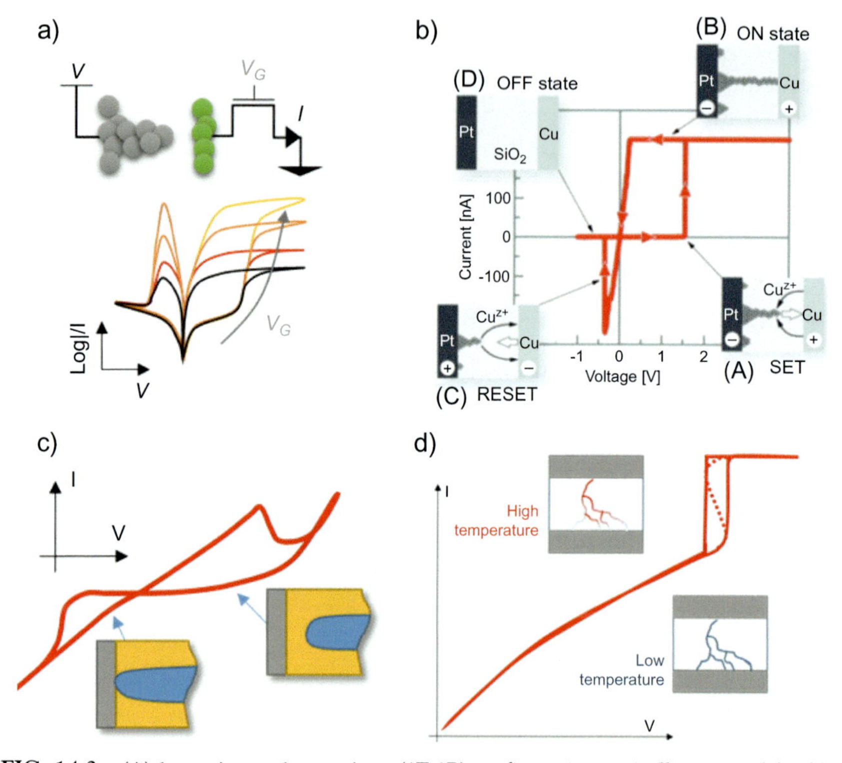

FIG. 14.3 (A) 1 transistor—1 memristor (1T-1R) configuration typically required for filamentary devices, which also allows the control of the value of the low resistance state, as depicted in the current–voltage characteristics. Current–voltage characteristics of a ECM (B) and of a VCM (C) devices together with the corresponding representation of filament configuration. (D) Temporary resistance switch of a threshold switching device. *Panel B Reprinted with permission from Ref. C. Schindler, G. Staikov, R. Waser, Electrode kinetics of Cu–SiO2-based resistive switching cells: overcoming the voltage-time dilemma of electrochemical metallization memories, Appl. Phys. Lett. 94 (2009) 072109. https://doi.org/10.1063/1.3077310. Copyright (2009) American Institute of Physics.*

In ***Electrochemical Metallization*** (ECM) devices [9,10], Cu and Ag-based (active) electrodes, when positively polarized release Cu and Ag ions in insulating layer. The cations drift into the oxide film, deposit and accumulate onto the opposite (inert) electrode giving rise to CF that eventually shorts the two electrodes, as shown in Fig. 14.3B [11]. Such *Cation-based filamentary* devices are also known as ***Conductive Bridge Random Access Memories*** (CBRAMs), ***Programmable Metallization Cells*** (PMC) and ***Atomic Switches***. Oxide-based ECM devices have been discussed in details in Chapters 1–2. Recently, a similar device concept has been proposed, namely the ***diffusive memristor***, in which both the electrodes are inert and Ag *diffusive* ions are provided by incorporating Ag nanoparticles in the insulator matrix [12].

In ***anion-based filamentary devices*** or filamentary ***Valence Change Memories*** (VCM) [9,13], the conductive filaments are locally high concentration of oxygen vacancies, $V_O^{\bullet\bullet}$. The voltage application promotes the drift of $V_O^{\bullet\bullet}$, as activated by self-heating through Joule effect. $V_O^{\bullet\bullet}$ defects can either be produced by a combination of temperature and electric field, or can already be present in fresh devices, especially at the interface between an oxide and a reactive electrode. Details can be found in Chapter 2. A typical current–voltage characteristic of filamentary VCM with the corresponding filament configuration is reported in Fig. 14.3C.

Finally, narrow current paths made of reduced oxides favors a high local temperature raise that triggers temporary effects like ***insulator-to-metal phase transitions*** [14,15] or ***threshold switching*** [16–18] events in oxides, like NbO_2. These phenomena typically lead to temporary resistance changes, sometimes associated to ***Negative Differential Resistance*** (NDR) when the device is current driven. Fig. 14.3D depicts a temperature-activated threshold-switching event (straight line) and the corresponding NDR in current mode (dashed line) in a filamentary device.

It is worth specifying that a different class of filamentary devices exist, namely Thermo-Chemical Memories [9,19]. Their operation requires only one polarity, which simplifies their integration in CMOS circuitry. Despite this fact, unipolar devices are usually characterized by poor endurance characteristics that prevent their used for computation [19]. For this reason, such unipolar devices will not be considered in the present chapter.

The class of **non-filamentary** devices is much less diversified than the filamentary one. Usually, the concentration of mobile oxygen vacancies at one metal/oxide interface modulates the overall device resistance through the change of the width or of the height of the interface barrier. Since the conduction pertains to the entire device cross-section, the temperature raise is more distributed and, therefore, locally less intense than in filamentary devices. Since thermal runaway is absent, resistance transitions are smooth and a current limitation is not actually necessary. Furthermore, the resistance of non-filamentary devices scales with their size, which has

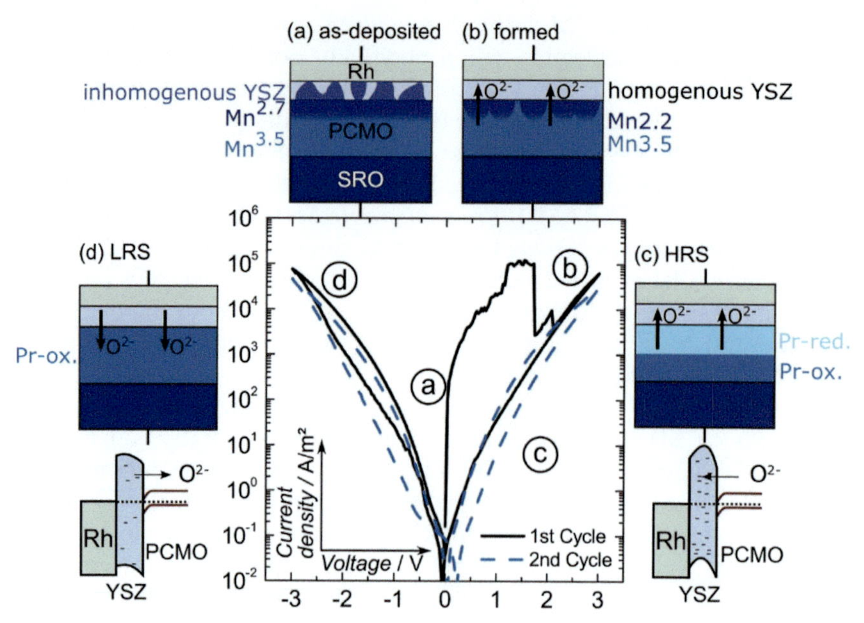

FIG. 14.4 Current–voltage characteristic of a non-filamentary device together with a representative evolution of the interfacial oxygen content and energy barrier modulation. *Reprinted from Ref. B. Arndt, F. Borgatti, F. Offi, M. Phillips, P. Parreira, T. Meiners, S. Menzel, K. Skaja, G. Panaccione, D.A. MacLaren, R. Waser, R. Dittmann, Spectroscopic indications of tunnel barrier charging as the switching mechanism in memristive devices, Adv. Funct. Mater. 27 (2017) 1702282. https://doi.org/10.1002/adfm.201702282 under CC-BY 4.0 license.*

bivalent consequences on energy containment. On one side, high fabrication efforts are required to increase resistance values and reduce energy consumption; on the other side, the perspective of energy scaling of non-filamentary device can be clearly addressed through a geometrical scaling roadmap. Fig. 14.4 reports a representative current–voltage characteristic of a non-filamentary device together with a sketch of the involved microscopic processes [20].

14.2.2 State retention and volatile effects

The previous sub-section presents the various mechanisms possibly responsible of reversible resistance changes of a memristive device. In the following, we will see that different kind of computation can be performed depending on the volatile or the non-volatile nature of the resistance states. Therefore, in this sub-section, it is useful to discuss the stability of the resistance states obtained according to the switching mechanisms described above. First, borrowing the terminology from the field of memory application, non-volatile memristors are ideally those devices whose two or more states can be programmed once and read many times for a period of time in the scale of years. In reality, the programmed states are always subjected to retention loss processes and resistance value degradation or fast drop with time [8,21–23]. In practice, non-volatile memristors are considered those whose retention is stable for months or years at least on average. Volatile memristors are those characterized by only one highly stable resistance state that is modified by voltage application. The subsequent voltage release produces the immediate restoration of the initial resistance value. Of course, the resistance restoration occurs with a certain fast dynamics that can support some novel computational paradigms.

Considering the switching mechanisms presented above and considering the Fig. 14.2, devices based on ion migration, being them ECM- or VCM-based, can display non-volatile state retention, even though the resistance values are affected by temperature-activated deterioration processes [21,22,24–27]. Indeed, both ECM [26,28,29] and filamentary VCM devices [8,22], at least at single device level, demonstrated to accomplish the industrial target for stand-alone non-volatile memory, i.e., the separation of high and low resistance states at 85 °C after 10 years.

Despite this general picture, in recent years, researchers have investigated temporary switching effects on much shorter timescales, roughly speaking below the *second* time scale. Metallic CFs in ECM devices can undergo dissolution, resulting in recovery of a high resistance value. Such mechanism is facilitated when CFs are especially thin, thanks to the use of a low value of the SET compliance current [24,25,30]. Similar short-term dynamics has been observed also in filamentary [31] and non-filamentary [32] VCM devices. Additionally, the class of diffusive memristors has been recently intentionally developed to display short-term state retention and resistance dynamics, as shown in Fig. 14.5A [12].

As discussed above, a peculiar feature of filamentary devices is the threshold switching phenomenon [17,33], which manifests as a temporary lowering of the device resistance, as shown in Fig. 14.5B [34]. In addition, temperature- or electric field-driven IMT produce a temporary rapid decrease of the resistance. In all cases, the back-switching to the stable resistance state is obtained, as soon as the voltage drop over the device is released, according to the parasitic capacitances present in the device structure [15,17].

14.2.3 Arrangement of devices into crossbars

Organization into crossbar structures is one of the main advantages of memristive devices compared to conventional memory technology. Crossbar structures host memristive devices at the intersection nodes of orthogonal word and bit lines and guarantee extremely high density

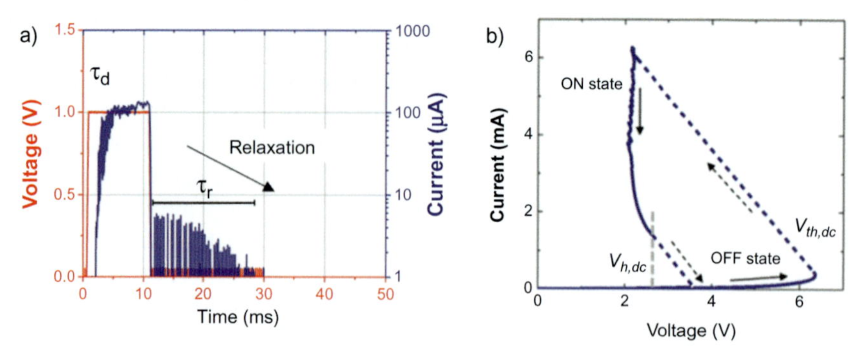

FIG. 14.5 Examples of volatile resistance switching effects: (A) switching and relaxation of a Ag-based diffusive memristor device. (B) Volatile threshold switching effect in a TaO_x based device. *Panel A Reprinted from Ref. R. Midya, Z. Wang, S. Asapu, X. Zhang, M. Rao, W. Song, Y. Zhuo, N. Upadhyay, Q. Xia, J.J. Yang, Reservoir computing using diffusive memristors, Adv. Intell. Syst. 1 (2019) 1900084. https://doi.org/10.1002/aisy.201900084 under CC-BY 4.0 license; Panel B Reprinted with permission from Ref. A.A. Sharma, I.V. Karpov, R. Kotlyar, J. Kwon, M. Skowronski, J.A. Bain, Dynamics of electroforming in binary metal oxide-based resistive switching memory, J. Appl. Phys. 118 (2015) 114903. https://doi.org/10.1063/1.4930051. Copyright (2015) American Institute of Physics.*

packing and possible 3D stacking of memory bits, as sketched in Fig. 14.6A. Beyond this opportunity, the arrangement in crossbars has an additional advantage for neuromorphic applications. Indeed, as depicted in Fig. 14.6B, crossbars can implement vector–matrix multiplication only by the exploitation of Kirchhoff's law of current summation. Once a vector $\boldsymbol{V}=\{V_n\}$ of n voltage values is applied to the word line of a crossbar containing $(n \times m)$ devices described by a matrix of conductances $\boldsymbol{G}=\{G_{ij}\}$, the vector of the resulting currents $\boldsymbol{I}=\{I_m\}$ read from the m bit lines is given by the dot product $\boldsymbol{I}=\boldsymbol{V}\bullet\boldsymbol{G}$. While the main works related to vector–matrix multiplication will be discussed in Section 14.4, here we would like to point out the main technical issues related to crossbar operation.

The design of crossbar matrices of memristive devices has to face the so-called sneak path issue. As sketched in Fig. 14.6C [8], applying the voltage to the bit and word lines corresponding to a specific device in a crossbar (green device between blue and orange lines) comprising just memristive devices, there is a high probability that a current will flow also in unselected devices (yellow ones), which produces a reading error or unwanted bit programming [8,35]. Selectors connected to each memristive device are able to avoid this problem. The most conventional selector device is a transistor that actively select which device in the array [8,35]. Of course, a transistor coupled to each memristive device in the array impact the ultimate bit density of the array both because the transistor footprint is larger than that of a memristive device and because of the additional current lines needed to control the gate and body terminals of

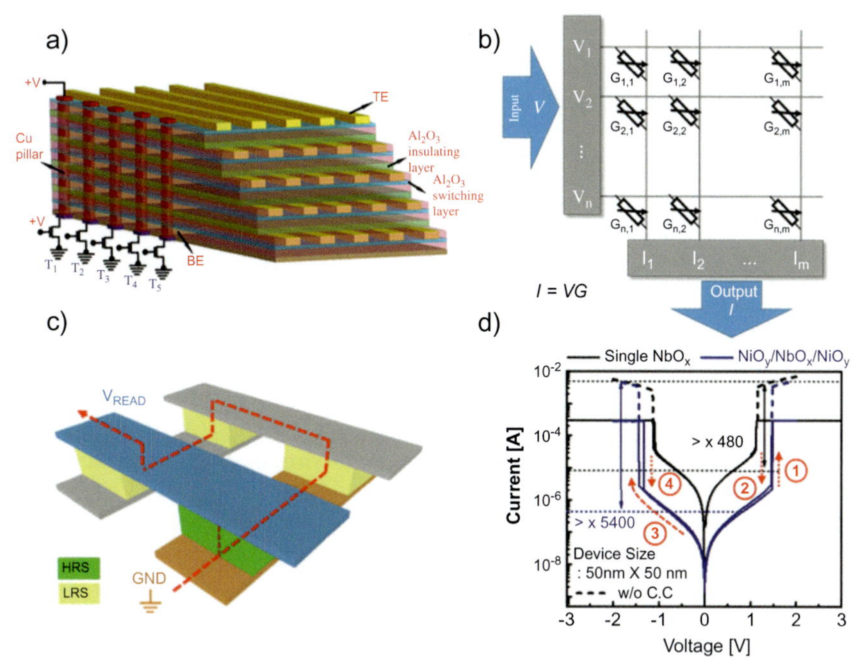

FIG. 14.6 (A) Sketch of a 3D stacking of crossbar arrays. (B) Implementation of dot product in a crossbar matrix. (C) Representation of the sneak path issue of passive crossbar array; (D) typical operation of a passive selector device. *Panel A Reprinted from Ref. S. Maikap, R. Panja, D. Jana, Copper pillar and memory characteristics using Al2O3 switching material for 3D architecture, Nanoscale Res. Lett. 9 (2014) 366. https://doi.org/10.1186/1556-276X-9-366 under CC-BY 4.0 license; Panel D Reprinted from Ref. J. Park, T. Hadamek, A.B. Posadas, E. Cha, A.A. Demkov, H. Hwang, Multi-layered NiOy/NbOx/NiOy fast drift-free threshold switch with high Ion/Ioff ratio for selector application, Sci. Rep. 7 (2017) 1–8. https://doi.org/10.1038/s41598-017-04529-4 under CC-BY 4.0 license; Panel C Reprinted with permission from Ref. H.-Y. Chen, S. Brivio, C.-C. Chang, J. Frascaroli, T.-H. Hou, B. Hudec, M. Liu, H. Lv, G. Molas, J. Sohn, S. Spiga, V.M. Teja, E. Vianello, H.-S.P. Wong, Resistive random access memory (RRAM) technology: from material, device, selector, 3D integration to bottom-up fabrication, J. Electroceram. 39 (2017) 21–38. https://doi.org/10.1007/s10832-017-0095-9 Copyright (2017) Springer.*

the transistors. The alternative that allows maintaining a dense $4F^2$ packing (with F being the so-called feature size) is the use of passive selectors. These elements assume high resistance values at the low voltages corresponding to the voltage drops of unselected cells and the low resistances for relatively high reading and programming voltages, as shown in Fig. 14.6D [36]. Devices of this kind take advantage conductance non-linearity of diode-like conduction mechanisms or volatile switching phenomena due to threshold switching, thermal runaway or phase transitions [35,37]. There are interesting examples in which the diode-like conduction is implemented in the memristive device itself, which, thus, does not need any additional selector element but, in turn, can be considered a self-selecting memristive device. Another example of passive memristive element is the so-called complementary resistance switching (CRS) device that can be realized by the anti-serial connection of two bipolar devices or just within a single device. The explanation of CRS and its use for passive crossbars can be found in Refs. [37,38].

14.3 Advanced memristor functionalities useful for neuromorphic applications

Resistance switching devices have been considered as candidates for non-volatile (or even the so-called universal) memory applications, at the beginning of their intensive research in the early 2000s. Once assimilated to the class of memristive devices, novel or advanced functionalities have been investigated with much more emphasis within the field of neuromorphic computing. For instance, neural accelerators applications renewed the interest toward multilevel operation. In the recent literature, many studies demonstrated that memristive devices can actively implement some computational tasks locally and collectively. For instance, they display a plastic change of resistance, i.e., an integrative response to sequences of voltage spikes and/or sensitive to their relative timing. In addition, the phenomenon of negative differential resistance has been recently spotlighted for the realization of oscillators in view of the emulation of collective neural dynamics. The following Sections 14.3.1–14.3.4 will deal with all these novel or advanced functionalities at the device level. System level works will be examined in depth in the following Section 14.4.

14.3.1 Multilevel operation

Multilevel operation is the property of a memristive device to settle into more than two stable resistance states. Therefore, it is a useful property when associated to non-volatile retention. According to the simple description drawn in Section 14.2, non-volatile switching is obtained through the modulation either of the CF dimension or of the width and height of an interface conduction barrier. Those quantities can be modulated to some extent by tuning the driving voltage or the compliance current during SET operations, giving rise to multiple resistance states.

In filamentary devices (either VCM and ECM type), the abrupt CF formation can be interrupted deliberately at precise resistance values through an external current limitation during SET (cf. Fig. 14.7A; Ref. [39]). In VCM devices, the dissolution of the CF occurs gradually. In this case, the maximum applied voltage, as well as the time interval of the voltage stimulation, governs the HRS resistance value (see the tuning of the

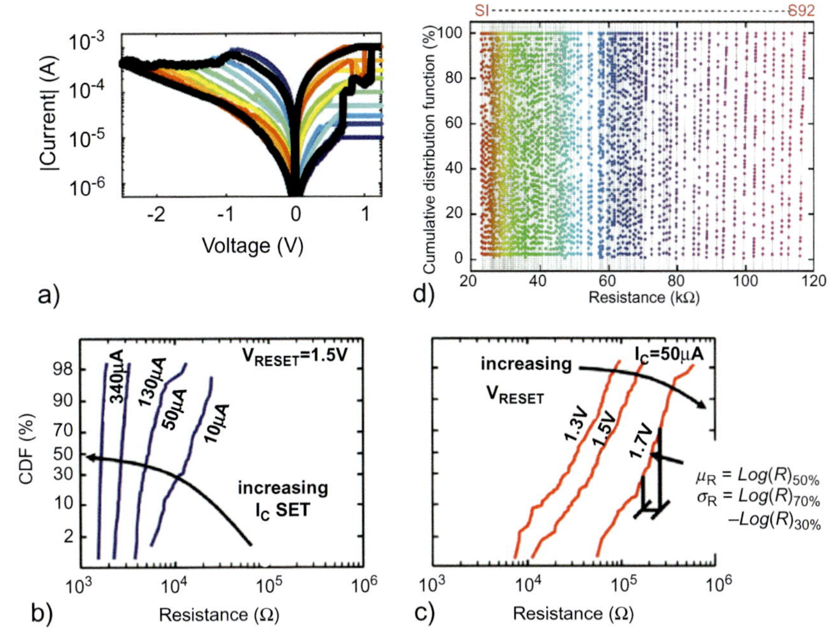

FIG. 14.7 (A) Representative multilevel programming of a filamentary HfO_2-based memristor obtained by limiting the current during the SET operation at progressively increasing values (positive voltages) and by tuning the RESET stop voltage (negative voltages). Representative distributions of resistance values belonging to distinct memory states programmed through the tuning of the compliance current (B) and RESET stop voltage (C) in a 16 kbit array of 1T-1 HfO_2-based RRAM. (D) Resistance cumulative distribution of 92 levels obtained through a program and verify technique on a single TiO_2-based device. *Panel A Re-arranged with permission from Ref. S. Brivio, S. Spiga, Stochastic circuit breaker network model for bipolar resistance switching memories, J. Comput. Electron. 16 (2017) 1154–1166. https://doi.org/10.1007/s10825-017-1055-y. Copyright (2018) Springer; Panel C Reprinted with permission from Ref. D. Garbin, E. Vianello, Q. Rafhay, M. Azzaz, P. Candelier, B. DeSalvo, G. Ghibaudo, L. Perniola, Resistive memory variability: a simplified trap-assisted tunneling model, Solid-State Electron. 115 (2016) 126–132. https://doi.org/10.1016/j.sse.2015.09.004. Copyright (2016) Elsevier; Panel D Reprinted under CC-BY license from Ref. S. Stathopoulos, A. Khiat, M. Trapatseli, S. Cortese, A. Serb, I. Valov, T. Prodromakis, Multibit memory operation of metal-oxide bilayer memristors, Sci. Rep. 7 (2017) 17532. https://doi.org/10.1038/s41598-017-17785-1.*

RESET operation for negative voltages in Fig. 14.7A; Ref. [39]). Conversely, in ECM devices the dissolution of the CF usually occurs in an abrupt fashion, which prevents the programming of more than two resistance states with a RESET operation. Generally, in non-filamentary devices, both SET and RESET processes occur gradually as a function of the applied ramped voltage and they can be both parceled out through the modulation of the maximum applied voltage or the time interval of voltage application.

From the conventional memory application point of view, *resistance bits*, or *resistance levels*, must be precisely distinguishable. An overlap in the distributions of the corresponding resistance values due to cycle-to-cycle, device-to-device variability, reading noise or retention degradation would result in a read failure. All the mentioned sources of noise severely affect the resistance distributions and prevent the identification of *resistance bits*, without the implementation of error correction codes. As an example, Fig. 14.7B–C reports the cumulative distributions of resistance levels obtained from 16kb filamentary RRAM arrays by tuning the current compliance (panel B) and the reset stop voltage (panel C) [40]. As said, the scientific results are rapidly making progress, though: Stathoupoulos et al. [41] were able to demonstrate 92 levels in a single TiO_2-based device. The result was obtained by a program and verify algorithm, which is usually adopted to overcome the variability of filamentary devices [42–45].

Area-dependent devices suffer less than filamentary ones from intrinsic variability and random telegraph noise (RTN). Furthermore, the power consumption can be reduced without severe constraints on the resistance window, as they can achieve very high OFF resistances [46]. Both features facilitate the multilevel operation of area-dependent devices.

It is usually considered that neuromorphic systems can tolerate to some extent device unreliability. However, the robustness of a neuromorphic hardware to device non-idealities depends a lot on the computing architecture, operation and task and, therefore, no prescription concerning device reliability has ever been formulated in a quantitative manner. In contrast to conventional memory application, neuromorphic systems that aim at providing an acceleration of deep learning algorithms and those designed to work in real-time typically cannot rely on time-consuming conventional error correction codes [47].

The multilevel operation as described above is mainly useful for neuromorphic hardware as it will be described in Section 14.4.

14.3.2 Memristor plasticity

Plasticity is a functionality of biological synapses that has been demonstrated in memristive devices, feeding the expectations toward spiking neural networks comprising memristive synapses (please refer to Section 14.4.2. Plasticity is considered as the property of the device to change its own resistance as a function of time in response to a repeated stimulation. Biological plasticity can be emulated in memristive devices by two ideally different operations. The first one corresponds to the analog operation, which is intended to be the *progressive or gradual change of resistance* over a continuum of resistance values. It is driven by sequences of identical spikes and results from the switching dynamics of the devices.

The second operation exploits the device *stochasticity*, i.e., the non-deterministic, binary switching between two resistance states. It is worth pointing out that there is an overlap between analog and stochastic operation as they both rely on the change of the atomic configuration. Nevertheless, the two following sub-sections are dedicated to these two ideally distinct programming operations.

14.3.2.1 Analog plasticity

An example of analog plasticity is reported in Fig. 14.8A and B, which shows the current increase and decrease as a function of trains of identical positive and negative pulses, respectively [48–51]. SET and RESET are usually called potentiation and depression operations for this synaptic electronic application [50]. The resistance evolution as reported in Fig. 14.8A and B is intended as plasticity because the i-th spike brings the device to a resistance value (R_i) that is a function of the previous resistance value, i.e., $R_i = f(R_{i-1})$. In plastic operation, the history of the stimulation of the device determines its final resistance value. In contrast, in standard memory operation one pulse is desired to bring the device into a precise resistance value independently from the previous programming history.

Fig. 14.8A and B shows an additional interesting detail: the analog operation is obtained by the proper tuning of the strength of the programming pulses (combination of pulse voltage and time width). Too long (or too high [48]) pulses result in abrupt conductance change. Frascaroli et al. [48], indeed, pointed out one of the main, and often neglected, issues of memristive plasticity, i.e., the limited programming window in which it is actually available. According to this observation, operational schemes for analog switching are usually considered *weak*, or equivalently *soft*,

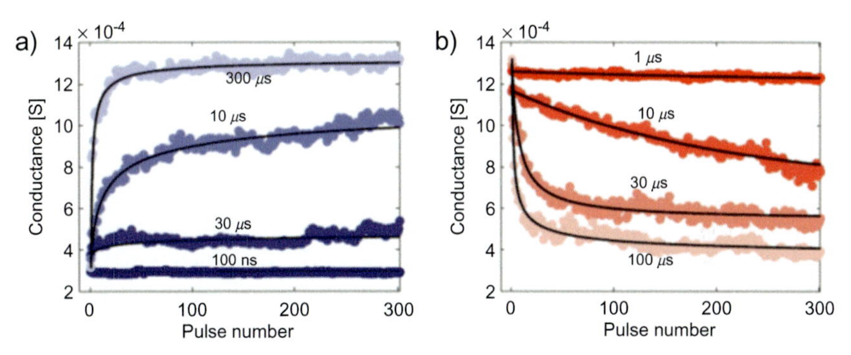

FIG. 14.8 Analog current dynamics of a HfO_2-based filamentary memristor stimulated by train of identical pulses for conductance potentiation (A) and depression (B). *Reprinted under CC-BY license from Ref. J. Frascaroli, S. Brivio, E. Covi, S. Spiga, Evidence of soft bound behaviour in analogue memristive devices for neuromorphic computing, Sci. Rep. 8 (2018) 7178. https://doi.org/10.1038/s41598-018-25376-x.*

programming conditions. Fig. 14.8A and B also shows that the conductance evolution is usually non-linear. Non-linear conductance plastic dynamics is considered to be a problem for machine learning accelerators, especially those implementing algorithms based on the backpropagation of the error [47,52,53], which has been driving huge research efforts toward the optimization of devices featuring linear conductance evolution [54–58]. On the other hand, recent studies pointed out that nonlinear conductance dynamics can be beneficial for spiking neural networks implementing biologically plausible learning rules [59–61].

It is worth mentioning that analog dynamics are subjected to severe cycle-to-cycle and device-to-device variability [53,58,59,62], intuitively because of the employed weak programming conditions. Furthermore, Brivio et al. [63] showed that trains of identical pulses delivered to analog filamentary devices stimulate random telegraphic resistance jumps, which constitutes an additional noise source for this kind of applications. Similar observations have been reported by Stathopoulos et al. [64] and by Doevenspeck et al. [65]

Considering the physical processes involved in the switching operation, as briefly summarized in Section 14.2, gradual resistance changes occur only if fast self-accelerated processes, like thermal runaway and abrupt filament formation, are inhibited. In filamentary devices, there are few tricks available to obtain this scope. The introduction of additional switching layers and tunneling barriers at one metal/oxide interface locally modifies the conduction and thermal dissipation properties [54,66]. Another solution is to harness the competitive switching of opposite interface, typical of the complementary resistance switching devices, to mitigate the thermal runaway process, in particular during the filament formation [49,51,67]. Contrarily to filamentary devices, in non-filamentary ones, the nearly ideal suppression of thermal runaway greatly favors the realization of analog dynamics [56,68,69].

14.3.2.2 Stochastic plasticity

Stochastic switching is a prerogative of filamentary devices, rather than non-filamentary ones. Indeed, CF formation and disruption is driven by stochastic processes at the nanoscale, e.g., defect generation or hopping from site to site, each of which individually produces a macroscopic resistance change. Therefore, the resistance transition in an ideal binary device, displaying only two levels with well-separated distributions, occurs with a certain probability when programming with weak pulses. In several recent works [70–73], it has been proposed to exploit this kind of operation for a different implementation of plasticity for neural applications. Indeed, trains of identical weak programming pulses drive the resistance change after a certain number of pulses on average. Therefore, a history-dependent and plastic operation on average is realized. An example of non-deterministic switching of a $Pt/HfO_x/TiO_x/HfO_x/TiO_x/TiN$ device is reported in Fig. 14.9. Indeed, Fig. 14.9A shows that, on different trials, the resistance transition from HRS to LRS is achieved after a different pulse number. An endurance test at weak programming conditions allows determining the average SET switching probability as shown in Fig. 14.9B [70]. In the reported case, the authors adopted a $HfO_x/TiO_x/HfO_x/TiO_x$ multilayer device to increase the resistance window. In this manner, they found stochastic SET transition between almost ideally binary states and gradual analog RESET dynamics [70]. Endurance tests performed on 1T-1R vertical $TiN/HfO_2/Ti/TiN$ RRAMs revealed that a correlation

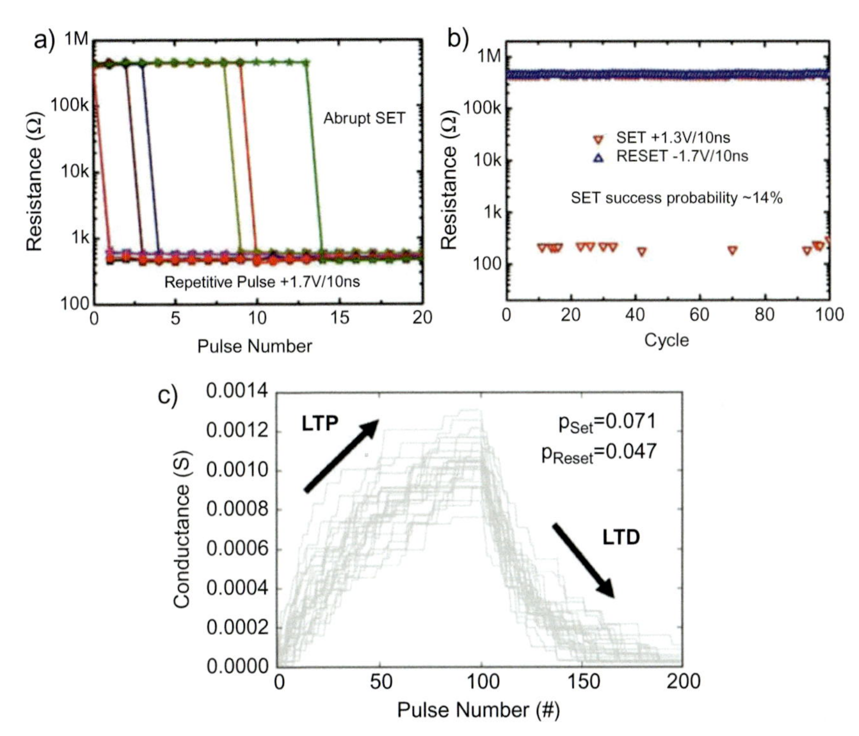

FIG. 14.9 (A) Example of stochastic operation of a binary RRAM under stimulation with trains of identical weak pulses; (B) Result of an endurance test with weak SET pulses. (C) Example of analog operation obtained with a parallel of 20 binary devices programmed in a stochastic manner. *Panels A and B are reprinted under the CC-BY license from Ref. S. Yu, B. Gao, Z. Fang, H. Yu, J. Kang, H.-S.P. Wong, Stochastic learning in oxide binary synaptic device for neuromorphic computing, Front. Neurosci. 7 (2013) 186. https://doi.org/10.3389/fnins.2013.00186; Panel C Reprinted under the CC-BY license from Ref. T. Werner, E. Vianello, O. Bichler, D. Garbin, D. Cattaert, B. Yvert, B. De Salvo, L. Perniola, Spiking neural networks based on OxRAM synapses for real-time unsupervised spike sorting, Front. Neurosci. 10 (2016). https://doi.org/10.3389/fnins.2016.00474.*

exists among the HRS and LRS values over tens of cycles and, thus, that the variability in the resistance states is not purely stochastic in the investigated system [74,75]. An alternative manner to analyze the stochastic device operation is to monitor the time required by a specific voltage value to drive the resistance transition. With this methodology, Gaba et al. found that wait times collected in an ensemble of programming trials follow a Poisson distribution in Ag/(amorphous Si)/Si devices [73], which indicates that the switching events are independent and completely uncorrelated, differently from Ref. [74].

The parallel arrangement of binary devices programmed in a stochastic way has been proposed by some authors [40,72,76] in order to endow the overall resulting circuit with the ability to go over many conductance values (given by the total resistance of all the device resistances in parallel, see Fig. 14.9C [76]). Bill and Legenstein [77] named this solution *compound synapse*. Obviously, the acquired property of gradual synaptic weight change comes at the expenses of the increase of the area occupancy.

14.3.3 Computing with spike timing and spike rate

The previous sections deal with implementations for storing memory bits and for keeping track of occurring events (plasticity). Advanced functionality is investigated to dress the memristive devices with the concepts of local computing. One example is the local coding of the rate and of the timing properties of the occurring events that stimulate the memristive device itself. The processing of the incoming events based on their rate and timing is believed to be a fundamental cognitive function and, therefore, has important applications in bio-inspired neuromorphic computing. Among such cognitive functions, in the recent literature, Spike Timing Dependent Plasticity (STDP) plays a primary role. According to the STDP prescription, the timing occurrence of spikes on two neurons (pre- and post-neurons) leads to a weight change of the synapse connecting the neurons [78]. If the pre-neuron contributes to the firing of the post-neuron, on average the pre-neuron fires always before the post-neuron. Vice versa is true if the post-neuron firing is completely independent of the pre-neurons. STDP prescribes a weight increase if the synapse is stimulated by the pre-neuron first and the post-neuron afterwards, within a certain time window. A weight decrease occurs in the opposite case. Therefore, STDP is a realization of local computing with timing and it has been demonstrated to be also sensitive on the spike rate in biological synapses [78,79]. In the following, we describe the recently proposed device level STDP implementations. In several literature reports, STDP has been implemented in plastic memristive devices, either stochastic [70–72,80] or analog [81–86], by delivering long (possibly pseudo-triangular-shaped) overlapping pulses to the two device terminals, as shown in the top panel of Fig. 14.10A [68]. The overlapping of such shaped pulses results in voltage drop on the device that depends on the relative pulse arrival timing and realizes a spike timing-dependent resistance change, see bottom panel of Fig. 14.10A [68]. In the stochastic version of STDP, the probability of switching assumes increasing finite values with increasing the delay-time window [70–72,80]. Such pair-wise STDP implementation is based on a clever programming of plasticity, as defined above. On the downside, such implementations require programming pulses as long as the timescales the system is to be sensitive to. Furthermore, it does not take into account pulse rates because it relies only on spike couples [59]. On the other hand, pioneering works have demonstrated that advanced cognitive functions are intrinsic to single devices, as we will describe in the following.

To explain the effect of the rate or of the timing of programming pulses delivered to a memristive device, let us first consider that a programming

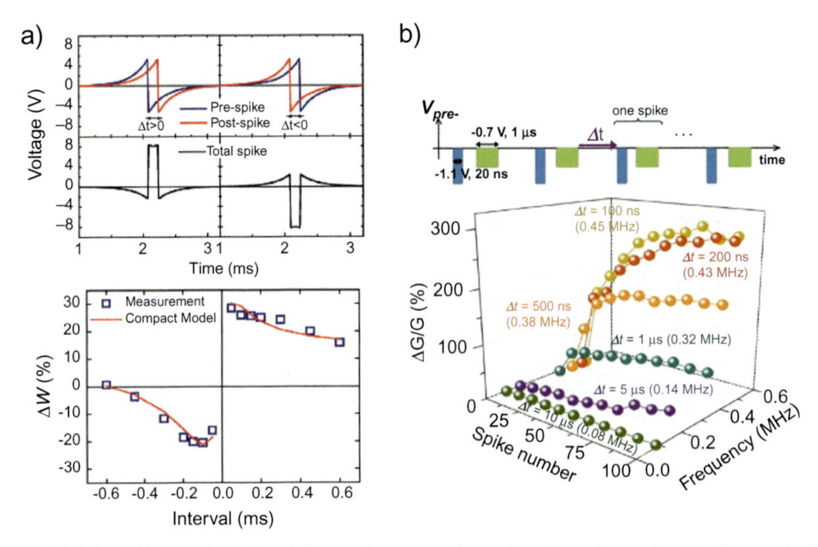

FIG. 14.10 (A) STDP obtained through temporal overlapping of couple of pulses as in the top panel, the weight change as a function of the delay among pulses is shown in the bottom panel. (B) PPF obtained with pre-heating low voltage and a programming high voltage spikes as reported in the top panel: the closer the two spikes, the more and the faster the conductance changes as shown in the bottom panel. *Panel A Reprinted under the CC-BY license from Ref. Y.-F. Wang, Y.-C. Lin, I.-T. Wang, T.-P. Lin, T.-H. Hou, Characterization and modeling of nonfilamentary Ta/TaOx/TiO2/Ti analog synaptic device, Sci. Rep. 5 (2015) 10150. https://doi.org/10.1038/srep10150; Panel B Reprinted (adapted) with permission from Ref. S. Kim, C. Du, P. Sheridan, W. Ma, S. Choi, W.D. Lu, Experimental demonstration of a second-order memristor and its ability to biorealistically implement synaptic plasticity, Nano Lett. 15 (2015) 2203–2211. https://doi.org/10.1021/acs.nanolett.5b00697. Copyright (2015) American Chemical Society.*

voltage spike, when isolated from previous or successive spikes, is able to drive a certain resistance change, ΔR_{alone}. It has been experimentally demonstrated that a preliminary conditional spike can prepare the ground for a subsequent programming spike so that the resistance change driven by the paired pulses (ΔR_{pair}) is larger than the one of a single pulse, ΔR_{alone}. This effect is named Pulse Pair Facilitation (PPF). In general, for this effect to take place, the conditional pulse and the programming pulse must be at a minimum time separation. The closer they are, the larger the difference $\Delta R_{pair} - \Delta R_{alone}$. This effect realizes a resistance change that depends on the spike timing or equivalently on the spike rate. As said, some pioneering works have experimentally observed PPF and attributed them to diverse physical mechanisms, depending on the employed material stack. Kim et al. [87] observed PPF for a filamentary $Pd/Ta_2O_{5-x}/TaO_y/Pt$ device, as shown in Fig.14.10B. The authors propose that one conditional spike (−1.1 V, 20 ns pulse in Fig. 14.10B) raises the temperature of the device, which slowly decays back to room temperature with a characteristic time of the order of 1 μs. A following spike (−0.7 V, 1 μs pulse in Fig. 14.10B) takes advantage of the pre-heating effect if it occurs within roughly 1 μs after the previous spike. The PPF is ascribed to a temperature effect because it does not depend on the voltage polarity of the conditional heating pulse [87]. Other studies demonstrates that summing up in short time volatile switching events results in a non-volatile storage of a conductance change, as representatively shown in Fig. 14.5B [30,31,60,88,89]. Volatile effects comprise both temporary ionic dynamics in filamentary VCM devices [31,88] or filament instabilities in ECM cells [30,60].

14.3.4 Negative differential resistance enabling oscillation

In the previous sub-sections, we discussed about the research aimed at localizing some computational tasks in individual devices. In this subsection, we describe research trends aimed at building nonlinear elements which can be used to implement sort of neuronal units [14]. The series connection of an element featuring NDR with a fixed resistor and a capacitor displays an oscillatory behavior when stimulated by a fixed voltage [34,90–92]. Let us consider the representative I—V curve of an NDR device operated in voltage mode (see Section 14.2) and the circuit reported in Fig. 14.11A. With an initial $V_{DD} = 0$ V, the memristor assumes an HRS value larger than that of the load resistor (R_{load}). In this case, the applied voltage V_{DD} drops mostly over the NDR device. Therefore, a V_{DD} value larger than the turn ON voltage, $V_{TH,ON}$, produces the switch of the NDR device to the LRS state with $R \ll R_{load}$, which causes most of the voltage to drop on the load resistor in a time interval governed by the capacitor C and the I—V hysteresis. The release of the voltage from the device, down to values

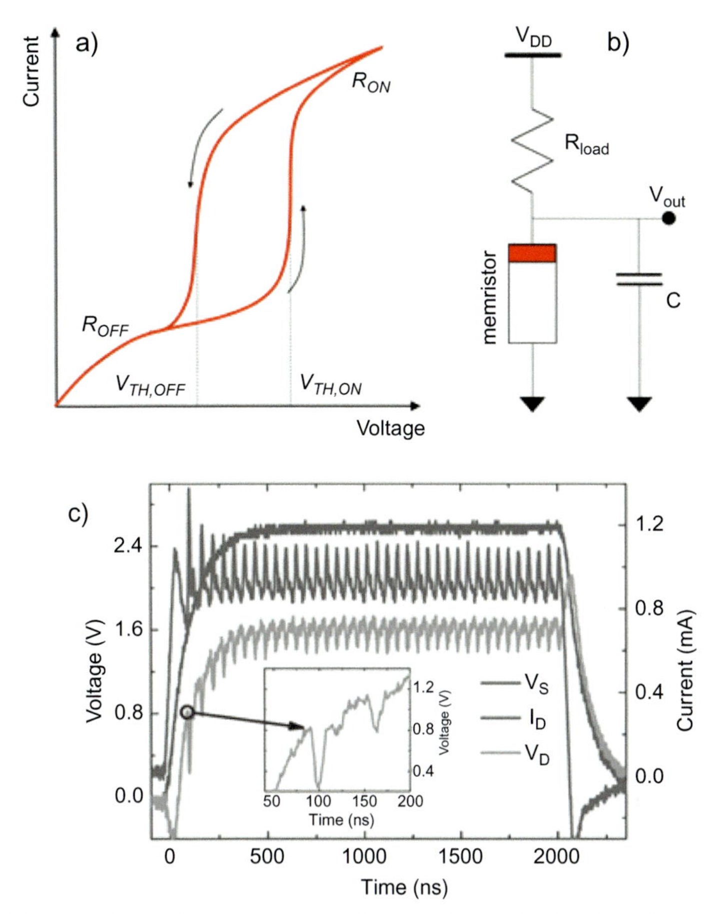

FIG. 14.11 (A) Sketch of the I—V curve as a function of applied voltage for a device showing NDR when programmed in current mode. (B) Oscillator circuit. (C) Experimental data for voltage applied to the oscillator, current through the NDR device and output voltage. *Reprinted with permission from Ref. S. Li, X. Liu, S.K. Nandi, D.K. Venkatachalam, R.G. Elliman, High-endurance megahertz electrical self-oscillation in Ti/NbOx bilayer structures, Appl. Phys. Lett. 106 (2015) 212902. https://doi.org/10.1063/1.4921745. Copyright (2015) American Institute of Physics.*

below the turn OFF voltage, $V_{TH,OFF}$, lets it switch back to the HRS state, which completes an entire oscillation cycle. The process continues in this way and the circuit oscillates as long as the voltage is applied to the structure, as the experimental result reported in Fig. 14.11C [93]. The circuit scheme in Fig. 11A is named Pearson-Anson relaxation oscillator after the scientists who first applied it to a Neon discharge lamp [94]. Further, the use of an external discrete capacitor may be avoided in the presence of parasitic capacitances that play the same role [34,92].

14.4 Overview of neuromorphic concepts and system prototypes

In this section, we review the status of the realization of neuromorphic computing machines comprising MO_x-based memristive elements. In particular, we review the works that have some connections with experimental data and real devices and circuits. Conversely, a more theoretical approach to memristive neural circuits can be found in Ref. [95].

As already mentioned in the introduction, there are proposed and realized systems that aim at accelerating neural networks and machine learning algorithms; others that take advantage from the inspiration from biological concepts. In the following sections we deal with all these systems.

14.4.1 Acceleration of neural networks and machine learning algorithms

Neural networks are probably the most important machine learning algorithm. They can be classified into two classes of architectures: feedforward neural networks, or multilayer perceptron (MLPs), and recurrent neural networks (RNNs). In feedforward neural networks, there are no feedback connections in which outputs of a node are feed back into itself (Fig. 14.12A) [4]. When many intermediate (hidden) layers separate input and output layers (leftmost and rightmost layers in Fig. 14.12A [4]), the network is called deep neural network (DNN). RNNs contain feedback connection loops enabling information to persist since the input at each step is composed of the data at that step in conjunction with the network output obtained at the previous step (Fig. 14.12C [96]). They are the natural architecture to use for sequential or temporal data. In particular, the well-known long-short-term memory (LSTM) RNN has recently found extensive application in text and speech recognition tasks. Fully connected neural networks (FCNN, Fig. 14.12A [4]) and convolutional neural networks (CNN, Fig. 14.12B [71]) are two example of feedforward neural networks. In FCNN, each neuron in one layer has directed connections (synaptic weights) to all the neurons of the subsequent layer. The number of weights and operations is directly proportional to the dimensions of the layers. On the other hand, CNN is composed of one or more convolutional layers, pooling or sub-sampling layers, and fully connected output layers (classification module). In a convolutional layer, a small set of synaptic weights constituting a kernel allows subsequent network layers to extract spatially localized features. CNN can achieve superb classification accuracy for image processing at much lower weight count than FCNN. All these classes of networks are trained using supervised learning and error

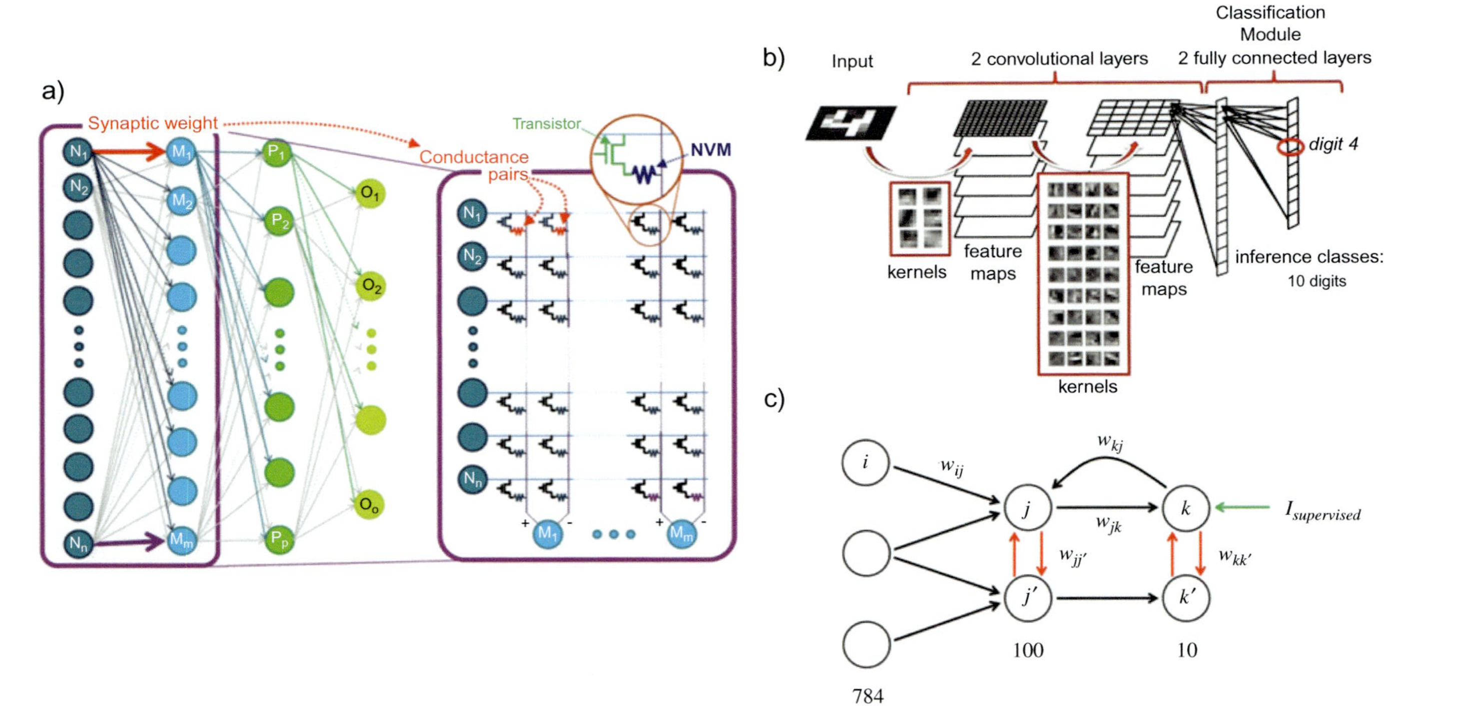

FIG. 14.12 Different Network Topology: (A) (Deep) Feedforward Neural Network. (B) Convolutional Neural Network (CNN) used for handwritten digits recognition (MNIST database). (C) Recurrent Neural Network. *Panel A Reprinted from Ref. G.W. Burr, R.M. Shelby, A. Sebastian, S. Kim, S. Kim, S. Sidler, K. Virwani, M. Ishii, P. Narayanan, A. Fumarola, L.L. Sanches, I. Boybat, M. Le Gallo, K. Moon, J. Woo, H. Hwang, Y. Leblebici, Neuromorphic computing using non-volatile memory, Adv. Phys. X. 2 (2017) 89–124. https://doi.org/10.1080/23746149.2016.1259585 under CC-BY 4.0 license; Panel B Adapted with permission from Ref. D. Garbin, O. Bichler, E. Vianello, Q. Rafhay, C. Gamrat, L. Perniola, G. Ghibaudo, B. DeSalvo, Variability-tolerant Convolutional Neural Network for Pattern Recognition applications based on OxRAM synapses, in: Electron Devices Meeting (IEDM), 2014 IEEE International, 2014: pp. 28.4.1-28.4.4. https://doi.org/10.1109/IEDM.2014.7047126. Copyright IEEE 2014; Panel C Reprinted from Ref. V. Demin, D. Nekhaev, Recurrent spiking neural network learning based on a competitive maximization of neuronal activity, Front. Neuroinform. 12 (2018). https://doi.org/10.3389/fninf.2018.00079 under CC-BY 4.0 license.*

backpropagation. The neuron inputs and outputs are real-valued numbers, processed on synchronous time steps.

Computation for DNNs includes both *training* and *inference*. During the training phase the synaptic weights are optimized on a training dataset (i.e., labeled dat(A), while during inference the already-learned network is used for classification or prediction on new, previously unseen 'test' data. Both training and inference require large number of *multiply and accumulate* (MAC) operations and memory accesses, leading to large power consumption when computing with conventional processors due to heavy communication between main memory and processors. Arrays of analog memristors are ideally suited to store the synaptic weights (the device conductance implementing the synaptic weights) and implement the MAC operations for both training and inference. The multiplication is performed at every memory point by Ohm's law, and current summation is done using kirchoff's current law, as discussed in Section 14.2.3. Since the memristors are used as a computing element, data movement is avoided. In addition, the MAC operations can be performed in parallel by applying analog input voltages to multiple word lines and reading the analog bit line current [97–99]. However, the analog results typically need to be converted into a digital format. Therefore, there is area and energy overhead for the analog-to-digital converter (ADC) and digital-to-analog converter (DAC). Another major difficulty of these networks is updating the conductance states in a reliable manner during the weight update process (training phase). The conductance modifications when a train of pulses is applied in nanoscale devices are stochastic, nonlinear, and asymmetric as well as of limited granularity. This has led to reduced classification accuracies when the training is performed on-chip compared with the off-chip training based on software simulations (i.e., the synaptic weight are defined by software simulation and just written one time on-chip). Several methods to improve the precision of the memristor based synaptic devices have been proposed in the literature: synaptic devices composed of multiple memristive elements [71,100,101], more complex synaptic architectures that augments PCM devices with a capacitor [98].

A fully connected neural network implemented entirely in integrated hardware is presented in Ref. [102]. Synaptic weights are stored by programming TaO_x- based memristors in multivalued resistance levels. The test chip, 4M synapses integrated into 180nm CMOS, results in 90.8% recognition rate (off-chip training) for the hand written digit classification problem (MNIST database). A small-scale fully connected neural network based on passive memristor crossbar array integrated with discrete CMOS components is presented in Ref. [103]. The network was trained both in-situ and ex-situ to perform classification of 4×4 pixel images.

Recently, a novel class of neural networks has been proposed, Binarized Neural Networks (BNN) [104,105]. In these neural networks, once trained, synapses as well as neurons assume only binary values meaning +1 or − 1. These neural networks have therefore limited memory requirements, and rely on highly simplified arithmetic (multiplications are replaced by one-bit XOR operations). Therefore, they are extremely attractive for realizing inference hardware. A hardware BNN based on HfO_2-based memristor devices has been recently proposed [106]. The network is based on a differential two resistors-two transistors (2T-2R) array, including all peripheral and sensing circuitry. Different hardware implementations of BNNs including memristive technology can be found in Refs [107,108].

The implementation of recurrent neural networks have been investigated recently, as well [109,110]. For instance, Guo et al. [109] simulate and build a hardware version of a small Hopfield network, i.e., a special form of RNN, performing the analog to 4-bit precision digital conversion. The hardware version is constituted by 16 wire-bonded TiO_x-based single devices programmed before the connection. Other investigations are performed at a purely theoretical level on generic RNN [111,112] and so-called Brain-State-in-a-Box models [113]. A first application of memristive technology to LSTM has also been recently investigated [114]. The already mentioned success of LSTM network is due to its capability of store and erase history observations, according to the required task. This operation, carried out by various LSTM cells, fundamentally gives a great advantage in learning compared to other RNN training algorithms [115]. Li et al [114] propose the implementation of an LSTM cell. They operate training and inference performed on a 128×64 1T-1R crossbar structure to demonstrate high speed and efficient computation in the LSTM scheme.

Implementations with memristive devices of other machine learning algorithms have been proposed in the literature. Such implementations take advantage of the vector–matrix multiplication that can be implemented in one step in memristive crossbars. Examples are the sparse coding algorithm demonstrated by Sheridan et al. [116] with a 25×20 subarray of the 32×32 WO_x-based memristive array and the principal component analysis algorithm simulated by Choi et al. [117] and implemented in a 9×2 forming-free Ta_2O_5-based memristor array by the same authors in later paper [118].

Alongside with training algorithms discussed so far, it is worth mentioning reinforcement learning as a way to train a DNN to a decision-making task and according a reward maximization approach. The success of AlphaGo demonstrated both the computational power of reinforcement learning [119] and its limitation in extreme energy requirements [120]. With the aim of containment of energy reduction, some steps of a reinforcement learning algorithm have been operated through a three-layer 1T-1R memristor crossbar, as reported in Ref. [121].

14.4.2 Spike-based brain-inspired architectures

Brain-inspired architectures based on spike coding of information comprise artificial emulators of synapses and neurons building brain-inspired networks operating through spike, or pulse, communication. In particular, we review various proposal of synaptic and neuronal architectures; routing synapse that are fundamental to achieve long-range neuron connections and finally full systems.

A large number of recent reports deals with proposal for *memristive synapses* [122,123]. Most of these units are conceived to connect neuron devices realized in conventional CMOS technologies. However, some proposals for the realization of *memristive neurons* exist in the literature [14,124–127]. The realization of tunable (plastic) synapses in memristive technologies may be considered the bottleneck for brain-inspired computing architectures. Indeed, the key factor determining the brain efficiency is its high synaptic interconnectivity. It is estimated that there are 10^4–10^5 per neurons [123]. Therefore, the realization of compact low-power synaptic units is already a crucial step for neuromorphic brain-inspired machines.

14.4.2.1 Memristive synapses

Plasticity is a suitable memristive functionality for electronic synapses in spike-based systems. In Section 14.3, we defined analog and stochastic versions of plasticity implementations. Ideally, a single memristive device, being an element with tunable resistance, can modulate the communications strength among other electronic units, as the neurons, thus emulating the function of synapses in biological brains. Many early studies report about simulated neuromorphic systems as proof of principle of the use of one memristor playing the role of one synapse [70,81–84,86,128,129]. In most cases, the authors adopt an STDP learning rule based on the temporal overlapping of long pulses, as described in Section 14.3. However, proposal for realistic synaptic architecture are those including a selecting transistor that avoids the sneak path problems in memristive arrays [59,76,130,131]. For instance, Fig. 14.13A reports a synaptic architecture comprising a 1T-1R device. The structure takes advantage of the transistor, beyond for selecting purposes, also to implement STDP. Indeed, the temporal overlapping of a long pulse delivered to the transistor gate (*PRE spike* in the figure) and a short pulse to the top memristive electrode (*Feedback spike* in the figure) drives synaptic potentiation or depression depending on their arrival timing [130]. The same architecture, as well as a variation including a 4T-1R device [132], can be used also for both spike-timing and spike-rate dependent learning rule [131]. Differently, Fig. 14.13B reports a 6T-1R structure that enables the activation of specific voltage values to drop on the memristive devices for various operations, i.e., long term potentiation (LTP), long term

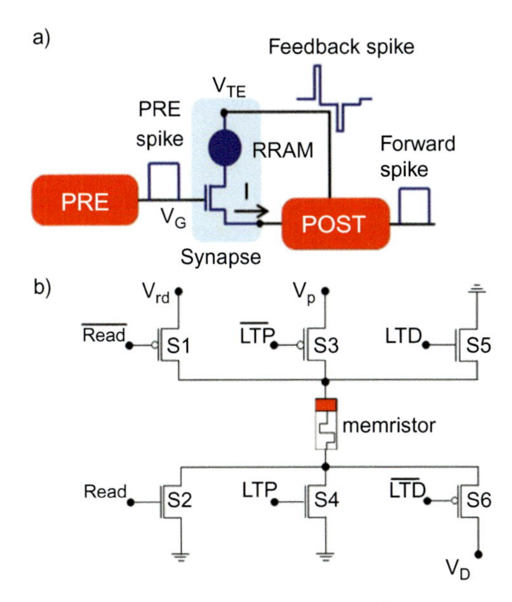

FIG. 14.13 (A) Synaptic architecture comprising 1T-1R devices and implementing STDP. (B) Versatile 6T-1R synaptic architecture possibly implementing various spike-based learning rules. *Panel A Reprinted from Ref. G. Pedretti, V. Milo, S. Ambrogio, R. Carboni, S. Bianchi, A. Calderoni, N. Ramaswamy, A.S. Spinelli, D. Ielmini, Memristive neural network for on-line learning and tracking with brain-inspired spike timing dependent plasticity, Sci. Rep 7 (2017). https://doi.org/10.1038/s41598-017-05480-0 under CC-BY 4.0 license; Panel B Reprinted from Ref. S. Brivio, D. Conti, M.V. Nair, J. Frascaroli, E. Covi, C. Ricciardi, G. Indiveri, S. Spiga, Extended memory lifetime in spiking neural networks employing memristive synapses with nonlinear conductance dynamics, Nanotechnology 30 (2019) 015102. https://doi.org/10.1088/1361-6528/aae81c under CC-BY 3.0 license.*

depression (LTD) and read [59]. The read operation corresponds to the application of a low voltage spike that delivers a synaptic current to the post-neuron. In this synapse scheme, the read operation is separated from LTP and LTD operations, which crucially prevents the feeding of the post-neuron with high programming currents. Indeed, programming voltages are higher than reading voltages and they result in much higher flowing current due to the typical nonlinear conduction mechanisms and unavoidable overshoot currents [133]. Therefore, programming current fed to the post-neurons can stimulate them by an improper amount, thus, invalidating the network operation. The 6T-1R synaptic architecture is compatible with mixed digital-analog sub-threshold CMOS neurons that control the 6T gate through a digital asynchronous, event-based circuitry [59]. The neuron-synapse realizes a spike-timing and spike-rate dependent plasticity rule [59], as well as, other possible learning algorithm including a probabilistic spike-based local version of the gradient descent algorithm, named delta rule [134]. The operation of such a synapse-neuron elemental block has been also demonstrated in hardware [135]. Many reports in the literature propose synaptic architectures comprising two memristive elements for each synaptic units. Doubling the memristive element with opposite orientation is used as a solution for improving the symmetry between LTP and LTD plasticity dynamics [136,137] and to implement positive (excitatory) and negative (inhibitory) synaptic weights [103,136,138–140]. Manu et al. [141] report on a 2-memristor-per-synapse architecture that provides the following functionalities: First, it scales of the sense current through the memristive elements; it applies an on-the-fly normalization of the resistance values allowing implementing positive and negative weights and it reduces the variability of the synapse induced by the memristive variability thanks to a push-pull programming of the two elements. Details can be found in Refs. [141,142].

In the proposals reviewed above, we considered the long-term version of plasticity, i.e., the long-term storage of synaptic weight updates. Short-term plasticity, i.e., temporary variation of the weight, in turn has an important computational role in encoding spatiotemporal spike patterns in brain-inspired computing [143,144]. CMOS implementations exist in which the temporal evolution is provided by capacitive charging and discharging, e.g., through the so call differential pair integrator circuit (DPI) [144]. The slow operation of the brain requires relatively large capacitances, even in case of circuits operating in the sub-threshold pA-fA range of transistors, that for scaling purposes must be shared among various synapses. Therefore, the emulation of short-term dynamics with nanoscale memristive devices allows enlarging the variety of timescales and extending their values in a single chip. The circuit in Fig. 14.14A is a modified version of a DPI circuit that includes memristive devices for the long term weight storage (green) and short term dynamics (light blue and pink) [143]. With such a scheme, even though the capacitor is shared among an array of synapses, a variety of timescales can be recovered through the adjustment of the charging current through the memristive elements. A very pragmatic, yet effective and reliable, implementation of short-term dynamics is the one reported in Fig. 14.14B, in which a presynaptic spike sets a non-volatile device that is afterwards reset back to its initial value through several partial RESET operations [145]. The evolution of the weight of such a volatile circuit block is shown in Fig. 14.14B. The described short-term is implemented in a simulated spiking network and demonstrate beneficial in improving the efficiency of the task of car detection in a highway [146].

The presented solutions to implement short-term dynamics in neuromorphic network are those that can be considered more robust. As already mentioned in Sections 14.2.2 and 14.3.3, memristive devices show a set of volatile and temporary effects that can exploited to implement short-term synaptic dynamics. Such effects have been investigated by some level of

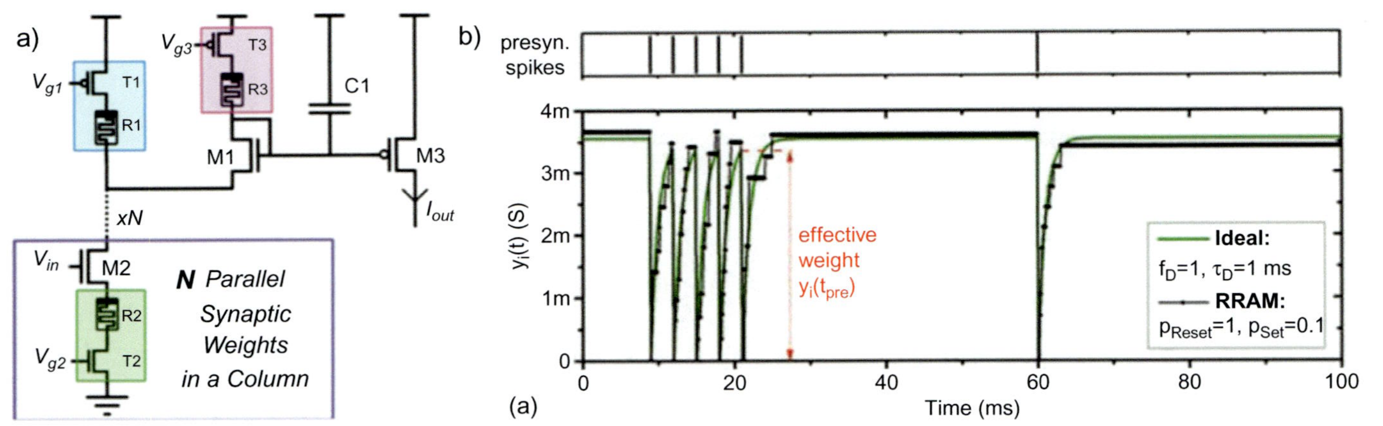

FIG. 14.14 (A) Representative short-term dynamics obtained from non-volatile HfO_2-based devices through a SET pre-spike and a sequence of gradual RESET operations [145] (B) Differential pair integrator circuit including non-volatile memristive devices modulating the capacitor charging. *Reprinted from Ref. T. Dalgaty, M. Payvand, F. Moro, D.R.B. Ly, F. Pebay-Peyroula, J. Casas, G. Indiveri, E. Vianello, Hybrid neuromorphic circuits exploiting non-conventional properties of RRAM for massively parallel local plasticity mechanisms, APL Mater. 7 (2019) 081125. https://doi.org/10.1063/1.5108663 under CC-BY 4.0 license.*

details in ECM devices based on chalcogenides [147], atomic switches [30] and, recently, diffusive memristors [12].

14.4.2.2 Memristive neurons

Memristive elements, besides being considered key-constituents for synaptic units, have being considered also to improve scalability [124] and the functionality of spiking neurons in tasks like, spike integration [124], spike firing [14], stochasticity [126] and intrinsic plasticity [127,143]. The simplest neuron model with some computational functionality is the Leak, Integrate and Fire (LIF) neuron. A LIF neuron integrates the input firing rate into an internal variable (usually associate to the membrane potential of a biological neuron). If the input firing rate is larger than the leakage rate, the membrane potential eventually overcomes a threshold for firing and a spike is generated at the neuron output. The CMOS design of LIF is relatively simple but it occupies a large area because of the capacitive contribution due to the integrative unit. pA-fA currents of sub-threshold analog circuits allow to use relatively small capacitors for biologically-plausible integration time constants [123]. It is worth pointing out that matching the time scales of the hardware operating 'speed' with those of the events to process allows an efficient real time processing of information. Indeed, conventional logic operating in the GHz range would require the storing pieces of signals to identify variation in slowing evolving sensor signals (with time scale longer than ms) or correlation in time among several signals. This operation reverts to von Neumann's bottleneck issues together with the associated energy inefficiency. Memristive versions of LIF neurons have been reported in which a capacitor is always needed to obtain integrative and spiking response. Fig. 14.15A [148] show a scheme of a neuron model (called neuristor), as first proposed by Pickett et al. [14], which includes two Mott memristive devices showing volatile switching due to thermally driven IMT transition. Two oscillatory units composed by capacitors and Mott elements implement the bio-realistic bi-phasic pulse firing shown in Fig. 14.15B (top) in response to a constant input current (Fig. 14.15B bottom). The variability of the output firing rate visible in Fig. 14.15B (top) is a consequence of memristive device variability and, interestingly, it corresponds, as well, to a Poissonian firing rate distribution similar to the biological counterpart [148]. Impressively, all these functionalities have been implemented in a simple neuronal unit that does not need discrete capacitors thanks to the leaky-integrative operation of an ECM memristive device in Ref. [124] or a diffusive memristor in Ref. [149].

Alternative implementations of neuronal units exist in the literature, which employ hybrid CMOS-memristor circuits to endow LIF neurons with additional functionality. Dalgaty et al. [143] introduce non-volatile memristive devices in a sub-threshold analog CMOS neuron that control

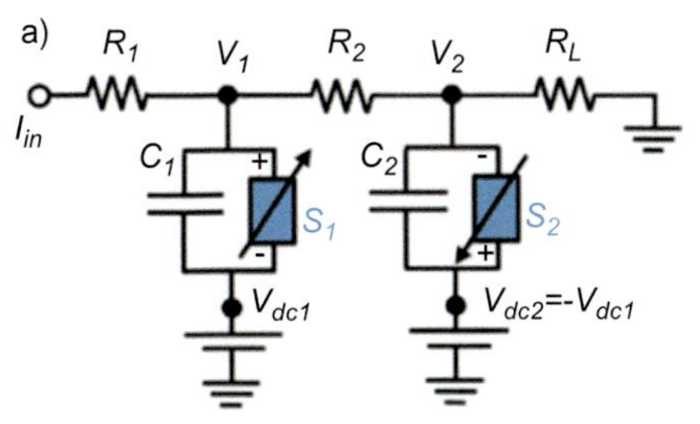

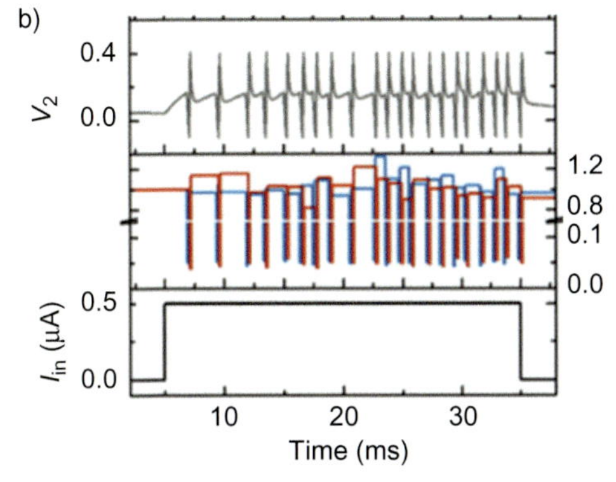

FIG. 14.15 (A) Circuit implementing the neuristor model comprising two Mott memristive devices (S_1 and S_2). (B) Simulated circuit operation from bottom to top: input current, memristor resistances and output voltage. *Reprinted from Ref. H. Lim, V. Kornijcuk, J.Y. Seok, S.-K. Kim, I. Kim, C.S. Hwang, D.S. Jeong, Reliability of neuronal information conveyed by unreliable neuristor-based leaky integrate-and-fire neurons: a model study, Sci. Rep. 5 (2015) 9776. https://doi.org/10.1038/srep09776 under CC-BY 4.0 license.*

the current flow toward the capacitors (in a somewhat similar way as in Fig. 14.14B) governing the integrative time constant and refractory period. This solution allows avoiding storing those parameters in external memory elements thus moving a further step forward toward in-memory computation. Furthermore, they also demonstrate that the neuron time constant can be adapted on-line to make a neuron fire around a target rate. This operation is known in biology as *intrinsic plasticity* and is considered an important phenomenon optimizing the trade-off between the volume of transmitted information and power consumption [127].

14.4.2.3 Routing solutions for reconfigurable networks

As already discussed, a SNN is composed of neurons and synapses. A third critical element for the hardware implementation of SNN is the routing method of the spikes (events) among the neurons. Routing can be realized either with dedicated routing, i.e., neurons are hardwired together through a set of synapses, or with synaptic lookup tables (LUTs). The dedicated routing schemes suffers from limited (or no) reconfigurability unless N^2 programmable synapses are used for N neurons. The LUT routing scheme addresses this issue by enabling any synapse in the synaptic array to be used for any arbitrary pair of neurons. When a neuron fires an event, the event is sent to a synaptic routing LUT storing the entire neuronal connections of the circuit, and then is transmitted to the appropriate neurons following the LUT. Most of spike-based reconfigurable neural networks use the Address Event Representation (AER) [150] as data representation and communication protocol [151]. AER representation allows for asynchronous communication between neurons: each neuron is assigned an address that is encoded as a digital word, and transmits their address to every neuron they are connected to as soon as they produce an event. LUTs can be efficiently implemented with Ternary Content Addressable Memories (TCAMs). In TCAM systems, a stored data is accessed by its content rather than by its address, therefore they allow to perform parallel search operation. The multi-core neuromorphic chip DYNAP-SEL [152] comprises multiple TCAM cells per neuron, to implement memory-optimized source-address routing schemes (see Ref. [153] for details). These TCAM cells are typically small in size (e.g., 22 bit in the DYNAP-SEL chip) and are programmed only at network configuration time. Conventional SRAM-based TCAM circuits are usually implemented with 16 transistors (16 T) [154]. This takes up valuable silicon real-estate in neuromorphic computing spiking neural network chips [152,153]. In order to increase storage density, memristor-based TCAM cells have been proposed to implement routing in multi-core [155]. Moreover, assuming future neuromorphic computing architectures of this type will have thousands of cores, the non-volatility feature of the proposed TCAM circuits will provide an additional crucial benefit, as it will require the user to upload all the configuration bits only the first time the network is configured, and will be able to skip this potentially time-consuming process every time the chip is reset or power-cycled.

14.4.2.4 Brain-inspired architectures and systems

In the literature, various brain-inspired computing architectures have been recently investigated. Many of the works present system level simulations relying on experimental data for memristive synapses and a mathematical formulation for neurons [70,77,81,83,156]. In other cases, the simulations of neurons unit are based on equation derived from analog sub-threshold neuromorphic circuits designed in various CMOS technology nodes [152,153,157]. Typically, the investigated architecture is a perceptron Spiking Neural Network with input and output layers directly all-to-all connected through plastic synapses. In some cases, a lateral inhibitory connection between output neurons implements a Winner-Take-All (WTA) process, according to which the first neuron firing prevents the other to fire for a certain period [83,156]. Usually, a STDP learning rule is implemented through temporal overlapping of pre- and post-synaptic spikes delivered to a memristive synapse, as in Fig. 14.10A, or to a 1T-1R structure [70,77,81,83,156], as in Fig. 14.13A, and provides an unsupervised learning process. Such kind of learning rule is thus purely dependent on couples of spikes and their timing. In other systems, the simulated learning rule is dependent on the pre-and post-firing rates and not on their timing, i.e., following the well-known Hebb's principle *"neurons that fire together wire together"* [132,158]. Combination of spike timing and rate sensitivity is obtained by Milo et al. [131] who introduce additional CMOS elements in the synapse architecture as already described the Section 14.3.2 and by Brivio et al. [59], who describe a generalized version of STDP implemented by the sub-threshold CMOS neuron circuits described above, which can also be operated in a semi-supervised manner with the help of a teacher neuron.

Despite the literature on memristive spiking neural network is dominated by feed forward networks, few examples exists of spiking Recurrent Neural Network [138,159]. For instance, Milo et al. [138,159] simulate a brain-inspired spiking RNN architecture comprising 1T-1R elements in the form of Hopfield network developing associative learning.

The investigation of spiking networks has recently moved some steps forward through mixed hardware-simulation experiments and full-hardware systems including microcontrollers. For instance, Serb et al. [160] used a memristive crossbar as a synaptic layer and simulated stochastic neurons to build STDP and Winner-Take-All networks. Conversely, Pedretti et al. [130] used Arduino Due for neuron units and 16 1T-1R synapses implementing STDP in hardware according to the implementation described with reference to Fig. 14.13A. Hansen et al. [158] demonstrated Hebb's rule in a network comprising a 16×16 memristive array and neurons programmed in an Arduino microcontroller. An impressive network implementation is demonstrated by Wang et al. [149]: they use non-volatile 1T-1R VCM devices, as plastic synapses and arrays of volatile diffusive memristors as LIF neuron. The cascade connection of synaptic and neuronal units is used as a WTA network. The unsupervised STDP learning protocol is, instead, simulated.

Despite neuromorphic systems, and in particular brain-inspired ones, have been conceived as analog systems in their original formulation, it is worth mentioning that digital machines can implement with versatility many neural and synaptic behaviors. For instance, Ref. [161] describes a digital neuromorphic processor that take advantage of memristive elements as synaptic units.

14.4.3 Non-spiking analog computing

In the previous section, we deal with computing concepts that borrow some cognitive processes from the brains and operate in the framework of information coding through spikes. However, other computational models exists that rely on the idea of a brain operating as a complex dynamical system where dynamical variables interact and synchronize [162]. Taking inspiration from this idea, the simplest building block that has been taken into account is a nano-oscillator. The final aim is to get a large amount of coupled oscillatory behavior giving rise to complex dynamical system responses useful for cognitive tasks. Relatively large systems have already been realized harnessing the technological maturity of magnetic memories to build spintronic oscillators [163]. The investigation in this field through memristive technology is still at an infancy level, given also the lower technology readiness level compared to magnetic memories. As already described in Section 14.3.3, memristor based oscillators relying on volatile resistance switching effects have been produced at a single-device level [90,91,164–166]. The coupling has been investigated in a hardware experiment involving two memristive oscillators by Li et al. [167]. Conversely, systems of memristor oscillators have demonstrated associative memory properties [168] and chaotic behavior useful for the solution of constrained optimization problems [166].

14.5 Conclusions and outlook

In summary, we described the salient features of different memristive technologies that are useful for neuromorphic computing and we reviewed the status of the research of neuromorphic systems including MO_x memristive devices distinguishing between various approaches. It is worth pointing out that, despite the tremendous advances in recent years with the (partial) realization of various hardware prototypes, much work is still required at different levels. Concerning the acceleration of machine learning algorithms, the crucial point is the development of devices with proper and reliable multilevel capability. On the system level, solutions to mitigate device non-ideality should be proposed and assessed to provide possibly a mid-term target. On the other side, brain-inspired systems are a step behind accelerators. Indeed, learning processes are still not well established, which does not allow a proper definition of the optimum device functionality and requirement. For this reason, we foresee that the most successful route for the realization of brain-inspired systems is a holistic approach aiming at matching functionality of devices and their specific reliability issues together with CMOS system design as a unique computational framework.

References

[1] D. Reinsel, J. Gantz, J. Rydning, The Digitization of the World from Edge to Core, 2018, p. 28.

[2] D.S. Jeong, K.M. Kim, S. Kim, B.J. Choi, C.S. Hwang, Memristors for energy-efficient new computing paradigms, Adv. Electron. Mater. 2 (2016) 1600090, https://doi.org/10.1002/aelm.201600090.

[3] H. Jeong, L. Shi, Memristor devices for neural networks, J. Phys. D Appl. Phys. 52 (2018), https://doi.org/10.1088/1361-6463/aae223, 023003.

[4] G.W. Burr, R.M. Shelby, A. Sebastian, S. Kim, S. Kim, S. Sidler, K. Virwani, M. Ishii, P. Narayanan, A. Fumarola, L.L. Sanches, I. Boybat, M. Le Gallo, K. Moon, J. Woo, H. Hwang, Y. Leblebici, Neuromorphic computing using non-volatile memory, Adv. Phys. X. 2 (2017) 89–124, https://doi.org/10.1080/23746149.2016.1259585.

[5] D.S. Jeong, C.S. Hwang, Nonvolatile memory materials for neuromorphic intelligent machines, Adv. Mater. 30 (2018) 1704729, https://doi.org/10.1002/adma.201704729.

[6] C. Mead, Neuromorphic electronic systems, Proc. IEEE 78 (1990) 1629–1636, https://doi.org/10.1109/5.58356.

[7] S. Brivio, G. Tallarida, E. Cianci, S. Spiga, Formation and disruption of conductive filaments in a HfO_2/TiN structure, Nanotechnology 25 (2014) 385705, https://doi.org/10.1088/0957-4484/25/38/385705.

[8] H.-Y. Chen, S. Brivio, C.-C. Chang, J. Frascaroli, T.-H. Hou, B. Hudec, M. Liu, H. Lv, G. Molas, J. Sohn, S. Spiga, V.M. Teja, E. Vianello, H.-S.P. Wong, Resistive random access memory (RRAM) technology: from material, device, selector, 3D integration to bottom-up fabrication, J. Electroceram. 39 (2017) 21–38, https://doi.org/10.1007/s10832-017-0095-9.

[9] R. Waser, R. Dittmann, G. Staikov, K. Szot, Redox-based resistive switching memories—nanoionic mechanisms, prospects, and challenges, Adv. Mater. 21 (2009) 2632–2663, https://doi.org/10.1002/adma.200900375.

[10] E. Ambrosi, P. Bartlett, A.I. Berg, S. Brivio, G. Burr, S. Deswal, J. Deuermeier, M. Haga, A. Kiazadeh, G. Kissling, M. Kozicki, C. Foroutan-Nejad, E. Gale, Y. Gonzalez-Velo, A. Goossens, L. Goux, T. Hasegawa, H. Hilgenkamp, R. Huang, S. Ibrahim, D. Ielmini, A.J. Kenyon, V. Kolosov, Y. Li, S. Majumdar, G. Milano, T. Prodromakis, N. Raeishosseini, V. Rana, C. Ricciardi, M. Santamaria, A. Shluger, I. Valov, R. Waser, R.S. Williams, D. Wouters, Y. Yang, A. Zaffora, Electrochemical metallization ReRAMs (ECM)—experiments and modelling: general discussion, Faraday Discuss. 213 (2019) 115–150, https://doi.org/10.1039/C8FD90059K.

[11] C. Schindler, G. Staikov, R. Waser, Electrode kinetics of Cu–SiO_2-based resistive switching cells: overcoming the voltage-time dilemma of electrochemical metallization memories, Appl. Phys. Lett. 94 (2009), https://doi.org/10.1063/1.3077310, 072109.

[12] Z. Wang, S. Joshi, S.E. Savel'ev, H. Jiang, R. Midya, P. Lin, M. Hu, N. Ge, J.P. Strachan, Z. Li, Q. Wu, M. Barnell, G.-L. Li, H.L. Xin, R.S. Williams, Q. Xia, J.J. Yang, Memristors with diffusive dynamics as synaptic emulators for neuromorphic computing, Nat. Mater. 16 (2017) 101–108, https://doi.org/10.1038/nmat4756.

[13] M. Aono, C. Baeumer, P. Bartlett, S. Brivio, G. Burr, M. Burriel, E. Carlos, S. Deswal, J. Deuermeier, R. Dittmann, H. Du, E. Gale, S. Hambsch, H. Hilgenkamp, D. Ielmini, A.J. Kenyon, A. Kiazadeh, A. Kindsmüller, G. Kissling, I. Köymen, S. Menzel, D.P. Asesio, T. Prodromakis, M. Santamaria, A. Shluger, D. Thompson, I. Valov, W. Wang, R. Waser, R.S. Williams, D. Wrana, D. Wouters, Y. Yang, A. Zaffora, Valence change ReRAMs (VCM)—experiments and modelling: general discussion, Faraday Discuss. 213 (2019) 259–286, https://doi.org/10.1039/C8FD90057D.

[14] M.D. Pickett, G. Medeiros-Ribeiro, R.S. Williams, A scalable neuristor built with Mott memristors, Nat. Mater. 12 (2013) 114–117, https://doi.org/10.1038/nmat3510.

[15] M.D. Pickett, J. Borghetti, J.J. Yang, G. Medeiros-Ribeiro, R.S. Williams, Coexistence of memristance and negative differential resistance in a nanoscale metal-oxide-metal system, Adv. Mater. 23 (2011) 1730–1733, https://doi.org/10.1002/adma.201004497.
[16] C. Funck, S. Menzel, N. Aslam, H. Zhang, A. Hardtdegen, R. Waser, S. Hoffmann-Eifert, Multidimensional simulation of threshold switching in NbO_2 based on an electric field triggered thermal runaway model, Adv. Electron. Mater. 2 (2016) 1600169, https://doi.org/10.1002/aelm.201600169.
[17] J.M. Goodwill, A.A. Sharma, D. Li, J.A. Bain, M. Skowronski, Electro-thermal model of threshold switching in TaO_x-based devices, ACS Appl. Mater. Interfaces 9 (2017) 11704–11710, https://doi.org/10.1021/acsami.6b16559.
[18] S. Slesazeck, H. Mähne, H. Wylezich, A. Wachowiak, J. Radhakrishnan, A. Ascoli, R. Tetzlaff, T. Mikolajick, Physical model of threshold switching in NbO_2 based memristors, RSC Adv. 5 (2015) 102318–102322, https://doi.org/10.1039/C5RA19300A.
[19] D. Ielmini, R. Bruchhaus, R. Waser, Thermochemical resistive switching: materials, mechanisms, and scaling projections, Phase Transit. 84 (2011) 570–602, https://doi.org/10.1080/01411594.2011.561478.
[20] B. Arndt, F. Borgatti, F. Offi, M. Phillips, P. Parreira, T. Meiners, S. Menzel, K. Skaja, G. Panaccione, D.A. MacLaren, R. Waser, R. Dittmann, Spectroscopic indications of tunnel barrier charging as the switching mechanism in memristive devices, Adv. Funct. Mater. 27 (2017) 1702282, https://doi.org/10.1002/adfm.201702282.
[21] S. Brivio, J. Frascaroli, S. Spiga, Role of Al doping in the filament disruption in HfO_2 resistance switches, Nanotechnology 28 (2017) 395202, https://doi.org/10.1088/1361-6528/aa8013.
[22] J. Frascaroli, F.G. Volpe, S. Brivio, S. Spiga, Effect of Al doping on the retention behavior of HfO_2 resistive switching memories, Microelectron. Eng. 147 (2015) 104–107, https://doi.org/10.1016/j.mee.2015.04.043.
[23] X. Li, H. Wu, B. Gao, N. Deng, H. Qian, Short time high-resistance state instability of TaO_x-based RRAM devices, IEEE Electron Device Lett. 38 (2017) 32–35, https://doi.org/10.1109/LED.2016.2630044.
[24] J. Guy, G. Molas, E. Vianello, F. Longnos, S. Blanc, C. Carabasse, M. Bernard, J.F. Nodin, A. Toffoli, J. Cluzel, P. Blaise, P. Dorion, O. Cueto, H. Grampeix, E. Souchier, T. Cabout, P. Brianceau, V. Balan, A. Roule, S. Maitrejean, L. Perniola, B. De Salvo, Investigation of the physical mechanisms governing data-retention in down to 10nm nano-trench Al_2O_3/CuTeGe conductive bridge RAM (CBRAM), in: Electron Devices Meeting (IEDM), 2013 IEEE International, 2013, pp. 30.2.1–30.2.4, https://doi.org/10.1109/IEDM.2013.6724722.
[25] G. Molas, J. Guy, M. Barci, F. Longnos, G. Palma, E. Vianello, P. Blaise, B.D. Salvo, L. Perniola, Conductive Bridge RAM (CBRAM): functionality, reliability and applications, in: 2015 International Conference on Solid State Devices and Materials, 2015, pp. 1142–1143.
[26] C. Nail, G. Molas, P. Blaise, G. Piccolboni, B. Sklenard, C. Cagli, M. Bernard, A. Roule, M. Azzaz, E. Vianello, C. Carabasse, R. Berthier, D. Cooper, C. Pelissier, T. Magis, G. Ghibaudo, C. Vallée, D. Bedeau, O. Mosendz, B.D. Salvo, L. Perniola, Understanding RRAM endurance, retention and window margin trade-off using experimental results and simulations, in: 2016 IEEE International Electron Devices Meeting (IEDM), 2016, pp. 4.5.1–4.5.4, https://doi.org/10.1109/IEDM.2016.7838346.
[27] B. Traoré, P. Blaise, E. Vianello, H. Grampeix, S. Jeannot, L. Perniola, B. De Salvo, Y. Nishi, On the origin of low-resistance state retention failure in HfO_2-based RRAM and impact of doping/alloying, IEEE Trans. Electron Devices 62 (2015) 4029–4036, https://doi.org/10.1109/TED.2015.2490545.
[28] J.R. Jameson, P. Blanchard, C. Cheng, J. Dinh, A. Gallo, V. Gopalakrishnan, C. Gopalan, B. Guichet, S. Hsu, D. Kamalanathan, D. Kim, F. Koushan, M. Kwan,

K. Law, D. Lewis, Y. Ma, V. McCaffrey, S. Park, S. Puthenthermadam, E. Runnion, J. Sanchez, J. Shields, K. Tsai, A. Tysdal, D. Wang, R. Williams, M.N. Kozicki, J. Wang, V. Gopinath, S. Hollmer, M. Van Buskirk, Conductive-bridge memory (CBRAM) with excellent high-temperature retention, in: Electron Devices Meeting (IEDM), 2013 IEEE International, 2013, pp. 30.1.1–30.1.4, https://doi.org/10.1109/IEDM.2013.6724721.

[29] D. Jana, S. Roy, R. Panja, M. Dutta, S.Z. Rahaman, R. Mahapatra, S. Maikap, Conductive-bridging random access memory: challenges and opportunity for 3D architecture, Nanoscale Res. Lett. 10 (2015) 188, https://doi.org/10.1186/s11671-015-0880-9.

[30] T. Ohno, T. Hasegawa, T. Tsuruoka, K. Terabe, J.K. Gimzewski, M. Aono, Short-term plasticity and long-term potentiation mimicked in single inorganic synapses, Nat. Mater. 10 (2011) 591–595, https://doi.org/10.1038/nmat3054.

[31] R. Berdan, E. Vasilaki, A. Khiat, G. Indiveri, A. Serb, T. Prodromakis, Emulating short-term synaptic dynamics with memristive devices, Sci. Rep. 6 (2016) 18639, https://doi.org/10.1038/srep18639.

[32] Z.-H. Tan, R. Yang, K. Terabe, X.-B. Yin, X.-D. Zhang, X. Guo, Synaptic metaplasticity realized in oxide memristive devices, Adv. Mater. 28 (2016) 377–384, https://doi.org/10.1002/adma.201503575.

[33] U. Celano, L. Mirabelli, L. Goux, K. Opsomer, W. Devulder, F. Crupi, C. Detavernier, M. Jurczak, W. Vandervorst, Tuning the switching behavior of conductive-bridge resistive memory by the modulation of the cation-supplier alloys, Microelectron. Eng. 167 (2017) 47–51, https://doi.org/10.1016/j.mee.2016.10.018.

[34] A.A. Sharma, I.V. Karpov, R. Kotlyar, J. Kwon, M. Skowronski, J.A. Bain, Dynamics of electroforming in binary metal oxide-based resistive switching memory, J. Appl. Phys. 118 (2015) 114903, https://doi.org/10.1063/1.4930051.

[35] A. Chen, Memory selector devices and crossbar array design: a modeling-based assessment, J. Comput. Electron. 16 (2017) 1186–1200, https://doi.org/10.1007/s10825-017-1059-7.

[36] J. Park, T. Hadamek, A.B. Posadas, E. Cha, A.A. Demkov, H. Hwang, Multi-layered $NiO_y/NbO_x/NiO_y$ fast drift-free threshold switch with high I_{on}/I_{off} ratio for selector application, Sci. Rep. 7 (2017) 1–8, https://doi.org/10.1038/s41598-017-04529-4.

[37] R. Aluguri, T. Tseng, Overview of selector devices for 3-D stackable cross point RRAM arrays, IEEE J. Electron Devices Soc. 4 (2016) 294–306, https://doi.org/10.1109/JEDS.2016.2594190.

[38] E. Linn, R. Rosezin, C. Kügeler, R. Waser, Complementary resistive switches for passive nanocrossbar memories, Nat. Mater. 9 (2010) 403–406, https://doi.org/10.1038/nmat2748.

[39] S. Brivio, S. Spiga, Stochastic circuit breaker network model for bipolar resistance switching memories, J. Comput. Electron. 16 (2017) 1154–1166, https://doi.org/10.1007/s10825-017-1055-y.

[40] D. Garbin, E. Vianello, Q. Rafhay, M. Azzaz, P. Candelier, B. DeSalvo, G. Ghibaudo, L. Perniola, Resistive memory variability: a simplified trap-assisted tunneling model, Solid-State Electron. 115 (2016) 126–132, https://doi.org/10.1016/j.sse.2015.09.004.

[41] S. Stathopoulos, A. Khiat, M. Trapatseli, S. Cortese, A. Serb, I. Valov, T. Prodromakis, Multibit memory operation of metal-oxide bilayer memristors, Sci. Rep. 7 (2017) 17532, https://doi.org/10.1038/s41598-017-17785-1.

[42] A. Fantini, G. Gorine, R. Degraeve, L. Goux, C.Y. Chen, A. Redolfi, S. Clima, A. Cabrini, G. Torelli, M. Jurczak, Intrinsic program instability in HfO_2 RRAM and consequences on program algorithms, in: 2015 IEEE International Electron Devices Meeting (IEDM), 2015, pp. 7.5.1–7.5.4, https://doi.org/10.1109/IEDM.2015.7409648.

[43] F.M. Puglisi, C. Wenger, P. Pavan, A novel program-verify algorithm for multi-bit operation in HfO_2 RRAM, IEEE Electron Device Lett. 36 (2015) 1030–1032, https://doi.org/10.1109/LED.2015.2464256.

[44] K. Higuchi, T.O. Iwasaki, K. Takeuchi, Investigation of verify-programming methods to achieve 10 million cycles for 50 nm HfO_2 ReRAM, in: 2012 4th IEEE International Memory Workshop, IEEE, Milan, 2012, pp. 1–4, https://doi.org/10.1109/IMW.2012.6213665.

[45] J.J. Ryu, B.K. Park, T.-M. Chung, Y.K. Lee, G.H. Kim, Optimized method for low-energy and highly reliable multibit operation in a HfO_2-based resistive switching device, Adv. Electron. Mater. 4 (2018) 1800261, https://doi.org/10.1002/aelm.201800261.

[46] Y.J. Ho, K.K. Min, S.S. Ji, S.J. Yeong, Y.K. Jean, K.D. Eun, P.T. Hyung, K.Y. Jae, S. Xinglong, H.C. Seong, Pt/Ta_2O_5/HfO_{2-x}/Ti resistive switching memory competing with multilevel NAND flash, Adv. Mater. 27 (2015) 3811–3816, https://doi.org/10.1002/adma.201501167.

[47] S. Agarwal, S.J. Plimpton, D.R. Hughart, A.H. Hsia, I. Richter, J.A. Cox, C.D. James, M.J. Marinella, Resistive memory device requirements for a neural algorithm accelerator, Neural Networks (IJCNN), 2016 International Joint Conference On, IEEE, 2016, pp. 929–938. http://ieeexplore.ieee.org/abstract/document/7727298/.

[48] J. Frascaroli, S. Brivio, E. Covi, S. Spiga, Evidence of soft bound behaviour in analogue memristive devices for neuromorphic computing, Sci. Rep. 8 (2018) 7178, https://doi.org/10.1038/s41598-018-25376-x.

[49] S. Brivio, E. Covi, A. Serb, T. Prodromakis, M. Fanciulli, S. Spiga, Experimental study of gradual/abrupt dynamics of HfO_2-based memristive devices, Appl. Phys. Lett. 109 (2016) 133504, https://doi.org/10.1063/1.4963675.

[50] E. Covi, S. Brivio, M. Fanciulli, S. Spiga, Synaptic potentiation and depression in Al:HfO_2-based memristor, Microelectron. Eng. 147 (2015) 41–44, https://doi.org/10.1016/j.mee.2015.04.052.

[51] S. Brivio, E. Covi, A. Serb, T. Prodromakis, M. Fanciulli, S. Spiga, Gradual set dynamics in HfO_2-based memristor driven by sub-threshold voltage pulses, in: Memristive Systems (MEMRISYS) 2015 International Conference On, 2015, pp. 1–2, https://doi.org/10.1109/MEMRISYS.2015.7378383.

[52] T. Gokmen, Y. Vlasov, Acceleration of deep neural network training with resistive cross-point devices: design considerations, Front. Neurosci. 10 (2016), https://doi.org/10.3389/fnins.2016.00333.

[53] P.-Y. Chen, B. Lin, I.-T. Wang, T.-H. Hou, J. Ye, S. Vrudhula, J. Seo, Y. Cao, S. Yu, Mitigating effects of non-ideal synaptic device characteristics for on-chip learning, in: Computer-Aided Design (ICCAD), 2015 IEEE/ACM International Conference On, IEEE, 2015, pp. 194–199.

[54] J. Woo, K. Moon, J. Song, S. Lee, M. Kwak, J. Park, H. Hwang, Improved synaptic behavior under identical pulses using AlO_x/HfO_2 bilayer RRAM array for neuromorphic systems, IEEE Electron Device Lett. 37 (2016) 994–997, https://doi.org/10.1109/LED.2016.2582859.

[55] Z. Wang, M. Yin, T. Zhang, Y. Cai, Y. Wang, Y. Yang, R. Huang, Engineering incremental resistive switching in TaO_x based memristors for brain-inspired computing, Nanoscale 8 (2016) 14015–14022, https://doi.org/10.1039/C6NR00476H.

[56] K. Moon, A. Fumarola, S. Sidler, J. Jang, P. Narayanan, R.M. Shelby, G.W. Burr, H. Hwang, Bidirectional non-filamentary RRAM as an analog neuromorphic synapse, part I: Al/Mo/$Pr_{0.7}Ca_{0.3}MnO_3$ material improvements and device measurements, IEEE J. Electron Devices Soc. 6 (2018) 146–155, https://doi.org/10.1109/JEDS.2017.2780275.

[57] J.W. Jang, S. Park, G.W. Burr, H. Hwang, Y.H. Jeong, Optimization of conductance change in $Pr_{1-x}Ca_xMnO_3$-based synaptic devices for neuromorphic systems, IEEE Electron Device Lett. 36 (2015) 457–459, https://doi.org/10.1109/LED.2015.2418342.

[58] D. Lee, K. Moon, J. Park, S. Park, H. Hwang, Trade-off between number of conductance states and variability of conductance change in $Pr_{0.7}Ca_{0.3}MnO_3$-based synapse device, Appl. Phys. Lett. 106 (2015) 113701, https://doi.org/10.1063/1.4915924.

[59] S. Brivio, D. Conti, M.V. Nair, J. Frascaroli, E. Covi, C. Ricciardi, G. Indiveri, S. Spiga, Extended memory lifetime in spiking neural networks employing memristive synapses

with nonlinear conductance dynamics, Nanotechnology 30 (2019), https://doi.org/10.1088/1361-6528/aae81c, 015102.

[60] S. La Barbera, D.R.B. Ly, G. Navarro, N. Castellani, O. Cueto, G. Bourgeois, B. De Salvo, E. Nowak, D. Querlioz, E. Vianello, Narrow heater bottom electrode-based phase change memory as a bidirectional artificial synapse, Adv. Electron. Mater. 4 (2018) 1800223, https://doi.org/10.1002/aelm.201800223.

[61] S. Brivio, D.R.B. Ly, E. Vianello, S. Spiga, Non-linear memristive synaptic dynamics for efficient unsupervised learning in spiking neural networks, Front. Neurosci. 15 (2021) 580909.

[62] S. Kim, M. Lim, Y. Kim, H.-D. Kim, S.-J. Choi, Impact of synaptic device variations on pattern recognition accuracy in a hardware neural network, Sci. Rep. 8 (2018) 2638, https://doi.org/10.1038/s41598-018-21057-x.

[63] S. Brivio, J. Frascaroli, E. Covi, S. Spiga, Stimulated ionic telegraph noise in filamentary memristive devices, Sci. Rep. 9 (2019) 6310, https://doi.org/10.1038/s41598-019-41497-3.

[64] S. Stathopoulos, A. Serb, A. Khiat, M. Ogorzałek, T. Prodromakis, A memristive switching uncertainty model, IEEE Trans. Electron Devices 66 (2019) 2946–2953, https://doi.org/10.1109/TED.2019.2918102.

[65] J. Doevenspeck, R. Degraeve, A. Fantini, P. Debacker, D. Verkest, R. Lauwereins, W. Dehaene, Low voltage transient RESET kinetic modeling of OxRRAM for neuromorphic applications, in: 2019 IEEE International Reliability Physics Symposium (IRPS), IEEE, Monterey, CA, USA, 2019, pp. 1–6, https://doi.org/10.1109/IRPS.2019.8720555.

[66] J. Woo, A. Padovani, K. Moon, M. Kwak, L. Larcher, H. Hwang, Linking conductive filament properties and evolution to synaptic behavior of RRAM devices for neuromorphic applications, IEEE Electron Device Lett. 38 (2017) 1220–1223, https://doi.org/10.1109/LED.2017.2731859.

[67] S. Brivio, J. Frascaroli, S. Spiga, Role of metal-oxide interfaces in the multiple resistance switching regimes of $Pt/HfO_2/TiN$ devices, Appl. Phys. Lett. 107 (2015), https://doi.org/10.1063/1.4926340, 023504.

[68] Y.-F. Wang, Y.-C. Lin, I.-T. Wang, T.-P. Lin, T.-H. Hou, Characterization and modeling of nonfilamentary $Ta/TaO_x/TiO_2/Ti$ analog synaptic device, Sci. Rep. 5 (2015) 10150, https://doi.org/10.1038/srep10150.

[69] J. Park, M. Kwak, K. Moon, J. Woo, D. Lee, H. Hwang, TiO_x-based RRAM synapse with 64-levels of conductance and symmetric conductance change by adopting a hybrid pulse scheme for neuromorphic computing, IEEE Electron Device Lett. 37 (2016) 1559–1562, https://doi.org/10.1109/LED.2016.2622716.

[70] S. Yu, B. Gao, Z. Fang, H. Yu, J. Kang, H.-S.P. Wong, Stochastic learning in oxide binary synaptic device for neuromorphic computing, Front. Neurosci. 7 (2013) 186, https://doi.org/10.3389/fnins.2013.00186.

[71] D. Garbin, E. Vianello, O. Bichler, Q. Rafhay, C. Gamrat, G. Ghibaudo, B. DeSalvo, L. Perniola, HfO_2-based OxRAM devices as synapses for convolutional neural networks, IEEE Trans. Electron Devices 62 (2015) 2494–2501, https://doi.org/10.1109/TED.2015.2440102.

[72] M. Suri, D. Querlioz, O. Bichler, G. Palma, E. Vianello, D. Vuillaume, C. Gamrat, B. DeSalvo, Bio-inspired stochastic computing using binary CBRAM synapses, IEEE Trans. Electron Devices 60 (2013) 2402–2409, https://doi.org/10.1109/TED.2013.2263000.

[73] S. Gaba, P. Sheridan, J. Zhou, S. Choi, W. Lu, Stochastic memristive devices for computing and neuromorphic applications, Nanoscale 5 (2013) 5872–5878, https://doi.org/10.1039/C3NR01176C.

[74] G. Piccolboni, G. Molas, D. Garbin, E. Vianello, O. Cueto, C. Cagli, B. Traore, B.D. Salvo, G. Ghibaudo, L. Perniola, Investigation of cycle-to-cycle variability in HfO_2-based OxRAM, IEEE Electron Device Lett. 37 (2016) 721–723, https://doi.org/10.1109/LED.2016.2553370.

[75] Y. Nishi, U. Böttger, R. Waser, S. Menzel, Crossover from deterministic to stochastic nature of resistive-switching statistics in a tantalum oxide thin film, IEEE Trans. Electron Devices 65 (2018) 4320–4325, https://doi.org/10.1109/TED.2018.2866127.
[76] T. Werner, E. Vianello, O. Bichler, D. Garbin, D. Cattaert, B. Yvert, B. De Salvo, L. Perniola, Spiking neural networks based on OxRAM synapses for real-time unsupervised spike sorting, Front. Neurosci. 10 (2016), https://doi.org/10.3389/fnins.2016.00474.
[77] J. Bill, R. Legenstein, A compound memristive synapse model for statistical learning through STDP in spiking neural networks, Front. Neurosci. 8 (2014) 412, https://doi.org/10.3389/fnins.2014.00412.
[78] G. Bi, M. Poo, Synaptic modifications in cultured hippocampal neurons: dependence on spike timing, synaptic strength, and postsynaptic cell type, J. Neurosci. 18 (1998) 10464–10472.
[79] P.J. Sjöström, G.G. Turrigiano, S.B. Nelson, Rate, timing, and cooperativity jointly determine cortical synaptic plasticity, Neuron 32 (2001) 1149–1164, https://doi.org/10.1016/S0896-6273(01)00542-6.
[80] S. Ambrogio, S. Balatti, V. Milo, R. Carboni, Z.Q. Wang, A. Calderoni, N. Ramaswamy, D. Ielmini, Neuromorphic learning and recognition with one-transistor-one-resistor synapses and bistable metal oxide RRAM, IEEE Trans. Electron Devices 63 (2016) 1508–1515, https://doi.org/10.1109/TED.2016.2526647.
[81] S. Yu, B. Gao, Z. Fang, H. Yu, J. Kang, H.-S.P. Wong, A low energy oxide-based electronic synaptic device for neuromorphic visual systems with tolerance to device variation, Adv. Mater. 25 (2013) 1774–1779, https://doi.org/10.1002/adma.201203680.
[82] E. Covi, S. Brivio, A. Serb, T. Prodromakis, M. Fanciulli, S. Spiga, HfO_2-based memristors for neuromorphic applications, in: 2016 IEEE International Symposium on Circuits and Systems (ISCAS), 2016, pp. 393–396, https://doi.org/10.1109/ISCAS.2016.7527253.
[83] E. Covi, S. Brivio, A. Serb, T. Prodromakis, M. Fanciulli, S. Spiga, Analog memristive synapse in spiking networks implementing unsupervised learning, Front. Neurosci. 10 (2016) 482, https://doi.org/10.3389/fnins.2016.00482.
[84] E. Covi, S. Brivio, J. Frascaroli, M. Fanciulli, S. Spiga, (Invited) analog HfO_2-RRAM switches for neural networks, ECS Trans. 75 (2017) 85–94.
[85] I.-T. Wang, Y.-C. Lin, Y.-F. Wang, C.-W. Hsu, T.-H. Hou, 3D synaptic architecture with ultralow sub-10 fJ energy per spike for neuromorphic computation, in: Electron Devices Meeting (IEDM), 2014 IEEE International, 2014, pp. 28.5.1–28.5.4, https://doi.org/10.1109/IEDM.2014.7047127.
[86] M. Prezioso, F. Merrikh Bayat, B. Hoskins, K. Likharev, D. Strukov, Self-adaptive spike-time-dependent plasticity of metal-oxide memristors, Sci. Rep. 6 (2016) 21331, https://doi.org/10.1038/srep21331.
[87] S. Kim, C. Du, P. Sheridan, W. Ma, S. Choi, W.D. Lu, Experimental demonstration of a second-order memristor and its ability to biorealistically implement synaptic plasticity, Nano Lett. 15 (2015) 2203–2211, https://doi.org/10.1021/acs.nanolett.5b00697.
[88] C. Du, W. Ma, T. Chang, P. Sheridan, W.D. Lu, Biorealistic implementation of synaptic functions with oxide memristors through internal ionic dynamics, Adv. Funct. Mater. 25 (2015) 4290–4299, https://doi.org/10.1002/adfm.201501427.
[89] T. Chang, S.-H. Jo, W. Lu, Short-term memory to long-term memory transition in a nanoscale memristor, ACS Nano 5 (2011) 7669–7676, https://doi.org/10.1021/nn202983n.
[90] A. Ascoli, S. Slesazeck, H. Mähne, R. Tetzlaff, T. Mikolajick, Nonlinear dynamics of a locally-active memristor, IEEE Trans. Circuits Syst. Regul. Pap. 62 (2015) 1165–1174, https://doi.org/10.1109/TCSI.2015.2413152.
[91] S.K. Nandi, S. Li, X. Liu, R.G. Elliman, Temperature dependent frequency tuning of NbO_x relaxation oscillators, Appl. Phys. Lett. 111 (2017) 202901, https://doi.org/10.1063/1.4999373.

[92] X. Liu, S. Li, S.K. Nandi, D.K. Venkatachalam, R.G. Elliman, Threshold switching and electrical self-oscillation in niobium oxide films, J. Appl. Phys. 120 (2016) 124102, https://doi.org/10.1063/1.4963288.

[93] S. Li, X. Liu, S.K. Nandi, D.K. Venkatachalam, R.G. Elliman, High-endurance megahertz electrical self-oscillation in Ti/NbO_x bilayer structures, Appl. Phys. Lett. 106 (2015) 212902, https://doi.org/10.1063/1.4921745.

[94] S.O. Pearson, H.S.G. Anson, The neon tube as a means of producing intermittent currents, Proc. Phys. Soc. Lond. 34 (1921) 204, https://doi.org/10.1088/1478-7814/34/1/341.

[95] R. Kozma, R.E. Pino, G.E. Pazienza (Eds.), Advances in Neuromorphic Memristor Science and Applications, Springer Netherlands, Dordrecht, 2012, https://doi.org/10.1007/978-94-007-4491-2.

[96] V. Demin, D. Nekhaev, Recurrent spiking neural network learning based on a competitive maximization of neuronal activity, Front. Neuroinform. 12 (2018), https://doi.org/10.3389/fninf.2018.00079.

[97] G.W. Burr, R.M. Shelby, C. di Nolfo, J.W. Jang, R.S. Shenoy, P. Narayanan, K. Virwani, E.U. Giacometti, B. Kurdi, H. Hwang, Experimental demonstration and tolerancing of a large-scale neural network (165,000 synapses), using phase-change memory as the synaptic weight element, in: Electron Devices Meeting (IEDM), 2014 IEEE International, 2014, pp. 29.5.1–29.5.4, https://doi.org/10.1109/IEDM.2014.7047135.

[98] S. Ambrogio, P. Narayanan, H. Tsai, R.M. Shelby, I. Boybat, C. di Nolfo, S. Sidler, M. Giordano, M. Bodini, N.C.P. Farinha, B. Killeen, C. Cheng, Y. Jaoudi, G.W. Burr, Equivalent-accuracy accelerated neural-network training using analogue memory, Nature 558 (2018) 60, https://doi.org/10.1038/s41586-018-0180-5.

[99] M. Prezioso, F. Merrikh-Bayat, B.D. Hoskins, G.C. Adam, K.K. Likharev, D.B. Strukov, Training and operation of an integrated neuromorphic network based on metal-oxide memristors, Nature 521 (2015) 61–64, https://doi.org/10.1038/nature14441.

[100] I. Boybat, M.L. Gallo, S.R. Nandakumar, T. Moraitis, T. Parnell, T. Tuma, B. Rajendran, Y. Leblebici, A. Sebastian, E. Eleftheriou, Neuromorphic computing with multi-memristive synapses, Nat. Commun. 9 (2018) 1–12, https://doi.org/10.1038/s41467-018-04933-y.

[101] S. Agarwal, R.B.J. Gedrim, A.H. Hsia, D.R. Hughart, E.J. Fuller, A.A. Talin, C.D. James, S.J. Plimpton, M.J. Marinella, Achieving ideal accuracies in analog neuromorphic computing using periodic carry, in: 2017 Symposium on VLSI Technology, 2017, pp. T174–T175, https://doi.org/10.23919/VLSIT.2017.7998164.

[102] R. Mochida, K. Kouno, Y. Hayata, M. Nakayama, T. Ono, H. Suwa, R. Yasuhara, K. Katayama, T. Mikawa, Y. Gohou, A 4M synapses integrated analog ReRAM based 66.5 TOPS/W neural-network processor with cell current controlled writing and flexible network architecture, in: 2018 IEEE Symposium on VLSI Technology, 2018, pp. 175–176, https://doi.org/10.1109/VLSIT.2018.8510676.

[103] F.M. Bayat, M. Prezioso, B. Chakrabarti, H. Nili, I. Kataeva, D. Strukov, Implementation of multilayer perceptron network with highly uniform passive memristive crossbar circuits, Nat. Commun. 9 (2018) 2331, https://doi.org/10.1038/s41467-018-04482-4.

[104] M. Courbariaux, I. Hubara, D. Soudry, R. El-Yaniv, Y. Bengio, Binarized Neural Networks: Training Deep Neural Networks With Weights and Activations Constrained to +1 or -1, 2016, ArXiv:1602.02830 [Cs] http://arxiv.org/abs/1602.02830. accessed July 29, 2019.

[105] M. Rastegari, V. Ordonez, J. Redmon, A. Farhadi, XNOR-Net: ImageNet Classification Using Binary Convolutional Neural Networks, Springerprofessional.De, 2016. https://www.springerprofessional.de/en/xnor-net-imagenet-classification-using-binary-convolutional-neur/10709306. accessed November 22, 2019.

[106] M. Bocquet, T. Hirztlin, J.-O. Klein, E. Nowak, E. Vianello, J.-M. Portal, D. Querlioz, In-memory and error-immune differential RRAM implementation of binarized deep

neural networks, in: 2018 IEEE International Electron Devices Meeting (IEDM), 2018, pp. 20.6.1–20.6.4, https://doi.org/10.1109/IEDM.2018.8614639.
[107] P. Huang, Z. Zhou, Y. Zhang, Y. Xiang, R. Han, L. Liu, X. Liu, J. Kang, Hardware implementation of RRAM based binarized neural networks, APL Mater. 7 (2019), https://doi.org/10.1063/1.5116863, 081105.
[108] X. Sun, S. Yin, X. Peng, R. Liu, J. Seo, S. Yu, XNOR-RRAM: a scalable and parallel resistive synaptic architecture for binary neural networks, in: 2018 Design, Automation Test in Europe Conference Exhibition (DATE), 2018, pp. 1423–1428, https://doi.org/10.23919/DATE.2018.8342235.
[109] X. Guo, F. Merrikh-Bayat, L. Gao, B.D. Hoskins, F. Alibart, B. Linares-Barranco, L. Theogarajan, C. Teuscher, D.B. Strukov, Modeling and experimental demonstration of a hopfield network analog-to-digital converter with hybrid CMOS/memristor circuits, Front. Neurosci. 488 (2015), https://doi.org/10.3389/fnins.2015.00488.
[110] S.G. Hu, Y. Liu, Z. Liu, T.P. Chen, J.J. Wang, Q. Yu, L.J. Deng, Y. Yin, S. Hosaka, Associative memory realized by a reconfigurable memristive Hopfield neural network, Nat. Commun. 6 (2015) 7522, https://doi.org/10.1038/ncomms8522.
[111] G.M. Tornez Xavier, F. Gómez Castañeda, L.M. Flores Nava, J.A. Moreno Cadenas, Memristive recurrent neural network, Neurocomputing 273 (2018) 281–295, https://doi.org/10.1016/j.neucom.2017.08.008.
[112] G. Zhang, Y. Shen, Q. Yin, J. Sun, Global exponential periodicity and stability of a class of memristor-based recurrent neural networks with multiple delays, Inform. Sci. 232 (2013) 386–396, https://doi.org/10.1016/j.ins.2012.11.023.
[113] M. Hu, H. Li, Q. Wu, G.S. Rose, Hardware realization of BSB recall function using memristor crossbar arrays, in: DAC Design Automation Conference 2012, 2012, pp. 498–503.
[114] C. Li, Z. Wang, M. Rao, D. Belkin, W. Song, H. Jiang, P. Yan, Y. Li, P. Lin, M. Hu, N. Ge, J.P. Strachan, M. Barnell, Q. Wu, R.S. Williams, J.J. Yang, Q. Xia, Long short-term memory networks in memristor crossbar arrays, Nat. Mach. Intell. 1 (2019) 49, https://doi.org/10.1038/s42256-018-0001-4.
[115] S. Hochreiter, J. Schmidhuber, Long Short-Term Memory (2006), https://doi.org/10.1162/neco.1997.9.8.1735.
[116] P.M. Sheridan, F. Cai, C. Du, W. Ma, Z. Zhang, W.D. Lu, Sparse coding with memristor networks, Nat. Nanotechnol. 12 (2017) 784–789, https://doi.org/10.1038/nnano.2017.83.
[117] S. Choi, P. Sheridan, W.D. Lu, Data clustering using memristor networks, Sci. Rep. 5 (2015), https://doi.org/10.1038/srep10492.
[118] S. Choi, J.H. Shin, J. Lee, P. Sheridan, W.D. Lu, Experimental demonstration of feature extraction and dimensionality reduction using memristor networks, Nano Lett. 17 (2017) 3113–3118, https://doi.org/10.1021/acs.nanolett.7b00552.
[119] D. Silver, A. Huang, C.J. Maddison, A. Guez, L. Sifre, G. van den Driessche, J. Schrittwieser, I. Antonoglou, V. Panneershelvam, M. Lanctot, S. Dieleman, D. Grewe, J. Nham, N. Kalchbrenner, I. Sutskever, T. Lillicrap, M. Leach, K. Kavukcuoglu, T. Graepel, D. Hassabis, Mastering the game of Go with deep neural networks and tree search, Nature 529 (2016) 484–489, https://doi.org/10.1038/nature16961.
[120] Another Way Of Looking At Lee Sedol vs AlphaGo Jacques Mattheij, 2016. https://jacquesmattheij.com/another-way-of-looking-at-lee-sedol-vs-alphago/. accessed November 27, 2019.
[121] Z. Wang, C. Li, W. Song, M. Rao, D. Belkin, Y. Li, P. Yan, H. Jiang, P. Lin, M. Hu, J.P. Strachan, N. Ge, M. Barnell, Q. Wu, A.G. Barto, Q. Qiu, R.S. Williams, Q. Xia, J.J. Yang, Reinforcement learning with analogue memristor arrays, Nat. Electron. 2 (2019) 115–124, https://doi.org/10.1038/s41928-019-0221-6.
[122] D. Kuzum, S. Yu, H.-S.P. Wong, Synaptic electronics: materials, devices and applications, Nanotechnology 24 (2013) 382001, https://doi.org/10.1088/0957-4484/24/38/382001.

[123] G. Indiveri, B. Linares-Barranco, R. Legenstein, G. Deligeorgis, T. Prodromakis, Integration of nanoscale memristor synapses in neuromorphic computing architectures, Nanotechnology 24 (2013) 384010, https://doi.org/10.1088/0957-4484/24/38/384010.

[124] Y. Zhang, W. He, Y. Wu, K. Huang, Y. Shen, J. Su, Y. Wang, Z. Zhang, X. Ji, G. Li, H. Zhang, S. Song, H. Li, L. Sun, R. Zhao, L. Shi, Highly compact artificial memristive neuron with low energy consumption, Small 14 (2018) 1802188, https://doi.org/10.1002/smll.201802188.

[125] J.H. Yoon, Z. Wang, K.M. Kim, H. Wu, V. Ravichandran, Q. Xia, C.S. Hwang, J.J. Yang, An artificial nociceptor based on a diffusive memristor, Nat. Commun. 9 (2018) 417, https://doi.org/10.1038/s41467-017-02572-3.

[126] M. Al-Shedivat, R. Naous, G. Cauwenberghs, K.N. Salama, Memristors empower spiking neurons with stochasticity, IEEE J. Emerg. Sel. Topics Circuits Syst. 5 (2015) 242–253, https://doi.org/10.1109/JETCAS.2015.2435512.

[127] T. Dalgaty, M. Payvand, B. De Salvo, J. Casas, G. Lama, E. Nowak, G. Indiveri, E. Vianello, Hybrid CMOS-RRAM neurons with intrinsic plasticity, in: 2019 IEEE International Symposium on Circuits and Systems (ISCAS), 2019, pp. 1–5, https://doi.org/10.1109/ISCAS.2019.8702603.

[128] D. Querlioz, O. Bichler, C. Gamrat, Simulation of a memristor-based spiking neural network immune to device variations, in: The 2011 International Joint Conference on Neural Networks (IJCNN), 2011, pp. 1775–1781, https://doi.org/10.1109/IJCNN.2011.6033439.

[129] D. Querlioz, O. Bichler, P. Dollfus, C. Gamrat, Immunity to device variations in a spiking neural network with memristive nanodevices, IEEE Trans. Nanotechnol. 12 (2013) 288–295, https://doi.org/10.1109/TNANO.2013.2250995.

[130] G. Pedretti, V. Milo, S. Ambrogio, R. Carboni, S. Bianchi, A. Calderoni, N. Ramaswamy, A.S. Spinelli, D. Ielmini, Memristive neural network for on-line learning and tracking with brain-inspired spike timing dependent plasticity, Sci. Rep. 7 (2017), https://doi.org/10.1038/s41598-017-05480-0.

[131] V. Milo, G. Pedretti, R. Carboni, A. Calderoni, N. Ramaswamy, S. Ambrogio, D. Ielmini, Demonstration of hybrid CMOS/RRAM neural networks with spike time/rate-dependent plasticity, in: 2016 IEEE International Electron Devices Meeting (IEDM), 2016, pp. 16.8.1–16.8.4, https://doi.org/10.1109/IEDM.2016.7838435.

[132] V. Milo, G. Pedretti, R. Carboni, A. Calderoni, N. Ramaswamy, S. Ambrogio, D. Ielmini, A 4-transistors/1-resistor hybrid synapse based on resistive switching memory (RRAM) capable of spike-rate-dependent plasticity (SRDP), IEEE Trans. Very Large Scale Integr. VLSI Syst. 26 (2018) 2806–2815, https://doi.org/10.1109/TVLSI.2018.2818978.

[133] S. Ambrogio, V. Milo, Z. Wang, S. Balatti, D. Ielmini, Analytical modeling of current overshoot in oxide-based resistive switching memory (RRAM), IEEE Electron Device Lett. 37 (2016) 1268–1271, https://doi.org/10.1109/LED.2016.2600574.

[134] M. Payvand, L.K. Muller, G. Indiveri, Event-based circuits for controlling stochastic learning with memristive devices in neuromorphic architectures, in: 2018 IEEE International Symposium on Circuits and Systems (ISCAS), 2018, pp. 1–5, https://doi.org/10.1109/ISCAS.2018.8351544.

[135] E. Covi, R. George, J. Frascaroli, S. Brivio, C. Mayr, H. Mostafa, G. Indiveri, S. Spiga, Spike-driven threshold-based learning with memristive synapses and neuromorphic silicon neurons, J. Phys. D Appl. Phys. 51 (2018) 344003, https://doi.org/10.1088/1361-6463/aad361.

[136] O. Krestinskaya, A.P. James, L.O. Chua, Neuro-memristive circuits for edge computing: a review, IEEE Trans. Neural Netw. Learn. Syst. (2019) 1–20, https://doi.org/10.1109/TNNLS.2019.2899262.

[137] A.M. Sheri, H. Hwang, M. Jeon, B. Lee, Neuromorphic character recognition system with two PCMO memristors as a synapse, IEEE Trans. Ind. Electron. 61 (2014) 2933–2941, https://doi.org/10.1109/TIE.2013.2275966.
[138] V. Milo, D. Ielmini, E. Chicca, Attractor networks and associative memories with STDP learning in RRAM synapses, in: 2017 IEEE International Electron Devices Meeting (IEDM), 2017, pp. 11.2.1–11.2.4, https://doi.org/10.1109/IEDM.2017.8268369.
[139] L. Deng, G. Li, N. Deng, D. Wang, Z. Zhang, W. He, H. Li, J. Pei, L. Shi, Complex learning in bio-plausible memristive networks, Sci. Rep. 5 (2015), https://doi.org/10.1038/srep10684.
[140] C.H. Bennett, S.L. Barbera, A.F. Vincent, J.O. Klein, F. Alibart, D. Querlioz, Exploiting the short-term to long-term plasticity transition in memristive nanodevice learning architectures, in: 2016 International Joint Conference on Neural Networks (IJCNN), 2016, pp. 947–954, https://doi.org/10.1109/IJCNN.2016.7727300.
[141] M.V. Nair, L.K. Muller, G. Indiveri, A differential memristive synapse circuit for on-line learning in neuromorphic computing systems, Nano Futures 1 (2017), https://doi.org/10.1088/2399-1984/aa954a, 035003.
[142] M. Payvand, M.V. Nair, L.K. Müller, G. Indiveri, A neuromorphic systems approach to in-memory computing with non-ideal memristive devices: from mitigation to exploitation, Faraday Discuss. 213 (2019) 487–510, https://doi.org/10.1039/C8FD00114F.
[143] T. Dalgaty, M. Payvand, F. Moro, D.R.B. Ly, F. Pebay-Peyroula, J. Casas, G. Indiveri, E. Vianello, Hybrid neuromorphic circuits exploiting non-conventional properties of RRAM for massively parallel local plasticity mechanisms, APL Mater. 7 (2019), https://doi.org/10.1063/1.5108663, 081125.
[144] C. Bartolozzi, G. Indiveri, Synaptic dynamics in analog VLSI, Neural Comput. 19 (2007) 2581–2603, https://doi.org/10.1162/neco.2007.19.10.2581.
[145] T. Werner, E. Vianello, O. Bichler, A. Grossi, E. Nowak, J.-F. Nodin, B. Yvert, B. DeSalvo, L. Perniola, Experimental demonstration of short and long term synaptic plasticity using OxRAM multi k-bit arrays for reliable detection in highly noisy input data, in: 2016 IEEE International Electron Devices Meeting (IEDM), IEEE, 2016, pp. 16.6.1–16.6.4.
[146] E. Vianello, T. Werner, O. Bichler, A. Valentian, G. Molas, B. Yvert, B. De Salvo, L. Perniola, Resistive memories for spike-based neuromorphic circuits, in: Memory Workshop (IMW), 2017 IEEE International, IEEE, 2017, pp. 1–6.
[147] S. La Barbera, D. Vuillaume, F. Alibart, Filamentary switching: synaptic plasticity through device volatility, ACS Nano 9 (2015) 941–949, https://doi.org/10.1021/nn506735m.
[148] H. Lim, V. Kornijcuk, J.Y. Seok, S.K. Kim, I. Kim, C.S. Hwang, D.S. Jeong, Reliability of neuronal information conveyed by unreliable neuristor-based leaky integrate-and-fire neurons: a model study, Sci. Rep. 5 (2015) 9776, https://doi.org/10.1038/srep09776.
[149] Z. Wang, S. Joshi, S. Savel'ev, W. Song, R. Midya, Y. Li, M. Rao, P. Yan, S. Asapu, Y. Zhuo, H. Jiang, P. Lin, C. Li, J.H. Yoon, N.K. Upadhyay, J. Zhang, M. Hu, J.P. Strachan, M. Barnell, Q. Wu, H. Wu, R.S. Williams, Q. Xia, J.J. Yang, Fully memristive neural networks for pattern classification with unsupervised learning, Nat. Electron. 1 (2018) 137–145, https://doi.org/10.1038/s41928-018-0023-2.
[150] S.-C. Liu, T. Delbruck, G. Indiveri, A. Whatley, R. Douglas, Event-Based Neuromorphic Systems: Liu/Event-Based, John Wiley & Sons, Ltd, Chichester, UK, 2015, https://doi.org/10.1002/9781118927601.
[151] K.A. Boahen, Point-to-point connectivity between neuromorphic chips using address events, IEEE Trans. Circuits Syst. II. 47 (2000) 416–434, https://doi.org/10.1109/82.842110.
[152] N. Qiao, G. Indiveri, Scaling mixed-signal neuromorphic processors to 28 nm FD-SOI technologies, in: 2016 IEEE Biomedical Circuits and Systems Conference (BioCAS),

IEEE, Shanghai, China, 2016, pp. 552–555, https://doi.org/10.1109/BioCAS.2016.7833854.

[153] S. Moradi, N. Qiao, F. Stefanini, G. Indiveri, A scalable multicore architecture with heterogeneous memory structures for dynamic neuromorphic asynchronous processors (dynaps), IEEE J. Emerging Sel. Top. Circuits Syst. 12 (2018) 106–122, https://doi.org/10.1109/TBCAS.2017.2759700.

[154] I. Hayashi, T. Amano, N. Watanabe, Y. Yano, Y. Kuroda, M. Shirata, S. Morizane, K. Hayano, K. Dosaka, K. Nii, H. Noda, H. Kawai, A 250-MHz 18-Mb full ternary CAM with low voltage match line sense amplifier in 65nm CMOS, in: 2012 IEEE Asian Solid State Circuits Conference (A-SSCC), 2012, pp. 65–68, https://doi.org/10.1109/IPEC.2012.6522628.

[155] D.R.B. Ly, B. Giraud, J.-P. Noel, A. Grossi, N. Castellani, G. Sassine, J.-F. Nodin, G. Molas, C. Fenouillet-Beranger, G. Indiveri, E. Nowak, E. Vianello, in-depth characterization of resistive memory-based ternary content addressable memories, in: 2018 IEEE International Electron Devices Meeting (IEDM), 2018, pp. 20.3.1–20.3.4, https://doi.org/10.1109/IEDM.2018.8614603.

[156] D. Querlioz, O. Bichler, A.F. Vincent, C. Gamrat, Bioinspired programming of memory devices for implementing an inference engine, Proc. IEEE 103 (2015) 1398–1416, https://doi.org/10.1109/JPROC.2015.2437616.

[157] E. Chicca, F. Stefanini, C. Bartolozzi, G. Indiveri, Neuromorphic electronic circuits for building autonomous cognitive systems, Proc. IEEE 102 (2014) 1367–1388, https://doi.org/10.1109/JPROC.2014.2313954.

[158] M. Hansen, F. Zahari, H. Kohlstedt, M. Ziegler, Unsupervised Hebbian learning experimentally realized with analogue memristive crossbar arrays, Sci. Rep. 8 (2018) 1–10, https://doi.org/10.1038/s41598-018-27033-9.

[159] V. Milo, E. Chicca, D. Ielmini, Brain-inspired recurrent neural network with plastic RRAM synapses, in: 2018 IEEE International Symposium on Circuits and Systems (ISCAS), IEEE, Florence, 2018, pp. 1–5, https://doi.org/10.1109/ISCAS.2018.8351523.

[160] A. Serb, J. Bill, A. Khiat, R. Berdan, R. Legenstein, T. Prodromakis, Unsupervised learning in probabilistic neural networks with multi-state metal-oxide memristive synapses, Nat. Commun. 7 (2016) 12611, https://doi.org/10.1038/ncomms12611.

[161] Y. Kim, Y. Zhang, P. Li, A reconfigurable digital neuromorphic processor with memristive synaptic crossbar for cognitive computing, J. Emerg. Technol. Comput. Syst 11 (2015) 38:1–38:25, https://doi.org/10.1145/2700234.

[162] G. Buzsáki, Rhythms of the Brain, Oxford University Press, 2006, https://doi.org/10.1093/acprof:oso/9780195301069.001.0001.

[163] M. Romera, P. Talatchian, S. Tsunegi, F.A. Araujo, V. Cros, P. Bortolotti, J. Trastoy, K. Yakushiji, A. Fukushima, H. Kubota, S. Yuasa, M. Ernoult, D. Vodenicarevic, T. Hirtzlin, N. Locatelli, D. Querlioz, J. Grollier, Vowel recognition with four coupled spin-torque nano-oscillators, Nature 563 (2018) 230, https://doi.org/10.1038/s41586-018-0632-y.

[164] A.A. Sharma, Y. Li, M. Skowronski, J.A. Bain, J.A. Weldon, High-frequency TaO_x-based compact oscillators, IEEE Trans. Electron Devices 62 (2015) 3857–3862, https://doi.org/10.1109/TED.2015.2475623.

[165] A.A. Sharma, T.C. Jackson, M. Schulaker, C. Kuo, C. Augustine, J.A. Bain, H.S.P. Wong, S. Mitra, L.T. Pileggi, J.A. Weldon, High performance, integrated 1T1R oxide-based oscillator: stack engineering for low-power operation in neural network applications, in: 2015 Symposium on VLSI Technology (VLSI Technology), 2015, pp. T186–T187, https://doi.org/10.1109/VLSIT.2015.7223672.

[166] S. Kumar, J.P. Strachan, R.S. Williams, Chaotic dynamics in nanoscale NbO_2 Mott memristors for analogue computing, Nature 548 (2017) 318–321, https://doi.org/10.1038/nature23307.

[167] S. Li, X. Liu, S.K. Nandi, D.K. Venkatachalam, R.G. Elliman, Coupling dynamics of Nb/Nb_2O_5 relaxation oscillators, Nanotechnology 28 (2017) 125201, https://doi.org/10.1088/1361-6528/aa5de0.
[168] T.C. Jackson, A.A. Sharma, J.A. Bain, J.A. Weldon, L. Pileggi, Oscillatory neural networks based on TMO nano-oscillators and multilevel RRAM cells, IEEE J. Emerg. Sel. Topics Circuits Syst. 5 (2015) 230–241, https://doi.org/10.1109/JETCAS.2015.2433551.

Index

Note: Page numbers followed by *f* indicate figures and *t* indicate tables.

R

S

T

Printed in the United States
by Baker & Taylor Publisher Services